CHEMICAL, ELECTROKINETIC, THERMAL, AND BIOENGINEERING METHODS

Ground Improvement Case Histories

CHEMICAL, ELECTROKINETIC, THERMAL, AND BIOENGINEERING METHODS

Edited by

BUDDHIMA INDRARATNA

Professor and Research Director, Centre for Geomechanics and Railway Engineering, Faculty of Engineering and Information Sciences, University of Wollongong, NSW 2522, Australia

JIAN CHU

Professor and James M. Hoover Chair in Geotechnical Engineering, Iowa State University, USA
Nanyang Technological University, Singapore

CHOLACHAT RUJIKIATKAMJORN

Associate Professor, Centre for Geomechanics and Railway Engineering, Faculty of Engineering and Information Sciences, University of Wollongong, NSW 2522, Australia

ELSEVIER

AMSTERDAM • BOSTON • HEIDELBERG • LONDON
NEW YORK • OXFORD • PARIS • SAN DIEGO
SAN FRANCISCO • SINGAPORE • SYDNEY • TOKYO

Butterworth-Heinemann is an imprint of Elsevier

Butterworth Heinemann is an imprint of Elsevier
The Boulevard, Langford Lane, Kidlington, Oxford OX5 1GB, UK
225 Wyman Street, Waltham, MA 02451, USA

British Library Cataloguing in Publication Data
A catalogue record for this book is available from the British Library

Library of Congress Cataloging-in-Publication Data
A catalog record for this book is available from the Library of Congress

ISBN: 978-0-08-100191-2

For information on all Butterworth Heinemann publications visit our website at http://store.elsevier.com/

DEDICATION

This Book is dedicated to several thousands who lost their lives in the catastrophic earthquake that hit Nepal on 25 April 2015 causing devastating and widespread destruction.

Appropriate Ground Improvement will be crucial during the reconstruction of affected regions including the surrounds of Kathmandu.

CONTENTS

CONTRIBUTORS

Toshihisa Adachi
Geo-Research Institute, Osaka, Japan

Hifumi Aoki
Japan Railway Construction Agency, Yokohama-shi, Japan

Arul Arulrajah
Swinburne University of Technology, Melbourne, Australia

Laura Banasiak
Centre for Geomechanics and Railway Engineering, Faculty of Engineering and Information Sciences, University of Wollongong, NSW, Australia

Dennes T. Bergado
Asian Institute of Technology, Bangkok, Thailand

Tejo Bheemasetti
Department of Civil Engineering, The University of Texas at Arlington, Arlington, Texas

Myint Win Bo
DST Consulting Engineers Inc., Thunder Bay ON, Canada

Don Cameron
School of Natural and Built Environments, University of South Australia, Adelaide, South Australia

J.-C. Chai
Institute of Lowland Technology, Saga University, Saga, Japan

Jian Chu
Professor and James M. Hoover Chair in Geotechnical Engineering, Iowa State University, USA; Nanyang Technological University, Singapore

Nicolas Denies
Belgian Building Research Institute, Brussels, Belgium

Mike Fabius
DST Consulting Engineers Inc., Thunder Bay ON, Canada

Behnam Fatahi
Centre for Built Infrastructure Research, School of Civil and Environmental Engineering, University of Technology Sydney (UTS), Sydney, Australia

Behzad Fatahi
Centre for Built Infrastructure Research, School of Civil and Environmental Engineering, University of Technology Sydney (UTS), Sydney, Australia

Ernst Friedlaender
Keller Ground Engineering, NSW, Australia

Stephanie Glendinning
Newcastle University, Newcastle upon Tyne, UK

Jia He
Nanyang Technological University, Singapore

Suksun Horpibulsuk
Swinburne University of Technology, Melbourne, Australia, and Suranaree University of Technology, Nakhon Ratchasima, Thailand

Noël Huybrechts
Belgian Building Research Institute, Brussels, and KU Leuven, Leuven, Belgium

Buddhima Indraratna
Professor and Research Director, Centre for Geomechanics and Railway Engineering, Faculty of Engineering and Information Sciences, University of Wollongong, NSW, Australia

Volodymyr Ivanov
Nanyang Technological University, Singapore

Pitthaya Jamsawang
King Mongkut's University of Technology North Bangkok, Bangkok, Thailand

Colin J.F.P. Jones
Newcastle University, Newcastle upon Tyne, UK

Richard Kelly
Coffey Geotechnics Pty Ltd., and ARC Centre of Excellence in Geotechnical Science and Engineering, University of Newcastle, NSW, Australia

Gunnar Keppel
School of Natural and Built Environments, University of South Australia, Adelaide, South Australia

Hadi Khabbaz
Centre for Built Infrastructure Research, School of Civil and Environmental Engineering, University of Technology Sydney (UTS), Sydney, Australia

Jayantha Kodikara
Monash University, Victoria, Australia

Lyesse Laloui
Ecole Polytechnique Fédérale de Lausanne, EPFL, and Soil Mechanics Laboratory, Lausanne, Switzerland

John Lamont-Black
Electrokinetic Ltd., Newcastle upon Tyne, UK

Eng Choy Lee
Infra Tech Pty Ltd., Burwood, NSW, Australia

Serge Leroueil
Department of Civil and Water Engineering, Laval University, Québec City, Québec, Canada

Glenys Lugg
Manildra Group, Nowra, NSW, Australia

Maryam Maeimi
Nanyang Technological University, Singapore

Maria Esther Soares Marques
Military Institute of Engineering, Rio de Janeiro, Rio de Janeiro, Brazil

N. Miura
Institute of Soft Ground Engineering, Co. Ltd., Saga, Japan

Theva Muttuvel
Coffey Geotechnics Pty Ltd., NSW, Australia

Mathieu Nuth
Ecole Polytechnique Fédérale de Lausanne, EPFL, and Soil Mechanics Laboratory, Lausanne, Switzerland

Muditha Pallewattha
Centre for Geomechanics and Railway Engineering, Faculty of Engineering and Information Sciences, University of Wollongong, NSW, Australia

Udeshini Pathirage
Centre for Geomechanics and Railway Engineering, Faculty of Engineering and Information Sciences, University of Wollongong, NSW, Australia

Aravind Pedarla
Department of Civil Engineering, The University of Texas at Arlington, Arlington, Texas

Teoh Yaw Poh
Building and Construction Authority, Singapore

Anand J Puppala
Department of Civil Engineering, The University of Texas at Arlington, Arlington, Texas

Amnart Rittirong
Department of Civil and Environmental Engineering, The University of Western Ontario, London, Ontario, Canada

Julie Shang
Department of Civil and Environmental Engineering, The University of Western Ontario, London, Ontario, Canada

Takeshi Tamura
Kyoto University, Kyoto, Japan

Masaru Tateyama
Railway Technical Research Institute, Tokyo, Japan

Fumio Tatsuoka
Tokyo University of Science, Chiba, Japan

Panich Voottipruex
King Mongkut's University of Technology North Bangkok, Bangkok, Thailand

Stacey Vorwerk
School of Natural and Built Environments, University of South Australia, Adelaide, South Australia

Laurent Vulliet
Ecole Polytechnique Fédérale de Lausanne, EPFL, and Soil Mechanics Laboratory, Lausanne, Switzerland

Kenji Watanabe
Railway Technical Research Institute, Tokyo, Japan

Dazhi Wen
Geotech & Tunnel Consult, Singapore

Ing Hieng Wong
MITIC Engineers Pte Ltd., Singapore

Shifan Wu
Iowa State University, USA

Shuwang Yan
Geotechnical Research Institute, Tianjin University, Tianjin, China

Richard Yeo
ARRB Group Ltd., Vermont South, Victoria, Australia

FOREWORD

There is hardly any country in the world that does not encounter problematic soil either due to natural formation or caused by manmade activities for land use and construction at a given location or at any given time. Moreover, with increasing population in many countries, it is almost inevitable that even the most marginal land and often difficult terrains are also utilised for infrastructure development. In Civil Engineering practice, without advanced ground improvement techniques, such developments over formations that consist of highly compressible estuarine clays, dispersive and erodible soils, landfills weak residual soils, historic landslide debris such as colluvium or highly weathered and fractured rock mass would not be possible. With many countries now embracing high speed rail, offshore land reclamation, deep excavations and tunnels among other construction challenges, detailed review and mandatory budgeting for optimal ground improvement techniques have become the norm in every infrastructure project.

This series of 3 Books on Ground Improvement Case Histories edited by Professors Buddhima Indraratna, Jian Chu and Cholachat Rujikiatkamjorn offer a significant expansion of the very popular previous book published more than a decade ago. This concerted effort of three world-renowned ground improvement experts is to be commended for an exceptionally beneficial contribution to geotechnical engineering. The contents are based on detailed knowledge acquired through many years of painstaking field observations and R & D studies by many internationally reputed academics and practitioners. The balance of theory and practice as well as the depth of the subject matter of about 75 Chapters is undeniable.

Numerous advances in almost every ground improvement technique have been made in the past decade, and these 3 books together provide methodically compiled new case histories plus expanded or revised Chapters from the previous book where warranted. In this series, the first two books comprehensively cover the entire range of physical methods of ground improvement while the third book is a significant contribution covering chemical, thermal, electrokinetic and biological methods of ground improvement, which are rapidly developing multi-disciplinary fields.

The Ground Improvement Case Histories in this standalone three book compilation offer an outstanding package for practitioners, researchers, academics and students, culminating in the understanding of concepts and the delivery of competitive ground breaking solutions in almost every facet of ground improvement. Undoubtedly, it is indeed the only comprehensive compilation of Ground Improvement Case Histories available today and should be on the shelf of every geotechnical engineer.

Serge Varaksin

Conseiller Scientifique/Scientific advisor
Vice-Chairman TC-211 ISSMGE
APAGEO-ZA de Gomberville – rue Salvador Allende
78114 Magny les Hameaux – France

PREFACE

It has been almost ten years since the first book on Ground Improvement - Case Histories (Indraratna and Chu) was published by Elsevier. Since then there has been a substantial number of major projects and case studies reported in various parts of the world in all areas of Ground Improvement.

Ground improvement has been both a science and art, with significant developments observed through ancient history. From the use of straw as blended infill with soils for additional strength during the ancient Roman civilizations, and the use of lime mortar for the construction of the Great Wall in China, the concepts of using additives and reinforcement for soil improvement and the more recent use of electrokinetics and thermal modifications of soils have come a long way. The selection and use of the most appropriate ground improvement technique can not only improve considerably the design and performance of foundations and earth structures, including embankments, cut slopes, roads, railways and tailings dams, but also save costs. Ground improvement works have become increasingly challenging when more and more problematic soils and marginal land have to be utilized for infrastructure development.

The third Book of this Ground Improvement Series, ***Chemical, Electrokinetic, Thermal and Bioengineering***, is divided into 3 major themes:

1. Chemical Admixtures, Grouting and Deep Mixing (12 Chapters)
2. Electrokinetic, Thermal and Explosion Based Techniques (6 Chapters)
3. Bio-Engineering (4 Chapters)

This Book contains a collection of specialised chapters written by invited experts from both academia and practice in various areas of ground improvement, illustrating the basic concepts and the applications of different ground improvement techniques via real-life projects. Case histories from many countries ranging from Asia, America, Australia and Europe are addressed insightfully and in detail. As a whole, different soils or ground materials are discussed by about 60 authors and co-authors including slurry, soft clays, sands, tailings and waste materials. Various ground improvement techniques that seek to enhance the engineering properties of soil to achieve the desired project goals are also presented. Most chapters provide an overview of the specific technology followed by applications, and in some cases, comprehensive back analysis through numerical modelling is also presented.

Environmental and economic advantages are also addressed as warranted. An extensive list of references is provided at the end of each chapter.

We are most grateful for all authors and co-authors whose efforts when put together have produced this comprehensive volume of well-described case histories, which reflect a balanced international expert view on ground improvement. The support and inspiration received from the members of ISSMGE-TC211 over many years is gratefully acknowledged. In particular, the Editors wish to thank Serge Varaksin, Prof Sven Hansbo, Vasantha Wijeyakulasuriya, Prof Dennes Bergado, Prof A. S. Balasubramaniam, Dr Myint Win Bo, Prof John Carter, Prof Robert Lo, Geoff McIntosh, Prof Shuwang Yan, Prof Hanlong Liu, Prof Serge Leroueil, Prof Harry Poulos, Dr Richard Kelly, Dr Jay Amaratunga, Babak Hamidi, Prof Norbert Morgenstern, Prof Sarah Springman, Prof Jean-Louis Briaud, Prof Anand Puppala, Prof Pedro Pinto, Prof Marcio Almeida, Prof Colin Leung, Prof Maosong Huang, Prof Roger Frank, Prof Noël Huybrechts, Prof Kerry Rowe and Dr Venu Raju among others for their support, technical suggestions and encouragement on various occasions.

The dedicated help of Mr. Ken McCombs (Elsevier) for his enthusiastic assistance during publishing of this Ground Improvement Case Histories Series is appreciated. The assistance of Bill Clayton and Manori Indraratna during the final copy editing and proof reading of various Chapters is also appreciated.

Buddhima Indraratna

Jian Chu

Cholachat Rujikiatkamjorn

PART ONE

Chemical Admixtures, Grouting, and Deep Mixing

CHAPTER 1

Cement/Lime Mixing Ground Improvement for Road Construction on Soft Ground

J.-C. Chai[1], N. Miura[2]
[1]Institute of Lowland Technology, Saga University, Saga, Japan
[2]Institute of Soft Ground Engineering, Co. Ltd., Saga, Japan

1.1 INTRODUCTION

In lowland areas of soft clay deposits, to reduce embankment load-induced settlement and construction cost, normally low embankment (<3.0 m height) road is constructed (Fujikawa et al., 1996; Miura et al., 2000; Miura, 2002). However, the low embankment road can be easily affected by traffic load, causing the differential settlement and the cracking of pavement. Then periodic repairing (overlayering) is required during the lifetime of the road. This kind of periodical repairing activity influences not only the traffic but also the costs.

There are some options for reducing the traffic-load-induced settlement, differential settlement, and the cracking of pavement—namely cement/lime mixing improvement of soft clayey subsoil, reinforcing the base course and/or subgrade of a road, using light-weighted fill material to reduce embankment load, and combinations of these methods. Practically, to select an improvement method, not only technical availability but also economic feasibility must be considered. To find an economic and effective way of constructing low embankment roads on soft subsoil, a test section approximately 1 km long with different ground improvement techniques was constructed on a soft Ariake clay deposit in Saga, Japan, and monitored for a period of more than 1 year. This chapter first reports the field conditions, ground improvement construction, and the field monitoring results. Then, the methods for predicting the traffic-load-induced permanent settlement of a low embankment road are briefly reviewed, and the predicted values by a recommended method are compared with the field data of the test road. Finally, the concept of total cost (construction cost and maintenance cost) is introduced, and some possible ways to reduce the total cost are briefly discussed.

Chemical, Electrokinetic, Thermal, and Bioengineering Methods
http://dx.doi.org/10.1016/B978-0-08-100191-2.00001-0

1.2 SUBSOIL CONDITION AND TEST SECTION CONSTRUCTION

The test section was located at Kawazoe Machi, Saga, Japan. At the site, nine piezocone penetration tests and three boreholes were made within 800 m length to investigate the subsoil conditions. The measured cone tip resistances (q_t) are given in Fig. 1.1 (Miura, 2002). The estimated soil strata are also indicated in Fig. 1.1. The total thickness of soft deposit is approximately 20 m, and it can be subdivided into five alluvial clay layers (A_{c1}–A_{c5}) and two thin alluvial sand layers (A_{s1} and A_{s2}). The softest layer is located at approximately 0 m elevation with a q_t value of 0.1–0.2 MPa. On the basis of the piezocone test results, the subsoil has been divided into four zones, I–IV, as indicated in Fig. 1.1. Zones I and IV are similar where sand layer A_{s1} is thin. In zone II, there is no A_{c3} layer, and the sand layer A_{s1} is relatively thicker. In zone III, the thickness of A_{s1} layer varies from that of zones II and IV. Generally, zones II and III are stronger. The three boreholes were located at No. 1, No. 6, and No. 8 locations as shown in Fig. 1.1. The soil properties from laboratory tests with soil samples recovered from the three boreholes are summarized in Fig. 1.2 (Miura, 2002). The natural water contents of the clay were 80–130% and slightly higher than the corresponding liquid limits. Within approximately 10 m depth, the compression index (C_c) of the soil was generally 1.0–2.0. The groundwater level was approximately 1.0 m below the ground surface.

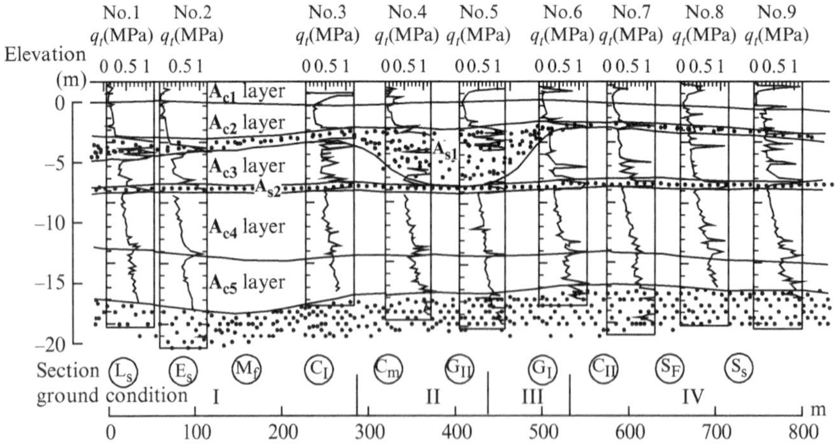

Figure 1.1 Estimated soil profile based on piezocone test results. *(Source: After Miura, 2002; reproduced with the permission of the International Association of Lowland Technology).*

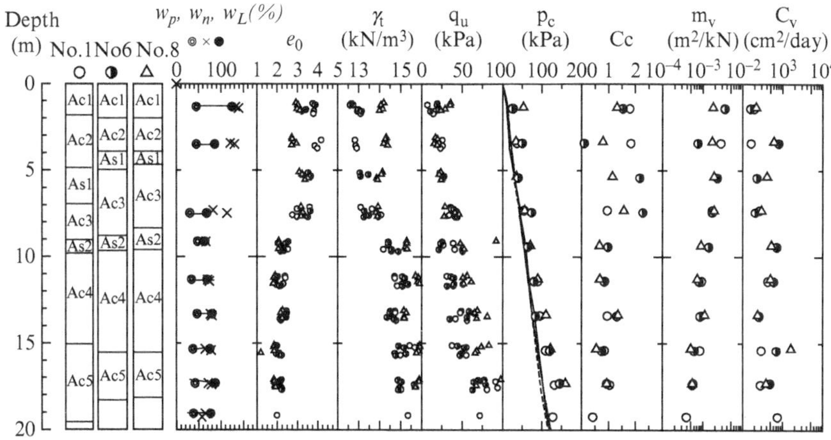

Figure 1.2 Soil properties at test section site. *(Source: After Miura, 2002; reproduced with the permission of the International Association of Lowland Technology).*

The approximately 1-km-long test section had 10 subsections. Each subsection had a different ground improvement method. The location of each subsection is indicated in Fig. 1.1, and the cross section of each subsection is illustrated in Fig. 1.3 (Miura, 2002). The Ascon in the figure means asphalt–concrete. The amounts of admixtures added and the resulting unconfined compressive strengths (q_u) from laboratory tests using the field samples are summarized in Table 1.1.

In M_f subsection, 0.3-m-thick base course was treated by air-bubbled cement (lightweight with a resulting unit weight of 13.1 kN/m^3) and

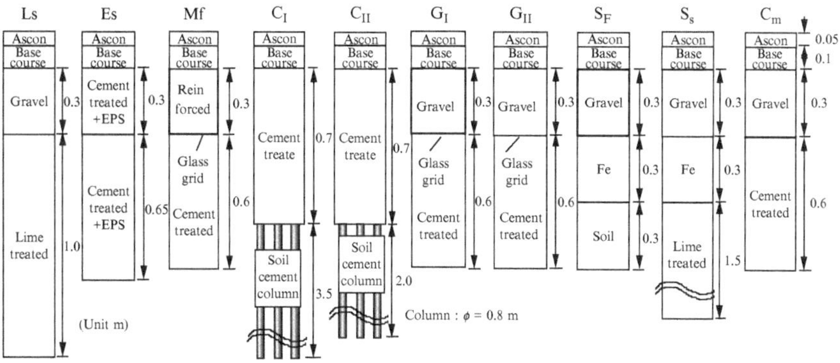

Figure 1.3 Subgrade and base course structures of 10 kinds of test road. *(Source: After Miura, 2002; reproduced with the permission of the International Association of Lowland Technology).*

Table 1.1 Summary of ground improvement methods

Subsection symbol	Subsection name	Base/ subgrade	Thickness (m)	Amount of admixture used (kN/m^3)	Unconfined compressive strength, q_u (kPa)
L_s	Lime surface treatment	Subgrade	1.00	0.96	640
E_s	Light-weighted fill material[a] and cement mixing	Base	0.30	2.26	310
		Subgrade	0.65	2.26	110
M_f	Multifunctional treatment	Base	0.30	3.38	1750
		Subgrade	0.60	1.10	1980
C_I	Soil–cement column–slab method I	Base	0.70	1.37	2440
		Subgrade	3.5 column	2.00 column	2390
C_{II}	Soil–cement column–slab method II	Base	0.7	1.37	1400
		Subgrade	2.0	1.40	1810
G_I	Geogrid + cement treatment method I	Subgrade	0.6	1.10	580
G_{II}	Geogrid + cement treatment method II	Subgrade	0.6	1.10	580
S_F	Special lime treatment + decomposed granite replacement	Subgrade	0.3	—	680
		Subgrade	0.3 decomposed granite replacement		
S_s	Special lime treatment	Subgrade I	0.3	—	720
		Subgrade II	1.5	0.69	320
C_m	Cement surface treatment	Subgrade	0.6	1.10	1410

[a]Expanded polystyrol (EPS) beads.

reinforced by a geogrid. The layout of the geogrid is illustrated in Fig. 1.3. The strength of the geogrid was 98 kN/m, and failure strain was 5%. The nominal grid size was 25 × 25 mm, and the unit weight was 170 g/m^2. Then a 0.6-m-thick subgrade was treated with cement, and the amount of cement used was 1.10 kN/m^3. For two soil–cement column–slab subsections (C_I and C_{II}), the diameter of the column was 0.8 m and the area replacement ratio of the soil–cement column was 36%. The difference between those two geogrid plus the cement treatment subsections (G_I and G_{II}) results in the layout of the geogrid. In G_n, the two ends of the geogrid were connected, but in G_I they were not. In subsection E_s, expanded polystyrol (EPS) beads were mixed in a 0.95-m-thick layer, and the resulting unit weight of the treated layer was 14.9–15.1 kN/m^3. The methods for constructing surface cement–lime mixing and soil–cement column are as follows:

Surface mixing method: There are two methods of mixing admixture with surface soil, namely dry mixing and wet mixing. The machine used for dry mixing was modified from a backhoe. The modification was made by replacing the normal bucket of a backhoe with one having several open slits as shown in Fig. 1.4. In the case of wet mixing, the admixture and the soil were mixed in a mixer and poured on the ground surface. The field tests (two test yards with a length of 5 m, width of 3 m, and thickness of 0.5 m) showed that the scatter of the strength of the samples from dry mixing yard was larger than that from wet mixing. However, the average strength of the dry mixed sample was higher, and dry mixing was easier

Figure 1.4 Modified backhoe bucket.

to operate in the field (Fujikawa, 1996). Therefore, for the test road, the dry mixing method was adopted.

Soil–cement column construction method: The soil–cement column was constructed by Teno-column method. The method of mixing cement slurry with clayey soil by a mixer is shown in Fig. 1.5. The mixer consists of a mixing wing and a free wing. The function of the free wing is to prevent the co-rotation of the soil with the mixing wing and achieve a better mixing.

Average embankment thickness of the test section was approximately 0.75 m (including pavement). The total width of the test section was 10 m, in which 7.5 m was for traffic and 2.5 m for sidewalk. Figure 1.6 shows a typical cross section of the road.

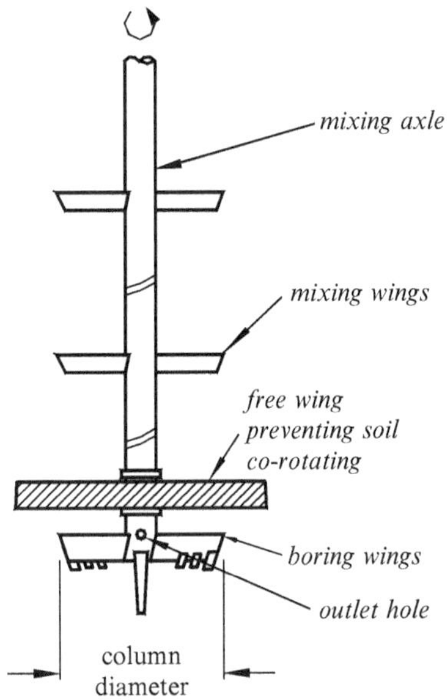

Figure 1.5 Sketch of Teno-column mixer. *(Source: After Shen, 1998; reproduced with the permission of S.-L. Shen).*

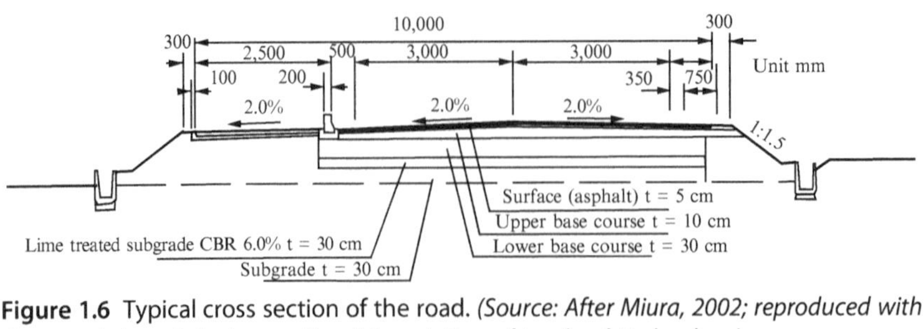

Figure 1.6 Typical cross section of the road. *(Source: After Miura, 2002; reproduced with the permission of the International Association of Lowland Technology).*

1.3 PERFORMANCE EVALUATIONS

1.3.1 Surface settlement

The settlement was measured after the completion of the road, and the settlement during the construction was not monitored. Approximately 100 days after completion of the construction, the road was opened to traffic. Figures 1.7(a) and 1.7(b) show the measured settlement variation of each subsection. The field data are from the Civil Engineering Department, Saga Prefecture, Japan (1996). Figure 1.7(a) shows the data of the relative stronger subsoil zones (zones II and III), and Fig. 1.7(b) shows the data of the relative weaker subsoil zones (zones I and IV). It can be seen that the subsoil condition had an obvious effect on settlement, and zones II and III had a smaller settlement than zones I and IV. Note that before the road opened to traffic, in all subsections except subsections S_F and S_s, the embankment (including pavement) load-induced consolidation settlement was almost finished. The reason why the embankment load-induced settlement for S_F and S_s was not finished when the road opened to traffic is not clear. In later discussion, we consider that the settlement after the road opened to traffic was the traffic-load-induced settlement.

Comparing lime treatment with cement treatment: Subsections L_s, S_F, and S_s were treated with lime, and the improvement depth varied from 0.9 to 2.1 m. These three subsections were all in weaker subsoil zones—subsection L_s in zone I and subsections S_F and S_s in zone IV. Compared with other subsections in zones I and IV that have the same or less improvement depth (E_s and M_f), the settlements of lime-treated subsections are larger (Fig. 1.7b). Therefore, generally, the lime treatment was not as effective as cement treatment for soft Ariake clay. However, when comparing C_{II} (in zone IV with cement treatment and improvement depth of 2.7 m) with L_s (in zone I with lime treatment and improvement depth of

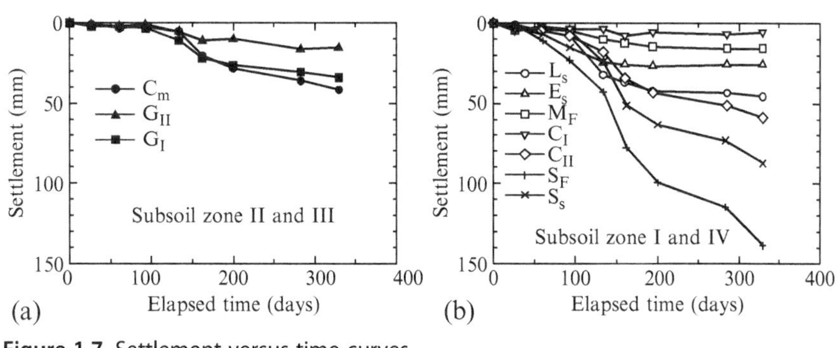

Figure 1.7 Settlement versus time curves.

1.3 m), they had almost the same amount of settlement (Fig. 1.7b). This means that not only the admixtures used but also the quality of treatment is an important factor.

Effect of improvement depth: C_I and C_{II} used the similar improvement method, but C_I had a deeper improvement depth and smaller settlement (Fig. 1.7b). The major difference between C_I and C_{II} was the settlement after opened for traffic (i.e., the traffic-load-induced settlement). Subsection C_{II} might have a poor improvement quality as discussed previously. Comparison of C_I (4.2-m improvement depth) and C_m (0.9-m improvement depth) emphasizes the effect of improvement depth (Figs. 1.7a and 1.7b). Also, S_F and S_s had a similar improvement technique, but S_s had a deeper improvement depth and a smaller settlement (Fig. 1.7b). Normally, traffic load has a limited influence depth (Fujikawa et al., 1996; Chai and Miura, 2002; Miura, 2002). If the improvement depth exceeds this influence depth, the traffic-load-induced settlement can be substantially reduced. This is discussed in detail later.

Effect of strength of base course/subgrade: The importance of the strength/stiffness of the improved layer was already mentioned when comparing C_I and C_{II} and also C_{II} and L_s. This point can be further demonstrated by comparing the subsections M_f and G_I. These two subsections had the similar subgrade improvement method and improvement depth. G_I was in the relatively stronger subsoil zone, zone III, but the improved subgrade had an unconfined compressive strength (q_u) of 580 kPa (Table 1.1). M_f is in zone I, and improved subgrade had a q_u value of 1980 kPa (Table 1.1). As a result, M_f had less settlement than G_I (Figs. 1.7a and 1.7b).

Effect of geogrid reinforcement: C_m, G_I, and G_{II} were in zones II and III and had the same improvement depth, but G_I and G_{II} had a geogrid reinforcement later and C_m did not. Also, C_m had higher subgrade strength than G_I and G_{II} (Table 1.1). However, G_I and G_{II} had less settlement, which is considered an effect of geogrid. The difference between G_I and G_{II} is that for G_{II}, the reinforcement was connected at two ends, but in the case of G_I it was not. G_{II} had a smaller settlement than G_I (Fig. 1.7b), but the subsoil of subsection G_{II} was stronger than that of subsection G_I. Thus, the effect of different layouts of reinforcement is inconclusive. Miura et al. (1990) suggested that a layer of geogrid can have an equivalent effect of increase in thickness of improved layer by 0.1 m.

Effect of light-weighted material: E_s and M_f were constructed with light-weighted materials. Subsection M_f also had a layer of geogrid reinforcement and a higher subgrade strength; therefore, the effect of light-weighted material cannot be identified. For E_s, EPS beads were mixed with a 0.95-m-thick layer below the base course. Although the strength of the treated layer was relatively lower (Table 1.1), the settlement was relatively smaller, and it is considered as an effect of light-weighted material. The settlement curve of E_s shows that after being opened to traffic, the settlement increment rate is relatively higher, which may indicate that light-weighted material can reduce embankment load-induced settlement but has an insignificant effect on the traffic-load-induced settlement.

1.3.2 Deflection measurement results

Benkelman beam tests (BBTs) and falling weight deflection (FWD) tests were conducted to evaluate the deflection behavior of each subsection. The tests were conducted twice—immediately after construction and 1 year after construction. The results are depicted in Figs. 1.8 and 1.9 for BBT and FWD, respectively (Miura, 2002). In Fig. 1.8, D_0 is the deflection directly under the loading point, and D_{150} is the deflection at a point 1.5 m away from the loading point in the direction along the road. D_B in Fig. 1.9 is the calculated deflection at the base of subgrade directly under the loading point. The calculation was made under the assumption that the load was transmitted from the loading plate at the surface to the base of subgrade with a spreading angle of 45° (Miura, 2002). For FWD test, with a load spread angle of 45°, the deflection at x m from the edge of the loading plate was mainly influenced by the properties of the soil layers below x m depth rather than the layers above. Let us denote the thickness of base course and subgrade as H_b. It is further assumed that the value of D_B will be the surface deflection at distance H_B from the edge of the loading plate (Fujikawa, 1996).

Generally, tendency is the same as settlement measurement. The thicker the improved layer and higher the strength/stiffness of the base course and the subgrade, the smaller the deflection. C_I and M_f subsections had smaller settlements (Fig. 1.7b) and smaller deflections (Figs. 1.8 and 1.9). However, there are some contradictions with regard to settlement measurement. Subsections C_{II} and S_s had a relatively larger settlement (Fig. 1.7b) but a

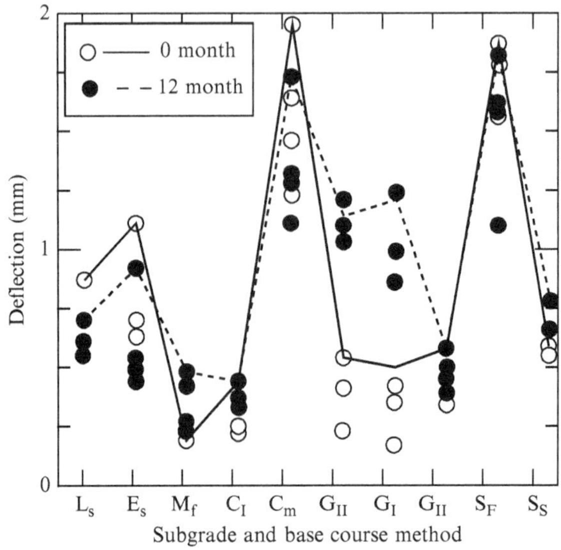

Figure 1.8 Deflection from Benkelman beam test. *(Source: After Miura, 2002; reproduced with the permission of the International Association of Lowland Technology).*

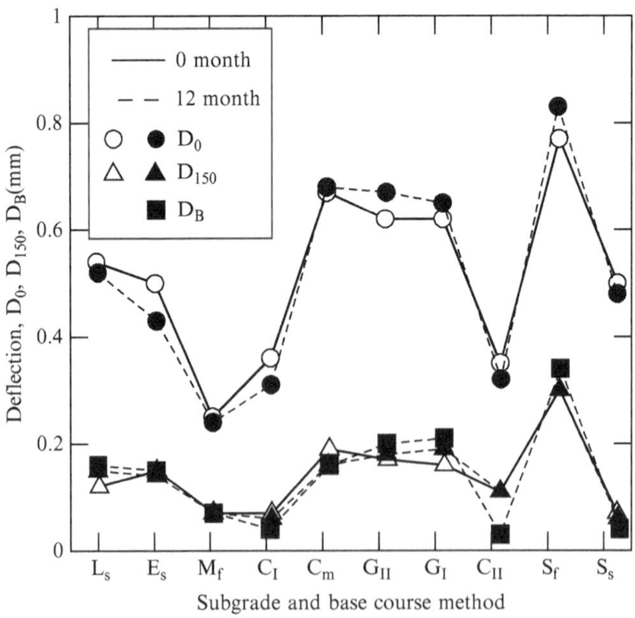

Figure 1.9 Results of falling weight diflectometer (FWD) test. *(Source: After Miura, 2002; reproduced with the permission of the International Association of Lowland Technology).*

smaller deflection (Figs. 1.8 and 1.9). The reason for this is not clear. Both C_{II} and S_s had a relatively thicker improved layer, and it might contribute to the smaller deflection measurement. From Fig. 1.9, the fact that subsections C_I, C_{II}, and S_s had smaller deflections at the base of subgrade (D_B) supports this argument. Another point is that subsection C_m had a modest settlement, which is consistent with FWD measurement but different from the BBT result. BBT resulted in the largest deflection of subsection C_m; the reason is not clear.

Comparing the two measurements—immediately after construction and 1 year after construction—there was no significant difference between the two except for subsections G_I and G_{II} in the case of BBT results. For G_I and G_{II}, BBT results indicate that 1 year after construction, the measurement was approximately three times that immediately after construction. Again, the reason is not clear; it is possibly due to the deterioration of reinforcement effect, such as creep deformation of the geogrid, etc.

Based on the previous discussion, it can be stated that both the improvement depth and the strength/stiffness of the improved layer have a significant influence on the traffic-load-induced settlement and deflection of a road. The test results show that an improvement depth of more than 1.3 m from the bottom of the base course is required to substantially reduce the traffic-load-induced settlement and the deflection of a road. If the improvement depth is approximately 1.0 m, an unconfined compressive strength of more than approximately 1000 kPa for the improved layer is desirable.

For low embankment road, the traffic-load-induced settlement is the main part of the post-construction settlement and therefore it controls the maintenance cost. In the following section, the methods for predicting the traffic-load-induced settlement are discussed, and some subsections of the test road reported in this chapter are analyzed.

1.4 PREDICTING THE TRAFFIC-LOAD-INDUCED PERMANENT SETTLEMENT

1.4.1 Brief review on existing methods

Several factors affect the traffic-load-induced deformation, namely (1) the strength and deformation characteristics of soft subsoil; (2) the properties of pavement, base course, and subgrade of a road; and (3) the magnitude and number of applications of traffic load. Any useful settlement prediction method should consider the previously mentioned factors directly or indirectly. The existing methods can be divided into three groups: numerical methods, equivalent static loading methods, and empirical equations.

Hyodo et al.'s (1996) dynamic analysis method: Theoretically, explicit simulation of the response of subsoil under repeated load is preferable. A method proposed by Hyodo et al. (1996) combines dynamic numerical analysis (two-dimensional (2D)) and dynamic triaxial test results to predict the traffic-load-induced deformation. However, the methods of explicit simulation are difficult to use in engineering practices. The response of subsoil under traffic load is a three-dimensional (3D) problem, and the number of load applications is usually extremely large.

Fujikawa et al.'s (1996) equivalent static load method: On the basis of field measurement, Fujikawa et al. (1996) proposed a method to estimate the distribution of the traffic-load-induced consolidation settlement in the subsoil. In Fujikawa et al.'s method, a triangular distribution of traffic-load-induced stress increments is assumed—that is, maximum at ground surface and linearly decreased to zero at a depth of influence. Fujikawa et al.'s method can calculate only the final settlement; the variation of the settlement with time cannot be predicted. Furthermore, the traffic-load-induced stress is not calculated explicitly; the properties of pavement, base course, and subgrade and the behavior of subsoil under repeated load are not considered.

Li and Selig's (1996) empirical method: A number of empirical equations have been proposed to predict the permanent deformation of cohesive soil under repeated load. Among them, the power equation proposed by Monismith et al. (1975) has been widely used. Li and Selig (1996) proposed a method to determine the constants in the power equation, in which the magnitude of traffic load applications and the strength of subsoil are directly included in the equation, and the physical state of subsoil is considered indirectly. Li and Selig (1998) showed some successful applications of the equation to predict the settlement of cohesive soils under train loading. The values of the constants suggested by Li and Selig seem to be applicable to compacted cohesive soils but may be less suitable for natural clay deposit. Also, the effect of initial static deviator stress in subsoil is not considered.

Chai and Miura's (2002) empirical method: Chai and Miura's (2002) method is a modification of Li and Selig's (1996) method. The main modifications are that it (1) considers the effect of initial deviator stress and (2) links the constant a in Li and Selig's (1996) equation to the compression index (C_c) of subsoil. This constant mainly controls the magnitude of the traffic-load-induced settlement. This method was used to analyze some subsections of the test road in Saga, Japan, and a brief description is given here.

The equation for calculating the cumulative plastic strain (ε_p) of soft cohesive soil under repeated loading is as follows:

$$\varepsilon_p = a \left(\frac{q_d}{q_f} \right)^m \left(1 + \frac{q_s}{q_f} \right)^n N^b \tag{1.1}$$

where q_d is the traffic-load-induced dynamic deviator stress; q_f is the static failure deviator stress of soil; N is the number of repeated load applications; q_s is the initial static deviator stress; and a, b, m, and n are the constants. The methods for determining the variables and constants in Eq. (1.1) are as follows:

Dynamic deviator stress q_d: To estimate the value of q_d, the 3D traffic load transfer mechanism and the characteristics of multilayer foundation must be considered. Burmister's (1945) multilayer elastic solution is considered suitable for this purpose. Using numerical techniques, the nonlinearity of subsoil behavior can be considered. In calculation, equivalent uniform distributed loads over circular areas simulate the tire loads of a truck. A relationship between the magnitude of the tire load and the radius of the equivalent loading area reported by Uchida (1988) is recommended (Fig. 1.10). If there are no measured data, the traffic load amplification factor of 1.0 is suggested (considering the traffic load as a static load).

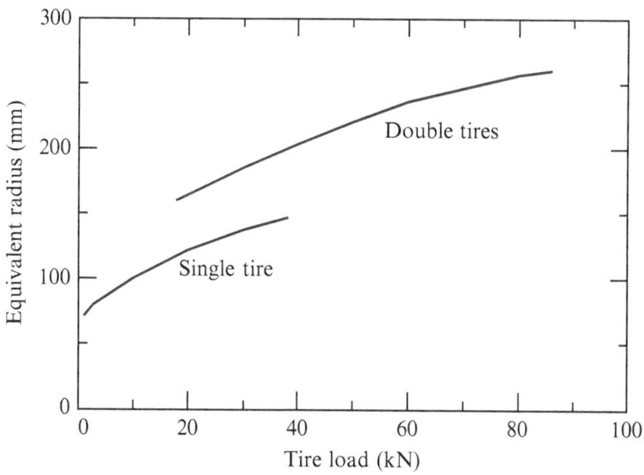

Figure 1.10 Relationship between the magnitude of the tire load and the equivalent radius of loading area. *(Source: Data from Uchida, 1988).*

Static failure deviator stress q_f: $q_f = 2S_u$, where S_u is the undrained shear strength of soft subsoil and can be measured by field vane shear test or determined by the following empirical equation (Ladd, 1991):

$$S_u = S\sigma'_v(\text{OCR})^{m1} \tag{1.2}$$

where σ'_v is the effective vertical stress, OCR is the overconsolidation ratio, and S and $m1$ are constants.

Initial deviator stress q_s: After a road embankment construction, the static stress distribution in subsoil can be calculated by one of the following two methods:

2D finite element (FE) analysis

Hand calculation: After embankment construction (assume the consolidation is finished), the vertical stress in the subsoil is calculated as follows:

$$\sigma'_v = \sum \gamma_i h_i + \Delta\sigma'_v \tag{1.3}$$

where $\Sigma\gamma_i h_i$ is the effective stress due to gravitational force of subsoil (below water level, buoyancy unit weight should be used), and $\Delta\sigma'_v$ is the embankment-load-induced stress increment, which can be estimated by Osterberg's (1957) method. For calculating the horizontal stress, the following equation is suggested to estimate the horizontal earth pressure coefficient, K_o (Mayne and Kulhawy, 1982):

$$K_o = (1 - \sin\ \phi')(\text{OCR})^{\sin\phi'} \tag{1.4}$$

where ϕ' is the the effective stress internal friction angle of subsoil.

Constants b and m: Currently, the values suggested by Li and Selig (1996) are recommended (Table 1.2). The parameter b controls the incremental rate of plastic strain with the number of repeated load applications. Li and Selig showed that b is not sensitive to the magnitude of dynamic deviator stress. It is mainly influenced by soil type. The parameter m influences both the magnitude and the distribution of plastic strain with depth. Because q_d/q_f is < 1.0 (and not a failure problem), the larger the m value, the faster the decrease in ε_p with depth.

Constant a: The parameter a influences the magnitude of plastic strain. The traffic-load-induced deformation mainly consists of two parts, namely dynamic consolidation and shear deformation. The amount of dynamic consolidation deformation is directly related to the compression index (C_c) of soil. Also, C_c is one of the parameters affecting the

Table 1.2 Values of constants *a*, *b*, and *m* (suggested by Li and Selig, 1996)

Soil Type	*a*	*b*	*m*
CH (high plasticity clay)	1.2	0.18	2.4
CL (low plasticity clay)	1.1	0.16	2.0
MH (elastic silt)	0.84	0.13	2.0
ML (silt)	0.64	0.10	1.7

magnitude of shear deformation. Therefore, it is rational to relate a with the compression index (C_c) of subsoil:

$$a = \alpha C_c \tag{1.5}$$

where α *is* a constant. On the basis of back-calculated results, Chai and Miura (2002) proposed that $\alpha = 8.0$.

Constant n: n is a parameter controlling the degree of the effect of initial deviator stress. Chai and Miura (2002) suggested that $n = 1.0$.

1.4.2 Comparing predicted settlements with the field measurements

In the analyses, only trucks were considered. This is because in comparison, the weight of passenger cars is 1/10 to 1/30 of that of heavy trucks. The effect of a passenger car on the permanent deformation of a road may be approximately 1/100 to 1/900 of that of a heavy truck. Another factor is that there are many types of trucks, and the number of axles and the distance between the axles are different. For simplicity, a load distribution pattern as shown in Fig. 1.11 was adopted. According to Japanese standard, the rear axle supports 80% and the front axle supports 20% of the total load (Uchida, 1988). The adopted values of constants in Eq. (1.1) were $\alpha = 8.0$, $b = 0.18$, $m = 2.0$, and $n = 1.0$ (Chai and Miura, 2002).

Among 10 subsections, 4 (C_I, C_m, M_f, and G_I) were analyzed to investigate the effect of improvement depth (comparing C_I and C_m) and the strength/stiffness of subgrade (comparing M_f and G_I):

Traffic intensity and the parameters adopted: The total traffic intensity was approximately 2500 cars/day (two-way), most of which were passenger cars. Trucks comprised approximately 6% of the total traffic (150 trucks/day) (Chai and Miura, 2002). Because the road is relatively narrow, the two-way traffic was used for calculating the traffic-load-induced permanent deformation. The trucks that passed this road were mainly light-weighted ones, and heavy trucks were very rare. In the analysis, it was assumed that the average weight of the truck was 100 kN/truck and

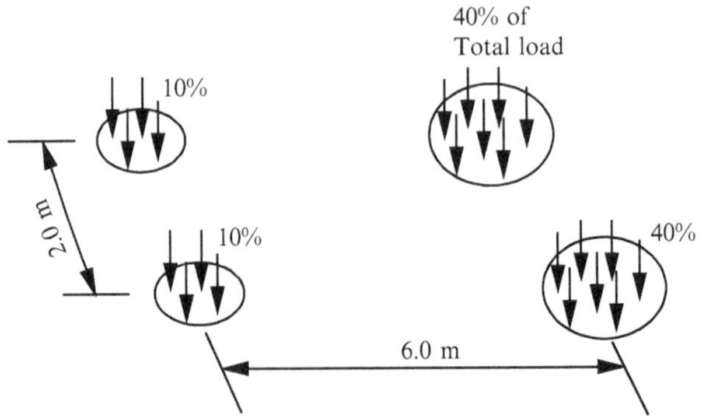

Figure 1.11 Load distribution pattern of a truck. *(Source: After Chai and Miura, 2002).*

equivalent radii of tire/road contact areas were 200 mm for back tires (double wheels) and 100 mm for front tires (Fig. 1.10). For analyzing the traffic-load-induced stress in subsoil, the adopted Young's modulus (E) and thickness (H) of each layer are summarized in Table 1.3. Poisson's ratios adopted were 0.2 for pavement and base course, 0.25 for subgrade, and 0.4 for soft subsoil. Modulus for cement-treated layers was estimated as 100 times the unconfined compressive strength (q_u) (Kitazume, 1996). Modulus for natural subsoil was assumed to be approximately 200 times the undrained shear strength (S_u) (Fujikawa, 1996). For the four subsections considered, the calculated dynamic deviator stresses were 3–5 kPa just below the subgrade, and they reduced to approximately 1.0 kPa at a depth of approximately 6 m.

Traffic-load-induced settlement: The calculated traffic-load-induced settlements are compared with measured data in Figs. 1.12 and 1.13. The calculation predicted the field data reasonably well. Note that for these four

Table 1.3 Thickness and Young's modulus of each subsoil layer of C_m, C_I, M_f, and G_I subsections[a]

	Pavement base course		Subgrade base course		Upper subsoil		Lower subsoil[b]	
Subsection	***H*** **(m)**	***E*** **(MPa)**	***H*** **(m)**	***E*** **(MPa)**	***H*** **(m)**	***E*** **(MPa)**	***H*** **(m)**	***E*** **(MPa)**
C_m	0.45	35	0.6	140	3.5	3	2.5	2.5
C_I	0.15	35	0.7	244	3.5	86.4	2.5	2.5
M_f	0.45	175	0.6	198	2.0	3	4.0	2.5
G_I	0.45	35	0.6	58	2.0	3	4.0	2.5

[a] H is the layer thickness, and E is the Young's modulus.
[b] Below the lower subsoil is a semi-infinite half-space.

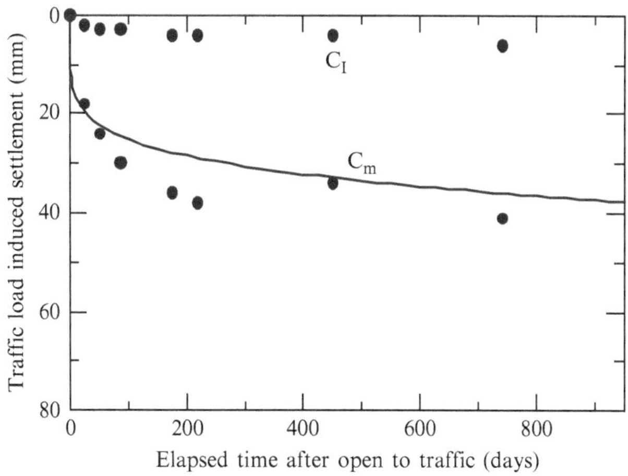

Figure 1.12 Comparing the traffic-load-induced settlement of subsections C_I and C_m. *(Source: Modified from Chai and Miura, 2002).*

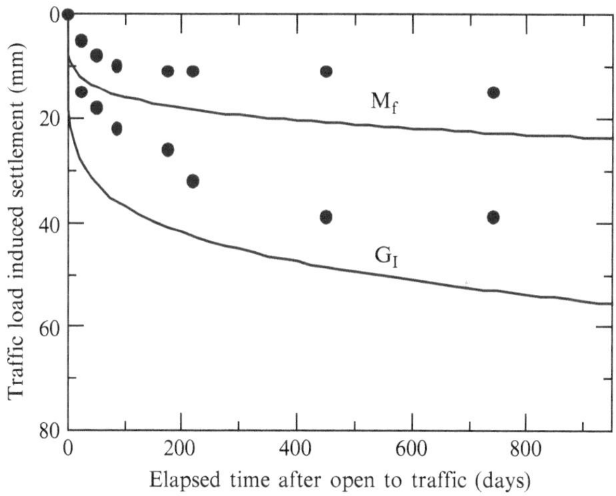

Figure 1.13 Comparing the traffic-load-induced settlement of subsections G_I and M_f. *(Source: Modified from Chai and Miura, 2002).*

subsections, we were able to get the measured data more than 2 years after the road opened to traffic. In Fig. 1.12, subsections C_m and C_I are compared, which had different depth of improvement. For C_I, soil–cement columns (36% replacement ratio by area) improved the upper subsoil. Dynamic deviator stress below the columns was small, and the traffic-load-induced settlement was much smaller than C_m. In Fig. 1.13, M_f and G_I are compared. These two subsections had a similar subgrade structure, but the strength and stiffness of the cement-treated layer were different; G_I had a lower strength and stiffness. Settlement of G_I was more than twice that of M_f. This case history indicates that an increase in the thickness and stiffness of improved subgrade is efficient for reducing the traffic-load-induced permanent settlement of soft subsoil.

Distribution of traffic-load-induced plastic strain with depth: The calculated plastic strain distribution with depth is depicted in Fig. 1.14 for 1 year after opening to traffic condition. It indicates that for the case considered, the significant influence depth of traffic load is approximately 6 m below the base of the embankment (pavement and base course). Also, the plastic strain reduced very quickly within the upper 2 m.

1.5 TOTAL COST OF LOW EMBANKMENT ROAD CONSTRUCTION ON SOFT SUBSOIL

1.5.1 Concept of total cost

In the case of a road, the total cost mainly consists of construction cost (initial cost) and maintenance cost. It is generally true that the initial cost and maintenance cost are reversely related—that is, higher initial cost will require less maintenance cost, and vice versa. The maintenance cost can be divided into direct and indirect costs. The direct cost is the cost for repairing the road, and the indirect cost is the effect of repairing the road on social activities. For a low embankment road on soft ground, the repair cost is directly related to

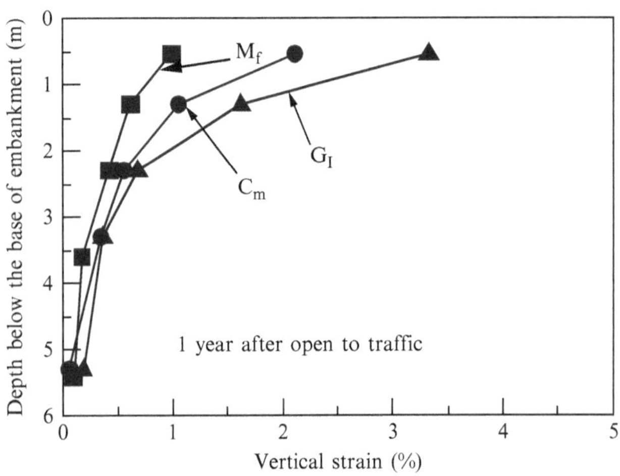

Figure 1.14 Calculated plastic strain distribution with depth. *(Source: Modified from Chai and Miura, 2002).*

residual settlement and can be divided into two types again—namely the cost for repairing the differential settlement and the cost for repairing the cracks and ruts:

Repairing of differential settlement: Along a road, the bridges and box culverts are generally supported by piles, which penetrate into stiffer layers and therefore have less settlement than the road section directly on soft deposit. The differential settlement will influence traffic and have to be repaired. Figure 1.15 shows an example of the differential settlement in Saga, Japan. In Japan, it is required that (1) when the different settlement reaches 30 mm, it must be repaired (Japan Road Association (JRA), 1983) and (2) the repaired road must satisfy the required slope for a vehicle speed at 60 km/h (JRA, 1984). Figure 1.16 shows the scheme/sequence of repairing.

Repairing the cracks and ruts: For a low embankment road on soft subsoil, if the subgrade is not strong enough, traffic load can cause cracks and ruts in pavement. These must be repaired by overlaying of a road with a layer of asphalt.

Figure 1.15 Example of differential settlement in Saga, Japan.

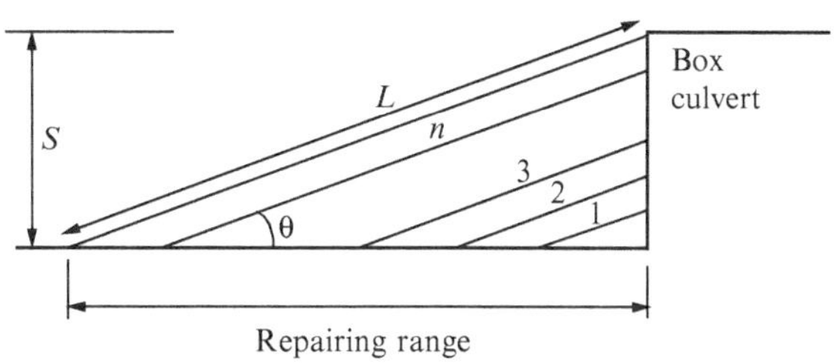

Figure 1.16 Repairing of differential settlement. *(Source: After Miura, 2002; reproduced with the permission of the International Association of Lowland Technology).*

As shown in Fig. 1.17, a lower initial cost implies a higher residual settlement and higher maintenance cost. Ideally, there is an optimum initial investment that will result in the lowest total cost.

The indirect cost includes the social influence of blocking a road for repairs and the uncomfortable feeling of driving on a road with differential settlement. However, it is not easy to evaluate quantitatively.

1.5.2 Traffic intensity definition in Japan and estimated direct repairing cost

In Japan, traffic intensity is divided into four groups: A- to D-traffic (Uchida, 1988). The traffic volumes are as follows:

A-traffic: 100–250 vehicles/day/one-way
B-traffic: 250–1000 vehicles/day/one-way
C-traffic: 1000–3000 vehicles/day/one-way
D-traffic: >3000 vehicles/day/one-way

Generally, for A- and B-traffic, the width of the road is approximately 7.0 m (one lane in each direction), and for C- and D-traffic, the width of the road is 24 m (two or three lanes in each direction).

To repair the different settlement between the crossroad structures (box culverts) and the adjacent road sections, there are different requirements and of course different costs. The strictest requirement is to satisfy the preferable

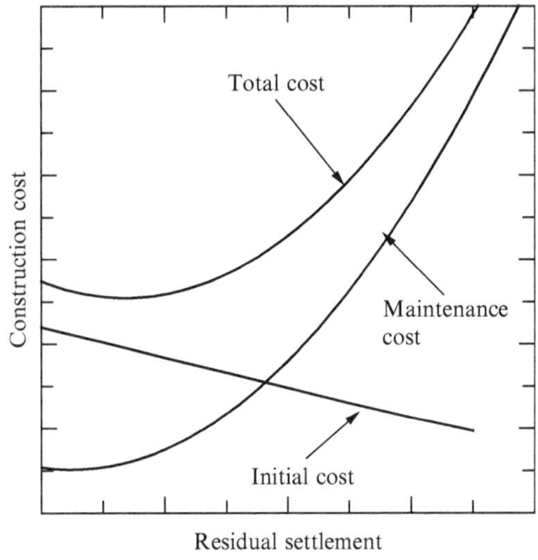

Figure 1.17 Concept of total cost. *(Source: After Miura, 2002; reproduced with the permission of the International Association of Lowland Technology).*

radius (slope) of curvature in longitudinal direction of a road with a value of θ in Fig. 1.16 of 0.64° for the speed of 60 km/h (this is called the best method). The next less strict requirement is to satisfy the minimum radius of curvature with a value of θ in Fig. 1.16 of 0.77° (will is called the second best method), and the least strict requirement is to have minimum overlay with a θ angle of 2.86° (this is called the conventional method). As mentioned previously, when the different settlement reaches 30 mm, repair will be carried out. Using a unit price of 1400 JP Yen/m^2/30 mm, which was the price in Japan in 1995 (Miura, 2002), and assuming that the residual settlement equals the differential settlement (no residual settlement for box culvert structure), Miura (2002) calculated the repair cost per box culvert as a function of residual settlement (Fig. 1.18). It can be seen that repair cost increased rapidly with the increase of residual settlement. Also, the difference between the best method and the second best method is not significant, and the difference between the conventional method and the best and the second best method is substantial.

1.5.3 An example calculation of total cost

For the 10 ground improvement methods adopted for the test section, the initial construction cost has been estimated as in Fig. 1.19 based on the actual price in Japan in 1995. The cost differences between C- and D-traffic and

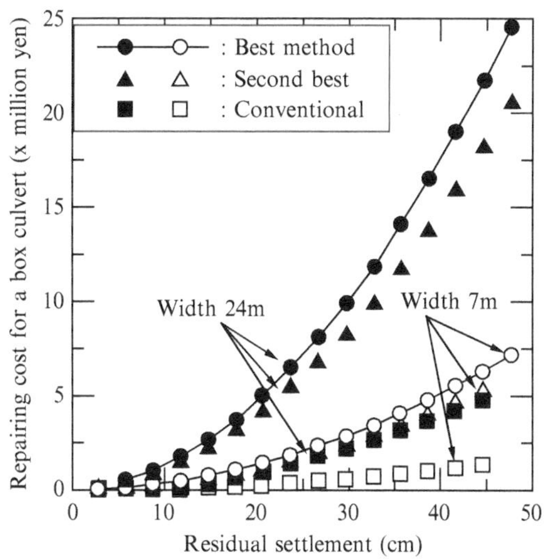

Figure 1.18 Repairing cost per box culvert structure. *(Source: After Miura, 2002; reproduced with the permission of the International Association of Lowland Technology).*

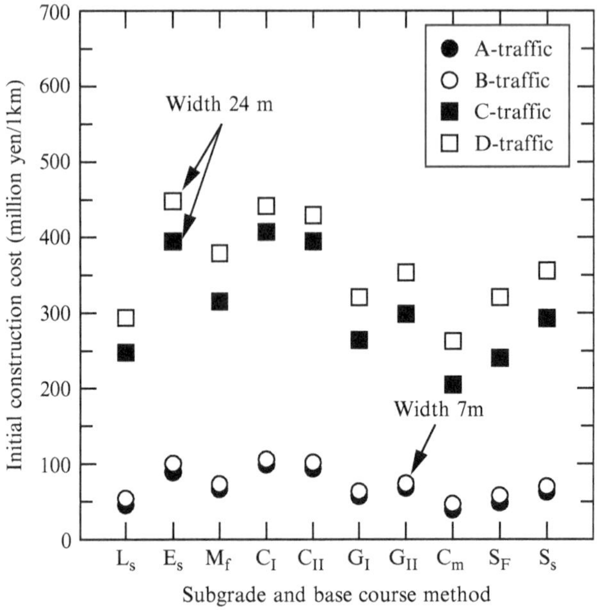

Figure 1.19 Initial construction cost for each method. *(Source: After Miura, 2002; reproduced with the permission of the International Association of Lowland Technology).*

between A- and B-traffic are due to the different requirement on pavement and base course structure (JRA, 1994). Considering ground improvement schemes C_I and C_m (highest and lowest initial construction costs as shown in Fig. 1.19), the initial construction cost for a 7-m-wide road (B-traffic) is approximately 100 and 50 million JP Yen/km, respectively.

Figure 1.12 shows that C_I had almost no residual settlement (traffic-load-induced) and C_m had approximately 40 mm. Considering a service time of 20 years, this differential settlement may reach approximately 50 mm or more. From Fig. 1.18, for a 7-m-wide road (the width of the test road was 7.5 m), the repair cost per box culvert is approximately 0.2 million JP Yen for a residual settlement of approximately 50 mm. Further assuming 10 box culverts per kilometer, the repair cost will be approximately 2 million JP Yen/km. For the test road, during an approximately 2-year period, no obvious ruts were observed and the overall overlay was not conducted and it was not considered to be performed in the future. Therefore, the total cost will be 100 and 52 million JP Yen for C_I and C_m ground improvement methods, respectively; thus, the C_m method is cheaper. Here, the social cost of blocking the road for repair is not considered.

Now let us consider a D-traffic and 24-m-wide road. From Fig. 1.19, the initial construction costs are approximately 260 million JP Yen/km for the C_m method and approximately 440 million JP Yen/km for the Q method. To obtain a realistic estimation of residual (or traffic-load-induced) settlement, two other case histories—Saga airport road and the test section at national road No. 34, Hyogo Machi, Saga, Japan—are briefly introduced.

Saga airport road

Saga airport road was constructed from 1990 to 1992 with a total length of 9 km. At the site, there is an approximately 20-m-thick, highly compressible and highly sensitive Ariake clay deposit. The total thickness of the road embankment was approximately 1.1 m, as shown in Fig. 1.20. The material for subgrade was decomposed granite (it is called *masado* in Japan). Although the lime (instead of cement) was used for stabilizing subgrade, generally the road structure was close to C_m of the test road. The total width of the road is 20 m, in which 11 m (two lanes) is for traffic and 4.5 m on each side is sidewalk.

Saga airport road is the main access road from Saga city to Saga airport. It was used for transporting construction materials (fill materials and others) during the airport construction. The traffic intensity of heavy trucks (200 kN/truck) was approximately 400 trucks/day (one-way). Settlements were monitored for more than 4 years at three monitoring stations. The measured residual (mostly the traffic-load-induced) settlement was 0.2–0.3 m (Chai and Miura, 2002). In addition to the repeated repairing of differential settlement between box culvert structures and adjacent road sections, there was an overall overlay approximately 2½ years after

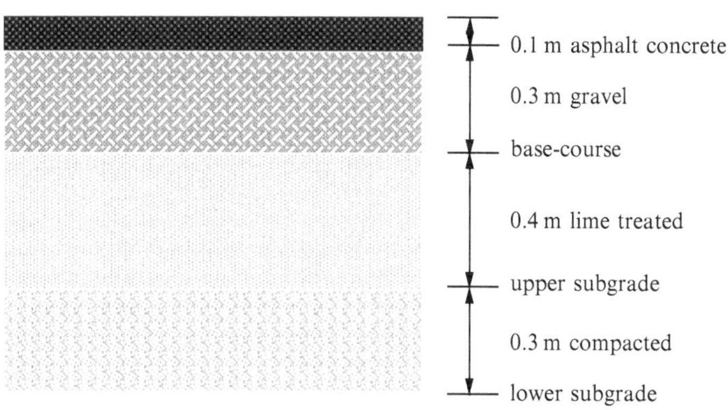

Figure 1.20 A typical cross section of Saga airport road.

construction was complete. If considering a service time of 20 years and a more intensive traffic condition, the residual settlement can be estimated to reach approximately 0.4 m or more.

Test section at national road No. 34, Japan

At Hyogo Machi, Saga, Japan, national road No. 34 was 12–14 m wide, and it had been decided to expand the road to four lanes. To further confirm the effect of some ground improvement methods, a test section 135 m long with three different ground improvement methods was constructed in 2001 and opened to traffic in March 2003. One subsection used the soil–cement slab–column system, and its cross section is shown in Fig. 1.21. The ground improvement scheme was close to Q of the test road. The area improvement ratio by soil–cement column was 12.6%. The amount of the cement used was 1.67 kN/m^3 of soil, and the designed unconfined compressive strength of the column was 600 kPa. At the site, the thickness of soft Ariake clay is approximately 10 m. Traffic intensity has been approximately 16,000 cars/day, including approximately 2800 trucks/day (one-way). One year after opening to traffic, the measured settlement (traffic-load-induced) was 4 mm (Motohara et al., 2004). This test section can be considered as maintenance-free.

Considering a road 24 m wide and residual settlement of 0.4 m (C_m ground improvement scheme in Saga area), the maintenance cost for

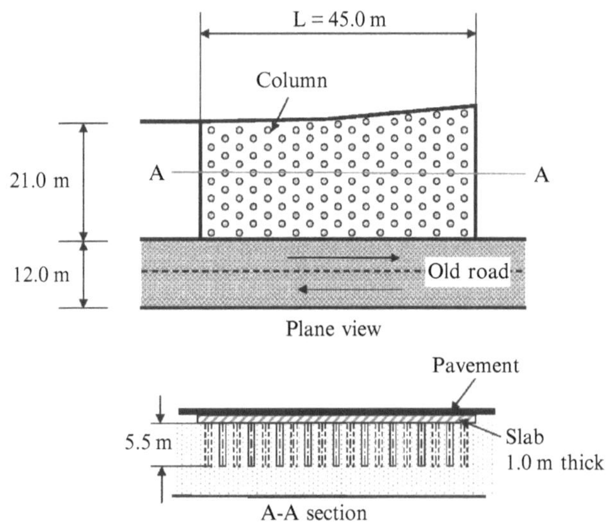

Figure 1.21 Plane and cross-sectional view of ground improvement scheme at Hyogo Mach, Saga, Japan.

differential settlement between box culverts (10 box culverts/km) and adjacent road sections will be approximately 150 million JP Yen/km (Fig. 1.18). The cost of just one overall overlay (30 mm thick) will be approximately 34 million JP Yen/km. The total maintenance cost will be 184 million JP Yen/km. This added to the initial construction cost of approximately 260 million JP Yen/km totals 444 million JP Yen/km. If we adopt a ground improvement scheme of C_I, it is almost maintenance-free and the total cost will be 440 million JP Yen/km. It is almost the same as the total cost of the ground improvement scheme of C_m. When considering the social impact of blocking the road for repair, the ground improvement scheme of C_I should be preferable.

1.6 CONCLUSION

A road test section approximately 1 km long on soft Ariake clay deposit with 10 different ground improvement methods was described in this chapter, and the field monitored results on settlements during a period of more than 1 year and BBT and FWD test results were reported. At the test site, the soft clayey layer was approximately 18 m, and the height of the road embankment was approximately 0.75 m (including the pavement). Both the settlement and the deflection test results indicate that an increase in the strength/stiffness of the base course and the subgrade and/or an increase in the improvement depth of the subgrade of a lower embankment road on soft subsoil can reduce the traffic-load-induced settlement and the deflection of the road significantly. Also, the field data indicate that cement treatment was more effective than lime treatment of soft Ariake clay, and a layer of geogrid in the base course had a positive effect on reducing deflection of the test road.

For a low embankment road, the traffic-load-induced settlement consists of a larger part of residual settlement and controls the maintenance cost of a road. The methods for predicting the traffic-load-induced settlement can be divided into three groups—namely numerical simulation methods, equivalent static load methods, and empirical equations. Comparison with the measurements of some subsections of the test road reported in this chapter shows that Chai and Miura's (2002) empirical equation, which considers the number of the traffic load application, the strength and compression index of soft subsoil, and the traffic-load-induced stress distribution in the ground, is a useful tool for predicting the traffic-load-induced settlement.

The concept of total cost (construction cost and maintenance cost) was introduced, and conceptually there is an optimum combination of initial construction cost and maintenance cost resulting in the lowest total cost. It was demonstrated that for a low embankment road on soft subsoil with intensive traffic (D-traffic in Japan), improving the subsoil by the soil–cement slab–column system to a depth of 4 or 5 m from the base of a road embankment is attractive in terms of the total cost.

REFERENCES

Burmister, D.M., 1945. The general theory of stresses and displacements in layered system I. J. Appl. Phys. 16, 89–94.

Chai, J.C., Miura, N., 2002. Traffic-load-induced permanent deformation of road on soft subsoil. J. Geotech. Geoenviron. Eng., ASCE 128 (11), 907–916.

Civil Engineering Department, Saga Prefecture, 1996. A Study on Predicting the Differential Settlement and Rational Design Method for Road on Soft Subsoil. Research Report, Saga Prefecture, Japan, p. 77, (in Japanese).

Fujikawa, K., 1996. On Optimistic Design of Low Embankment Road on Soft Subsoil by Considering the Traffic-Load-Induced Settlement. Dr. of Engineering Dissertation, Saga University, p. 199, (in Japanese).

Fujikawa, K., Miura, N., Beppu, I., 1996. Field investigation on the settlement of low embankment road due to traffic load and its prediction. Soils Found. J. Geotech. Soc. 36 (4), 147–153 (in Japanese).

Hyodo, M., Yasuhara, K., Murata, H., 1996. Deformation analysis of the soft clay foundation of low embankment road under traffic loading. In: Proc. 31st Symp. Japanese Soc. Soil Mech. Found. Eng, pp. 27–32 (in Japanese).

Japan Road Association (JRA), 1983. Road Maintenance and Repairing Instructions (in Japanese).

Japan Road Association (JRA), 1984. Road Structure Manual—Instructions and Practice, pp. 286–297 (in Japanese).

Japan Road Association (JRA), 1994. General Outline of Asphalt Pavement (in Japanese).

Kitazume, M., 1996. Deep mixing method. Found. Eng. Equip. 24 (7), 14–19 (in Japanese).

Ladd, C.C., 1991. Stability evaluation during stage construction. J. Geotech. Eng., ASCE 117 (4), 541–615.

Li, D., Selig, E.T., 1996. Cumulative plastic deformation for fine-grained subgrade soils. J. Geotech. Eng., ASCE 122 (12), 1006–1013.

Li, D., Selig, E.T., 1998. Method for railway track foundation design: II. Application. J. Geotech. Geoenviron. Eng., ASCE 124 (4), 316–322.

Mayne, P.W., Kulhawy, F.H., 1982. Ko-OCR relationships in soils. J. Geotech. Eng. Div., ASCE 108 (6), 851–872.

Miura, N., 2002. A field study concerning total cost of road construction on soft ground. In: Hayashi, S., Araki, H., Hokao, K. (Eds.), Proceedings of the International Symposium on Lowland Technology 2002. Institute of Lowland Technology, Saga University, Japan, pp. 1–10.

Miura, N., Fujikawa, K., Sakai, A., Hamatake, A., 2000. Consideration of the total costs for construction and maintenance of pavements on soft ground based on test road performance. J. Civ. Eng. Jpn. Soc. Civil Eng. 659 (III-52), 253–263, in Japanese.

Miura, N., Sakai, A., Taesiri, Y., Yamonouchi, T., Yasuhara, K., 1990. Polymer grid reinforced pavement on soft clay ground. Geotex and Geomembranes 9, 99–123.

Monismith, C.L., Ogawa, N., Freeme, C.R., 1975. Permanent deformation characteristics of subsoil due to repeated loading, Transp. Res. Rec. No. 537. Transportation Research Board, Washington, DC, pp. 1–17.

Motohara, K., Chai, J.-C., Miura, N., 2004. Prediction of traffic-load-induced permanent settlement of low embankment road on soft ground. In: Proc. 59th Annual Meeting Japanese Soc. Civil Eng, pp. 963–964, in Japanese.
Osterberg, J.O., 1957. Influence values for vertical stresses in semi-infinite mass due to embankment loading. In: Proc. 4th Int. Conf. Soil Mech. Found. Eng, 1.
Shen, S.-L., 1998. Behavior of Deep Mixing Columns in Composite Clay Ground. Dr. of Engineering dissertation, Saga University, p. 247.
Uchida, I., 1988. Road Engineering. Morikita, Tokyo (in Japanese).

CHAPTER 2

Full-Scale Tests on Stiffened Deep Cement Mixing Piles Including Three-Dimensional Finite Element Simulation

Pitthaya Jamsawang[1], Dennes T. Bergado[2], Panich Voottipruex[1]
[1]King Mongkut's University of Technology North Bangkok, Bangkok, Thailand
[2]Asian Institute of Technology, Bangkok, Thailand

2.1 INTRODUCTION

Deep cement mixing (DCM) pile is widely used to improve the engineering properties of thick deposits of soft ground. DCM piles can effectively reduce settlements of full-scale embankments (Bergado et al., 1999; Lai et al., 2006). However, they have variable strength and stiffness, specifically flexural strength (Petchgate et al., 2003, 2004), leading to low bearing capacity and large settlements (Wu et al., 2005). Consequently, DCM piles are not suitable for medium to high design loads (Dong et al., 2004). Liu et al. (2007) introduced geogrid-reinforced and cast-in-place concrete pile for supported embankment over soft clay. Dong et al. stated that concrete or cast-in-situ pile is deemed uneconomical as a friction pile for this purpose because much of the strength of pile materials has not been utilized when the low-capacity ground fails. The DCM pile can be subjected to both vertical and horizontal forces induced by the embankment loads.

Thus, the stiffened deep cement mixing (SDCM) pile is more suitable than DCM pile because SDCM pile has higher strength and stiffness and can sustain higher bending moment and resist higher lateral loads. SDCM pile employs a precast reinforced concrete pile inserted at the center of DCM pile. The concrete core, which is a precast concrete pile, takes most of the load and gradually transmits it to the surrounding soil–cement through the interface between the concrete core pile and the DCM pile. This chapter presents a series of full-scale tests consisting of SDCM piles and DCM piles in soft Bangkok clay under axial compression load, lateral load, and pullout interface. Moreover, a 5-m-high embankment on stiffened deep cement

Chemical, Electrokinetic, Thermal, and Bioengineering Methods
http://dx.doi.org/10.1016/B978-0-08-100191-2.00002-2

mixing pile was constructed on soft Bangkok clay improved with both SDCM and DCM piles in order to compare their performances under working stress level.

2.2 FULL-SCALE LOAD TESTS ON SDCM AND DCM PILES

Full-scale pile load test studies of soft clay foundation improved by SDCM piles and DCM piles installed by jet mixing method were performed. A series of the full-scale tests consisted of SDCM piles and DCM piles under axial compression load and lateral and pullout interface between concrete and deep cement mixing under axial tensile load. Tests of the SDCM piles and DCM piles under axial compression load and lateral load were conducted to study their load-bearing capacities and the effect of section area and length of concrete core on bearing capacities, whereas the pullout interface tests were conducted to study interface resistance between concrete and deep cement mixing.

2.2.1 Test site and subsurface investigation

The test site was located at the northern part of the Asian Institute of Technology (AIT) campus in Klong Luang, Pathumthani, Thailand. The soil profile and soil properties of the subsoil in the uppermost three layers at the AIT campus are presented in Fig. 2.1. The uppermost 10 m can be divided into four layers. The weathered crust forms the uppermost layer consisting of weathered clay, with the uppermost 2.0 m underlain by a soft

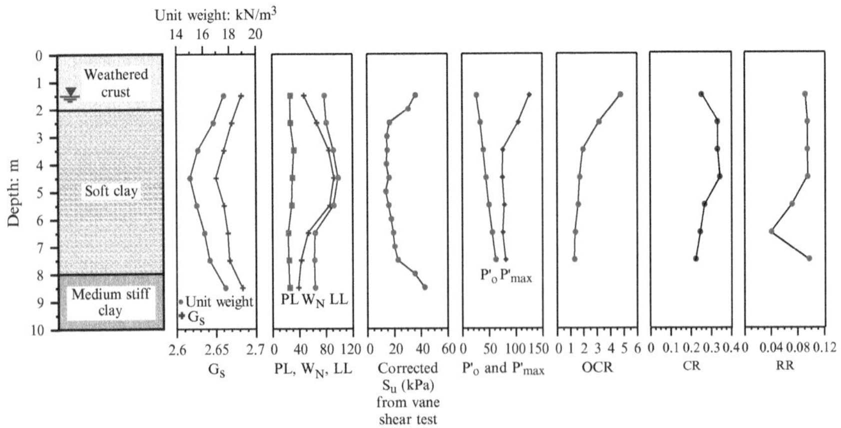

Figure 2.1 Subsoil profile and relevant parameters.

clay layer down to approximately 6-m depth. The medium stiff clay layer was found at 8- to 10-m depth. Below this layer is the stiff clay layer. The undrained shear strength obtained from field vane test of the soft clay was 16 or 17 kPa. Underlying the soft clay layer is the medium stiff clay layer, with a strength of more than 30 kPa.

2.2.2 Concrete core pile

The prestressed concrete pile was proposed to bear compression and lateral loads. The prestressed concrete pile was selected as the stiff core because it has high strength and stiffness, and it is also cheaper than steel pile. The selected prestressed concrete piles were square shaped with dimension of 0.18 and 0.22 m in cross section and 4.0 m and 6.0 m in length, as shown in Fig. 2.2.

2.2.3 Installation of deep mixing piles by jet grouting

A total of 27 deep mixing piles were installed down to the bottom of the soft clay layer, as shown in Fig. 2.3, with 2.0-m spacing in a square pattern by jet

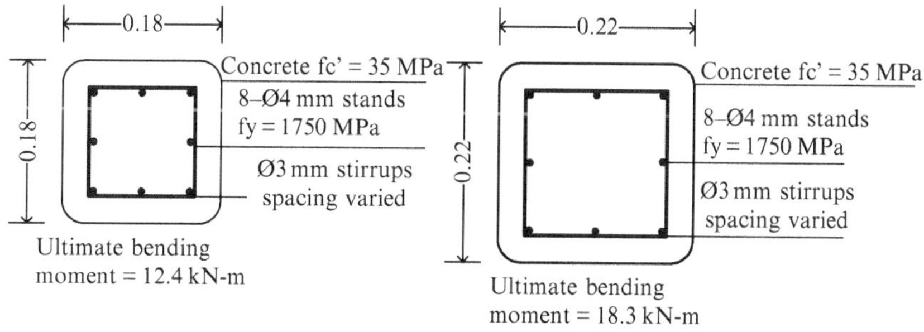

Figure 2.2 Details of prestressed concrete piles.

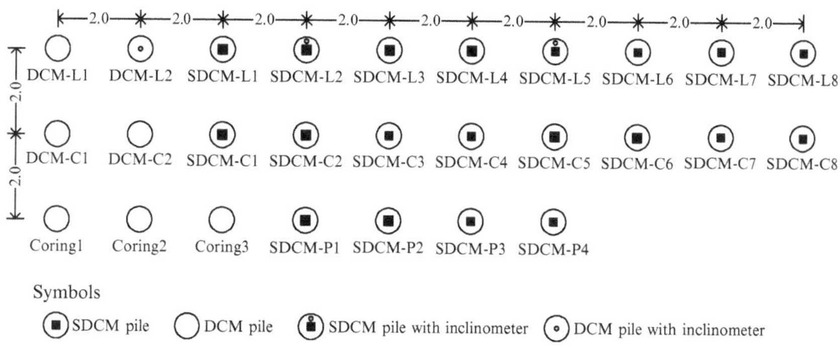

Figure 2.3 Pile test layout.

Figure 2.4 Construction of SDCM pile.

mixing method employing a jet pressure of 22 MPa. The water:cement ratio (W:C) of the cement slurry and the cement content employed for the construction of deep mixing piles were 1.5 and 150 kg (m^3 of soil), respectively. Each deep mixing pile had a diameter of 0.6 m and length of 7.0 m. Twenty of the deep mixing piles were utilized as SDCM piles. Each SDCM pile was constructed by insertion of prestressed concrete pile immediately after deep cement mixing had completed. The prestressed concrete pile was sunk by its weight without any pushing force because of very low fiction between prestressed concrete pile and cement–admixed clay due to the high water content produced by jet mixing. The construction of SDCM pile is shown in Fig. 2.4. All deep mixing piles were allowed to cure in situ for 80 days after completion of installation; three deep mixing piles installed to represent all deep mixing piles were cored by drilling machine to obtain unconfined compressive strength and modulus of elasticity of undisturbed samples in laboratory, and a full-scale pile load test program was performed.

2.2.4 Axial compression load tests

Figure 2.5 shows the schematic setup for applying axial compression loads to the test pile using a hydraulic jack acting against a platform. The platform was composed of steel sheets and concrete boxes with a total weight of 500 kN supported by upper cross H beams. The reaction beam or test beam was support by lower cross H beams supported by a concrete box that distributed the total weight of the platform to the surrounding soil. The vertical load was applied to the test pile through a 600-kN capacity hydraulic jack. A 500-kN capacity proving ring was inserted between the jack and the reaction beam

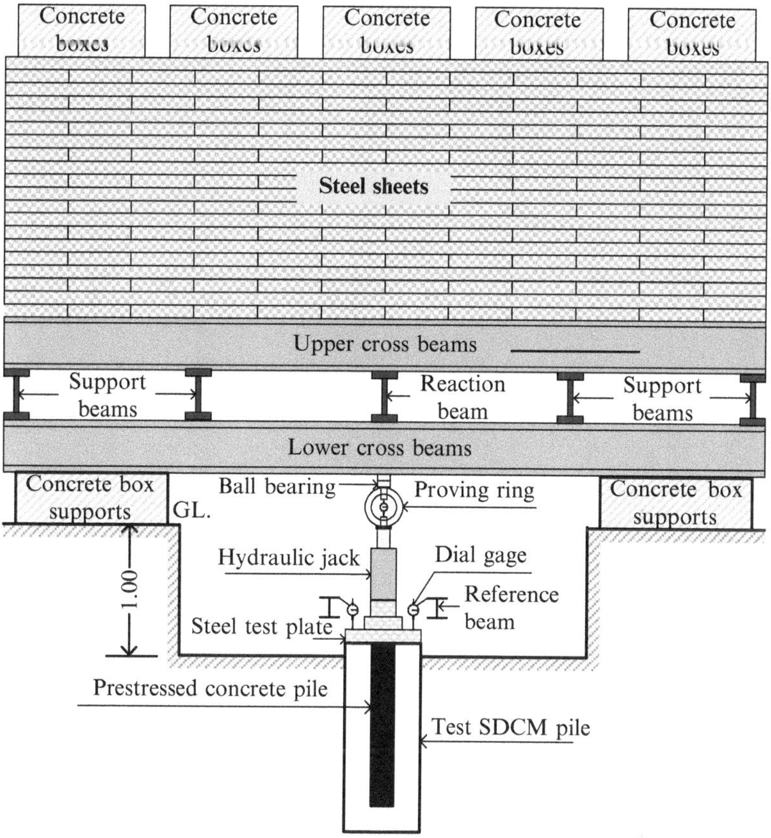

Figure 2.5 Schematic setup for pile under axial compression load.

to measure the applied load. The ball bearing was inserted between the proving and the reaction beam to ensure the vertical direction of the applied loads from the hydraulic jack. The vertical settlement of the test pile under the applied load was measured using two dial gages, which were connected to two reference beams placed on both sides of the jack. The axial compression tests were performed in accordance with ASTM D1143 (1994a). The load was applied in increments of 10 kN. Each load increment was maintained for 5 min. The load was applied until continuous vertical displacements occurred at a slight or no increase in load.

2.2.5 Lateral load tests

Figure 2.6 shows the schematic setup for applying the lateral loads to the SDCM pile using a hydraulic jack acting against the sides of an excavation.

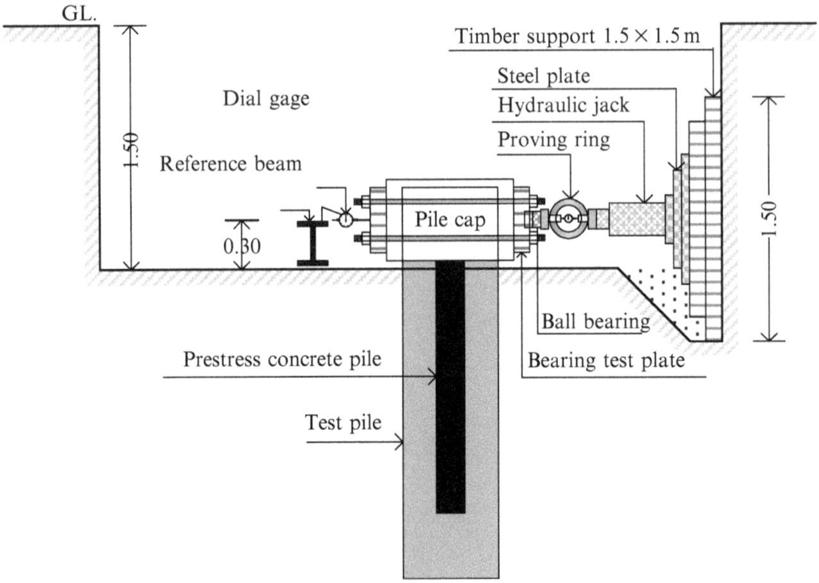

Figure 2.6 Schematic setup for pile under lateral load.

The base excavation was performed around the test pile at a depth of 1.5 m from the original ground level to provide enough area on the side of the excavation for the necessary reactive capacity to the maximum anticipated lateral test loads. A concrete pile cap with a height of 0.4 m was placed on the pile to prevent local failure on the pile head, and the load was applied 0.3 m from the base excavation level. The hydraulic jack, with a capacity of 600 kN, was used for applying lateral loads. The loads were read from proving ring with capacity of 100 kN, and the lateral displacement at the load application level was read from one dial gage connected to reference beams. The ball bearing was inserted between the proving and the reaction beam to adjust the horizontal direction of the applied loads from the hydraulic jack. Thick timber sheets were use as support to distribute the load from the hydraulic jack to the side of the excavation. The lateral load tests were performed in accordance with ASTM D3966 (1995). The load was applied in increments of 0.5 and 1 kN for DCM pile and SDCM pile, respectively. Each load increment was maintained for 10 min. The load was applied until continuous lateral displacements occurred at a slight or no increase in load. The inclinometer casings were installed in DCM-L2, SDCM-L2, and SDCM-L5 to measure the lateral displacement with depth caused by the lateral load. The inclinometer casing consisted of 70-mm- and 6.10-m-long plastic tube sections with four grooves to align the torpedo sensor in the required directions of measurement.

2.2.6 Pullout interface shear test

Figure 2.7 shows the schematic setup for applying axial tensile loads to the test pile using a hydraulic jack with a capacity of 600 kN acting between the test beam and the reaction frame. The length of the cores embedded in the SDCM was 1 m, with another 1 m protruding out of the SDCM to perform the pullout tests. During the preparation for the test, clay cement over

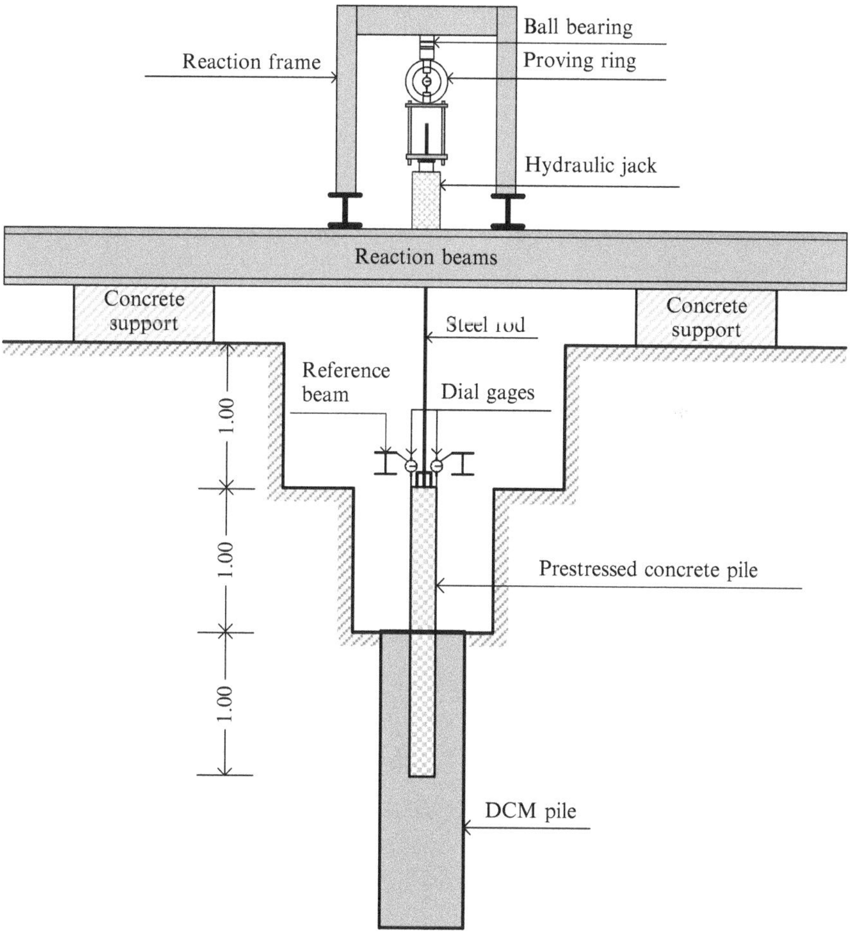

Figure 2.7 Schematic setup for pile under tensile load.

the top 1 m of the SDCM was removed, leaving 1 m of the core embedded in the DCM pile. A steel rod was connected to the test pile and the reaction frame in order to pull the prestressed concrete core pile from the DCM. The test beam was supported by the concrete box that distributed the total load to the surrounding soil. The vertical settlement of the test pile under the applied load was measured using two dial gages, which were connected to two reference beams placed on both sides of the jack. The pullout interface tests were performed in accordance with ASTM D3689 (1994b). The load was applied in increments of 5 kN. Each load increment was maintained for 5 min. The load was applied until continuous vertical displacements occurred at a slight or no increase in load.

2.3 FULL-SCALE LOAD TEST RESULTS

2.3.1 Unconfined compression tests on deep cement mixing core samples

To obtain the in situ strength of the DCM pile in the test field, three DCM piles were constructed (Fig. 2.3) so that core samples could be extracted for unconfined compression tests in the laboratory to determine unconfined compressive strength, q_u, and modulus of elasticity corresponding at 50% unconfined compressive strength, E_{50}. Unconfined compressive tests were performed on samples with a diameter of 50 mm and height of 100 mm. The values are scattered over the entire depth without any clear trend of the influence of the depth on the values of unconfined compressive strength and modulus of elasticity (Fig. 2.8). The values of unconfined compressive strength ranged from 500 to 1500 kPa, with an average value of 900 kPa, whereas the modulus of elasticity ranged from 50,000 to 150,000 kPa, with an average value of 90,000 kPa, indicating an empirical relationship of $E_{50} = 100\ q_u$. The correlation ratio of E_{50}/q_u obtained from field core samples ranged from 60 to 150.

2.3.2 Axial compression tests

Two axial pile load tests on DCM piles (DCM-C1 and DCM-C2) and eight axial pile load tests on SDCM piles (SDCM-C1 to SDCM-C8) varying the length and the sectional area of prestressed concrete core pile as listed in Table 2.1 were conducted in accordance with ASTM D1143-94 (1994a). The load was applied in increments of 10 kN. Each load increment was maintained for 5 min. The load was applied until continuous settlements occurred at a slight or no increase in axial load. Figure 2.9 shows the axial

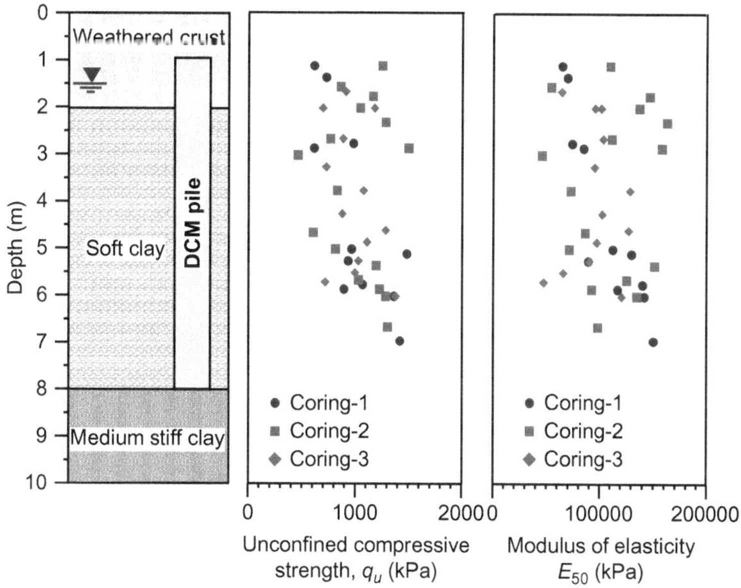

Figure 2.8 Field test results on DCM piles.

Table 2.1 Comparison of ultimate bearing capacities (Q_{ult}) in axial compression tests

No.	L_{DCM} (m)	D_{DCM} (m)	L_{core} (m)	Core size (m × m)	L_{core}/L_{DCM}	A_{core}/A_{DCM}	Q_{ult} (kN)
DCM-C1	7.0	0.60	—	—	—	—	200
DCM-C2	7.0	0.60	—	—	—	—	140
SDCM-C1	7.0	0.60	6.0	0.22 × 0.22	0.85	0.17	320
SDCM-C2	7.0	0.60	6.0	0.22 × 0.22	0.85	0.17	310
SDCM-C3	7.0	0.60	6.0	0.18 × 0.18	0.85	0.11	300
SDCM-C4	7.0	0.60	6.0	0.18 × 0.18	0.85	0.11	300
SDCM-C5	7.0	0.60	4.0	0.22 × 0.22	0.57	0.17	280
SDCM-C6	7.0	0.60	4.0	0.22 × 0.22	0.57	0.17	270
SDCM-C7	7.0	0.60	4.0	0.18 × 0.18	0.57	0.11	270
SDCM-C8	7.0	0.60	4.0	0.18 × 0.18	0.57	0.11	260

compression load versus settlement relationships for all 10 test piles. The ultimate bearing capacities of all test piles and corresponding settlements are tabulated in Table 2.1 and were determined by the slope tangent method at the point of intersection of the initial and final tangents to the load settlement curve following Butler and Hoy (1977).

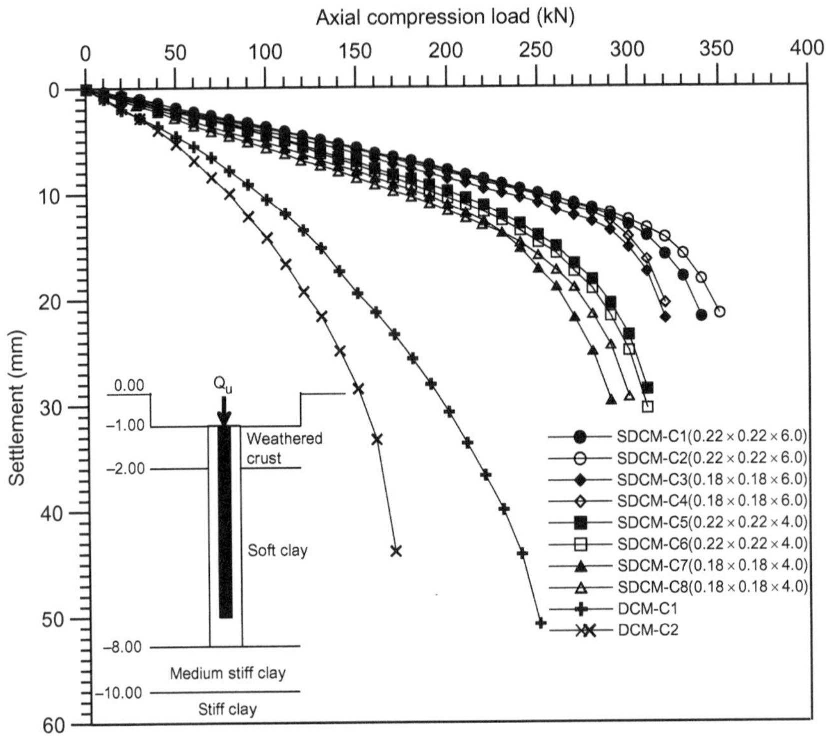

Figure 2.9 Axial load–settlement curves from field tests.

Axial compression test on DCM piles

The ultimate bearing capacities of DCM piles DCM-C1 and DCM-C2 were 200 and 140 kN, respectively. For DCM-C1, the load was applied until continuous vertical displacements occurred between loads of 240 and 250 kN and no increase in load after a load of 250 kN. After the test, the failure mode of DCM-C1 was observed to be pile failure by excavation of 1-m depth around the pile head. Cracks were found around the pile head due to high stress concentration at the pile head. During testing, it was found that the pile DCM-C2 failed suddenly after increasing the axial from 160 to 170 kN. The test procedure was stopped, and the failure mode of DCM-C2 was observed. The pile failure of DCM-C2 took place at a depth approximately 0.50 m from the pile head due to low shear strength and low quality of this part of DCM pile. The major difference in the ultimate bearing capacities of DCM-C1 and DCM-C2—as much as 60 kN—confirmed the low quality that commonly occurs on DCM piles resulting in its low bearing capacity (Petchgate et al., 2003).

Axial compression pile load test on SDCM with 0.18-m square core pile
The ultimate bearing capacities of SDCM piles with 4-m-long and 0.18-m square concrete core (SDCM-C7 and SDCM-C8) were 270 and 260 kN, respectively. The ultimate bearing capacities of SDCM-C7 and SDCM-C8 and the load settlement curves were not significantly different, and they are much more consistent than those of DCM piles. Therefore, the average ultimate bearing capacity of SDCM pile with 0.18-m square and 4-m-long core pile is 265 kN and higher than those of DCM-C1 and DCM-C2 by approximately 1.3 and 1.9 times, respectively. This implies that concrete core can increase the bearing capacity. The load settlement curve of SDCM pile shows linear behavior, whereas the load settlement curves of DCM piles show nonlinear behavior. Moreover, the settlements for SDCM pile were less than that for DCM pile at the same load. This implies that concrete core can increase the stiffness of the SDCM pile, causing more linear behavior and thus reducing the settlements. The ultimate bearing capacities of the SDCM piles with 6-m-long concrete core (SDCM-C3 and SDCM-C4) were the same at 300 kN and higher than those of DCM-C1 and DCM-C2 by approximately 1.5 and 2.1 times, respectively.

The ultimate bearing capacity of SDCM pile with 6-m-long core pile is higher than that of the 4-m-long core pile by 35 kN. This implies that the longer concrete core can increase the bearing capacity. The load settlement curve of SDCM with 6-m-long core pile shows further range of linear behavior than SDCM with 4-m-long core pile resulting in the increase in ultimate bearing capacity. Moreover, the settlements for SDCM pile with 6-m-long core pile were less than that for SDCM pile with 4-m-long core pile at the same load. This implies that the longer concrete core pile can transfer the load from the top part to the bottom part more efficiently than the shorter core pile transferring the load directly to the unimproved part of DCM, as shown in Fig. 2.10.

Axial compression pile load test on SDCM with 0.22-m square core pile
The ultimate bearing capacities of the SDCM piles with 4-m long and 0.22-m square concrete core (SDCM-C3 and SDCM-C4) were 280 and 270 kN, respectively. The average ultimate bearing capacity was 275 kN and higher than those of DCM-C1 and DCM-C2 by approximately 1.4 and 2.0 times, respectively. The load settlement curves of SDCM pile with 0.22-m square and 4-m-long core pile were quite similar to that of SDCM with 0.18-m square and 4-m-long core pile. However, the average ultimate bearing capacity of the former was slightly higher than the latter by 10 kN,

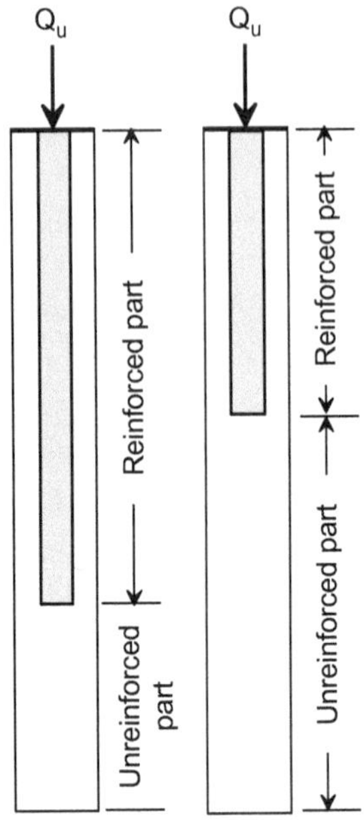

Figure 2.10 SDCM piles with different length of concrete core pile.

with associated lower settlements. This implies that section area of concrete core slightly affects the ultimate bearing capacity and settlement of SDCM pile with the same length.

The load settlement curves of SDCM pile with 0.22-m square and 4-m-long core pile were quite different from that of SDCM with 0.18-m square and 6-m long core pile. The average ultimate bearing capacity of the former was lower than the latter by 25 kN. The load settlement curves of SDCM with 0.22-m square and 4-m-long core pile showed shorter range of linear behavior and also lower stiffness than the 0.18-m square and 6-m-long pile. This implies that length of concrete core affects both the ultimate bearing capacity and settlement, which is more dominant than the section area of concrete core pile.

The ultimate bearing capacities of the SDCM piles with 0.22-m square and 6-m-long concrete core (SDCM-C1 and SDCM-C2) were 320 and 310 kN, respectively. The average ultimate bearing capacity of SDCM pile with 0.22-m square and 6-m-long core pile is 315 kN and higher than those of DCM-C1 and DCM-C2 by approximately 1.6 and 2.2 times, respectively.

The load settlement curves of SDCM with 0.22-m square and 6-m-long core pile were slightly different compared to that of SDCM with 0.18-m square and 6-m-long core pile. The average ultimate bearing capacity of the former was slightly higher than the latter by 15 kN. This confirms that section area of concrete core slightly affects the ultimate bearing capacity and settlement of SDCM pile with the same length of concrete core pile.

The load settlement curves of SDCM with 0.22-m square and 6-m-long core pile were quite different from that of SDCM with 0.22-m square and 4-m long core pile. The average ultimate bearing capacity of the former was also higher than the latter by 40 kN. The load settlement curves of SDCM with 0.22-m square and 6-m-long core pile showed further range of linear behavior and also higher stiffness than the 0.22-m square and 4-m-long pile. This confirms that length of concrete core affects both the ultimate bearing capacity and settlement, and the longer concrete core pile can have higher bearing capacity and stiffness than the shorter core pile, as shown in Fig. 2.10. Therefore, the concrete core increases quality control and increase bearing capacity of DCM pile by factors of 1.4–2.2. The length of concrete core affects both the ultimate bearing capacity and settlement, and the length is more dominant than section area of concrete core pile.

Mode of failure in axial load tests

Lorenzo (2005) and Petchgate et al. (2003) suggested that the ultimate bearing capacity of individual DCM piles can be obtained depending on the mode of failure using Eqs. (2.1) and (2.2):

$$Q_{\mathrm{ult}}^{\mathrm{soil}} = \left(\pi D_{\mathrm{DCM}} L_{\mathrm{DCM}} + 9\pi {D_{\mathrm{DCM}}}^2/4\right) c_{\mathrm{u}} (\text{Soil failure}) \tag{2.1}$$

$$Q_{\mathrm{ult}}^{\mathrm{pile}} = \frac{\pi}{4} (D_{\mathrm{DCM}})^2 (q_{\mathrm{u,DCM}}) (\text{Pile failure}) \tag{2.2}$$

The ultimate bearing capacity in case of soil failure using Eq. (2.1) was calculated as 320 kN, whereas that in case of pile failure using Eq. (2.2) depends on unconfined compressive strength of DCM piles as shown in Fig. 2.11. The ultimate bearing capacities of DCM-C1 and DCM-C2 indicate that the unconfined compressive strengths of DCM-C1 and DCM-C2 are 700 and 500 kPa, respectively, which do not reach the average (900 kPa). Moreover, to obtain ultimate bearing capacity in case of soil failure (320 kN), the unconfined compressive strength of DCM pile should be higher than 1100 kPa. It can be considered that the ultimate bearing capacities of SDCM piles with the core length of 6 m (SDCM 1 to SDCM 4) can almost reach ultimate bearing capacity in case of soil failure, whereas SDCM piles with the core length of 4 m (SDCM 5 to SDCM 8) can be considered as pile failure. Based on post-test observations, no damage occurred in all

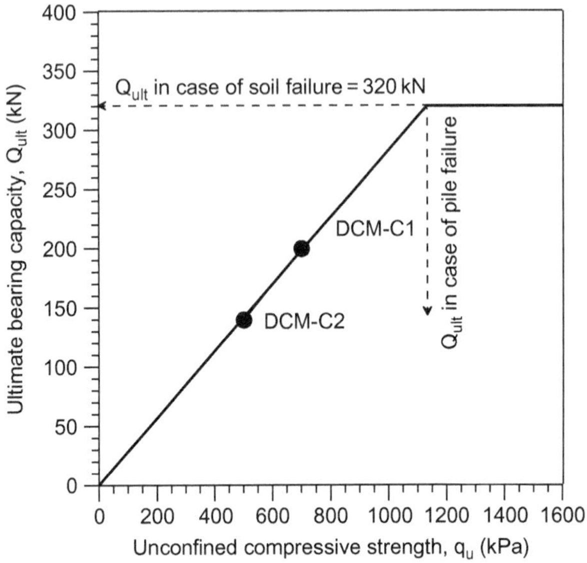

Figure 2.11 Ultimate bearing capacity of DCM pile and unconfined compressive strength.

concrete core piles inserted in DCM piles in the SDCM piles, implying that pile body failure can be considered.

2.3.3 Lateral load tests

Two lateral load tests on DCM piles (DCM-L1 and DCM-L2) and eight lateral load tests on SDCM piles (SDCM-L1 to DCM-L8) varying the length and the sectional area of prestressed concrete core pile as listed in Table 2.2 were conducted in accordance with ASTM D3966-95 (1995). The lateral load tests were performed to measure the ultimate lateral bearing capacity and to obtain the relationship between lateral load and lateral displacement of the test piles. Moreover, the lateral displacement versus depth was obtained by inclinometers to study the influence of depth of surrounding clay acting on the test pile. The loads were applied in increments of 1 and 0.5 kN for SDCM pile and DCM pile, respectively. Each load increment was maintained for 10 min. The load was applied until continuous lateral displacements occurred at a slight or no increase in load. Figure 2.12 shows the lateral loads versus lateral displacements relationships for all 10 test piles. The ultimate bearing capacities of all test piles are tabulated in Table 2.2.

Table 2.2 Comparison of ultimate bearing capacities (P_{ult}) in lateral load tests

No.	L_{DCM} (m)	D_{DCM} (m)	L_{core} (m)	Core size (m × m)	L_{core}/L_{DCM}	A_{core}/A_{DCM}	P_{ult} (kN)
DCM-L1	6.5	0.60	—	—	—	—	3.5
DCM-L2	6.5	0.60	—	—	—	—	2.5
SDCM-L1	6.5	0.60	5.5	0.2 × 0.22	0.85	0.17	46
SDCM-L2	6.5	0.60	5.5	0.22 × 0.22	0.85	0.17	45
SDCM-L3	6.5	0.60	3.5	0.22 × 0.22	0.54	0.17	44
SDCM-L4	6.5	0.60	3.5	0.22 × 0.22	0.54	0.17	43
SDCM-L5	6.5	0.60	5.5	0.18 × 0.18	0.85	0.11	35
SDCM-L6	6.5	0.60	5.5	0.18 × 0.18	0.85	0.11	34
SDCM-L7	6.5	0.60	3.5	0.18 × 0.18	0.54	0.11	33
SDCM-L8	6.5	0.60	3.5	0.18 × 0.18	0.54	0.11	33

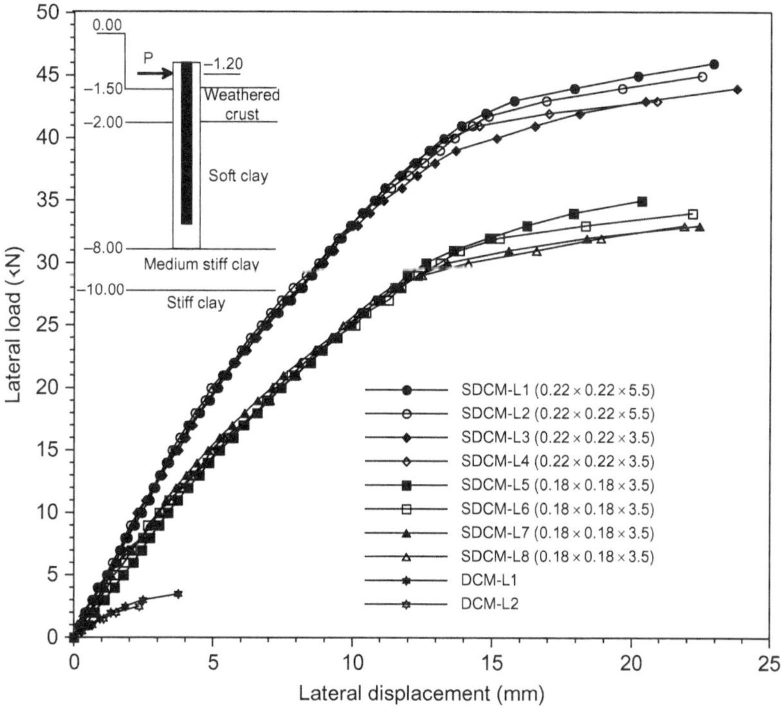

Figure 2.12 Lateral load–lateral displacement curves from field tests.

Lateral load test on DCM piles

The ultimate lateral loads of the DCM piles (DCM-L1 and DCM-L2) were 3.5 and 2.5 kN, respectively, with average ultimate lateral loads of 3.0 kN. Excavation after the test revealed cracks approximately 0.10 m below the base of the excavated pit in DCM-L1. The DCM-L2 cracked at the level of the base of the excavated pit. This means that the failure mode of two DCM piles was pile body failure due to bending moment induced by the lateral load.

Lateral load test on SDCM piles with 0.18-m square core pile

The ultimate lateral loads for the SDCM piles with 0.18-m square and 3.5-m-long concrete core pile (SDCM-L7 and SDCM-L8) were the same at 33 kN and much higher than that for DCM pile by 11 times due to much higher flexural strength of prestressed concrete core pile. The lateral load versus displacement curve shows nonlinear behavior after lateral load of 15 kN, approximately half of ultimate lateral loads, whereas the DCM pile behavior was nonlinear from the beginning. After the test, the surrounding soil was excavated to observe the failure mode of the SDCM piles. The cracks were located 0.46 and 0.50 m below the excavated base in SDCM-L7 and SDCM-L8, respectively. The ultimate lateral loads for the SDCM piles with 0.18-m square and 5.5-m-long concrete core pile (SDCM-L5 and SDCM-L6) were 35 and 34 kN, respectively. The average ultimate lateral load was 34.5 kN, which was similar to the average ultimate lateral load for SDCM piles with 3.5-m-long concrete core pile. The lateral load–lateral displacement curves and the locations of cracks for SDCM pile with 3.5-m-long and 5.5-m-long concrete core pile were similar, which means the overall behavior was the same. Consequently, increasing the length of the concrete core pile does not affect the ultimate lateral load and displacement of SDCM pile, which is quite different from that of SDCM pile under compression load.

Lateral load test on SDCM piles with 0.22-m square core pile

The ultimate lateral loads for the SDCM piles with 0.22-m square and 3.5-m-long concrete core pile (SDCM-L3 and SDCM-L4) were 44 and 43 kN, respectively. The average ultimate lateral load was 43.5 kN and much higher than the corresponding value for DCM pile by 14.5 times, which is higher than that for SDCM piles with 0.18-m square concrete core pile by approximately 1.3 times due to a larger section area causing higher flexural resistance. The lateral load versus displacement curve shows non-linear behavior after lateral load of 20 kN, approximately half of the ultimate lateral loads, and also shows higher flexural stiffness than that of the SDCM piles with 0.18-m square core pile. After the test, the surrounding soil was excavated to observe the failure mode of the SDCM piles. The crack locations were approximately 0.55 and 0.60 m below the base of the excavated pit in SDCM-L3 and SDCM-L4, respectively.

The ultimate lateral loads for the SDCM piles with 0.22-m square and 5.5-m-long concrete core pile (SDCM-L1 and SDCM-L2) were 46 and 45 kN, respectively. The average ultimate lateral load was 45.5 kN, which is similar to the average ultimate lateral load for SDCM piles with 3.5-m-long concrete core pile. After the test, the surrounding soil was excavated to observe the failure mode of the SDCM piles. The crack locations were approximately 0.62 m and 0.58 m below the excavated base in SDCM-L1 and SDCM-L2, respectively. The lateral load versus displacement curves and the locations of cracks for SDCM pile with 3.5-m and 5.5-m-long concrete core pile were similar, which means the overall behavior was the same. Consequently, increasing the length of the core does not affect the ultimate lateral load and displacement of SDCM pile, whereas the section area of concrete core affected the results, which is different from SDCM pile under compression load in which the length of core pile affected ultimate bearing capacity and settlement.

Lateral displacement versus depth curves

Figures 2.13 to 2.15 show the lateral displacement versus depth (below excavated base) curves for DCM-L2, SDCM-L2, and SDCM-L1, respectively, obtained from the inclinometer data at different load intervals. These curve shapes are typical of free-head pile where moment at the loading point is zero (Broms, 1964). According to the lateral displacement versus depth curves for DCM-L2 as shown in Fig. 2.13, the lateral displacement was measured at a load interval of 0.5 kN until failure occurred. The lateral displacement approached zero at a depth of 0.5 m (1 DCM pile diameter) below the excavated base. Thus, the lateral movement developed only within the depth of 1 to 2 DCM pile.

The lateral displacement versus depth curves for SDCM-L2 are shown in Fig. 2.14. The lateral displacement was measured at load intervals of 5 kN until failure occurred. The lateral displacement approached zero at a depth of 2.0 m (3 DCM pile diameter) below the excavated base. Thus, the lateral displacement developed only at a depth of 3 DCM pile diameter, and the maximum lateral displacement occurred at the excavated base. Similarly, the lateral displacement versus depth curve for SDCM-L1 is shown in Fig. 2.15. The lateral displacement was measured at a load interval of 5 kN until failure. The lateral displacement approached zero at a depth of 2.0 m (3 DCM pile diameter) below the excavated base. Thus, the influence

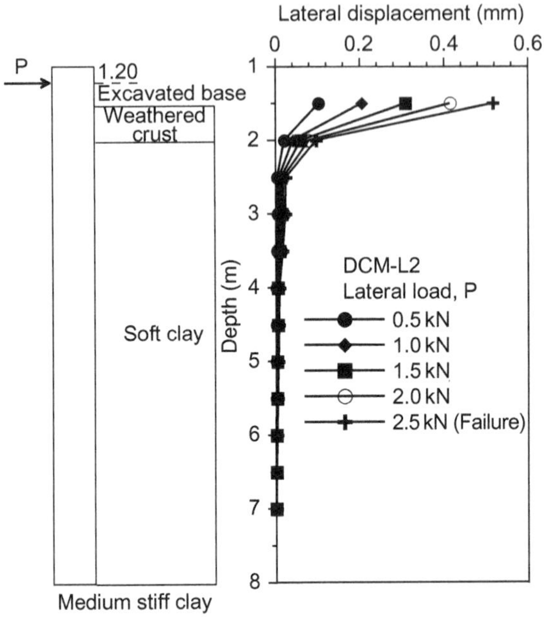

Figure 2.13 Lateral displacement versus depth curves of DCM-L2.

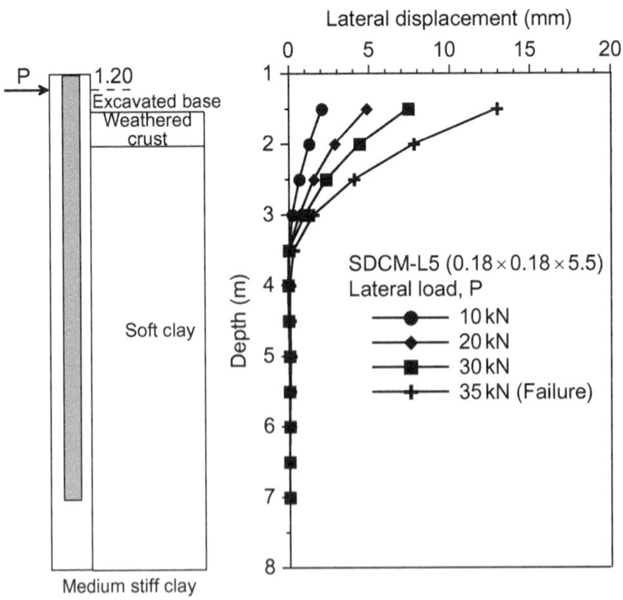

Figure 2.14 Lateral displacement versus depth curves of SDCM-L2.

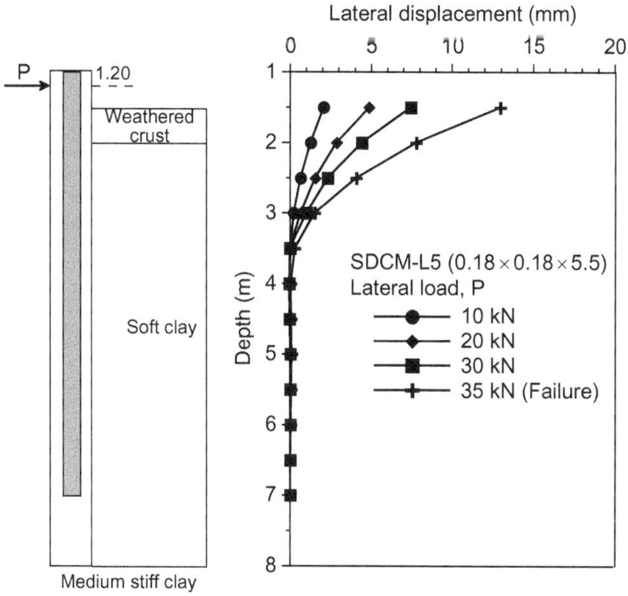

Figure 2.15 Lateral displacement versus depth curves of SDCM-L1.

zone of surrounding clay on DCM pile is only approximately 1 DCM pile diameter. On the other hand, the influence zone of SDCM pile is 3 DCM pile diameter. Consequently, the stiffness of the SDCM pile affected the lateral displacement to greater depths.

Damage characteristics of DCM and SDCM piles

The locations of plastic hinge referring to maximum bending moment in the DCM and SDCM piles were at and below the excavated level as inferred from the crack locations. All failures in the DCM (DCM-L1 and DCM-L2) piles occurred near the ground surface due to the breaking of the pile body approximately 0.10 m below the excavated base and at the excavated base for DCM-L1 and DCM-L2, respectively. The maximum bending moments can be calculated from Fig. 2.16. The flexural strength, σ_f, can be calculated from following relationship (Petchgate et al., 2004):

$$M_{max} = P_{ult} y_{crack} \tag{2.3}$$

$$\sigma_f = \frac{M_{max}(D_{DCM}/2)}{I_{DCM}} \tag{2.4}$$

The maximum moments calculated from Eq. (2.4) are 1.4 and 0.75 kN-m for DCM-L1 and DCM-L2, respectively. The flexural strengths

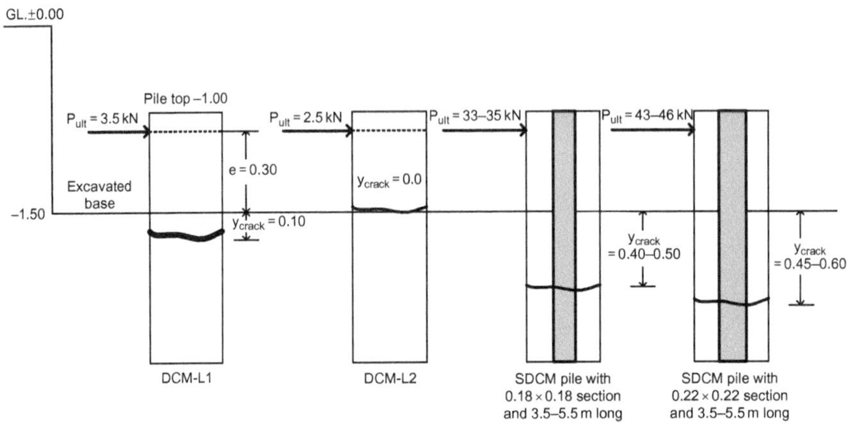

Figure 2.16 Modes of failure of DCM and SDCM piles under lateral load.

from back-analysis are 60 and 30 kPa for DCM-L1 and DCM-L2, respectively, corresponding to 7% and 4% of average unconfined compressive strength (900 kPa). The locations of the maximum bending moments and cracks in the SDCM were deeper than those for the DCM piles. Inspection of SDCM-L1 to SDCM-L8 piles revealed no damage above the excavated base, and failure resulted from the breaking of the DCM and cracking of prestressed concrete core pile. The locations of these cracks were found at deeper depths varying from 0.40 to 0.60 m below the excavated base.

2.3.4 Pullout interface test

Four pullout interface tests between prestressed concrete core pile and surrounding DCM were performed. Two pullout interface tests were conducted on 0.22-m square prestressed concrete core piles (SDCM-P1 and SDCM -P2) and another two on 0.18-m square prestressed concrete core piles (SDCM-P3 and SDCM –P4). The length of the cores embedded in the SDCM was 1 m for all tests. Figure 2.17 shows the tension load versus vertical displacement relationships. The interface shear stress was determined by dividing the pullout force by the surface area of concrete core pile embedded in the DCM pile. The maximum tension loads were 165, 155, 135, and 120 kN for the test piles SDCM-P1, SDCM-P2, SDCM-P3, and SDCM-P4, respectively, as tabulated in Table 2.3. Subsequently, the interface shear strengths ($\tau_{interface}$) were calculated from maximum tensile load divided by surface area of concrete core embedded in DCM pile equal to 188, 176, 188, and 167 kPa for SDCM-P1, SDCM-P2, SDCM-P3, and SDCM-P4, respectively, with an average value of 179 Pa. Consequently,

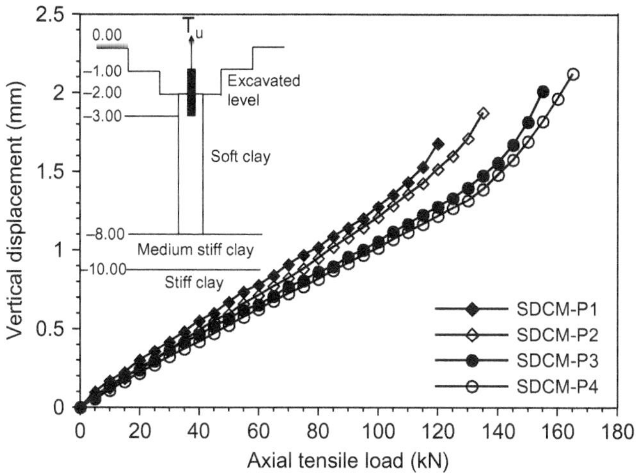

Figure 2.17 Tensile load–vertical displacement curves from field tests.

Table 2.3 Comparison of maximum tensile load-bearing capacities (T_{ult}) in pullout interface tests

No.	D_{DCM} (m)	L_{core} (m)	Core size (m × m)	Interface area (m^2)	T_{ult} (kN)	τ_{inter} (kPa)
SDCM-P1	0.60	1.0	0.22 × 0.22	0.88	165	188
SDCM-P2	0.60	1.0	0.22 × 0.22	0.88	155	176
SDCM-P3	0.60	1.0	0.18 × 0.18	0.72	135	188
SDCM-P4	0.60	1.0	0.18 × 0.18	0.72	120	167

the strength reduction factor for interfaces (R_{inter}) defined by Brinkgreve and Broere (2006) was calculated as follows:

$$R_{inter} = \tau_{interface}/c_{soil} \tag{2.5}$$

where R_{inter} was calculated as 179/450 = 0.40. This value is within the range of pullout interface test results on concrete core pile and cement–admixed clay performed in the laboratory ranging from 0.36 to 0.44.

The interface shear strengths between concrete core pile and the surrounding DCM for axial compression tests can be calculated as 518, 778, 634, and 950 kN for 0.18-m square with 4- and 6-m-long as well as 0.22-m square with 4- and 6-m-long concrete core piles, respectively, which are much higher than axial ultimate bearing capacities. Therefore, no slippage occurred between the interfaces during the axial load tests.

2.4 FULL-SCALE BEHAVIOR OF SDCM PILES UNDER EMBANKMENT LOADING

The objective of this study was to investigate the behavior of the improved ground by SDCM piles compared to DCM piles under vertical and horizontal loading conditions induced by a full-scale embankment. Field observations were performed during and after embankment construction. The observed results were analyzed and discussed. The project site was located near the site of the full-scale load test.

2.4.1 Concrete core pile

The reinforced concrete core pile was proposed to be the plug inserted in the DCM pile to form the SDCM pile. A prestressed reinforced concrete pile was selected as the core pile because it has high strength and stiffness, and it is also cheaper than steel pile. The selected prestressed reinforced concrete pile was square section with 0.22-m width and 6.0-m length.

2.4.2 Construction of foundation improvement

Prior to embankment construction, the foundation subsoil was improved with SDCM and DCM piles. The DCM piles were installed in situ by a jet mixing method employing a jet pressure of 22 MPa. Both SDCM and DCM piles were installed at 2.0-m spacing, as shown in Fig. 2.18. The W:C ratio of the cement slurry and the cement content employed for the construction of deep mixing were 1.5 and 150 kg/m^3 of soil, respectively. Each deep mixing pile had a diameter of 0.6 m and a length of 7.0 m, penetrating down to the bottom of the soft clay layer, as shown in section view of the embankment (Fig. 2.18b).

2.4.3 Field instrumentation

To observe the behavior of SDCM and DCM piles under embankment loading as well as the behavior of the test embankment, various instruments were installed at the site. Figures 2.18(a) and 2.18(b) show the instrumentation in plan and cross-section views, respectively. Five surface settlement plates were installed to a depth of 1.0 m from the original ground. The surface settlement plates were installed on the top of the SDCM pile (S10) and the DCM pile (S11), on the surrounding clay of SDCM pile (S1) and the DCM pile (S7), and on the unimproved clay (S4). The installations were carried out after excavation of 1.0-m depth, the base level of the test embankment. Six subsurface settlement gauges were also installed at

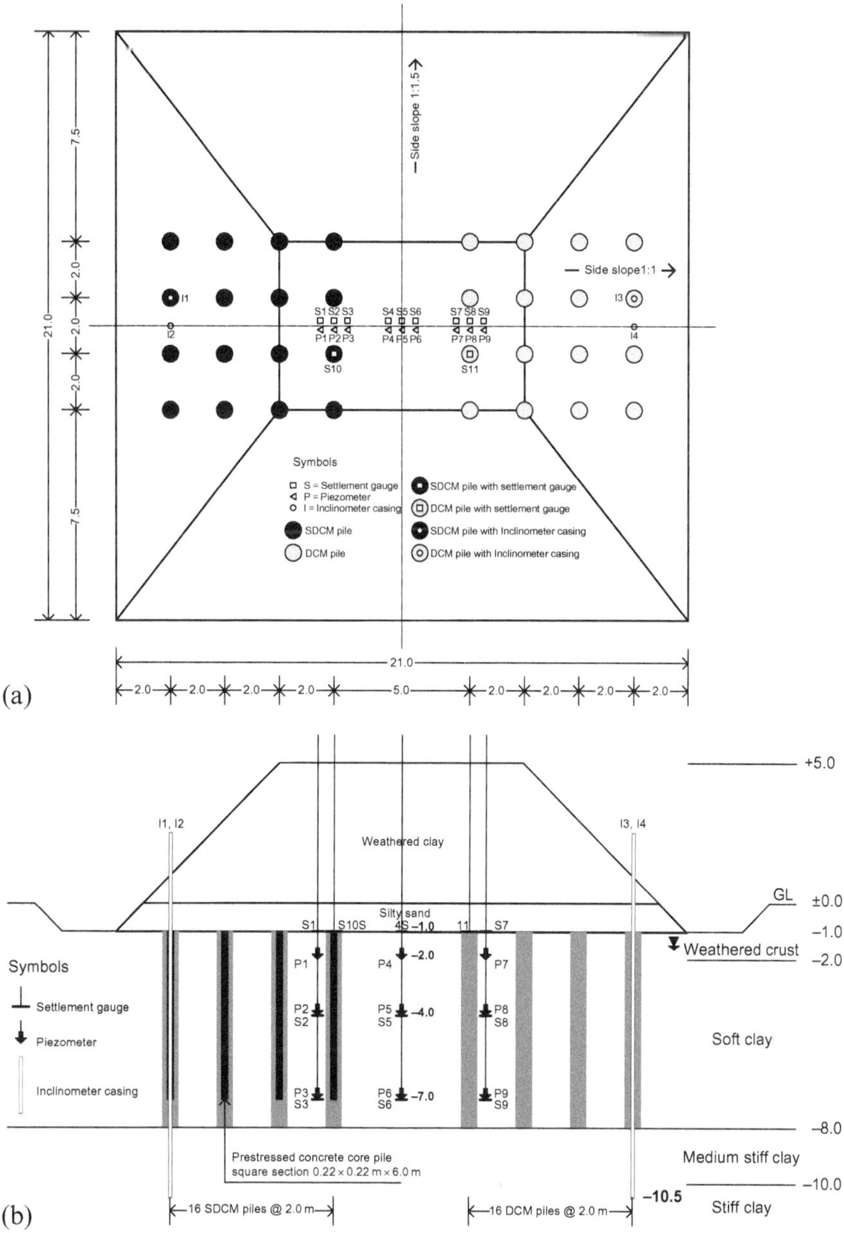

Figure 2.18 (a) Plan view of test embankment. (b) Cross section of test embankment.

4.0- and 7.0-m depths from the original ground. The subsurface settlement gauges were installed in the surrounding clay of SDCM pile (S2 and S3), DCM pile (S7 and S8), and the unimproved clay (S5 and S6). Three piezometers were installed at 4.0-m depths from the original ground in the surrounding clay of SDCM pile (P1), DCM pile (P3), and the unimproved clay (P2). Four vertical inclinometer casings were installed. The vertical inclinometer casings were installed in SDCM pile (I1), DCM pile (I3), the surrounding clay of SDCM pile (I2), and the surrounding clay of DCM pile (I4).

2.4.4 Construction of embankment

The undrained shear strength of the uppermost 2-m-thick weathered crust layer is significantly higher than that of the soft clay layer below. In order to obtain larger deformation characteristics of the improved ground for both SDCM and DCM piles, a 1-m depth of weathered crust was excavated. Accordingly, prior to embankment construction, a 21-m-wide, 40-m-long, and 1-m-deep foundation of soil was removed. After excavation of the trench, the area covered by the test embankment was backfilled to 1.0-m depth of compacted silty sand. The silty sand was compacted to a unit weight of 17 kN/m^3 by a vibratory roller. The area around the instruments was compacted by a vibratory hand compactor. Then, a 5-m-high embankment was constructed with end slopes of 1:1 and side slopes of 1:1.5, with base dimensions of 21×21 m and top dimensions of 9×6 m (Fig. 2.18). The embankment material mostly consisted of weathered clay excavated from the area near the test embankment. The weathered clay was compacted at 0.30-m lift thickness to a density of 16 kN/m^3. The embankment construction was completed within 30 days from June 26, 2006, to July 25, 2006.

2.5 FULL-SCALE EMBANKMENT TEST RESULTS

2.5.1 Settlement behavior of SDCM and DCM piles improved soft clay foundation

Figure 2.19 shows the settlements on top of SDCM pile, DCM pile, surrounding clay of SDCM pile, surrounding clay of DCM pile, and the surface of unimproved clay obtained from surface settlement plate S1, S7, S10, S11, and S4, respectively, during and after construction up to 570 days of full embankment loading. After embankment construction, the total settlements on SDCM pile, DCM pile, adjacent clay of SDCM pile, adjacent clay of DCM pile, and unimproved clay were 92, 149, 136, 184, and 186 mm,

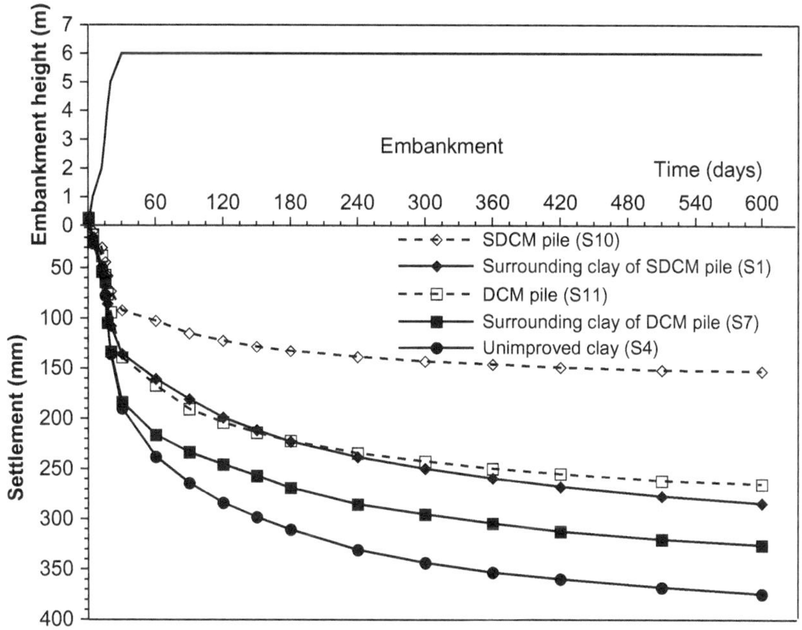

Figure 2.19 Comparison of surface settlements.

respectively. The maximum settlements on SDCM pile, DCM pile, adjacent clay of SDCM pile, adjacent clay of DCM pile, and unimproved clay were approximately 153, 265, 284, 326, and 374 mm, respectively, at 570 days after embankment construction. Approximately 50% of the total settlement occurred during the 30 days of construction of the test embankment. The settlement on surrounding clay of SDCM pile was slightly less than that on surrounding clay of DCM pile. Thus, the embankment load was transferred to the SDCM pile more efficiently than to the DCM pile. Moreover, the settlement on SDCM pile was 40% less than the corresponding value on DCM pile due to its higher stiffness. In addition, the settlements on top of the SDCM piles were almost 60% less than that of unimproved ground, whereas those of DCM piles were only 30% less than that of unimproved ground.

2.5.2 Excess pore water pressure

Figures 2.20, 2.21, and 2.22 show the comparisons of excess pore water pressures between unimproved clay and surrounding clays of DCM and SDCM piles at 2, 4, and 7 m, respectively. The measured excess pore water pressure at 2-, 4-, and 7-m depths demonstrated similar trends. The

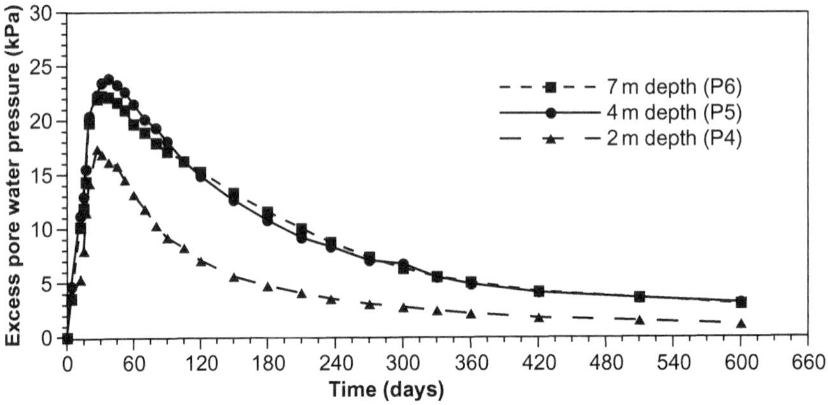

Figure 2.20 Comparison of excess pore water pressure in unimproved clay.

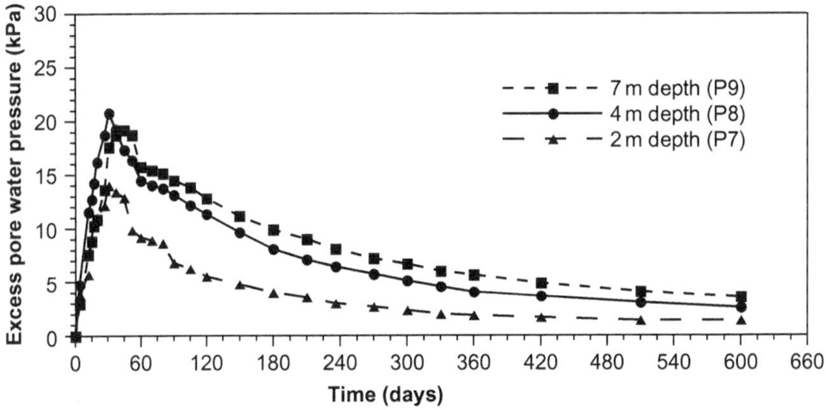

Figure 2.21 Comparison of excess pore water pressure in surrounding clay of DCM pile.

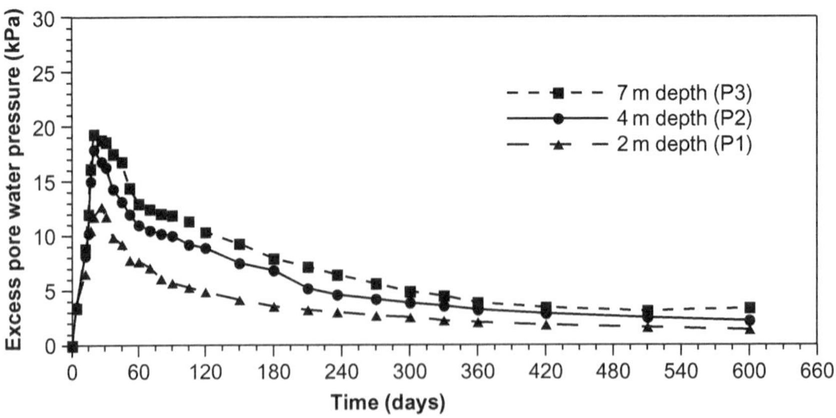

Figure 2.22 Comparison of excess pore water pressure in surrounding clay of SDCM pile.

measured pore pressures were highest in the unimproved clay, followed by the surrounding clay of DCM pile and lowest in the surrounding clay of SDCM pile at any depth, implying that the embankment load was transferred to DCM and SDCM piles. The SDCM pile carried more embankment load than the DCM pile; consequently, lower excess pore water pressures were measured in adjacent clays of the SDCM piles.

2.5.3 Lateral movement behavior of SDCM and DCM piles improved soft clay foundation

Lateral movement profiles of SDCM and DCM piles

The comparison of lateral movement profiles of SDCM and DCM piles after embankment construction and 90 and 570 days after construction is plotted in Fig. 2.23. The measured maximum lateral movements at a depth of 1.0 m,

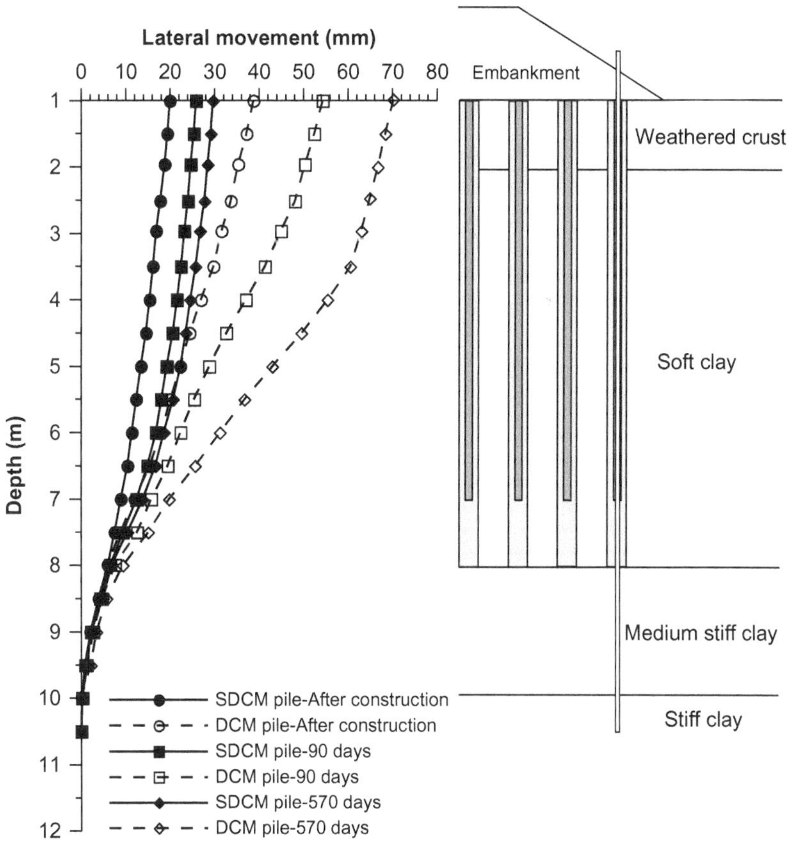

Figure 2.23 Comparison of lateral movement profiles between SDCM and DCM piles.

or at the pile head, immediately after embankment construction and after 570 days for SDCM pile were 19.9 and 29.6 mm, respectively, whereas the corresponding values for DCM pile were 38.7 and 70.2 mm, respectively. Thus, approximately 65% and 55% of the total lateral movements occurred immediately after construction of the test embankment for SDCM and DCM piles, respectively. Due to its higher flexural stiffness, the lateral movement in SDCM piles was 60% less than that in DCM piles.

Lateral movement profiles of surrounding clay adjacent to SDCM and DCM piles

The comparison of lateral movement profiles of adjacent clay of SDCM and DCM piles immediately after construction and 90 and 570 days after construction of the test embankment is plotted in Fig. 2.24. The measured maximum lateral movements at a depth of 1.0 m (excavation base level) immediately after embankment construction and after 570 days for surrounding clay of

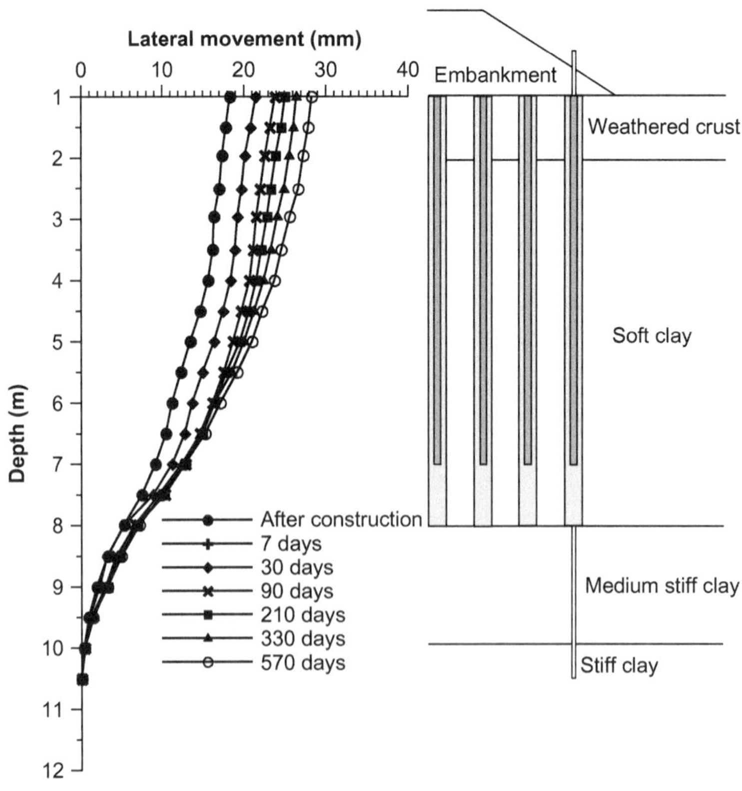

Figure 2.24 Lateral movement profiles of surrounding clay of SDCM pile.

SDCM pile were 18.3 and 28.3 mm, respectively, whereas those for surrounding clay of DCM pile were 36.0 and 67.3 mm, respectively. Thus, approximately 65% and 50% of the total lateral movements occurred during the construction of the test embankment for adjacent clays of SDCM piles and DCM piles, respectively, immediately after construction and 570 days after construction. Consequently, the SDCM pile can reduce the magnitude of lateral movement 60% and increase the stability on the improved foundation.

Rate of lateral movement

The comparison of maximum lateral movements at the top of the pile or the surface (excavation base level) for surrounding clay with time during and after embankment construction is shown in Fig. 2.25, which shows that the rates of movements were significantly faster during construction and then became slower after embankment construction. The rate of the maximum lateral movement can be divided into three stages for SDCM pile, DCM pile, surrounding clay of SDCM pile, and surrounding clay of DCM pile; namely (1) during embankment construction (0–45 days), (2) at transition period (45–90 days), and (3) during consolidation (90–600 days). The average rates of the lateral movements were calculated by dividing the amount of lateral movements measured by the time interval. The average rates during embankment construction stage were 0.52, 0.46, 1.02, and

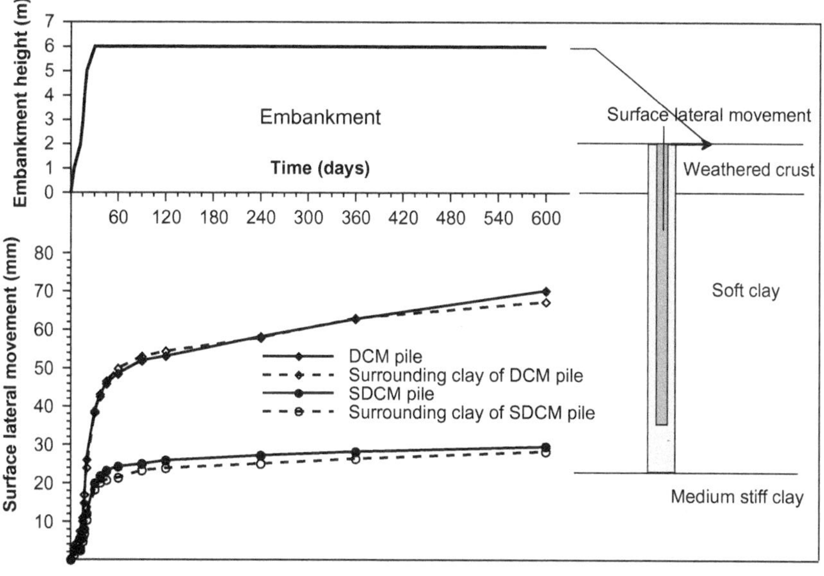

Figure 2.25 Comparison of surface lateral movements.

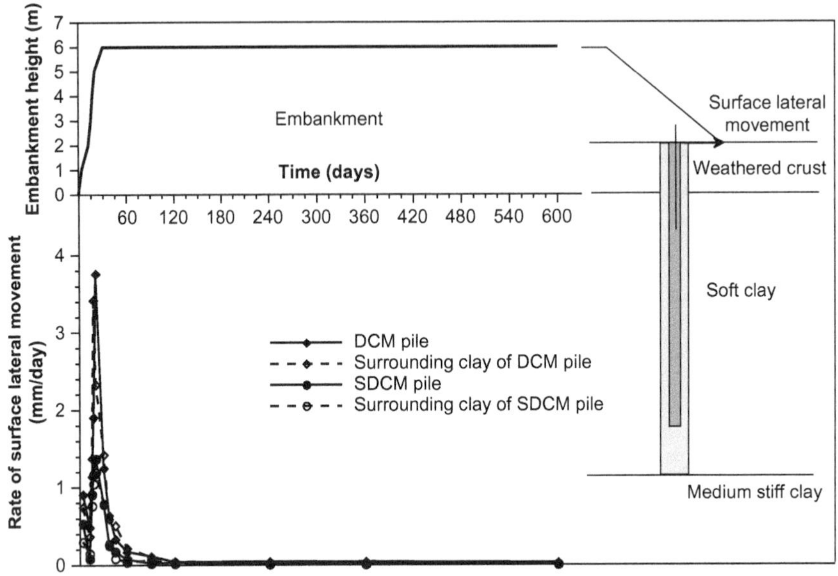

Figure 2.26 Comparison of rate of surface lateral movements.

1.04 mm/day for SDCM pile, surrounding clay of SDCM pile, DCM pile, and surrounding clay of DCM pile, respectively. The average rates of lateral movements in the consolidation stage were 0.01, 0.01, 0.04, and 0.03 mm/day for SDCM pile, surrounding clay of SDCM pile, DCM pile, and surrounding clay of DCM pile, respectively.

The same data are also plotted in terms of rate of surface lateral movement with time, as shown in Fig. 2.26. The maximum rates are 1.37, 3.76, 1.20, and 3.43 mm/day for SDCM pile, DCM pile, surrounding clay of SDCM pile, and surrounding clay of DCM pile, respectively, occurring during embankment construction. Thus, the SDCM pile reduced the rates of lateral movements by approximately 64% during construction and the transition period as well as by approximately 70% during the consolidation stage.

Ratio of lateral movement to surface settlement

The ratio of maximum lateral movement to maximum surface settlement was used as the criterion for safety control during embankment construction (Moh and Lin, 2006). According to Indraratna et al. (1992), the ratio of lateral movement to surface settlement is defined as the maximum lateral displacement at the toe to the maximum settlement at the center line of an embankment. By studying the performance of a test embankment load to failure, Indraratna et al. concluded that a small ratio value is necessary to maintain the stability of the embankment over the subsoil. Chai (1992) showed that when an embankment approached failure, the lateral movement to settlement ratio increased rapidly. At the embankment failure, the value was larger than 0.5. In this study, the lateral movement to settlement ratios of SDCM pile, surrounding clay of SDCM pile, DCM pile, and surrounding clay of DCM pile were determined by the data obtained from the surface lateral movements and surface settlements, respectively.

Figure 2.27 shows the measured lateral movement:settlement ratio versus time curve during embankment construction until 570 days after construction for SDCM pile, DCM pile, surrounding clay of SDCM pile, and surrounding clay of DCM pile. The measured SDCM pile, surrounding clay of SDCM pile, DCM pile, and surrounding clay of DCM pile show a similar trend in which the ratios decreased from an embankment height of 1 m to a height of 2 m and then increased until the embankment height was 6 m and decreased with time until at 240 days or 210 days after construction

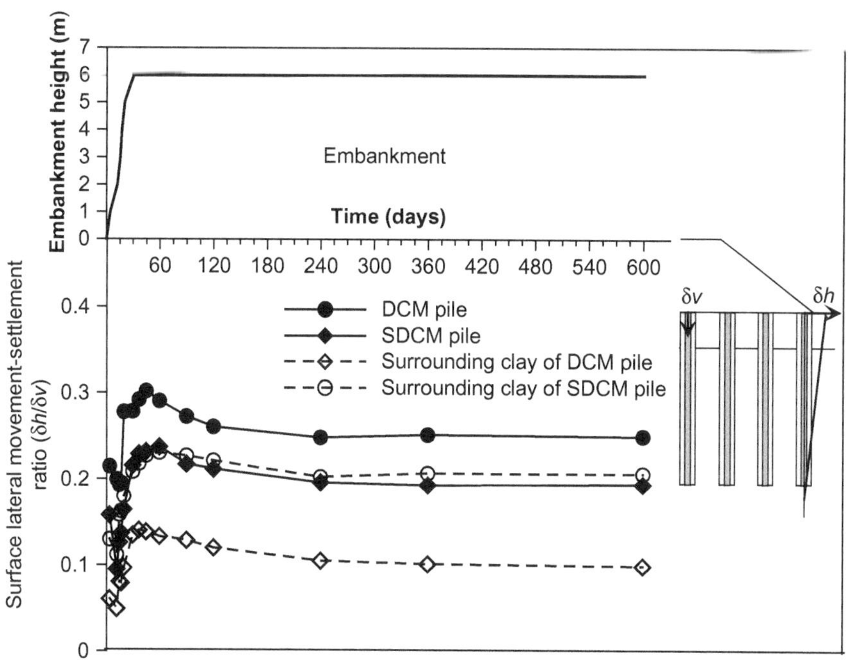

Figure 2.27 Comparison of surface lateral movement:settlement ratio.

and became constant, implying that both vertical settlements and lateral movements increased slowly at similar rate. The maximum ratios are 0.23, 0.14, 0.30, and 0.23 for the lateral movement to settlement ratios of SDCM pile, surrounding clay of SDCM pile, DCM pile, and surrounding clay of DCM pile, respectively, occurring during embankment construction to 30 days after embankment construction. The ratios became constant at 240 or 210 days after construction and were 0.19, 0.10, 0.25, and 0.21 for SDCM pile, surrounding clay of SDCM pile, DCM pile, and surrounding clay of DCM pile Thus, the maximum ratios are still less than 0.5, and the ratios of SDCM pile and surrounding clay of SDCM pile are less than those of DCM pile and surrounding clay of DCM pile. Therefore, the SDCM pile increased stability and safety during construction.

Figure 7.28 shows the relationship between measured lateral movements and vertical settlements of SDCM pile, DCM pile, surrounding clay of SDCM pile, and surrounding clay of DCM pile during embankment construction until 570 days after construction. The average values of the ratios are 0.20, 0.11, 0.27, and 0.21 for SDCM pile, surrounding clay of SDCM pile, DCM pile, and surrounding clay of DCM pile, respectively. Therefore,

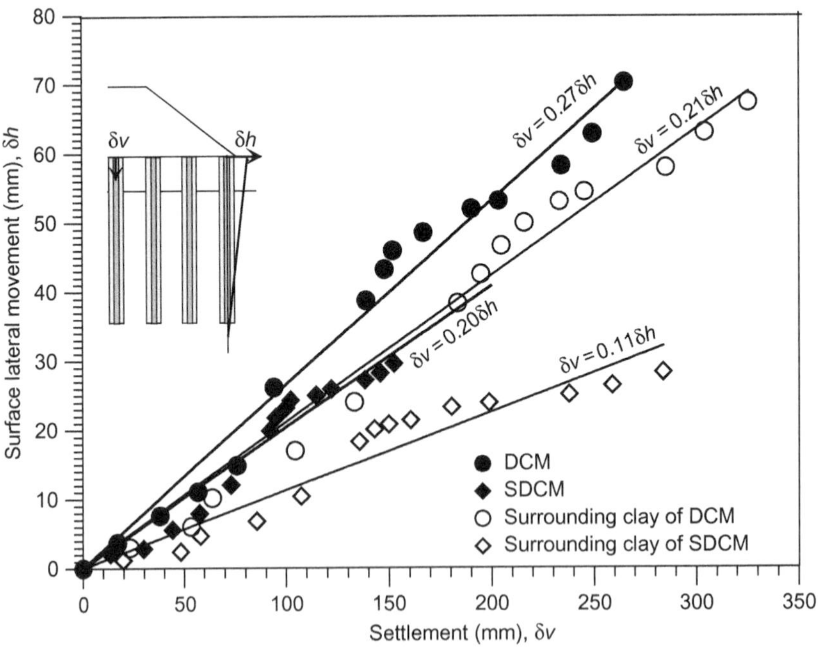

Figure 2.28 Comparison of surface lateral movement and settlement.

the lateral movement to settlement ratio can be reduced by SDCM pile, especially in the surrounding clay. These ratios can be reduced by as much as 50% compared to the corresponding values in the DCM piles.

2.6 THREE-DIMENSIONAL NUMERICAL SIMULATION OF FULL-SCALE SDCM PILES UNDER EMBANKMENT LOADING

In this section, the details of the method used in three-dimensional (3D) finite element simulation of the results of full-scale embankment load on SDCM and DCM piles as outlined in Section 2.5 are presented. The parameters significantly affecting the consolidation settlements and lateral movements of SDCM piles are sectional area of concrete core pile, A_{core}, and length of concrete core pile, L_{core}. The numerical simulation was accomplished through finite element analysis using PLAXIS 3D foundation software version 1.6. The best fit of surface settlement versus time curves was performed in order to obtain the back-analyzed parameters to be utilized in the following numerical simulations for analyzing the effects of A_{core} and L_{core} on the settlement and lateral movement behavior of SDCM pile under lateral loading.

2.6.1 Three-dimensional finite element simulation

Finite element discretization

Finite element simulations were performed using the PLAXIS 3D foundation software version 1.6 developed by Brinkgreve and Broere (2006). The software was used in the simulation of the full-scale pile load tests in Section 5. The 3D finite element model consisted of the embankment, SDCM pile, DCM pile, and foundation soil as illustrated in Fig. 2.29, which shows the finite element mesh consisting of more than 28,950 elements.

The in situ stresses in the foundation soil were generated by the K_o procedure. Then, the 32 DCM piles were constructed from a depth of 1 m to 8 m. After construction of all DCM piles, the 16 prestressed concrete core piles were inserted into the center of the 16 DCM piles to form the 16 SDCM piles. Then, excavation of the topmost weathered crust layer was performed to 1-m depth. The embankment was divided into six layers following the field construction and was placed on top of DCM and SDCM improved foundation. After the placement of a compacted fill layer, plastic calculation analysis was used to simulate the plastic deformation and consolidation calculation analysis was used to simulate the consolidation deformation that occurred during construction in each layer of embankment

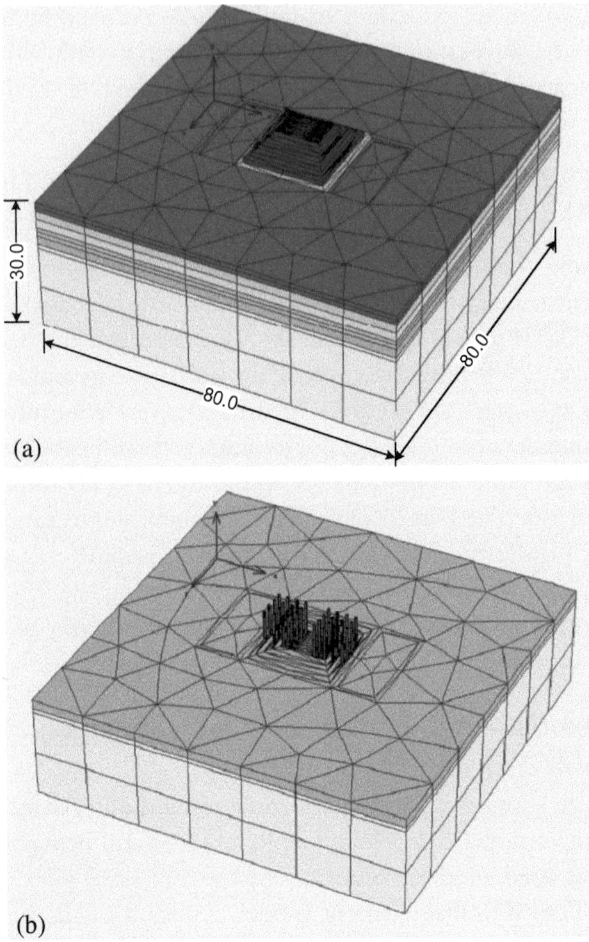

Figure 2.29 (a) Portion of 3D finite element model of test embankment. (b) Portion of 3D finite element model of SDCM and DCM piles.

material from 1 to 6 m. After completion of the full height of the embankment, the consolidation calculation analysis was used to simulate the consolidation process until 600 days.

Material models and parameters

Backfill soil materials

The backfill soil materials used in the embankment were compacted silty sand and weathered clay. The elastic, perfectly plastic Mohr–Coulomb model (MCM) was used to represent the embankment materials. The

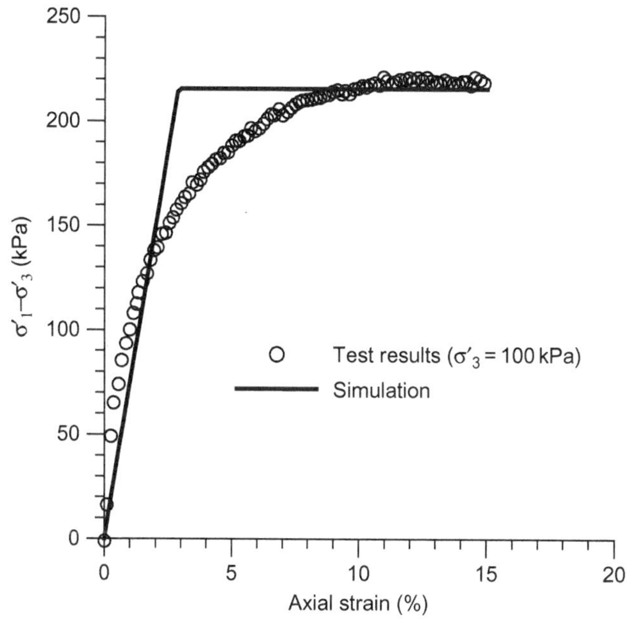

Figure 2.30 Comparisons of simulated and tested stress–strain curves from CID tests on backfill material.

effective cohesions, friction angles, and deformation characteristics of the backfill materials were determined from back-analyses by fitting stress–strain curve between simulated and laboratory test results obtained from consolidated drained triaxial (CID) tests using representative specimens at actual densities in the embankment materials as shown in Fig. 2.30.

Prestressed concrete core pile

The concrete core pile was modeled as "massive" volume element with interfaces at its periphery. Failure of the concrete core pile can occur in the case of high load. Thus, the MCM was selected to model the behavior of the prestressed concrete core pile. The parameters were obtained from back-analyses by fitting compressive stress–strain curve between observed and simulated results as shown in Fig. 2.31.

DCM pile

The MCM was selected to model the behavior of DCM piles. The MCM assumed a fixed yield surface in a principal stress space, which is fully defined by model parameters. For any stress state that is below the yield surface, the

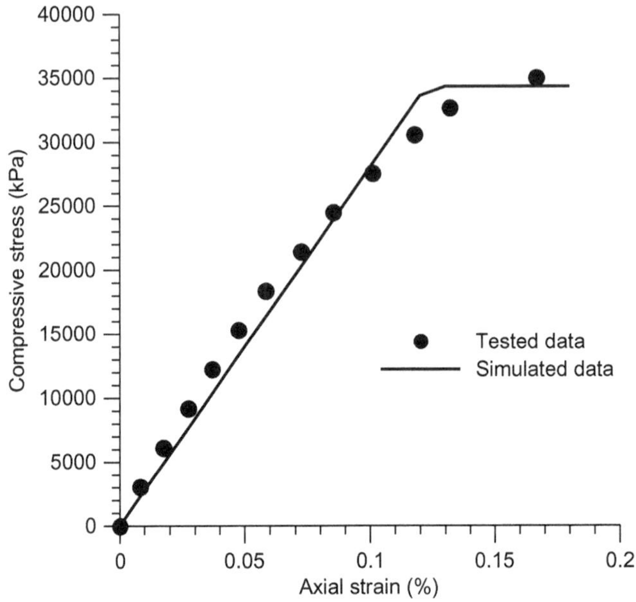

Figure 2.31 Comparison of tested and simulated compressive stress and axial strain of prestressed concrete core pile.

material behavior is purely elastic and all strains are reversible, similar to the elastic model. The drained cohesion and modulus of elasticity were varied to obtain the back-analyzed parameters.

Weathered crust

At the construction site, the uppermost layer in the soil profile layer from 0.0- to 2.0-m depth consisted of the heavily overconsolidated weathered crust. The elastic, perfectly plastic model with a constant value of Poisson's ratio has been used for this soil (Chai, 1992). In this study, the elastic, perfectly plastic MCM was also used for this type of soil. The strength and modulus of elasticity parameters were obtained from the existing data on Bangkok clay by Lai et al. (2006).

Soft clay

The soft clay layer with a depth from 2 to 8 m was modeled by the soft soil model (SSM). The SSM is a Cam clay-type model developed by Brinkgreve and Broere (2006). The SSM parameters were determined based on the 1D consolidation test. The modified compression index, λ^*, was found to be

CR/2.303, where CR is the compression ratio. The CR values obtained from 1D consolidation tests as shown in Fig. 2.1 ranged from 0.22 to 0.33, corresponding to λ^* of 0.10 to 0.15. The ratio of virgin compression to recompression, κ^*/λ^*,is commonly in the range of 0.2–0.4 (Britto and Gunn, 1987). Based on the results of previous investigations (Chai, 1992; Bergado et al., 1995, 2002; Alfaro et al., 1997), κ^*/λ^* was assumed to be 0.2, in agreement with the recommendation by Vermeer and Brinkgreve (1995). The effective friction angle of the soft clay layer was assumed to be 23°, as suggested by Balasubramaniam et al. (1978) for soft Bangkok clay. The permeability of clays is one of the most difficult parameters to determine. The representative vertical permeability used in this study was based on the adjusted oedometer test results. The vertical permeability input parameter for the finite element method (FEM) is two times the oedometer test data (Chai and Bergado, 1993). The horizontal permeability input data was assumed to also be two times the corresponding input vertical values (Bergado et al., 2000).

Medium stiff clay and stiff clay

The MCM was used for modeling the behavior of the medium stiff clay layer. There were no direct measurements of the moduli of the soils at the site. However, the undrained modulus, E_u, can be given in terms of the field vane shear strength, $C_{u(VST)}$, such as $E_u = \alpha C_{u(VST)}$. Balasubramaniam and Brenner (1981) reported the value of α to be 250 for Bangkok clay. For Bangkok area, the value of drain modulus, E', was suggested by Parnploy (1985) as $E' = 0.57\ E_u$. The drained strength parameters were obtained from data on Bangkok clay obtained by Balasubramaniam et al. (1978). The parameters used in this study are shown in Table 2.4.

2.6.2 Back-analysis of full-scale SDCM and DCM piles under embankment loading

Surface settlement versus time curves

The surface settlements versus time comparisons between observed and simulated data from FEM analysis are plotted in Fig. 2.32. The settlements from the FEM analysis agreed well with the observed data considering both settlement magnitudes and settlement rates. The parametric study on the effect of length and sectional area of concrete core pile on the surface settlement of SDCM pile was performed by varying the lengths of concrete core piles from 4 to 7 m using the sectional areas of 0.22- and 0.30-m square concrete core pile as shown in Fig. 2.33. The length of concrete

Table 2.4 Soil models and parameters used in 3D FEM simulation SDCM and DCM piles under embankment loading

Material	γ (kN/m^3)	$k_x = k_z$ (m/day)	k_v (m/day)	E'_{ref} (kPa)	ν'	λ^*	κ^*	c' (kPa)	ϕ' (deg)	Over consolidation ratio (OCR)
Subsoil										
Weathered crust	17	0.0020	0.0010	2,500	0.25			10	23	
Soft clay	15	0.0008	0.0004			0.15	0.030	4	23	1.5
Medium stiff clay	18	0.0004	0.0002	5,000	0.25			10	25	
Stiff clay	19	0.0040	0.0020	9,000	0.25			30	26	
Embankment fill										
Weathered clay	16			3,000	0.25			10	26	
Silty sand	17			7,500	0.30			8	29	
Foundation										
Concrete core pile	24			2.8×10^7	0.15					
DCM pile (with interface elements, $R_{inter} = 0.4$)	15	0.0240	0.0120	30,000	0.33			200	30	

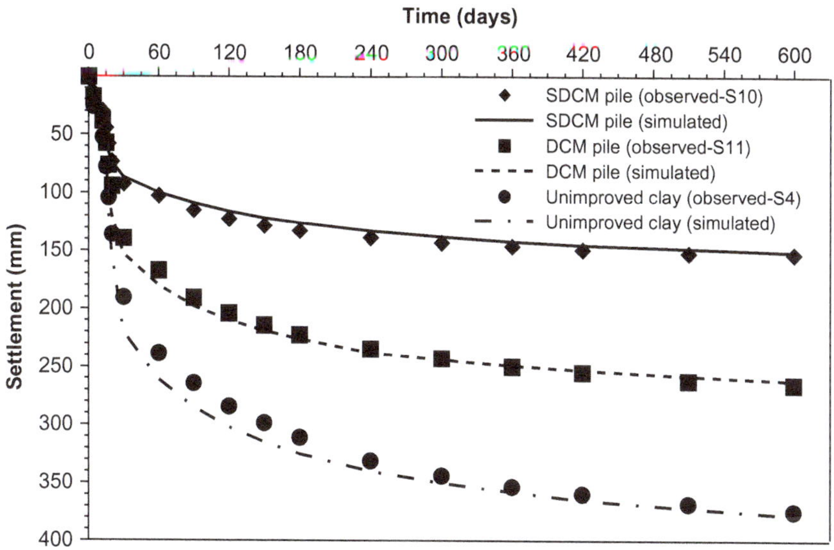

Figure 2.32 Comparison of observed and simulated surface settlements.

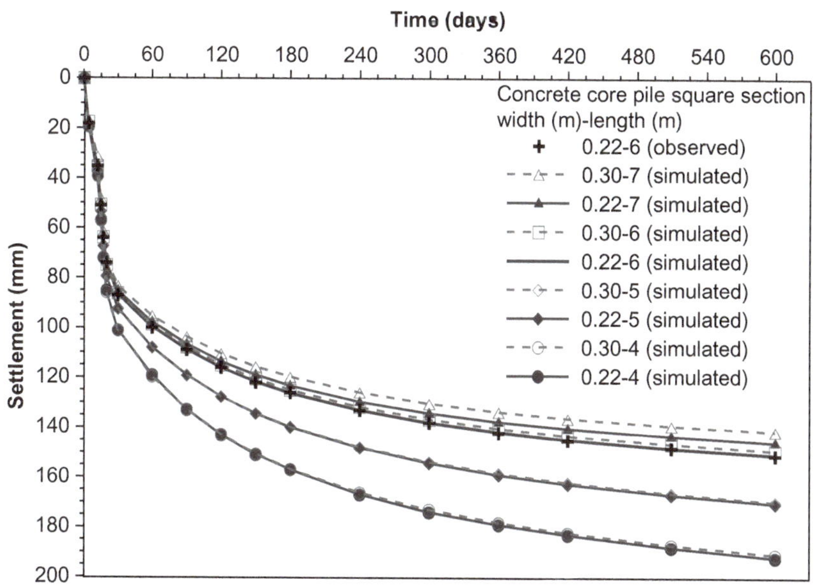

Figure 2.33 Effect of length and sectional area of concrete core pile on settlement–time curves of SDCM pile from simulation.

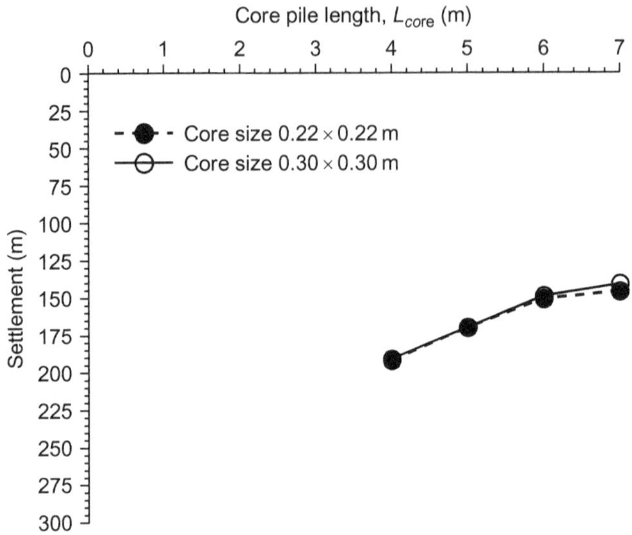

Figure 2.34 Effect of length and sectional area of concrete core pile on ultimate settlement of SDCM pile.

core pile significantly affected the settlement versus time curves. The longer the concrete core, the smaller the settlement. On the other hand, the sectional areas of the concrete core pile only slightly affected the settlements, similar to the observed data from full-scale test on SDCM pile under compression tests. Figure 2.34 shows the effect of length and sectional areas of the concrete core pile on the ultimate settlement of SDCM pile. The settlement of SDCM pile decreased linearly with increasing length from 0 m ($L_{core}=0$ m) to the core pile length of 6 m ($L_{core}=6$ m). In contrast, the settlement decreased only slightly from $L_{core}=6$ m to $L_{core}=7$ m. Thus, the settlement reduction is proportional to the increase in length of the core pile. There was minimal effect in the settlements caused by the section areas of the concrete core pile.

Excess pore water pressure

The measured and simulated excess pore water pressure data during and after construction at 2-, 4-, and 7-m depths are plotted in Figs. 2.35, 2.36, and 2.37, respectively. The simulated results overestimated the excess pore water pressures, especially in early stages of the construction of the embankment. This trend is expected because the pore pressures were measured using the slow responding open standpipe piezometers. However,

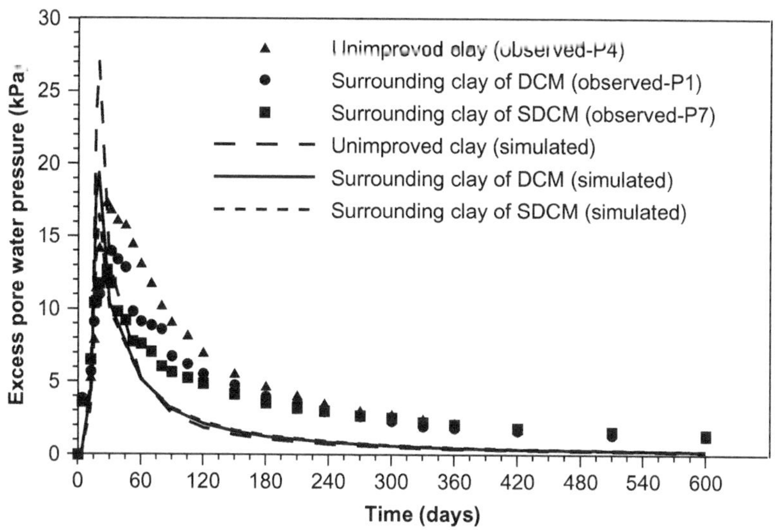

Figure 2.35 Excess pore water pressure at 2-m depth from element simulation.

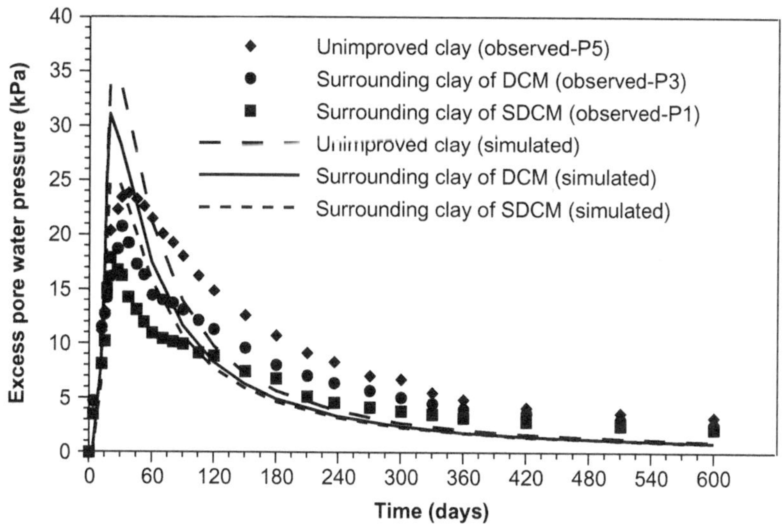

Figure 2.36 Excess pore water pressure at 4-m depth from simulation.

the prediction of the dissipation rate of pore water pressures during the consolidation process seems to be better. In addition, the overestimation can be explained by the undrained condition applied in the staged construction function of PLAXIS 3D foundation software. In reality, both consolidation and drainage occurred gradually, and the excess pore

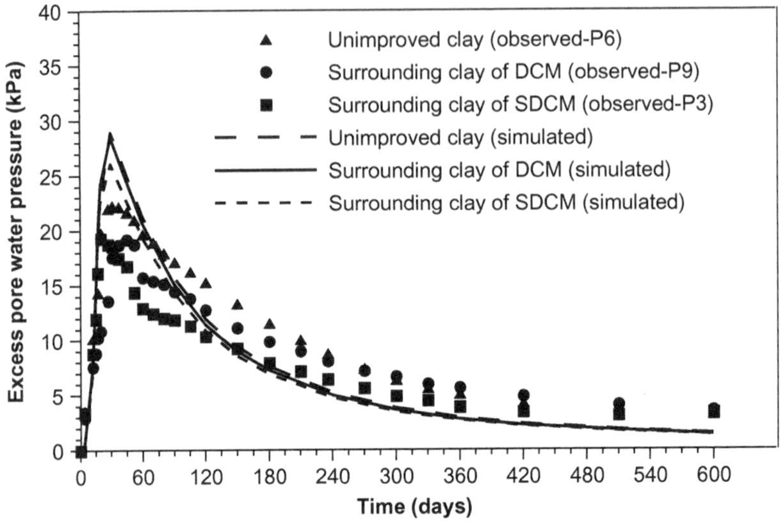

Figure 2.37 Excess pore water pressure at 7-m depth from simulation.

pressures subsequently decreased during each stage of construction. Furthermore, the simulated results were in agreement with the measured data, which indicated that the highest excess pore pressures were registered at the unimproved foundation and decreasing at the DCM and SDCM improved foundations.

Lateral movements

Comparisons between FEM simulated results and the measured lateral displacement profiles immediately after construction and 90 and 570 days after construction are plotted in Figs. 2.38 and 2.39 for SDCM pile and DCM pile, respectively. The FEM results for lateral movements are in good agreement with the observed data. There were slight underestimations of the simulated results for the lateral displacements in SDCM and DCM pile from 3- to 8-m depth. The observed lateral movement profiles of SDCM pile had similar shape with those of DCM pile obtained from the simulations immediately after construction until 570 days after construction. The results of the effect of length and section area of concrete core pile on surface lateral movement of SDCM pile are illustrated in Fig. 2.40. It can be concluded that both the length and the sectional area of concrete core pile affected the lateral movement of SDCM pile under embankment loading.

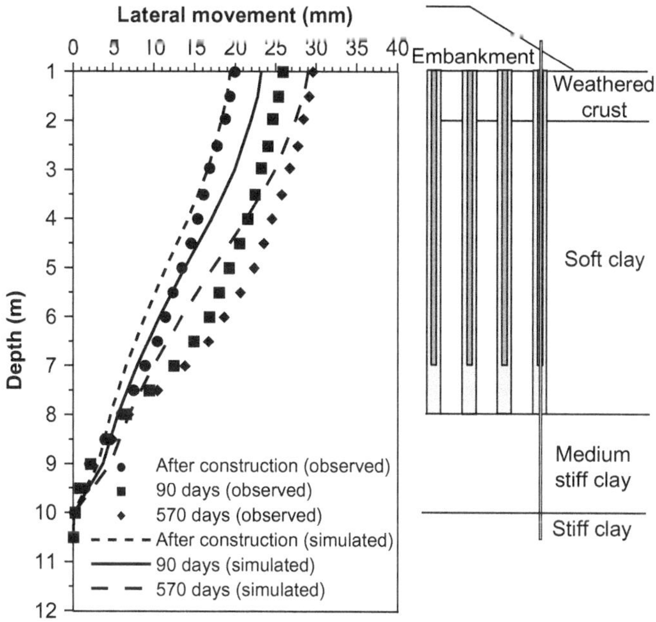

Figure 2.38 Lateral movement profiles of SDCM pile from simulation.

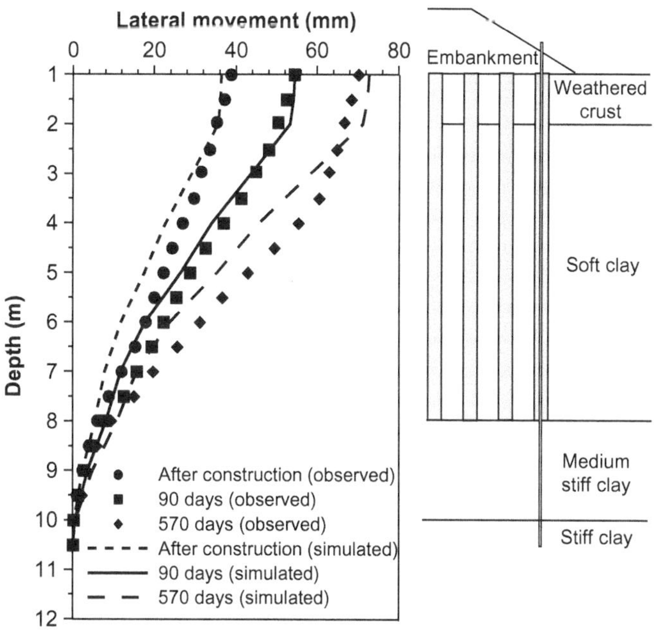

Figure 2.39 Lateral movement profiles of DCM pile from simulation.

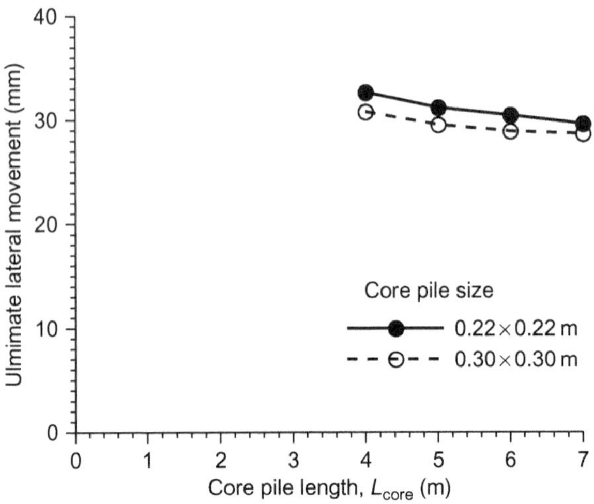

Figure 2.40 Effect of sectional area and length of concrete core pile on maximum lateral movement of SDCM pile.

2.7 CONCLUSION

The ultimate bearing capacities of DCM piles are significantly lower—as much as 60 kN or 30%—compared to that of SDCM pile due to the low quality that commonly occurs on DCM piles. The ultimate bearing capacities of SDCM piles with a length ratio of 0.57 and area ratios of 0.11 and 0.17 can improve by factors of 1.3 or 1.4 compared to DCM piles. The ultimate bearing capacities of SDCM piles with a length ratio of 0.85 and area ratios of 0.11 and 0.17 can improve by factors of 1.9–2.3 compared to DCM piles. Thus, the length of concrete core affects both the ultimate bearing capacity and settlement of the cross-sectional area.

The mode of failure of DCM pile was fail due to its unconfined compressive strength, which did not exceed 1100 kPa in order to attain soil failure. The SDCM pile with a length ratio of 0.57 failed by DCM pile body below the concrete core pile tip. On the other hand, the mode of failure of SDCM pile with a length ratio of 0.85 was soil failure. Thus, the effective length ratio of SDCM pile should be larger than 0.85.

The average lateral ultimate bearing capacities of SDCM piles with an area ratio of 0.11 and length ratios of 0.54 and 0.85 could be improved by a factor of 11 compared to those of DCM piles. The average lateral ultimate bearing capacities of SDCM piles with an area ratio of 0.17 and length ratios of 0.54 and 0.85 could be improved by a factor of 15 compared to those of DCM piles. The sectional area of concrete core affects both the ultimate bearing capacities and lateral movements, whereas the length only slightly affects them.

During the lateral load tests, the DCM pile cracked near the excavation due to the bending moment. In the SDCM pile, the cracks were located deeper down their length. This difference in the location of the cracks was due to the low stiffness and low flexural strength of the DCM pile that cannot transfer moment to deeper depths. On the other hand, the SDCM pile, which has higher stiffness and higher flexural, can transfer moment to deeper depths. The influence zone of surrounding clay on DCM pile is only approximately 1 DCM pile diameter, whereas that on SDCM pile is 3 DCM pile diameter. The stiffness of the SDCM pile affected the lateral displacement into greater depths.

The strength reduction factor for interfaces, R_{inter}, obtained from full-scale pullout interface test was 0.40, which was in agreement with the laboratory measured value, implying that the shear strength at the interface between concrete core and surrounding DCM is strong enough to prevent any slippage during axial compression tests.

The SDCM pile with an area ratio and length of 0.17 and 0.85, respectively, has effectively reduced the settlements by as much as 40% and lateral movements by as much as 60% induced by embankment loading compared to the DCM pile. The SDCM pile can reduce the magnitude of lateral movement by 60%, and the SDCM and DCM piles were confirmed to move laterally together with their adjacent grounds.

The SDCM pile reduced the rates of lateral movements by approximately 64% during construction and 70% during the consolidation stage compared to the DCM pile. Moreover, the results indicated that both SDCM and DCM piles moved laterally together with their surrounding ground.

The measured lateral movement to settlement ratios were 0.23, 0.14, 0.30, and 0.23 for the SDCM pile, surrounding clay of SDCM pile, DCM pile, and surrounding clay of DCM pile, respectively, during embankment construction until 30 days after embankment construction. The ratios tended to be constant at 240 or 210 days after construction and amounted to 0.19, 0.10, 0.25, and 0.21 for SDCM pile, surrounding clay of SDCM pile, DCM pile, and surrounding clay of DCM pile, respectively. Thus, this implies that the maximum ratios are still less than the failure value of 0.5. Also, the ratios of SDCM pile and surrounding clay of SDCM pile are less than those of DCM pile and surrounding clay of DCM pile, which means that SDCM pile increased the stability and safety during construction. The average values of the ratios are 0.20, 0.11, 0.27, and 0.21 for SDCM pile, surrounding clay of SDCM pile, DCM pile and surrounding clay of DCM pile, respectively. Therefore, the lateral movement to settlement ratio can be reduced by SDCM pile, especially in the surrounding clay, by as much as 50%.

The effectiveness of SDCM ground improvement employing jet mixing technique in improving thick deposit of soft clay for foundation support of an embankment has been confirmed from field observations. The implementation of this ground improvement technique enabled the following improvements in the engineering performance of the SDCM improved soft ground: (1) increased bearing capacity and lateral resistance of the soft clay foundation, (2) increased stability of the soft clay foundation, and (3) reduced compressibility of the improved foundation.

The length of concrete core pile significantly affected the settlement of the SDCM pile. The longer the concrete core, the smaller the settlement. The settlement of SDCM pile decreased linearly with increased amounts of pile core length from 0 m ($L_{core}=0$ m) to 6 m ($L_{core}=6$ m). On the other hand, the sectional areas of the core pile only slightly affected the settlements. The lengths of concrete core pile were varied from 4 to 7 m. The magnitudes of lateral movement of SDCM with 0.22-m square concrete core pile are larger than those of SDCM with 0.30-m square concrete core pile, which means that the section area of concrete core pile affected the lateral movement as expected, similar to results of the lateral load full-scale test and simulation on SDCM pile under lateral load. Both the lengths and sectional areas of concrete core pile affected the lateral movements of the SDCM pile under embankment loading.

REFERENCES

Alfaro, M.C., Hayashi, S., Miura, N., Bergado, D.T., 1997. Deformation of reinforced wall/embankment system on soft clay foundation. Soils Found. 37 (4), 33–46.

ASTM D1143-94, 1994. Standard Test Method for Piles Under Static Axial Compression Load. Annual Book of ASTM Standards.

ASTM D3689-94, 1994. Standard Test Method for Piles Under Static Axial Tensile Load. Annual Book of ASTM Standards.

ASTM D3966-95, 1995. Standard Test Method for Piles Under Lateral Loads. Annual Book of ASTM Standards.

Balasubramaniam, A.S., Brenner, R.P., 1981. Consolidation and settlement of soft clay. In: Brand, E.W., Brenners, R.P. (Eds.), Soft Clay Engineering. Elsevier, New York.

Balasubramaniam, A.S., Hwang, Z.M., Uddin, W., Chaudhry, A.R., Li, Y.G., 1978. Critical state parameters and peak stress envelopes for Bangkok clays. Q. J. Eng. Geol. 11, 219–232.

Bergado, D.T., Chai, J.C., Miura, N., 1995. FE analysis of grid reinforced embankment system on soft Bangkok clay. Comput. Geotechnics 17, 447–471.
Bergado, D.T., Long, P.V., Srivinasa Murthy, B.R., 2002. A case study of geotextile-reinforced embankment of soft ground. Geotext. Geomembranes 20 (6), 343–365.
Bergado, D.T., Ruenkrairergsa, T., Taesiri, Y., Balasubramaniam, A.S., 1999. Deep soil mixing to reduce embankment settlement. Ground Improvement J. 3 (3), 1–18.
Bergado, D.T., Teerawattanasuk, C., Youwai, S., Voottipruex, P., 2000. FE modeling of hexagonal wire reinforced embankment on soft clay. Can. Geotech. J. 37, 1209–1226.
Brinkgreve, R.B., Broere, W., 2006. PLAXIS 3D Foundation Version 1.6 Manual. Balkema, A. A., Rotterdam, Brookfield, The Netherlands.
Britto, A.M., Gunn, M.J., 1987. Critical State Soil Mechanics via Finite Elements. Horwood, Chichester, UK.
Broms, B.B., 1964. Lateral resistance of piles in cohesive soils. J. Soil Mech. Found. Div 90 (SM2), 27–61.
Butler, H.D., Hoy, H.E., 1977. The Texas Quick Load Method for Foundation Load Testing—User's manual. Federal Highway Administration report No. FHWA-IP-77-8.
Chai, J.C., 1992. Interaction Behavior between Grid Reinforcement and Cohesive Frictional Soils and Performance of Reinforced Wall/Embankment on Soft Ground. D. Eng. Dissertation No. GT-91-1Asian Institute of Technology, Bangkok, Thailand.
Chai, J.C., Bergado, D.T., 1993. Some techniques for FE analysis of embankment on soft ground. Can. Geotech. J. 30, 710–719.
Dong, P., Qin, R., Chen, Z., 2004. Bearing capacity and settlement of concrete-cored DCM pile in soft ground. Geotech. Geological Eng. 22 (1), 105–119.
Indraratna, B., Balasubramanlam, A.S., Balachandran, S., 1992. Performance of test embankment constructed to failure on soft marine clay. J. Geotech. Eng. 118 (1), 12–33.
Lai, Y.P., Bergado, D.T., Lorenzo, G.A., Duangchan, T., 2006. Full-scale reinforced embankment on deep jet mixing improved ground. Ground Improvement J. 10 (4), 153–164.
Liu, H.L., Ng, C.W.W., Fei, K., 2007. Performance of a geogrid-reinforced and pile-supported highway embankment over soft clay: Case study. J. Geotech. Geoenviron. Eng. 133 (2), 1483–1493.
Lorenzo, G.A., 2005. Fundamentals of Cement-Admixed Clay in Deep Mixing and Its Behavior as Foundation Support of Reinforced Embankment on Subsiding Soft Clay Ground. D. Eng. Dissertation, Asian Institute of Technology, Thailand.
Moh, Z.C., Lin, P.C., 2006. Geotechnical history of the development of the Suvarnabhumi International Airport. In: Proc. Int. Symp. Geotechnical Aspects Second Bangkok International (Suvarnabhumi) Airport in Thailand, May 15–16. Miracle Grand Convention Hotel, Bangkok, Thailand.
Parnploy, U., 1985. Deformation analysis and settlement prediction of Bangna-Bangpakong Highway (Section 1). M. Eng. Thesis, Asian Institute of Technology, Bangkok, Thailand.
Petchgate, K., Jongpradist, P., Jamsawang, P., 2004. Field flexural behavior of soil–cement pile. In: Proc. Fifth Symp. 2004 on Soil/Ground Improvement and Geosynthetics. King Mongkut's University of Technology, Thonburi, Thailand, pp. 85–90.
Petchgate, K., Jongpradist, P., Panmanajareonphol, S., 2003. Field pile load test of soil–cement pile in soft clay. In: Proc. Int. Symp. 2003 on Soil/Ground Improvement and Geosynthetics in Waste Containment and Erosion Control Applications, December 2–3, 2003. Asian Institute of Technology, Thailand, pp. 175–184.
Vermeer, P.A., Brinkgreve, R.B.J., 1995. Finite Element Code for Soil and Rock Analysis. Balkema, Rotterdam, The Netherlands.
Wu, M., Zhao, X., Dou, Y.M., 2005. Application of stiffened deep cement mixed pile in ground improvement. In: Proc. Int. Conf. on Deep Mixing Best Practices and Recent Advances, Stockholm, Sweden, pp. 463–468.

CHAPTER 3

Undersea Tunnel

Effect of Drainage and Grouting

Toshihisa Adachi[1], **Takeshi Tamura**[2]
[1]Geo-Research Institute, Osaka, Japan
[2]Kyoto University, Kyoto, Japan

3.1 INTRODUCTION

The most important problem in the construction of undersea tunnels and openings in grounds under high water pressure is how to take proper measures to meet the water conditions. It is well-known that drain holes or drain tunnels can improve the stability of the tunnel facing as well as the surrounding ground. An experiment was carried out at a test tunnel of the Seikan Tunnel in order to investigate what kind of effects are caused to the ground by the changes in water pressure brought about by the opening and/or the closing of drain holes. As a result, it was proved that the earth pressure acting on the supports increased and that the radius of the tunnel contracted when drainage from the drain holes was stopped.

On the other hand, when constructing a deep undersea tunnel, a very thick lining system is required to counteract the high water pressure that builds up behind the lining if water inflow to the tunnel opening is not allowed. Therefore, in the case of a highly permeable ground, the grouting method is applied to decrease the amount of inflow by reducing the permeability of the ground and by letting the water enter the tunnel in order to avoid the buildup of high water pressure behind the lining. Grouting is not only time-consuming and costly but also it largely affects the progress of the tunnel excavation. Therefore, it becomes another important task to properly determine the size of the grouting zone. In this study, the mechanical interaction between the surrounding ground and the existing water pressure is theoretically investigated in order to clarify the effectiveness of drain holes and to determine the optimum extent of grouting in undersea tunnel construction projects. First, the problems are analyzed by idealizing that the ground is an elasto-plastic body, defined so as to satisfy the Mohr–Coulomb type of plastic yield criterion in terms of the effective stress with

Chemical, Electrokinetic, Thermal, and Bioengineering Methods
http://dx.doi.org/10.1016/B978-0-08-100191-2.00003-4

the nonassociated flow rule, and that the region which includes the drain holes and the grouted zone can be treated as being in the plane strain state of a thick wall cylinder. Second, the effectiveness of the drain holes is discussed by comparing the experimental results with the theoretical solutions. Finally, the optimum extent for the grouted area is predicted by assuming the Seikan Tunnel as a model.

There are previous excellent studies by Shimokawachi (1971), Sakurai (1971), and Kudo (1971) concerning the mechanical interaction between the surrounding ground and the existing water with high pressure. Taking the water pressure into account and analyzing with the Mohr–Coulomb type of plastic yield criterion, Shimokawachi discussed the design principle for tunnel linings. Using a visco-elasto-plastic model, Sakurai also solved the problem of tunnel linings in terms of the effective stress. Meanwhile, Kudo proposed a method to determine the proper size of the grouted zone under the assumption that no plastic region will occur in a grouted zone. The current study, the purpose of which is similar to that of the previously mentioned studies, is characterized by more general constitutive relations based on the nonassociated flow rule and by a more practical subject, such as the estimation of the effect of drainage and the determination of the optimum extent of grouting (Adachi and Tamura, 1978a; Fujita et al., 1982; Adachi, 1986).

3.2 ELASTO-PLASTIC ANALYSIS OF THE GROUND SURROUNDING A TUNNEL

3.2.1 Assumptions for the analysis

Figure 3.1 shows a tunnel built at a depth of H_2 m from the sea bottom of which the depth is H_1 m. The first problem is to investigate how the ground behaves when drainage is stopped from the drain holes, which are bored apart from the tunnel wall after the tunnel is excavated as shown in Fig. 3.1(a), and the support system is installed under draining conditions. That is, it is necessary to confirm the fact that draining the leaking water from the drain holes stabilizes the surrounding ground. The second problem is to presume the optimum extent of the grouted zone in order to stabilize the ground as shown in Fig. 3.1(b). To investigate the effect of drainage and to establish the optimum extent of grouting, an idealized boundary value problem is theoretically analyzed, based on the following assumptions (Adachi and Tamura, 1978a,b):

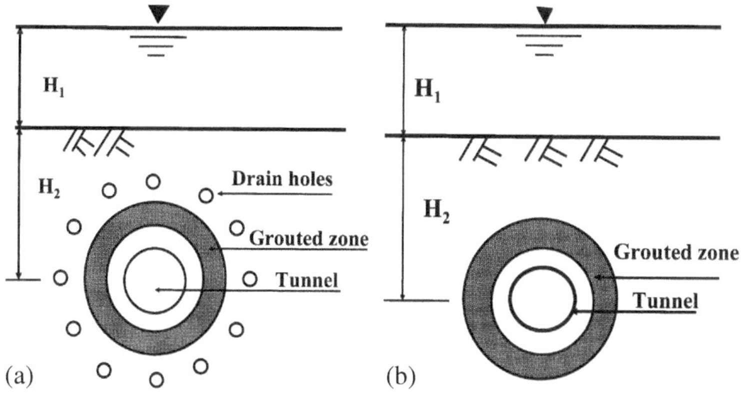

Figure 3.1 Schematic view of the undersea tunnel (a) drainage and (b) grouting problems.

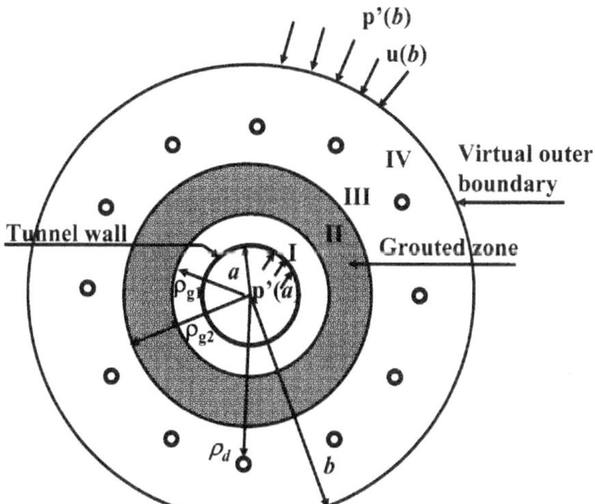

Figure 3.2 Analytical conditions of the thick wall cylinder problem.

1. The problem is idealized as a plane strain thick wall cylindrical problem, the geometrical and boundary conditions of which are shown in Fig. 3.2. In the figure, a is the tunnel radius, b is the virtual outside radius of the thick wall cylinder, ρ_{g1} is the inner radius of the grouted area, ρ_{g2} is the outer radius of the grouted area, ρ_d is the distance from the tunnel axis to the drain holes, $p'(b)$ is the effective earth pressure acting on the outside boundary, $u(b)$ is the water pressure at the outside boundary, and $p'(a)$ is the effective pressure acting on the tunnel wall, namely the reaction by the tunnel support structures.

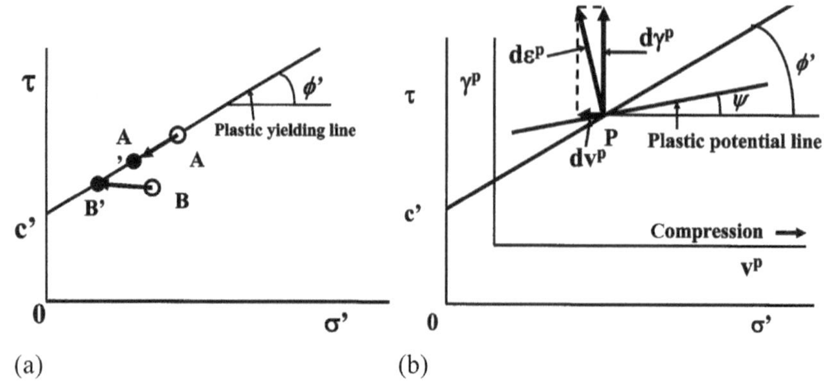

Figure 3.3 Mohr–Coulomb type of (a) yield conditions and (b) plastic potential conditions.

2. The ground is of an elasto-plastic material governed by a Mohr–Coulomb type of yield condition in terms of the effective stress, as illustrated in Fig. 3.3(a). The plastic constitutive equations are given by the nonassociated flow rule in the theory of plasticity. The plastic potential function is also given as a Mohr–Coulomb type of criterion, as shown in Fig. 3.3(b). The ground improved by grouting is assumed to be of the same elasto-plastic material as the original ground.
3. Darcy's law governs the movement of the interstitial water. No coupling between the ground water movement and the ground deformations is considered.
4. The coefficient of permeability for the original ground, k_0, that for the grouted ground k_g, and that for the plastic region developed in the grouted ground, k_{gp}, can differ from each other. Based on typical experiences in the Seikan Tunnel, the water cutoff efficiency by grouting is improved by 10^{-2} in terms of the coefficient of permeability, namely $k_g/k_0 = 1/100$.

The effective stress, σ'_{ij}, which was originally defined by Terzaghi, is equal to the value of the total stress, σ_{ij}, minus water pressure, u:

$$\sigma'_{ij} = \sigma_{ij} - u\delta_{ij} \tag{3.1}$$

The effect of drainage from the drain holes on the stabilization of the surrounding ground for the tunnel excavation can be explained in terms of the effective stress as follows. When the drainage is stopped—that is, the drain holes are closed after the tunnel is built—effective stress states A and B at two locations in the ground might change to states A′ and B′ on

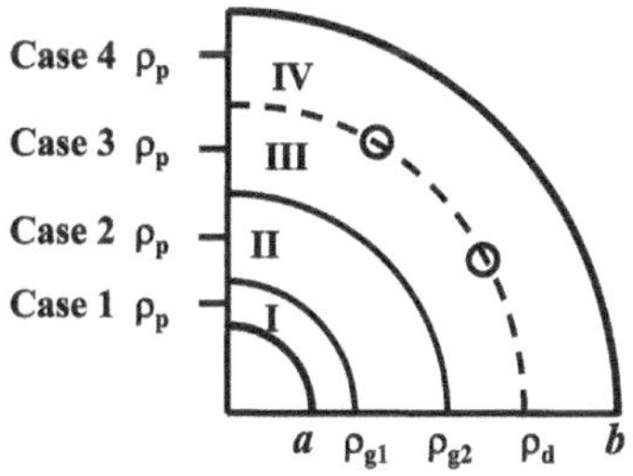

Figure 3.4 Zones I–IV and various cases.

the plastic yielding line, respectively, because the water pressure increases in the ground. This results in an increase in the plastic region and the ground becomes unstable.

For convenience in analyzing the problem, four zones, namely Zones I—IV, are defined for the surrounding ground, as shown in Figures 3.2 and 3.4. In other words, Zone I is between the tunnel wall and the grouted area ($\rho_{g1} \geq r \geq a$), Zone II is the grouted area ($\rho_{g2} \geq r \geq \rho_{g1}$), Zone III is between the drain holes and the grouted area ($\rho_d \geq r \geq \rho_{g2}$), and Zone IV is outside the drain holes ($b \geq r \geq \rho_d$). Based on where the elasto-plastic boundary is located in the previously defined four zones, the problem was analyzed for five cases. For example, Case I is defined when the elasto-plastic boundary, $r=\rho_p$, lies in Zone I ($\rho_{g1} \geq r \geq a$). Cases II–IV are similarly defined. In addition, Case V should be defined to give the ground deformation in the original state of stress before the tunnel excavation.

3.2.2 Parameter for the effect of drainage and pore water pressure distribution

The pore water pressure distribution is obtained from an analysis of the steady-state problem of seepage with the boundary conditions, namely $u(b)$ and $u(a)$. In order to evaluate the effect of drainage by drain holes, a parameter for the effect of drainage, $m_d = Q_d/Q$, is defined as the value of the ratio of the amount of drainage, Q_d, by drain holes against the total amount of inflow, Q, into the inside of the virtual outside boundary—that is, $r=b$. For example, $m_d=0$ denotes no drainage by drain holes. On the other hand, $m_d=1$ corresponds to the case in which the total amount of inflow, Q, is completely drained through the drain holes. Besides m_d, $m_{d'}$ is defined as $m_{d'}=1 - u(r_d)/u(b)$, which is useful in an analysis when the water pressure, $u(r_d)$, is given. It is important to recognize that $m_{d'}=1$ indicates $m_d=1$, however, $m_{d'}=0$ does not correspond to $m_d=0$.

The coefficient of permeability for the original ground, k_0, that for the grouted area, k_g, that for the plastic region developed in the original ground, k_{0p}, and that for the plastic region developed in the grouted area, k_{gp}, can differ from each other. For convenience, the following ratios for the coefficients of permeability are introduced: $n_g = k_0/k_g$, $n_{0p} = k_0/k_{0p}$, and $n_{gp} = k_g/k_{gp}$.

The governing equations for the underground water flow for zone α—that is, α = Zones I, II, III, and IV—are as follows:

$$\text{Darcy's law} \quad v_{r\alpha} = -\frac{k_\alpha}{w}\frac{du_\alpha(r)}{dr} \tag{3.2}$$

$$\text{Equation of continuity} \quad -Q_\alpha = 2\pi r v_{r\alpha} \tag{3.3}$$

where $v_{r\alpha}$, k_α, $u_\alpha(r)$, w, and Q_α denote the velocity of the underground water, the permeability coefficient, the underground water pressure, the density of water, and the amount of inflow water, respectively. Eliminating $v_{r\alpha}$ from Eqs. (3.2) and (3.3), the next equations are obtained:

$$r\frac{du_\alpha(r)}{dr} = \frac{wQ_i}{2\pi k\alpha} = q_\alpha \quad \text{(constant value)} \tag{3.4}$$

Integrating Eq. (3.4) with the following boundary conditions, namely

$$u_I(a) = u(a) \text{ and } u_{IV}(b) = u(b)$$

the underground water pressure distribution for each zone can be obtained. The calculated results for only Cases 1 and 5 are given here. However, solutions for the other cases can be similarly obtained.

1. Case 1 ($a \leq \rho_p \leq \rho_{g1}$) (the plastic zone located inside of the unimproved original ground)

   Zone I_p ($a \leq r \leq \rho_p$) (in the unimproved original ground, the plastic state)

$$r\frac{du_{Ip}(r)}{dr} = \frac{wQ_{Ip}}{2\pi k_{0p}} = \frac{w(Q - Q_d)}{2\pi k_{0p}} = n_{0p}(1 - m_d)q = q_{IP} \tag{3.5.1a}$$

$$u_{Ip}(r) = (1 - m_d)qn_{0p} \ln\frac{r}{a} + u(a) \tag{3.5.1b}$$

   Zone I ($\rho_p \leq r \leq \rho_{g1}$) (in the unimproved original ground, the elastic zone)

$$r\frac{du_I(r)}{dr} = \frac{wQ_I}{2\pi k_0} = \frac{w(Q - Q_d)}{2\pi k_0} = (1 - m_d)q = q_I \tag{3.5.2a}$$

$$u_{\mathrm{I}}(r)=(1-m_{\mathrm{d}})q\left[\ln\frac{r}{\rho_{\mathrm{p}}}+n_{0\mathrm{p}}\ln\frac{\rho_{\mathrm{p}}}{a}\right]+u(\mathrm{a}) \qquad (3.5.2\mathrm{b})$$

Zone II ($\rho_{\mathrm{g1}}\leq r\leq\rho_{\mathrm{g2}}$) (in the grouted area, the elastic state)

$$r\frac{\mathrm{d}u_{\mathrm{II}}(r)}{\mathrm{d}r}=\frac{w\mathrm{Q}_{\mathrm{II}}}{2\pi k_{\mathrm{g}}}=\frac{w(\mathrm{Q}-\mathrm{Q}_{\mathrm{d}})}{2\pi k_{\mathrm{g}}}=n_{\mathrm{g}}(1-m_{\mathrm{d}})q=q_{\mathrm{II}} \qquad (3.5.3\mathrm{a})$$

$$u_{\mathrm{II}}(r)=(1-m_{\mathrm{d}})q\left[n_{\mathrm{g}}\ln\frac{r}{\rho_{\mathrm{g1}}}+\ln\frac{\rho_{\mathrm{g1}}}{\rho_{\mathrm{p}}}+n_{0\mathrm{p}}\ln\frac{\rho_{\mathrm{p}}}{a}\right]+u(\mathrm{a}) \qquad (3.5.3\mathrm{b})$$

Zone III ($\rho_{\mathrm{g2}}\leq r\leq\rho_{\mathrm{d}}$) (between the outside boundary of the grouted area and the drain holes, the elastic state)

$$r\frac{\mathrm{d}u_{\mathrm{III}}(r)}{\mathrm{d}r}=\frac{w\mathrm{Q}_{\mathrm{III}}}{2\pi k_0}=\frac{w(\mathrm{Q}-\mathrm{Q}_{\mathrm{d}})}{2\pi k_0}=(1-m_{\mathrm{d}})q=q_{\mathrm{III}} \qquad (3.5.4\mathrm{a})$$

$$u_{\mathrm{III}}(r)=(1-m_{\mathrm{d}})q\left[\ln\frac{r}{\rho_{\mathrm{g2}}}+n_{\mathrm{g}}\ln\frac{\rho_{\mathrm{g2}}}{\rho_{\mathrm{g1}}}+\ln\frac{\rho_{\mathrm{g1}}}{\rho_{\mathrm{p}}}+n_{0\mathrm{p}}\frac{\rho_{\mathrm{p}}}{a}\right]+u(\mathrm{a}) \qquad (3.5.4\mathrm{b})$$

Zone IV ($\rho_{\mathrm{d}}\leq r\leq b$) (outside the drain holes, the elastic state)

$$r\frac{\mathrm{d}u_{\mathrm{IV}}(r)}{\mathrm{d}r}=\frac{w\mathrm{Q}_{\mathrm{IV}}}{2\pi k_0}=\frac{w(\mathrm{Q})}{2\pi k_0}=q=q_{\mathrm{IV}} \qquad (3.5.5\mathrm{a})$$

$$u_{\mathrm{IV}}(r)=q\left\{\ln\frac{r}{\rho_{\mathrm{d}}}+(1-m_{\mathrm{d}})\left[\ln\frac{\rho_{\mathrm{d}}}{\rho_{\mathrm{g2}}}+n_{\mathrm{g}}\ln\frac{\rho_{\mathrm{g2}}}{\rho_{\mathrm{g1}}}+\frac{\rho_{\mathrm{g1}}}{\rho_{\mathrm{p}}}+n_{0\mathrm{p}}\ln\frac{\rho_{\mathrm{p}}}{a}\right]\right\}+u(\mathrm{a}) \qquad (3.5.5\mathrm{b})$$

in which

$$q=[u(\mathrm{b})-u(\mathrm{a})]\bigg/\left\{\ln\frac{b}{\rho_{\mathrm{d}}}+(1-m_{\mathrm{d}})\left[\ln\frac{\rho_{\mathrm{d}}}{\rho_{\mathrm{g2}}}+n_{\mathrm{g}}\ln\frac{\rho_{\mathrm{g2}}}{\rho_{\mathrm{g1}}}+\ln\frac{\rho_{\mathrm{g1}}}{\rho_{\mathrm{p}}}+n_{0\mathrm{p}}\ln\frac{\rho_{\mathrm{p}}}{a}\right]\right\} \qquad (3.5.6)$$

and

$$m'_{\mathrm{d}}=[1-u(\rho_{\mathrm{d}})/u(b)]$$

$$=\ln\frac{b}{\rho_{\mathrm{d}}}\bigg/\left\{\ln\frac{b}{\rho_{\mathrm{d}}}+(1-m_{\mathrm{d}})\left[\ln\frac{\rho_{\mathrm{d}}}{\rho_{\mathrm{g2}}}+n_{\mathrm{g}}\ln\frac{\rho_{\mathrm{g2}}}{\rho_{\mathrm{g1}}}+\ln\frac{\rho_{\mathrm{g1}}}{\rho_{\mathrm{p}}}+n_{0\mathrm{p}}\ln\frac{\rho_{\mathrm{p}}}{a}\right]+u(a)\right\} \qquad (3.5.7)$$

2. Case 5 (before the tunnel excavation)

Because the tunnel has not been excavated, no groundwater flow has taken place inside the drain holes. In this case, the following solutions are given:

Zones I, II, and III (a more inner area than the drain holes)

$$r\frac{du_\alpha(r)}{dr}=\frac{wQ_\alpha}{2\pi k_\alpha}=0=q_\alpha \quad (\alpha=\text{I, II, and III}) \tag{3.6.1a}$$

$$u_\alpha(r)=\text{constant}=u_{\text{IV}}(\rho_\text{d})=u(\rho_\text{d}) \quad (\alpha=\text{I, II, and III}) \tag{3.6.1b}$$

Zone IV ($\rho_\text{d}\leq r\leq b$) (a more outer part than the drain holes)

$$r\frac{du_{\text{IV}}(r)}{dr}=\frac{wQ_{\text{IV}}}{2\pi k_{\text{IV}}}=q=q_{\text{IV}} \tag{3.6.2a}$$

$$u_{\text{IV}}(r)=m_\text{d}q\ \ln\frac{r}{b}+u(b) \tag{3.6.2b}$$

in which

$$q=[u(\text{b})-u(\rho_\text{d})]/m_\text{d}\ \ln\frac{b}{\rho_\text{d}} \tag{3.6.3}$$

and

$$m_\text{d}'=m_\text{d}q\ \ln\frac{b}{\rho_\text{d}}\Big/u(\text{b}) \tag{3.6.4}$$

The groundwater pressure in each zone, $u_\alpha(r)$, is given by the previous discussion. However, the radius of the plastic zone, ρ_p, is still an unknown parameter. It can be determined in the stress analysis discussed in the next section.

3.2.3 Determination of stress levels and displacements by an elasto-plastic analysis

Because this problem is handled under axisymmetric plane strain conditions, the displacement and the strain components are given as follows:

$$u_r=u_r(r), u_\vartheta=0,\ u_z=0 \tag{3.7.1a}$$

$$\varepsilon_r=\partial u_r(r)/\partial r, \varepsilon_\vartheta=u_r(r)/r,\ \varepsilon_\vartheta=0,\ \varepsilon_{\vartheta z}=\varepsilon_{zr}=\varepsilon_{r\vartheta}=0 \tag{3.7.1b}$$

in which u_r, u_ϑ, and u_z are the displacement components in the radial, tangential, and axial directions, respectively, and ε_r, ε_ϑ, and ε_z denote the strain components in each direction.

The next relation gives the compatibility equation under these conditions:

$$\frac{\partial}{\partial r}(r\varepsilon_{\vartheta}) = \varepsilon_r \tag{3.8}$$

Assuming the infinitesimal displacement and strain conditions, the total strain components, ε_{ij}, are given simply by the sum of the elastic components, $\varepsilon_{ij}^{\mathrm{E}}$, and the plastic components, $\varepsilon_{ij}^{\mathrm{P}}$:

$$\varepsilon_{ij} = \varepsilon_{ij}^{\mathrm{E}} + \varepsilon_{ij}^{\mathrm{P}} \tag{3.9}$$

Using the deviatoric stress, S_{ij}, and the deviatoric strain, e_{ij}, the following constitutive equations are assumed to apply:

$$e_{ij}^{\mathrm{E}} = \frac{1}{2G} S_{ij} \tag{3.10.1}$$

$$e = \frac{1}{3K} S' = \frac{1}{3K}(S - u_{\alpha}(r)) \tag{3.10.2}$$

$$\dot{\varepsilon}_{ij}^{\mathrm{P}} = \Lambda \frac{\partial g}{\partial \sigma'_{ij}} \tag{3.10.3}$$

in which G and K are the shear modulus and the bulk modulus, respectively; g is the plastic potential function; Λ is the proportional parameter; and the deviatoric stress, S_{ij}, the mean stress, S, the deviatoric strain, e_{ij}, and the mean strain, e, are defined as follows:

$$S_{ij} = \sigma_{ij} - S\delta_{ij}, \quad S = \frac{1}{3}\sigma_{kk} \tag{3.11.1}$$

$$e_{ij} = \varepsilon_{ij} - e\delta_{ij}, \quad e = \frac{1}{3}\varepsilon_{kk} \tag{3.11.2}$$

The equilibrium equation under axisymmetric plane strain conditions is expressed as

$$\frac{\partial \sigma_r}{\partial r} + \frac{\sigma_r - \sigma_{\vartheta}}{r} = \frac{\partial \sigma'_r}{\partial r} + \frac{\partial u_{\alpha}(r)}{\partial r} + \frac{\sigma'_r - \sigma'_{\vartheta}}{r} = 0 \tag{3.12}$$

where σ_r and σ_{ϑ} are the radial stress and the tangential stress, respectively, and σ'_r and σ'_{ϑ} are their effective stress expressions. In this chapter, the positive values for both stress and strain are defined to stand for the compressive state.

1. Solutions for the elastic regions

   In the elastic analysis, such parameters as e and ξ are used, and they are defined as

$$e=\frac{1}{3}(\varepsilon_r+\varepsilon_\vartheta+\varepsilon_z)=\frac{1}{3}(\partial u_r/\partial r+u_r/r) \quad (3.13.1)$$

$$\xi=\frac{1}{3}(\varepsilon_r-\varepsilon_\vartheta)=\frac{1}{3}(\partial u_r/\partial r-u_r/r) \quad (3.13.2)$$

   Using e and ξ, the constitutive equations can be expressed as follows:

$$S_r=2Ge_r=G(e+3\xi),\quad \sigma_r'=(G+3K)e+3G\xi \quad (3.14.1)$$

$$S_\vartheta=2Ge_\vartheta=G(e-3\xi),\quad \sigma_\vartheta'=(G+3K)e-3G\xi \quad (3.14.2)$$

$$S_z=2Ge_z=-2Ge,\quad \sigma_z'=(3K-2G)e \quad (3.14.3)$$

$$S'=3Ke,\quad \sigma_{rz}=0 \quad (3.14.4)$$

where S_r, S_ϑ, and S_z are the deviatoric components of σ_r, σ_ϑ, and σ_z, respectively.

   Introducing Eq. (3.13) into the compatibility equation (Eq. 3.8) and introducing Eq. (3.14) into the equilibrium equation (Eq. 3.12), the following equations can be obtained for Zone α. In other words,

$$\frac{\partial e}{\partial r}-\frac{\partial \xi}{\partial r}=2\frac{\xi}{r} \quad \text{(Compatibility equation)} \quad (3.15)$$

$$(G+3K)\frac{\partial e}{\partial r}+3G\left(\frac{\partial \xi}{\partial r}+2\frac{\xi}{r}\right)+\frac{\partial u_\alpha(r)}{\partial r}=0 \quad \text{(Equilibrium equation)} \quad (3.16)$$

   From Eqs. (3.15) and (3.16), two individual differential equations for e or ξ are obtained, respectively:

$$(4G+3K)\frac{\partial e}{\partial r}+\frac{\partial u_\alpha(r)}{\partial r}=0 \quad (3.17.1)$$

$$\frac{\partial \xi}{\partial r}+2\frac{\xi}{r}+\frac{1}{(4G+3K)}\frac{\partial u_\alpha(r)}{\partial r}=0 \quad (3.17.2)$$

   Integrating the previous equations gives the following relations for e and ξ in elastic Zone α:

$$e=\frac{1}{(4G+3K)}[C_{1\alpha}-u_\alpha(r)] \quad (3.18)$$

Introducing the relations in Eq. (3.18) into the relations in Eqs. (3.13) and (3.14) yields the stress levels and the displacement in Zone α, as follows:

$$\sigma'_r = \frac{G+3K}{4G+3K}[C_{1\alpha} - u_\alpha(r)] + 3G\left[\frac{C_{2\alpha}}{r^2} - \frac{q_\alpha}{2(4G+3K)}\right] \tag{3.19.1}$$

$$\sigma'_\vartheta = \frac{G+3K}{4G+3K}[C_{1\alpha} - u_\alpha(r)] - 3G\left[\frac{C_{2\alpha}}{r^2} - \frac{q_\alpha}{2(4G+3K)}\right] \tag{3.19.2}$$

$$\sigma'_z = \frac{3K-2G}{4G+3K}[C_{1\alpha} - u_\alpha(r)] \tag{3.19.3}$$

$$u_r = \frac{3}{2}\left[\frac{C_{1\alpha} - u_\alpha(r)}{4G+3K} - \frac{C_{2\alpha}}{r^2} + \frac{q_\alpha}{2(4G+3K)}\right] \tag{3.19.4}$$

where the constants of the integration, $C_{1\alpha}$ and $C_{2\alpha}$, will be determined in order to satisfy the boundary conditions given in the next section, and $u_\alpha(r)$ and q_α are the groundwater pressure and the amount of water flow in Zone α.

2. Solutions for the plastic regions

In the current study, the following Mohr–Coulomb types of plastic yield function, f, and plastic potential function, g, are used:

$$f = (\sigma'_\vartheta - \sigma'_z) - (\sigma'_\vartheta + \sigma'_r)\sin\phi' - 2c'\cos\phi' = 0 \tag{3.20.1}$$

$$g = (\sigma'_\vartheta - \sigma'_r) - (\sigma'_\vartheta + \sigma'_r)\sin\psi \tag{3.20.2}$$

where c' is the cohesive strength based on the effective stress, ϕ' is the friction angle based on the effective stress, and ψ is the plastic potential angle, as shown in Figure 3.3(b).

Because the plastic yield condition in Eq. (3.20.1) must be satisfied in the plastic state, eliminating σ'_ϑ in the equilibrium equation (Eq. 3.12) by using the plastic yield condition in Eq. (3.20.1) results in the following differential equation expressed only by σ'_r and q_α:

$$\frac{\partial\sigma'_r}{\partial r} - \frac{2\sin\phi'}{1-\sin\phi'}\frac{\sigma'_r}{r} + \frac{1}{r}\left[q_\alpha - \frac{2c'\cos\phi'}{1-\sin\phi'}\right] = 0 \tag{3.21}$$

The stress values, σ'_r and σ'_ϑ, can be obtained by integrating the previous equation as follows:

$$\sigma'_r = D_{1\alpha} r^{2\sin\phi'/(1-\sin\phi')} + \frac{1-\sin\phi'}{2\sin\phi'}\left[q_\alpha - \frac{2c'\cos\phi'}{1-\sin\phi'}\right] \tag{3.22}$$

$$\sigma'_\vartheta = \frac{1+\sin\phi'}{1-\sin\phi'}\sigma'_r + \frac{2c'\cos\phi'}{1-\sin\phi'} \tag{3.23}$$

Based on the plane strain conditions as well as the constitutive equations (Eqs. 10.3 and 20.2), the plastic strain in the z-direction, $\varepsilon_z^{\mathrm{P}}$, never takes place. Therefore, $\varepsilon_z = \varepsilon_z^{\mathrm{E}} = 0$ is satisfied, and σ'_z is given as follows:

$$\sigma'_z = \frac{3K-2G}{6K+2G}(\sigma'_r + \sigma'_\vartheta) = \upsilon(\sigma'_r + \sigma'_\vartheta) \tag{3.24}$$

where υ is Poisson's ratio.

Using Eqs. (3.10.3) and (3.20.2) to determine the displacement in the plastic region, $u_r(r)$, gives the following relations:

$$\dot{\varepsilon}_r^{\mathrm{P}} = -\Lambda(1+\sin\psi), \dot{\varepsilon}_\vartheta^{\mathrm{P}} = \Lambda(1-\sin\psi), \text{and } \dot{\varepsilon}_z^{\mathrm{P}} = 0 \tag{3.25}$$

Integrating Eq. (3.25) with the initial conditions—that is, $\varepsilon_r^{\mathrm{P}} = \varepsilon_\vartheta^{\mathrm{P}} = \varepsilon_z^{\mathrm{P}} = 0$ —yields the following relation:

$$\varepsilon_r^{\mathrm{P}} = -\frac{1+\sin\psi}{1-\sin\psi}\varepsilon_\vartheta^{\mathrm{P}}, \quad \varepsilon_z^{\mathrm{P}} = 0 \tag{3.26}$$

This relation shows the normality rule; that is, the plastic strain increment vector, $\dot{\varepsilon}^{\mathrm{P}}$, is perpendicular to the plastic potential line, as shown in Figure 3.3(b).

The plastic volumetric strain, $v^{\mathrm{P}} = 3e^{\mathrm{P}} = \varepsilon_r^{\mathrm{P}} + \varepsilon_\vartheta^{\mathrm{P}} + \varepsilon_z^{\mathrm{P}}$, is expressed by

$$v^{\mathrm{P}} = -\frac{2\sin\psi}{1-\sin\psi}\varepsilon_\vartheta^{\mathrm{P}} \tag{3.27}$$

Under the condition of $0 \le \psi \le 90°$, plastic volumetric strain v^{P} is always negative; that is, plastic volume expansion occurs.

Using the relations in Eqs. (3.9) and (3.25) in the compatibility equation (Eq. 3.8) yields the next equation:

$$r\frac{\partial\varepsilon_\vartheta^{\mathrm{P}}}{\partial r} + r\frac{\partial\varepsilon_\vartheta^{\mathrm{E}}}{\partial r} = \left(\varepsilon_r^{\mathrm{E}} - \varepsilon_\vartheta^{\mathrm{E}}\right) - \frac{2}{1-\sin\psi}\varepsilon_\vartheta^{\mathrm{P}} \tag{3.28}$$

Integrating the previous equation gives the following solution:

$$\varepsilon_\vartheta^{\mathrm{P}} + \varepsilon_\vartheta^{\mathrm{E}} = r^{2/(1-\sin\psi)}\left[H_\alpha = \int r^{\frac{1+\sin\psi}{1-\sin\psi}}\left(\varepsilon_r^{\mathrm{E}} - \varepsilon_\vartheta^{\mathrm{E}}\right)\mathrm{d}r + \frac{2}{1-\sin\psi}\int r^{\frac{1+\sin\psi}{1-\sin\psi}}\varepsilon_\vartheta^{\mathrm{E}}\,\mathrm{d}r\right] \tag{3.29}$$

The following relations are always satisfied between the elastic strain values, ε_r^E and ε_ϑ^E, and the stress values, σ'_r and σ'_ϑ; namely

$$\varepsilon_r^E - \varepsilon_\vartheta^E = \frac{1}{2G}(\sigma'_r - \sigma'_\vartheta) \tag{3.30.1}$$

$$\varepsilon_\vartheta^E = \frac{1}{4G(3K+G)}\left[(3K-4G)\sigma'_\vartheta - (3K-2G)\sigma'_r\right] \tag{3.30.2}$$

Using the plastic yield condition in Eq. (3.20.1) gives the following relations:

$$\varepsilon_r^E - \varepsilon_\vartheta^E = -\frac{1}{2G}\left[\frac{2\sin\phi'}{1-\sin\phi'}\sigma'_r + \frac{2c'\cos\phi'}{1-\sin\phi'}\right] \tag{3.31.1}$$

$$\varepsilon_\vartheta^E = \frac{1}{2G(3K+G)}\left[\frac{3G+(3K+G)\sin\phi'}{1-\sin\phi'}\sigma'_r + \frac{(3K+4G)c'\cos\phi'}{1-\sin\phi'}\right] \tag{3.31.2}$$

Integrating Eq. (3.28) with the previous relations gives the next solutions for the strain and the displacement:

$$\varepsilon_\vartheta = r^{2/(1-\sin\psi)}\left[H_\alpha + AF_\alpha(r) + BF(r)\right] \tag{3.32.1}$$

$$u_r(r) = r\varepsilon_\vartheta = r^{(1+\sin\psi)/(1-\sin\psi)}\left[H_\alpha + AF_\alpha(r) + BF(r)\right] \tag{3.32.2}$$

where H_α is a constant of the integration, and $F_\alpha(r)$, $F(r)$, A, and B are given as follows:

$$F_\alpha(r) = \left[\begin{array}{l} \dfrac{D_{1\alpha}}{\dfrac{1+\sin\psi}{1-\sin\psi}+\dfrac{1+\sin\phi'}{1-\sin\phi'}}r^{\left(\frac{1+\sin\psi}{1-\sin\psi}+\frac{1+\sin\phi'}{1-\sin\phi'}\right)} \\ +\dfrac{1-\sin\phi'}{2\sin\phi'}\left(q_\alpha - \dfrac{2c'\cos\phi'}{1-\sin\phi'}\right)\dfrac{1-\sin\psi}{2}r^{\left(\frac{2}{1-\sin\psi}\right)} \end{array}\right] \tag{3.33.1}$$

$$F(r) = \frac{1-\sin\psi}{2}r^{2/(1-\sin\psi)} \tag{3.33.2}$$

$$A = \left[\frac{3G+(3K+G)\sin\phi'}{G+(3K+G)(1-\sin\psi)(1-\sin\phi')} - \frac{\sin\phi'}{G(1-\sin\phi')}\right] \tag{3.33.3}$$

$$B = \left[\frac{(3G+4G)c'\cos\phi'}{G(3K+G)(1-\sin\psi)(1-\sin\phi')} - \frac{c'\cos\phi'}{G(1-\sin\phi')}\right] \tag{3.33.4}$$

where $D_{1\alpha}$ and H_α are the constants of integration, and $F_\alpha(r)$ becomes definite if $D_{1\alpha}$ and q_α are given.

3.2.4 Stress and displacement distributions in each zone

Only the solutions for Cases 1 and 5 are given here; however, the solutions for the other cases can be similarly obtained.

1. Case 1 ($a \leq \rho_p \leq \rho_{g1}$)

   Zone I_p ($a \leq r \leq \rho_p$) (inside the grouted area, in the plastic state)

$$\sigma'_r = D_{1I_p} r^{2\sin\phi'/(1-\sin\phi')} + \frac{1-\sin\phi'}{2\sin\phi'}\left[q_{I_p} - \frac{2c'\cos\phi'}{1-\sin\phi'}\right] \tag{3.34.1}$$

$$\sigma'_\vartheta = \frac{1+\sin\phi'}{1-\sin\phi'}\sigma'_r + \frac{2c'\cos\phi'}{1-\sin\phi'} \tag{3.34.2}$$

$$\sigma'_z = \frac{3K-2G}{6K+2G}\left(\sigma'_r + \sigma'_\vartheta\right) \tag{3.34.3}$$

$$u_r = r^{(1+\sin\psi)/(1-\sin\psi)}\left[H_{1p} + AF_{1p} + BF\right] \tag{3.34.4}$$

   Zones I, II, III, and IV ($\rho_p \leq r \leq b$)

$$\sigma'_r = \frac{G+3K}{4K+3G}[C_{1\alpha} - u_\alpha(r)] + 3G\left[\frac{C_{2\alpha}}{r^2} - \frac{q_\alpha}{2(4G+3K)}\right] \tag{3.35.1}$$

$$\sigma'_\vartheta = \frac{G+3K}{4G+3K}[C_{1\alpha} - u_\alpha(r)] - 3G\left[\frac{C_{2\alpha}}{r^2} - \frac{q_\alpha}{2(4G+3K)}\right] \tag{3.35.2}$$

$$\sigma'_z = \frac{3K-2G}{4K+3G}[C_{1\alpha} - u_\alpha(r)] \tag{3.35.3}$$

$$u_r = \frac{3}{2}\left[\frac{C_{1\alpha}}{4G+3K} - \frac{C_{2\alpha}}{r^2} + \frac{q_\alpha}{2(4G+3K)}\right]r \tag{3.35.4}$$

where $\alpha = $ I, II, III, and IV. There are 11 unknowns: D_{1Ip} and H_{1Ip} in the plastic region; C_{1I}, C_{1II}, C_{1III}, C_{1IV}, C_{2I}, C_{2II}, C_{2III}, and C_{2IV} in the elastic region; and the radius of the plastic region, ρ_p. These unknowns are determined based on the following 11 boundary conditions:

$$\sigma'_r(a) = p'(a) \text{ at } r = a$$
$$\sigma'_r(b) = p'(b) \text{ at } r = b$$

The individual continuity of $\sigma_r'(r)$ and $u_r(r)$ at

$$r=\rho_{\rm d},\rho_{\rm g2},\rho_{\rm g1},\rho_{\rm p}$$

The stress levels in the elastic region must satisfy the plastic yield condition in Eq. (3.20.1) at $r=\rho_{\rm p}$.

The determined unknowns are given here:

$$C_{1\rm I}=C_{1\rm II}=C_{1\rm III}=C_{1\rm IV} \tag{3.36.1}$$

$$C_{1\rm IV}=u_{\rm I}\left(\rho_{\rm p}\right)+\frac{(4G+3K)}{(G+3K)(1-\sin\phi')}\left[\begin{array}{l} D_{1\rm I_p}\rho_{\rm p}^{\frac{2\sin\phi'}{1-\sin\phi'}} \\ +\dfrac{1-\sin\phi'}{2\sin\phi'}\left(q_{\rm Ip}-\dfrac{2c'\cos\phi'}{1-\sin\phi'}\right) \\ +c'\cos\phi' \end{array}\right] \tag{3.36.2}$$

$$D_{1\rm I_p}=\left[p'(a)-\frac{1-\sin\phi'}{2\sin\phi'}\left(q_{\rm I_p}-\frac{2c'\cos\phi'}{1-\sin\phi'}\right)\right]a^{-\frac{2\sin\phi'}{1-\sin\phi'}} \tag{3.36.3}$$

$$C_{2\rm III}=C_{2\rm IV}-\frac{1}{2(4G+3K)}\rho_{\rm d}^2(q_{\rm IV}-q_{\rm III}) \tag{3.36.4}$$

$$C_{2\rm II}=C_{2\rm III}-\frac{1}{2(4G+3K)}\rho_{\rm g_2}^2(q_{\rm III}-q_{\rm II}) \tag{3.36.5}$$

$$C_{2\rm I}=C_{2\rm II}-\frac{1}{2(4G+3K)}\rho_{\rm g_1}^2(q_{\rm II}-q_{\rm I}) \tag{3.36.6}$$

$$C_{2\rm IV}=b^2\left\{\frac{q_{\rm IV}}{2(4G+3K)}+\frac{1}{3G}\left[p'(b)-\frac{G+3K}{4G+3K}(C_{1\rm IV}-u(b))\right]\right\} \tag{3.36.7}$$

$$H_{\rm I_p}=\frac{3}{2}\left[\frac{C_{1\rm I}-u_{\rm I}\left(\rho_{\rm p}\right)}{4G+3K}-\frac{C_{2\rm I}}{\rho_{\rm p}^2}+\frac{q_{\rm I}}{2(4G+3K)}\right]\rho_{\rm p}^{2/(1-\sin\psi)} -\left[AF_{\rm I_p}\left(\rho_{\rm p}\right)+BF\left(\rho_{\rm p}\right)\right] \tag{3.36.8}$$

The radius of plastic region $\rho_{\rm p}$ can be obtained by analyzing the following equation:

$$\frac{C_{2\rm I}}{\rho_{\rm p}^2}-\frac{q_{\rm I}}{2(4G+3K)}+\frac{(G+3K)}{3G(4G+3K)}\left[C_{1\rm IV}-u_{\rm I}\left(\rho_{\rm p}\right)\right]\sin\phi'+\frac{c'\cos\phi'}{3G} \tag{3.37}$$

Because $C_{2\mathrm{I}}$, $C_{1\mathrm{IV}}$, and $u_{\mathrm{I}}(\rho_{\mathrm{p}})$ in the equation are functions of ρ_{p}, Eq. (3.37) is only a function of ρ_{p}.

The analysis of this problem starts by solving Eq. (3.37).

2. Case 5

Because the stress levels and the displacement in the ground before the tunnel excavation are also given by Eq. (3.19), the unknown parameters need to be determined based on the boundary conditions.

In this case, it is enough to discuss two regions, namely Zone IV and the other zones (Zones I, II, and III).

Zones I, II, and III ($0 \le r \le \rho_{\mathrm{d}}$)

$$\sigma'_r = \frac{G+3K}{4G+3K}\left[C_{1\mathrm{III}} - u_{\mathrm{IV}}(\rho_{\mathrm{d}})\right] + 3G\frac{C_{2\mathrm{III}}}{r^2} \tag{3.38.1}$$

$$\sigma'_\vartheta = \frac{G+3K}{4G+3K}\left[C_{1\mathrm{III}} - u_{\mathrm{IV}}(\rho_{\mathrm{d}})\right] - 3G\frac{C_{2\mathrm{III}}}{r^2} \tag{3.38.2}$$

$$\sigma'_z = \frac{3K-2G}{4G+3K}\left[C_{1\mathrm{III}} - u_{\mathrm{IV}}(\rho_{\mathrm{d}})\right] \tag{3.38.3}$$

$$u_r(r) = \frac{3}{2}\left[\frac{C_{1\mathrm{III}} - u_{\mathrm{IV}}(\rho_{\mathrm{d}})}{4G+3K} - \frac{C_{2\mathrm{III}}}{r^2}\right] r \tag{3.38.4}$$

Zone IV ($\rho_{\mathrm{d}} \le r \le b$)

$$\sigma'_r = \frac{G+3K}{4G+3K}\left[C_{1\mathrm{IV}} - u_{\mathrm{IV}}(r)\right] + 3G\left[\frac{C_{2\mathrm{IV}}}{r^2} - \frac{q_{\mathrm{IV}}}{2(4G+3K)}\right] \tag{3.39.1}$$

$$\sigma'_\vartheta = \frac{G+3K}{4G+3K}\left[C_{1\mathrm{IV}} - u_{\mathrm{IV}}(r)\right] - 3G\left[\frac{C_{2\mathrm{IV}}}{r^2} - \frac{q_{\mathrm{IV}}}{2(4G+3K)}\right] \tag{3.39.2}$$

$$\sigma'_z = \frac{3K-2G}{4G+3K}\left[C_{1\mathrm{IV}} - u_{\mathrm{IV}}(r)\right] \tag{3.39.3}$$

$$u_r(r) = \frac{3}{2}\left[\frac{C_{1\mathrm{IV}} - u_{\mathrm{IV}}(r)}{4G+3K} - \frac{C_{2\mathrm{IV}}}{r^2} + \frac{q_{\mathrm{IV}}}{2(4G+3K)}\right] r \tag{3.39.4}$$

Four unknowns—$C_{1\mathrm{III}}$, $C_{1\mathrm{IV}}$, $C_{2\mathrm{III}}$, and $C_{2\mathrm{IV}}$—can be determined by the following boundary conditions:

$$\sigma'_r(b) = p'(b) \text{ at } r = b$$

The individual continuity of $\sigma'_r(r)$ and $u_r(r)$ at $r = \rho_{\mathrm{d}}\sigma'_r(0)$ and the others have their own definite values at r = 0

The determined unknowns are as follows:

$$C_{1\mathrm{III}} = C_{1\mathrm{IV}} \tag{3.40.1}$$

$$C_{1\mathrm{IV}} = u(b) + \frac{4G+3K}{G+3K}\left[p'(b) - \frac{3G}{2(4G+3K)}\left(\frac{\rho_{\mathrm{d}}^2 - b^2}{b^2}\right) q_{\mathrm{IV}}\right] \tag{3.40.2}$$

$$C_{2\mathrm{III}} = 0 \tag{3.40.3}$$

$$C_{2\mathrm{IV}} = \frac{\rho_{\mathrm{d}}^2}{2(4G+3K)} q_{\mathrm{IV}} \tag{3.40.4}$$

When the displacement given in Eqs. (3.38.4) and (3.39.4), $u_r(r)$, is denoted as $u_{r0}(r)$, the displacement that takes place due to tunneling is given by $[u_r(r) - u_r0(r)]$.

3.3 EFFECT OF DRAINAGE

First, the results of the opening and closing tests on the drain holes, which were based on the obtained theoretical solutions and conducted at a test tunnel of the Seikan Tunnel, are considered in this section. Second, the relationship between the effect of drainage and the mechanical properties of the ground are studied. Finally, an effective arrangement for the drain holes is discussed in connection with the strength characteristics of an unimproved ground by grouting.

3.3.1 Experiments on the effect of drainage

The experiment was carried out at the test tunnel in order to investigate the kind of effect caused to the mechanical behavior of the surrounding ground when the groundwater pressure is changed by opening or closing the drain holes. The test tunnel, 3.2 m in diameter and 25 m in length, is located at a depth of 240 m from the sea surface with a soil cover of 200 m, as shown in Fig. 3.5. The tunnel was excavated after improving the ground by grouting with an extent of $\rho_{\mathrm{g}} = 5.3$ m and by draining from the drain holes. After the support system was installed and the earth pressure acting on the support system reached a steady state, the drainage from the drain holes was stopped. The changes in earth pressure and the convergence during this process were measured in the fractured zone near the 22-m position, as shown in Fig. 3.6. The convergence changes were measured at five different locations with respect to A, B, and C directions, as also indicated in Fig. 3.6.

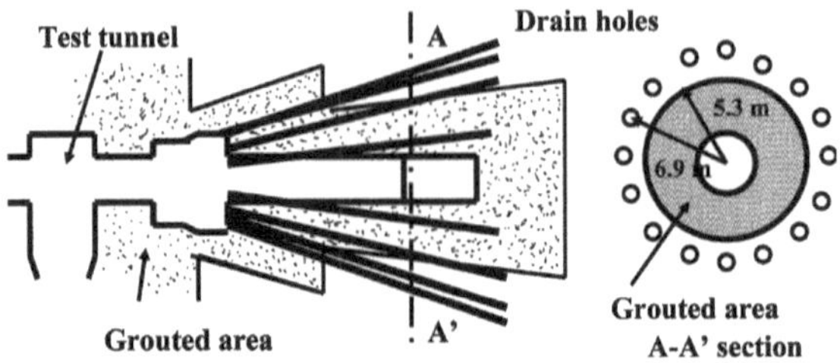

Figure 3.5 Arrangement of the test tunnel.

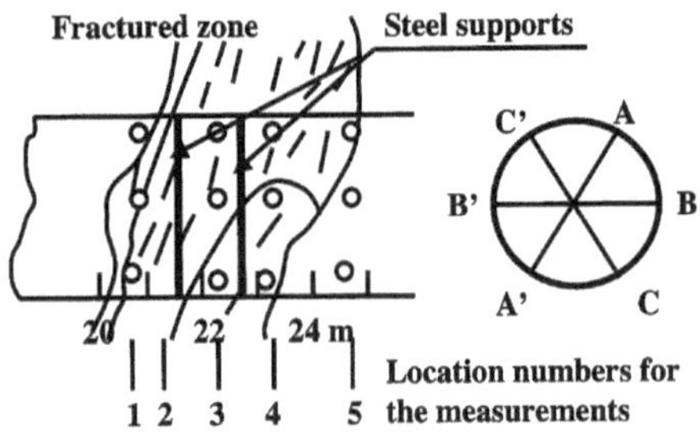

Figure 3.6 Locations for measuring the convergence and the earth pressure on the steel supports.

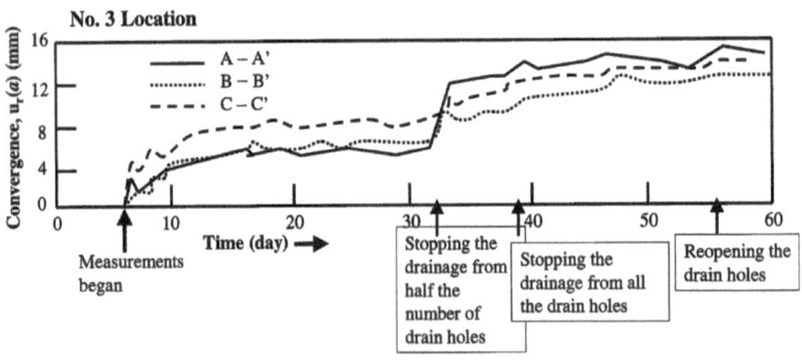

Figure 3.7 Convergence and time curves.

Figure 3.7 shows the results at Location 3 that were measured after the face was advanced approximately half a tunnel diameter beyond Location 2. Figure 3.8 illustrates the well-known relation between the tunnel convergence and the tunnel advancement that was based on the theory of elasticity. From Fig. 3.8, 40% of the convergence at a certain location has taken place when the tunnel face reaches the location, and 90% of the convergence has

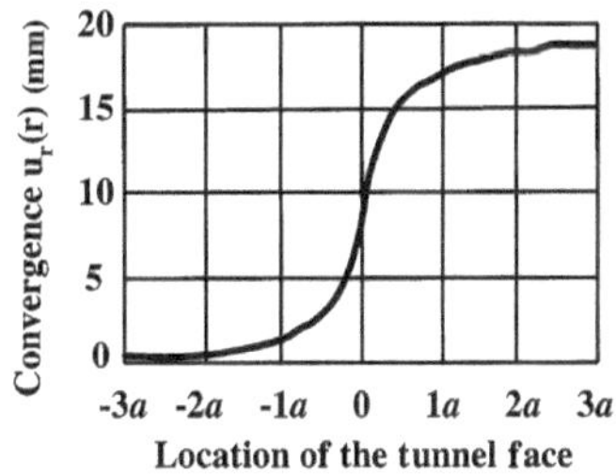

Figure 3.8 Convergence due to the tunnel face advancement.

occurred when the face advances for half a diameter beyond the location. According to the previous discussion, the measured results shown in Fig. 3.7 should be interpreted as only 10% of the total convergence took place due to tunneling. Thus, the total convergence can be thought of 10 times the measured value—that is, 60–70 mm. In Fig. 3.7, it is recognized that the tunnel diametrical convergences increase by 6 to 7 mm when the drainage is stopped from the drain holes. This increase in convergence must be due to the plastic behavior because the convergence never returns to the original value even after the drainage from the drain holes resumes. Corresponding to the increase in convergence, the earth pressure acting on the support system increases from 0.04 to 0.1 MPa.

In order to discuss the test results based on the theoretical analysis, the following analysis is conducted with the given boundary conditions and material parameters. In this analysis, the inner radius of the grouted zone is assumed to be $\rho_{g1} = 0$ m—that is, not a doughnut shape in the grouted area. The virtual outer radius of the domain, b, is 40 m. The groundwater pressure at b, u(b), is 2.4 MPa, and the effective pressure acting at b, p'(b), is 2 MPa when considering an overburden of 200 m and a submerged unit weight of the ground material of 1.0 t/m^3.

The question arises, however, of how to take into account the hanging effect on the fractured zone that is sandwiched between neighboring sound hard rock grounds. For this question, Fig. 3.9 schematically illustrates the sandwiched effect caused by the hard grounds. If the whole ground is hard, the displacement is δ_1, whereas the displacement should be δ_2 when the whole ground is fractured. In an actual case, measured displacement δ is affected by the sandwiched effect. In order to take into account the sandwiched effect, the effective earth pressure acting on the external boundary, p'(b), is assumed to reduce to one-third to one-fifth of the actual acting pressure—that is, 0.7, 1.0, and 1.36 MPa. The reduction rate should be

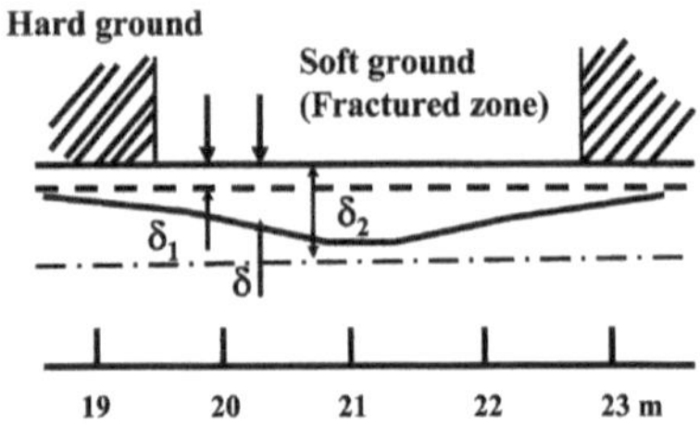

Figure 3.9 Sandwiched effect by the hard grounds.

determined by the deformation constants of both grounds, and in these grounds the reduction rate is predicted to be one-third to one-fourth. From the results of the mechanical tests on the ground material, the material parameters, $E = 3 \times 10^2$ MPa, $\upsilon = 0.4$, and $\phi' = 30°$ are used in the analyses. Particularly for cohesive strength, c', three values, namely 0.4, 0.45, and 0.5 MPa, are used in the analyses. The coefficient of permeability for the grouted area is 1/100 that for the original ground, and the coefficients for the plastic zones that developed due to tunneling becomes five times those for the original ground as well as for the grouted area. The analytical conditions are summarized in Table 3.1. m'_d is applied as the drainage effect parameter. The analytical results are given in Figs. 3.10 and 3.11. Figure 3.10

Table 3.1 Conditions for the analysis of drainage problem 1

Tunnel radius (m)	a	1.6
Radius of virtual outer boundary (m)	b	40
Radius of grouted area (m)	ρ_g	5.3
Location of drain holes (m)	ρ_d	6.9
Effective pressure acting on the tunnel wall (MPa)	$p'(a)$	0, 0.02, 0.04, 0.06, 0.08, 0.1
Effective earth pressure acting on the outer boundary (MPa)	$p'(b)$	0.7, 1.0, 1.36
Water pressure at the outer boundary (MPa)	$u(b)$	2.4
Young's modulus (MPa)	E	300
Poisson's ratio	υ	0.4
Cohesive strength (MPa)	c'	0.4, 0.45, 0.5
Internal friction angle (°)	ϕ'	30°
Plastic potential parameter (°)	ψ	0°
Coefficient of permeability		
Original ground (cm/s)	k_0	k_0
Grouted area (cm/s)	k_g	$k_0/100$
Plastic zone in original ground (cm/s)	k_{0p}	$5k_0$
Plastic zone in grouted area (cm/s)	k_{gp}	$5k_g$

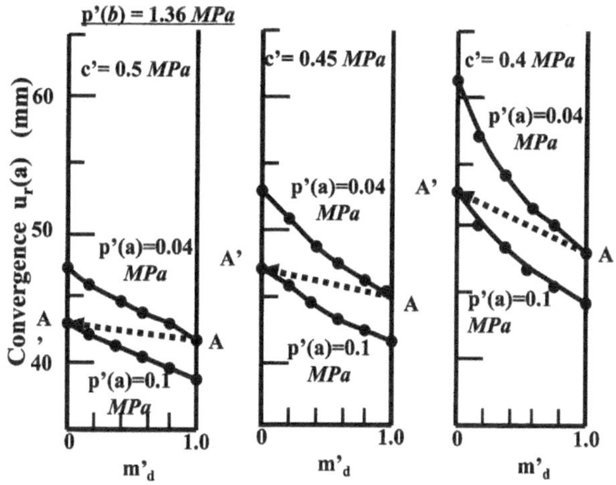

Figure 3.10 Effect of drainage: convergence versus drainage effect parameter.

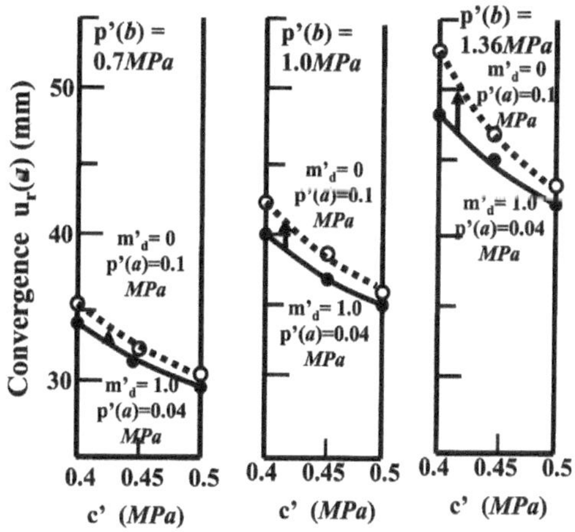

Figure 3.11 Effect of drainage: convergence versus cohesive strength.

shows the relationship between the tunnel convergence due to tunneling and drainage effect parameter, m'_d ($m'_d = 0$ denotes no drainage by drain holes, whereas $m'_d = 1.0$ denotes perfect drainage by drain holes). The state denoted by A in the figure corresponds to that prior to stopping the drainage—that is, $m'_d = 1.0$ and $p'(a) = 0.04$ MPa. The state expressed by A′ represents the state after stopping the drainage, namely $m'_d = 0$ and $p'(a) = 0.1$ MPa. It is clearly seen in the figure that the convergence increases from state A to state A′ by stopping the drainage.

To evaluate the test results more quantitatively, the analytical results are given as the relationship between the tunnel convergence and the cohesive strength of the ground, c'. As illustrated in Fig. 3.11, the solid lines are for the state prior to stopping the drainage (i.e., $m'_d = 1.0$ and $p'(a) = 0.04$ MPa), whereas the dotted lines are for the state after stopping the drainage (i.e., $m'_d = 0$ and $p'(a) = 0.1$ MPa). In this figure, the increase in convergence due to stopping the drainage is expressed by the change from the solid lines to the dotted lines. The experimental results—that is, a 5- to 7-mm increase in convergence by stopping the drainage—can be explained well by this analysis for the case in which $p'(b) = 1.36$ MPa and $c' = 0.4$ MPa.

3.3.2 Effect of drainage and the mechanical characteristics of the ground

In order to investigate how the effect of drainage is changed by the mechanical properties of the ground, an analysis is carried out under the conditions listed in Table 3.2. Figure 3.12 shows how the relationship between the drainage effect parameter, m'_d, and the radius of the plastic region, ρ_p, changes with cohesive strength, c', for the fixed value of the internal friction angle, $\phi' = 30°$. On the other hand, how the tunnel convergences change

Table 3.2 Conditions for the analysis of drainage problem 2

Tunnel radius (m)	a	1.6
Radius of virtual outer boundary (m)	b	20
Radius of grouted area (m)	ρ_g	5.3
Location of drain holes (m)	ρ_d	6.9
Effective pressure acting on the tunnel wall (MPa)	$p'(a)$	0.01
Effective earth pressure acting on the outer boundary (MPa)	$p'(b)$	1.77
Water pressure at the outer boundary (MPa)	$u(b)$	2.4
Young's modulus (MPa)	E	200
Poisson's ratio	υ	0.4
Cohesive strength (MPa)	c'	Varied
Internal friction angle (°)	ϕ'	30°, 35°, 40°
Plastic potential parameter (°)	Ψ	0°
Coefficient of permeability		
Original ground (cm/s)	k_0	k_0
Grouted area (cm/s)	k_g	$k_0/10$
Plastic zone in original ground (cm/s)	k_{0p}	$2k_0$
Plastic zone in grouted area (cm/s)	k_{gp}	$2k_g$

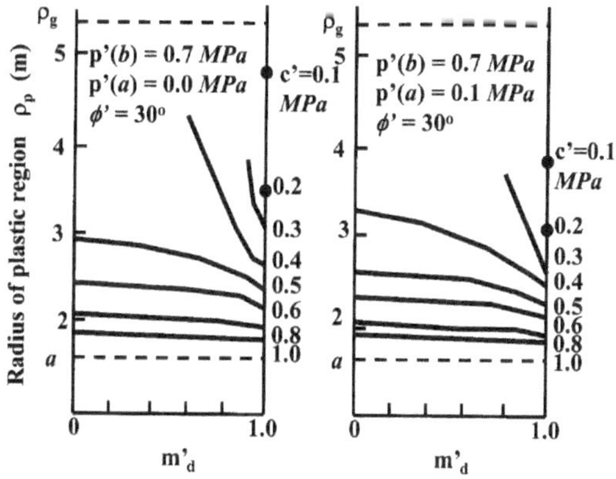

Figure 3.12 Drainage effect: extent of plastic region versus drainage effect parameter.

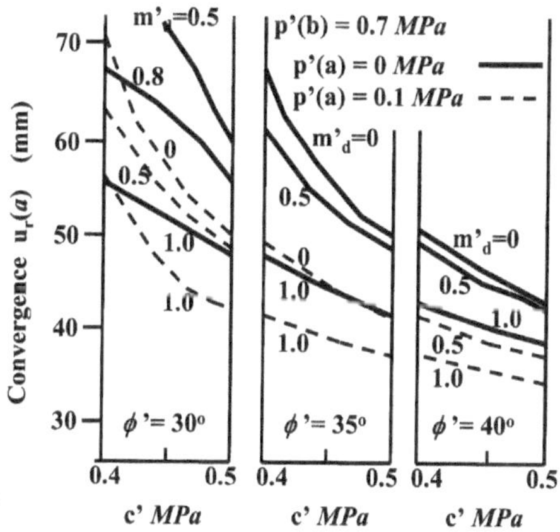

Figure 3.13 Drainage effect: convergence versus strength parameters of ground, c' and ϕ'.

with cohesive strength, c', as well as internal friction angle, ϕ', is shown in Fig. 3.13. The following conclusions are derived from these analyses:

1. The extent of the developed plastic region decreases with an increase in the material strength—that is, an increase in c' and ϕ'.
2. The extent of the plastic region and the tunnel convergence both increase when the parameter for the drainage effect, m'_d, decreases from 1.0 to 0—that is, by stopping the drainage.
3. When cohesive strength c' is large enough, the developed plastic region becomes very small and the drainage becomes ineffective. This means that the drainage is ineffective if the ground is thought to show elastic behavior.

3.3.3 Drainage effect for the case of an unimproved ground

An attempt was made to find the most effective arrangement for the drain holes in the case of an unimproved ground. The analytical conditions are given in Table 3.3. Figure 3.14 gives the results obtained under the most severe conditions—that is, $p'(b)=1.36$ MPa and $p'(a)=0$ MPa. Because m_d' is an index giving the water pressure, $u(\rho_d)$, at $r=\rho_d$, $m_d=1.0$ means that the water pressure is zero at $r=\rho_d$, and $m_d=0.4$ means that $u(\rho_d)=0.6u(b)$. Because the case for $m_d=0$ cannot be analyzed, the value $m_d'=0.02$ is used. This means that $u(\rho_d)=0.98u(b)$ in this state, namely almost the same amount of high water pressure, $u(b)$, is acting at $r=\rho_d$. At any rate, Fig. 3.14 shows that instability of the tunnel occurs when high water pressure acts on a certain portion of the surrounding ground. For example, in a ground with $c'=0.4$ MPa, it is necessary to reduce the water pressure, $u(\rho_d)$, below $0.8u(b)$ because tunnels become unstable when water pressure higher than $0.8u(b)$ (corresponding to $m_d'<0.2$) acts on approximately $\rho_d=4.8$ $m=3a$ (three times the tunnel radius). On the other hand, even in the case of $\rho_d=3.2$ $m=2a$ (twice the tunnel radius), tunneling can be done stably, when the water pressure, $u(\rho_d)$, can be reduced to less than $0.4u(b)$—that is, $m_d'>0.6$. In connection with the discussion in the next section about

Table 3.3 Conditions for the analysis of drainage problem 3

Tunnel radius (m)	a	1.6
Radius of virtual outer boundary (m)	b	40
Radius of grouted area (m)	ρ_g	3.2, 4.8, 6.9, 9.6
Location of drain holes (m)	ρ_d	6.9
Effective pressure acting on the tunnel wall (MPa)	$p'(a)$	0, 0.04, 0.1, 0.2, 0.3
Effective earth pressure acting on the outer boundary (MPa)	$p'(b)$	0.7, 1.0, 1.36
Water pressure at the outer boundary (MPa)	$u(b)$	2.4
Young's modulus (MPa)	E	300
Poisson's ratio	υ	0.4
Cohesive strength (MPa)	c'	0.3, 0.4, 0.5
Internal friction angle (°)	ϕ'	30°
Plastic potential parameter (°)	ψ	0°
Coefficient of permeability		
Original ground (cm/s)	k_0	k_0
Grouted area (cm/s)	k_g	k_0
Plastic zone in original ground (cm/s)	k_{0p}	$5k_0$
Plastic zone in grouted area (cm/s)	k_{gp}	k_{0p}

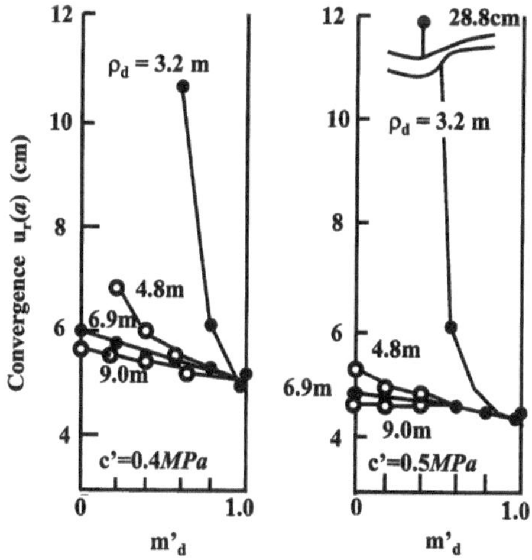

Figure 3.14 Drainage effect: convergence versus drainage effect parameter and location of drain holes.

the optimum extent of grouting, note that a stable ground is guaranteed under the conditions of $c' = 0.5$ MPa and $p'(b) = 1.36$ MPa, even if a very high water pressure, as much as $u(b)$, is acting at the location, $\rho_d = 4.8\ m = 3a$ (three times the tunnel radius).

3.4 OPTIMUM EXTENT OF GROUTING

In the previous section, the effect of drainage was discussed, comparing the test results with the analytical solutions. As already mentioned in the introduction, because a costly water-pressure-resisting tunnel lining would be required to prevent the introduction of leakage water into the inside of the tunnel, it was eventually decided that the deep undersea tunnel should be constructed by the conventional mountain tunneling method—that is, by introducing leakage water into the inside of the tunnel and then using a proper cutoff method to reduce the amount of water inflow. However, cutoff by grouting is both time-consuming and costly, and it greatly affects the progress of the tunneling. Furthermore, if the extent of the grouting is insufficient, the surrounding ground becomes unstable because high water pressure acts in the ground close to the tunnel wall. In any case, it is very important to establish a method for estimating the optimum extent of grouting.

3.4.1 Effect of cohesive strength c' on the optimum extent of grouting

In order to investigate the relationship between the cohesive strength of the ground material and the optimum extent of grouting, an analysis is carried out for a tunnel that is located at a depth of 240 m from the sea surface with an earth covering of 100 m. The analytical conditions are summarized in Table 3.4. Figure 3.15 shows how the relationship between the tunnel convergence and the extent of grouting is affected by the cohesive strength of ground material c'. In the figure, the dotted lines correspond to the case in which a complete drainage is conducted. Although the inner pressure, $p'(a)$, by the tunnel support system is zero right after the tunnel excavation, some inner pressure due to the so-called three-dimensional restriction effect can be taken into account in the plane strain tunneling analysis. From the results given in Fig. 3.15, when inner pressure is not anticipated, at least 5.8 m (3.625 times the tunnel radius) of the extent of the grouting, ρ_g, is required for a stable excavation in a ground with $c' = 0.4$ MPa, whereas a stable excavation is possible under $\rho_g = 4.8$ m (3 times the tunnel radius) in the case of a ground with $c' = 0.5$ MPa. The desirable extent of the grouting is approximately 3 or 4 times the tunnel radius. There is a question of how to excavate

Table 3.4 Conditions for the analysis of the grouting problem

Tunnel radius (m)	a	1.6
Radius of virtual outer boundary (m)	b	20
Radius of grouted area (m)	ρ_g	4.3, 4.8, 5.3, 5.8
Location of drain holes (m)	ρ_d	6.9
Effective pressure acting on the tunnel wall (MPa)	$p'(a)$	0, 0.02, 0.04, 0.06, 0.08, 0.1
Effective earth pressure acting on the outer boundary (MPa)	$p'(b)$	1.36
Water pressure at the outer boundary (MPa)	$u(b)$	2.4
Young's modulus (MPa)	E	300
Poisson's ratio	υ	0.4
Cohesive strength (MPa)	c'	0.35, 0.4, 0.45, 0.5, 0.55
Internal friction angle (°)	ϕ'	30°
Plastic potential parameter (°)	ψ	0°
Coefficient of permeability		
Original ground (cm/s)	k_0	k_0
Grouted area (cm/s)	k_g	$k_0/100$
Plastic zone in original ground (cm/s)	k_{0p}	$5k_0$
Plastic zone in grouted area (cm/s)	k_{gp}	$5k_g$

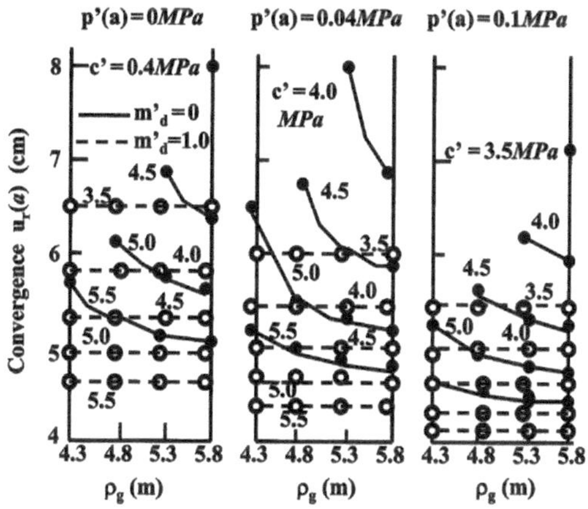

Figure 3.15 Effect of grouting: convergence versus extent of grouting.

a stable tunnel when the ground is not so strong—for example, $c' = 0.35$ MPa. In this case, the drainage is applied together with 5.8 m of the extent of grouting, ρ_g. Namely, a stable excavation can be conducted by using complete drainage at $r = \rho_d$, even under the conditions of $c' = 0.35$ MPa and $p'(a) = 0$ MPa, as seen in Fig. 3.15(left). After the excavation, 0.1 MPa of the inner pressure $p'(a)$ is applied by installing supports prior to stopping the drainage, and a stable tunnel can be constructed as seen in Fig. 3.15(right).

3.4.2 Optimum extent of the grouting

The optimum extent of the grouting can be predicted by this analysis. For example, Fig. 3.16 gives the results for the case of the main tunnel of the Seikan Tunnel using the analytical conditions shown in Table 3.5.

In Fig. 3.16, changes in tunnel convergence, $u_r(a)$, and in the amount of leaking inflow water, Q, with an increase in the extent of the grouting are given. It can be seen from the figure that if the extent of the grouting is insufficient, the tunnel convergence increases so much that instability of the tunnel takes place. This means that the grouting should be extensive enough to eliminate the high groundwater pressure area from the ground close to the tunnel wall. Furthermore, insufficient grouting makes the ground more unstable than the case in which no grouting has been applied. In the case shown in Fig. 3.16—that is, obtained under the conditions given in

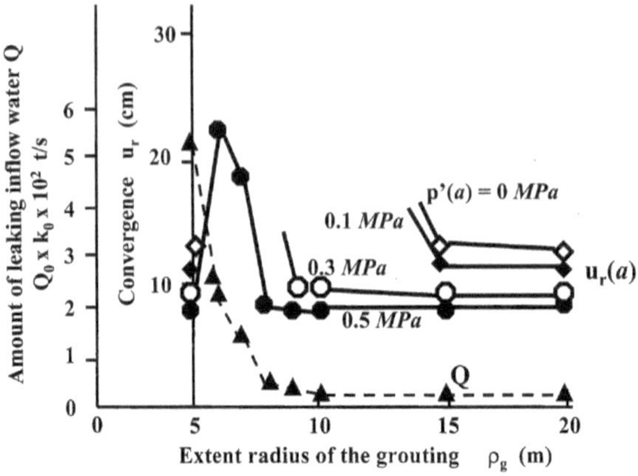

Figure 3.16 Optimum extent of grouting: changes in convergence and the amount of leaking inflow water with an increase in the extent of the grouting.

Table 3.5 Conditions for the analysis of the optimum extent of grouting

Tunnel radius (m)	a	5.0
Radius of virtual outer boundary (m)	b	40
Radius of grouted area (m)	ρ_g	Varied
Location of drain holes (m)	ρ_d	
Effective pressure acting on the tunnel wall (MPa)	$p'(a)$	Varied
Effective earth pressure acting on the outer boundary (MPa)	$p'(b)$	1.36
Water pressure at the outer boundary (MPa)	$u(b)$	1.9
Young's modulus (MPa)	E	300
Poisson's ratio	υ	0.4
Cohesive strength (MPa)	c'	0.5
Internal friction angle (°)	ϕ'	30°
Plastic potential parameter (°)	ψ	0°
Coefficient of permeability		
Original ground (cm/s)	k_0	k_0
Grouted area (cm/s)	k_g	$k_0/100$, $k_0/10$
Plastic zone in original ground (cm/s)	k_{0p}	$5k_0$
Plastic zone in grouted area (cm/s)	k_{gp}	$5k_g$

Table 3.5—the extent of the grouting should be more than three times the tunnel radius. When the ground is weak, such as the fractured zone shown in Fig. 3.17, the grouting extent reaches up to six times the tunnel radius.

The amount of seeping inflow water decreases naturally with an increase in the extent of grouting, as shown in Fig. 3.16.

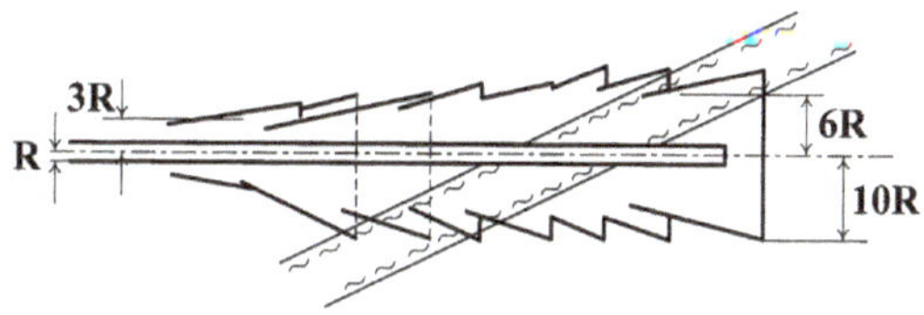

Figure 3.17 Extent of the grouting depending on the strength of the ground—that is, c' and ϕ'.

3.5 CONCLUSION

In this study, the important subjects related to drainage and grouting during undersea tunneling were discussed. The mechanical interaction between the surrounding ground and the existing water pressure was theoretically investigated in order to clarify the effectiveness of drain holes and to determine the optimum extent of grouting with the Seikan Undersea Tunnel in mind. The effectiveness of drain holes was verified by comparing the experimental results with the theoretical solutions. The optimum extent of grouting was established theoretically, and this method had to be applied to the construction of the Seikan Tunnel.

It is noteworthy that the extent of grouting depends on the ground quality and seepage conditions. If the ground quality was average and the seepage volume was small, grouting was carried out to a depth of three times the tunnel radius. If the ground was weak or the seepage volume was large, grouting was undertaken to six times the tunnel radius (Fujita et al., 1982).

On January 27, 1983, the pilot tunnel was driven through. In addition to the continuation of work to discover any important factors that might govern the undersea tunnel driving method, attempts were made to measure the earth pressure that had developed on the tunnel lining. The main tunnel was driven through on March 10, 1985.

At the end of March 1985, the Japanese Society of Civil Engineers (JSCE) Geotechnical Committee on Seikan Tunnel was disbanded by pointing out that the distinctive feature of the Seikan Tunnel construction is the development of the proper cutoff method by grouting (Adachi, 1986).

This study was carried out after the first author participated in the JSCE Geotechnical Committee on the Seikan Tunnel from 1971 to 1985.

ACKNOWLEDGMENTS

We express our sincere appreciation to the chairman of the JSCE Geotechnical committee, late Professor Shunzo Okamoto, as well as to all the members of the committee and to Japan Railway Construction Public Corporation.

REFERENCES

Adachi, T., 1986. Geotechnical report on the Seikan Tunnel. Tunnell. Underground Space Technol. 1 (3/4), 351–355.

Adachi, T., Tamura, T., 1978a. Effectiveness of drain holes and proper dimensioning of grouting zone for tunneling in high water pressure acting ground. Proc. Japan Soc. Civ. Eng. 280, 87–98 (in Japanese).

Adachi, T., Tamura, T., 1978b. Undersea tunnel—Effect of drainage and grouting. In: Proc. Symp. Soil Reinforcing and Stabilising Techniques, Sydney, Australia, pp. 513–529.

Fujita, M., Nishimura, T., Kitamura, A., 1982. Constructing the undersea section of the Seikan Tunnel. Tunnels Tunnelling 14 (8), 31–33.

Kudo, A., 1971. Optimum Extent of Grouting Based on the Elastic Stability. Report of the JSCE Geotechnical Committee on the Seikan Tunnel (in Japanese).

Sakurai, S., 1971. Tunneling in the Visco-Elastic Ground in Terms of Effective Stress. Report of the JSCE Geotechnical Committee on the Seikan Tunnel (in Japanese).

Shimokawachi, M., 1971. Design Concept of Undersea Tunnel Lining in the Elasto-Plastic Ground in Terms of Effective Stress. Report of the JSCE Geotechnical Committee on the Seikan Tunnel (in Japanese).

CHAPTER 4

Use of Jet Grouting in Deep Excavations

Dazhi Wen
Geotech & Tunnel Consult, Singapore

4.1 INTRODUCTION

Jet grouting is an increasingly used technique for in situ soil improvement in many types of soil. It is a general term used to describe various grouting techniques in which high-pressure air, water, and cementing grout are injected into the ground at high velocity. Depending on the system used, the in situ soil may be mixed with the grout, partly mixed and partly removed, or completely replaced. Typically, a single tube system is used to mix the grout with in situ soil and form a grouted pile in the ground. A double tube system injects high-velocity cementing grout and compressed air simultaneously. By introducing the compressed air, the double tube system is able to produce jet grouted piles (JGPs) approximately twice as large as the piles by the single tube system. The in situ soil is lifted up to the ground surface by the compressed air. A triple tube system injects air, water, and grout simultaneously into the drilled hole. The injection of both water and compressed air permits more soil to be removed from the cutting area, and the in situ soil can be replaced completely by grout during the process of triple tube jet grouting works. Because the injection of grout and air/water is separate, the size of the grouted pile by the triple tube system is usually larger for the same soil type than the size produced by single tube or double tube systems. Typically, the diameter of a JGP by the triple tube system in Singapore marine clay is 1600–2000 mm. Figure 4.1 shows the typical arrangement of air, water, and grout jets in the three grouting systems.

The application of the jet grouting technique in geotechnical engineering falls generally into three categories:

- Strengthening of ground as excavation support or underpinning support
- Temporary or permanent stabilization of soil
- Groundwater control

Chemical, Electrokinetic, Thermal, and Bioengineering Methods
http://dx.doi.org/10.1016/B978-0-08-100191-2.00004-6

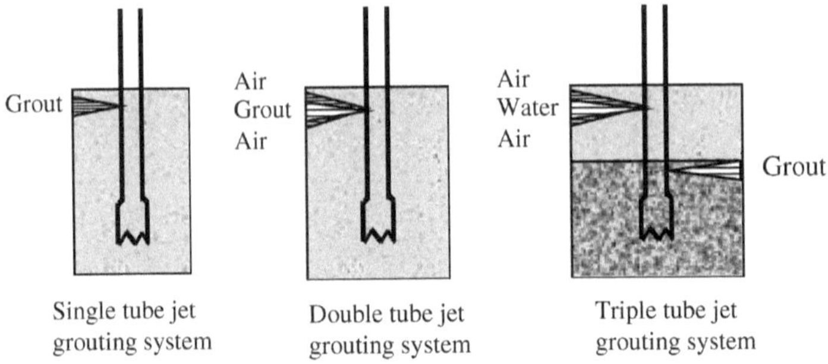

Figure 4.1 Typical arrangement of jets.

For the application in excavation, the JGPs are typically installed at the base of the excavation. The improved soil layer acts as a temporary strut to reduce the deflection of the retaining walls. For underpinning, the JGPs are installed under the structures, sometimes at an angle from the ground surface to enhance the bearing capacity of the original ground. Examples of temporary stabilization can be found at launching or receiving shafts for tunnel boring machines. For groundwater seepage control, JGPs are installed at the base of embankment to cut off water flow.

4.2 QUALITY CONTROL FOR JET GROUTING WORKS

Quality assurance and quality control are critical components of a successful jet grouting program to ensure that subsurface soils are consistent with design assumptions and that the design parameters are met or exceeded throughout the project.

4.2.1 Quality assurance

Quality assurance begins with a full-scale trial on site. The trial is to establish operation parameters for subsequent working pile installation to achieve the design geometry and the quality and strength characteristics of the JGPs. The operating parameters include air, water, and/or grout flow and pressure together with monitor rotation and withdrawal speed. Typical grouting parameters for jet grouting in the Singapore marine clay during the construction of the Mass Rapid Transit (MRT) North East Line (NEL) projects in Singapore are listed in Table 4.1.

The targeted diameter of the JGPs was generally 1.6–2.0 m in the NEL projects. The purpose of the jet grouting was to strengthen the very soft to

Table 4.1 Typical grouting parameters used in MRT NEL projects in Singapore

Operating parameters	Units	Triple tube grouting parameters
Water pressure	MPa	35–45
Air pressure	MPa	0.7–1.5
Grout pressure	MPa	7–11
Water flow rate	l/min	75–150
Air flow rate	m^3/min	1
Grout flow rate	l/min	62–105
Lifting speed or withdrawal speed	min/m	8–10
Rotation speed	rev/min	5–10

soft marine clays in deep excavations and in launching and receiving shafts of tunnel boring machines.

To verify the geometry of the trial piles, indirect methods are often used. This is especially so for applications in deep excavations because exposure of the trial piles is impractical. The indirect methods include coring, cone penetration tests (CPTs), standard penetration tests (SPTs), or cross-hole geophysical tests. Retrieved core samples are laboratory tested to confirm that satisfactory unconfined compressive strengths and stiffness (Young's modulus) are achieved.

The preproduction quality assurance measures form the basis for quality control during production grouting. It may be assumed that in comparable soil conditions, the same jet grouting parameters produce the same JGP dimension, properties, and spoil return. For the triple tube jet grouting works at Race Course Road during the construction of the NEL projects in Singapore, the density of spoil return of 123 samples ranges from 1.01 to 1.52 g/cm^3, with an average of 1.31 g/cm^3 (Shirlaw et al., 2003). The density of grout typically ranges from 1.51 to 1.56 g/cm^3, with an average of 1.53 g/cm^3.

4.2.2 Quality control during production

The minimum quality control consists of reporting of operation parameters and observing the spoil return. The technology of computerized data collection system for all jet grouting parameters is available along with continuous real-time observation.

The selected operation parameters should preferably be automatically controlled and monitored throughout construction (Table 4.2). Reduced flow or increased withdrawal speed will produce smaller jet grouted geometry.

Table 4.2 Quality control inspection items

Drilling	Location, drilling angle and drilling depth
Batching	Preparation of grout slurry for consistency in cement content and properties (density measurement)
Jetting	Operating parameters (lift speed, rotation rate, and pressure and flow of water, air, and grout), observing spoil returns with density measurement
Documentation	Documentation for each element constructed; construction times and correlation to any sampling performed
Sampling and testing	Retrieval of representative samples for laboratory testing

In addition to the quality control inspection items for JGP production, additional project-specific quality control measures are generally required. This can be in the form of coring for laboratory tests to establish the unconfined compressive strength of the cores or in situ pumping tests to determine the permeability of the grouted mass. Permeability of the treated ground can also be verified by installing piezometers on both sides of the treated ground.

The frequency of quality control tests is not fully standardized. Various standards and manuals have different recommendations. The designer will have to specify the testing regime based on his or her design and how critical the JGPs are in the overall system. A common method is to specify a frequency of quality control testing based on the volume of treated soil. For example, it can be specified that one core per 1000 m^3 treated soil through the full depth of the JGPs should be taken and a minimum of three unconfined compressive strength tests should be carried out with strain measurement to verify strength and stiffness of the installed JGPs. The samples for the tests can be taken from the top, middle, and bottom of the core. The cores should be fully logged, and total core recovery or rock quality designation can also be used as a guide to determine the quality of the JGP mass. The location of cores is often taken at the overlapping areas of the JGPs because these areas are typically the weak areas.

Grouting sequences and documentation form an important element in the quality control system. A systematic approach is typically adopted to ensure that all JGPs are installed as required by design and no gaps are left in the grouted mass. One example of a systematic approach is shown in Fig. 4.2 used in the jet grouting works at Race Course Road in Singapore. The grouting rigs were mounted on a steel truss supported on rails founded on diaphragm walls. The truss beam could move along the longitudinal direction of the diaphragm walls that acted as temporary supports during the subsequent excavation for the construction of the cut-and-cover tunnels.

Figure 4.2 Schematic approach in grouting sequences.

4.3 PROPERTIES OF JET GROUTED PILES

The strength and stiffness properties of JGPs are often obtained from unconfined compressive strength (UCS) tests on core samples taken from the grouted soil. In situ pressuremeter tests can also be conducted to determine the in situ Young's modulus of the grouted mass. Other tests, such as SPTs and CPTs, are used to verify the quality of the grouting and to correlate with UCS tests and in situ pressuremeter tests to obtain the strength and stiffness of the JGPs for design purposes.

The design undrained shear strength for JGPs in deep excavations in Singapore marine clay is typically set at $c_u = 300$ kPa, with an equivalent UCS, q_u, of 600 kPa. The Young's modulus is typically set at 120 MPa—that is, $400c_u$ or $200q_u$. This is believed to be a lower bound value of the JGP mass. Actual UCS tests on cores taken from the JGPs generally show much higher UCS, typically exceeding 1000 kPa, even for cores taken from overlapping areas of the JGPs. Table 4.3 shows some of the UCS test results on JGP core samples at Race Course Road in Singapore. The cores were taken within the middle-third diameter of the JGPs and were tested at 14 days.

The distribution of 28-day UCS for samples taken from Race Course Road is summarized in Fig. 4.3. Similar results from another site at Clarke Quay are shown in Fig. 4.4 (Shirlaw et al., 2000a). These results demonstrate that the typical design undrained shear strength, c_u, of 300 kPa is a lower bound value. An upper bound value can be as high as 10,000 kPa at Race Course Road.

The failure of the JGP core samples at UCS tests indicates a brittle failure mode at a typical strain range of 0.5–1.5% (Fig. 4.5), especially for the

Table 4.3 Test results of JGP at Race Course Road

No.	Bulk density (kN/m^3)	Unconfined compressive strength at 14 days (kPa)	No.	Bulk density (kN/m^3)	Unconfined compressive strength at 14 days (kPa)
1	19.3	6280	7	14.8	978
2	18.8	6050	8	15.4	879
3	14.4	3390	9	14.2	876
4	18.6	4900	10	15.1	780
5	14.3	1491	11	14.9	667
6	14.8	1290			

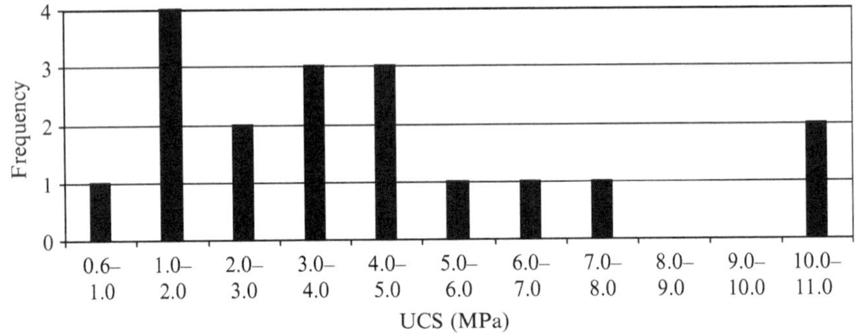

Figure 4.3 Distribution of 28-day UCS at Race Course Road.

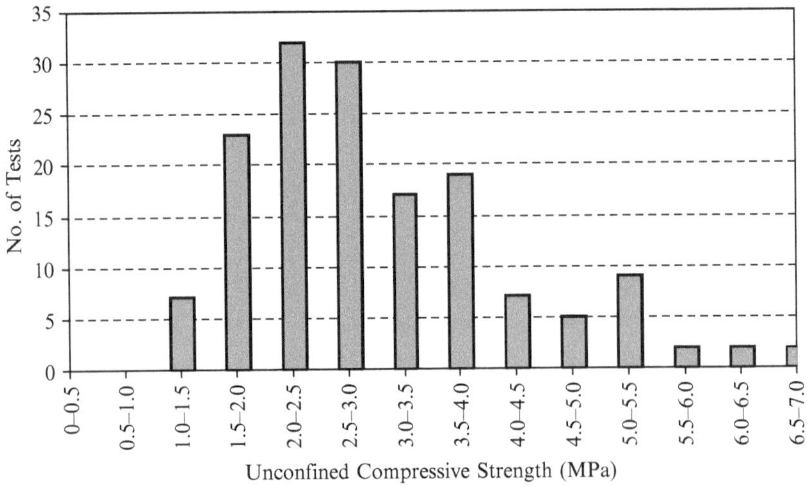

Figure 4.4 Twenty-eight-day UCS at Clarke Quay. *(Source: After Shirlaw et al. (2000a)).*

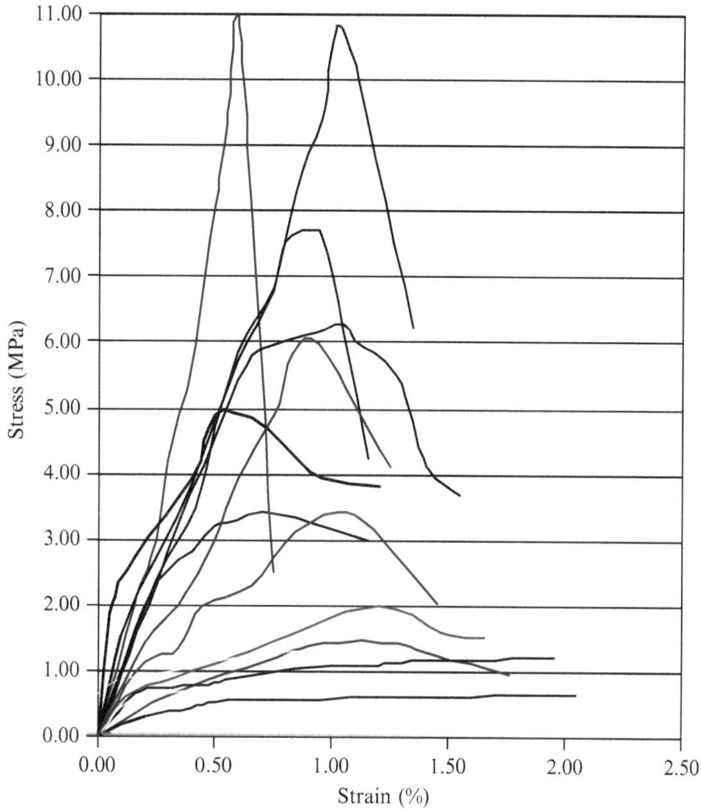

Figure 4.5 Stress–strain curves of JGP core samples in UCS tests.

samples showing very high UCS values. This is in contrast to the in situ soft clay, which fails under compression in plastic mode. It is interesting to note that when the UCS is relatively low in the range of 1.0 MPa, the brittleness is less remarkable.

What should also be noted is the density of the JGP samples. The treated ground does not gain in density compared with that of the in situ soil. JGPs will have similar density of the in situ soil. When checking for uplift, the density of the JGPs should not be taken higher than that of the in situ soft clay. The samples at Race Course Road showed densities typically close to the density of the in situ marine clay, which is generally in the range of 14–16 kN/m^3, as shown in Fig. 4.6. Similar results at Clarke Quay are given in Fig. 4.7.

The ratio of Young's modulus of the JGP core samples to the UCS strength is typically taken as $E/q_u = 200$. The results at Clarke Quay confirm that the ratio for design purposes is reasonable (Fig. 4.8).

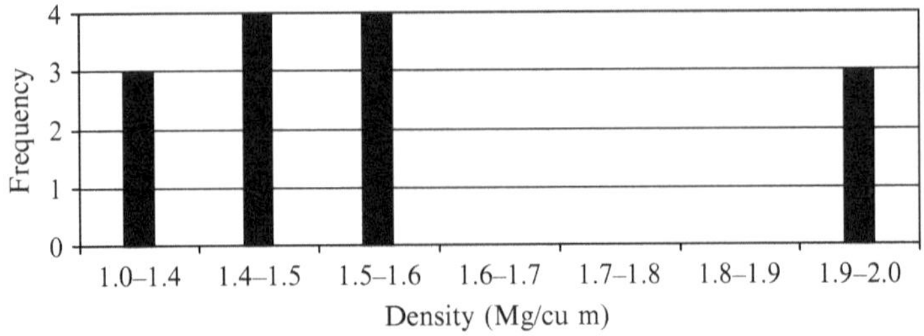

Figure 4.6 Density of JGPs at Race Course Road.

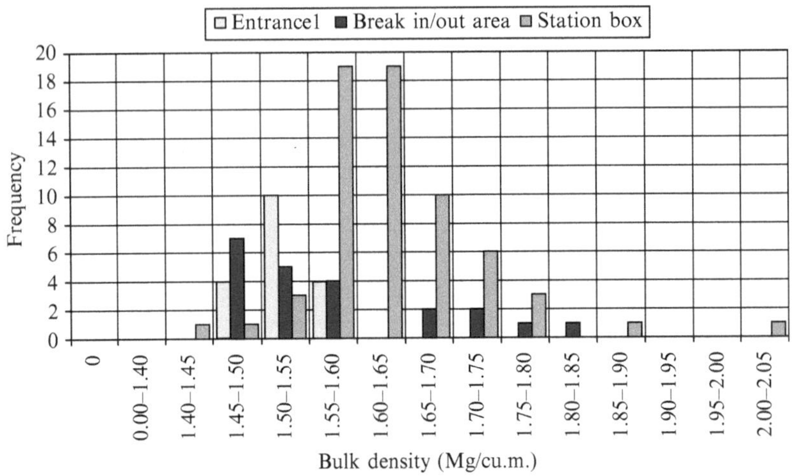

Figure 4.7 Density of JGPs at Clarke Quay.

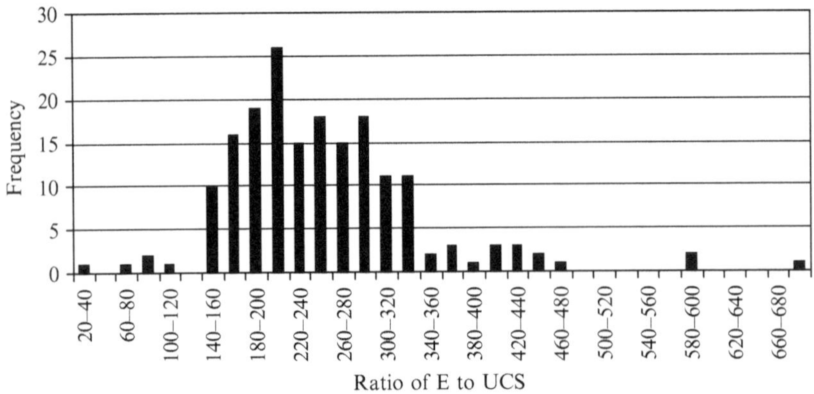

Figure 4.8 Ratio of Young's modulus to UCS at Clarke Quay.

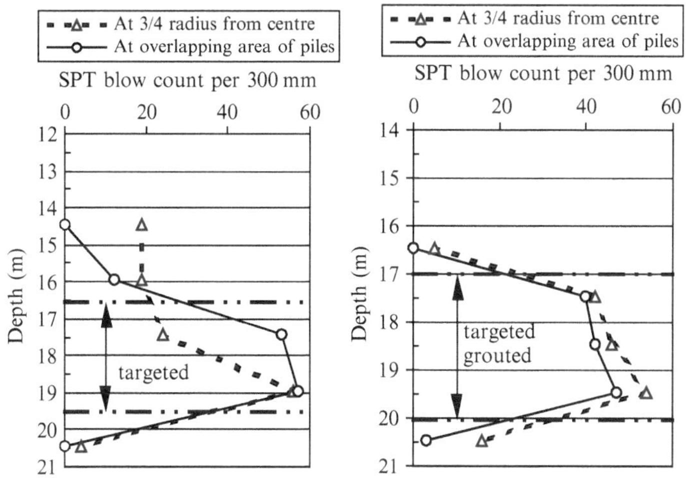

Figure 4.9 SPT blow counts versus depth at Race Course Road at two locations.

The SPT blow counts at two locations at Race Course Road 20 days after the grouting at three-fourths of the radius of the pile from the center and at the overlapping areas of three piles are shown in Fig. 4.9. The SPT blow counts increased from the original 0 blow counts to 24–57 blows. SPT blow counts can therefore give a clear indication of the success of the grouting program.

4.4 DESIGN ISSUES FOR JET GROUTED PILES

It is a common practice to design JGPs as a layer of extremely strong soil. In Singapore, the use of JGPs at the base of the excavation to treat the very soft to soft marine clay is very popular. As shown in Fig. 4.5, the stress and strain curves for JGPs and the natural occurring marine clay are very different. The JGP samples reach peak strength typically at 0.5–1.5% of strain, and the failure tends to be brittle. The incompatibility of strains at failure of the natural occurring soft clay and JGPs should be considered. Although the behavior of JGP slab as a mass of treated ground may be different from that of a solid JGP core, because there may be gaps and pockets of untreated soft clay in the JGP mass, the design should take into consideration the issue of strain incompatibility by verifying the strains in the JGPs at various stages of the excavation. Where the strains in the JGPs are outside the typical range, close monitoring should be carried out, and where necessary, additional measures should be taken to prevent a brittle failure of the JGP slabs during excavation.

In addition to verifying the strains in the JGPs at different stages of the excavation during design, all failure mechanisms should also be checked. Shirlaw et al. (2000b) identified all the mechanisms for JGP slabs at the base of excavation for two cases (Fig. 4.10):

- With sufficient retaining wall embedment to hard stratum at Race Course Road to control inward movement of the wall
- No wall embedment into hard stratum at Clarke Quay to control basal stability for the excavation

The JGP slabs may not necessarily be at the base of the excavation. They can also be constructed above the base of excavation prior to commencement of excavations to further limit the deflection of the walls (Fig. 4.11). These layers are sometimes called sacrificial JGP layers because they will be removed stage by stage during the excavation.

Traditionally, geotechnical design parameters are selected conservatively—that is, design values are selected toward the lower bound values. For the design of the sacrificial JGP slabs, it is considered necessary to carry out sensitivity analysis for the upper bound values. This is to cater for the worst loading

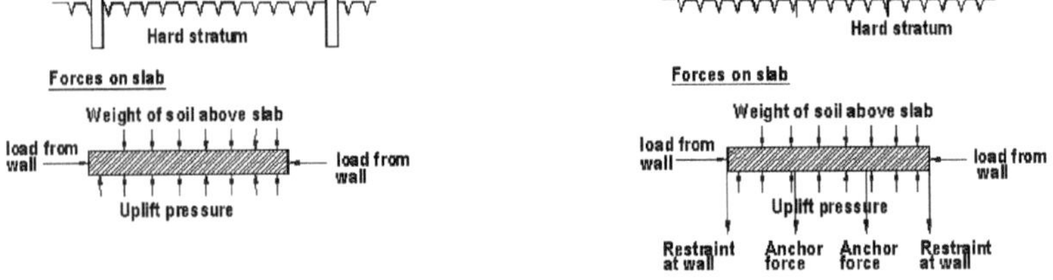

Potential causes of failure

a) Hydraulic uplift causing slab to heave and bend. Critical pressure could be on base of slab or on interface between high and low permeability strata below slab.
b) Swelling of soft clay causing slab to heave and bend.
c) Axial force on slab causing failure.
d) Defect in slab causing failure due to axial forces.

Potential causes of failure

a) Hydraulic uplift (in sands) causing failure or slab in bending or shear.
b) Undrained base heave pressure (in clay) causing failure of slab in bending or shear.
c) Axial force on slab causing failure.
d) Defect in slab causing failure due to the other listed mechanisms.
e) Insufficient capacity of restraints at walls of anchors.
f) Lack of bond to transfer load from slab to walls and anchors.

Figure 4.10 Design issues for JGP slabs for deep excavations. *(Source: after Shirlaw et al. (2000b)).*

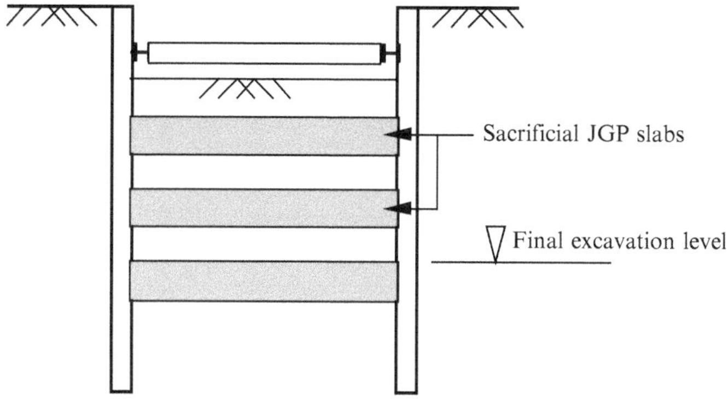

Figure 4.11 Use of sacrificial JGP slabs in deep excavations.

condition for the struts above the sacrificial JGP layer. When the JGP slab is removed, the load taken by the JGP slab will be taken over by the struts above the JGP slab. If a lower bound value is used, this may underestimate the loading for the struts immediately above the sacrificial JGP slab.

4.5 CONSTRUCTION ISSUES FOR JET GROUTED PILES

The main problem associated with jet grouting in Singapore marine clay is heave. At Race Course Road, localized heave was reported to be in excess of 300 mm (Maguire and Wen, 1999). During the early phases of the construction of MRT lines, the maximum measured heave was 1600 mm (Shirlaw et al., 2000a). Large lateral ground movement has also been recorded at Race Course Road (Fig. 4.12). Similar lateral ground movement was also reported by Wong and Poh (2000).

During the jet grouting process, a large volume of grout, air, and water is injected into the ground. Surplus material produced by the injection of these materials is expelled to the surface via the annulus between the drilling rods and the surrounding marine clay. If this passage is restricted or blocked, the pressure inside the cavity created by the high-pressure jetting will be built up. When the pressure exceeds the cavity expansion pressure, both lateral and vertical ground movement will be initiated.

The hydrodynamic pressure created by the high-pressure jetting exists only within a zone of influence, which is approximately 300 times that of the nozzle diameter (Covil and Skinner, 1994). For a nozzle diameter of 2 mm, the zone of influence is only 0.6 m. Therefore, the jet pressure does not control the pressure within the cavity. It is believed that the movement of the ground (both lateral and vertical) is not a direct result of the high-pressure jetting adopted in the jet grouting works. The ground movement

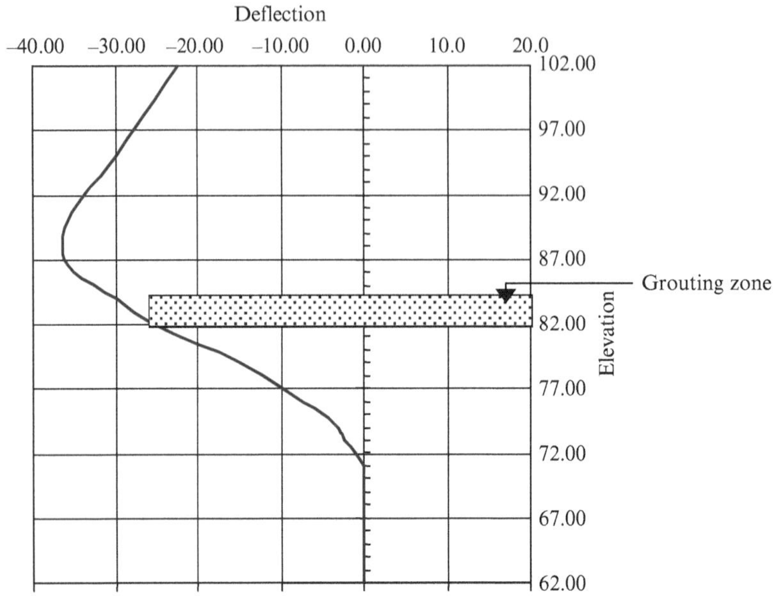

Figure 4.12 Lateral diaphragm wall movement due to jet grouting works.

during jet grouting works is a result of the pressure built up to expel the sludge to the ground surface (Fig. 4.13). When the annulus between the grouting tube and the ground is blocked, pressure can be locked into the ground, causing the ground to move both horizontally and vertically.

On the basis of the understanding of the cause of the ground movement during jet grouting, the following measures can be taken to reduce the

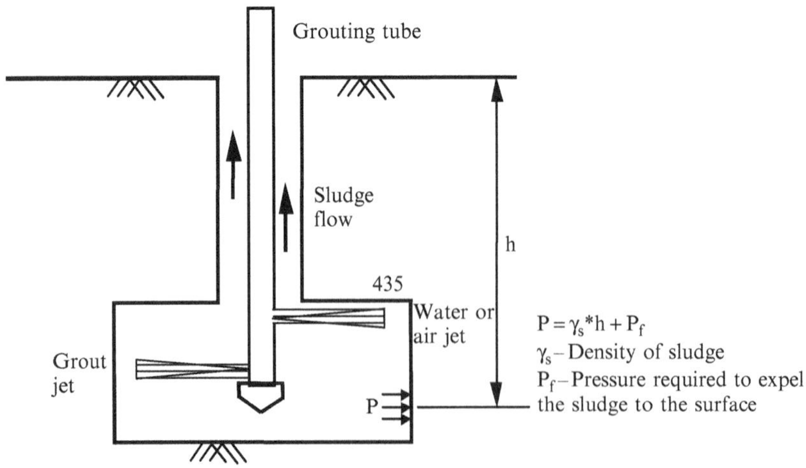

Figure 4.13 Pressure required to expel the sludge to surface.

pressures to expel the sludge to the surface, and thus to reduce the ground movement:

- Reducing the density of the sludge
- Increasing the size of the annulus by drilling a larger hole
- Using casings to prevent blockage of the sludge flow

Shirlaw et al. (2003) showed that the diameter of the casing should be 200 mm or larger for it to be effective. This may explain why the use of 100- and 150-mm casing at Race Course Road was of limited success (Maguire and Wen, 1999). Other measures implemented at Race Course Road included pregrouting to condition the soft clays and the use of pressure relief holes. These measures proved to be more effective. All these measures were undertaken to ensure free flow of the sludge and to reduce the pressure locked into the ground.

Another issue that should be carefully examined is the layout of the JGPs next to the retaining walls. Gaps between the retaining walls and the JGP slab should not exist to ensure the effectiveness of the JGP slab in reducing the inward movement of the walls and in resisting the uplift force. At Clarke Quay station, shear connectors were provided at the sheet piles to provide adequate restraints at the interface between the JGPs and the sheet piles. Because the connection is critical, the spacing of the JGPs closest to the sheet pile wall was significantly reduced. As shown in Fig. 4.14, there would be shadows at the interface between the JGPs and the sheet piles if the standard spacing of the JGPs were kept at 1.4 m center to center. After reducing the

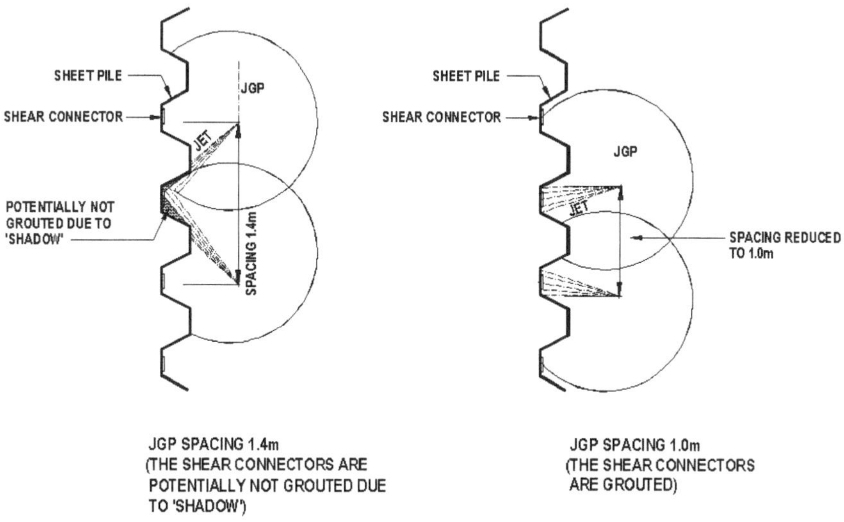

Figure 4.14 Layout of JGPs closest to sheet piles. *(Source: after Shirlaw et al. (2000a)).*

spacing to 1.0 m center to center, the interface would be fully grouted in the critical area of the shear connection to the sheet piles.

4.6 CONCLUSION

There are three stages in quality assurance and quality control for jet grouting works: a full-scale trial to establish the operating parameters, monitoring of the parameters during working pile installation, and verification tests after the installation. It can be assumed that the same jet grouting parameters would produce JGPs of the same dimension and characteristics.

In the design of JGP slab in deep excavations, it is necessary to check the strains of the JGP slab at different stages of the excavation. All failure mechanisms should be considered for the JGP slabs. Sensitivity analyzes are necessary to obtain the worse loading conditions of the struts when sacrificial JGP slabs are used. Density of the JGPs is typically similar to that of the in situ soft clay and should be used when checking uplift forces.

Heave and horizontal movement of the ground during grouting can be controlled if the pressures in the cavity created by the high-pressure jetting can be controlled. There are a few measures to prevent the buildup of pressure in the cavity, including the use of a 200-mm or larger diameter polyvinyl chloride casing, increasing the size of the annulus of the grout tube and the ground by drilling larger holes, and reduction of sludge density. The layout of the JGPs closest to the retaining walls should be planned carefully to ensure that no gaps exist at the interface of the wall and the JGPs.

REFERENCES

Covil, C.S., Skinner, A.E., 1994. Jet grouting—A review of some of the operating parameters that form the basis of the jet grouting process. In: Bell, A.L. (Ed.), Proc. Grouting in the Ground, Telford, London.

Maguire, C., Wen, D., 1999. Practical aspects of jet grouting in Singapore marine clay. In: Proc. Int. Conf. Rail Transit, Singapore.

Shirlaw, J.N., Wen, D., Kheng, H.Y., Osborne, N., 2003. Controlling heave during jet grouting on marine clay. In: Proc. RTS Conference, Singapore.

Shirlaw, J.N., Wen, D., Nadarajah, P., Yoon, S.I., Sugawara, S., 2000a. Construction issues related to jet grouted slabs at the base of excavations. In: Proc. Tunnels and Underground Structures—An International Conference, November 26–29, 2000, Singapore.

Shirlaw, J.N., Wen, D., Nadarajah, P., Yoon, S.I., Sugawara, S., 2000b. Design issues related to jet grouted slabs at the base of excavations. In: Proceedings of Tunnels and Underground Structures—An International Conference, November 26–29, 2000, Singapore.

Wong, I.H., Poh, T.Y., 2000. Effects of jet grouting on adjacent ground and structures. J. Geotech. Geoenviron. Eng. 126 (3), 247–256.

CHAPTER 5

A Case History of Jet Grouting in Marine Clay

Ing Hieng Wong[1], Teoh Yaw Poh[2]
[1]MITIC Engineers Pte Ltd., Singapore
[2]Building and Construction Authority, Singapore

5.1 INTRODUCTION

In Singapore, for deep excavations next to sensitive structures such as Mass Rapid Transit (MRT) structures and old buildings associated with thick soft soil deposits, the use of diaphragm walls alone often will result in movements that exceed the design movement criteria specified by the authorities or clients. Consequently, soil improvement methods such as jet grouting are usually required to improve the strength and deformation characteristics of the soft soils before the start of excavation to reduce these ground movements.

Examples of deep excavations in Singapore associated with soft soils improved by jet grouting include the Newton MRT station (Gaba, 1990), Singapore Arts Center, Bugis Junction (Sugawara et al., 1996), Grandlink Square, and United Engineers Square (Khoo et al., 1997). All these case histories involved jet grouting in marine clay except for the last one, which involved both organic and marine clays.

The process of jet grouting involves the cutting of soil by a mixture of water and air under high pressure and the placement of grout, also under high pressure. Hence, the process of jet grouting may tend to displace the adjacent soil away from the grouted zone. Thus, the process of jet grouting may have some effects on the retaining structures, adjacent soils, and nearby structures.

The performance of a field jet grout trial and the production grouting carried out during the basement construction for the Singapore Post Centre is presented and evaluated herein. Emphasis is given to the effects of jet grouting on movements of diaphragm walls, adjacent soils, and structures and on changes in lateral earth pressure and water and piezometric levels in adjacent soils.

Chemical, Electrokinetic, Thermal, and Bioengineering Methods
http://dx.doi.org/10.1016/B978-0-08-100191-2.00005-8

5.2 JET GROUTING SYSTEM

Jet grouting is a process in which the erosion of soil by a very high-pressure fluid jet followed by the injection of grout improves the physical characteristics of the ground. Jet grouting systems have been developed throughout the years and can be broadly classified into three groups, namely single, double, and triple tube systems. The triple tube system, which was used for this project, erodes the soil by the independent action of a water jet, shrouded by air to achieve maximum efficiency, and grout is injected through a separate orifice beneath the erosive jet. The triple tube system was adopted for this project because of its versatility and because it allowed relatively large grouted soil columns to be formed. The details of the jet grouting process, its application, and the influence of jet grouting operation parameters have been reported by many, such as Bell (1993) and Covil and Skinner (1994).

5.3 PROJECT DESCRIPTION

The construction of the basement for the 14-story Singapore Post Centre involved a deep braced excavation in soft marine clay supported by diaphragm walls. The excavation was 9–12.2 m deep (with localized excavation levels up to 14.7 m), 87–104 m wide, and 241 m long. The plan layout for the basement excavation and the locations of the monitoring instruments are shown in Fig. 5.1. The combined thickness of the soft soil

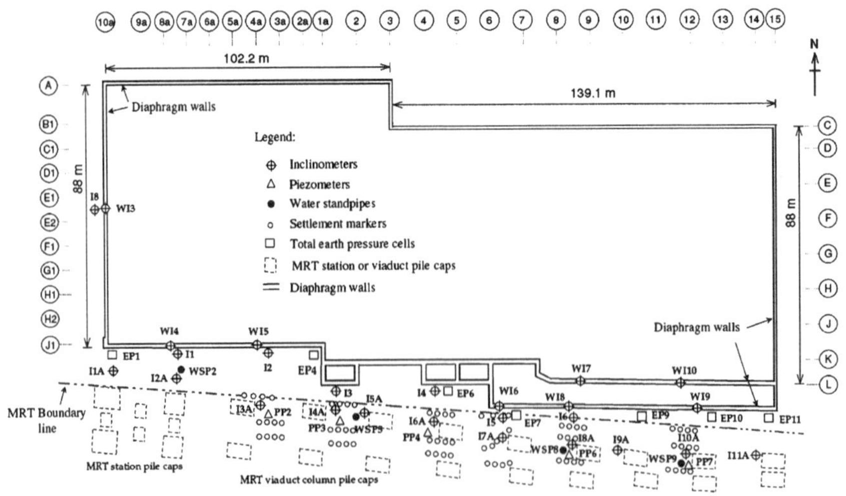

Figure 5.1 Plan layout for the basement excavation and the locations of the monitoring instruments.

layers ranged from 10 to 25 m. The site was close to an existing MRT station and viaduct. The excavation thus had to comply with MRT requirements that movements induced in the MRT structures be limited to 15 mm and the lowering of the groundwater level be limited to 1 m.

In view of the presence of the thick layers of soft marine clay at the site and the strict regulatory requirements, the basement excavation was supported by a 1.2- and a 0.8 m-thick diaphragm wall for the boundary along and away from the MRT station and viaduct, respectively. Prior to the start of the basement excavation, the soft to medium stiff marine clay layer below the bottom of the excavation level, over the entire basement area, was treated with jet grout. The objective was to use the treated soil mass, with enhanced strength and deformation characteristics, to act as an internal strut below the bottom of the excavation level in order to reduce movements caused by the basement excavation.

A trial jet grout program was carried out to confirm the grouting parameters and assess the effects of the trial grouting on adjacent ground and diaphragm wall panels. Following the trial grout program, the jet grouting parameters were modified and the production grouting was executed based on these modified parameters.

5.4 SUBSURFACE CONDITIONS

A longitudinal cross section showing the soil profile along the excavation boundary next to the MRT station and viaduct is presented in Fig. 5.2. The ground surface level varied from reduced level (RL) 103.2–103.6 m. The groundwater level was at approximately 1.5 m below the ground surface.

The subsurface soils consisted of a very loose to loose clayey silty sand or soft silty clay fill layer overlying a soft to medium stiff marine clay layer. The thickness of the fill layer varied from 2 to 5.5 m, with standard penetration test (SPT) *N* values ranging from 2 to 10 and an average *N* value of 4. The thickness of the soft to medium stiff marine clay layer varied from 11.5 to 25.5 m, with an average thickness of 15.5 m. The marine clay was underlain by a medium stiff to very stiff silty clay layer or a soft soil sediment that was in turn underlain by a medium dense or a dense clayey sandy silt or clayey silty sand layer.

The soft soil sediments generally consisted of pockets of soft to medium stiff marine clay (lower marine clay layer), soft sandy clay, soft silty clay, and loose clayey silty sand. Directly below the dense sandy soils were very dense

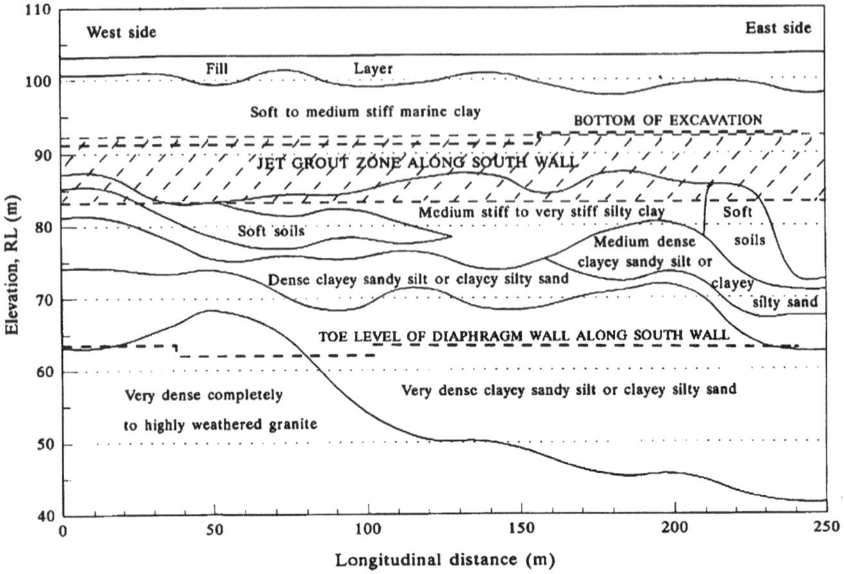

Figure 5.2 Soil profile along the excavation boundary next to the MRT station and viaduct.

sandy soils of similar nature, followed by a very dense to hard, completely to highly weathered granite.

The average total unit weight of the marine clay layer was 15.1 kN/m^3. The undrained shear strength of the marine clay obtained from the field vane shear tests (corrected value) ranged from 14 to 42 kPa, with an undrained strength ratio, s_u/σ'_{vo}, ranging from 0.20 to 0.26. The results from the consolidated undrained triaxial tests with pore pressure measurements indicated that the effective friction angle ranged from 20 to 24°, with an average value of 22°, and the value of the effective cohesion was negligible. Results of Menard pressuremeter tests carried out during the site investigation showed that the soil modulus from the first unload–reload cycle ranged from 2.8 to 4.9 MPa, with soil modulus over undrained shear strength ratio ranging from 170 to 373. The water content varied from 42% to 69%, with an average value of 59%. It is a high plasticity clay, with an average liquid limit and plastic limit of 92% and 39%, respectively, giving an average plasticity index of 53%. The void ratio ranged from 1.82 to 2.11, with an average value of 1.93. The average compression index, C_c, and recompression index, C_r, were 0.82 and 0.16, respectively, giving an average value of plastic volumetric strain ratio, Λ, of 0.80. The coefficient of permeability of the marine clay was 6×10^{-10} m/s.

The stiff intermediate clay layer had a total unit weight of 18 kN/m^3, water content of 31%, liquid limit of 60%, and plastic limit of 22%. The medium dense to very dense clayey sandy silt or clayey silty sand layer belongs to the Old Alluvium formation (*Geology of the Republic of Singapore*, 1976). The total unit weight of the very dense clayey sandy silt or clayey silty sand ranged from 20 to 21 kN/m^3. The effective friction angle generally ranged from 30° to 40°, with an average value of 35°. The average coefficient of permeability of the hard sandy silty clay was 1×10^{-9} m/s. The soil modulus obtained from the first unload–reload cycle from the Menard pressuremeter tests conducted during the site investigation ranged from 35.9 to 63.6 MPa, 100.2 to 257.3 MPa, and 235.5 to 455 MPa, with a soil modulus over undrained shear strength ratio ranging from 246 to 758, 394 to 953, and 545 to 1517, for the medium dense, dense, and very dense clayey sandy silt or clayey silty sand, respectively.

5.5 JET GROUT TRIAL

5.5.1 Description of jet grout trial

In the design, jet grouting was proposed to form a mat as an internal strut beneath the base of the excavation to control the lateral movements of the diaphragm walls during excavation for the basement of the Singapore Post Centre. However, there was concern that the grouting process itself, within the excavation area, might cause the diaphragm walls to deflect outward beyond acceptable limits. In order to study the effects of jet grouting on the diaphragm walls and adjacent ground, a field jet grouting trial with a plan area of 6.4 × 6.75 m was carried out. The jet grouting trial also served to verify the jet grouting operating parameters in order to achieve the required strength and stiffness of the grouted soil mass.

A comprehensive set of instrumentation was installed in the field trial area to monitor the effects of grouting on the diaphragm walls and the surrounding soils. After the completion of the jet grouting trial, four boreholes were drilled through the grout columns and core samples were obtained and tested to assess the quality, strength, and stiffness of the improved soil mass. Pressuremeter tests were also conducted. The targeted undrained shear strength and Young's modulus of elasticity of the grouted soil mass were 300 kPa and 150 MPa, respectively. The targeted grout strength of 300 kPa was selected to provide sufficient strength to restrain lateral soil movements but not too large as to impede the installation of bored piles and barrettes after the completion of jet grouting.

The field trial was located next to the north wall boundary of the proposed basement, approximately 110 m away from the MRT viaduct. Figure 5.3 shows the plan view of the layout of the jet grout trial, diaphragm wall panels, and monitoring instruments. Figure 5.4 shows the cross section of the jet grout trial. Three diaphragm wall panels, Panels 3–5, each with wall thickness of

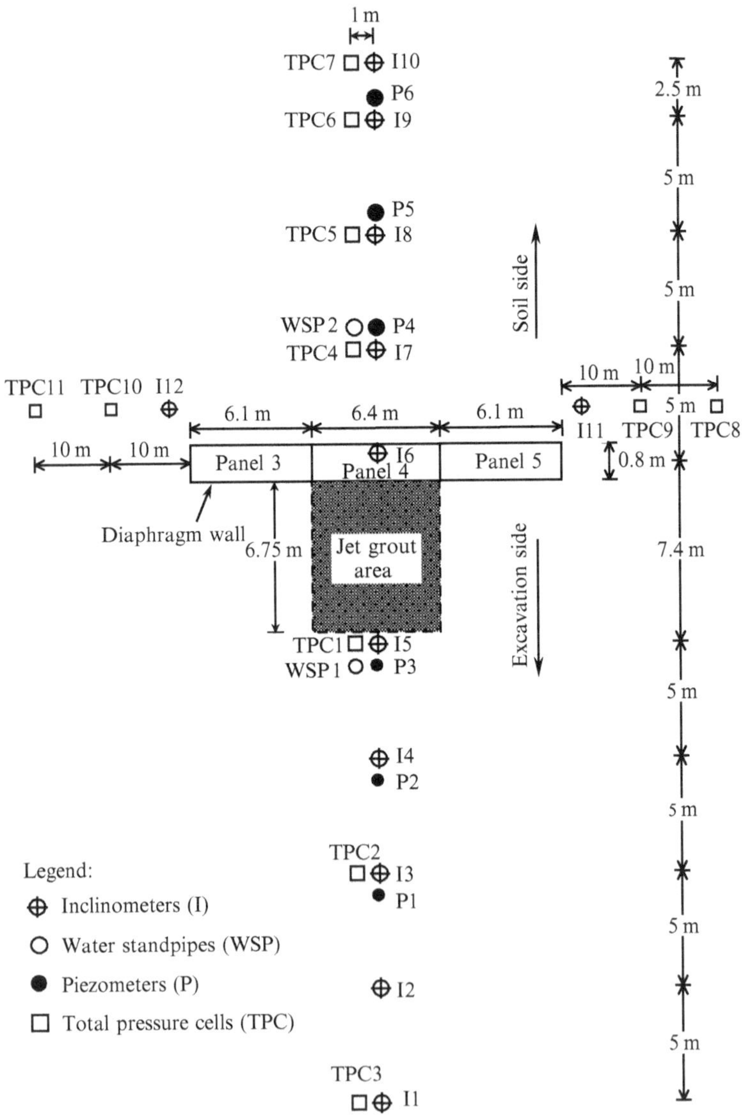

Figure 5.3 Plan layout of the jet grout trial, diaphragm wall panels, and monitoring instruments.

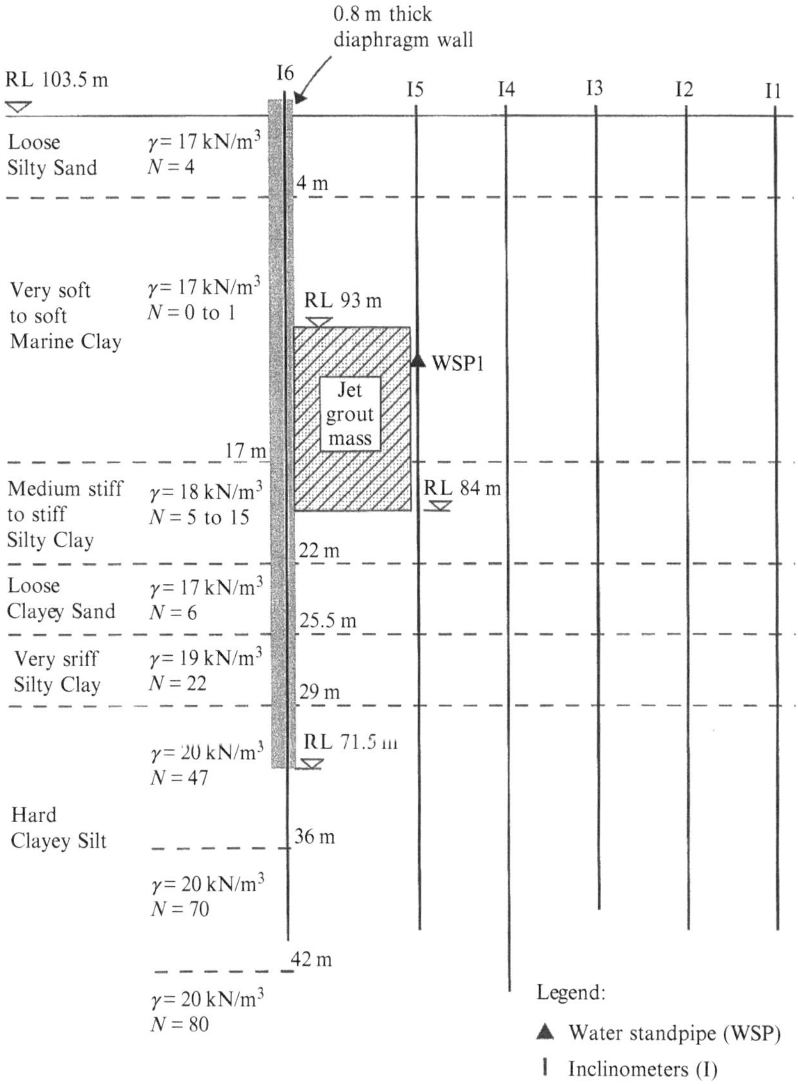

Figure 5.4 Soil profile and cross section through the jet grout trial.

0.8 m and panel length of 32 m, were constructed prior to the start of jet grouting. The diaphragm wall panels were socketed into a hard clayey sandy silt layer. A total of 18 jet grout columns numbered as C1–C18 were installed next to the middle diaphragm wall panel (Panel 4) to form a 6.4-m-wide and 6.75-m-long jet grouted area.

5.5.2 Construction sequences

Figure 5.5 shows the layout of jet grout columns for the jet grouting trial. The target diameter of the columns was 1.8 m, and the columns were formed in a triangular grid pattern spaced 1.55 and 1.35 m apart, center to center, parallel and perpendicular to the wall, respectively. Each column was 9 m long and was formed upward, from 19.5 to 10.5 m below the ground surface. As shown in Fig. 5.4, 6.5 m of the jet grout column was in the soft to medium stiff marine clay layer, whereas the remaining 2.5 m was in the medium stiff to very stiff silty clay layer. The sequence of jet column installation followed the columns' numbers starting with C1 and ending with C18. Each day, only two columns were formed. Figure 5.5 shows that no two adjacent columns were formed within the same day in order to allow time for an earlier formed adjacent column to set.

The details of the construction sequence of the jet grout columns were as follows:

1. Construction of three 0.8-m-thick diaphragm wall panels (June 6, 1994)
2. Installation of monitoring instruments (May 23–June 27, 1994)
3. Reestablishing the initial readings of monitoring instruments (July 4, 1994)
4. Construction of jet grout columns C1–C18, with two columns formed per day (July 5–13, 1994)

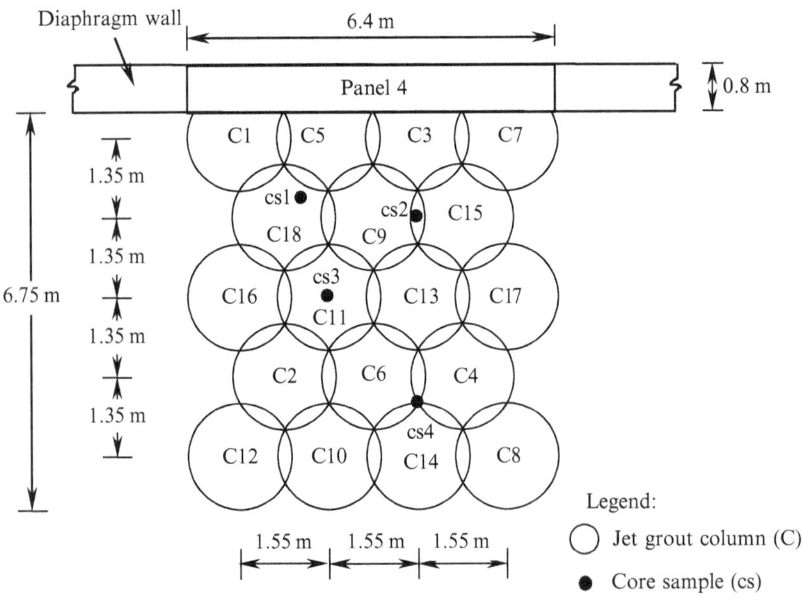

Figure 5.5 Layout of jet grout column for the jet grout trial.

The jet grout columns were formed by using a triple fluid jetting system. The technique involved using air-shrouded water jetting to penetrate and cut the soil mass. A separate injection nozzle was used for the placement of grout. A jet grout column was formed in the following stages: First, a 150-mm-diameter borehole was drilled to the required base depth of the proposed jet grout column by using a self-drilling machine. Maintenance of the stability of the borehole was important because the waste slurry generated during the jet grouting was required to reach the ground surface through the hole. With self-drilling, borehole stability was maintained primarily by the presence of the grout pipes and jetting monitor in the hole. During the short delay between drilling and the start of jetting, the drill flushing medium was kept flowing to maintain the stability of the borehole.

Once jetting started, stability was maintained by the escaping waste slurry. Jetting began where the high-pressure water jet shrouded by air was used to erode the ground locally, thus enabling the placement of the grout slurry. The process was followed by lifting and rotating the jetting pipes at a constant rate as grout was simultaneously injected to create a column of treated soil until it reached the required length. The treated soil thus consisted of a mixture of the injected grout and the eroded soil. Waste generated from the jetting (excess water, soil fines, and grout) under pressure traveled up the annulus between the grout pipes and the borehole sides for collection and disposal at the ground surface. Approximately 80–90% of the soil to be grouted flowed to the surface as waste.

The borehole was frequently checked, and the mechanical rotary head was adjusted upward or downward as necessary to ensure waste flow to the ground surface. At the required depth, the cutting medium was switched off, and the grouting was continued for a time to ensure all the cut material was flushed to the surface. Then the grouting was halted, and the rods were removed from the ground. The jet grouting operation parameters used during the jet grouting trial are summarized in Table 5.1.

5.5.3 Instrumentation and monitoring program

The monitoring instruments consisted of one wall inclinometer, 11 soil inclinometers, six pneumatic piezometers, two water standpipe observation wells, and 11 total earth pressure cells. The locations of these instruments are shown in Fig. 5.3. Most of these instruments were installed after the construction of the diaphragm walls, but some were installed before the construction of the diaphragm walls. However, the initial readings for these instruments were retaken on July 4, 1994, one day prior to the start of

Table 5.1 Parameters used in the jet grout trail

Operation parameter	Range
Water injection pressure	40–45 MPa
Water flow rate	60–70 l/min
Grout injection pressure	8–12 MPa
Grout flow rate	80–90 l/min
Compressed air pressure	0.7–1.0 MPa
Rod withdrawal rate	110–125 mm/min
Rod rotation rate	8–9 rpm
Water:cement ratio	1:1
Cement content	750 kg/m^3 grout

the jet grouting trial. Hence, the effects of construction of the diaphragm walls are not reflected in the results presented herein. The tips of the inclinometers were located at least 4 m below the toe of the diaphragm wall. The earth pressure cells and piezometers were installed at RL 88.5 m, which was the mid-elevation of the jet grout columns. In accordance with regulatory requirements, water standpipes were installed to monitor changes in the groundwater table. The standpipes were installed to a depth of 12 m, and except for the top 1.0–1.5 m, they were perforated throughout their length.

The monitoring program covered the entire process of the jet grouting trial and ended on the fifth day after the last jet grout column was formed. Readings of instruments were taken after the completion of each jet grout column. The monitoring results are summarized and discussed in the following sections.

5.5.4 Performance of jet grout trial

The effects of jet grouting on the lateral movements of the diaphragm wall panels and adjacent soil mass, changes in lateral earth pressures, and piezometric and water levels in the adjacent soils are presented and evaluated in this section.

Lateral movements

Figure 5.6 shows the plot of the maximum lateral movement profiles for the various inclinometers during the jet grouting trial. The column number (in parenthesis) indicated next to each inclinometer number in the figure legend corresponds to the jet grout column installation that caused the maximum lateral movement in the inclinometer. Figure 5.6 shows that the maximum lateral soil movements in the excavation (front) side of the jet grouted area (inclinometer I1–I4) decreased with increasing distance away from the jet

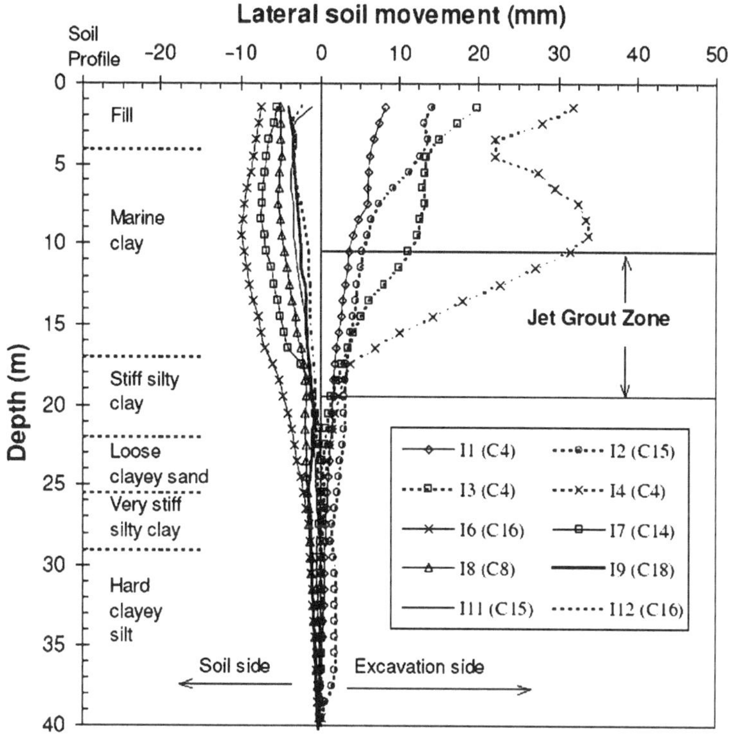

Figure 5.6 Effects of jet grout trial on lateral soil movements of adjacent soils.

grouted area. Inclinometer I5 located adjacent to the jet grouted mass was damaged during the jet grout trial. The monitoring results for inclinometers I7–I9 located at the soil (rear) side of the jet grouted area also show a similar movement trend, with the movements decreasing with distance, away from the jet grout area. During the jet grouting trial, the diaphragm wall was pushed a maximum of 10 mm away from the jet grout area, toward the soil side, as indicated by inclinometer I6. The deflected shape of the wall inclinometers suggested that the loose silty sand fill layer acted as a restraint against the outward movement of the wall, causing a change in curvature of the wall.

Figure 5.7 shows the maximum lateral movements versus distance from the grouted mass. This figure indicates that the maximum lateral movements decreased with distance from the jet grouted mass. Also, the maximum free field lateral soil movements in the excavation side of the jet grouted mass were much larger than the corresponding lateral movements behind the wall. The maximum lateral free field soil movement at 5.25 m in front of the jet grouted mass was 35 mm, decreasing to 9.5 mm at 20.25 m from

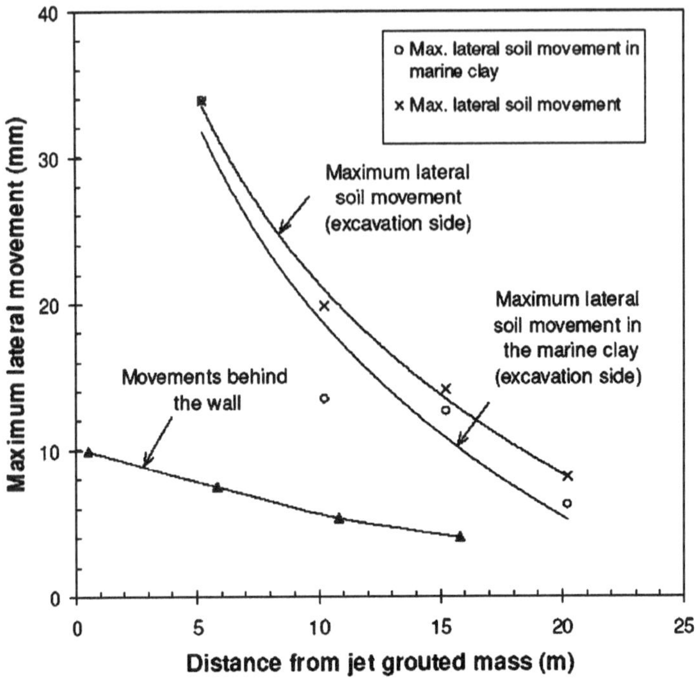

Figure 5.7 Lateral movements versus distance from jet grouted mass for jet grout trial.

the jet grouted mass. The lateral movements behind the wall decreased from 10 mm at 0.5 m to 4 mm at 15.8 m. These results suggest that the diaphragm wall provided considerable restraint, thereby reducing the lateral movements of the soil behind the wall induced by jet grouting.

The lateral soil movement profiles on the excavation side of the jet grout trial where inclinometer I3 was located at a distance of 10 m from the jet grouted mass, are shown in Fig. 5.8. The movements in the fill increased at a faster rate than the corresponding soil movements below the fill. These movements are typical of the inclinometers on the excavation side. This movement trend was different from the observed movement trend at the soil side. This may be due to the presence of the diaphragm wall panels, which provided significant restraint, thereby reducing movements behind the wall but causing larger upward and outward movements at the excavation side.

Bending moments

As described previously, jet grouting caused the wall to deflect, which resulted in a change in wall curvature and hence a change in wall-bending moments. Typical bending moment profiles plotted against depth are shown

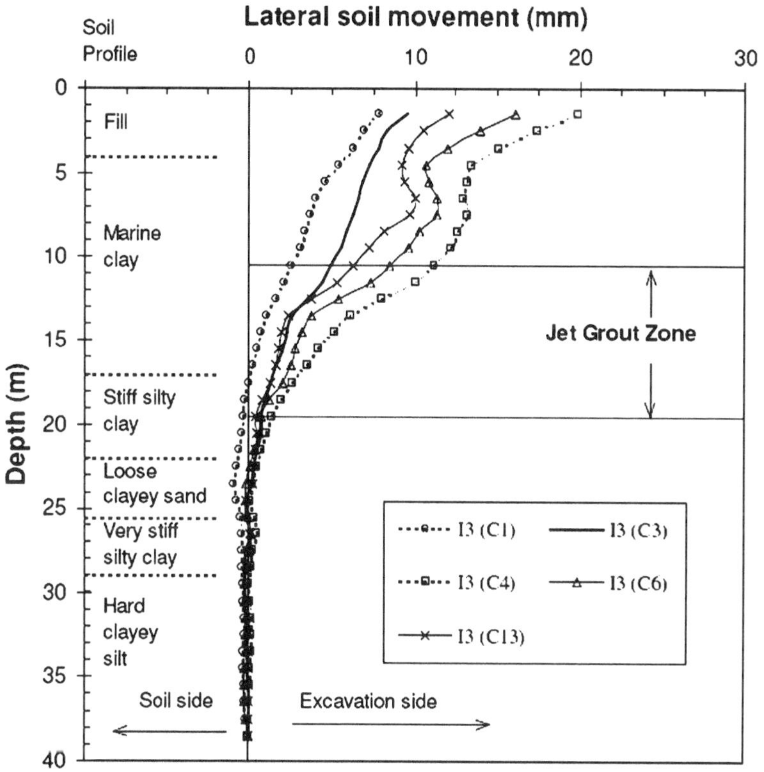

Figure 5.8 Development of lateral soil movements for inclinometer I3 during jet grout trial.

in Fig. 5.9. The variation of the maximum negative and positive wall-bending moments with time during the jet grout trial are shown in Fig. 5.10. These bending moments were deduced from readings of wall inclinometer I6. The wall-bending moment is defined as positive when the tension face of the wall is on the excavation side.

The jet grouting caused the wall to deflect backward (toward the soil side). The deflected shape of the wall (as shown by inclinometer I6 in Fig. 5.6) suggests that the fill layer provided restraint against the backward movement, resulting in a change in the curvature of the wall and hence a maximum negative bending moment at a depth of 7.5–9 m below the ground surface, slightly above the top of the jet grouted zone. Below the jet grout zone, the backward movements were restrained by the presence of the stiff soil layer, resulting in a change in wall curvature and hence a maximum positive bending moment at a depth of 18.5–23.5 m, near the bottom

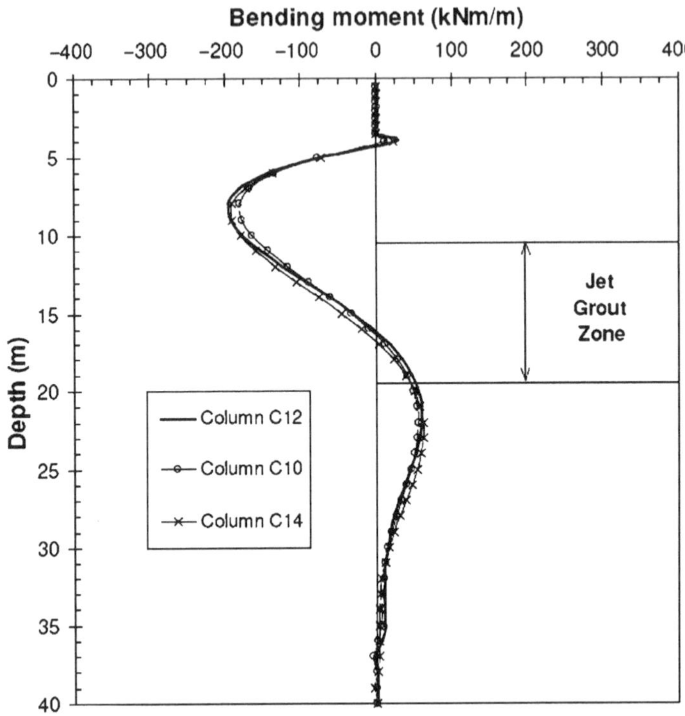

Figure 5.9 Typical bending moment profiles plotted against depth in Panel 4 during jet grout trial.

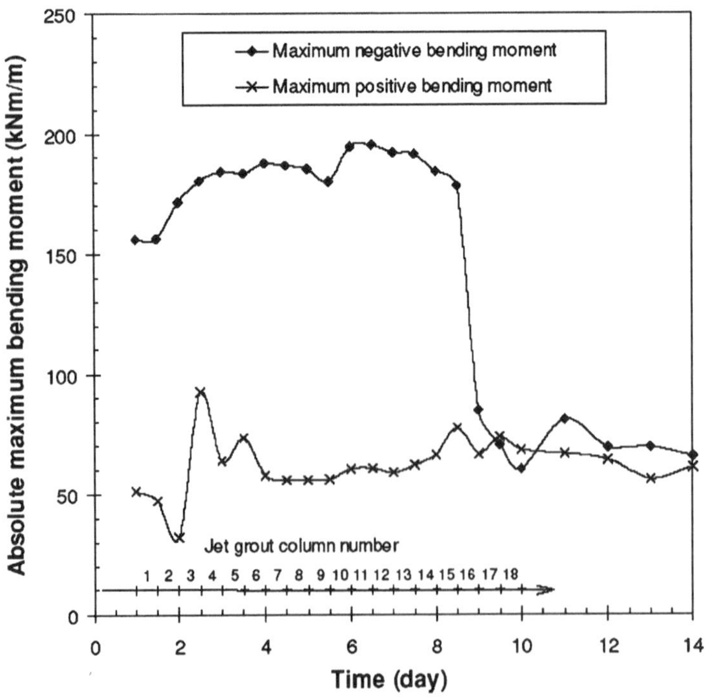

Figure 5.10 Maximum wall-bending moment in Panel 4 during jet grout trial.

of the jet grouted zone. Figure 5.10 shows that the negative wall-bending moment varied from 156 to 195 kN m/m during the jet grouting of the first 16 columns but decreased drastically thereafter and finally stabilized at approximately 70 kN m/m.

The deflection profile in Figure 5.6 shows that the wall also rotated in addition to bending. A closer study of the monitored wall deflections revealed that during the jet grouting of column C17, the top of the wall moved by an additional 3 mm, from −3.3 to −6.3 mm. This relatively large wall top movement released some of the wall restraint, resulting in a large reduction in the negative bending moment. Figure 5.10 also shows smaller variations in positive bending moments from 32.5 to 92.5 kN m/m during the jet grout trial.

Lateral earth pressures

The effects of the jet grout trial on the lateral earth pressures in the soil adjacent to the jet grouted area are shown in Fig. 5.11. This figure shows that the

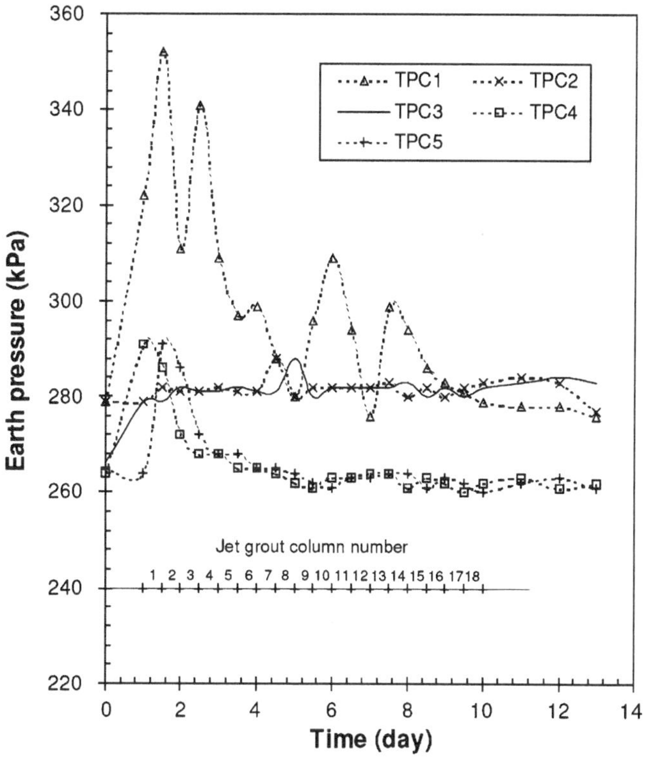

Figure 5.11 Effects of the jet grout trial on the lateral earth pressures in the adjacent soil.

jet grouting caused a large increase in lateral earth pressure at TPC1 and some increase in lateral earth pressure at TPC2–TPC5. The monitoring results showed minimal variations in the lateral earth pressures at TPC6–TPC11; thus, those results are not presented here. The lateral earth pressure at TPC1 located close to the jet grouted mass was highly variable throughout the jet grouting, with a maximum increase of lateral earth pressure of 73 kPa. Generally, there was a slight increase in the lateral earth pressure at TPC2 and TPC3 during the process of jet grouting.

The lateral earth pressure for TPC4 and TPC5 installed behind the diaphragm wall panel increased from approximately 270 kPa to approximately 290 kPa during the installation of the first two jet grout columns. However, it decreased to the original level after the fourth column was installed. Thereafter, there were only minimal variations in the lateral earth pressure. These results suggest that the jet grouting trial caused only a temporary increase in lateral earth pressures. The results also suggest that the wall caused the lateral forces to be distributed, thereby resulting in a smaller increase in lateral pressures behind the wall than in front of the wall.

Water and piezometric levels

The monitoring results indicated some minor variations of the groundwater level in the excavation side, whereas the groundwater level behind the wall remained almost constant during the jet grouting. Readings of the piezometric level during the jet grouting trial are shown in Fig. 5.12. This figure shows that the process of jet grouting caused a large increase in piezometric level in piezometer P3 located immediately in front of the jet grout area. The maximum increase of piezometric level was 7 m. Piezometers P2 and P4 also experienced an increase in the piezometric level during the jet grouting trial. However, for piezometers P1, P5, and P6, located farther away from the jet grout area, the variations in the piezometric levels were minimal. The incremental pore pressure (Δu) versus the incremental total soil pressure (ΔP) has been plotted in Fig. 5.13. This figure suggests an approximately linear relationship between Δu and ΔP. Figures 5.11 and 5.12 show that the magnitude of the increase in Δu and ΔP was largest at the beginning of the jet grouting trial and then subsequently decreased with time. Figure 5.12 shows that at the end of the jet grouting trial, the piezometric levels for all piezometers returned to their initial values, except for P3, which was slightly higher than its initial value. These results show that the jet grouting trial caused only a temporary increase in the piezometric level.

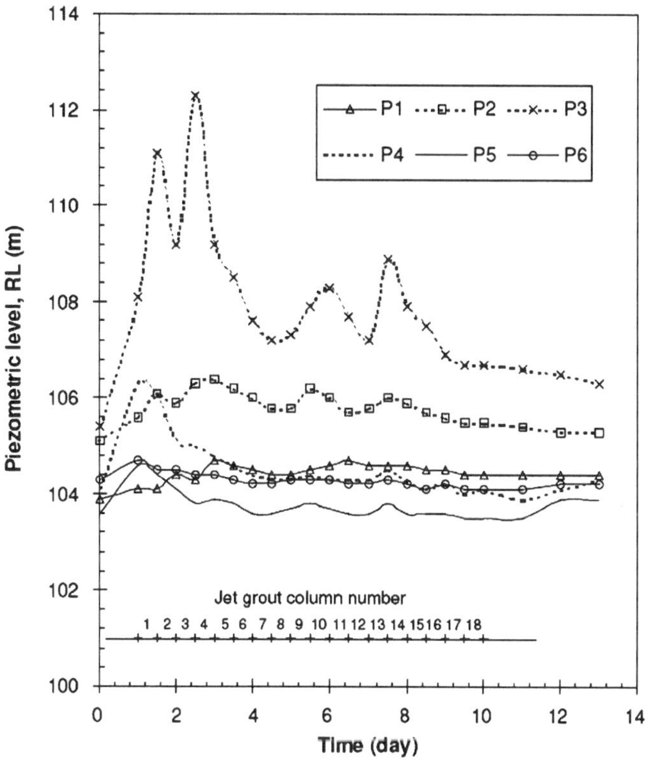

Figure 5.12 Readings of the piezometric level during the jet grouting trial.

Jet grout formation

For jet grout columns formed within the soft to medium stiff marine clay, core samples obtained using a triple fluid core barrel indicated that a treated jet grout mass was properly formed with a total core recovery of 70–100%. However, within the medium stiff to very stiff silty clay layer, the core samples showed that the jet grout columns were not properly formed or not formed at all. Over this zone, the core samples consisted of stiff silty clay or silty clay with only traces of jet grout. These results indicate that the use of a single set of jet grout parameters was not efficient for forming a jet grout column in layered soils with large differences in soil properties between the soil layers. The jet grout parameters as shown in Table 5.1 were suitable for forming a proper jet grout column within the soft-to-medium stiff marine clay layer. However, the use of the same parameters failed to form a jet grout column in the medium stiff clay layer. Hence, for the medium stiff to very stiff silty clay layer, a different set of jet grout parameters

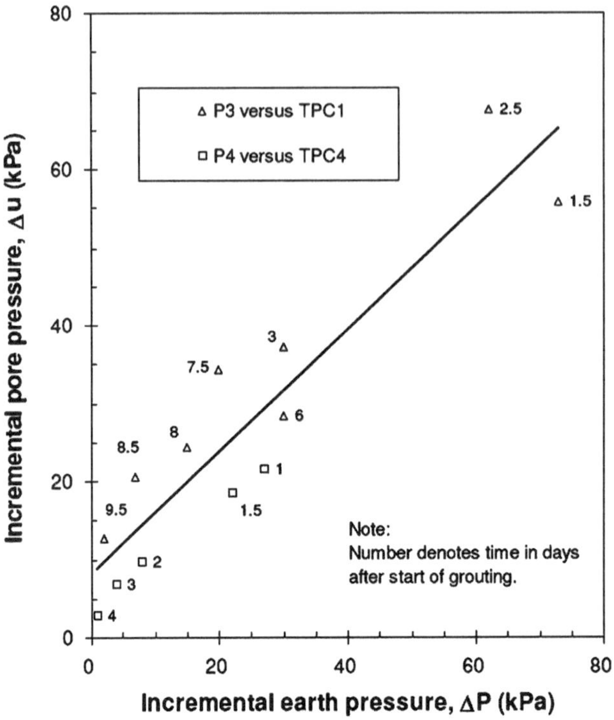

Figure 5.13 Incremental pore pressure (Δu) versus the incremental total soil pressure (ΔP).

with more energy per unit lift as suggested by Bell (1993) would be more suitable.

The results from unconfined compression tests on the core showed that the average undrained shear strength and Young's modulus of elasticity of the jet grouted mass were 3070 kPa and 257 MPa, respectively. Results from pressuremeter tests indicated a higher stiffness for the jet grouted mass. The modulus of elasticity obtained from the first unloading–reloading curve ranged from 176 to 519 MPa. These test results indicated that the actual grout strength was an order of magnitude higher than the target strength, whereas the measured Young's modulus of elasticity was approximately 70% higher than the target value.

5.6 PRODUCTION JET GROUTING

The target diameter of the columns was 1.8 m, and the columns were formed in a triangular grid pattern spaced 1.55 and 1.35 m apart, center to center, parallel and perpendicular to the long axis of the excavation, respectively. Prior to the production grouting, a field jet grout trial was carried out at the site to assess the effects of the installation sequence of the jet grout columns and the effects of jet grouting on the diaphragm walls and adjacent ground. The results of the jet grout trial suggested that the jet grout columns in the soft clay were properly formed and the strength and stiffness were met when the operation parameters shown in Table 5.1 were adopted in the grouting program. Hence, in the production grouting, the operation parameters for jet grouting in marine clay were the same as that used in the jet grout trial. The results of the jet grout trial revealed that a higher energy input was required for grouting in the medium stiff to very stiff silty clay. Hence, a slower rod withdrawal rate of approximately half that for the marine clay was used for this layer, with other operation parameters remaining unchanged. The thickness of the jet grout mass varied across the site and is also shown in Fig. 5.14. A 9-m-thick, 10-m-wide jet grout mass was formed along the excavation boundary next to the MRT station and viaduct. For the rest of the basement area, the thickness of jet grout mass ranged from 3 to 4 m. The jet grout layers were generally located within the soft to medium stiff marine clay, with some portions extending into the medium stiff to very stiff silty clay. The total treated soil area was approximately 22,000 m^2, with a total volume of 87,000 m^3.

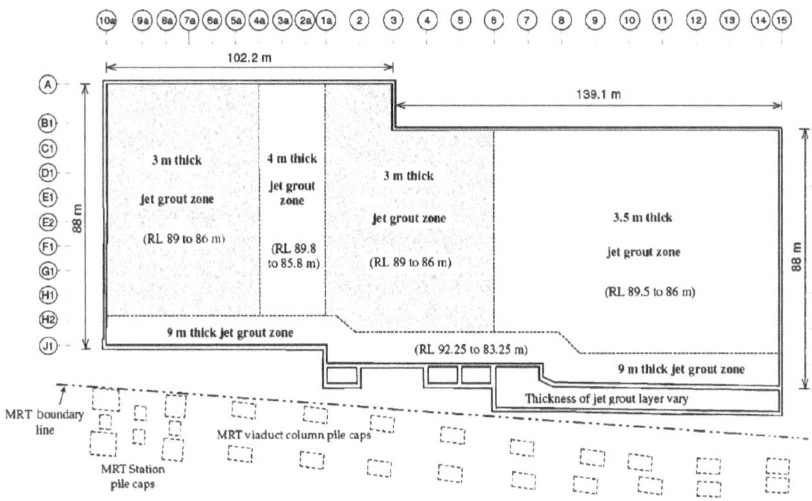

Figure 5.14 Plan layout of thickness of jet grout layer for Singapore Post Centre.

5.6.1 Instrumentation and monitoring program

Various instruments were installed to monitor the effects of the production grouting on the behavior of the diaphragm walls, adjacent soils, and structures. These instruments included wall and soil inclinometers, soil settlement markers, pneumatic piezometers, water standpipe observation wells, total earth pressure cells, tilt plates, and three-dimensional (3D) prism survey points. These instruments were installed prior to the construction of the diaphragm walls except for the wall inclinometers, which were installed during the wall construction. The initial readings of the wall inclinometers were taken 3 or 4 weeks after the wall panel was constructed to allow the concrete of the wall to fully harden. The locations of these instruments are shown in Fig. 5.1. The behavior monitored included movements of the diaphragm wall panels and adjacent soils, changes in lateral earth pressures, changes in piezometric and water levels, and the tilting and movement of the MRT station and viaduct columns. The tips of the inclinometers were installed into the very dense clayey silt or silty sand layer. The total earth pressure cells were installed at RL 87.5 m, except for EP4, which was installed at RL 84.5 m. These earth pressure cells were installed at 1.5–2 m, behind the diaphragm walls. The pneumatic piezometers were installed in the marine clay, with their tips at RL 91.5–97.5 m, at a distance between 9 and 20.5 m behind the wall. The tips of the water standpipes were installed at RL 90.2–91 m, located 7–19.5 m behind the wall panel. Soil settlement markers were installed at between 3.3 and 29.7 m behind the walls, along the MRT pile caps. Tilt plates were mounted on the MRT station and viaduct columns to measure their tilting, whereas the 3D prism survey points were installed on the MRT station and viaduct columns to monitor movements resulting from the production grouting and subsequent excavation.

The monitoring program covered the entire period of the production grouting and subsequent basement excavation. Readings of instruments were taken twice a week, except for the soil and wall inclinometers, which were monitored daily. The monitoring results are summarized and discussed in the following sections.

5.6.2 Construction sequences

The construction of the diaphragm wall panels began at the intersection of the east and north wall boundaries and progressed toward the south and west simultaneously. After the completion of the installation of the panels along

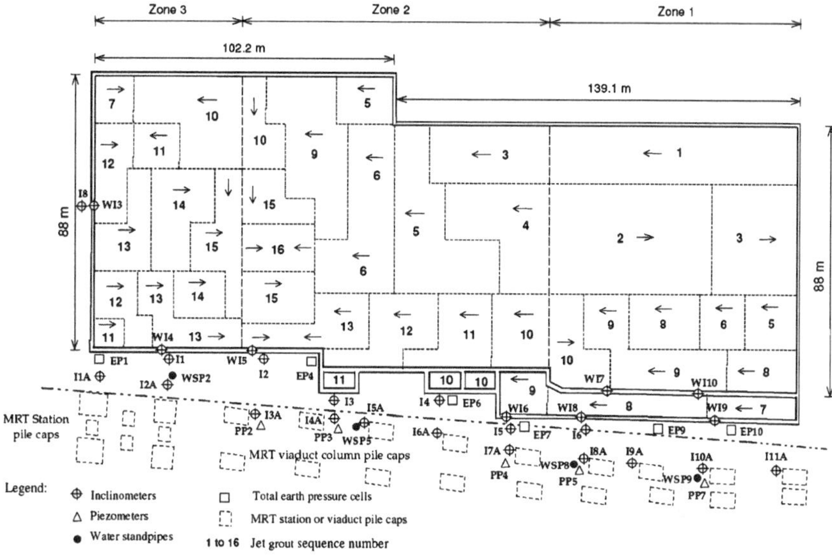

Figure 5.15 Details of jet grout sequence during production grouting for Singapore Post Centre.

the east wall and most of the north wall in zone 1 (Fig. 5.15), the jet grouting work started at the intersection of the east and north walls. Subsequently, the construction of the diaphragm wall panels at the north, south, and west wall boundaries was carried out concurrently with the jet grouting work. Construction of the south wall boundary started with the construction of the outer wall of the double wall at the intersection between the south and east wall boundary. This was followed by construction of the inner wall of the double wall from east to west.

After the diaphragm wall panels were constructed, generally there was a minimum clear distance between the constructed diaphragm wall panels and the grouted area. This was to provide enough time for the concrete in the wall panels to harden. For zones 1–3, the minimum clear distance between the constructed wall and grouted area was approximately 30, 20, and 17 m, respectively. In terms of time, there was an allowance of at least 2 weeks after a wall panel was cast before any jet grouting immediately next to it could proceed. With reference to the grid lines in Fig. 5.1, the details of the diaphragm wall construction sequences are as follows:

1. Construction of the east wall (grid lines C–L) and north wall (grid lines 15–6)
2. Construction of the north wall (grid lines 6–3)

3. Construction of the outer wall of the south double wall (grid lines 15–6) and north wall (grid lines 3–1a)
4. Construction of the inner wall of the south double wall (grid lines 15–6) and the north wall (grid lines 1a–10a), and a portion of the west wall (grid lines A–C1)
5. Construction of a portion of the south wall (grid lines 6–4, 2a–10a) and a portion of the west wall (grid lines J1–F1)
6. Construction of the rest of the south wall (grid lines 4–2a) and the west wall (grid lines F1–C1)

5.6.3 Production jet grout sequence

The production grouting was divided into three zones, namely zones 1–3, as shown in Fig. 5.15. The zones were separated by temporary sheet pile walls. The jet grout columns were installed row by row, in a direction away from the walls. Generally, jet grouting began at the north wall and progressed toward the south wall and from east to west. With this grouting sequence, the south boundary or boundary along the MRT station and viaduct was the last boundary to be grouted. The details of the jet grout sequence are shown in Fig. 5.15, with numbers indicating the progression of grouting area and arrows indicating the direction of the jet grout work.

A jet grout column was formed in the following steps: First, a 150-mm-diameter borehole was drilled to the required base depth of the proposed jet grout column. Maintenance of the stability of the borehole was important because the waste slurry generated during the jet grouting was required to reach the ground surface through the hole. Borehole stability was maintained primarily by the presence of the grout pipes and jetting monitor in the hole. During the short delay between drilling and the start of jetting, the drill-flushing medium was kept flowing to maintain the stability of the borehole. Once jetting started, stability was maintained by the escaping waste slurry. Jetting began where the high-pressure water jet shrouded by air was used to erode the ground locally, thus enabling placement of the grout slurry. Then, grouting began by lifting and rotating the jetting pipes at a constant rate as grout was simultaneously injected to create a column of treated soil until it reached the required height.

The treated soil thus consisted of a mixture of the injected grout and the eroded soil. The waste generated from the jetting (excess water, soil fines, and grout) under pressure traveled up the annulus between the grout pipes and the borehole sides for collection and disposal at the ground surface. Approximately 80–90% of the soil within the grouted zone was displaced, and most of it flowed to the surface as waste. The borehole was frequently checked, and the mechanical rotary head was adjusted upward or downward as necessary to ensure waste flow to the ground surface. At the required column height, the cutting jet was switched off, and the grouting was continued for a time to ensure that all the cut material was flushed to the surface. Then the grouting was halted, and the rods were removed from the ground.

After the completion of each zone, core samples were obtained and tested. Most of the cores were taken at the intersection of the jet grouted columns. The core samples indicated that the jet grout columns were properly formed and achieved their target diameter of 1.8 m. The rod withdrawal rate used in the production grouting for the marine clay ranged from 110 to 125 mm/min, as indicated in Table 5.1. Stroud (1994) reported similar results, where 1.8-m-diameter jet grow columns were formed using a similar rod withdrawal rate reported herein. Unconfined compression tests on the core samples showed that the average undrained shear strength and Young's modulus of elasticity of the jet grouted mass were 2.3 and 173 MPa, respectively.

5.6.4 Effects of production grouting on diaphragm walls

Field observation

The lateral movements and the bending moments in the diaphragm wall panels along the south and west boundaries as a result of the jet grouting are summarized in this section. The south walls were 40–41.5 m deep, as shown in Fig. 5.2. Note that generally the wall panels were constructed only approximately 17–30 m ahead of the nearest jet grout area. Also, the initial readings of the inclinometers in the walls were taken only 3 or 4 weeks after the wall panel was constructed. Hence, only partial effects of jet grouting are reflected in these results.

Lateral wall movements induced by the jet grouting are shown in Fig. 5.16. Negative or backward movement refers to movement toward the soil side, away from the jet grouted area. The diaphragm wall movements as a result of the jet grouting were predominantly backward, away from the jet grouted area. The maximum lateral movements ranged from 9.7 to 36.4 mm. The figure also shows that the lateral wall movements increased from east to west. In the area of the double wall, grouting was first performed on the south side of inclinometers WI7 and WI10 and resulted in movements toward the north—that is, opposite the direction of movement shown in Fig. 5.16. When subsequent grouting occurred in subzones 9 and 10 (Fig. 5.15), north of this wall, the direction of the movements was

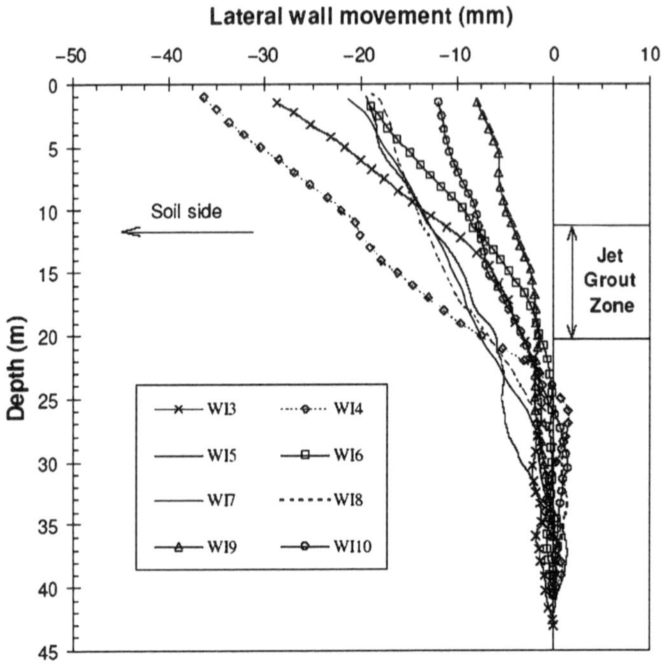

Figure 5.16 Effects of production grouting on lateral wall movement of diaphragm walls.

reversed. However, these movements were impeded by the previously jet grouted zone on the south side. This explains why the movements of WI7 and WI10 were smaller than those of the other inclinometers.

Figure 5.15 shows that inclinometers WI3 and WI4 were located in an area where grouting was carried out in a confined area—that is, the soil adjacent to this area had already been treated. Review of the measurements for WI3 and WI4 showed that a significant portion of the movement occurred during the grouting of subzones 14–16. The confinement by the surrounding subzones, which were previously grouted, probably resulted in considerable displacement of the soft soils (under essentially undrained conditions) and is the likely reason why WI3 and WI4 showed larger movements.

Similar backward movements have been reported by others, including Sugawara et al. (1996) and Khoo et al. (1997). The maximum wall movements shown in Fig. 5.16 are comparable to the reported movements for the aforementioned two sites, which had similar soft soil conditions.

Bending moments in a diaphragm wall were estimated from the deflected shape of the walls as described by Poh et al. (1997). The wall bending

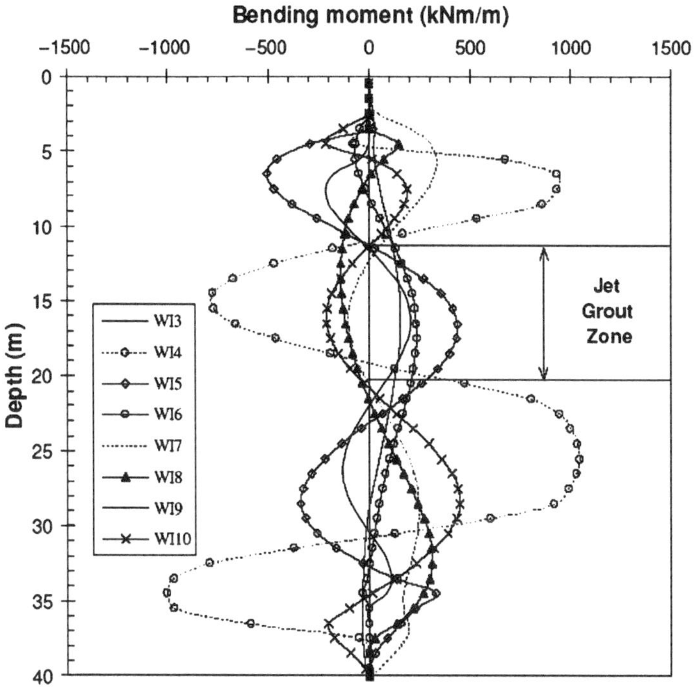

Figure 5.17 Effects of production grouting on bending moments in diaphragm walls.

moment is defined as positive when the tension face of the wall is on the excavation side. Figure 5.17 shows the deduced bending moment profiles of the diaphragm walls corresponding to the wall movements shown in Fig. 5.16, caused by the production grouting. The production grouting induced both positive and negative bending moments in the walls, depending on the shape or the curvature of the wall. As expected, the maximum bending moments varied in proportion to wall stiffness. The thicker, 1.2-m wall experienced maximum moments ranging from 333 to 1043 kN m/m and −103 to −1005 kN m/m. The maximum bending moment for the 0.8-m-thick wall panel (west wall) containing WI3 was relatively small because its stiffness was only 30% of that of the 1.2-m-thick south wall. These results suggest that a stiffer wall attracts larger bending moments, as has been reported by Wong and Poh (1995) and Poh et al. (1997), and that the process of the production grouting may cause significant bending moments in the diaphragm walls. If the reinforced concrete diaphragm wall is not properly designed, these bending moments may exceed the yield moment of the wall.

Verification by numerical analysis

Finite element analysis has been carried out to study the effects of jet grouting on the wall movements, with the finite element mesh and simplified soil profile shown in Fig. 5.18. The computer program used is PLAXIS. The soil is represented by six-node solid elements and the diaphragm wall by beam elements. In the analysis, the pressure from the grouting process acting on the wall is taken to be 43 kPa, which is the maximum incremental lateral earth pressure reported in production grouting.

The computed deflections of the diaphragm wall are shown in Fig. 5.19. The maximum computed deflection of 25 mm is at the top of the wall. The shape of the deflections is broadly similar to, although not identical with, the measured shapes. The upper part of the marine clay and the overlying fill layer are soft to medium stiff in strength, resulting in the cantilever shape of the wall deflections.

Typical results of the net lateral wall deflections at the end of excavation are presented in Fig. 5.20 to illustrate the effectiveness of the jet grout strut in controlling movements caused by the subsequent basement excavation. Figure 5.20 shows that wall inclinometers WI3 and WI10 recorded relatively small deflections, with the maximum lateral deflections ranging from 11 to 20 mm, which suggests that the jet grout strut was effective in controlling wall deflections during basement excavation.

Currently in Singapore, the deflections and resulting internal forces induced in the diaphragm walls by jet grouting are not specifically taken

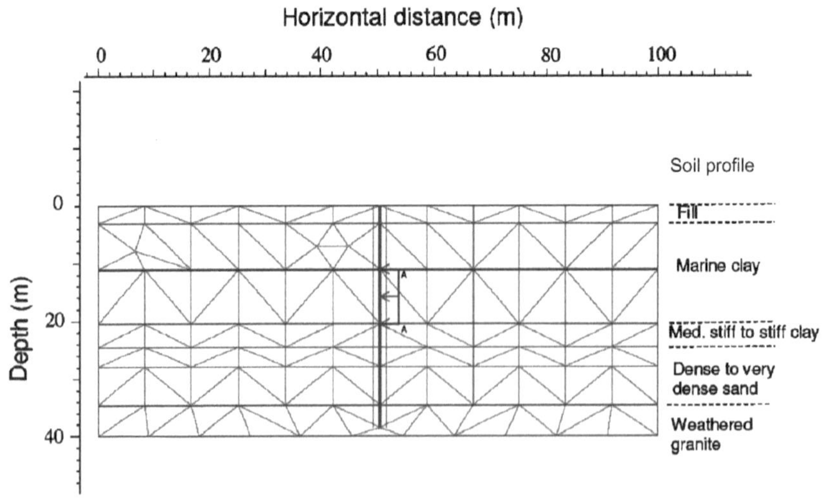

Figure 5.18 Finite element mesh used in the analysis of the effect of jet grouting.

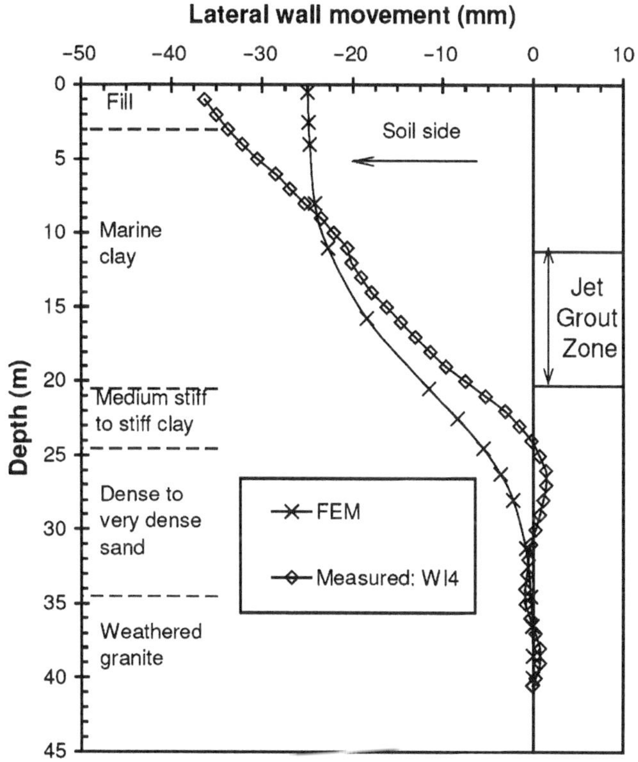

Figure 5.19 Lateral wall deflections computed by finite element analysis.

account of in the design. One of the reasons why is that, as pointed out herein, during the ensuing excavation for the basement structure, the deflections and forces would be expected to be reversed. This benign attitude is changing toward requiring that the ground movements induced by jet grouting be kept to a minimum.

Recent projects involving jet grouting in marine clay with the use of an external casing during the jet grouting operation to ensure a free return of sludge flow to the ground surface have resulted in substantially smaller lateral wall deflections (with maximum lateral deflections of a few millimeters at the end of grouting) (Poh, 2003). The grouting parameters adopted in these projects were similar to those used in the Singapore Post Centre. The use of such temporary external casing would help to prevent excessive movement likely caused by the jet grouting process; however, it slows the grouting process and thus increases cost.

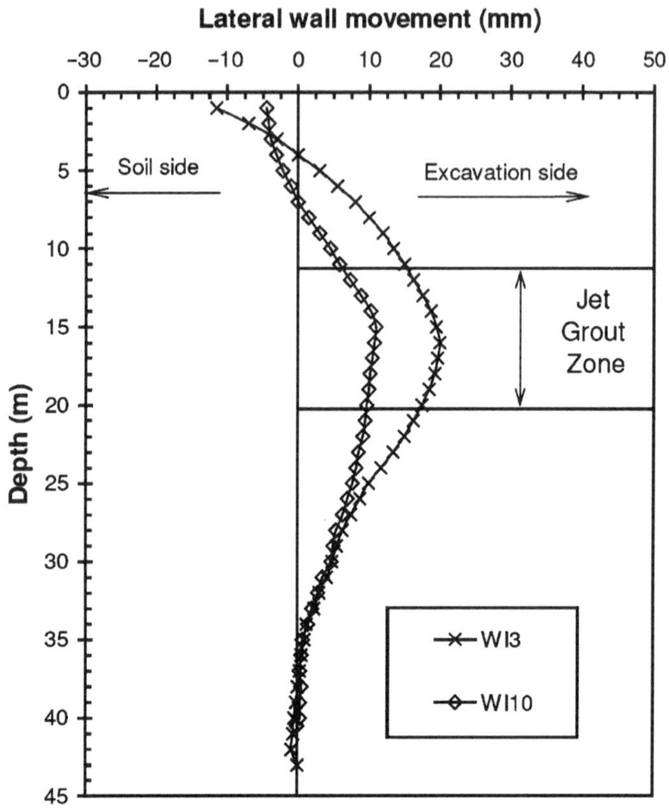

Figure 5.20 Typical results of total net lateral wall deflections at the end of excavation.

5.6.5 Effects of production grouting on adjacent ground response

The readings from the soil inclinometers, soil settlement markers, total pressure cells, piezometers, and water standpipes taken during the production grouting are presented herein. These results are used to evaluate adjacent ground responses as a result of production grouting. Soil inclinometers were monitored daily and were able to capture the effects of the construction of the wall panels. Thus, the effects of jet grouting on the lateral movements could be separated from the movements arising from the construction of the wall panels. Other monitoring instruments were monitored only twice per week; therefore, the effects of jet grouting and the construction of wall panels could not be assessed separately.

Lateral soil movements

Figure 5.21 shows the lateral soil movements caused by the production grouting recorded by the seven inclinometers next to the wall panels. The maximum lateral soil movement was 53.6 mm (inclinometer I8), occurring at the mid-section of the west wall boundary. Along the south wall boundary, the maximum lateral soil movements ranged from 35.3 to 51.0 mm (inclinometers I1–I6). In general, soil movements below the jet grouted zone were insignificant except for I8. The lateral soil movements above the grouted zone generally followed a triangular wedge pattern, except in the top few meters in the fill layer, where the deflection profiles were uniform, curved, or bent backwards. The deflected movement profiles shown in Fig. 5.21 suggest that significant shear deformation occurred over the depth interval corresponding to the jet grouted zone. The magnitudes of the lateral soil movements shown in Fig. 5.21 were considered large for the zone next to the MRT station and viaduct columns. These movements were

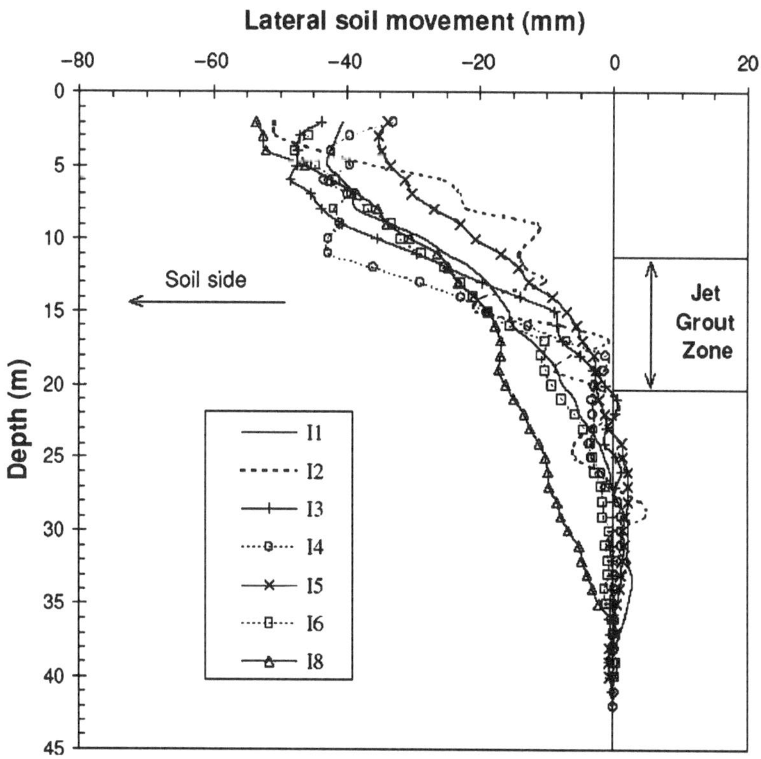

Figure 5.21 Lateral soil movements caused by the production grouting.

slightly larger than the corresponding wall movements because the soil movements reflected the total cumulative effects caused by the production grouting, whereas the wall movements showed only the effects of production grouting after the walls were in place.

The effects of the production grouting on the lateral soil movements at some distance behind the diaphragm walls are shown in Fig. 5.22. This figure shows the monitoring results for the 11 inclinometers installed near the column pile caps along the MRT station and viaduct. These inclinometers were located 9.5–20.5 m behind the wall panels, except for inclinometers I1A and I2A, which were located slightly closer to the wall. The maximum lateral soil movements generally ranged from 19.8 to 32.8 mm, except for inclinometer I11A, for which the maximum lateral soil movement was only 13.5 mm. The movement patterns shown in Fig. 5.22 also suggest that shear deformation occurred at a depth corresponding to the jet grouted zone, whereas movements below the jet grouted zone were insignificant.

The net movements (i.e., excluding the effects of wall construction) presented in Figs. 5.21 and 5.22 have been replotted as maximum lateral soil

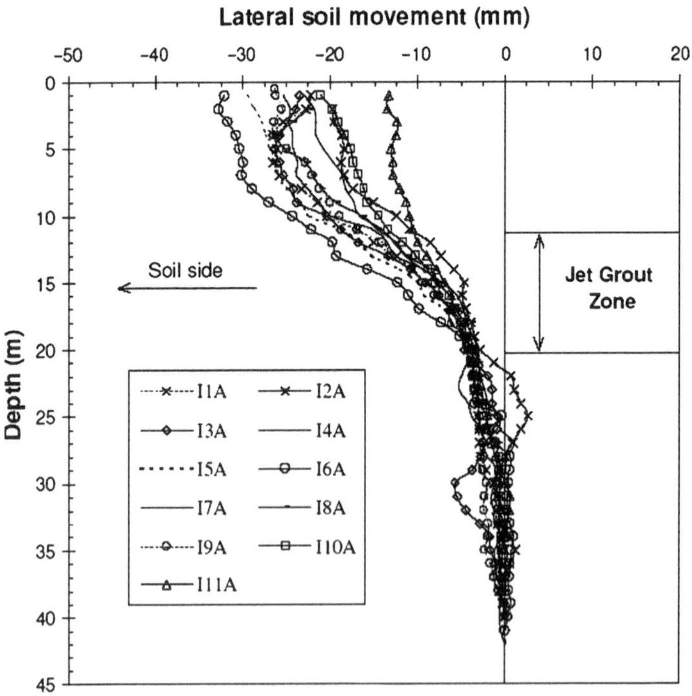

Figure 5.22 Lateral soil movements at some distance behind wall caused by the production grouting.

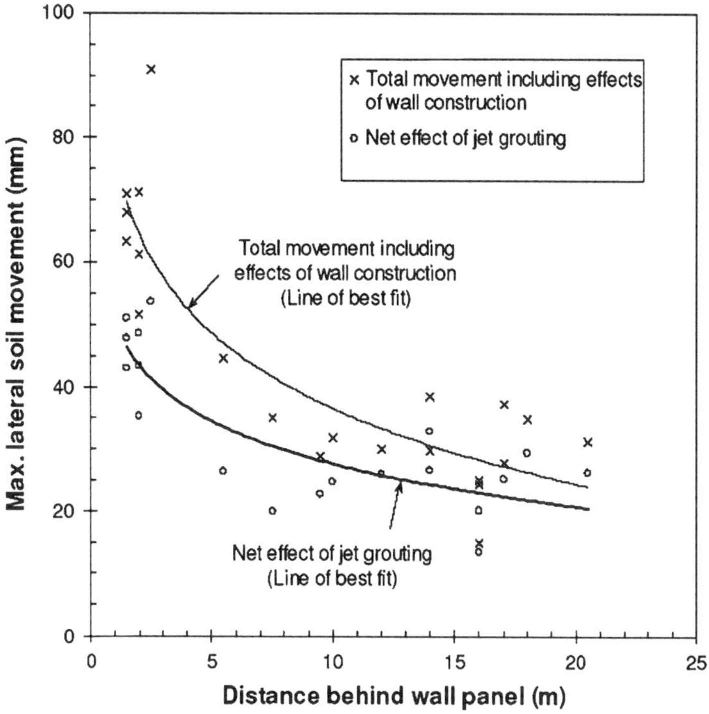

Figure 5.23 Maximum lateral soil movements versus distance behind the wall panel for the production grouting.

movements versus distance behind the wall panel and are shown in Fig. 5.23. The maximum lateral soil movements decreased significantly within the first 5 m from the wall. Beyond this distance, the decrease was more gradual. At a distance of 20 m behind the wall, the maximum lateral soil movement was approximately 21 mm. These results suggest that the zone of influence of the jet grouting extended more than 20 m behind the wall.

The total maximum lateral soil movements including the effects of construction of wall panels versus distance behind wall panels are also plotted in Fig. 5.23. Readings from the soil inclinometers show that the construction of the diaphragm wall panels also caused the soft to medium stiff marine clay to move backward, to the soil side. These backward soil movements are likely due to the process of concreting of the wall panel. Figure 5.23 shows that immediately adjacent to the walls, the process of the wall construction caused an average maximum backward movement (total movement minus movement due to jet grouting) of 23 mm. These movements decreased with distance away from the wall panels. Nevertheless, Fig. 5.23 shows that the effects of the construction of the wall panels extended to a considerable distance. At a distance of 20 m behind the wall panels, the average maximum backward lateral soil movement induced by the construction of wall panels was approximately 4 mm. Poh and Wong (1998) provide more details of the movements caused by diaphragm wall construction. The results described by Poh and Wong were obtained from a trial section near the intersection of the north and east walls.

Soil heave

The maximum soil heave recorded by settlement markers during the production grouting is plotted with distance behind wall panel in Fig. 5.24. The settlement markers were located 3.5–29.5 m behind the walls, along the MRT viaduct columns' pile caps. The results presented in Fig. 5.24 include the effects of the construction of the wall panels. This figure shows that the maximum soil heave caused by the production grouting was generally less than 15 mm, with a majority of the data being less than 10 mm, except for three groups of markers for which the soil heave ranged from 16 to 24 mm. These three groups of markers (shown by circular symbols

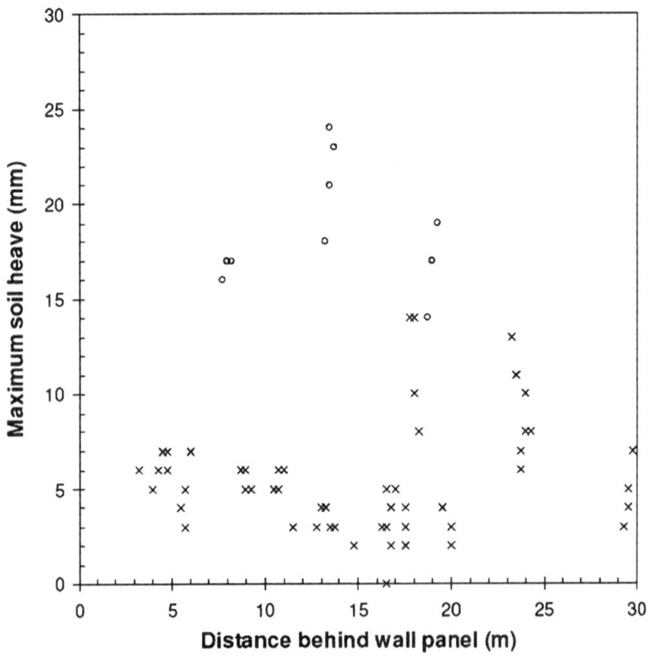

Figure 5.24 Maximum soil heave recorded by settlement markers during the production grouting.

in Fig. 5.24) were located between grid lines 1a and 2 (Fig. 5.1), where the walls were the last to be constructed. Hence, this larger soil heave was likely due to the longer time that the ground there was unrestrained by the presence of the walls while grouting in adjacent areas was in progress.

Lateral earth pressures

The readings for the total earth pressure cells during the production grouting are shown in Fig. 5.25. These cells were installed at RL 87.5 m, except for EP4 (TPC4), which was installed at RL 84.5 m. The readings included the effects of the construction of the wall panels. Because the readings were taken only twice per week, the effects of construction of the wall panels could not be separated from the effects of grouting. Figure 5.25 shows that the lateral earth pressures generally reached a peak value and then decreased. The maximum increase in the total earth pressure ranged from 10 to 43 kPa.

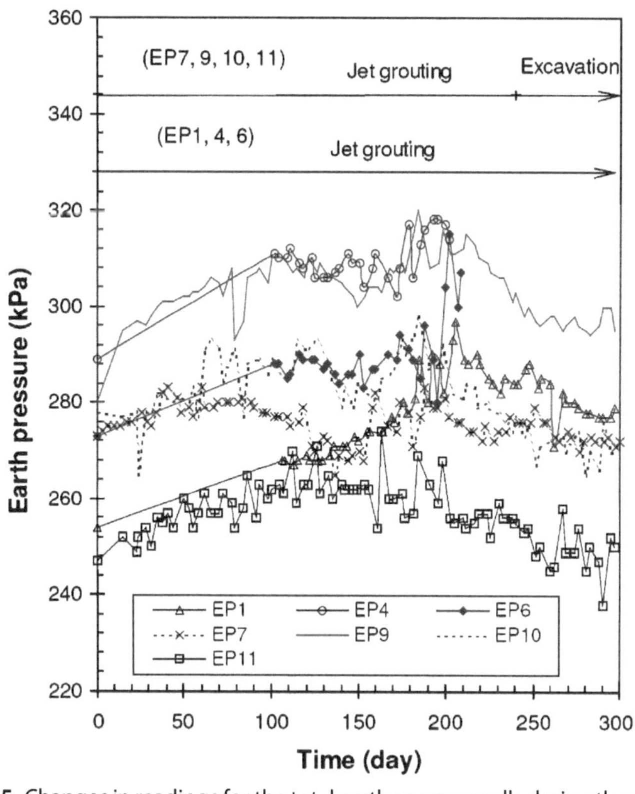

Figure 5.25 Changes in readings for the total earth pressure cells during the production grouting.

That represented an increase of 3.7–16.9% from the initial values. Figure 5.25 also shows that the earth pressures started to decrease before the start of basement excavation. These results indicate that the production grouting caused only a temporary increase in the lateral earth pressures in the adjacent soils. This may likely be due to the dissipation of excess pore pressure, which is indicated in the subsequent section on piezometric water levels.

Piezometric and water levels

Figure 5.26 shows the readings from some typical piezometers during the production grouting. Piezometers PP2–PP4, PP6, and PP7 were installed in the marine clay, with their tip level ranging from RL 91.5 to 97.5 m at 20.5, 9, 12.5, 11, and 15 m behind the wall, respectively. Water standpipes with their tips at RL 90.2–91 m were installed 7–19.5 m behind the wall panel. The piezometric readings included the effects of construction of

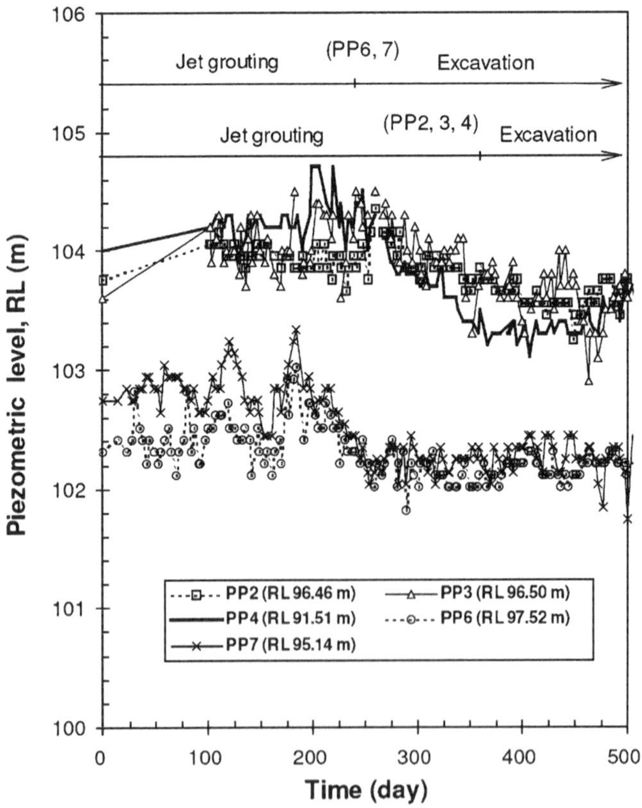

Figure 5.26 Changes in piezometric levels during the production grouting.

the wall panels. Figure 5.26 shows that the piezometric levels increased slightly during the production grouting. Then they started to decrease before the start of basement excavation. Only minor changes in groundwater levels were recorded in the water standpipes during the production grouting. Generally, the water level fluctuations were within 0.5 m of the original level.

5.6.6 Effects of production grouting on adjacent structures

The existing MRT station and viaduct are sensitive structures and are located close to the south wall boundary of the basement excavation. Hence, the columns of the MRT station and viaduct along the south wall boundary were instrumented and monitored closely during the production grouting. The monitoring instruments included tilt plates and 3D prism survey points for monitoring tilting and movements of the columns. These instruments were monitored only twice weekly, and hence the effects of the construction of wall panels could not be separated from the effects of the jet grouting. Thus, the results presented herein reflect the combined effects of jet grouting and wall construction.

The MRT viaduct columns are 1.75 m in diameter, spaced 25 m apart. The pile foundations of the MRT viaduct columns typically consisted of six 1.0-m-diameter bored piles at 3-m centers, with pile caps 7.8 m long, 4.8 m wide, and 2.0 m thick. The tips of the bored piles ranged from RL 62.9 to 69.0 m. These piles were socketted 12.0–13.5 m into the medium-dense to very dense clayey sandy silt or clayey silty sand (Old Alluvium). At the west end of the site, where the MRT station is located, the columns are 1.9 m in diameter. The pile foundations consisted of nine 1.0-m-diameter bored piles, spaced at 2.5- to 3.5-m centers, with pile caps 7.0 × 7.0 and 7.0 × 9.0 m and 2 m thick. The pile toe level ranged from RL 73.5 to 77.5 m, with their toes socketted 14.0–18.0 m into the medium-dense to very dense clayey sandy silt or clayey silty sand (Old Alluvium). Because the station columns are supported by heavier and closer spaced foundations, they are more rigid than the viaduct columns.

The readings from the tilt plates and 3D prism survey points are shown in Fig. 5.27. Positive tilt values shown in the figure correspond to tilting of the column away from the diaphragm walls. Positive lateral and vertical movements refer to movements away from the diaphragm walls and heave or upward movements, respectively. Figure 5.27 shows that the maximum tilting of the MRT columns ranged from 20 to 92 s. These maximum tilts had a trapezoidal shape in a plan view, with the largest tilt occurring within the

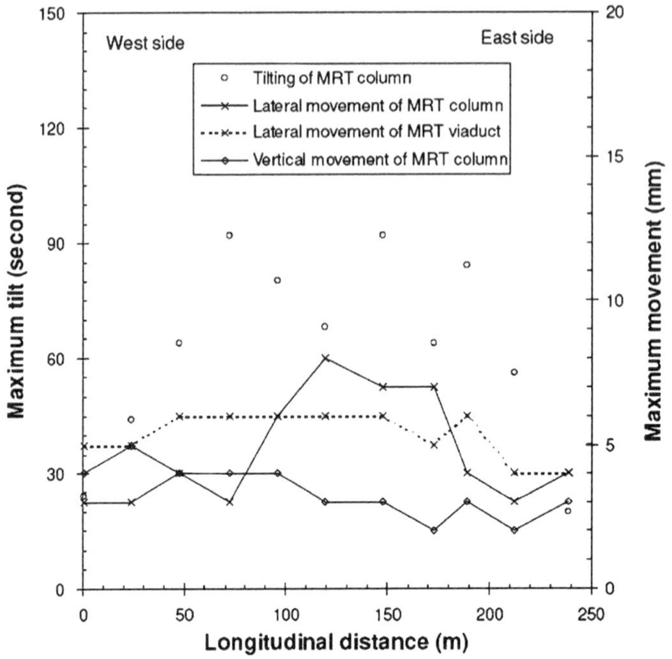

Figure 5.27 Maximum tilts and movements of MRT columns along MRT Station and viaduct.

middle zone and smallest tilt occurring near the two ends of the grouted zone. This trend was due to the geometry or 3D effects because the first and last MRT columns were located along the east and west wall boundary lines, respectively. In addition, at the two ends (especially the east end), the columns are farther from the diaphragm walls. Although the MRT columns were supported on fairly rigid pile foundations, the largest maximum column tilt induced by the grouting reached 89% of the allowable tilt (103 s). These columns tilted away from the grouted zone (excavation), and the readings during the subsequent basement excavation showed that the basement excavation caused the columns to tilt back toward the excavation side, thus offsetting the initial tilt.

The specifications by the authorities permit a maximum movement of 15 mm for any MRT structure. As shown in Fig. 5.27, the maximum lateral movement of the MRT columns induced by grouting ranged from 3 to 8 mm. These movements were away from the grouted zone (excavation), with the largest movement occurring within the middle section (between 90 and 170 m) along the south wall boundary where the columns were closest (approximately 11–14 m) to the diaphragm walls. As with the tilting, the smaller movements occurred near the two side wall boundaries. The lateral movements of the MRT viaducts were more uniform, with the maximum lateral movement ranging from 4 to 6 mm. As shown in Fig. 5.22, the corresponding maximum lateral movement of the adjacent soil next to the MRT column pile caps generally ranged from 19.8 to 32.8 mm. In comparison, the largest maximum lateral movement of the MRT structures was only 8 mm, much smaller than the adjacent soil movement. This smaller movement trend was due to the MRT columns being supported on pile foundations. The larger soil movements were in the upper 10 m, where the soils were generally soft to medium stiff. Hence, the forces induced in the pile foundations were likely to be small. Figure 5.27 shows that the MRT columns also experienced some vertical movements during the production grouting. The vertical upward movements of these columns ranged from 2 to 5 mm, with the movements decreasing from west to east. Subsequent readings showed that the basement excavation caused the columns to settle by approximately 1–6 mm.

The MRT column movements shown in Fig. 5.27 have been replotted against the distance behind the wall panel and are shown in Fig. 5.28. The lateral movements of the MRT columns decreased with distance away from the wall panel. However, the results show vertical movements of several millimeters even at distances up to 20 m from the walls. Because of the upward resistance of the pile foundations to ground heave, there is no clear trend in variation of foundation heave with distance from the diaphragm walls.

The monitoring results shown in Figs. 5.16, 5.21, 5.22, and 5.27 indicate that the production grouting caused the diaphragm walls, adjacent soils, and structures to move away from the jet grouted zone. However, as long as these movements were monitored and controlled well within their allowable limits, they would eventually help the performance of the diaphragm walls and adjacent structures during the subsequent basement excavation. These backward movements were responsible for minimizing the ultimate positive movements induced by the subsequent basement excavation.

5.7 CONCLUSIONS

The performance of the jet grout trial and production grouting during the basement construction for the Singapore Post Centre have been evaluated to study the effects of production grouting on the diaphragm walls, adjacent

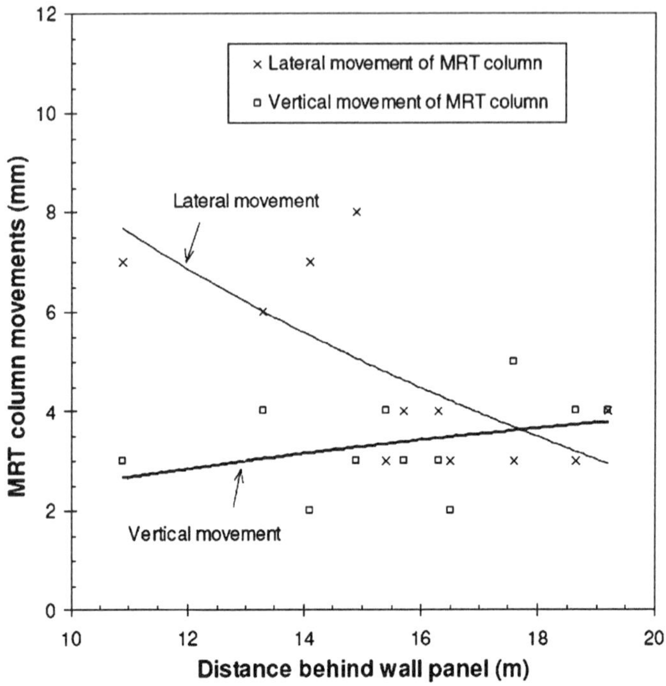

Figure 5.28 Maximum tilts and movements of MRT columns along MRT Station and viaduct.

soils, and nearby structures. The following summarizes the main points discussed in this chapter:

1. The maximum lateral soil movements on both sides of the jet grout area decreased with increasing distance away from the jet grout area. During the jet grout trial, the diaphragm wall was pushed a maximum of 10 mm away from the jet grout area, toward the soil side. The production grouting caused the retaining diaphragm walls and adjacent soils next to walls to move 9.7–36.4 and 35.3–53.6 mm, respectively, away from the jet grout area. The measured soil movements were considerably larger than the corresponding wall movements because they reflected the cumulative effects caused by the production grouting as well as the construction of the diaphragm walls, whereas the wall movements reflect merely the effects of jet grouting after the walls were in place. For soils located 9.5–20.5 m behind the diaphragm walls, the outward lateral movements ranged from 13.5 to 32.8 mm, with the movements decreasing with distance away from the diaphragm walls. Soil heave induced by production

grouting ranged from 2 to 24 mm, with the majority being less than 10 mm.

2. The results show that the forward lateral movement in front of the jet grout area may be influenced by the confining effects of the basement wall boundaries or the jet grout mass completed earlier. These effects explain why the lateral wall and soil movements behind the wall from the production grouting were significantly more than those observed during the jet grout trial.
3. The maximum free field lateral soil movements in the excavation side of the jet grouted mass were much larger than the corresponding lateral movements behind the wall. The maximum lateral free field soil movement at 5.25 m in front of the jet grouted mass was 35 mm, decreasing to 9.5 mm at 20.25 m from the jet grouted mass. The lateral movements behind the wall decreased from 10 mm at 0.5 m to 4 mm at 15.8 m. These results suggest that the diaphragm wall provided considerable restraint, thereby reducing the lateral movements of the soil behind the wall induced by jet grouting.
4. The jet grout trial caused a large increase (maximum of 73 kPa) in lateral earth pressure in the zone immediately next to the grouted area and some increase in lateral earth pressure farther away in the free field side. The production grouting caused some increase in the total lateral earth pressure in adjacent soils immediately behind the diaphragm walls, with a peak increase of 3.7–16.9%, and then started to decrease before the start of basement excavation. These results suggest that the increase in the total lateral earth pressure caused by the production grouting was temporary.
5. The monitoring results indicated some minor variations of the groundwater level in the excavation side, whereas the groundwater level behind the wall remained almost constant during the jet grouting. The maximum increase of piezometric level immediately in front of the jet grout area was 7 m. Piezometers P2 and P4 also experienced some increase in the piezometric level during the jet grout trial. However, the increase in the piezometric level farther away from the jet grout area was minimal.
6. The induced bending moments in diaphragm wall ranged from −195 to 92.5 kN m during the jet grout trial. During the production grouting, the induced wall-bending moments ranging from −787 to 1043 kN m/m. These results suggest that the process of the production grouting itself may cause significant bending moment in the diaphragm walls. If the reinforced diaphragm wall is not properly designed, these bending moments may exceed the yield moment of the wall.

7. For jet grout columns formed within the soft to medium stiff marine clay, core samples obtained indicated that a treated jet grout mass was properly formed with a total core recovery of 70–100%. However, within the medium stiff to very stiff silty clay layer, the core samples showed that the jet grout columns were not properly formed. These results indicate that the use of a single set of jet grout parameters was not efficient for forming a jet grout column in layered soils with large differences in soil properties between the soil layers.
8. The production grouting caused the MRT station and viaduct columns (6.5–15 m away from the grouted zone) to tilt 20–90 s, away from the jet grout area. Subsequent readings showed that basement excavation caused the columns to tilt back, toward the excavation side, thus offsetting the initial tilt. The maximum lateral movements of the MRT station and viaduct columns induced by the production grouting ranged from 3 to 8 mm, away from the grouted zone. The corresponding vertical upward movement ranged from 2 to 5 mm. Subsequent readings showed that the basement excavation caused the columns to settle.
9. Typical results of the net lateral wall deflections at the end of excavation suggested that the jet grout strut is effective in controlling movements caused by the subsequent basement excavation. For example, wall inclinometers WI3 and WI10 recorded relatively small deflections, with the maximum lateral deflections ranging from 11 to 20 mm. In Singapore, the deflections and resulting internal forces induced in the diaphragm walls by jet grouting are not specifically taken into account in the design. One of the reasons why is that during the ensuing excavation for the basement structure, the deflections and forces would be expected to be reversed. This benign attitude is changing toward requiring that the ground movements induced by jet grouting be kept to a minimum. However, in the event excessive movements are observed or anticipated from jet grouting works, it is recommended that this effect be checked or designed for. Recent projects involving jet grouting in marine clay use an external casing during the jet grouting operation to ensure a free return of sludge flow to the ground surface. This has resulted in substantially smaller lateral wall deflections. The grouting parameters adopted in these projects were similar to those reported herein. The use of such temporary external casing would help to prevent excessive movement likely caused by the jet grouting process; however, it slows the grouting process and thus increases costs.

NOTATION

C_c = compression index

C_r = recompression index

N = standard penetration test (SPT) blow count

S_u = undrained shear strength of clay

Λ = plastic volumetric strain ratio = $(C_c - C_r)/C_c$

σ_{v0}' = vertical effective overburden stress

REFERENCES

Bell, A.L., 1993. Jet grouting. In: Moseley, E.D. (Ed.), Ground Improvement. Chapman & Hall, Glasgow, pp. 149–174.

Covil, C.S., Skinner, A.E., 1994. Jet grouting—A review of some of the operating parameters that form the basis of the jet grouting process. In: Bell, A.L. (Ed.), Proc. Grouting in the Ground. Telford, London, pp. 605–629.

Gaba, A.R., 1990. Jet grouting at Newton Station, Singapore. In: Proc. 10th Southeast Asian Geotechnical Conference. Chinese Institute of Civil and Hydraulic Engineering, Taipei, pp. 77–79.

Geology of the Republic of Singapore. Public Works Dept, Singapore.

Khoo, K.S., Orihara, K., Egi, F., Arii, T., Yamamoto, K., 1997. Excavation with soil berm improved by jet grout piles. In: Tan, Y.C. et al., (Ed.), Proc. 3rd Asian Young Geotechnical Engineers Conference. Photoplates, Singapore, pp. 189–196.

Poh, T.Y., 2003. Movements associated with jet grouting and mitigation measures. In: 12th Asian Geotechnical Conference on Soil Mechanics and Geotechnical Engineering, Singapore.

Poh, T.Y., Wong, I.H., 1998. Effects of construction of diaphragm wall panels on adjacent ground—Field trial. J. Geotech. Geoenviron. Eng., ASCE 124 (8), 749–756.

Poh, T.Y., Wong, I.H., Chandrasekaran, B., 1997. Performance of two propped diaphragm walls in stiff residual soils. J. Perf. Constr., Facilities, ASCE 11 (4), 190–199.

Stroud, M.A., 1994. Report on Session 5 and 6: Jet grouting and soil mixing. In: Bell, A.L. (Ed.), Proc. Grouting in the Ground. Telford, London, pp. 539–560.

Sugawara, S., Shigenawa, S., Gotoh, H., Hosoi, T., 1996. Large-scale jet grouting for prestrutting in soft clay. In: Proc. 2nd Int. Conf. on Ground Improvement Geosystems: Grouting and Deep Mixing. Balkema, Rotterdam, The Netherlands, pp. 353–356.

Wong, I.H., Poh, T.Y., 1995. Design of diaphragm walls for excavations in stiff clay. In: Proc. Bengt B. Broms Symp. Geotech. Engrg. World Scientific, Singapore, pp. 429–446.

CHAPTER 6

Performance Evaluation of Road Pavements Stabilized In Situ

Jayantha Kodikara[1], Richard Yeo[2]
[1]Monash University, Victoria, Australia
[2]ARRB Group Ltd., Vermont South, Victoria, Australia

6.1 INTRODUCTION

The road network of a developed country is a valuable asset that underpins its economy, freight transport, and people's standard of living. Australia has a vast road network consisting of more than 810,000 km of roads, approximately 70% forming the rural network and approximately 30% the urban network (Land Transport, 2000). The relatively small population of Australia means that there is a relatively small funding base for road infrastructure compared to other developed nations. For example, the length of road per capita is approximately 280 m in the United States, 130 m in Great Britain, and 90 m in Japan in comparison to 450 m in Australia. Australia also has the largest road freight compared to the measured per capita or per unit gross national product of any OECD country. The growth of road freight in Australia has outpaced the growth of other road traffic and gross domestic product, and with this growth, more use of heavier vehicles (such as B-doubles), higher axle loads, and higher tire pressures is also taking place. This situation presents a considerable concern because the heavy vehicle traffic has substantially higher potential to cause damage to the road pavement infrastructure. On the other hand, increased wear and tear incurred to expensive freight trucks and associated road safety issues are also of concern. Under these circumstances, the cost-effective management of Australia's vital but aging road infrastructure needs the use of science and engineering at the current-state-of-knowledge, innovation, and highly skilled staff.

With the aging national road infrastructure and increasing environmental concerns, the rehabilitation of old degraded roads has become increasingly important. Rehabilitation may be achieved by reconstruction, overlays, or in situ stabilization. Reconstruction provides a brand new pavement but may be too costly and environmentally unfriendly due to the wastage of existing road materials and the use of new quarry and other manufactured

Chemical, Electrokinetic, Thermal, and Bioengineering Methods
http://dx.doi.org/10.1016/B978-0-08-100191-2.00006-X

products, which will exert pressure on new quarries and increase levels of greenhouse gas emissions. Overlays involve placement of a newly constructed layer on the old pavement. Although relatively less costly in the short term, this technique may not always extend the design life of the pavement as required. In situ stabilization involves the fragmentation of the old road and subsequent stabilization of the pavement with cementitious, bituminous, or other binders. This technique is an environmentally prudent approach because it reuses the existing road materials, and the stabilizing binder can be derived from waste by-products of the coal industry (e.g., fly ash) and the iron smeltering industry (e.g., slag).

Unbound road pavements with thin bituminous surfacings account for 95% of the sealed road network in Australia. This component of the road network is likely to be the most affected by the increasing traffic loads and aging. These pavements are particularly suited to in situ stabilization and are therefore increasingly rehabilitated in situ. Specialized equipment has been developed that can recycle pavements in deep lifts of up to approximately 400 mm. The percentage of stabilizing binder content used is much less than that for concrete and varies according to requirements determined through a mix design process and experience. On the basis of a survey of local municipalities, Chakrabarti et al. (2002) reported that GP (general Portland or general purpose) cement is the most commonly used binder, and it is used at an application rate of approximately 3% (by weight) in Victoria, 2.5–5% in Queensland, 4–5% in South Australia, and 3–6% in New South Wales. Other cementitious binders used include blends containing some mixture of two or more components of GP cement, slag, lime, bitumen, and fly ash. In addition, foamed bitumen is also used in in situ stabilization.

Crushed igneous and metamorphic rocks are commonly preferred as unbound pavement materials due to their high strength, and where economically available, they are preferred as *traditional* materials. Naturally occurring materials are also used in road construction, although they may have less strength than traditional materials. These materials would generally be considered as *marginal* materials (Yeo and Newman, 1998). Alluvial, colluvial, or chemically formed gravel, clayey gravel, sandstone, limestone, tuff, and scoria are examples of marginal materials. In most cases, they may have just satisfied the structural requirements of pavement material, but due to their proximity, they are cheaper than the cost of cartage of traditional materials from more distant sites. Therefore, the quality of unbound materials encountered in old degraded pavements can be wide ranging, and in situ stabilization needs to cater for these variations.

Considering that in situ stabilization is approximately 35–50% of the cost of reconstruction, it is not surprising that the use of this technology is on the rise primarily due to the potential for substantial cost savings. Environmental benefits provide an added bonus. Despite the assorted attractiveness of this method, there are a range of issues that require investigation in order to make this method viable in the long term. Otherwise, we may be creating another potential problem for the future, where rehabilitated pavements become substandard with potential to deteriorate prematurely. The main issue of concern is the long-term performance of these pavements. The long-term performance requirements must be catered for by effective material mix design and rational structural pavement design methods together with appropriate maintenance. In the past several decades, there has been a concerted effort to investigate these issues in the local scene. This chapter presents results of two case studies along with some advancements that followed in in situ stabilization technology. The focus of the chapter is limited to stabilization of unbound pavements using cementitious additives only.

This chapter places special focus on the use of activated slag as a potential binder in place of GP cement. It is well-established that if slag is activated by an alkali (e.g., lime, NaOH, or sodium silicate), then the resulting alkali activated slag (AAS) becomes a powerful stabilizer (Collins and Sanjayan, 2002). In addition, the use of slag can produce substantial environmental benefits. Approximately 3.1 million tons of slag is produced annually in Australia by the steel manufacturing industry as a waste product (in contrast to 6 million tons pa of Portland cement). The production of GP cement is very energy intensive, consuming 4–7 MJ of fossil fuel energy per kilogram, and it releases approximately 1 ton of carbon dioxide (CO_2) for manufacture of each ton of Portland cement, contributing to 5% of global greenhouse gas emissions. Although pavement rehabilitation uses only approximately 3–6% (by weight) of Portland cement compared to concrete's 10–16%, the amount of cement to be utilized in pavements would be substantial considering the large amounts of pavements to be rehabilitated in Australia. When in situ pavement rehabilitation is more widely used, as predicted by many, the binder use can become comparable to the usage in concrete (estimated to be approximately 300 tons/km). Therefore, the use of binders involving slag and fly ash, which are basically waste products of other industries, can potentially contribute significantly to reduce CO_2 emissions.

6.2 CASE STUDIES

Two case studies are discussed, namely the Cooma trial and the Dandenong trial. The Cooma trial was conducted from May to October 1994 at a site on the former Monaro Highway at Numeralla River, approximately 20 km north of Cooma. This study particularly examined the feasibility of deep-lift recycling of granular pavements to satisfy the structural design requirements of heavily trafficked rural pavements. Deep-lift recycling was considered because the structural pavement design required a thick depth of stabilized material to carry the anticipated traffic, and previous attempts to achieve thick layers using multiple thin lifts had resulted in subsequent delamination of layers, pumping of fines, and excessive cracking (Kadar, 1987). New construction techniques were introduced for deep-lift recycling from pilot stage to full-scale test stage in cooperation with road agencies and the industry. For the Cooma trial, the pavement base was constructed of imported weathered granite (Bunyan gravel) of subbase quality similar to that in flexible pavements in southeastern New South Wales (NSW). The study was undertaken by the Australian Road Research Board (ARRB) through funding provided by Road Traffic Authority (RTA) of NSW. A complete description of the trial was presented by Jameson et al. (1995).

The Dandenong trial was conducted by ARRB under the sponsorship of Austroads (the body representing Australian and New Zealand road authorities) between February 1996 and April 1997 at a special site at Dandenong, Victoria (approximately 30 km south of Melbourne). The main objective of this trial, within the context of the current chapter, was to compare the performance of high-quality crushed unbound rock (i.e., traditional material) with that of a marginal material stabilized using different techniques. The techniques included stabilization with a slow setting cementitious binder (a slag/lime blend) and a bitumen/cement binder. A control section of unstabilized marginal material was also included in the trial. The full details of the trial and associated references can be found in Moffatt et al. (1998).

6.3 COOMA TRIAL

6.3.1 Laboratory mix design

Laboratory mix design trials were undertaken to select the binder types and application rates to be used with Bunyan gravel. Bunyan gravel had 45% passing through 0.425-mm sieve and 26% passing through 0.075-mm sieve and a plasticity index (PI) of 10 for the fines content, signifying that fines comprised predominantly nonclay minerals. The standard maximum dry density (MDD) and optimum moisture content (OMC) were 2.08 t/m^3 and 8.6%, respectively. The material gave a soaked California bearing ratio (CBR) (%) of 38.

Because the relatively low working time available with GP cement was identified as a problem, emphasis was placed on the use of blends of lime, slag, and fly ash, which normally provide relatively high working time (i.e., slow setting blends). Commonly, unconfined compressive strength (UCS) is used as the key indicator for mix design. UCS can be obtained on specimens prepared in the laboratory using gyratory, dynamic, or static compaction. More common is the use of dynamic compaction of a mix in a standard Proctor mold to either 100% standard or 95% modified compaction at respective OMC. In this instance, 100% standard compaction at OMC was used. Another aspect that needs to be considered at the mix design stage is the use of an appropriate curing regime for the specimens. Ideally, specimens should be cured under the same conditions (i.e., mainly the temperature and the humidity) that would prevail in the stabilized pavement in the field. However, this is not practical because (1) we would not know exactly the conditions that would exist long term in the field pavement and (2) trial mix design results are needed in a short space of time (within several weeks and often earlier) in contrast to the curing time that can run into months or years under field conditions. Under these circumstances, it is common practice to carry out accelerated curing at an elevated temperature along with moist curing at room temperature (at approximately 23 °C). In this instance, therefore, specimens were subjected to 7 days of accelerated curing at 65 °C prior to testing. Table 6.1 shows the results obtained. It was considered that all binders provided adequate strength, and the binder of 85% slag/15% lime was chosen for the trial. This binder is a slow-setting binder and allows at least 8 hours of working time. This working time is required to ensure sufficient time for compaction of the deep-lift layer (single lifts up to 400 mm thick) and adequate time for final trimming of the pavement to ensure ride quality requirements are met.

Table 6.1 UCS results of the trial mix design for 28 days of curing (Cooma trial)

% Binder by weight	Lime/fly ash/ground slag (30/50/20) (MPa)		Ground slag/lime (50/50) (MPa)		Ground slag/lime (85/15) (MPa)	
	Specimen 1	*Specimen 2*	*Specimen 1*	*Specimen 2*	*Specimen 1*	*Specimen 2*
4	6.4	5.7	4.5	4.0	5.4	3.6
5	6.4	6.5	5.6	5.6	6.6	7.0
6	6.8	7.2	6.0	6.0	6.5	6.8

6.3.2 Trial pavement construction

Test description and layout

Figure 6.1(a) shows the layout of four test pads (Pads 2–5). In addition, another test pad (Pad 1) was constructed next to the main test bed. The construction was similar to that for Pads 2–4, and it was considered that this pad was affected by some deficiency in the construction process used. Therefore, it is not discussed further. Pads 2, 4, and 5 were stabilized in situ. Pad 3 was left in an unbound state to provide a control section for comparison with the stabilized pavements.

Figure 6.1(b) shows a cross section of Pads 2–5. The pavement structure was placed on the natural subgrade and, from bottom to top, included 200 mm of sand/rubble material, a minimum of 250 mm of imported clay subgrade, and the pavement base. The imported subgrade was a clayey silt from a cutting south of Bredo in southeastern NSW. As can be seen from Fig. 6.1(b), the thickness of the gravel placed over the subgrade varied depending on the pavement section.

Stabilization process

The pads were constructed on April 6 and 7, 1994. Each pad was constructed 4.6 m wide, done in two parts each 2.4 m in width (designated *a* and *b*) with an overlap of the paths of both binder spreader and the recycler. Table 6.2

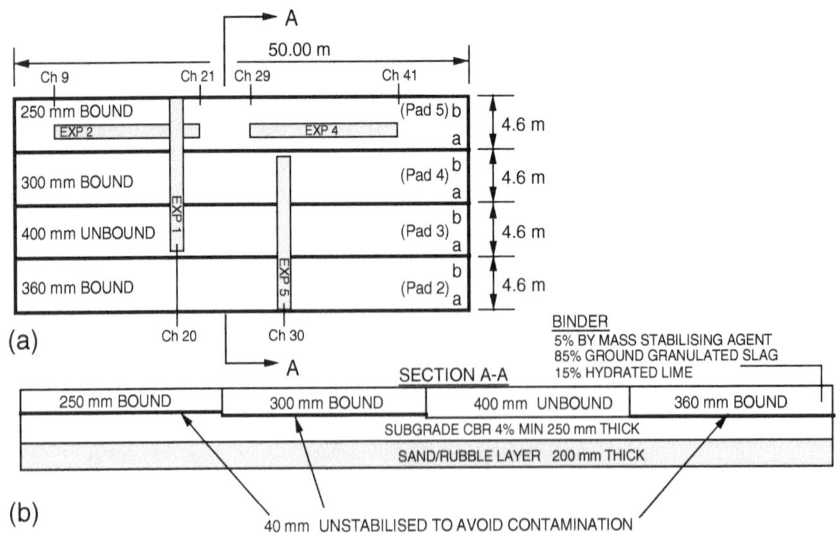

Figure 6.1 Test pad layout and cross sections in the Cooma trial. *(Source: Modified from Jameson et al. (1995)).*

Table 6.2 Depths of stabilization in the Cooma Trial

Pad	Nominal depth of stabilization (mm)	Actual depth of stabilization (mm)
2a, 2b	360	350–380
4a, 4b	300	260
5a, 5b	250	220–250

shows the nominal depths of stabilization along with the actual depths of stabilization achieved after construction, inferred from coring. A 40-mm additional depth of unbound (unstabilized) gravel was provided below the stabilized section to avoid mixing of subgrade clay with the gravel during stabilization.

The application of the binder (85% slag/15% lime) at a rate of 5% by dry mass of the host material was undertaken in two runs of 19 kg/m^2 each for the 360-mm beds. Mixing (using a CMI RS500 Recycler) and moisture application was also carried out in two runs. Subsequent compaction was undertaken with a 34-ton static pad foot roller, and the pavement was trimmed with a grader and then finished with a smooth drum and rubber-tire rollers. The pavement was moist-cured until a primerseal was applied within 48 h. A final 14/7-mm single/double seal was also applied before full-scale load performance testing with the accelerated loading facility (ALF).

6.3.3 Field testing

Test variables

The main test variable was the moisture state of the subgrade. As shown in Fig. 6.1, Experiments 1 and 2 were carried out on pads with relatively stiff subgrade (CBR = 15%), in which the water content was not altered by artificial wetting. For Experiments 4 and 5, however, the clay subgrade was soaked by filling trenches adjacent to the test section with water up to 100 mm above the bottom of the subgrade. The sand/rubble layer provided below the clay subgrade allowed the spread of water to other parts of subgrade. The water contents were measured by time domain reflectrometer gauges and direct grab samples underneath core holes drilled for dynamic cone penetrometer testing.

Laboratory testing of field samples

The overall subgrade comprised natural subgrade, the sand/rubble layer, and the imported subgrade layer. The clayey silt used for the imported subgrade had a liquid limit of 35% and a PI of 16. The soaked and unsoaked CBR of

Table 6.3 Soaked and unsoaked CBR of subgrade materials

Material	Unsoaked CBR		Soaked CBR	
	95% Standard compaction	*100% Standard compaction*	*95% Standard compaction*	*100% Standard compaction*
Imported clay	15	45	1.5	2
Sand/ rubble			13	
Natural subgrade	30	60	20	40

subgrade materials are given in Table 6.3. For the imported clay subgrade, which served as the primary subgrade, a dry density of 1.85 t/m^3 was obtained at OMC corresponding to standard MDD. These conditions were similar to those in Experiments 1 and 2 prior to the subgrade being wet up.

During the construction of the test pads, loose samples were taken from each stabilized bed and were compacted using 100% standard compactive effort. After curing the specimens in cylindrical molds at laboratory room temperature (15–25 °C) and subsequently soaking for 4 h, UCS tests were performed. The results are given in Table 6.4. Pad 4 was stabilized to a lower depth and hence contained a higher than expected binder percentage. This is reflected in the higher UCS values measured. The lower results obtained for Pad 5 were unexpected and were partly attributed to lower specimen densities.

Field cores were obtained from the stabilized test pavements at various times, and the specimens prepared from these cores were subjected to UCS testing. Figure 6.2 shows a typical result in which the UCS results showed a strong dependence on the dry density of the material. It was also apparent that the increase in UCS with time was dependent on the field

Table 6.4 UCS test results (in MPa) of laboratory compacted specimens prepared from stabilized gravel obtained during construction

Pad No.	Depth of stabilization (mm)	Accelerated curing (7 days)	Days of curing at ambient temperature			
			56 Days	*84 Days*	*127 Days*	*154 Days*
2	360	6.4	3.6	4.1	5.3	5.3
4	300	7.9	4.6	6.9	8.0	6.6
5	250	3.9	1.7	2.2	2.2	2.4

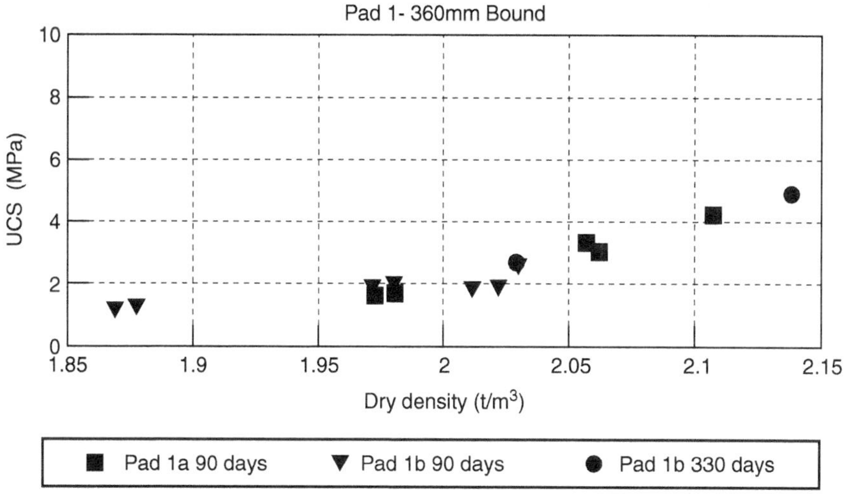

Figure 6.2 The relationship of UCS with dry density for field cores recovered from the Cooma trial. *(Source: Modified from Jameson et al. (1995)).*

curing environment. For instance, the specimens cured during cold winter months (at ages 60–120 days) did not show a substantial variation in strength. However, the cores tested after 330 days of curing (the period that spanned the hot summer months) generally showed an increase in strength. Furthermore, 6–7% difference in dry density was observed between top and bottom halves of the 360-mm-deep layer. This density difference typically gave rise to the UCS of the top half, approximately 1.6–1.8 times the UCS of the bottom half.

Resilient moduli were also determined from field cores. The cores obtained at an age of 60–65 days gave mean resilient modulus in the range 7.2–11.1 GPa, whereas the cores obtained at approximately 120 days gave moduli in the range 10.3–17 GPa. This indicated a possible increase in modulus along with the increase in strength. However, the resilient moduli obtained from laboratory-molded specimens were much lower than those obtained from field cores.

Full-scale performance testing using ALF

Full-scale performance testing was undertaken using a mobile ALF. The ALF can test road sections applying repeated cycles of full-scale wheel loading to a designated pavement structure. Figure 6.3 shows a picture of the machine, as located in the current indoor test site at Dandenong, Victoria. The total length of the machine is 29 m, and for a particular station of the

Figure 6.3 The accelerated loading facility (ALF) at the indoor test site, Dandenong, Victoria.

ALF, the length of test pavement where wheel loads are applied is 12 m. The rolling wheel load, which travels at a constant speed of 20 km/h, can be varied in 10-kN increments between 40 and 80 kN. Either single or dual wheels can be used, and the lateral distribution of the trafficking can be varied to replicate normal traffic wander or to a narrow distribution. A narrow distribution of loading can be used to concentrate the wheel loading and thereby to accelerate the pavement deterioration. Simulating normal highway applications, the wheel loads are applied in one direction only, with the wheels lifted off the pavement surface on the return cycle. Table 6.5 summarizes the experiments carried out using ALF.

Performance monitoring during the accelerated testing

A transverse permanent deformation of the test sections was determined using a profilometer after application of the load cycles each day. Surface cracking was also measured and was recorded as total crack length per square meter of pavement. The fatigue life was defined as the loading cycles giving 1.0 m/m^2 of surface cracking.

Typical permanent deformations measured for Experiment 1 (with unsoaked subgrade) are shown in Fig. 6.4. The permanent deformation was calculated from the change in profile from the start of the experiment. It is clear that stabilized pavements have performed in a superior manner compared to the unbound pavement. The rate of degradation (rate of increase of mean deformation with respect to loading cycles) is higher for the unbound pavement (Pad 3). In Experiment 5, the unbound material failed after only 180 cycles of 40-kN loading, whereas the stabilized pavements performed satisfactorily after 148 kcycles of loading. Figure 6.5 shows

Table 6.5 Loading cycles using the ALF for the Cooma trial

Experiment No.	Pad No.	Subgrade condition	Cycles of 40 kN ALF loading (kcycles)	Cycles of 50 kN ALF loading (kcycles)	Cycles of 60 kN ALF loading (kcycles)	Cycles of 80 kN ALF loading (kcycles)	Total cycles of ALF loading (kcycles)
1	3,4,5	Not soaked	46.7		56.7		
2	5	Not soaked		12.4		163.6	176.0
4	5	Soaked		11.5		195.9	207.4
5	2,3,4	Soaked		9.5		138.2	147.7

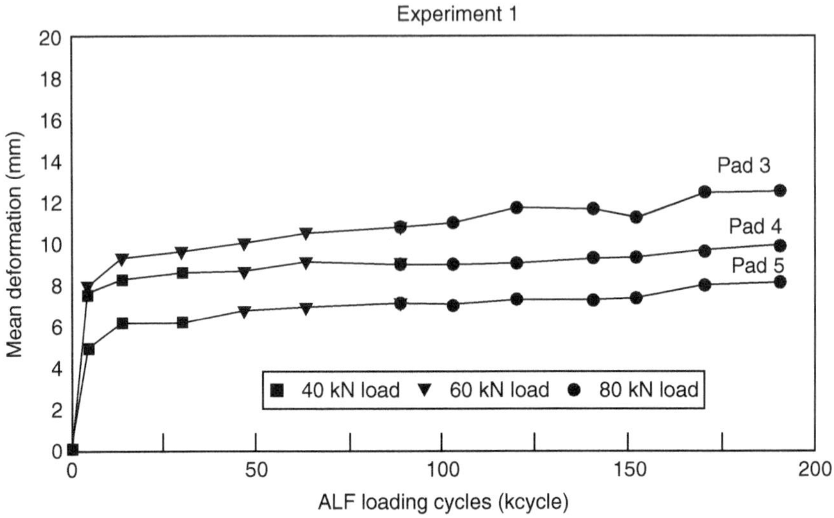

Figure 6.4 The mean permanent deformation measured (for unsoaked subgrade) at different ALF cycles during the Cooma trial. *(Source: Jameson et al. (1995)).*

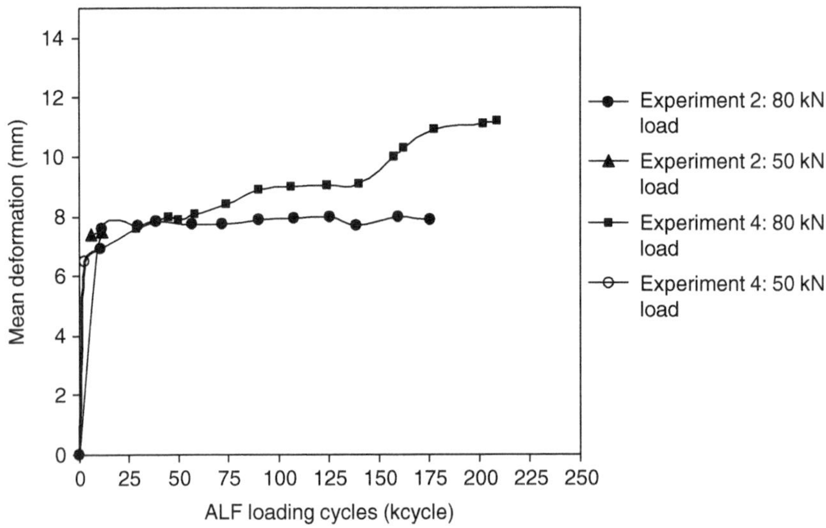

Figure 6.5 Variation of mean deformation with ALF for the unsoaked (Experiment 2) and soaked (Experiment 4) subgrade. *(Source: Jameson et al. (1995)).*

a comparison of the influence of stiffness of subgrade in Experiments 2 and 4 (with no-soaking and soaking, respectively) using Pad 5. No surface cracking was observed in Experiment 2 (and also in Experiment 5) after approximately 200 kcycles. In contrast, the pavement with soaked subgrade

(Experiment 4) failed after 169 kcycles, and the rate of pavement deformation was generally higher.

The main conclusion drawn from the accelerated loading testing was that all the stabilized pavements, including those with the low strength (soaked with CBR of 4%) subgrade, had fatigue lives at least twice the loading estimated for the Monaro Highway (i.e., 5.3×10^6 equivalent standard axle loadings) over a design period of 20 years. Consequently, it was suggested that pavement recycling by deep-lift stabilization could be used for pavements carrying moderate rural arterial traffic.

6.4 DANDENONG TRIAL

6.4.1 Laboratory mix design

The traditional unbound crushed rock was an angular, rough, hard, blue-gray rock from a rhyolite acid igneous source. It had a maximum particle size of 20 mm and PI of almost zero for the fines content. As noted previously, this material was used as a benchmark traditional material and was not cementitiously stabilized. The crushed rock complied with the VicRoads (Victorian Road Authority) classification for Class 1 base. The material gave a modified MDD of 2.2 t/m^3 at OMC of 6.8% and unsoaked and soaked CBR (%) values greater than 100.

The marginal unbound material used was a sandstone rubble from Wollen Rises Quarry, Victoria. It is often used as a subbase material by VicRoads Northern Region. The aggregate was considered as irregular, rough, and medium soft. The material gave a modified MDD of 1.82 t/m^3 and OMC of 12%, and unsoaked and soaked CBR (%) of 80 and 56, respectively. The PI of the fines was typically 6–8.

Only the sandstone was to be tested and stabilized, and prominence was given to a slow-setting binder. In line with the Cooma trial, a slow-setting binder comprising 85% slag/15% lime was chosen, and a target UCS of 5 MPa was specified. Laboratory specimens were prepared with 5% and 6% binder contents using static compaction, and they were cured using two different techniques: moist curing in a 100% humidity environment for 28 days or accelerated curing in a sealed plastic bag at a temperature of 65 °C for 7 days. Unfortunately, none of the specimens achieved UCS greater than 2 MPa, and many were less than 1.5 MPa. This was despite the accelerated curing giving higher UCS values than the moist curing. It was concluded that more time was necessary to develop strength with the slow-setting slag/lime binder in the stabilized sandstone, but a binder content of 5% was selected in line with the Cooma trial. In addition to UCS testing, resilient modulus was also measured using repeated load triaxial (RLT) tests. They were performed on 96% modified compaction at OMC using dynamic compaction for specimens prepared with 5% binder content. The tests were conducted at a mean normal stress of 400 kPa and an inverse stress ratio of 0.4 (i.e., the ratio of static confining stress to the maximum vertical stress applied to the specimen), which simulated the condition close to the surface of the stabilized pavement. For specimens cured under normal curing, there was a tendency for the modulus to increase with time, giving a value of 1540 and 3920 MPa at 28 and 113 days, respectively. For specimens for which curing was accelerated, a modulus of 2620 MPa was obtained at 7 days.

6.4.2 Trial pavement construction

Test description

This chapter describes only three pavements tested in the overall trial: the unbound crushed rock and sandstone pavements and the pavement constructed of stabilized sandstone with 85% slag/15% lime. All the pavements were constructed at the ALF site in Dandenong, Victoria (Fig. 6.3), in a manner similar to that described for the Cooma trial. The cross sections of the test pavements are illustrated in Fig. 6.6.

The imported subgrade used in the pavement construction was a residual clay of Silurian siltstone (sometimes interbedded with sandstone), which forms the bedrock for Melbourne and surrounding region. This rock outcrops to the surface in the east and dips below approximately 30–40 m at the city center. The soil featured a liquid limit of 16 and a PI of 4–5. The lower than expected typical plasticity for this soil may indicate the presence of quartz sand particles in the fines content. The standard MDD was 2.08 t/m^3 at a corresponding OMC of 8.6%. The unsoaked and soaked CBR of the clay were measured to be 10% and 3%, respectively.

Stabilization process

The test pavements were constructed on December 6, 1995. During stabilization, the binder was spread at an application rate of 18 kg/m^2 (5% by dry mass of the host material) using a special-purpose spreader. The stabilization was carried out by an RS500 stabilizer linked to a water truck, which measured the flow of water into the mixing chamber. The mixing was carried

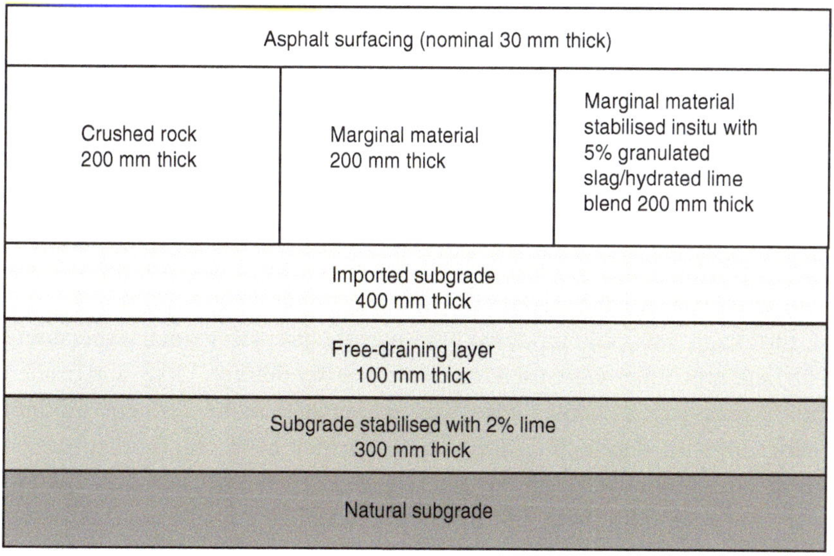

Figure 6.6 Cross sections of test pavements. *(Source: Modified from Moffatt et al. (1998)).*

out down to 200-mm depth in a single pass, and as in the Cooma trial, two runs were used to generate 4-m-wide test pavement strips. The mixed material was compacted using an 11-ton vibrating padfoot roller and a 10-ton vibrating smooth drum roller. The stabilized material was compacted to 98% modified MDD. After compaction, the surface was finished off with a multitire roller. It was intended to apply primerseal after a curing period of 48 h. However, the priming was not applied until approximately 37 days, by which time parts of the pavement had dried out and experienced significant shrinkage cracking.

6.4.3 Field testing

Test variables

The main test objective was to evaluate the performance of unbound sandstone and stabilized sandstone with respect to the performance of traditional unbound crushed rock material.

Full-scale performance testing and laboratory testing of field specimens

Full-scale accelerated performance testing was carried out between May 1996 and March 1997 using the ALF noted previously. Although many tests including some industry-funded tests (on contract) were performed, this chapter describes information only relevant to unbound crushed rock and unbound and slag/lime stabilized sandstone pavements. The tests were performed using 50- and 80-kN dual-wheel loading and 50-kN single-wheel loading. All experiments used a single tire pressure of 690 kPa. Unlike in the Cooma trial, no measure was taken to change the subgrade conditions; therefore, measures close to the as-constructed state could be considered to be prevalent during testing.

Cores were retrieved from the test pavements after the construction and during ALF loading. Specimens were prepared from the top, middle, and bottom sections of the pavement, where possible. However, the presence of shrinkage cracking hindered consistently preparing intact specimens. The field specimens were used for measuring dry density, UCS, and resilient modulus testing. Clearly, there was evidence that the resilient modulus decreased with depth, indicating that this may be a common feature of the in situ stabilization process. There was no clear evidence that modulus decreased with trafficking, but this might have been due to the lack of data collected on trafficked surfaces. As for the Cooma trial, there was also evidence that the dry density, UCS, and resilient modulus increased with age when not subjected to traffic loading.

Figure 6.7 shows the influence of age and the depth on resilient modulus. This graph also shows the laboratory-measured values (at 7, 28, and 113 days of curing) during mix design. It is evident that field values are significantly higher than those for the laboratory mixed specimens. However, the resilient modulus of the specimen cured at 113 days measured in the laboratory appears to be similar to those measured in cored specimens at a similar age. The UCS values obtained from laboratory molded specimens at 7 and 113 days of moist curing were 1.1 and 2.5 MPa, in comparison with the values of 2.5 and 4.0 MPa obtained from the field cores after 69 and 489 days, respectively. Therefore, the resilient modulus and UCS values from laboratory-prepared specimens can give indications of the field UCS values, provided the curing conditions and times are representative. The figure also highlights that a reduction in stiffness occurred with depth primarily due to a reduction in mixing efficiency during the stabilization process. This may also happen from mixing of subgrade material with similar adverse effects, as discussed further in Section 6.5.2.

Performance monitoring during the accelerated testing

As for the Cooma trial, transverse permanent deformation of the test sections was determined using a profilometer at various stages during the ALF trafficking. In addition, the measurement of surface deflection bowls using falling

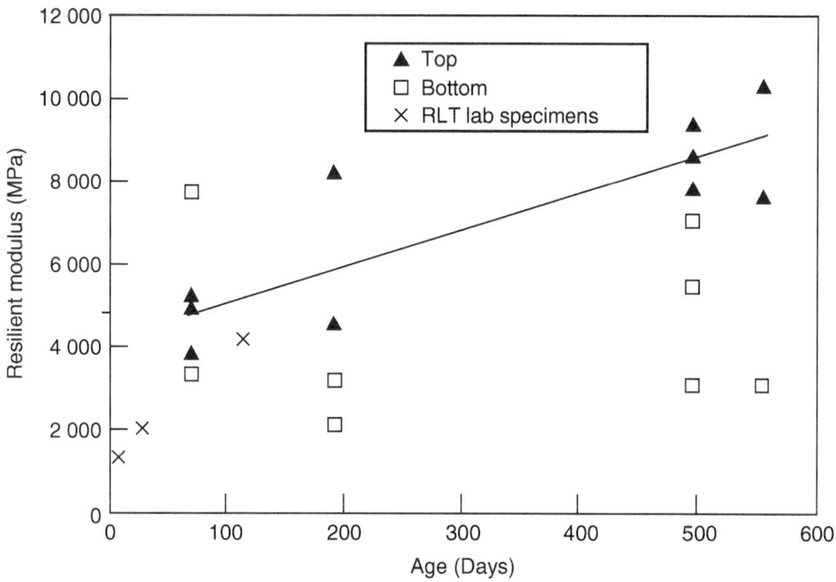

Figure 6.7 Comparison of resilient modulus of slag/lime cores with sample age for laboratory-prepared and field-retrieved specimens (the linear regression line shown is for field specimens obtained from the top half of the pavement). *(Source: Moffatt et al. (1998)).*

weight deflectometer (FWD) and surface cracking was also carried out at various loading cycles. Figure 6.8 shows mean deformation obtained for the stabilized sandstone pavement at various load cycles. It is clear that part of the test section experienced excessive deformation. This section was affected by shallow shrinkage cracking that occurred due to the delay in the application of primerseal. It was inferred that during the application of traffic loading, the surficial shrinkage cracking coalesced to form almost an unbound layer below the asphalt surface, which subsequently underwent premature cracking and excessive deformation. This highlighted the importance of applying a primerseal immediately after construction to minimize the possibility of shrinkage cracking due to rapid drying of the stabilized material. As can be seen from Fig. 6.8, the section that appeared to have not been affected by this problem performed satisfactorily even up to 55 kcycles of 80 kN of dual-wheel loading. In contrast, the unbound crushed rock and sandstone pavements underwent severe cracking and deformation even well below this level of loading cycles, highlighting the benefits of in situ stabilization.

Observation trenches were excavated post ALF trafficking to examine the layers and measure the deformation of each layer due to traffic loading.

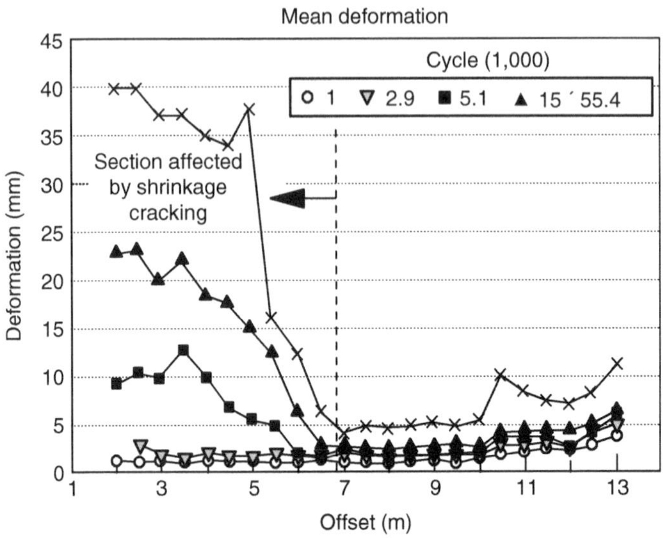

Figure 6.8 Variation of mean deformation of stabilized sandstone pavements at various ALF cycles of 80 kN dual-wheel loading. *(Source: Modified from Moffatt et al. (1998)).*

It was inferred that much of the deformation in the unbound crushed rock occurred within the crushed rock layer. Minimal permanent deformation was observed in the subgrade. In the unbound sandstone pavement, some permanent subgrade deformation had occurred, but the major contribution to the surface deformation had come from the deformation of the unbound sandstone. As noted previously, the primary deformation in the stabilized sandstone had resulted from the movement of asphalt due to the breaking down of the surficial layer of stabilized sandstone, influenced by shrinkage cracking. Figure 6.9 shows a comparison of average deformation of unbound and stabilized sandstone pavements under the 80-kN dual-wheel loading. Linear regression lines for each pavement type are also shown in this figure. It is clear that a substantially lower deterioration rate can be achieved using in situ stabilization if properly administered. Moffatt et al. (1998) summarized this result using two indicators: relative cost—the cost of stabilized pavement is two times the cost of unbound pavement; and relative performance—the performance (the rate of average increase in deformation per 80-kN load cycle) of stabilized pavement is 9.5 times better than that of a unbound sandstone pavement. This statement, however, was developed on the basis of a number of assumptions involving material cost and the pavement test configuration.

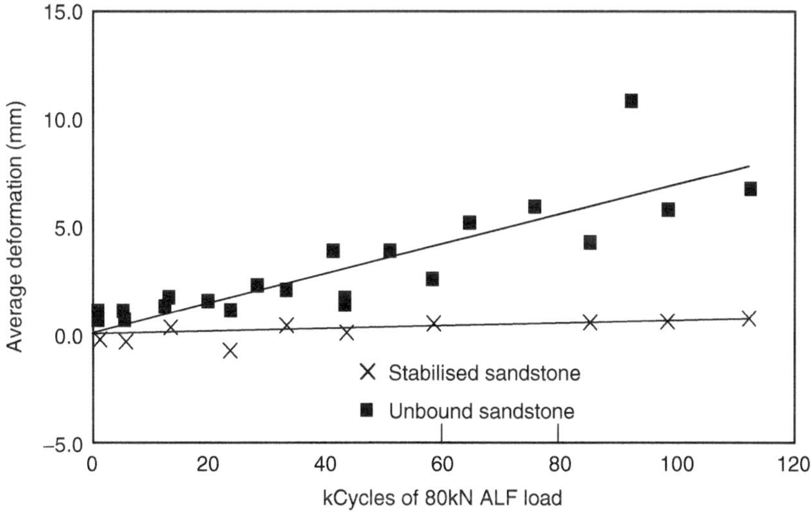

Figure 6.9 Comparison of average deformation of unbound and stabilized sandstone pavements under 80 kN dual-wheel loading. *(Source: Modified from Moffatt et al. (1998)).*

6.5 RECENT RESEARCH ON STABILIZED PAVEMENT MATERIALS

6.5.1 General

Previous sections presented case studies of in situ stabilization of unbound pavements using the cementitious binder comprising 85% slag/15% lime blend. As noted previously, this can be considered as a slow-setting binder in comparison to traditional GP cement. The main practical advantage of such a binder is that it can provide more working time during construction. The case studies highlighted the effectiveness of this binder for road stabilization, and the knowledge gained should be considered in pavement construction and stabilization practice. However, a number of issues need ongoing investigation so that this technology can be advanced further, including the development of reliable mix design methodologies; rational design of stabilized pavement bases taking into account traffic and environmental loading; and the issues of construction practices, such as material variability, mixing efficiency, and weather influences. Development of a thorough scientific understanding of the stabilized material behavior utilizing various binders including GP cement is essential to address these issues. The following sections provide brief accounts of relatively recent knowledge gathered on related topics.

6.5.2 Mix design

The decision to stabilize an old unbound pavement in situ will depend on a number of considerations: whether there is sufficient material available in the pavement; if not, whether more suitable unbound material can be added to supplement the existing material; how nonuniform the existing material is along the pavement; and the state of the subgrade and the drainage conditions and current elevation of the existing pavement. Once the decision is made, undertaking of mix design follows. The objectives of mix design are to select a suitable binder and an application rate and provide input parameters to the structural design of the stabilized pavement.

In some cases, the selected binder may not meet the capillary rise and the other optional tests but meet the UCS requirements; therefore, engineering judgment may override the results from these tests to use the specific binder and content.

The selection of a suitable binder and an application rate must consider the potential problems that stabilized pavement may experience during its construction and operation. These include (1) lack of durability, (2) lack of sufficient working time, (3) potential for premature fatigue cracking, (4) potential for excessive shrinkage cracking, (5) potential for thermal cracking, (6) erodibility, and (7) moisture susceptibility (Foley and Austroads Stabilisation Expert Group, 2001a; the authors also belong to this group). In addition, other environmental problems, such as acid generation, chemical leaching, and salt crystallization, may lead to premature deterioration (Kodikara et al., 2004). Generally, these problems are created by traffic loading, environmental loading, or a combination of both loading types. It may be cost-prohibitive or impractical to undertake detailed investigation of all these factors in a project. However, past experience and a general understanding of the behavior of stabilized materials can be applied to detect areas that require further scrutiny. A number of reports have been developed by Austroads to guide the mix design process (Foley and Austroads Stabilisation Expert Group, 2001b). Figure 6.10 shows a flowchart produced to assist the mix design process.

6.5.3 Strength and stiffness of stabilized materials

Unconfined compressive strength

As can be seen from Fig. 6.10, UCS is usually used as a key parameter in mix design and for indirect assessment of stiffness and some of the other properties, including durability. In situ stabilized pavements may be considered under two categories: lightly bound (UCS between 1.5 and 3 MPa), generally used

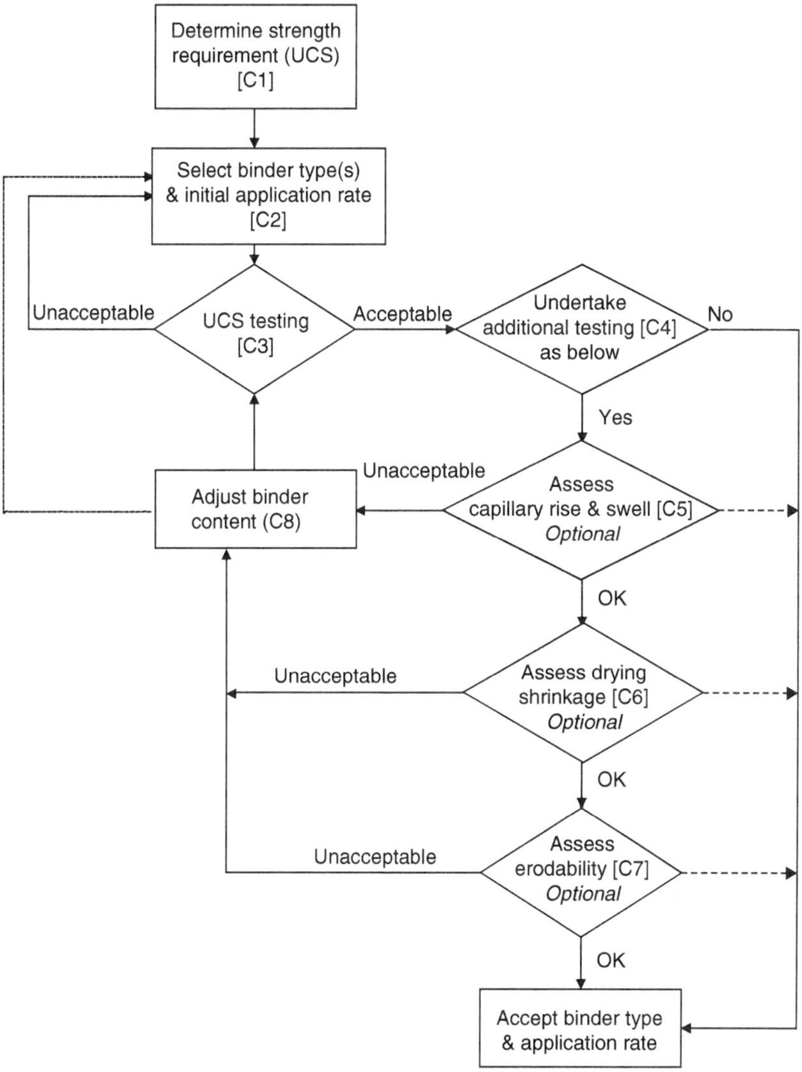

C1 Appropriate UCS design values for lightly bound pavement s are 2.0 MPa < UCS < 5.0 MPa

C2 In some cases the parent insitu pavement material may need an improvement in grading, and any additional material needs to be included in the laboratory samples and test program.

C3 Use AS 1141.51 for the determination of the UCS using standard compaction and either standard or accelerated curing. May consider and compare rate of strength gain between 7 and 28 days under standard curing.

C4 Where a new material is proposed for use without any previous knowledge of its performance in the proposed environment, then additional testing (i.e. capillary rise, shrinkage, erodibility) may be required to reduce the likelihood of moisture sensitivity, unacceptable cracking and erodibility.

C5 Capillary rise and swell assessment to be carried out to AS1141.

C6 As a rough indication of drying shrinkage use the linear shrinkage test (Standards Australia 1995) to discriminate between binder(s)/content(s). Some road authorities will use an equivalent density obtained using Modified compactive effort.

C7 Consider erodibility testing performed on a sample at 3% lower density as erodibility is sensitive to the degree of compaction.

C8 If there is no improvement in the stabilised material after increasing the binder content, it is suggested that another binder type is selected.

Figure 6.10 Flowchart for lightly bound material design. *(Source: After Foley and Austroads Stabilisation Expert Group (2001a)).*

for light to medium traffic, and heavily bound (UCS > 3 MPa), generally used for heavy traffic. This UCS is normally obtained by testing specimens with a diameter of 105 mm and height of 115.5 mm prepared with 100% standard compactive energy or 95% modified compactive energy at respective OMCs (Standards Australia, AS 1141.51, 1996). Curing for 28 days under a controlled environment with relative humidity (RH) more than 90% (generally close to 100%) and temperature between 21 °C and 24 °C is commonly used. Note that the length to diameter ratio used in this testing is close to 1. This appears to have been adopted mainly to take advantage of the preparation of specimens in the standard compaction mold. As noted by Sherwood (1993), the length to diameter ratio can influence UCS results, and specimens with length to diameter ratio of 1 can produce UCS values approximately 1.25 times that of specimens with a length to diameter ratio of 2 (as typically used in soil and rock testing). In addition, consideration must also be given to rate of loading because higher loading rates can increase the measured UCS (Jalinous, 1986; Chakrabarti and Kodikara, 2005).

The UCS of stabilized materials increase at a diminishing rate as the binder content is increased (Sherwood, 1993; Chakrabarti and Kodikara, 2003a, 2005). Figure 6.11 shows the development of UCS with the binder content for crushed basaltic rock stabilized with three cementitious binders, namely GP cement, GB cement, and AAS. GB cement and AAS are proprietary products manufactured locally in Melbourne. GB cement is a blend of blast furnace slag (approximately 50%) and Portland cement (approximately 50%). AAS is blast furnace slag activated by alkaline clinker from the cement-forming process. Table 6.6 shows the chemical compositions of the binders. Figures 6.11(a) and 6.11(b) show that in comparison to GP cement, both GB cement and AAS can be considered as slow-setting binders. It is also clear that after approximately 28 days, all three binders have given rise to similar strengths. As noted previously, an important benefit of slow-setting binders is the extra working time available during construction. For example, internal research undertaken by VicRoads indicated that materials stabilized with 3% GP cement produced a 30% reduction in UCS due to compaction with a 12-h delay after mixing, whereas similar specimens compacted with AAS did not show any appreciable decrease after a 12-h delay. The increase in UCS with increasing density was apparent from testing in the field trials. This has been further confirmed by laboratory testing involving a wide range of materials (Symons, 1999).

The results at 28 days of curing indicate that crushed basaltic rock can be stabilized quite readily to achieve sufficiently high UCS, potentially to carry heavy traffic. Note that this material is a traditional material and not a

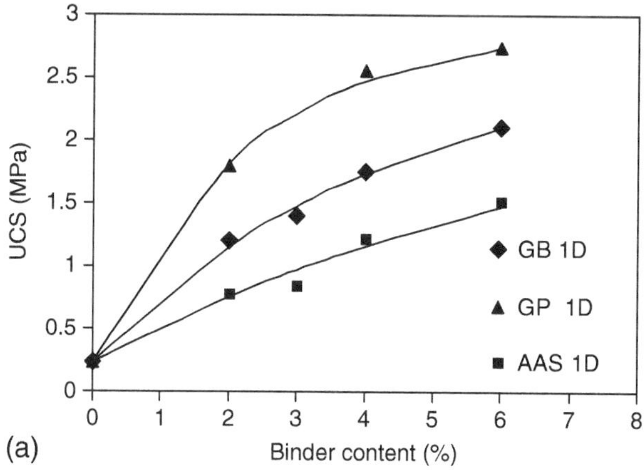

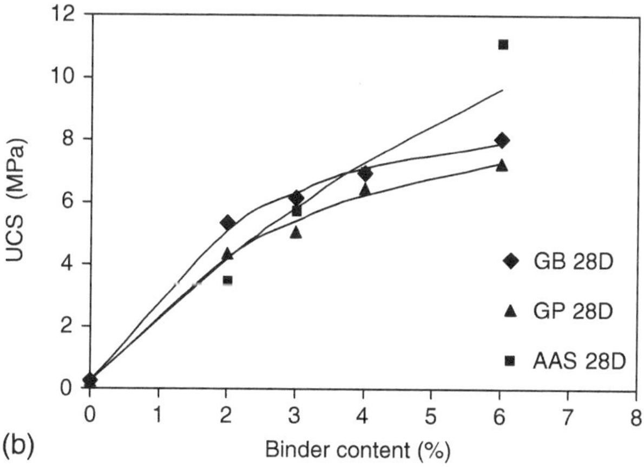

Figure 6.11 Development of UCS of crushed basaltic rock stabilized with GP cement, GB cement and AAS: (a) 1 day curing; (b) 28 day curing. *(Source: Chakrabarti, 2004).*

Table 6.6 Chemical compositions and loss of ignition (LOI) of binders

Material	**X-ray fluorescence spectroscopy (%)**								
	SiO_2	Al_2O_3	Fe_2O_3	*CaO*	*MgO*	SO_3	Na_2O	K_2O	*LOI*
GP cement	21.31	4.72	3.30	62.81	1.30	2.85	0.19	0.49	3.3
GB cement	23.88	7.27	2.51	57.36	2.76	3.01	0.14	0.42	2.7
AAS	22.49	7.97	1.63	44.15	3.56	4.04	0.71	4.53	10.9

marginal material. The increase in UCS with curing time was also evident from the field results of Cooma and Dandenong ALF trials.

Although not desirable, it is sometimes possible to have substantial quantities of clayey soil in the pavement material. It may either be present in the old pavement material or happen during the field mixing process either intentionally or unintentionally. Although the stabilization of clay soils with cementitious binders can be performed, the percentage of binder required is substantially higher than that required for rock aggregates. Figures 6.12 and 6.13 show the influence of artificially mixing high-plastic clay to crushed rock during stabilization. The high-plastic clay used in these experiments is the residual basaltic soil available in western regions of Melbourne, and it is known to display high shrink/swell properties when subjected to moisture fluctuations. It is clear that adding 6% and 15% clay substantially reduces the strength development. However, it may be possible to develop the required UCS strength by adding substantially higher percentages of binder content than that required for rock materials alone. Therefore, this issue needs to be considered carefully during mix design and even during budgeting of the stabilization projects.

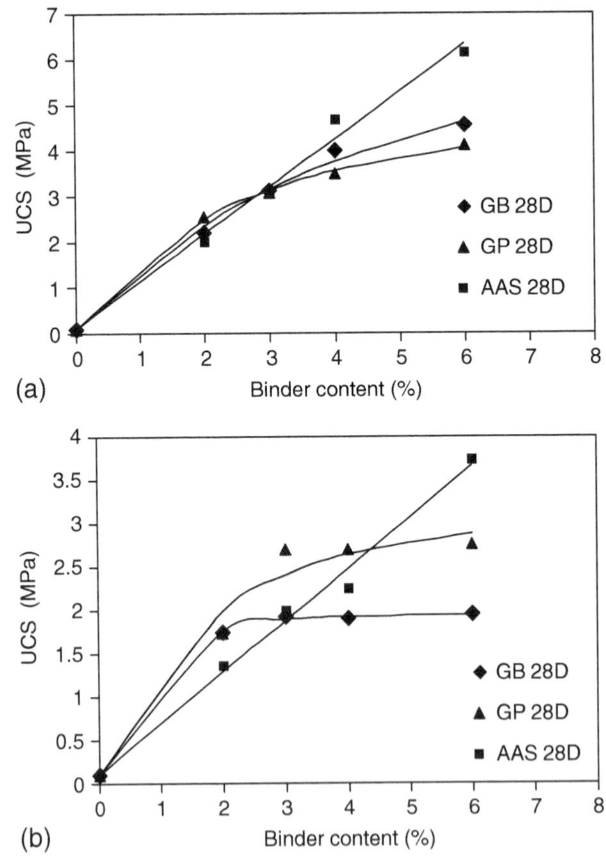

Figure 6.12 Influence of plastic clay soil on UCS development: (a) with 6% clay; (b) with 15% clay. *(Source: Chakrabarti (2004)).*

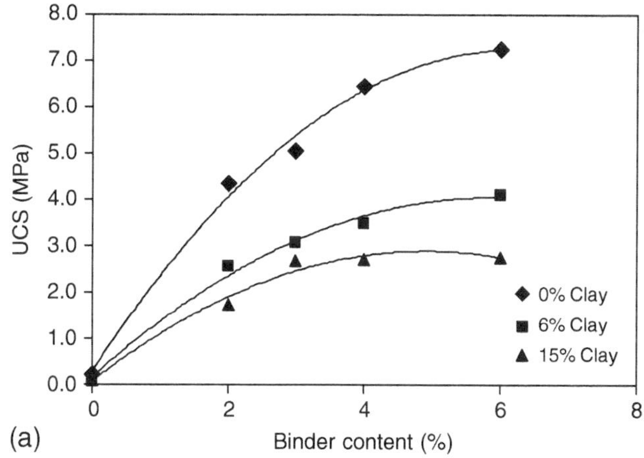

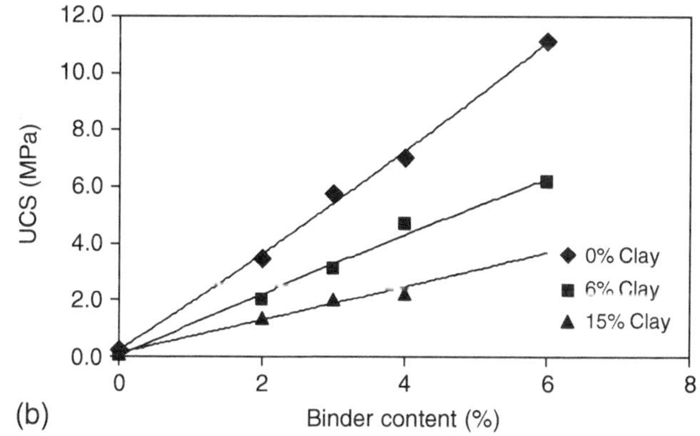

Figure 6.13 Influence of clay on UCS at 28 days curing: (a) GP cement; (b) AAS. *(Source: Chakrabarti (2004)).*

Tensile strength of stabilized materials

The common methods used for measuring tensile strength of materials can be categorized as *direct* (Wallace, 1998; Hannant et al., 1999; Tang and Graham, 2000) and *indirect* (Brazilian) tensile test (IDT) (Frydman, 1964), flexure beam test (Bofinger, 1970), and hollow cylinder test (Al-Hussaini, 1981). For stabilized pavement materials, the IDT and beam flexure tests are commonly used. Usually, the indirect methods assume that the material behaves in linear elastic manner (up to failure) to compute the tensile strength. The IDT test typically uses a cylindrical specimen 100 or 150 mm in diameter and 60 or 85 mm in height. The specimen is loaded diametrically until failure is reached. The failure load is used to compute the tensile strength of the material. In the flexure beam test, a compacted beam specimen is simply supported at the edges and is loaded to failure either centrally or at two locations at one-third of the beam length from each end. The tensile strength can be estimated assuming linear elastic behavior and pure bending of beams.

Direct methods, in contrast, measure the tensile strength without the assumption of linear elasticity and can give the tensile stress–strain behavior of the material in a more realistic sense. Nonetheless, the direct methods face challenges of effectively holding the specimen during testing and generating consistent test results. In both direct and indirect testing, it is important that the material is tested in the direction where tensile stresses and failure is expected. For instance, the tensile strength in the direction of compaction may be seriously affected by layering during compaction, particularly when dynamic compaction is used to prepare specimens.

The first author and his students developed a test method that is specifically suited for stabilized materials (Chakrabarti, 2004; Nahlawi et al., 2004). Figure 6.14 shows a picture of a specimen during testing. A special procedure is followed for preparing the specimens, but it involves compaction of stabilized material into the mold using a standard compaction hammer. After curing the specimen for the required period, it is tested under direct tension in a test rig. Therefore, the tensile strength is measured perpendicular to the direction of compaction. Internal keys are provided on the walls of the split molds to provide restraint to the specimen during tension. These keys are in addition to the provision of a neck, where the failure is forced to take place, as shown in Fig. 6.14. The tensile strain is estimated by a high-precision gauge placed across the neck section. A major advantage of the method is that it can be used to measure the development of tensile

Figure 6.14 Arrangement of stabilized material during direct tensile testing.

strength of stabilized materials with curing time. Nahlawi et al. (2004) presented results where the same equipment was used for measuring tensile strength of plastic clay at the liquid limit. Figure 6.15 shows typical results obtained for crushed basaltic rock stabilized with GP and GB cements at various curing times. Also shown is the repeatability of the test method specimens with 3% GB cement cured for 7 days. It can be seen that the stabilized materials develop substantial tensile strength and stiffness during curing, but the failure tensile strain is reduced as a result. Usually, experimental results have indicated that the tensile failure strain is approximately 10–50 μs for stabilized materials. The tensile strength at 28 days curing was measured typically highest for AAS and then GB cement and lowest for GP cement for 3% and 4% binder contents, but this relativity was not apparent for binder content of 2%. Typical values at 28 days curing ranged between 0.55 and 0.65 MPa for binder content of 3%.

Stiffness of stabilized materials

The stiffness of stabilized materials is characterized by Young's modulus or resilient modulus measured on recoverable (i.e., elastic) component of the strain. It can be measured by various means, including compressive or tensile loading in unconfined state, compressive loading using RLT apparatus,

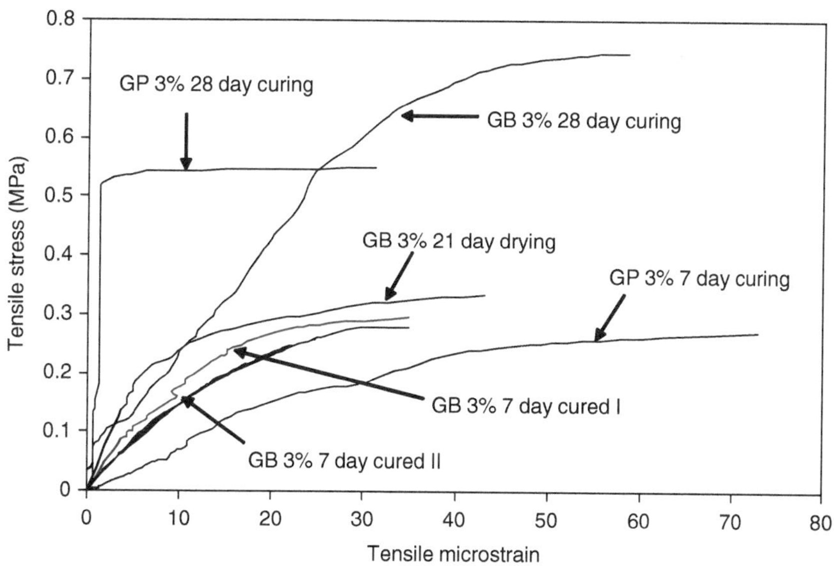

Figure 6.15 Typical tensile stress–strain results for crushed basaltic rock stabilized with GP and GB cements. *(Source: Chakrabarti (2004)).*

IDT, flexural testing, and dynamic testing (e.g., FWD). In mechanistic pavement design (see Section 6.5.4), material modulus is the key input parameter, not the strength. Therefore, it is important to understand its relationship to stress system, stress level, stress repetition, strength, and environmental factors such as moisture level and temperature.

The stress and strain system generated within the specimen or pavement can vary depending on the type of the test used. Therefore, significantly different modulus values may result from different tests. The RLT test generally operates in compressive stresses and strains due the lateral confinement provided. UCS and IDT tests apply compressive stresses, but failure may take place due to lateral tensile strains, and usually no confining stress is applied during these tests. In the IDT test, the lateral strain measured diametrically perpendicular to the direction of loading is used to compute the modulus, assuming linear elastic material behavior and the Poisson's ratio for the material. Flexural modulus is obtained by beam flexure test and, therefore, both tensile and compressive stresses and strains are involved. A measure of tensile modulus can also be obtained from the direct tensile test, as noted previously. Generally, dynamic tests give (approximately 20%) higher modulus values than continuous loading (sometimes referred to as static) or repeat loading tests (Williams, 1986; Alderson, 1999). Therefore, a difference in modulus values back-calculated from FWD testing and laboratory resilient modulus testing can be expected. However, the inaccuracies in interpretation of field results can also contribute to this difference.

Figure 6.16 shows results of RLT testing of a typical cemented material plotted as the inverse stress ratio (i.e., the ratio of static confining stress to the maximum vertical stress applied to the specimen) against the mean stress (average of all three principal normal stresses). The mean stress can be viewed as the overall confining stress, and the stress ratio signifies the deviator stress inducing shear strains. It can be seen that at a given mean stress, the modulus decreases as the stress ratio decreases (i.e., when deviator stress increases). Similar effect is seen if the confining stress is decreased. The stabilized material needs to be characterized for repeated loading to determine the fatigue life, which is normally characterized by flexural beam testing (Austroads, 2004). As shown in Fig. 6.17, the IDT test can also be used to measure the variation of resilient modulus with load repetition. However, it is apparent that depending on the method employed, modulus values ranging from 1 or 2 GPa to tens of GPa can be measured. Figure 6.17 highlights that damage in stabilized materials occurs early and increases steadily with the load cycles. This damage evolution has given rise to significant

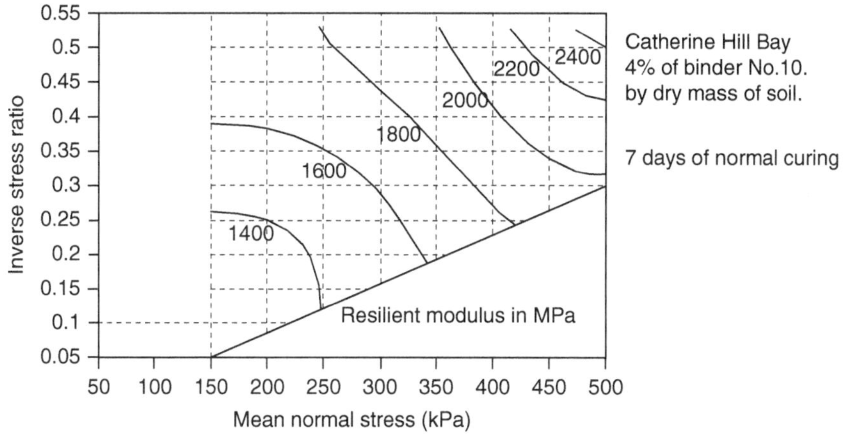

Figure 6.16 Typical results of resilient modulus of a cemented material from RLT testing. *(Source: Symons and Poli (1996)).*

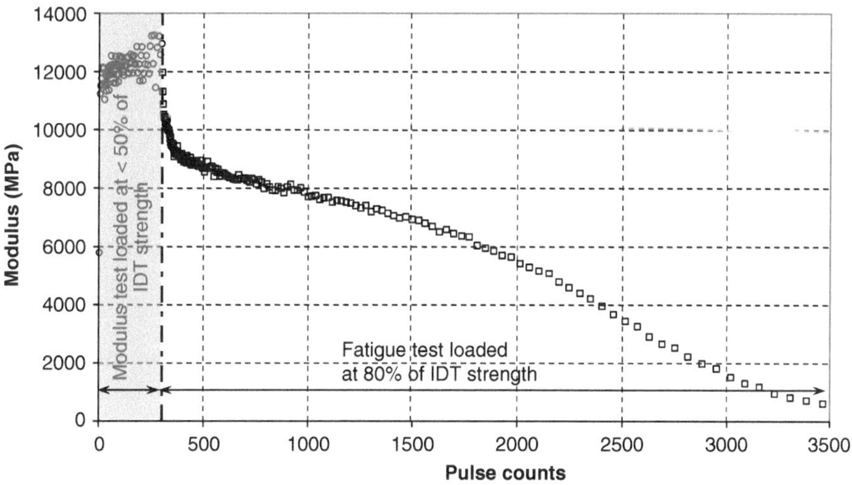

Figure 6.17 Typical resilient modulus variation under repeated indirect tensile loading for cemented materials.

reduction in modulus with load cycles. Therefore, the applicability of linear elastic fracture mechanics (LEFM) to characterize fatigue behavior of these materials is questionable.

A correlation between UCS (at 28 days moist curing) and flexural modulus is used in the *Austroads Pavement Design Guide* (Austroads, 2004). In the absence of measured data, a ratio of flexural modulus to UCS between 1000 and 1500 is suggested. However, experimental evidence appears to indicate that a reduction of UCS may result in a smaller proportion of reduction in modulus (Otte, 1978). Assuming linear elastic behavior, this may mean that when the UCS is reduced, the failure strain may not increase by the same proportion or vice versa. This behavior has important implications for not only the structural fatigue failure behavior but also the shrinkage crack development, as discussed in the next section. The influence of moisture on the resilient modulus has also been observed. For instance, Jameson et al. (1992) reported the moduli measurement of 3% cement treated crushed rock from a trial after 27 months from an area that was not subjected to traffic loading. Cores were tested in (field) dry state or wet state (after soaking for 1 week in water). The results gave average values in the ranges 7–14.3 GPa for dry cores and 5.2–11 GPa for wet cores.

6.5.4 Drying shrinkage of stabilized materials

Drying free shrinkage can be considered as a material property for a given environmental condition (Kodikara and Chakrabarti, 2001). The measurement of true shrinkage, however, should be such that no restraints, either internal or external, are placed on the specimen during shrinkage. External restraints can come from friction at the specimen boundaries, whereas internal shrinkage can arise from nonuniform shrinkage within the specimen. Therefore, practically, it is only possible to approximate the free shrinkage condition in the laboratory. Following the approach used in concrete technology, Chakrabarti and Kodikara (2003b) presented a method for free shrinkage assessment of stabilized materials. Stabilized specimens measuring 75 mm (wide) × 75 mm (high) × 280 mm (long) were compacted using Standard Proctor hammer. Two-gauge studs were placed at the middle of the end sections during compaction to facilitate shrinkage measurement. Specimens in duplicate per set were cured for 24 h at 90% or higher RH and air temperature of 21–24 °C. Subsequently, specimens were dried in a controlled environment with 50% RH and air temperature of 22 °C. Shrinkage measurements at various time intervals were obtained by using a special measuring apparatus equipped with a digital micrometer with an accuracy of 0.001 mm, as shown in Fig. 6.18(a). A referenced bar was used to initialize the reading prior to shrinkage measurement (Fig. 18b). Shrinkage strain was computed as the total shrinkage divided by the original length presented as units of microstrain.

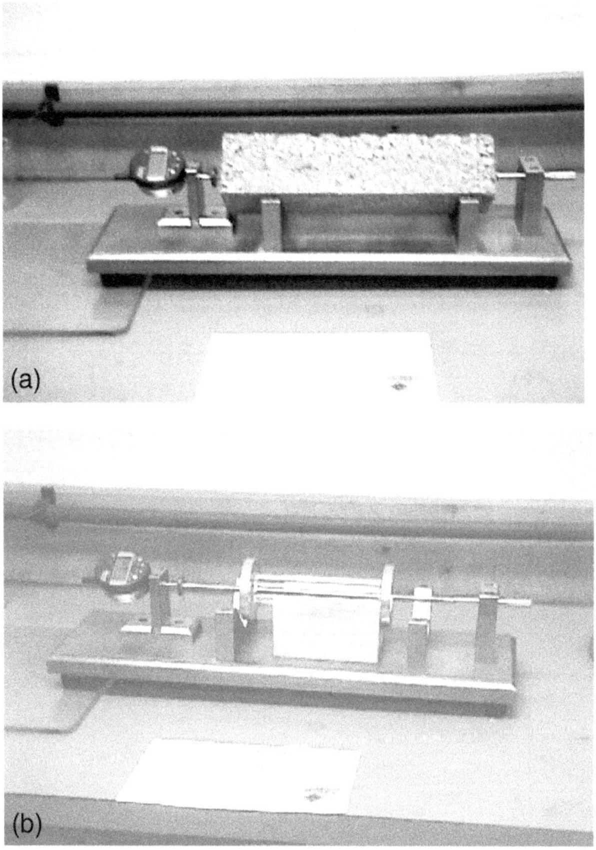

Figure 6.18 (a) Free shrinkage measurement; (b) Reference bar for length initialization.

Figure 6.19 shows typical shrinkage values measured for specimens stabilized with GP, GB, and AAS at 2%, 3%, and 4% binder contents, respectively. For all binders, shrinkage develops at a relatively rapid rate at the beginning, and the rate of shrinkage reduces with time. For GP and GB cements, the shrinkage stabilized to an ultimate (equilibrium) value after approximately 21 days, whereas AAS requires a longer time duration to stabilize. These results indicate that the rate of shrinkage development is lowest for AAS. Figure 6.19 also indicates that for GP and AAS, the shrinkage slightly increases when the binder content is increased, whereas for GB cement, shrinkage tends to decrease marginally when the binder content is increased. However, supplementary test results show that shrinkage can increase again for further increases in GB content (Table 6.7).

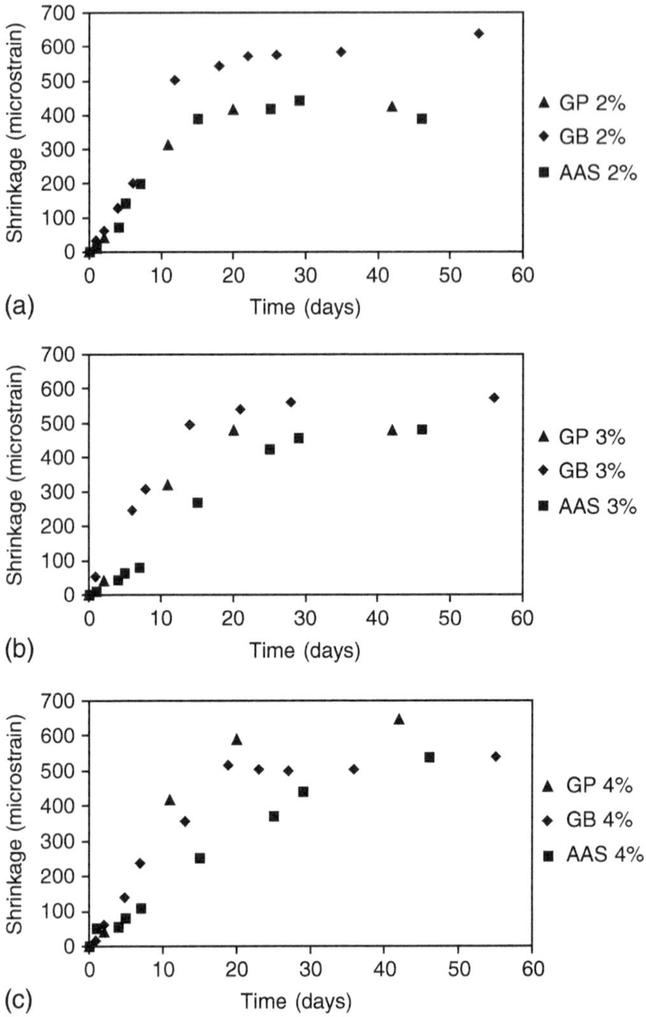

Figure 6.19 Drying shrinkage versus time relationship for various binder types with (a) binder content 2%, (b) binder content 3%, and (c) binder content 4%. *(Source: Chakrabarti (2004)).*

Table 6.7 Ultimate drying shrinkage for different types and quantities of binders at 50% RH and 23 °C

Binder type	**Shrinkage in microstrain**			
	2% Binder	*3% Binder*	*4% Binder*	*6% Binder*
GB	639	575	540	642
GP	425	478	645	684
AAS	391	472	542	581

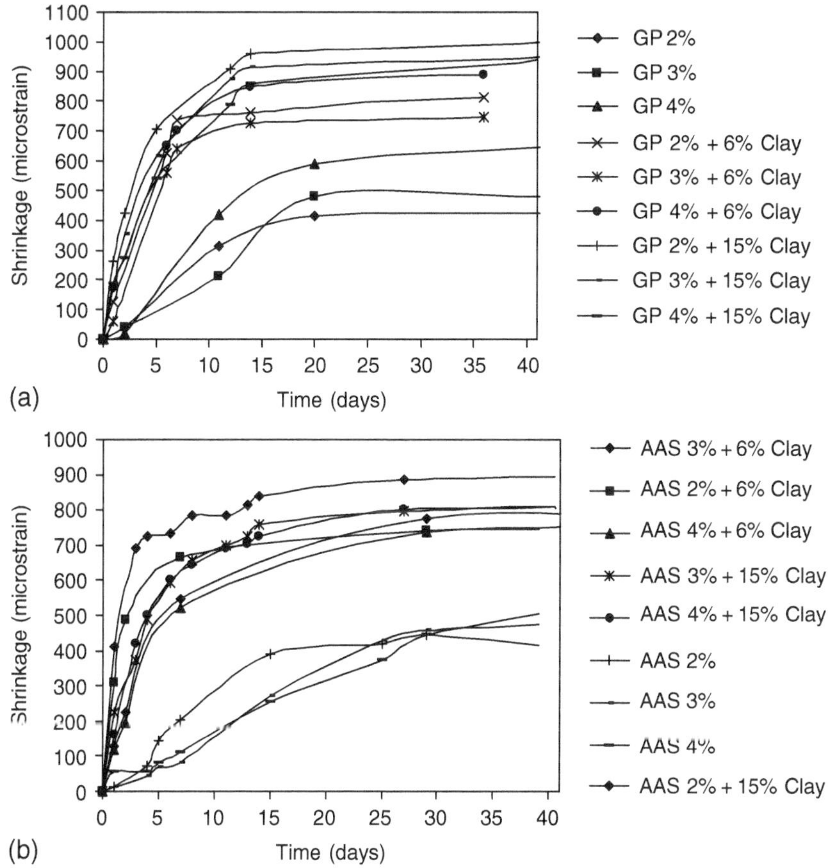

Figure 6.20 Shrinkage behavior with and without fines (a) for GP cement and (b) for AAS. *(Source: Chakrabarti (2004)).*

These differences arise from the complex interaction of binder and the particle matrix during drying shrinkage. Two major mechanisms in play are the suction generated due to capillarity and adsorbed water and the resistance provided by the particle matrix to material shrinkage. Figure 6.20 shows the influence of clay in the pavement material on drying shrinkage. It is clear that the presence of clay can substantially increase the shrinkage potential of the pavement material. Therefore, similar to the revelations on UCS, the likely presence of clay in the pavement material must be seriously considered during the design stage.

6.5.5 Design of stabilized pavement bases

Design for traffic loading

The design of bound stabilized pavements is based on the fatigue life due to flexure when subjected to traffic loading (Austroads, 2004). In Australia and New Zealand, the concept of standard axle repetition (SAR) is used to convert the likely traffic spectrum to the equivalent number of SARs. The standard axle is defined as a single axle with dual tires at each end, loaded with an axle load (at the middle) of 80 kN. According to Austroads, the fatigue life or allowable number of SARs is defined by

$$N = \mathrm{RF}\left[\frac{113{,}000/E^{0.84}{}_{\mathrm{flex}} + 191}{\mu\varepsilon}\right]^{12} \tag{6.1}$$

where E_{flex} (MPa) is the flexural modulus of the cement-treated material, and $\mu\varepsilon$ is the horizontal tensile strain at the bottom of the cemented layer expressed in microstrain (10^{-6}). RF is a reliability factor signifying the reliability of the design. For instance, an RF of 0.95 indicates that 95% of the pavements designed using this method should perform satisfactorily within the design life. The fatigue life model given in Eq. (6.1) has two factors overriding each other in terms of the material behavior. The first is the applied tensile strain, a decrease of which will increase the fatigue life. The second is the flexural modulus, an increase of which (at a given strain) will decrease the fatigue life. However, there is a stronger emphasis on the strain reduction than modulus increase; therefore, the overall tendency is to get longer fatigue life when the material is made stiffer. These trends are in accordance with the trends established with classical LEFM studies (Castell et al., 2000). Austroads (2004) stipulates that a second phase of pavement life may be possible when the cemented layer has approached design fatigue life. In the second phase, the pavement is treated primarily as an unbound layer without any cohesive properties. However, this two-level demarcation of performance may sometimes be too simplistic because there may be a transition period in which the material behaves as a cracked material before starting to display predominantly unbound behavior.

Jameson et al. (1995) compared the predicted and measured traffic lives for stabilized pavement experiments in the Cooma trial. They calculated the ratio of observed (the fatigue life was defined as the pavement cracking approached 1.0 m/m^2) and predicted (on the basis of Eq. (6.1)) fatigue lives for Experiments 3 (Pad 1) and 4 (Pad 5) as 0.034 and 5.1, respectively. The significant overestimation of the fatigue life in Experiment 1 was attributed to a number of factors, including the decrease of modulus with depth within the stabilized layer and saturation of the layer prior to trafficking. In other experiments of the Cooma trial, the actual fatigue lives were not determined because the trafficking was stopped before reaching the designated fatigued state. However, analysis of results indicated that predicted lives were less than the loading repetitions imposed during testing. It is considered generally that Eq. (6.1) provides a conservative estimate of the fatigue life of stabilized pavement.

It is apparent that the actual fatigue behavior is usually more complex than that depicted in Eq. (6.1). For instance, there may be a failure strain (as discussed previously) at which the allowable load repetition is unity. Similarly, there may be a lower limit for the strain, below which fatigue failure may not occur. More important, however, fatigue life for different load configurations and levels may be different, thereby generating errors with the use of the standard axle representation for all load types and levels. In addition, the effects of shrinkage cracking and other environmental influences can affect the tensile crack propagation influencing the fatigue life to traffic loading. On the other hand, the results presented previously highlighted that the damage of stabilized materials can occur progressively with continual reduction of modulus. This form of damage may be considered under the so-called residual strength models (Degrieck and Paepegem, 2001). Along with the environmental loading to be discussed in the next section, these are some of the challenges faced in relation to the future advancement of design of cementitiously stabilized pavement bases.

Design for environmental loading

Environmental loading for pavements can come from changes in moisture content due to water ingress or loss, temperature changes, and ingress or leaching of salts and other chemicals into or from the stabilized matrix. In addition to these, shrinkage and thermal cracking can accelerate the deterioration of stabilized materials, especially in the presence of excessive moisture. In the current practice, some of these issues are considered mainly in a qualitative sense during the mix design stage (e.g., see Fig. 6.9). It was highlighted that the modulus of the stabilized material depends on its moisture content. Therefore, the potential for ingress of water into the pavement system through rain or capillary flow from groundwater table or clogged up drains is important. The capillary rise tests (as shown in Fig. 6.10) may be used as an index test to assess the potential for water ingress, but Lee et al. (2004) showed that these tests should use specimens with a height comparable with the thickness of the prototype pavement base in order to obtain realistic results. Resulting saturation levels may be linked to the deterioration potential of pavements due to traffic loading. The extrapolation of laboratory results to field conditions invariably needs theoretical models, but simplification of these models for routine practice may be possible.

When a pavement loses moisture during in-service conditions, it will be subjected to restrained shrinkage, and shrinkage cracking may result. Currently, there are no established methods to incorporate this aspect into pavement mix design. On the basis of the research by the first author and coworkers, a conceptual approach is discussed here. The concept is based on shifting laboratory-measured ultimate shrinkage to shrinkage applicable under field conditions. Chakrabarti and Kodikara (2004) highlighted that under restrained shrinkage, the material can undergo creep strains relaxing the development of tensile strain. Therefore, the effective shrinkage strain to be considered for shifting to the field conditions would be the difference between the ultimate free shrinkage and the creep strain. Factors that need to be considered under field conditions include (1) climate, (2) pavement seal, (3) pavement thickness, (4) initial moisture content and time after construction, and (5) depth of groundwater table. A computer program capturing vapor and moisture flow may be able to characterize these factors. Once the effective field shrinkage strain (applicable to field conditions) is calculated, it can be compared with the failure tensile strain of the material for assessing the severity of the potential cracking in a particular stabilizing mix. This approach may be used to guide the mix design process to minimize the shrinkage cracking potential under field conditions. However, further research is necessary to develop a quantitative procedure to incorporate environmental loading into stabilized pavements.

6.6 CONCLUSION

In situ stabilization of degraded road pavements using cementitious binders is presented as a viable alternative to reconstruction and overlay. The quantity of binder depends on the design requirements of the pavement, but relatively small quantities of 3–6% may be suitable for pavements with light to medium traffic. Pavements with high traffic may be catered for by a higher level of binder content, by using deeper lift of recycling, or by a combination of both. The laboratory results as well as field testing highlight that pavements stabilized with slag blends can work effectively in comparison to the traditional binder, GP cement. In addition to the recycling of old pavement materials, this use has the advantage of using waste products in place of GP cement, reducing the pressure on global CO_2 emissions.

Accelerated loading can be used to assess the pavement performance under field conditions; however, to get maximum results, the tests should be properly planned, instrumented, and analyzed. The important aspects of the design of these pavements were highlighted, which include the design against traffic and environmental loading. Although the emphasis so far has been placed on traffic loading that can lead to fatigue failure of the materials, it was identified that the environmental loading in the forms of moisture influence and thermal and drying shrinkage can cause pavement to deteriorate and can exacerbate pavement deterioration due to traffic loading. A conceptual approach for incorporating environmental loading in the design of these pavements was outlined, but this should be followed by future research efforts. Other factors, such as erodibility, chemical attack due to salt crystallization, and potential chemical leaching and generation, are identified as important but not discussed in detail here. A comprehensive scientific basis required for the analysis of these issues is already available in other disciplines, such as in material science, geomechanics, and geoenvironmental engineering. Therefore, it is a matter for the pavement fraternity of road authorities, councils, industry, research bodies, and universities to participate in this research effort to advance this technology to make it an even more environmentally prudent and economically sustainable approach for pavement rehabilitation.

ACKNOWLEDGMENTS

The authors thank their respective organizations, Monash University and ARRB, for support in relation to this work. The first author thanks his graduate students, Dr. Srijib Chakrabarti, Mr. Hani Nahlawi, and Mr. Kok Yun Lee, for their contribution to the research related to this work. Furthermore, the first author acknowledges the support received from the Australian Research Council in the form of APA(I) linkage project in partnership with Chadwick Geotechnical Testing Pty. Ltd. to undertake some research reported here.

REFERENCES

Alderson, A., 1999. Structural Properties of Cement Stabilised Materials. Contract Report RC7101A, May, ARRB Transport Research Ltd, Vermont South, Victoria, Australia.

Al-Hussaini, M., 1981. Tensile properties of compacted soils. In: Young, R.N., Townsend, F.C. (Eds.), Laboratory Shear Strength of Soil. American Society for Testing and Materials, West Conshohocken, PA, pp. 207–225.

Austroads, 2004. Pavement Design—A Guide to the Structural Design of Road Pavements, Pavement Technology Series. Austroads, Sydney, Australia.

Bofinger, H.E., 1970. The Measurement of the Tensile Properties of Soil–Cement. Road Research Laboratory Report No. LR 365, Ministry of Transport, Crowthorne, Berkshire, UK.

Castell, M.A., Ingraffea, A.R., Irwin, L.H., 2000. Fatigue crack growth in pavements. J. Transp. Eng., ASCE 126 (4), 282–290.

Chakrabarti, S., 2004. Shrinkage Behaviour of Cementitiously Stabilised Crushed Basaltic Rocks as Applicable to Pavement Rehabilitation. PhD thesis, Department of Civil Engineering, Monash University, Victoria, Australia.

Chakrabarti, S., Kodikara, J.K., 2003a. Influence of reactive clay on crushed basaltic rock materials stabilised with cementitious additives. In: Australian Road Research Board Conference, Cairns, Australia, June, CD-ROM.

Chakrabarti, S., Kodikara, J.K., 2003b. Strength, shrinkage and capillary rise properties of cementitiously stabilised crushed basaltic rock. In: Transportation Research Record, USA, No. 1819, vol. 2, pp. 18–26.

Chakrabarti, S., Kodikara, J.K., 2004. Laboratory method for assessing shrinkage cracking potential in stabilised materials. In: Petit, C., Al-Qadi, I.L., Millen, A. (Eds.), Proc. Fifth Int. RILEM Conference, pp. 605–610.

Chakrabarti, S., Kodikara, J.K., 2005. Stabilization of crushed basaltic rocks and clay mixtures using cementitious additives. J. Mater. Civ. Eng. 17 (4), 432–439.

Chakrabarti, S., Kodikara, J.K., Pardo, L., 2002. Overview of pavement performance in Australia. Road Transport Res. 11 (3), 3–16.

Collins, F., Sanjayan, J.G., 2002. Development of novel alkali activated slag binders to achieve high early strength concrete for construction use. Aust. Civil Eng. Trans. CE44 (IEAust), 91–102.

Degrieck, J., Paepegem, W.V., 2001. Fatigue damage modelling of fibre-reinforced composite materials: review. Appl. Mech. Rev. 54 (4), 279–300.

Foley, G., Austroads Stabilisation Expert Group, 2001a. Mechanistic Design Issues for Stabilised Pavement Materials. Austroads Project Report NT&E9910, Austroads, Sydney, Australia.

Foley, G., Austroads Stabilisation Expert Group, 2001b. Effect of Design and Environmental Factors for Long-Term Performance of Stabilised Materials. Austroads Project NT&E910, Austroads, Sydney, Australia.

Frydman, S., 1964. The applicability of the Brazilian (indirect tension) test to soils. Aust. J. Appl. Sci. 15, 335–343.

Hannant, D.J., Branch, J., Mulheron, M., 1999. Equipment for tensile testing of fresh concrete. Mag. Concrete Res. 51 (4), 263–267.

Jalinous, M.A., 1986. Effect of Rate of Loading on the Relationship between the Modulus of Rapture and the Compressive Strength of Plain Concrete. MS thesis, Department of Civil Engineering, University of Massachusetts Lowell, Lowell, MA.

Jameson, G.W., Sharp, K.G., Yeo, R., 1992. Cement-Treated Crushed Rock Pavement Fatigue Under Accelerated Loading: The Mulgrave (Victoria) ALF trial, 1989/91. ARRB ARR 229, July 1992, ARRB Transport Research, Vermont South, Victoria, Australia.

Jameson, G.W., Dash, D.W., Tharan, Y., Veressy, N.J., 1995. Performance of Deep-Lift in Situ Pavement Recycling under Accelerated Loading: The Cooma ALF trial, 1994. Austroads Report No. APRG 11, ARRB Research Report No. ARR 265, p. 96.

Kadar, P., 1987. Accelerated Full Scale Testing of Thin CTB Pavements—Beerburrum, Queensland, 1986–87. Internal Report AIR 441-1, Australian Road Research Board, Vermont South, Victoria, Australia.

Kodikara, J.K., Chakrabarti, S., 2001. Shrinkage behaviour of cemented materials as applicable to in situ pavement stabilisation. In: 20th ARRB Conference, Melbourne, March, CD-ROM.

Kodikara, J.K., Sierakowski, J., Ranjith, P.G., 2004. A synthesis of mechanisms and factors affecting salt damage to road infrastructure: engineering salinity solutions. In: Dogramaci, S., Waterhouse, A. (Eds.), Proc. 1st National Salinity Engineering Conference. Institution of Engineers, Australia, pp. 101–106.

Land Transport, 2000. http://www.dotrs.gov.au/land/index.htm.

Lee, K.Y., Kodikara, J.K., Bouazza, A., 2004. Modelling and laboratory assessment of capillary rise in stabilised pavement materials. Transportation Res. Record 1868, 3–13.

Moffatt, M.A., Sharp, K.G., Vertessy, N.J., Johnson-Clarke, J.R., Vuong, B.T., Yeo, R.E. Y., 1998. The Performance of in Situ Stabilised Marginal Sandstone Pavements. Austroads Report No. APRG 22, ARRB Research Report No. ARR 322, p. 72.

Nahlawi, H., Chakrabarti, S., Kodikara, J.K., 2004. A direct tensile testing method for unsaturated geomaterials. Geotech. Test. J., ASCE 27 (4), 1–6.

Otte, E., 1978. A Structural Design Procedure for Cement-Treated Layers in Pavements. D. SC. (Eng.) thesis, University of Pretoria, South Africa.

Sherwood, P., 1993. Soil Stabilisation with Cement and Lime, TRL State of the Art Review. Transport Research Laboratory, HMSO, London.

Standards Australia, 1996. Unconfined Compressive Strength of Compacted Materials, AS 1141.51. Standards Australia, Sydney, Australia.

Symons, M.G., 1999. Shrinkage and Erodability of Cementitious Bound Soils, Research Report. Structural Materials and Assemblies Group, University of South Australia, Adelaide, South Australia.

Symons, M.G., Poli, D.C., 1996. Experimental Procedures for Soils Stabilised with Cementitious Binders, Research Report. Structural Materials and Assemblies Group, University of South Australia, Adelaide, South Australia.

Tang, G.X., Graham, J., 2000. A method for testing tensile strength in unsaturated soils. Geotech. Test. J. 23 (3), 377–381.

Wallace, K., 1998. The tensile response of an unbound granular pavement. Road Transport Res. J., ARRB 7 (3), 36–47.

Williams, R.I.T., 1986. Cement-Treated Pavement Materials, Design and Construction. Elsevier, London.

Yeo, R.E.Y., Newman, G., 1998. Stabilisation of Marginal Materials. VicRoads Report No. Tr110, Project No. 695, VicRoads, Kew, Victoria, Australia.

CHAPTER 7

Numerical Modeling of Clogging in a Permeable Reactive Barrier and Rejuvenation by Alkaline Fluid Injection in the Shoalhaven Floodplain, Australia

Udeshini Pathirage[1], Laura Banasiak[1], Buddhima Indraratna[1], Glenys Lugg[2]
[1]Centre for Geomechanics and Railway Engineering, Faculty of Engineering and Information Sciences, University of Wollongong, NSW, Australia
[2]Manildra Group, Nowra, NSW, Australia

7.1 INTRODUCTION

Acid sulfate soil (ASS) occupies more than 200,000 km^2 of the coastal Australian land. Acidic groundwater generated from ASS (mainly pyrite) is a major environmental and socioeconomic problem that occurs due to disturbances in the land and hydrological systems such as construction of deep flood mitigation drains and due to rainy and drought seasons. These changes can encourage pyrite oxidation in coastal areas, resulting in the generation of sulfuric acid in soil along with leaching out of aluminum (Al) and iron (Fe) from soil, which are toxic for living habitats (Dent, 1992; Indraratna et al., 1995; Regmi et al., 2009a). Subsequent transportation of this acidic water enriched with dissolved Al and Fe to nearby waterways has created unfavorable environment conditions.

A permeable reactive barrier (PRB) is a reactive material-filled trench installed in the subsurface to pass a contaminant plume through reactive media. This is a passive treatment that offers in situ solutions. During the residence time, reactions take place between contaminated groundwater and the reactive material. The reaction process can be via chemical precipitation, dissolution, sorption, and oxidation/reduction (Rumer and Ryan, 1995). PRBs have been identified as promising remediation technology, which can run for several years or decades (Gavaskar et al., 1998), exhausting the reactivity of the reactive material. Moreover, clogging and armoring

Chemical, Electrokinetic, Thermal, and Bioengineering Methods
http://dx.doi.org/10.1016/B978-0-08-100191-2.00007-1

(encrustation on the reactive surfaces) reduce the performance of PRBs due to secondary mineral precipitation (Wilkin et al., 2003; Li et al., 2005, 2006; Li and Benson, 2010). Clogging of the pore spaces decreases the porosity and hydraulic conductivity of the reactive media, which changes the hydraulic properties inside the PRB and reduces flow through the barrier (Mackenzie et al., 1999).

Golab et al. (2006) performed batch tests for 24 alkaline materials to identify suitable reactive material for this PRB and found that recycled concrete aggregates have the ability to remove Al and Fe efficiently from acidic groundwater and are capable of sustaining a near-neutral pH for a long time. With these promising results, a pilot-scale PRB was installed in the Lower Shoalhaven Floodplain area near the town of Bomaderry, Australia (approximately 100 km south of Sydney) in October 2006 (Fig. 7.1), where acidic groundwater generated from ASS terrain is a severe environmental problem. The PRB was filled with crushed recycled concrete ($d_{50}=40$ mm) in a $17.7 \times 1.2 \times 3.0$ m trench. Before the trench was filled with recycled concrete, it was lined with geotextile material (Fig. 7.1) to ensure that no fine sand or clay particles enter with groundwater and clog the barrier.

The field site was composed of 30 observation wells, 15 piezometers, and three data loggers installed inside, up-gradient and down-gradient of the PRB. A water level meter was used to calculate hydraulic gradients, phreatic surface variations, and permeability. Multiparameter field electrode probes were used to measure groundwater pH, oxidation reduction potential (ORP), and temperature. These parameters were measured onsite during monthly field visits since October 2006 by the researchers from the

Figure 7.1 (a) Pilot-scale PRB installed in the Shoalhaven Floodplain, NSW. (b) Installation of the PRB.

University of Wollongong (UOW). Furthermore, the data loggers recorded hourly data of pH, temperature, dissolved oxygen, and water pressure. Groundwater samples collected from each monitoring well were analyzed for concentrations of acidic cations: Fe, Al, major cations (Na^+, K^+, Ca^{2+}, Mg^{2+}, Al^{3+}, Fe^{3+}), and anions (Cl^-, $SO_4{}^{2-}$).

PRBs have proven to be a satisfactory engineering technology for acidic groundwater remediation despite the decrease in long-term efficiency due to mineral fouling. PRBs have been widely used for acid mine drainage (AMD) problems. For example, Blowes et al. (2003) reported a PRB filled with organic carbon-rich material such as wood chips, municipal compost, and paper mill pulp. The alkalinity generated from these organic materials could effectively raise the alkalinity and pH of the groundwater, wherein the only drawback was the armoring by metal sulfide precipitation and bacterial growth (Blowes et al., 2003). Another PRB used for AMD treatment was a limestone and red mud-filled PRB, in which the precipitation of heavy metals (Mn, Zn, Fe, Al, Cu, Co, $SO_4{}^{2-}$, and Ni) in their forms of hydroxides clogged the reactive media and decreased the reactivity (Komnitsas et al., 2004). The performance of PRBs has been hindered by mineral fouling wherein the pore space is reduced by mineral precipitation in the reactive media. Mineral fouling decreases the porosity and hydraulic conductivity of the reactive medium inside the PRB (Mackenzie et al., 1999; Rowe et al., 2000; Rowe, 2005; Jeen et al., 2007; Regmi et al., 2009b; Indraratna et al., 2010; Jeen et al., 2012; Pathirage et al., 2012). Changes in these hydraulic properties affect the flow paths and reactive times. Thus, it is important to understand the geochemistry of the remediation process and predict the performance to calculate the longevity of the PRB. Porosity and hydraulic conductivity reductions due to secondary mineral precipitation are higher at the entrance zone and lower in the middle and exit zones of the PRB (Liang et al., 2003; Wilkin et al., 2003; Li et al., 2006). Core samples collected from the PRBs can be used to calculate the volume of secondary minerals precipitated (Li et al., 2006). Moreover, stoichiometric relationship of the chemical reactions also can be used to calculate the volumetric change and hence the change in porosity and hydraulic conductivity.

The main aim of this research was to develop a numerical model to understand mineral dissolution and precipitation in a PRB in ASS terrain, with a contaminated transport model to simulate mineral accumulation and the associated changes in hydraulic parameters. First, a novel geochemical algorithm was developed to capture all the dominant reactions.

Thereafter, a transient groundwater flow model was developed to capture the change in hydraulic head with related changes in porosity and hydraulic conductivity. This head solution was used as the input solution for MODFLOW, which was used to couple the groundwater flow with contaminant transport in RT3D. The authors believe that the coupled model will be useful for geotechnical engineers and environmental scientists rehabilitating ASS terrains, especially in coastal Australia.

7.2 MATERIALS AND METHODS

7.2.1 Laboratory column experiments

Laboratory column experiments were carried out to validate the developed model. Twin columns were run in order to measure the pressure readings from one and to take samples from the other (Fig. 7.2). The reason for

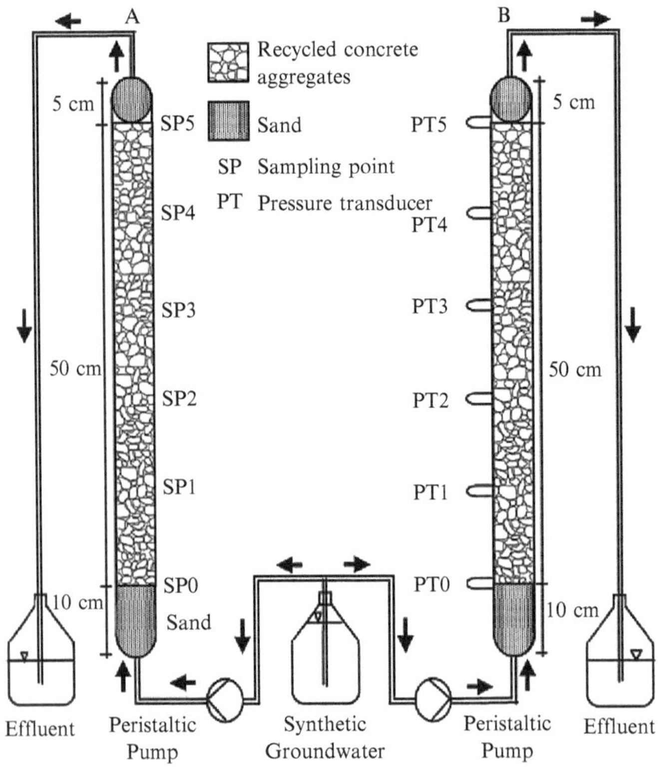

Figure 7.2 Schematic of the laboratory column experiments: A, Sampling column; B, pressure measuring column (Indraratna et al. (2014)).

running twin columns was to minimize the disturbance from sampling to the pressure readings (Johnson et al., 2005). A constant flow of 1.2 ml/min was supplied using a Masterflex peristaltic pump. Luer adapters were used to take samples every 100 mm along the sampling column (SC), and pressure transducers were used to measure the pressure in the pressure measuring column (PTC) at similar heights as those of SC. All other environmental conditions and input parameters were maintained similar for both columns. Recycled concrete collected from a demolished site was crushed, and uniform particle size (passed through a 4.75-mm sieve and retained on a 3.35-mm sieve) was chosen to fill the columns. The mineralogical analysis carried out by x-ray diffraction revealed the weight percentages as follows: Ca (57.3%), Fe (21.4%), Al (9.85%), Mg, (5.27%), Si (3.06%), and others (3.04%) (Regmi et al., 2011b; Pathirage et al., 2012). Acrylate transparent columns 650 mm in length with a 50-mm internal diameter were used, for which the first 100 mm and last 50 mm were placed with pure silica sand wrapped in geotextile material and the middle 500 mm was placed with crushed concrete particles. Synthetic acidic groundwater was prepared similar to the chemistry of acidic groundwater at the ASS terrains in southeast New South Wales (Regmi et al., 2009b; Pathirage et al., 2012) and fed into columns from bottom to top. Collected samples from the effluent and at sampling points were analyzed for alkalinity and Al, Fe, and Ca concentrations. Inductively coupled plasma–mass spectrometry (ICP-MS) was used to analyze for Al and Ca, and atomic absorption spectroscopy was used to analyze for Fe following the standard guidelines provided by the American Public Health Association (1998).

7.2.2 Geochemical algorithm

Transition state theory was used to develop the geochemical algorithm capturing all the dominant chemical reactions. Equation (7.1) gives the transition state theory that is well adopted in contaminant transport modeling (Yabusaki, 2001; Li et al., 2005; Mayer et al., 2006; Jeen et al., 2012). The alkalinity to neutralize the acidity in groundwater comes from the Ca-bearing minerals present within recycled concrete, which are $CaAl_2Si_2O_8$ (16.8% by weight), $CaCO_3$ (4.4% by weight), and $Ca(OH)_2$ (0.3% by weight) according to Regmi et al. (2011a). Twelve dominant mineral dissolution/precipitation reactions (Regmi et al., 2009b) are used in the geochemical algorithm, which is the first such algorithm developed for the remediation of acidic groundwater generated from ASS as listed in the appendix. Other mineral components given in Table 7.1 did not take

Table 7.1 Water chemistry of the influent solution prepared for column experiment simulating the water chemistry of the acidic groundwater in ASS terrain presented in Regmi et al. (2009a)[b]

pH	ORP (mV)	Acidity (mmol eq/l)[a]	mg/l							
			Na^+	K^+	Ca^{2+}	Mg^{2+}	Al^{3+}	Fe^{3+}	Cl^-	SO_4^{2-}
2.67	610	6.45	504.2	50.1	152.2	118.0	54.0	49	849.0	1450.0

[a] Acidity was measured equivalent with respect to $CaCO_3$.
[b] Indraratna et al. (2014).
ORP, oxygen reduction potential.

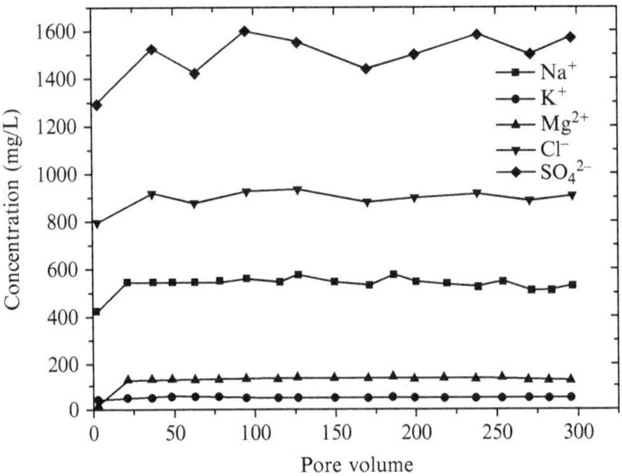

Figure 7.3 Other ions in the effluent that do not change significantly with time (Indraratna et al. (2014)).

part in the remediation process, as can be seen in Fig. 7.3, in which the concentrations of those ions did not change significantly with the pore volume (PV).

$$r = -k_{\text{eff}}\left(1 - \frac{\text{IAP}}{K_{\text{eq}}}\right) \tag{7.1}$$

The value for IAP/K_{eq} was taken by calculating the saturation indices (SI) by PHREEQC software using Eq. (7.2) (Regmi et al., 2011a; Walter et al., 1994):

$$\text{SI} = \log(\text{IAP}) - \log(K_{\text{eq}}) \tag{7.2}$$

PHREEQC software calculates the saturation indices by the influent chemistry of groundwater, such as the concentration of all ions, pH, alkalinity, and temperature. Saturation indices (SI) can be used to identify the dissolving and precipitating minerals because the SI values are less than 0 for dissolving minerals and more than 0 for precipitating minerals. The effective rate coefficient (k_{eff}) was taken by calibrating against a different data set and was assumed to be a spatially homogeneous and time-independent parameter during the simulation (Li et al., 2006).

7.2.3 Change of mineral quantity over time

The change in volume due to secondary mineral precipitation and the dissolution of Ca-bearing minerals in recycled concrete were calculated from their reaction kinetics and molar volume as suggested by Steefel and Lasaga (1994) in Eq. (7.3). Dissolved and precipitated minerals were assumed to be immobile within the system. The pore space occupied by each mineral was calculated from the relevant molar volume. Therefore, the associated porosity reductions with time can be calculated from Eq. (7.4):

$$\frac{\partial \Phi_k}{\partial t} = M_k R_k \tag{7.3}$$

$$n_t = n_0 - \sum_{k=1}^{N_m} M_k R_k t \tag{7.4}$$

$M_k R_k$ is a constant value for one time step. For the corresponding time step, a new value of R_k is substituted to Eq. 7.4 as calculated from Eqs. (7.1) and (7.3). The normalized Kozeny–Carmen equation (Eq. (7.5)) can be utilized to estimate the change in hydraulic conductivity at different time steps due to the dissolution and precipitation of minerals (Li et al., 2006; Pathirage et al., 2012):

$$K = K_0 \left[\frac{n_0 - \Delta n_t}{n_0}\right]^3 \Big/ \left[\frac{1 - n_0 + \Delta n_t}{1 - n_0}\right]^2 \tag{7.5}$$

7.2.4 Groundwater flow model and reactive transport flow model

MODFLOW and RT3D are finite different numerical codes that were used to simulate the contaminant transport of Ca, Al, and Fe in the column experiment and field PRB. In MODFLOW, groundwater flow is solved using the block-centered finite difference method (Harbaugh, 2005). Both column

Table 7.2 Experimental and model parameters[a]

Property	Experiment	Model (lab)	Model (field)
Flow	1.2 ml/min	1.2 ml/min	1.1×10^6 l/year
Initial porosity (n_0)	0.69	0.69	0.5
Initial hydraulic conductivity (K_0)	0.9565 m/day	0.9565 m/day	0.1 m/s
pH of influent	2.8	2.8	3.6

[a]Indraratna et al. (2014).

experiments and the field PRB were treated as confined aquifers with transient groundwater flow conditions. The recycled concrete-filled media was considered homogeneous and isotropic because a relatively uniform particle gradation was chosen and the particle angularity is assumed similar. Table 7.2 summarizes the experimental parameters and model inputs. The sides of the column were no-flow boundaries.

The governing equation for transient groundwater flow in one dimension is given by

$$\frac{\partial^2 h}{\partial x^2} = \nabla^2 h = \frac{S}{T}\left(\frac{\partial h}{\partial t}\right) \tag{7.6}$$

$$T = Kb \tag{7.7}$$

The Kozeny–Carmen equation (Eq. (7.5)) calculates the change in hydraulic conductivity due to mineral dissolution and precipitation. The head solution for governing Eq. (7.6) with the variations in hydraulic conductivity is given by

$$h = \left(\exp\left[-\frac{\mu^2 b K_0}{S\sum_{k=1}^{N_m} M_k R_k}\frac{(1-n_0)^2}{n_0^3}\left\{\alpha^2\left(1.5 + {}^1/_\beta\right) - 3(\alpha + \ln\beta)\right\}\right]\right)\cdot (C\sin\mu x + D\cos\mu x) \tag{7.8}$$

where

$$\alpha = n_0 + \sum_{k=1}^{N_m} M_k R_k t$$

$$\beta = 1 - n_0 - \sum_{k=1}^{N_m} M_k R_k t$$

μ, C, and D can be calculated by the following initial conditions:

$$h = h_1 \text{ at } x = 0 \text{ and } t = 0,$$
$$h = h_2 \text{ at } x = l \text{ and } t = 0,$$
$$\frac{\partial h}{\partial t} = 0 \text{ at } x = 0 \text{ and } t = 0$$

MODFLOW uses the solution from Eq. (7.8) as the starting head for every time step. Therefore, the head solution carries the change in porosity and hydraulic conductivity because MODFLOW cannot change these values automatically unless they are entered manually. For example, after the simulation was run for the first time step, the resulting values for porosity and hydraulic conductivity were updated in the second time step. Corresponding head for the next time step is now calculated using Eq. (7.8) as input for MODFLOW. Whereas MODFLOW simulated the groundwater flow for every time step, advection, diffusion, and dispersion (Eq. (7.9)) for contaminant transport were simulated using RT3D for every time step. RT3D (three-dimension multicomponent transport model) solves coupled partial differential equations using the implicit finite different method. In the current study, "user-defined module" in RT3D was used because the available seven predefined reaction modules did not cater for acidic water remediation due to ASS:

$$R_e \frac{\partial C}{\partial t} = D \frac{\partial^2 [C]}{\partial x^2} - u_b \frac{\partial [C]}{\partial x} - R_k M_k C \tag{7.9}$$

Once RT3D got the head solution from MODFLOW as the input, groundwater flow velocity (u_b) was then calculated using Eq. (7.10) for the subsequent time step; thus,

$$u_b = -\frac{K}{n} \frac{\partial h}{\partial x} \tag{7.10}$$

Therefore, MODFLOW and RT3D were run in tandem to get the concentrations of reactants at every time step. The kinetic reaction rate coefficients (k_{eff}) for Ca^{2+}, Al^{3+}, and total Fe (Fe^{2+} and Fe^{3+}) (Table 7.3) were obtained through calibration as stated in Li et al. (2006). Trial-and-error calibration was carried out using the data provided for the laboratory column experiments by Regmi et al. (2011b). The calibrated values for kinetic reaction rate coefficients are listed in Table 7.3. The developed model was validated using the data obtained from the laboratory column experiments carried out in this study. Both calibration and validation were carried out for to 40- to 190-PV range because this was the most important experiment zone in which a neutral pH was maintained and Al and Fe were

Table 7.3 Kinetic reaction rate coefficients (k_{eff}) for the mineral dissolution/precipitation that are calibrated values from the data provided by Regmi et al. (2011b)[b]

Mineral phase	Kinetic reaction rate coefficient (k_{eff}) (mol/l.s)	Kinetic reaction rate coefficient (k_{eff}) in literature (mol/l.s)[a]
Ca^{2+}	2.27×10^{-7}	(1×10^{-6})
Al^{3+}	6.86×10^{-8}	$(9.0 \times 10^{-7} - 1.0 \times 10^{-8})$
Total Fe (Fe^{2+} and Fe^{3+})	5.87×10^{-8}	$(1.0 \times 10^{-7} - 1.2 \times 10^{-8})$

[a] *Source:* Ouangrawa et al. (2009) and Jurjovec et al. (2004).
[b] Indraratna et al. (2014).

completely removed from the influent. The concentration of each substance at each cell in the finite difference grid was calculated for each time step.

7.2.5 Model application to the field PRB

One-dimensional reactive transport modeling was carried out through the centerline of the PRB. The total width of 1.2 m was discretized into 12 elements of 0.1 m. The governing equations used to predict the vertical flow of column experiments were adopted and assumed similar to the horizontal flow through centerline of the PRB. The same chemical reactions were assumed to occur in the PRB; thus, the geochemical algorithm used for column experiments was utilized for this study. The PRB was considered as a saturated flow domain with specified head boundaries where top, bottom, and lateral faces of the domain were no-flow boundaries. An average hydraulic gradient of 0.006 was used, which was similar to the field conditions observed from October 2006 to January 2012.

7.3 RESULTS AND DISCUSSION

The alkalinity generated from Ca-bearing minerals in recycled concrete elevated the pH inside the column. When reactions occurred between acidic groundwater and recycled concrete, the pH of the column effluent reduced from pH 9.6 to 8 within 15 PVs initially (Fig. 7.4), after which there was a slow reduction of pH from 7.9 (25 PV) to 7.5 (125 PV). Thereafter, a quick decrease of pH from 7.5 (125 PV) to approximately 6.8 (185 PV) was observed, followed by a fast depletion of pH from 6.8 (185 PV) to 4 (215 PV). Before test termination, there was another pH drop from 4 (215 PV) to 3.1 (295 PV). Indraratna et al. (2010) interpreted the reasons for these different pH declines, where the initial decline after 15 PV corresponded to

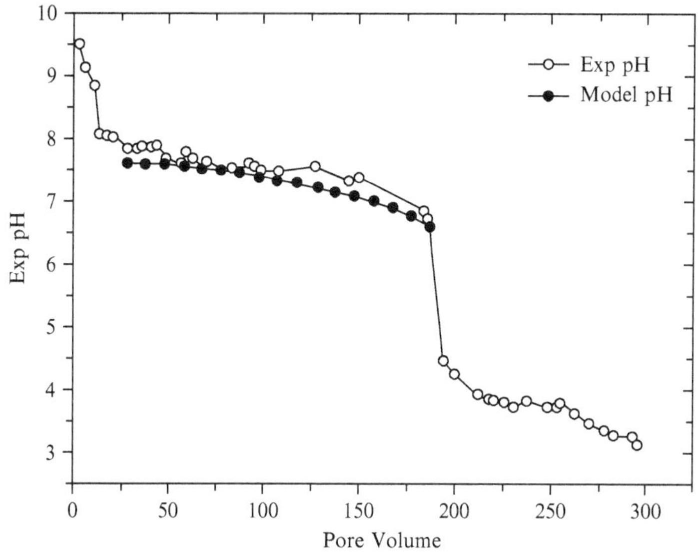

Figure 7.4 Predicted and experimental results of pH at the effluent (Indraratna et al. (2014)).

the exhaustion of carbonate alkalinity. The next decline of pH 6.8 to 4 was probably due to the OH^- present in the system having equilibrium throughout the exhaustion of carbonate minerals (Indraratna et al., 2010). The model results for the first pH plateau are given in Fig. 7.4. Moreover, the experimental data and model-predicted pH values along the height of the column are given in Fig. 7.5. At SP1, the rapid decrease in pH occurred at 25 PVs. This was due to the rapid exhaustion of reactive material and the fast neutralization process taking place at the entrance zone of the column. Another reason why there was no stepwise change in these sampling points (SP1, -2, -3, and -4) was probably because of the small number of sample collected at these ports to avoid disturbance to the water flow. In terms of removing Al from the influent, white precipitates were evident at the entrance zone of the column soon after the experiment was started (Fig. 7.6). Al was precipitated when the pH in the system was greater than 4.5. After effluent pH dropped to 4, Al was also witnessed in the effluent water, following an increase of Al concentration in the effluent (Fig. 7.6). In this manner, Fe was also precipitated out of solution when the pH was greater than 3.5. After 255 PV, the pH of the effluent decreased below pH 3.5, and a slight amount of Fe (<1 mg/l) was observed in the effluent (Fig. 7.7).

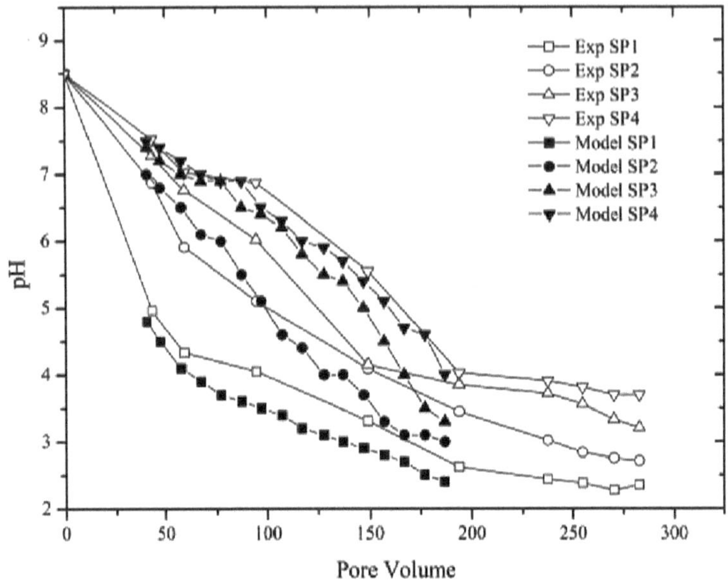

Figure 7.5 Predicted and experimental results of pH at the sampling points along the column (Indraratna et al. (2014)).

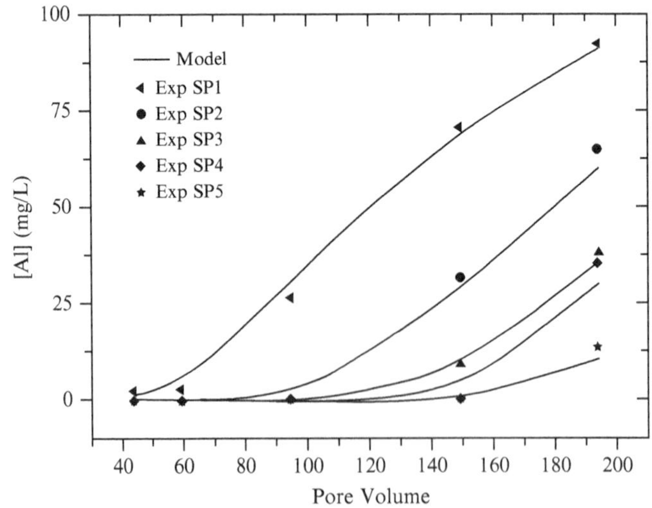

Figure 7.6 Calculated and measured Al^{3+} concentrations at sampling points along the column (Indraratna et al. (2014)).

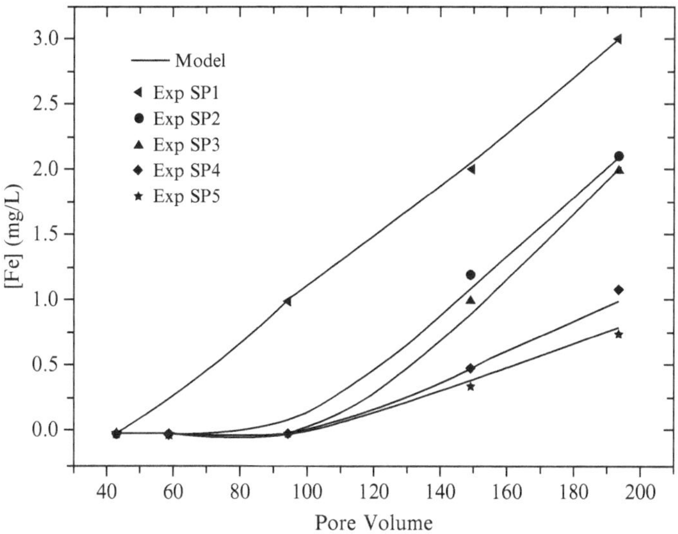

Figure 7.7 Calculated and measured total Fe (Fe^{2+} and Fe^{3+}) concentrations at sampling points along the column (Indraratna et al. (2014)).

The model predicted concentrations of Al^{3+}, total Fe, and Ca^{2+} (Figs. 7.6–7.8) that are compliant with the experimental data. In the geochemical algorithm, Fe^{2+} and Fe^{3+} are calculated individually and added later to match the experimental data. The secondary minerals, which precipitated out of solution ($Fe(OH)_3$, Fe(OOH), Fe_2O_3, $Fe(OH)_2$, $FeCO_3$, and $Al(OH)_3$), armored the reactive surfaces of recycled concrete particles and decreased the reactivity.

The calculated porosity values showed that the precipitated secondary minerals consequently reduced the porosity and hydraulic conductivity. Porosity calculation using a porosity meter (Trani and Indraratna, 2010) was not successful in this study because of the internal disturbance of the recycled concrete around the probe tip. Therefore, Eq. (7.6) was used to back-calculate some porosity values from the hydraulic conductivity data to confirm the porosity values calculated from the model (Table 7.4). The results are in agreement with the predicted porosity values from Eq. (7.5), which is evidence of the accuracy of the developed model.

Clogging is maximum at the entrance zone of the column. This behavior was found in previous studies carried out for groundwater treatment methods using zero-valent iron columns (Kamolpornwijit et al., 2004; Li et al., 2005), glass bead columns (Rowe et al., 2000), organic sediment columns (Bilek, 2006), and recycled concrete columns (Regmi et al., 2011b). The

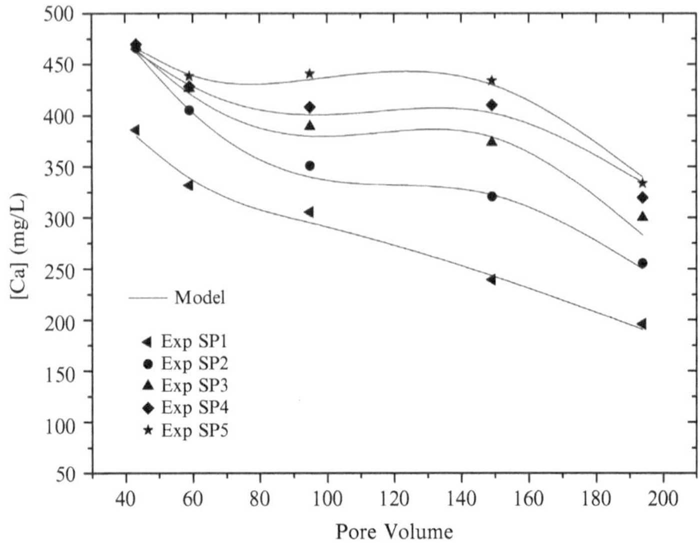

Figure 7.8 Ca^{2+} concentrations of model predicted values versus experimental values at sampling points along the column (Indraratna et al. (2014)).

Table 7.4 Comparison of porosities based on Kozeny–Carmen relationship with the model predictions (Eq. (7.5))[a]

Pore volume	Experimental K (m/day) based on Darcy's law	*n* back-calculated from Kozeny–Carmen equation (Eq. (7.6))	*n* predicted from geochemical model (Eq. (7.5))
43	0.957	0.690	0.690
59	0.919	0.685	0.687
95	0.808	0.673	0.679
149	0.682	0.656	0.668
194	0.628	0.648	0.663

[a]Indraratna et al. (2014).

porosity and hydraulic conductivity changes due to secondary mineral precipitation and dissolution were calculated from Eqs. 7.4 and 7.5 along the column at 100-mm intervals (Figs. 7.9 and 7.10). The maximum reductions in porosity and hydraulic conductivity were observed at Zone 1 (entrance zone) and decreased with distance (i.e., clogging in Zone 1 > Zone 2 > Zone 3, etc.).

The results obtained from model predictions to the field PRB were in good agreement with field measurements. Figures 7.11, 7.12, and 7.13 show the comparisons for pH, Al, and total Fe concentrations, respectively, for model predictions and field results for 2012, 6 years after PRB installation.

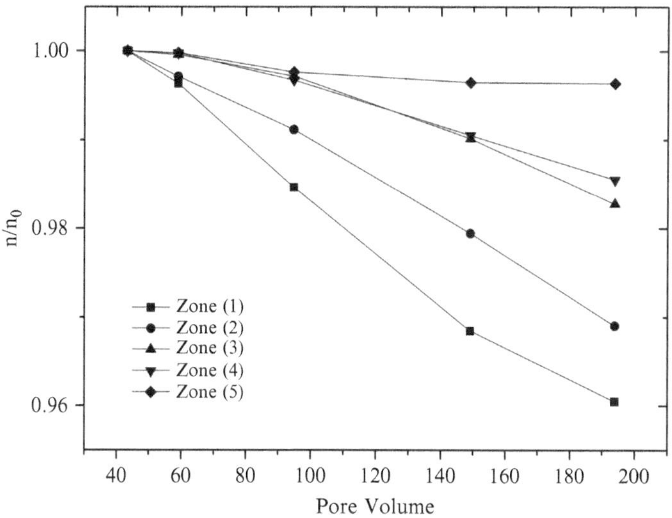

Figure 7.9 Normalized porosity values in Zone 1, SP1–SP2; Zone 2, SP2–SP3; Zone 3, SP3–SP4; Zone 4, SP4–SP5; and Zone 5, SP5–SP6 (Indraratna et al. (2014)).

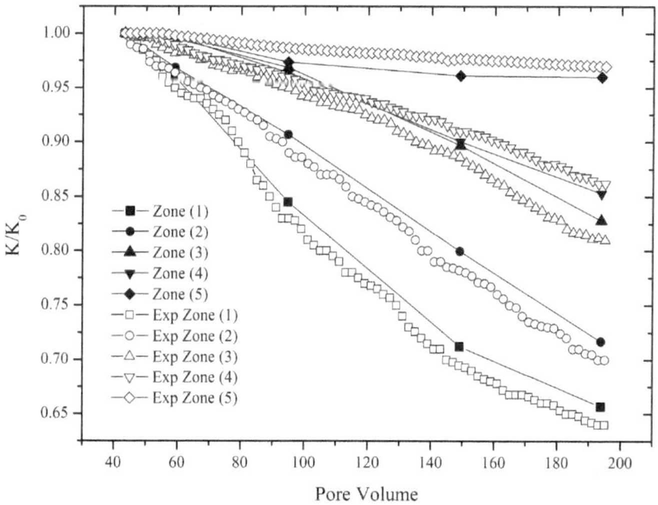

Figure 7.10 Experimental and predicted (normalized) hydraulic conductivity values in Zone 1, SP0–SP1; Zone 2, SP1–SP2; Zone 3, SP2–SP3; Zone 4, SP3–SP4; and Zone 5, SP4–SP5 (Indraratna et al. (2014)).

Up-gradient groundwater pH fluctuated between 3.2 and 4.1 (average of 3.6). Groundwater pH inside the PRB fluctuated from 6.7 to 7.4 (average of 7). Table 7.5 shows the model inputs, outputs, and averaged values of field measurements.

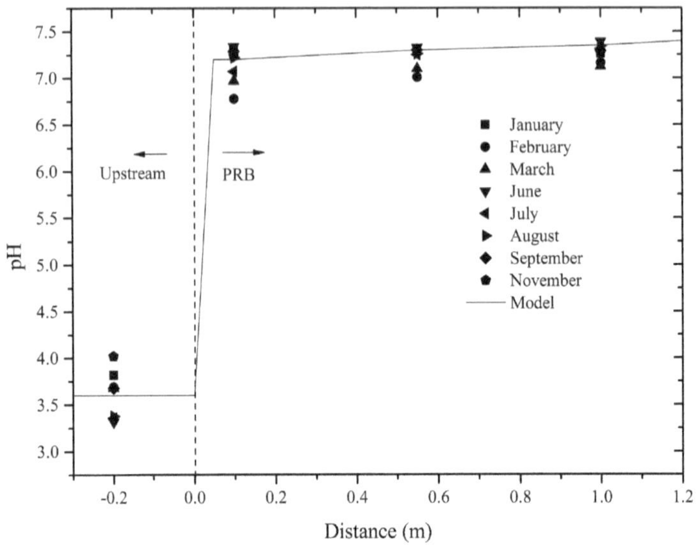

Figure 7.11 Measured and predicted pH values for field PRB for 2012 (Indraratna et al. (2014)).

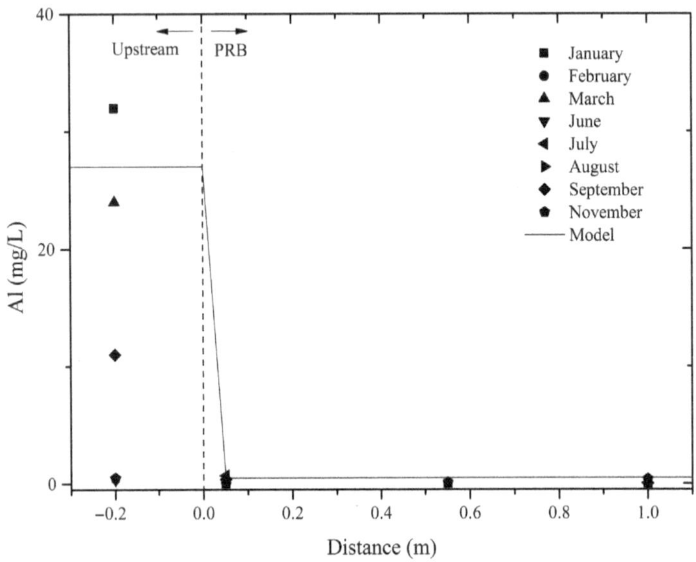

Figure 7.12 Measured and predicted Al concentrations for field PRB for 2012 (Indraratna et al. (2014)).

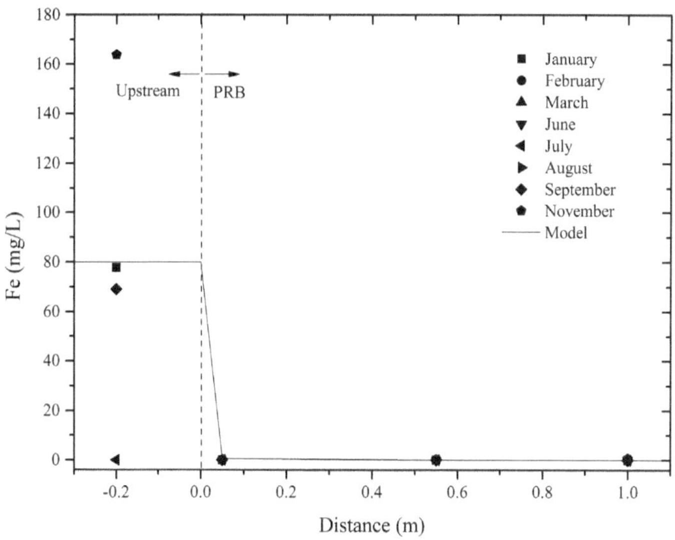

Figure 7.13 Measured and predicted Fe concentrations for field PRB for 2012 (Indraratna et al. (2014)).

Table 7.5 Model predicted and measured pH, Al, and total Fe concentrations in the field PRB[a]

Parameter	Input value	Averaged measured value inside the field PRB	Averaged model predicted value inside the field PRB
pH	3.6	7	7.3
[Al] (mg/l)	27	1	0.5
[Total Fe] (mg/l)	80	1	0

[a]Indraratna et al. (2014).

The model-predicted pH values and the concentrations of Al^{3+} and total Fe are in favorable trends with the observed pH and the concentrations of Al^{3+} and total Fe inside the PRB. The decrease in Al^{3+} and total Fe inside the PRB indicates that they have been precipitated as their oxy/hydroxides, which results in decreased porosity and hydraulic conductivity of the barrier. Nevertheless, the calculated reduction in hydraulic conductivity is only 3% for the period of October 2006 to October 2012. The small decrease in hydraulic conductivity over the 6 years is probably due to the use of coarse-grained recycled concrete aggregates ($d_{50} = 40$ mm) in the PRB, which delays total clogging of the barrier.

7.4 CONCLUSION

The current study focused on the alkalinity generation of Ca-bearing minerals in recycled concrete and the precipitation of secondary minerals during the neutralization process of acidic groundwater and recycled concrete. The resulting changes for the hydraulic properties such as porosity and hydraulic conductivity of crushed recycled concrete in a test column and a pilot-scale PRB were examined. Finite different codes MODFLOW and RT3D were used to simulate the groundwater flow and the reactive transport of major cations. A novel geochemical algorithm was developed to capture the reaction between recycled concrete and acidic groundwater and was fed into RT3D. The model-predicted concentrations of Ca^{2+}, Al^{3+}, and total Fe were in favorable agreement with the experimental and field data. Based on the results reported in this chapter, the following findings were concluded.

Secondary minerals precipitated out of solution were basically $Fe(OH)_3$, Fe(OOH), Fe_2O_3, $Fe(OH)_2$, $FeCO_3$, and $Al(OH)_3$), which formed in the pore space and on reactive surface of the recycled concrete. Clogging was most prominent at the entrance zone of the column; thus, the reduction in porosity and hydraulic conductivity was also most substantial at the entrance zone. The largest porosity reduction was found to be 4% at the entrance zone of the column, and it was 3% in the middle and 0.5% near the exit zone of the column. Therefore, the calculated hydraulic conductivity reductions using the Kozeny–Carmen equation were 34% near the entrance, 27% in the middle, and 4% near the exit zone of the column. The column experiments revealed that recycled concrete aggregates were a promising reactive material to neutralize the acidic groundwater and also for 99% removal of Al and Fe from solution. The performance of the PRB for the 6-year period was satisfactory, where the average pH was approximately 7 inside the PRB. However, the groundwater pH inside the PRB decreased slightly, similar to that in the column experiment, because of the effect of clogging from secondary minerals and due to the exhaustion of the alkalinity-generating minerals in recycled concrete.

7.5 REJUVENATION OF THE PERMEABLE REACTIVE BARRIER BY ALKALINE WASTEWATER INJECTION

Due to the slow decrease in PRB performance concerning the neutralization of acidity and removal of Al and Fe, management of the PRB is essential. Removal or replacement of the recycled concrete is one management option for restoring PRB performance; however, this is a substantial cost. Although criteria used for economic study of previous PRB sites have involved maintenance cost estimates assuming replacement of 25% of the reactive material every 10 years at sites with low precipitation potential and every 5 years for sites with high precipitation potential (Gavaskar et al., 1998), the pilot-scale PRB has been operating for more than 8 years and is performing relatively well. Thus, an alternative to material replacement is needed to increase the long-term performance of the PRB for remediating acidic groundwater.

A project between the UOW ASS research group, Manildra Group, Douglas Partners Pty Ltd., and South East Local Land Services was initiated to find an alternative to increase the long-term performance of recycled concrete in the PRB. Manildra Group's Shoalhaven Starch plant generates a water-based waste stream containing soluble protein and organic nutrients. Its water recovery plant (WRP) consists of an anaerobic digester bulk volume fermenter and a membrane bioreactor/reverse osmosis (RO). This process allows Manildra Group to generate biogas for energy from the organics in the wastewater and reuse 75% of wastewater produced on site. A novel approach to reuse this alkaline wastewater (the reject or retentate from the RO process) is to apply it to the PRB.

7.6 LONG-TERM COLUMN EXPERIMENTS

7.6.1 Reactive material, synthetic acidic groundwater, and alkaline wastewater

Large recycled concrete aggregates, which were from the same batch used in the pilot-scale PRB, were crushed such that what passed through a 4.75-mm sieve and was retained in a 3.35-mm sieve was used to ensure a uniform particle size in the column. The column influent (Table 7.6) was a synthetic acidic solution prepared matching average values of constituents in groundwater from the PRB site measured over 1½ years (Regmi et al., 2009b). Constituents of the wastewater collected from Manildra Group's WRP are given in Table 7.6.

7.6.2 Experimental protocol

The transparent acrylic column of 1.5 l capacity (L = 65 cm, I.D. = 5 cm) (Fig. 7.14) contained a 10-cm silica sand ball after the influent point followed by 50 cm of recycled concrete and another 5-cm sand ball. A geosynthetic membrane separated the sand zones from the inlet and outlet to prevent

Table 7.6 Alkaline wastewater chemistry

Parameter	**Alkaline wastewater**
pH	8.3–8.5
ORP (mV)	299.5
EC (mS/cm)	5.5–6.0
Alkalinity (mg/l as $CaCO_3$)	2460
Na^+ (mg/l)	696
K^+ (mg/l)	85
Ca^{2+} (mg/l)	41
Mg^{2+} (mg/l)	364
Al^{3+} (mg/l)	0.08
Total Fe (mg/l)	1.91
Cl^- (mg/l)	428
SO_4^{2-} (mg/l)	<1

EC, electrical conductivity; ORP, oxygen reduction potential.
Source: Courtesy of the Manildra Group.

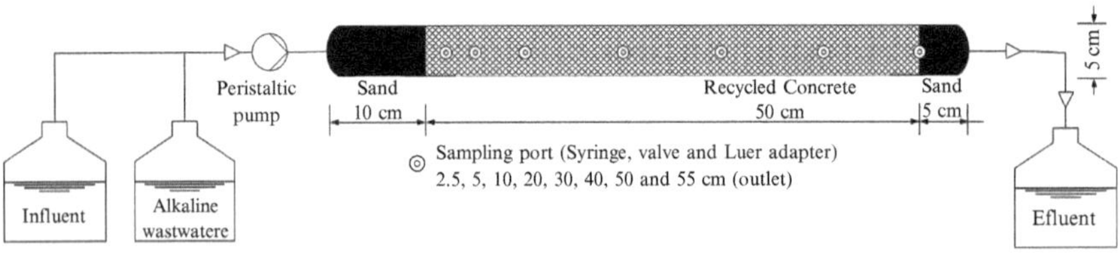

Figure 7.14 Column setup for long-term alkaline injection experiment (Banasiak et al. (2015)).

physical clogging by the sand. The injection process of pumping the alkaline wastewater through the inlet valve allowed the wastewater to pass through the entire column and enabled it to act in regions of concentrated chemical armoring. Table 7.7 summarizes the physical parameters of the column.

The synthetic acidic water was passed through the column at a flow rate of 1.2 ml/min using a Masterflex peristaltic pump. The retention time for

Table 7.7 Alkaline wastewater chemistry

Parameter	**Value**
Flow rate (ml/min)	1.2
Mass of concrete (kg)	1.2
Initial porosity (n_0)	0.71
Pore volume (l)	0.6924
Initial hydraulic conductivity (K_0 m/day)	0.880

Source: Courtesy of the Manildra Group.

one PV is 9.62 h, which is 10 times less than the retention time for a groundwater flow rate of 30 cm/day at the PRB field site. The column was flushed with 4 or 5 PVs of deionized (DI) water before the experiment was begun at room temperature.

Sampling ports (SP1–SP7), fitted with a luer adaptor to allow vacuum sampling during the experiment, were located at 2.5-, 5.0-, 10-, 20-, 30-, 40-, and 50-cm distance along the column from left to right. Effluent and water from the sampling ports were analyzed directly after the samples were collected for pH, electrical conductivity (EC), ORP, and temperature. Samples were then filtered through 0.45-μm cellulose acetate filter paper and stored at 4 °C before analysis. Alkalinity and other chemical analyses were performed following standard methods (American Public Health Association, 1998). Total Fe was measured using atomic absorption spectroscopy. Al^{3+} and Ca^{2+} were measured using ICP-MS (Southern Cross University Environmental Analysis Laboratory).

After the concrete's ANC was significantly reduced (when the effluent pH reached approximately 3), three trial wastewater injections were undertaken: T1, 10 PVs at 20 ml/min; T2, 10 PVs at 10 ml/min to increase reaction time between the wastewater and precipitates in the column; and T3, 20 PVs at 10 ml/min to investigate the effects of increased alkaline wastewater volume.

7.6.3 Changes in porosity and hydraulic conductivity

To calculate changes in column porosity due to Al and Fe precipitation during the experiment, their influent and effluent concentrations were plotted. Because a constant influent concentration was used, the volume of Al^{3+} and Fe retained within the column was obtained by subtracting the integrated data of the influent curve (using OriginPro 8.5) from the integrated data of the effluent curve. This was then multiplied by the column PV (0.6924 l) to give the mass (mg) of Al^{3+} and total Fe precipitated. Using the molar volume of the main precipitates formed within the column (gibbsite $Al(OH)_3$ 31.97 cm^3/mol; goethite FeOOH 20.33 cm^3/mol), the volume used by each mineral (V_P) was calculated. Column porosity (n_t) at different PVs with the change in precipitated minerals with time was calculated using Eq. (7.11):

$$n_t = n_0 - \left(\frac{V_P}{V_T}\right) \tag{7.11}$$

The normalized Kozeny–Carmen equation (Eq. (7.6)) was used to estimate hydraulic conductivity (*K*) with the change of precipitated minerals with time (Li et al., 2006; Pathirage et al., 2012; Indraratna et al., 2014).

7.6.4 Acid neutralization behavior during pre-alkaline injection period

The distinct plateaus in pH (Fig. 7.15(a)), similarly observed by Regmi et al. (2011b), can be attributed to three stages of buffering:

Stage 1: Dissolution of carbonate/bicarbonate alkalinity from the recycled concrete as indicated by the near-neutral plateau at pH 8.00–7.02

Stage 2: Precipitation/dissolution of Al hydroxide minerals at approximately pH 4.2

Stage 3: Precipitation/dissolution of Fe oxyhydroxide minerals at pH 3.4

The high initial effluent pH of 9.56 was attributed to the dissolution of Ca minerals portlandite ($Ca(OH)_2$) and ettringite ($(CaO)_6(Al_2O_3)(SO_3)_3.32H_2O$) from the concrete and the release of alkalinity. However, because the concrete contains a small amount of portlandite (0.3% weight; Regmi et al., 2011a) and the fact that dissolution of ettringite occurs at pH 10.7–9.5 (Alvarez-Ayuso and Nugteren, 2005), this pH was only maintained for 9 PVs. Following this, the pH decreased to 8–7.02 due to the dissolution of calcium aluminate hydrated compounds (C-A-H) such as anorthite ($CaAl_2Si_2O_8$) (Regmi et al., 2011a). The pH then decreased to 4.2 and continued until 212 PVs followed by a decrease to 3.4.

Low initial effluent ORP for SP1–SP3 (351–231 mV) indicated weak oxidizing conditions (Fig. 7.15(b)). ORP increased suddenly in Stage 1 for SP1–SP3 due to rapid reduction in alkalinity at the progressing acid front within the column corresponding to large decreases in pH. The effluent ORP increased to 500 mV between Stage 2 and Stage 3, indicating oxidizing conditions.

The low initial effluent EC (0.63 mS/cm) (Fig. 7.16) was due to flushing of the column with DI water. After $CaCO_3$ was released from the concrete, there was little variation in EC (4.5–5.8 mS/cm) in response to trends of Al^{3+} and total Fe armoring of the concrete. Slight variations in effluent EC during the experiment resulted from slight variations in the batches (60 l) of influent during the long-term experiment.

Alkalinity of the column effluent was due to the release of alkaline minerals from the recycled concrete and the influent chemistry. Before the depletion of alkalinity at 130 PVs (Fig. 7.17(a)), the pH was stable and

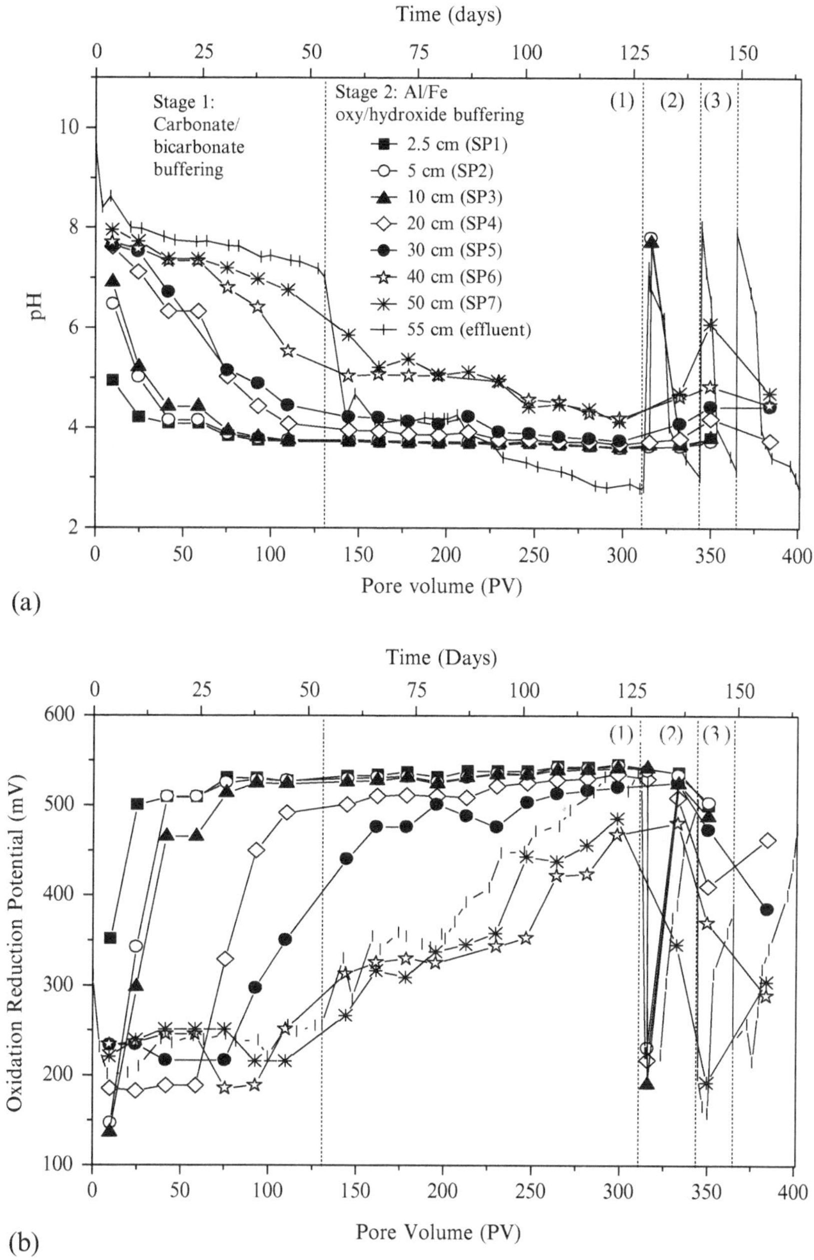

Figure 7.15 (a) pH and (b) oxidation reduction potential of the effluent and at sample points as a function of pore volume and time. Alkaline injection indicated by (1), (2) and (3) (Banasiak et al. (2015)).

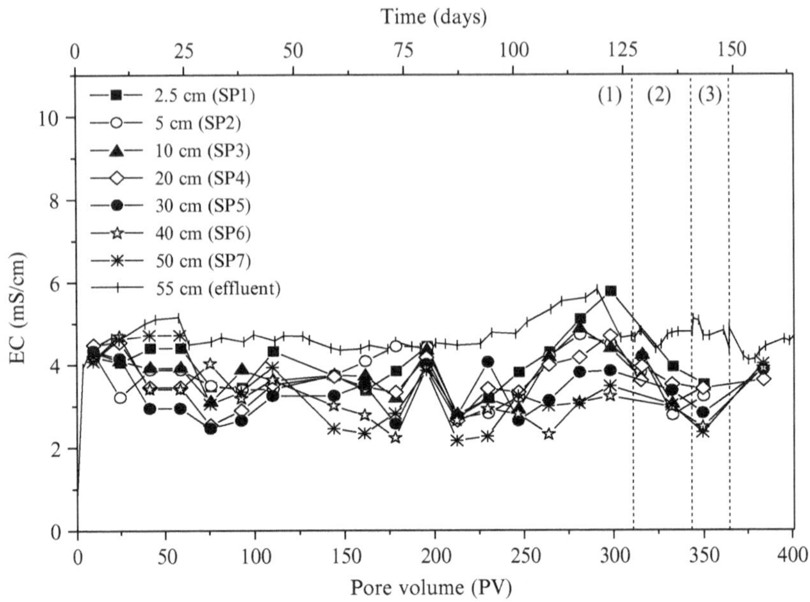

Figure 7.16 Electrical conductivity (mS/cm) of the effluent and at sample points as a function of pore volume and time. Alkaline injection indicated by (1), (2), and (3) (Banasiak et al. (2015)).

near-neutral (pH 7.2–7.3). Following this, there was a decrease in Ca^{2+} released from the recycled concrete (i.e., decrease in Ca^{2+} in the effluent from 466 mg/l at 130 PVs to 345 mg/l at 142 PVs) (Fig. 17(b)). The growth of white and orange precipitates inside the column decreased the reactivity of the recycled concrete and corresponded to the decrease in alkalinity and Ca^{2+} released from the concrete. These precipitates, attributed to Al and Fe oxy/hydroxides, respectively, initially formed near the column inlet and then further along the column as the flow of acidic influent continued.

7.6.5 Metal removal capacity during pre-alkaline injection period

Al^{3+} removal was approximately 100% until 130 PVs, whereby removal decreased to 38.1% (142 PVs; 56 days) (Fig. 7.18), coinciding with the depletion of alkalinity within the column effluent, and the decrease in pH from pH 7 to pH 4.2. Al^{3+} removal was due to the precipitation of Al in hydroxide form (gibbsite ($Al(OH)_3$) and boehmite (AlOOH)) as confirmed by analysis of the precipitates formed on the concrete during past column experiments (Regmi et al., 2011a) and reported in studies of field PRBs

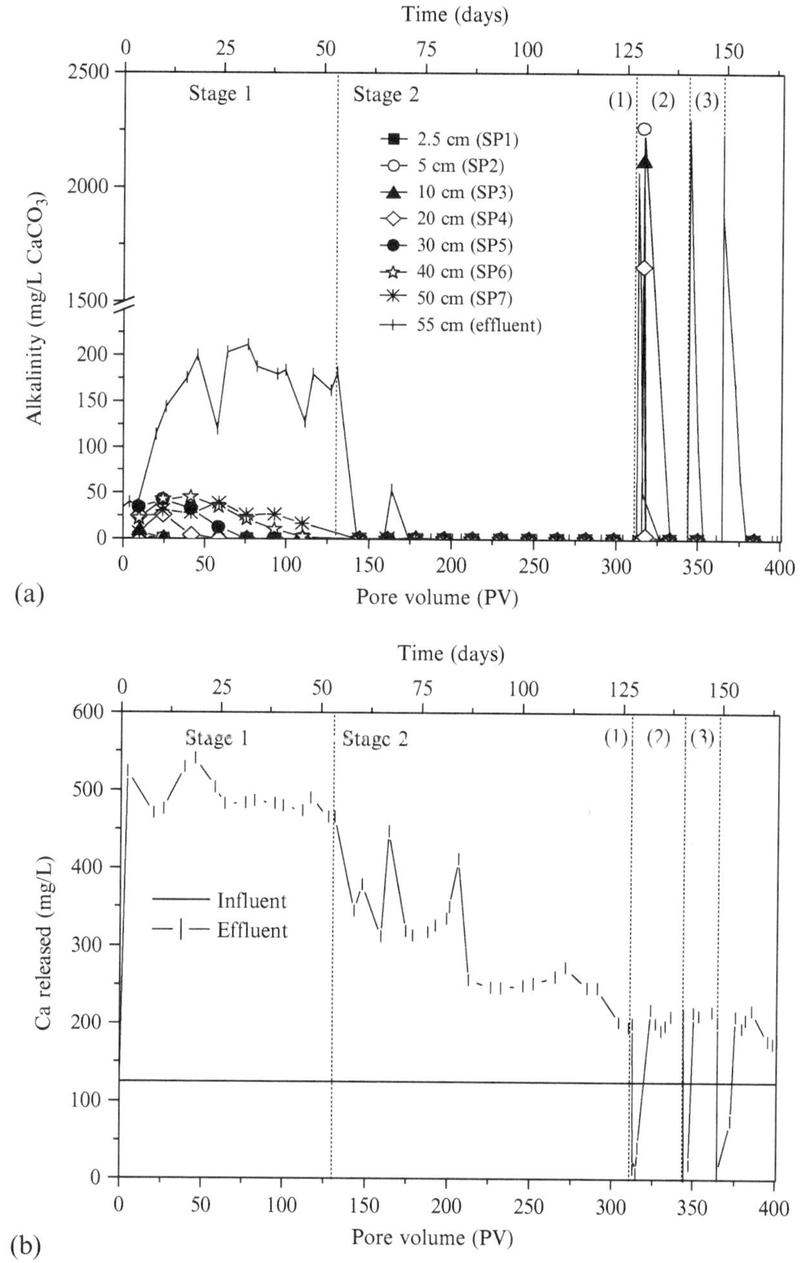

Figure 7.17 (a) Alkalinity and (b) calcium released from the concrete as a function of pore volume and time. Alkaline injection indicated by (1), (2), and (3) (Banasiak et al. (2015)).

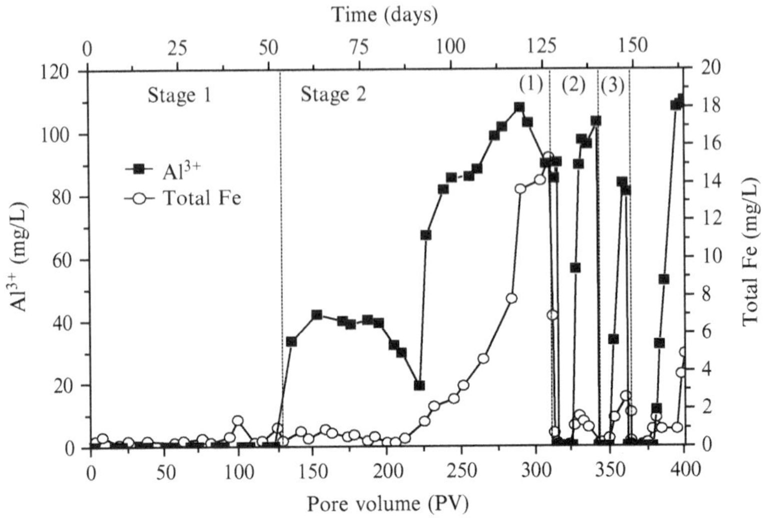

Figure 7.18 Concentration of Al^{3+} and total Fe as a function of pore volume and time. Alkaline injection indicated by (1), (2), and (3) (Banasiak et al. (2015)).

and column tests for acid mine drainage and other groundwater contamination (Blowes et al., 2003; Jurjovec et al., 2002). Fe removal was 99.6% until 212 PVs (93 days), after which it decreased to 89.8% just prior to the alkaline injection trial. Fe precipitated as Fe oxyhydroxide (goethite (FeOOH)). The flow of acidic influent through the column was stopped at 312 PVs (Day 128), when the effluent pH reached 2.83.

7.6.6 Changes in physical properties of the column

Chemical armoring of the concrete by precipitates was studied by observing changes in n/n_t and K/K_0 of the column (Fig. 7.19). Armoring of the concrete until the alkalinity was exhausted (212 PVs) reduced n/n_t of the column, which was similarly observed in other studies on PRBs (Kamolpornwijit et al., 2003; Wilkin et al., 2003; Li et al., 2006). The n/n_t decreased gradually from 0.710 to 0.703 (0.97% reduction), and K/K_0 decreased from 0.880 to 0.856 m/day (2.75%). However, the pores in the column were large enough that complete clogging did not occur.

7.6.7 Alkaline wastewater injection

Approximately 400 l of acidic influent was passed through the column before alkaline injection was attempted. The following trends in effluent pH (Fig. 7.15(a)) were observed after each alkaline injection trial: a sharp

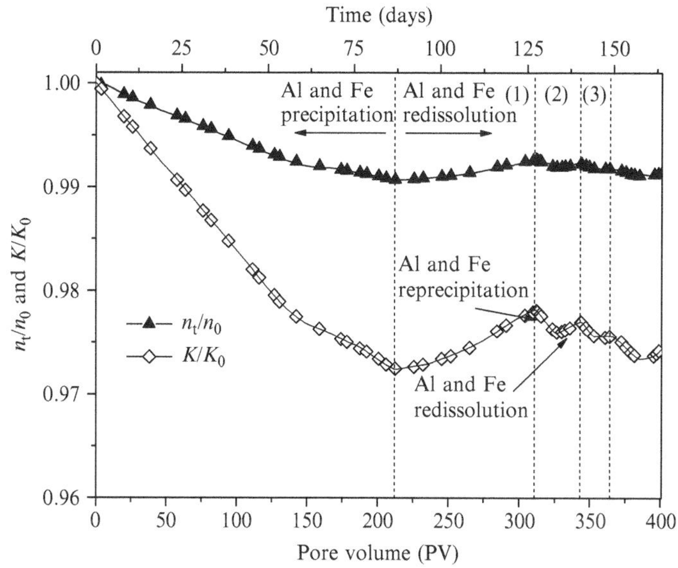

Figure 7.19 Normalized porosity (n_t/n_0) and hydraulic conductivity (K/K_0) within the column (Banasiak et al. (2015)).

but short increase in effluent pH (T1 at 310 PV, pH 2.8–7.2; T2 at 343 PV, pH 3.0–7.1; T3 at 364 PV, pH 3.2–6.7) indicating an increase in the concretes' ANC and also a steady reduction in effluent pH to values typical during pre-injection period 10 PVs after alkaline injection. This indicates the short-term temporary nature of the alkaline rejuvenation after significant chemical armoring of the recycled concrete has occurred. The maximum effluent pH in the post-alkaline injection period (pH 8.04) was similar to the maximum effluent pH in the pre-injection period (pH 8.00).

Within 3–10 PVs following alkaline injection, weak oxidizing conditions prevailed in the column, as shown by the reduction in effluent ORP (T1, 530–226 mV; T2, 501–167 mV; T3, 385–258 mV)) Fig. 7.15 (b)). These are ideal ORP values at which Al and Fe precipitation can occur. There was a return to strong oxidizing conditions within the column concluding the passing of 10–18 PVs of acidic influent. However, the ANC of the concrete benefited from multiple alkaline injection trials. Each rejuvenation attempt produced a slight reduction in the effluent ORP, as denoted by the steady reduction of peak ORP values from 537 mV after T1 to 311 mV after the third injection trial.

There was little variation in the EC of the effluent during the post-alkaline injection period (Fig. 7.16). The range in the effluent EC during the pre-injection period was 4.35–5.81 mS/cm, which was similar to that observed during the post-injection period (4.10–5.11 mS/cm). In addition, the EC of the alkaline wastewater (5.5–6 mS/cm) was very similar to the EC of the acidic influent and effluent during the pre-injection period.

The increase in alkalinity of the column effluent during the post-injection period is due to the chemistry of the alkaline wastewater—that is, alkalinity provided by the wastewater (2460 mg/l as $CaCO_3$)—and not due to the release of alkaline minerals from the concrete. Effluent alkalinity increased significantly 2 or 3 PVs after the injection trials ($CaCO_3$: T1, 0–2044 mg/l; T2, 0–2274 mg/l; T3, 2192 mg/l) and decreased rapidly within 3–6 PVs. (Fig. 7.17(a)). There was no significant difference in the concentration of Ca^{2+} released from the concrete before and after alkaline injection. Ca^{2+} released from the concrete decreased following injection and returned to the pre-injection concentration of 200 mg/l within 4–7 PVs (Fig. 7.17(b)).

The increase in effluent pH and decrease in ORP following each injection led to the precipitation of Al- and Fe-bearing minerals, as shown by the decrease in Al and Fe concentration in the effluent (T1, 90.4–0 mg/l; T2, 103.2–0 mg/l; T3, 80.9–0 mg/l) (Fig. 7.18(a)). When the ORP returned to 315–500 mV, the re-dissolution of Al and Fe precipitates occurred, as indicated by the increase in Al and Fe concentration after 9–11 PVs. The precipitation of Al- and Fe-bearing minerals after alkaline injection caused n/n_t and K/K_0 to slightly decrease (Fig. 7.19). However, due to the liberation of precipitates from within the pores in the column during the first alkaline injection trial, as indicated by precipitates in the effluent container, and the re-dissolution of Al and Fe minerals, n/n_t and K/K_0 increased again. Importantly, the removal of clogging within the pores eliminated undesired flow impedance, allowing the concrete to react with the acidic influent.

7.7 IMPLICATIONS OF WASTEWATER INJECTION FOR PILOT-SCALE PERMEABLE REACTIVE BARRIER

This study demonstrates that the reuse of alkaline wastewater has potential as a management tool for PRB technology in ASS terrain. It is evident that future rejuvenation attempts via injection of alkaline wastewater into a PRB should use trial injection at the initial stages of notable decreases in effluent pH (Stage 1) prior to exhaustion of carbonate/bicarbonate buffering. Further injection of the alkaline wastewater in Stage 2 could promote further alkaline buffering within the PRB. Although it is perceived that increasing alkalinity within the column in the initial stages of treatment may present a noticeable reduction in downstream pH levels for the PRB, enhanced precipitation of Al- and Fe-bearing minerals may increase the extent of chemical armoring, as indicated by the decrease in n/n_t and K/K_0 following the injection trials, and hasten the reduction in ANC of the concrete. However, due to the larger particle size used in the pilot-scale PRB ($d_{50}=40$ mm) compared to that used in the column test, and the fact that the decrease in n/n_t and K/K_0 within the column was marginal, any enhancement of chemical armoring is not expected to be significant prior to exhaustion of the ANC of the concrete.

Should this potential PRB management tool receive the necessary approval for field trials, a proposal will be developed for Manildra Group's WRP for the recycling of its alkaline wastewater for the remediation of local acidic groundwater and previous estuarine swamps. This would allow the complete rejuvenation process to be conducted "on site" using a waste material resource from within Manildra Group's Environmental Farm.

APPENDIX: GEOCHEMICAL ALGORITHM[1]

Bicarbonate buffering (maintenance of an almost neutral pH) and complete removal of Al and Fe from the solution:

$$Ca(OH)_2 + 2H^+ \rightarrow Ca^{2+} + 2H_2O$$

$$Ca(OH)_2 + CO_2(aq) \rightarrow CaCO_3 + H_2O$$

$$CaAl_2Si_2O_8 + 8H^+ \rightarrow Ca^{2+} + 2Al^{3+} + 2H_4SiO_4$$

$$CaCO_3 + 2H^+ \Leftrightarrow Ca^{2+} + H_2CO_3$$

$$CaCO_3 + H_2CO_3 \Leftrightarrow Ca^{2+} + 2HCO_3^-$$

$$CO_2(aq) + H_2O \Leftrightarrow H_2CO_3$$

Dissolution of Ca-bearing minerals and precipitation secondary minerals were assumed to follow the transition state theory with the following expressions:

$$\frac{d\left[m_{Ca(OH)_2}\right]}{dt} = \frac{1}{2}\frac{d[H^+]}{dt} = -\frac{d[Ca^{2+}]}{dt} = r_{1[Ca^{2+}]} = k_{[Ca^{2+}]}\left[\frac{a_{Ca^{2+}}a^2_{OH^-}}{K_{eq,Ca^{2+},OH^-}} - 1\right]$$

[1] Appendix from Indraratna et al. (2014).

$$\frac{d[m_{CaAl_2Si_2O_8}]}{dt}=\frac{1}{8}\frac{d[H^+]}{dt}=-\frac{d[Ca^{2+}]}{dt}=-\frac{1}{2}\frac{d[Al^{3+}]}{dt}$$

$$=r_{[Ca^{2+}Al^{3+}]}=k_{[Ca^{2+}Al^{3+}]}\left[\frac{a_{Ca^{2+}}a^2_{Al^{3+}}}{K_{eq,Ca^{2+}Al^{3+}}}-1\right]$$

$$\frac{d[m_{CaCO_3}]}{dt}=\frac{1}{2}\frac{d[H^+]}{dt}=-\frac{d[Ca^{2+}]}{dt}=-\frac{d[H_2CO_3]}{dt}$$

$$=r_{2[Ca^{2+}]}=k_{[Ca^{2+}]}\left[\frac{a_{Ca^{2+}}a_{CO_3^{2-}}}{K_{eq,Ca^{2+},CO_3^{2-}}}-1\right]$$

$$\frac{d[m_{CaCO_3}]}{dt}=\frac{d[H_2CO_3]}{dt}=-\frac{d[Ca^{2+}]}{dt}=-\frac{1}{2}\frac{d[HCO_3^-]}{dt}$$

$$=r_{3[Ca^{2+}]}=k_{[Ca^{2+}]}\left[\frac{a_{Ca^{2+}}a_{HCO_3^-}}{K_{eq,Ca^{2+},HCO_3^-}}-1\right]$$

$$Fe^{3+}+3H_2O\rightarrow Fe(OH)_{3(S)}+3H^+_{aq}$$

$$\frac{d[Fe^{3+}]}{dt}=-\frac{d\left[m_{Fe(OH)_3}\right]}{dt}=-\frac{1}{3}\frac{d[H^+]}{dt}=r_{1[Fe^{3+}]}$$

$$=k_{[Fe^{3+}]}\left[\frac{a_{Fe^{3+}}a^3_{OH^-}}{K_{eq,Fe^{3+},OH^-}}-1\right]$$

$$Fe^{3+}+2H_2O\rightarrow Fe(OOH)+3H^+_{aq}$$

$$\frac{d[Fe^{3+}]}{dt}=-\frac{d[m_{Fe(OOH)}]}{dt}=-\frac{1}{3}\frac{d[H^+]}{dt}=r_{2[Fe^{3+}]}$$

$$=k_{[Fe^{3+}]}\left[\frac{a_{Fe^{3+}}a_{OOH^{3-}}}{K_{eq,Fe^{3+},OOH^{3-}}}-1\right]$$

$$2Fe^{3+}+3H_2O\rightarrow Fe_2O_3+6H^+_{aq}$$

$$\frac{1}{2}\frac{d[Fe^{3+}]}{dt}=-\frac{d[m_{Fe_2O_3}]}{dt}=-\frac{1}{6}\frac{d[H^+]}{dt}=r_{3[Fe^{3+}]}=k_{[Fe^{3+}]}\left[\frac{a^2_{Fe^{3+}}a^3_{O^{2-}}}{K_{eq,Fe^{3+},O^{2-}}}-1\right]$$

$$Al^{3+}+3H_2O\rightarrow Al(OH)_{3(S)}+3H^+_{(aq)}$$

$$\frac{d[Al^{3+}]}{dt}=-\frac{d\left[m_{Al(OH)_3}\right]}{dt}=-\frac{1}{3}\frac{d[H^+]}{dt}=r_{[Al^{3+}]}=k_{[Al^{3+}]}\left[\frac{a_{Al^{3+}}a^3_{OH^-}}{K_{eq,Al^{3+},OH^-}}-1\right]$$

$$Fe^{2+}+2(OH)^-\leftrightarrow Fe(OH)_{2(S)}$$

$$\frac{d[Fe^{2+}]}{dt}=\frac{1}{2}\frac{d[OH^-]}{dt}=-\frac{d[m_{Fe(OH)_2}]}{dt}=r_{1[Fe^{2+}]}=k_{[Fe^{2+}]}\left[\frac{a_{Fe^{2+}}a^2_{OH^-}}{K_{eq,Fe^{2+},OH^-}}-1\right]$$

$$Fe^{2+}+CO_3^{2-}\leftrightarrow FeCO_{3(S)}$$

$$\frac{d[Fe^{2+}]}{dt}=\frac{d[CO_3^{2-}]}{dt}=-\frac{d[m_{FeCO_3}]}{dt}=r_{2[Fe^{2+}]}=k_{[Fe^{2+}]}\left[\frac{a_{Fe^{2+}}a_{CO_3^{2-}}}{K_{eq,Fe^{2+},CO_3^{2-}}}-1\right]$$

$$Ca^{2+}+CO_3^{2-}\leftrightarrow CaCO_{3(S)}$$

$$\frac{d[Ca^{2+}]}{dt}=\frac{d[CO_3^{2-}]}{dt}=-\frac{d[m_{CaCO_3}]}{dt}=r_{4[Ca^{2+}]}=k_{[Ca^{2+}]}\left[\frac{a_{Ca^{2+}}a_{CO_3^{2-}}}{K_{eq,Ca^{2+},CO_3^{2-}}}-1\right]$$

The overall reactive kinetics for each species in the algorithm are listed as follows:

$$\frac{d[Ca^{2+}]}{dt}=-r_{1[Ca^{2+}]}-r_{[Ca^{2+}Al^{3+}]}-r_{2[Ca^{2+}]}-r_{3[Ca^{2+}]}+r_{4[Ca^{2+}]}$$

$$\frac{d[Fe^{3+}]}{dt}=r_{1[Fe^{3+}]}+r_{2[Fe^{3+}]}+2r_{3[Fe^{3+}]}$$

$$\frac{d[Fe^{2+}]}{dt}=r_{1[Fe^{2+}]}+r_{2[Fe^{2+}]}$$

$$\frac{d[Al^{3+}]}{dt}=r_{[Al^{3+}]}-2r_{[Ca^{2+}Al^{3+}]}$$

$$\frac{d[H^+]}{dt}=2r_{1[Ca^{2+}]}+8r_{[Ca^{2+}Al^{3+}]}+2r_{2[Ca^{2+}]}-3r_{1[Fe^{3+}]}-3r_{2[Fe^{3+}]}-6r_{3[Fe^{3+}]}-3r_{[Al^{3+}]}$$

$$\frac{d[HCO_3^-]}{dt}=-r_{2[Ca^{2+}]}-2r_{3[Ca^{2+}]}+r_{2[Fe^{2+}]}+r_{4[Ca^{2+}]}$$

NOTATION

Φ_k = volume fraction of mineral
μ, C, D = constants
b = aquifer thickness (m)
C = aqueous species concentration (mg/l)
D = hydrodynamic dispersion coefficient (m)
h = head (m)

IAP = ion activity product (depends on the reaction)
K = hydraulic conductivity at time t (m/s)
K_0 = initial hydraulic conductivity (m/s)
k_{eff} = effective rate coefficient ($mol/m^{-3}{}_{bulk}/s$)
K_{eq} = solubility constant (depends on the reaction)
l = length of the column (m)
M_k = mineral molar volume (m^3/mol)
N_m = number of minerals
n_o = initial porosity
n_t = porosity at time t
R_e = retardation coefficient
r = reaction rate ($mol/m^{-3}{}_{bulk}/s$)
R_k = overall reaction rate for the mineral ($mol/m^{-3}{}_{bulk}/s$)
S = storage coefficient
SI = saturation index
T = transmissivity (m^2/s)
t = time (s)
v = seepage velocity (m/s)
V_P = volume occupied by each mineral

ACKNOWLEDGMENTS

The authors acknowledge funding from the Australian Research Council (ARC) and industry partners Southern Rivers Catchment Management Authority, Douglas Partners Pty Ltd., and Manildra Group. Support of Prof. Long Nghiem and Dr. Gyanendra Regmi is greatly appreciated. The authors acknowledge Bob Rowlan and Frank Crabtree (University of Wollongong) for their technical assistance during this study. Some of the technical content was reprinted with permission of Elsevier and the Institution of Civil Engineers, United Kingdom.

REFERENCES

Alvarez-Ayuso, E., Nugteren, H.W., 2005. Synthesis of ettringite: a way to deal with acid wastewaters of aluminium anodizing industry. Water Res. 39, 65–72.
American Public Health Association, 1998. Standard Methods for the Examination of Water and Wastewater, 20th ed. American Public Health Association, Washington, DC.
Banasiak, L.J., Indrarata, B., Lugg, G., Pathirage, P., McIntosh, G., Rendell, N., 2015. Permeable reactive barrier rejuvenation by alkaline wastewater. Environ. Geotech. 2, 45–55.

Bilek, F., 2006. Column tests to enhance sulphide precipitation with liquid organic electron donators to remediate AMD-influenced groundwater. Environ. Geol. 49, 674–683.

Blowes, D.W., Ptacek, C.J., Jambor, J.L., Weisener, C.G., 2003. The geochemistry of acid mine drainage. In: Holland, H.D., Turekian, K.K. (Eds.), Treatise on Geochemistry. Elsevier, Oxford, pp. 149–204.

Dent, D., 1992. Reclamation of acid sulphate soils. Adv. Soil Sci. 17, 79–122.

Gavaskar, A.R., Gupta, N., Sass, B.M., Janosy, R.J., O'Sullivan, D., 1998. Permeable Barriers for Groundwater Remediation: Design, Construction and Monitoring. Battelle Press, Columbus, OH.

Golab, A.N., Peterson, M.A., Indraratna, B., 2006. Selection of potential reactive materials for a permeable reactive barrier for remediating acidic groundwater in acid sulphate soil terrains. Q. J. Eng. Geol. Hydroge. 39 (2), 209–223.

Harbaugh, A.W., 2005. MODFLOW-2005: The U.S. Geological Survey modular ground-water model—The Ground-Water flow process. U.S. Geological Survey Techniques and Methods 6-A16.

Indraratna, B., Sullivan, J., Nethery, A., 1995. Effect of groundwater table on formation of acid sulphate soils. Mine Water Environ. 14, 71–84.

Indraratna, B., Regmi, G., Nghiem, L.D., Golab, A., 2010. Performance of a PRB for the remediation of acidic groundwater in acid sulfate soil terrain. J. Geotech. Geoenviron. Eng. 136 (7), 897–906.

Indraratna, B., Pathirage, P.U., Kerry Rowe, R., Banasiak, L.J., 2014. Coupled hydro-geochemical modelling of a permeable reactive barrier for treating acidic groundwater. Comp. Geotech. 55, 429–439.

Jeen, S.-W., Mayer, K.U., Gillham, R.W., Blowes, D.W., 2007. Reactive transport modeling of trichloroethene treatment with declining reactivity of iron. Environ. Sci. Technol. 41, 1432–1438.

Jeen, S.-W., Amos, R.T., Blowes, D.W., 2012. Modeling gas formation and mineral precipitation in a granular iron column. Environ. Sci. Technol. 46, 6742–6749.

Johnson, R.L., Tratnyek, P.G., Miehr, R., Thoms, R.B., Banstra, J.Z., 2005. Reduction of hydraulic conductivity and reactivity in zero-valent iron columns by oxygen and TNT. Ground Water Monit. Remed. 25, 129–136.

Jurjovec, J., Ptacek, C.J., Blowes, D.W., 2002. Acid neutralization mechanisms and metal release in mine tailings: a laboratory column experiment. Geochim. Cosmochim. Acta 66 (9), 1511–1523.

Jurjovec, J., Blowes, D.W., Ptacek, C.J., Mayer, K.U., 2004. Multicomponent reactive transport modeling of acid neutralization reactions in mine tailings. Water Resour. Res. 40 (11), W1120201–W1120217.

Kamolpornwijit, W., Liang, L., West, O.R., Moline, G.R., Sullivan, A.B., 2003. Preferential flow path development and its influence on long-term PRB performance: column study. J. Contam. Hydrol. 66 (3/4), 161–178.

Kamolpornwijit, W., Liang, L., Moline, L., Hart, T., West, O.R., 2004. Identification and quantification of mineral precipitation in Fe^0 filings from a column study. Environ. Sci. Technol. 38 (21), 5757–5765.

Komnitsas, K., Bartzas, G., Paspaliaris, I., 2004. Efficiency of limestone and red mud barriers: laboratory column studies. Miner. Eng. 17, 183–194.

Li, L., Benson, C.H., 2010. Evaluation of five strategies to limit the impact of fouling in permeable reactive barriers. J. Hazard. Mater. 181, 170–180.

Li, L., Benson, C.H., Lawson, E.A., 2005. Impact of mineral fouling on hydraulic behavior of permeable reactive barriers. Ground Water 43 (4), 582–596.

Li, L., Mergener, E.A., Benson, C.H., 2006. Modeling porosity reductions caused by mineral fouling in continuous-wall permeable reactive barriers. J. Contam. Hydrol. 83 (1/2), 89–121.

Liang, L., Sullivan, A.B., West, O.R., Moline, G.R., Kamolpornwijit, W., 2003. Predicting the precipitation of mineral phases in permeable reactive barriers. Environ. Eng. Sci. 20 (6), 635–653.
Mackenzie, P.D., Horney, D.P., Sivavec, T.M., 1999. Mineral precipitation and porosity losses in granular iron columns. J. Hazard. Mater. 68, 1–17.
Mayer, K.U., Benner, S.G., Blowes, D.W., 2006. Process-based reactive transport modeling of a permeable reactive barrier for the treatment of mine drainage. J. Contam. Hydrol. 85 (3/4), 195–211.
Ouangrawa, M., Molson, J., Aubertin, M., Bussiere, B., Zagury, G.J., 2009. Reactive transport modelling of mine tailings columns with capillarity induced high water saturation for preventing sulfide oxidation. Appl. Geochem. 24 (7), 1312–1323.
Pathirage, P.U., Indraratna, B., Nghiem, L.D., Banasiak, L.J., 2012. Armouring by precipitates and the associated reduction in hydraulic conductivity of recycled concrete aggregates used in a novel PRB for the treatment of acidic groundwater. Paper presented at the 11th Australia–New Zealand Conference on Geomechanics: Ground Engineering in a Changing World, Melbourne, Australia.
Regmi, G., Indraratna, B., Nghiem, L.D., 2009a. Effective remediation of groundwater in acid sulphate soil terrain. Aust. Geomech. 44, 33–40.
Regmi, G., Indraratna, B., Nghiem, L.D., 2009b. Long-term performance of a permeable reactive barrier in acid sulphate soil terrain. Water Air Soil Poll. 9 (5/6), 409–419.
Regmi, G., Indraratna, B., Nghiem, L.D., Banasiak, L.J., 2011a. Evaluating waste concrete for the treatment of acid sulphate soil groundwater from coastal floodplains. Desal. Water Treat. 32 (1–3), 126–132.
Regmi, G., Indraratna, B., Nghiem, L.D., Golab, A., Guruprasad, B., 2011b. Treatment of acidic groundwater in acid sulphate soil terrain using recycled concrete: column experiments. J. Environ. Eng. 137 (6), 433–443.
Rowe, R.K., 2005. Long-term performance of contaminant barrier systems. Geotechnique 55 (9), 631–678.
Rowe, R., Armstrong, M., Cullimore, D., 2000. Particle size and clogging of granular media permeated with leachate. J. Geotech. Geoenviron. Eng. 126 (9), 775–786.
Rumer, R.R., Ryan, M.E., 1995. Barrier Containment Technologies for Environmental Remediation Applications. Wiley, New York.
Steefel, C.I., Lasaga, A.C., 1994. A coupled model for transport of multiple chemical-species and kinetic precipitation dissolution reactions with application to reactive flow in single-phase hydrothermal systems. Am. J. Sci. 294 (5), 529–592.
Trani, L.D.O., Indraratna, B., 2010. Use of impedance probe for estimation of porosity changes in saturated granular filters under cyclic loading: calibration and application. J. Geotech. Geoenviron. Eng., ASCE. 136, 1469–1474.
Walter, A.L., Frind, E.O., Blowes, D.W., Ptacek, C.J., Molson, J.W., 1994. Modeling of multicomponent reactive transport in groundwater: model development and evaluation. Water Resour. Res. 30 (11), 3137–3148.
Wilkin, R.T., Puls, R.W., Sewell, G.W., 2003. Long-term performance of permeable reactive barriers using zero-valent iron: Geochemical and microbiological effects. Ground Water 41 (4), 493–503.
Yabusaki, S., 2001. Multicomponent reactive transport in an in situ zero-valent iron cell. Environ. Sci. Technol. 35 (7), 1493–1503.

CHAPTER 8

Improving Geotechnical Properties of Closed Landfills for Redevelopment Using Chemical Stabilization Techniques

A Case Study on Samples of a Landfill Site in Southwest of Sydney

Behnam Fatahi, Hadi Khabbaz, Behzad Fatahi
Centre for Built Infrastructure Research, School of Civil and Environmental Engineering, University of Technology Sydney (UTS), Sydney, Australia

8.1 INTRODUCTION

Landfilling is the most predominant method of disposing municipal solid wastes (MSW). With an expanding world and the remarkable growth of industry, the demand for larger and higher capacity landfills is rapidly increasing. As available space becomes scarce in urban areas, development on the top of or adjacent to old landfills has become increasingly common. Landfill design and construction technology has rapidly progressed during the past two decades in response to more strict controlling requirements. However, before any construction on top of closed landfills occurs, properties of decomposed waste materials should be improved as required. Developers are not interested in constructing over lands containing MSW due to uncertainties with regard to their engineering properties. Municipal solid wastes are also considered problematic due to their low shear strength, high amount of moisture content, and compressibility. There is a high demand to reuse these sites for redevelopment despite potential risks to human health and the environment.

The stress–strain response of MSW requires further study to identify the level of sustainability between shear strength and deformation along failure surfaces. This chapter presents the experimental results and numerical

Chemical, Electrokinetic, Thermal, and Bioengineering Methods
http://dx.doi.org/10.1016/B978-0-08-100191-2.00008-3

analyses for the use of fly ash and quicklime in improvement of the geotechnical properties of MSW. Compressibility, compressive strength, shear strength parameters, permeability, stiffness, and the brittleness index of MSW are examined.

8.2 BACKGROUND

8.2.1 Settlement of MSW

Bjarngard and Edgers (1990) noted that total settlement of MSW landfills can be between 25% and 50% of their original thickness. Some settlement occurs during filling in the short term, and a significant amount of settlement occurs in the long term under constant load. Landfill settlement is the result of load and biodegradation.

Sowers (1973) utilized primary and secondary compression equations of soils to determine settlements of solid waste landfills. The primary compression of MSW includes physical compression of particles and consolidation. Physical compression occurs soon after application of loads. Therefore, primary compression occurs within a period of a few days to a few weeks. The time-dependent secondary settlement, which includes all creep effects (i.e., the mechanical compression under a constant stress) and those relating to degradation (both chemical and biological), begins after the primary settlement is complete and generally continues for a longer period.

Table 8.1 presents the variation of compression ratio of MSW, taken from previously published articles. In addition, Bareither et al. (2012) presented the results of compression tests in 305-mm cells on fresh and decomposed waste. The results indicated that compression ratio (C'_c) values were in the range of 0.22 to 0.28 for the stress range between 25 and

Table 8.1 Compression ratio (C'_c) of MSW from the literature

Reference	C'_c
Anderson et al. (2004)	0.17–0.23
Hossain et al. (2003)	0.16–0.37
Marques et al. (2003)	0.073–0.132
Landva et al. (2000)[a]	0.17–0.24
Gabr and Valero (1995)[a]	0.2–0.23
Wall and Zeiss (1995)[a]	0.21–0.25
Sowers (1973)[a]	0.1–0.41

[a]Data presented by Landva et al. (2000).

100 kPa. Change in C'_c was related to the waste compressibility index, which was a function of water content, dry unit weight, and biodegradable organic waste content such as cardboard and food wastes. Chen et al. (2009) reported that C'_c) reduces with decreasing organic content. Kavazanjian et al. (1999) and Dixon et al. (2008) also reported that C'_c reduces by increasing incompressible waste materials.

Table 8.2 summarizes the recommended coefficient of secondary compression (C_α) values for different conditions of waste materials (Sharma and De, 2007). External loads were applied during or after closure of landfills. Due to different rates of degradation with time, using an average value of C_α for the whole period of secondary compression may not be reasonable. Hence, further investigations are required to determine the values of C_α according to MSW conditions.

8.2.2 Strength of MSW

Characterization of shear strength of MSW is an important stage in reliable stability analyses of landfills. A comprehensive laboratory testing program employing direct shear, triaxial was carried out by Bray et al. (2009) on municipal solid waste retrieved from a landfill in San Francisco. The results of direct shear tests showed that the shear strength of MSW was best characterized by a cohesion of 15 kPa and a friction angle of 36° at normal stress of 100 kPa, with the friction angle decreasing by 5° for every log cycle increase in the normal stress.

Reddy et al. (2009) presented the results of a series of direct shear tests executed on fresh MSW collected from the working face of Orchard Hills landfill, located in Illinois. According to their results, drained cohesion

Table 8.2 Recommended values of C_α parameters

C_α under external load, $C_{\alpha(EL)}$	
Overall range of C_α	0.01–0.07
With pretreatment using DDC	0.014
With pretreatment using RC	0.03
With pretreatment (surcharge)	0.045
C_α under self-weight, $C_{\alpha(SW)}$	
Fresh waste	0.014–0.06
Waste during active decomposition	0.1–0.34

DDC, deep dynamic compaction; RC, roller compaction.
Source: After Sharma and De (2007).

values and the drained friction angles were in the range of 31–64 kPa and 26–30°, respectively. The triaxial compression results obtained by Bray et al. (2009) showed that MSW shear strength mobilized at 5% strain can be characterized by a secant friction angle between 34° and 44° for confining stresses up to 400 kPa.

8.2.3 Water content of MSW

A wide range of data on the water content of MSW has been reported by researchers. Tchobanoglous et al. (1993) reported water content of fresh MSW in U.S. landfills between 15% and 40% on a wet weight basis, depending on local climatic conditions at the time of waste collection and waste composition. They noted that MSW with high a content of food waste or organic matter also had high water content—up to 74%.

Zornberg et al. (1999) measured an average water content of 28% for 80 samples of waste collected from depths down to 65 m from a landfill in Southern California. The estimated unit weight of the waste was 12–14 kN/m^3. A slight increase in water content and unit weight was observed with depth. Skhiri et al. (2006) determined the water content of MSW cores taken by drilling at different distances from a leachate injection well. After 1 year of leachate injection, the water content had increased from 27% to 33–39% at a depth of 6–8 m and a distance of 5 m from the leachate injection well.

8.2.4 Porosity of MSW

Beaven (2000) examined the drainable porosity of fresh and aged MSW in a 2-m-diameter compression cell. A reduction in drainable porosity with stress level of less than 200 kPa was reported. For both types of waste, the porosity was approximately 45% at stresses higher than 200 kPa. Zeiss and Major (1993) estimated waste porosities in a range of 47–58% for the waste compacted to densities of 0.17 to 0.3 $tons/m^3$.

8.2.5 Permeability of MSW

Decomposed municipal solid waste is recognized as a material with high moisture content. Many traditional or old landfills have no proper leachate collection or cover system. Jain et al. (2006) presented results from various investigations on the permeability of MSW, combining field and laboratory-based tests. The coefficient of permeability of MSW, reported in the literature, varies between approximately 1×10^{-3} and 1×10^{-9} m/s, although typically values are in the range 10^{-5} to 10^{-6} m/s.

Landva and Clark (1990) conducted large-scale permeability tests in pits at the surface of different landfills in Canada. Permeability was measured based on the rate of decrease in water level and the flow nets appropriate to each stage. They reported a coefficient of permeability ranging between 4×10^{-4} and 1×10^{-5} m/s. The unit weights of the waste from the pits were typically in the range 10–14 kN/m^3.

Hudson et al. (2009) studied the effect of anisotropy on the hydraulic conductivity of MSW. A wide range in the ratio of horizontal to vertical hydraulic conductivity, k_h:k_v, has been reported. Tests have been performed on 20-year-old waste and on fresh waste in the Pitsea cell. All tests showed higher horizontal than vertical hydraulic conductivities. The ratio was between 5 and 7 at low stress and approached 10 at higher stresses. Fresh sample showed slightly greater k_h:k_v ratios compared to the aged sample, which may be attributed to a slight breakdown in the anisotropic structure of waste with decomposition.

8.3 CHEMICAL STABILIZATION

8.3.1 Improvement techniques

The deep dynamic compaction (DDC) technique is a common ground improvement technique because of its economical application. Zekkos and Flanagan (2011) reviewed 64 case histories with respect to the effectiveness of DDC on MSW landfill sites. With DDC, large voids are reduced, and subsequently other techniques such as fly ash–lime grouting can further decrease the remaining smaller voids. In addition, the lime/fly ash slurry injection has a significant impact on protecting groundwater, neutralizing leachate. Several research works have been published on the chemical stabilization of problematic soils incorporating some waste products such as carpet fibers (Fatahi et al., 2012, 2013a, b). Replacement of cement with by-product materials such as fly ash can further reduce the expenses associated with stabilization. Many researchers have reported the application of fly ash in geotechnical projects (Kawasaki et al., 1981; Kehew, 1995; Kitazume et al., 2000; Hokmabadi et al., 2014a, b). Horpibulsuk et al. (2012) reported that from an economic and environmental standpoint, rich materials in slaked lime ($Ca(OH)_2$, calcium hydroxide) can be treated together with pozzolanic materials, such as fly ash, to develop a cementitious material. A series of laboratory permeability tests on decomposed MSW, using chemical agents, have been conducted to quantify the behavior of stabilized MSW samples.

8.3.2 Activation of fly ash with lime

Fly ash is a major by-product of the combustion of pulverized coal. Fly ash mainly consists of oxides of silicon, aluminum, iron, and calcium. According to ASTM C618 (ASTM International, 2012), fly ash is classified as either Class C or Class F. Class C fly ash is categorized as high-calcium fly ash, having cementitious properties due to including more than 20% free lime. Class F ashes are obtained from bituminous coals composing of alumino-silicate glass, with quartz, mullite, and magnetite. Quicklime (CaO) is produced by calcination of limestone ($CaCO_3$) at 1100 °C. High amounts of CO_2 (nearly 40%) are emitted during the calcination process (Rogbeck et al., 2008). After quicklime is exposed to water, it reacts and forms hydrated lime ($Ca(OH)_2$).

The reaction of fly ash with lime results in a quick improvement of the soil mechanical characteristics during stabilization. It decreases the moisture content of the soil, which has an immediate stabilizing effect; increases soil pH, which is required for secondary pozzolanic reactions; and produces heat, which speeds the chemical reactions (Janz and Johansson, 2002). Many activation techniques have been recommended by researchers to treat unstable soils with fly ash. The main aim of these attempts is to improve the reactivity of the pozzolan in order to increase the stability and mechanical characteristics of the integrated product. Curing at high temperatures (Shi and Day, 1993), alkali activation (Xie and Xi, 2001), and chemical activation (Shi and Day, 1995) are some of the approaches that have been examined to achieve that goal.

Extensive investigations have been performed to examine fly ash activation based on a chemical technique, particularly with utilization of lime. Brooks et al. (2011) reported that the addition of fly ash and lime to clay soils reduced plasticity characteristics and swell potential, and it improved the compressive strength of the soils. Kumar and Sharma (2004) illustrated that by mixing fly ash with expansive soil, the optimum water content of mixture was reduced and the maximum dry density increased. Huang and Cheng (1986) employed the fly ash and hydrated lime mixture to determine the optimum amount of hydrated lime for accelerating the reaction of fly ash at various ages. Shi (1999) examined the outcomes of hydrated lime and also quicklime mixed with pozzolan/cement and found that cement mixed with quicklime exhibited significantly higher strength than the cement mixed with hydrated lime. Gray and Lin (1972) showed that the strength gain ratio of lime-treated fly ash was highly dependent on the curing temperature.

8.3.3 Effect of fly ash and lime on settlement and strength of soils

Several researchers conducted experimental studies to treat and stabilize different types of problematic soils using lime and fly ash in combination or separately. Phanikumar and Sharma (2007) reported laboratory test data on the effect of fly ash content on the volume change of two expansive clay samples and one non-expansive clay. They reported that for 20% fly ash content, the compression index of both expansive and non-expansive clays decreased by approximately 50% and the coefficient of secondary consolidation of both clays decreased by 40%. Okoro et al. (2011) performed a study on the consolidation characteristics of two soil samples stabilized with Class C fly ash and lime. Class C fly ash and lime stabilization reduced the compressibility of soil due to new cementing products in the stabilized specimens. They concluded that both the compression index (C_c) and the swelling index (C_s) decreased due to fly ash–lime stabilization. The changes in C_c and C_s varied with the type of stabilizing agent. Tastan et al. (2011) presented the results of the tests performed on three different organic soils blended with fly ash at in different percentages and with two different water contents. They reported that the unconfined compressive strength of organic soils increased using fly ash, but the amount of increase depended on the type of soil and the characteristics of the fly ash. Furthermore, resilient moduli of organic soils also significantly improved. The authors attributed the increases in strength and stiffness to cementing caused by pozzolanic reactions, although the reduction in water content caused by the addition of dry fly ash solid also contributed to the gain in strength.

8.4 EXPERIMENTAL PROGRAM

8.4.1 Materials

Samples of 30-year-old decomposed municipal solid waste, utilized in this study, were taken from two test pits from the Bankstown landfill site southwest of Sydney. Test pit 1 (TP1) and test pit 2 (TP2) were excavated down to 4.6 and 4 m deep, respectively.

Depending on the size of the prepared specimen, limitations on the maximum recommended particle size exist. Typically, the largest particle size in a representative specimen should be one-sixth the specimen diameter. Therefore, the waste samples were sieved, and particles smaller than 9 mm were employed for tests. The decomposed waste material used for this research was soil-like material that contained mainly soil and degraded

Table 8.3 Chemical composition of the MSW samples used in this study

Element	Carbon	Hydrogen	Oxygen	Nitrogen	Sulfur	Ash
%	43.7	6.7	39.1	0.8	0.2	9.5

Table 8.4 Chemical composition of Eraring fly ash (class F)

Material	Al_2O_3	CaO	Fe_2O_3	K_2O	MgO	MnO	Na_2O	P_2O_5	SiO_2	TiO_2
%	25.50	2.27	3.92	1.24	0.69	0.08	0.52	0.33	64.20	0.97

organic waste such as cardboard, wood, and food. Table 8.3 present the general chemical properties of the used MSW. Considering the size of particles, the materials were classified as silty sand (SM) according to the Unified Soil Classification System. The specific gravity of solids (G_s) was 1.8. The grain size distribution showed that 22% of particles were in the range of gravel, 44% in the range of sand, and 34% were fine-grained material with a uniformity coefficient (C_u) of 58. Atterberg limits of the fine portion of material were 38% liquid limit (LL) and 28% plastic limit (PL), which yielded a plasticity index (PI) of 10%. The average organic content and natural moisture content of the samples were 35% and 22%, respectively. Table 8.4 provides a general chemical composition of Eraring fly ash (Class F), which was used for the current project. The specific gravity of the employed fly ash was 2.1. In this study, quicklime (calcium oxide) was used to improve the reactivity of the applied fly ash.

8.4.2 Mixing of materials

Treated waste samples smaller than 9 mm were prepared by mixing various contents of quicklime–fly ash but with a constant ratio of 1:3, as shown in Table 8.5. Following this preparation, the specimens were mixed thoroughly.

Table 8.5 Moisture content of MSW samples

FA (%)/QL (%)	MC before mixing (%)	MC just after mixing (%)	MC after 28 days curing (%)
5/1.7	21	19	18.5
10/3.3	21	18	16.5
15/5	21	16.5	15
20/6.7	21	15	13.5

FA, fly ash; MC, moisture content; QL, quicklime.

A mechanical mixer was used for the mixing of the waste material with fly ash and quicklime. Due to the naturally high amount of moisture content of MSW samples, which was wetter than the optimum moisture content, and the water absorption characteristics of fly ash and quicklime, the dry mixing method (i.e., not adding extra water to MSW samples during the mixing process) was applied. Furthermore, mixing of fly ash and quicklime with MSW specimens in their natural moisture content can simulate the field application condition. To achieve these purposes, certain contents of fly ash–quicklime were mixed with MSW samples. All tests were conducted in an air-conditioned laboratory with limited temperature fluctuation. Table 8.5 presents the moisture content of MSW samples by various amounts of fly ash–quicklime. After mixing of the material, the specimens were prepared for permeability tests to determine the coefficient of permeability of treated and untreated MSW samples for selected fly ash–quicklime contents.

8.4.3 Compaction tests

The maximum dry density and the optimum water content of the untreated MSW samples with particles smaller than 9 mm were measured with the application of 596 KJ/m^3 energy (Standards Australia, 2008). Then, the treated MSW samples were prepared by mixing MSW materials and fly ash–quicklime in various ranges. Again, the optimum water content and the maximum dry density of the treated MSW samples were determined. Based on the fact that MSW material in landfills is less compacted than normal soil layers, a series of revised compaction tests with the application of 357 KJ/m^3 energy (15 blows instead 25 blows per layer) were performed, and the revised maximum dry density values were used in sample preparation. Furthermore, by comparing the results of standard and revised compaction tests, the effect of the number of blows on the maximum dry density and the optimum moisture content could be observed. Table 8.6 demonstrates the

Table 8.6 Compaction test results

MSW mix No.	Fly ash (%)	Quicklime (%)	OMC (%)[a]	MDD (kg/m^3)[a]	OMC (%)	MDD (kg/m^3)
1	0	0	17.5	1132	15.7	1243
2	5	1.7	16.2	1151	14.5	1264
3	10	3.4	15.7	1167	14	1276
4	15	5	15.2	1175	13.6	1287
5	20	6.7	14.8	1182	13.3	1295

[a]Revised with 15 blows per layer.
MDD, maximum dry density; OMC, optimum moisture content.

compaction test results for untreated and treated MSW samples with various fly ash–quicklime content. Each test was repeated three times, and the average data were reported in this table. Note that all the prepared samples for this study were also compacted to achieve the revised maximum dry density. Referring to Table 8.6, with regard to the revised compaction tests, by mixing MSW samples with 26.7% fly ash–quicklime, the maximum dry density increased from 1132 kg/m^3 to 1182 kg/m^3 and the optimum moisture content decreased from 15.7% to 13.3%. Note that the increase in the unit weight and the reduction in the moisture content are two important factors in stabilization of the closed landfill sites.

8.4.4 Specimen preparation and testing procedure

After conducting the mixture process of fly ash–quicklime with waste materials, treated and untreated specimens, 50 mm in diameter and 100 mm in height, were prepared in three layers using the tamping technique to reach the desired dry density. The specimens were prepared to estimate permeability parameters during consolidation using the triaxial apparatus. Each test was repeated three times, and the average data were reported. After specimen preparation and curing for a period of 28 days, the specimens were set up in the equipment. In triaxial cell, all specimens were fully enveloped with a thin rubber membrane and sealed by rubber O-rings at the top and bottom ends. The test cell was assembled by placing the reinforced Perspex cylinder cell wall and the top assembly. Flexible tubings were attached from the panel to the base assembly ports. The space between the specimen and the cell wall was filled with the de-aired water by applying positive pressure to water in a static tank. The saturation process started after applying the back pressure to the top and bottom of the specimen. This was conducted by setting the confining pressure to 25 kPa and setting the water pressure to 20 kPa flowing into the top and bottom of the specimen. The specimen subjected to this condition left for a period of time until a B-value of 0.95 or higher was reached. This value was reached after approximately 1 day (20–24 h). Once the specimen was saturated, the consolidation process started. The confining pressure increased so that the difference between the confining pressure and the back pressure matched the desired effective consolidation stress. Because the thickness of waste in the landfill was 20 m and the average unit weight of decomposed municipal waste material was assumed to be approximately 15 kN/m^3, the maximum effective confining pressure of 300 kPa was chosen for the tests (i.e., 50, 100, 150, 200, and 300 kPa). A series of isotropic consolidation tests were performed. The pore pressure readings at specified times were recorded. These data were used to verify the completion of the consolidation process. The hydraulic conductivity was measured for flow parallel to the long axis of the sample (i.e., perpendicular to the direction of compaction). The permeate was Sydney tap water, having high quality with minimal impurities. More details are provided in Fatahi (2013).

8.5 LABORATORY RESULTS AND DISCUSSION

8.5.1 Stress–strain response

The average stress–strain responses of MSW specimens in the triaxial compression tests are depicted in Figs. 8.1–8.3 for untreated and treated specimens with various percentages of fly ash–quicklime. Based on these figures, increasing fly ash–quicklime content causes a significant change in the peak strength, residual response, stiffness, and brittleness of MSW specimens. For example, under effective confining pressure of 300 kPa, the peak principal stress difference of MSW specimens increased from 600 kPa (without treatment) to 1150 kPa (mixed with 26.7% fly ash–quicklime). Comparing the curves indicates that by increasing fly ash–quicklime content in MSW specimens, the shear strength and the stiffness significantly increased, whereas the residual strength experienced a 30% increase by adding fly ash–quicklime admixture (up to 13.3%) and remained relatively constant after any extra addition of fly ash–quicklime.

Figure 8.4 provides an additional analysis of the secant modulus of elasticity at failure stress (E_f) and at 50% of the failure stress ($E_{50\%}$) for untreated

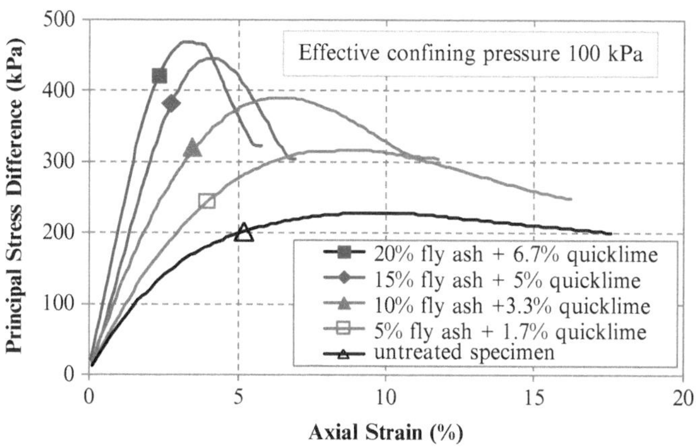

Figure 8.1 Stress–strain response under effective confining pressure of 100 kPa.

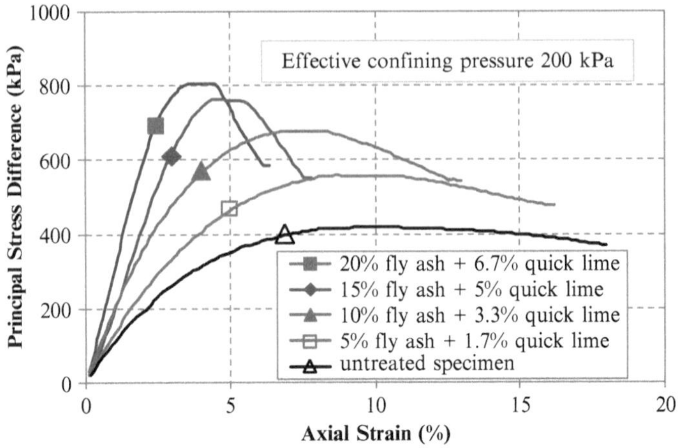

Figure 8.2 Stress–strain response under effective confining pressure of 200 kPa.

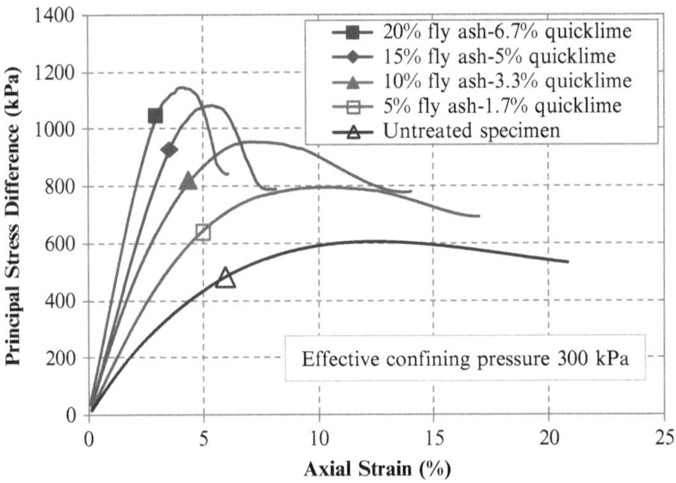

Figure 8.3 Stress–strain response under effective confining pressure of 300 kPa.

and treated MSW specimens under different effective confining pressures. The values of E_f and $E_{50\%}$ were calculated using Eqs. (8.1) and (8.2):

$$E_f = \frac{(\sigma_1' - \sigma_3')_{max}}{\varepsilon_f} \tag{8.1}$$

$$E_{50\%} = \frac{(\sigma_1' - \sigma_3')_{max}}{2 \times \varepsilon_{50\%}} \tag{8.2}$$

where ε_f is the axial strain when the principal effective stress difference is equal to $(\sigma_1' - \sigma_3')_{max}$ and $\varepsilon_{50\%}$ is the axial strain when the principal effective

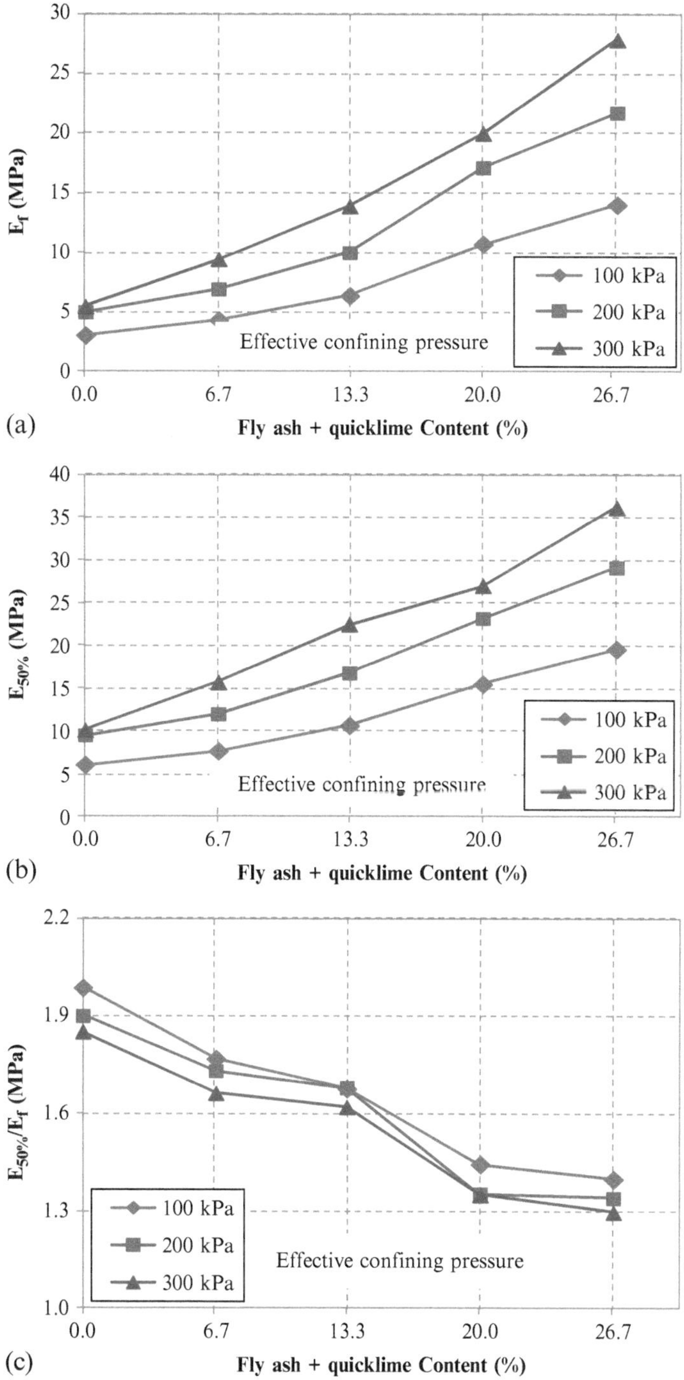

Figure 8.4 Variation of Young's modulus at failure stress (a) and at 50% failure stress (b) and the stiffness ratio (c) of untreated and treated MSW specimens. *(Source: After Fatahi and Khabbaz (2013a)).*

Table 8.7 Effective peak and residual strengths and elastic parameters for untreated and treated MSW in triaxial compression tests

FA (%)	QL (%)	Cohesion, C'_f (kPa)	Friction angle, φ'_f (°)	C'_{res} (kPa)	φ'_{res} (°)	E'_f (MPa)	E'_{50} (MPa)
0	0.0	11	29	11	27.0	3.1–5.5	6.1–12.1
5	1.7	20	33	11	31.0	4.3–9.4	7.7–15.7
10	3.3	26	36	17	33.0	6.4–13.9	10.7–21.5
15	5.0	29	38	17	33.5	10.7–20	15.5–27.0
20	6.7	30	39	16	34.5	14–27.9	19.6–36.1

stress difference is equal to $\frac{(\sigma'_1-\sigma'_3)_{max}}{2}$. It can be inferred from Fig. 8.4 and Table 8.7 that by increasing fly ash–quicklime content from 0 to 26.7%, both E_f and $E_{50\%}$ increase from a range of 3.1–5.5 to 14–27.9 MPa and from 6.1–12.1 to 19.6–36.1 MPa, respectively. At the same time, the ratio of $E_{50\%}$ to E_f decreases.

8.5.2 Shear strength parameters

It has been observed that by increasing fly ash–quicklime content from 0 to 26.7%, the peak friction angle increased from 29° to 39° and the cohesion intercept increased from 11 to 30 kPa. The peak and residual strength envelopes are presented in Figs. 8.5 and 8.6, respectively, and the corresponding parameters are also shown in Table 8.7. The results illustrate that the residual friction angle is also increased from 27° for untreated MSW specimens to 34.5° for treated MSW specimens with 26.7% fly ash–quicklime content, whereas the changes in the residual cohesion values were insignificant. Note that the friction angle depends on the confining pressure under which the MSW specimens were tested.

8.5.3 Brittleness

The dramatic decrease in strength after achieving peak values can be determined by the brittleness index, I_B, based on Eq. (8.3), defined by Bishop (1967):

$$I_B = \frac{q_f - q_u}{q_u} \tag{8.3}$$

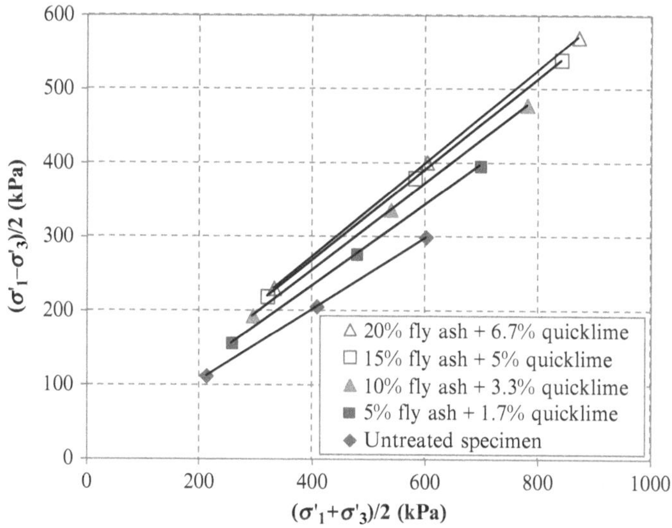

Figure 8.5 Peak strength envelopes of untreated and treated MSW specimens. *(Source: After Fatahi and Khabbaz (2013a)).*

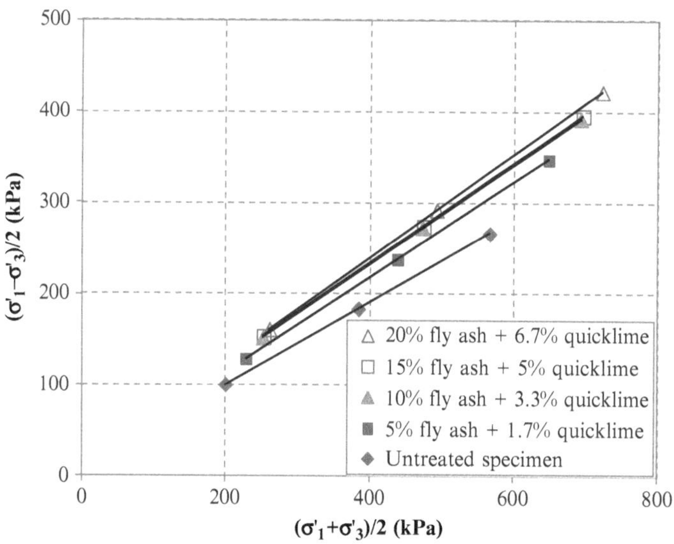

Figure 8.6 Residual strength envelopes of untreated and treated MSW specimens. *(Source: After Fatahi and Khabbaz (2013a)).*

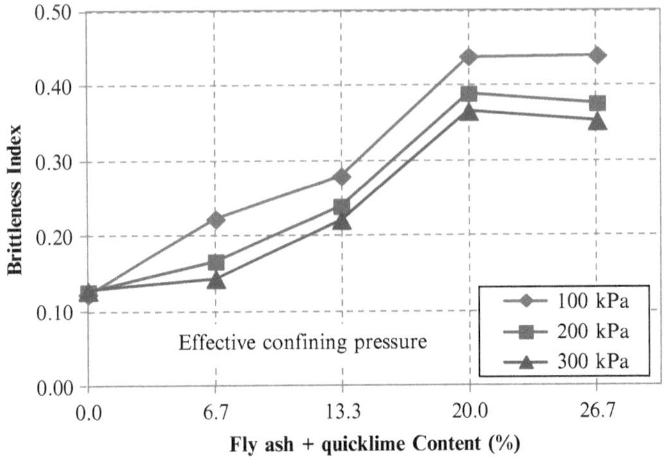

Figure 8.7 Effect of fly ash–quicklime contents on brittleness index of MSW specimens. *(Source: After Fatahi and Khabbaz (2013a)).*

where q_f and q_u are the peak deviator stress and the residual deviator stress, respectively. Figure 8.7 shows variation of the brittleness index for untreated and treated MSW specimens with various percentages of fly ash–quicklime. The results indicate that the brittleness index of MSW specimens treated by fly ash–quicklime up to 20% increased from 0.13 to 0.35 under an effective confining pressure of 300 kPa, whereas for 20–26.7% fly ash–quicklime content, the brittleness index was relatively constant. It was also observed that for specimens under higher effective confining pressure, the brittleness index had a relatively lower value.

8.5.4 Primary and secondary compression

Based on the consolidation tests performed, the effect of fly ash–quicklime content on primary compression of MSW specimens was investigated. Results presented in Fig. 8.8 indicate that addition of fly ash–quicklime reduced compressibility characteristics of MSW specimens. The large voids present in the untreated soil were reduced in the MSW treated with 20% fly ash and 6.7% quicklime. This is due to the microstructural development of cementing compounds in stabilized specimens. Table 8.8 shows the values of the compressibility index, C_c, for the MSW specimens treated with various fly ash–quicklime content. It reveals that the value of C_c decreases from 0.33 for untreated MSW specimens to 0.23 for treated specimens with 26.7% fly ash–quicklime content. As the fly ash–quicklime content increased, the

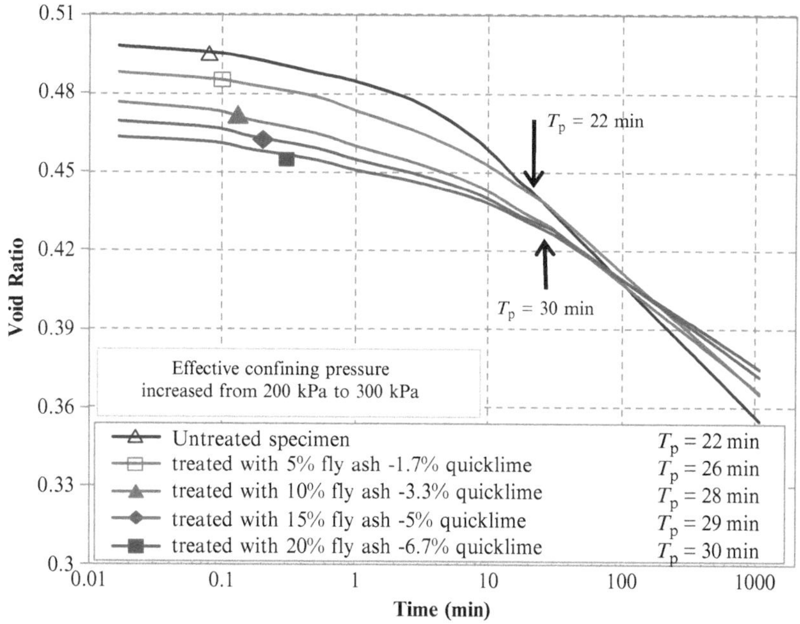

Figure 8.8 Primary and secondary compression of MSW specimens. *(Source: After Fatahi and Khabbaz (2013a)).*

Table 8.8 Primary and secondary compression indices

Fly ash/quicklime content (%)	C_c	C_α
0/0	0.33	0.052
5.0/1.7	0.29	0.045
10.0/3.3	0.28	0.040
15.0/5.0	0.24	0.036
20.0/6.7	0.23	0.033

treated specimens resisted the compressive loading better and consequently showed fewer compressibility characteristics.

Figure 8.8 also presents the time required for the end of primary consolidation and beginning of secondary consolidation (T_p), during which effective confining pressure increased from 200 to 300 kPa. It also demonstrates the secondary consolidation behavior of the untreated and treated MSW specimens with various fly ash–quicklime contents under an effective confining pressure of 300 kPa. It has been shown that changes in void ratio due to creep and sliding of particles after the end of primary consolidation were less in MSW specimens mixed with fly ash–quicklime. Table 8.8 shows that the value of the secondary compression index, C_α, for MSW specimens decreased from 0.052 to 0.033, whereas fly ash–quicklime content increased from 0 to 26.7%. This means that for any given time period, the volume change during secondary compression is more for untreated specimens than for those treated with fly ash–quicklime. Hence, fly ash–quicklime is effective in reducing volume change during primary consolidation as well as secondary consolidation. The reduction in compression index and secondary consolidation of fly ash–quicklime-treated MSW can also be attributed to the effect of cementation and pozzolanic activity of fly ash–quicklime.

To validate the reliability of test results, coefficient of variation (CV) values for the shear strength parameters, the primary and secondary compression indices, and the moisture content were calculated. Results indicated that CV values were 3–6% for the friction angle and the cohesion values; 4–8% for the primary and secondary compression indices; and 3–12% for moisture content, with the majority less than 10%. Thus, the distribution of the test results was categorized as low variance, which made the results from three tests reliable. Accordingly, in this study, the average of three measured or calculated results was used for the analysis.

8.5.5 Coefficient of consolidation

The consolidation process in the triaxial cell was used to determine several important soil parameters, such as the coefficient of consolidation, c_v, which governs the rate of consolidation under a change in applied pressure. Wray (1986) described graphical procedures for determining c_v. The following conventional equation for the log-of-time method has been used for the calculation of c_v:

$$c_v = \frac{0.196d^2}{t_{50}} \tag{8.4}$$

where d is the initial thickness of the specimen before a change in the applied pressure. The values of t_{50} reflect the time required for a 50% reduction in the thickness of the specimen to occur (or 50% reduction in sample volume), and the c_v values were obtained for each consolidation pressure applied.

Figure 8.9 shows the variation in the coefficient of consolidation with different contents of fly ash–quicklime under different effective confining pressures. These results confirmed that increasing the content of fly ash–quicklime in the specimen reduced the coefficient of consolidation. For example, under 200 kPa of average effective confining pressure, the c_v values were 1.6×10^{-5} and 1.2×10^{-5} m^2/s for the untreated specimens and the specimens mixed

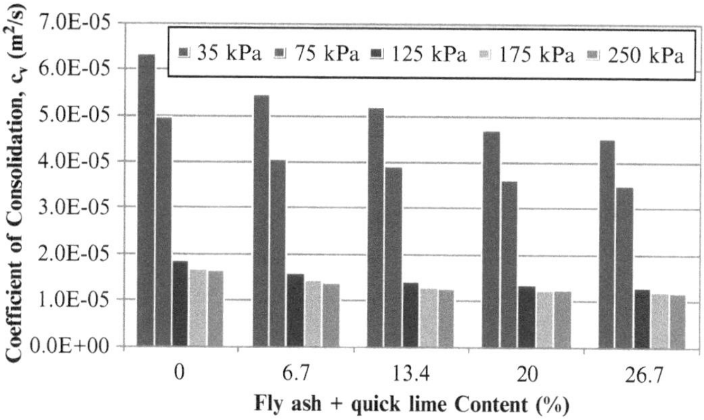

Figure 8.9 Coefficient of consolidation of MSW specimens for different fly ash–quicklime contents under various effective confining pressures.

with 26% fly ash–quicklime, respectively. In addition, with an increase in the confining pressure up to the preconsolidation pressure, the c_v value was reduced and changes were insignificant for higher confining pressures. Lower c_v at higher confining pressure may cause a decrease in compressibility/stiffness of the material. Fatahi and Khabbaz (2013a, 2013b) reported that preconsolidation pressure for specimens used in this study were in the range of 110–130 kPa, depending on fly ash quicklime content.

8.5.6 Hydraulic conductivity

Saturated hydraulic conductivity is another important parameter determined through consolidation testing in the triaxial apparatus. The flow rate of water through waste material depends on the magnitude of the hydraulic conductivity values. The saturated hydraulic conductivity for each load increment was calculated using Terzaghi's (1943) equation (cited by Das, 1983):

$$k = c_v . m_v . \gamma_w \tag{8.5}$$

where γ_w is the unit weight of water, c_v is the coefficient of consolidation, and m_v is the coefficient of volume compressibility, defined as the volume change per unit volume per unit increase in the effective stress. The value given by Eq. (8.5) represents the vertical component of the hydraulic conductivity tensor.

The hydraulic conductivity was measured under average effective confining pressures of approximately 35, 75, 125, 175, and 250 kPa. Figure 8.10 shows the coefficient of permeability of MSW specimens stabilized with different contents of fly ash–quicklime. These results show that by increasing

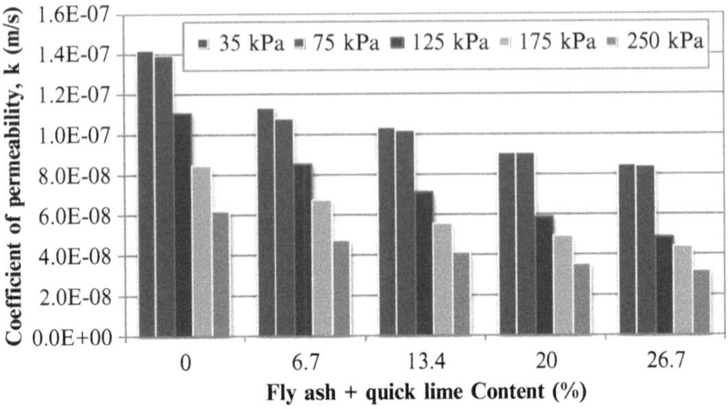

Figure 8.10 Coefficient of permeability of MSW specimens for different fly ash–quicklime contents under various effective confining pressures.

the content of fly ash–quicklime in the MSW specimen, the coefficient of permeability was reduced. The coefficient of permeability for an untreated specimen was 6.2×10^{-8} m/s and was reduced to 3.2×10^{-8} m/s for specimens mixed with 20% fly ash and 6.7% quicklime (under average confining pressure of 250 kPa). This parameter is 4.7×10^{-8}, 4.1×10^{-8}, and 3.5×10^{-8} m/s for MSW specimens mixed with 6.7%, 13.3%, and 20% fly ash–quicklime, respectively. Furthermore, Fig. 8.10 shows the effect of the effective confining pressure on the permeability of untreated and treated MSW specimens, where increasing the effective confining pressure up to the preconsolidation pressure caused no significant change in the coefficient of permeability. However, at higher confining pressures, the coefficient of permeability was reduced. This reduction can also be attributed to the conversion of soluble calcium hydroxide to cementitious compounds, a reduction in the bleed channels and void spaces, and an increase in the density. The large voids present in the untreated soil were reduced in MSW treated with 20% fly ash and 6.7% quicklime. This is due to the microstructural development of cementing compounds in stabilized specimens.

8.5.7 Permeability change index

The value of the permeability coefficient depends on the size, shape, and void distributions of particles. Taylor (1948) proposed an empirical linear relationship between log k and e. The relationship can be expressed as follows:

$$\log k = \log k_0 - \frac{e_0 - e}{c_k} \tag{8.6}$$

Table 8.9 Relationship between the void ratio and permeability

FA/QL content (%)	e_0	k_0 (m/s)	C_k	Relationship
0/0	0.598	1.4×10^{-7}	0.36	log k = 2.78e–8.52
5/1.7	0.593	1.1×10^{-7}	0.32	log k = 3.12e–8.77
10/3.3	0.575	1.0×10^{-7}	0.27	log k = 3.70e–9.07
15/5	0.558	9.0×10^{-8}	0.23	log k = 4.34e–9.41
20/6.7	0.531	8.4×10^{-8}	0.2	log k = 5.00e–9.73

where c_k is the permeability change index, and k_0 and e_0 are the initial values of permeability and the void ratio, respectively. This type of relationship has been commonly used to describe the variation of permeability with the void ratio. Based on a linear regression analysis, the relationship between the void ratio and permeability (*e*-log *k*) and c_k values is shown in Table 8.9 for different contents of fly ash–quicklime. The results show that increasing the content of fly ash–quicklime reduced the c_k value. For example, by mixing 26% of fly ash–quicklime, the c_k value was reduced from 0.36 (for untreated specimen) to 0.20, whereas the c_k values were 0.32, 0.27, and 0.23 for specimens with 6.7%, 13.6%, and 20% of fly ash–quicklime content, respectively.

8.6 FINITE ELEMENT MODELING

8.6.1 Model description

The finite element program PLAXIS version 9 was used to evaluate the settlement of stabilized landfill. The cross section utilized for the numerical analysis is presented in Fig. 8.11. Five layers of solid waste were considered for the landfill to estimate the effect of depth of stabilization on landfill settlement. All the dimensions for the model are given in Fig. 8.12. Accordingly, the solid waste properties need to be adjusted to consider various

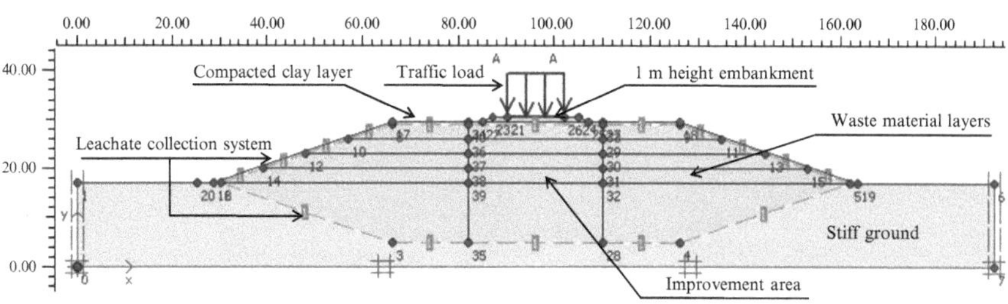

Figure 8.11 Cross section of the numerical model.

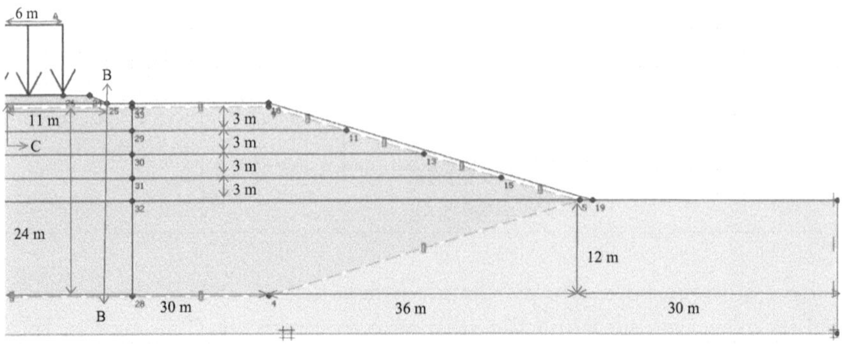

Figure 8.12 Dimensions of the model.

improvement options. The landfill is modeled as a two-dimensional plane strain model. More details are provided in Fatahi (2013).

8.6.2 Mesh generation and boundary conditions

The 15-noded triangle elements were employed in the modeling, and the mesh generation algorithm of PLAXIS version 9.0 was used in this study. It provides a fourth-order interpolation for displacements, and the numerical integration involves 12 stress points. The powerful 15-node element provides an accurate calculation of stresses and failure loads. The two vertical boundaries on both sides are free to move in a vertical direction, whereas the horizontal boundary at the base is considered to be fixed in both vertical and horizontal directions. The foundation soil was considered to be stiff soil, and its stability and deformation were not considered directly in this analysis. Cross sections of generated mesh are shown in Fig. 8.13.

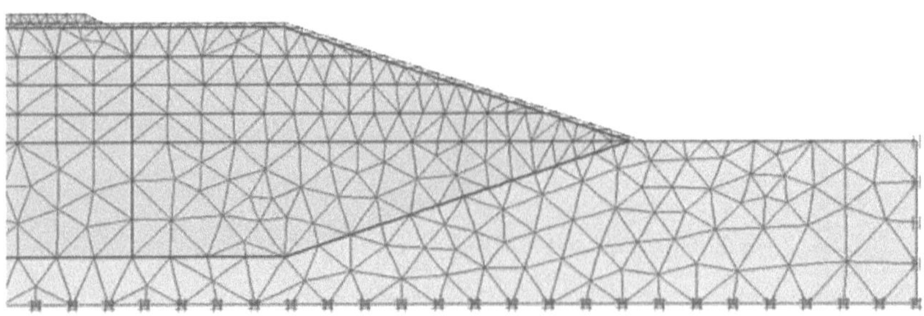

Figure 8.13 Cross section of generated mesh.

8.6.3 Adopted material models

All soils exhibit some creep, and primary compression is thus always followed by a certain amount of secondary compression. The secondary compression (e.g., during a period of 10 or 30 years) can be a certain percentage of the primary compression. For example, this is the case when constructing embankments over closed landfill sites. Indeed, the primary settlement of footings and embankments is usually followed by a substantial creep settlement in later years. In such cases, it is desirable to estimate the creep from finite element method (FEM) computations. The soft soil creep model was used for this analysis based on the following characteristics:

- Stress-dependent stiffness
- Distinction between primary loading and unloading–reloading
- Secondary compression
- Memory of preconsolidation stress
- Failure behavior according to the Mohr–Coulomb criterion

8.6.4 Material parameters

Municipal solid waste parameters used in FEM analysis for different contents of stabilizer are presented in Table 8.10. In addition, parameters used for compacted clay as a cover layer of closed landfill model and road embankment are summarized in Table 8.11.

Some of the important waste properties required for the analyses are unit weight, shear strength, permeability, primary compression index, and secondary compression index. These parameters were obtained from the results of the extensive laboratory tests performed on the MSW samples.

8.6.5 Analysis type

The settlement of the proposed landfill model in PLAXIS was computed using "consolidation analysis" as a calculation type and "staged construction" as a

Table 8.10 Parameters for soft soil creep model in FEM analysis

FA/QL content (%)	γ_{sat} (kN/m^3)	K (m/day)	c_r	c_c	c_α	Void ratio (e_0)	Cohesion, C (kN/m^2)	$\varphi°$
0/0	14.6	0.012	0.075	0.33	0.052	0.63	11	29
5/1.7	14.8	0.0092	0.07	0.29	0.045	0.61	20	33
10/3.3	15	0.0088	0.068	0.28	0.04	0.59	26	36
15/5	15.2	0.0088	0.064	0.24	0.036	0.57	29	38
20/6.7	15.4	0.0082	0.061	0.23	0.033	0.56	30	39

Table 8.11 Model parameters used for cover layer and road embankment

ID	γ_t (kN/m^3)	*E* (kPa)	ν	C (kPa)	$\varphi°$	$\psi°$
Cover	17.5	2.5E4	0.35	10	25	0
Road	18	6.5E4	0.3	5	30	5

loading input. The effects of various fly ash–quicklime contents and different depths of improvement were examined. The model used a 30-year-old closed landfill, and selected MSW materials were decomposed waste. In the first stage of modeling, the filling process of landfill was completed in four stages (each stage consisted of 4 m over 30 days). After filling the site with decomposed waste material, the waste was then covered with a 0.5-m layer of compacted clay. Next, the model evaluated the settlement of the landfill after 30 years of self-weight of MSW. In the next stage, a 1-m-thick road embankment was constructed on the closed landfill model over a 6-month period, and then a traffic loading of 20 kPa with a reduction factor of 0.5 was applied onto the road embankment. In the final step, the vertical displacement of the landfill model 10 and 20 years after applying the traffic load for different amounts of fly ash–quicklime and various depths of improvement was calculated.

8.6.6 Vertical settlement 10 and 20 years after applying the traffic load

The outcomes of the numerical estimation of the model are predicted in PLAXIS. Tables 8.12 and 8.13 show the vertical displacement of the landfill model 10 and 20 years after applying the traffic load, respectively, at the mid-point below the embankment. These data are based on different depths of treatment and various amounts of fly ash–quicklime. The results showed that

Table 8.12 Vertical displacement of the landfill model 10 years after applying traffic load

	Treatment depth (m)				
FA/QL (%)	3	6	9	12	24
Untreated	370	370	370	370	370
5/1.7	328	309	283	271	248
10/3.3	312	277	255	239	207
15/5	301	258	232	213	172
20/6.7	296	251	221	201	154

Source: After Fatahi and Khabbaz (2013b).

Table 8.13 Vertical displacements of the landfill model, 20 years after applying the traffic load

FA/QL (%)	Treatment depth (m) 3	6	9	12	24
Untreated	536	536	536	536	536
5/1.7	486	451	426	408	372
10/3.3	466	419	387	363	311
15/5	451	394	356	327	262
20/6.7	444	383	339	308	244

Source: After Fatahi and Khabbaz (2013b).

the settlement of the untreated landfill model was approximately 536 mm 20 years after applying traffic load. Moreover, these results indicate that treatment of MSW with fly ash–quicklime reduced vertical displacement of the model significantly, and this reduction was more significant for higher depths of improvement. It showed that 20 years after applying traffic load on 3 m of improved landfill with 26.7% fly ash–quicklime, the vertical settlement was reduced by 19% (from 536 to 444 mm), and this reduction for 6, 9, 12, and 24 m improved landfill was 29, 37, 43, and 55%, respectively.

Furthermore, Fig. 8.14 indicates the effects that the amount of fly ash–quicklime has on the vertical displacement versus time for 6-m improvement depth at the midpoint below the embankment. This figure

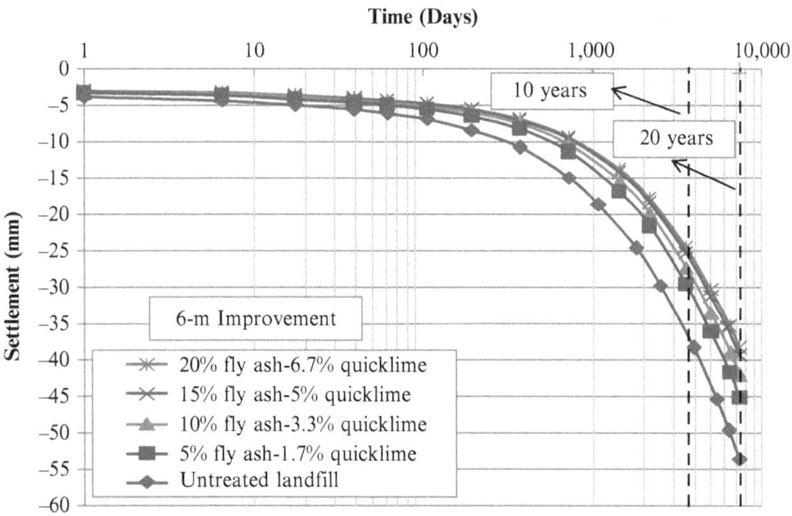

Figure 8.14 Vertical settlement versus time for 6-m improved landfill.

clearly indicates that increasing the amount of fly ash–quicklime reduced vertical settlement to a major extent.

8.7 CONCLUSION

During the past several decades, there has been a significant increase in the development of closed MSW landfills. A widespread experimental program was conducted to study the properties of old solid waste before and after chemical stabilization. The samples of decomposed MSW used in this study were collected from a landfill site southwest of Sydney; the site was closed approximately 30 years ago. Based on the results of the experimental and numerical analysis of this study, it was found that chemical stabilization can effectively increase the compressive strength, shear strength parameters, stiffness, and brittleness index while decreases the settlement and permeability of the MSW landfills. It can be beneficial and effective in redevelopment of closed landfill sites incorporating chemical treatments. The findings of this study may facilitate calculations of the bearing capacity and settlement as well as the slope stability analysis of chemically treated closed landfill sites.

REFERENCES

Anderson, E.O., Balanko, L.A., Lem, J.M., Davis, D.H., 2004. Field monitoring of the compressibility of municipal solid waste and soft alluvium. In: Proc 5th Int. Conf. on Case Histories in Geotechnical Engineering, New York.

ASTM International, 2012. Standard C618-12: Standard Specification for Coal Fly ash and Raw or Calcined Natural Pozzolan for Use in Concrete. ASTM International, West Conshohocken, PA.

Bareither, C.A., Benson, C.H., Edil, T.B., 2012. Compression behavior of municipal solid waste: immediate compression. J. Geotech. Geoenviron. Eng. 138 (9), 1047–1062.

Beaven, R.P., 2000. The Hydrogeological and Geotechnical Properties of Household Waste in Relation to Sustainable Landfilling. PhD dissertation, Queen Mary and Westfield College, University of London, London.

Bishop, A.W., 1967. Progressive failure—with special reference to the mechanism causing it. In: Proc. Geotechnical Conf. on Shear Strength Properties of Natural Soils and Rocks, Oslo, Norway, vol. 2, pp. 142–150.

Bjarngard, A., Edgers, L., 1990. Settlement of municipal solid waste landfills. In: Proc. 13th Annu. Madison Waste Conf., Madison, WI, pp. 192–205.

Bray, J.D., Zekkos, D., Kavazanjian Jr., E., Athanasopoulos, G.A., Riemer, M.F., 2009. Shear strength of municipal solid waste. J. Geotech. Geoenviron. Eng., ASCE 135 (6), 709–722.

Brooks, R., Udeyo, F.F., Takkalapelli, K.V., 2011. Geotechnical properties of problem soils stabilized with fly ash and limestone dust in Philadelphia. J. Mater. Eng., ASCE 23, 711–716.

Chen, Y.M., Zhan, L.T., Wei, H.Y., Ke, H., 2009. Aging and compressibility of municipal solid wastes. Waste Manage. 19, 86–95.

Das, B.M., 1983. Advanced Soil Mechanics. Hemisphere, Washington, DC.

Dixon, N., Langer, U., Gotteland, P., 2008. Classification and mechanical behavior relationships for municipal solid waste: Study using synthetic wastes. J. Geotech. Geoenviron. Eng. 134, 79–90.
Fatahi, B., 2013. Improving Geotechnical Properties of Closed Landfills for Redevelopment Using Fly Ash and Quicklime. Doctoral dissertation, University of Technology Sydney, Sydney, Australia.
Fatahi, B., Khabbaz, H., 2013a. Influence of fly ash and quicklime addition on behaviour of municipal solid wastes. J. Soils Sediments 13 (7), 1201–1212.
Fatahi, B., Khabbaz, H., 2013b. Predicting settlement of chemically stabilized landfills. Int. J. GEOMATE5, 700–705.
Fatahi, B., Khabbaz, H., Fatahi, B., 2012. Mechanical characteristics of soft clay treated with fibre and cement. Geosynthetics Int. 19 (3), 252–262.
Fatahi, B., Fatahi, B., Le, T., Khabbaz, H., 2013a. Small-strain properties of soft clay treated with fibre and cement. Geosynthetics Int. 20 (4), 286–300.
Fatahi, B., Le, T., Fatahi, B., Khabbaz, H., 2013b. Shrinkage properties of soft clay treated with cement and geofibers. Geotech. Geol. Eng. 31 (5), 1421–1435.
Gabr, M.A., Valero, S.N., 1995. Geotechnical properties of municipal solid waste. Geotech Test J ASTM 18 (2), 241–251.
Gray, D.H., Lin, Y.K., 1972. Engineering properties of compacted fly ash. J. Soil Mech. Found. Div. 98 (4), 361–380.
Hokmabadi, A.S., Fatahi, B., Samali, B., 2014a. Assessment of soil–pile–structure interaction influencing seismic response of mid-rise buildings sitting on floating pile foundations. Comput. Geotech. 55, 172–186.
Hokmabadi, A.S., Fatahi, B., Samali, B., 2014b. Seismic response of mid-rise buildings on shallow and end-bearing pile foundations in soft soil. Soils Found. 54 (3), 345–363.
Horpibulsuk, S., Phetchuay, C., Chinkulkijniwat, A., 2012. Soil stabilization by calcium carbide residue and fly Ash. J. Mater. Civil Eng. 24 (2), 184–193.
Hossain, M.S., Gabr, M.A., Barlaz, M.A., 2003. Relationship of compressibility parameters to municipal solid waste decomposition. J. Geotech. Geoenviron. Eng. 129 (12), 1151–1158.
Huang, S., Cheng, J., 1986. Kinetic of reaction in the system fly ash $Ca(OH)_2$–H_2O. J. Chinese Ceramic Soc. 14, 191–197.
Hudson, A.P., Beaven, R.P., Powrie, W., 2009. Assessment of vertical and horizontal hydraulic conductivities of household waste in a large scale compression cell. In: Cossu, R., Diaz, L.F., Stegman, R. (Eds.), Proc. 12th Waste Management and Landfill Symp. Sardinia 2009. CISA, S. Margherita di Pula, Italy, pp. 641–642 (Abstract). (Full paper CD only).
Jain, P., Powell, J., Townsend, T.G., Reinhart, D.R., 2006. Estimating the hydraulic conductivity of landfilled municipal solid waste using the borehole permeameter test. J. Environ. Eng., ASCE 132, 645–652.
Janz, M., Johansson, S., 2002. The function of different binding agents in deep stabilization. Svensk djupstabilisering report No. 9.
Kavazanjian Jr., E., Matasovic, N., Bachus, R.C., 1999. Large-diameter static and cyclic laboratory testing of municipal solid waste. In: Proc. 7th Int. Waste Management and Landfill Symp., Cagliari, Italy, pp. 437–444.
Kawasaki, T., Niina, A., Saitoh, S., Suzuki, Y., Honjo, Y., 1981. Deep mixing method using cement hardening agent. Proc. In: 10th Int. Conf. on Soil Mechanics and Foundation Engineering. Southeast Asian Geotechnical Society, Bangkok, pp. 721–724.
Kehew, E.A., 1995. Geology for Engineers and Environmental Scientists, second ed. Prentice Hall, Englewood Cliffs, NJ, pp. 295–302.
Kitazume, M., Yamazaki, H., Tsuchida, T., 2000. Recent soil admixture stabilization techniques for port and harbor constructions in Japan—deep mixing method, premix method, light-weight method. In: Proc. Int. Semin. on Geotechnics in Kochi, ISGK 2000, Kochi, Japan, pp. 23–40.

Kumar, P., Sharma, R.S., 2004. Effect of fly ash on engineering properties of expansive soils. J. Geotech. Geoenviron. Eng., ASCE 130, 764–767.
Landva, A.O., Clark, J.I., 1990. Geotechnics of waste fills. In: Landva, A.O., Knowles, G.D. (Eds.), Geotechnics of Waste Fills—Theory and Practice, ASTM Special Technical Publication. ASTM, Baltimore, MD.
Landva, A.O., Valsangkar, A.J., Pelkey, S.G., 2000. Lateral earth pressure at rest and compressibility of municipal solid waste. Can. Geotech. J. 37 (6), 1157–1165.
Marques, A.C.M., Filz, G.M., Vilar, O.M., 2003. Composite compressibility model for municipal solid waste. J. Geotech. Geoenviron. Eng. 129 (4), 372–378.
Okoro, C., Vogtman, J., Yousif, M., Agnaou, M., Khoury, N., 2011. Consolidation characteristics of soils stabilized with lime, coal combustion products and plastic waste. In: Proc. GeoFrontiers 2011, Dallas, TX.
Phanikumar, B., Sharma, R., 2007. Volume change behavior of fly ash-stabilized clays. J. Mater. Civil Eng. 19 (special issue), 67–74.
Reddy, K.R., Hettiarachchi, H., Parakalla, N.S., Gangathulasi, J., Bogner, J.E., 2009. Geotechnical properties of fresh municipal solid waste at Orchard Hills landfill, USA. Waste Manage. 29, 952–959.
Rogbeck, Y., Larsson, L., Åhnberg, H., Rodenstam, S., 2008. Kalkrik aska som bindemedel i markstabilisering. Statens geotekniska institut, Rapport Varia 590.
Sharma, H.D., De, A., 2007. Municipal solid waste landfill settlement postclosure perspectives. J. Geotech. Geoenviron. Eng. 133 (6), 619–629.
Shi, C., 1999. Studies on several factors affecting hydration and properties of lime–pozzolan cements. J. Mater. Civil Eng. 29 (4), 467–472.
Shi, C., Day, R.L., 1993. Acceleration of strength gain of lime–pozzolan cements by thermal activation. Cem. Concr. Res. 23 (4), 824–832.
Shi, C., Day, R.L., 1995. Acceleration of the reactivity of fly ash by chemical activation. Cem. Concr. Res. 25, 15–21.
Skhiri, N., Guerbois, M., Hebe, I., 2006. Technical and environmental performances assessment for a bioreactor. In: Lagerkvist, A. (Ed.),. ICLRS 2006—Abstract Proc. 4th Intercontinental Landfill Research Symp., Gallivare, Sweden, pp. 183–184.
Sowers, G.F., 1973. Settlement of waste disposal fills. In: Proc. 8th Int. Conf. Soil Mechanics and Foundation Engineering, Moscow.
Standards Australia, 2008. Australian Standard 1289: Methods of Testing soil for Engineering Purposes. Standards Australia, Sydney, Australia.
Tastan, E., Edil, T., Benson, C., Aydilek, A., 2011. Stabilization of organic soils with fly ash. J. Geotech. Geoenviron. Eng. 137 (9), 819–833.
Taylor, D.W., 1948. Fundamentals of Soil Mechanics. Wiley, New York.
Tchobanoglous, G., Theisen, H., Vigil, S., 1993. Integrated Solid Waste Management: Engineering Principles and Management Issues. McGraw-Hill, New York.
Terzaghi, K., 1943. Theoretical Soil Mechanics. Wiley, New York.
Wall, D.K., Zeiss, C., 1995. Landfill biodegradation and settlement. Journal of Geotechnical and Geoenvironmental Engineering, ASCE 121 (3), 214–224.
Wray, W.K., 1986. Measuring Engineering Properties of Soil. Prentice-Hall, Englewood Cliffs, NJ.
Xie, Z., Xi, Y., 2001. Hardening mechanisms of an alkaline-activated class F fly ash. Cem. Concr. Res. 31 (9), 1245–1249.
Zeiss, C., Major, W., 1993. Moisture flow through municipal solid waste. J. Environ. Syst. 22 (3), 211–231.
Zekkos, D., Flanagan, M., 2011. Case histories-based evaluation of the deep dynamic compaction technique on municipal solid waste sites. Geo-Frontiers, 529–538.
Zornberg, J.G., Jernigan, B.L., Sanglerat, T.R., Cooley, B.H., 1999. Retention of free liquids in landfills undergoing vertical expansion. J. Geotech. Geoenviron. Eng. 125, 583–594.

CHAPTER 9

Dry Soil Mixing for the Ballina Bypass Motorway Upgrade

Richard Kelly[1,2], Ernst Friedlaender[3], Theva Muttuvel[1]
[1]Coffey Geotechnics Pty Ltd., NSW, Australia
[2]ARC Centre of Excellence in Geotechnical Science and Engineering, University of Newcastle, NSW, Australia
[3]Keller Ground Engineering, NSW, Australia

9.1 INTRODUCTION

The Ballina Bypass project constructed a mostly four-lane dual-carriageway motorway over a length of approximately 12 km. The southern half of the route traversed a low-lying floodplain associated with the Richmond River. The northern half of the route climbed through rolling hills. The alignment of the Ballina Bypass is shown in Fig. 9.1.

In the floodplain, 18 bridges and three large concrete culverts were constructed along with embankments ranging from approximately 2 to 6 m in height on soft soils ranging in depth from 2 to 30 m. Historically on the existing highway, embankments had failed at heights on the order of 3 m and settlements on the order of 3.5 m had occurred over approximately 50 years.

The original concept design adopted prefabricated vertical drains coupled with staged construction to reduce post-construction settlement to acceptable values. The associated construction period was estimated at 13 years due to the periods of consolidation required during filling for staged construction. An improved concept design was subsequently developed incorporating a suite of ground improvement technologies that reduced the construction period to 2 years of early works followed by 4 years of construction.

The project was delivered as an Alliance between the Roads and Maritime Authority of New South Wales (RMS), Leighton Contractors, SMEC, AECOM, and Coffey. Keller Ground Engineering (KGE) was selected as the specialist ground improvement subcontractor to the Alliance.

The design settlement criteria were tight at bridge abutments in order to provide a smooth transition from the embankments to the structure, as well as to prevent soil displacements loading the abutment piles. Creating the smooth transition required a gradual increase in settlement moving away

Chemical, Electrokinetic, Thermal, and Bioengineering Methods
http://dx.doi.org/10.1016/B978-0-08-100191-2.00009-5

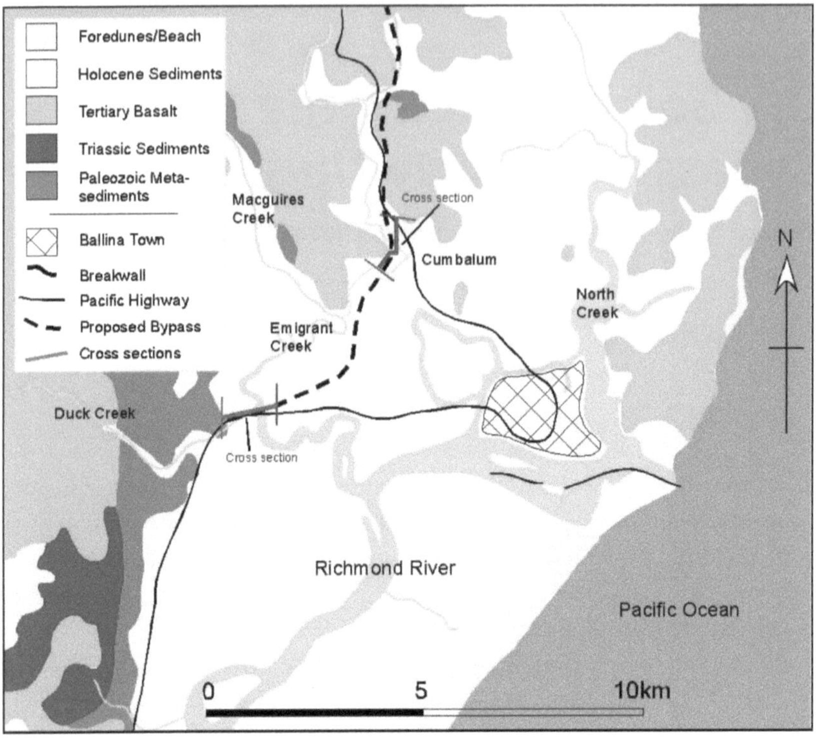

Figure 9.1 Alignment of the Ballina Bypass.

from the bridges. In addition, a diversion road needed to be constructed immediately so that mainline construction could be done offline to meet a tight delivery date.

A wide range of ground improvement technologies was considered during design development. Selection of the final mix of ground improvement technologies was based on optimizing a balance between cost, construction time, performance, and risk. Locations on the critical path, such as one of the diversion roads and a number of the bridge approach embankments, were treated with dry cement soil mixing (DSM). In particular, the structures construction program had multiple critical paths where several bridges were constructed in parallel, followed by the next group of two or three bridges and so on. Therefore, some embankments had to be constructed rapidly, whereas others had in excess of 12 months prior to bridge construction. Where time was available, surcharging without ground improvement or using prefabricated vertical drains was adopted. Where settlement needed to be controlled and less time was available, DSM, dynamic replacement, or stone columns were adopted.

DSM was the key technology adopted to achieve an optimal cost–performance–risk balance for the Ballina Bypass Pacific Highway Upgrade project. The use of DSM allowed embankments to be constructed within a 3-month period and reduced settlement on the order of 100 mm during construction and on the order of 10s of millimeters post construction. The DSM technique allowed construction of columns tapered in length for a gradual settlement transition away from embankments. DSM was also used in conjunction with other ground improvement methods such as prefabricated vertical drains to reduce settlement at intermediate cost.

The Ballina Bypass was the first major project in Australia for which DSM was used extensively. The challenging ground conditions created difficulties early in the project. These difficulties were progressively overcome, and a process was developed that delivered the required outcomes with certainty. This chapter reports on some of the challenges that were faced and overcome, and it presents the process that is now used routinely on Australian projects.

9.2 BACKGROUND INFORMATION

9.2.1 Dry soil mixing

The process of dry soil mixing is described in detail by Kitazume and Terashi (2013) and EuroSoilStab (1997) and is summarized here.

Figure 9.2 shows the DSM equipment in use on site. The binder is contained in the yellow pod on the left-hand side of the excavator-mounted mixing tool. The mixing tool is first rotated slowly into the ground until it reaches its target depth. Then the rotations are reversed in direction, and the revolutions per minute (rpm) are increased up to 100–150 rpm. The binder pod is pressurized with air, and the binder is transported from the pod via a tube to the top of the Kelly bar. The rate of extraction of the tool is controlled by the rate of binder flow in order to achieve the target binder content with an accuracy of ±10%. The process is optimized to maximize production while achieving conforming columns.

Although the type of soil and in situ conditions affect the column strength and stiffness, the strength and stiffness are a function of the ratio of natural water content of soils to cement and the quantity of mixing (measured as a blade rotation number) (Kitazume and Terashi, 2013). Experience shows, however, that adjacent DSM columns can have very different

Figure 9.2 Dry soil mix equipment.

strengths and stiffnesses even though they are installed in the same ground using nearly identical methods (Kitazume et al., 2001; Lee et al., 2005). Many factors contribute to the variability of column strength and stiffness, including the inherent variability of the ground, the moisture content, non-uniform feed rates and mixing of the cement, the soil mineralogy (clay content, clay mineralogy, and sulfate content), the cation exchange capacity, the specific surface area of the soil, and the type of cement used. The coefficient of variation of strength is reported to range between 0.3 and 0.8 (Filz and Navin 2010; Madhyannapu et al., 2010; Al-Naqshabandy and Larsson, 2013), in contrast to natural soil, which is reported to have a variance ranging between 0.1 and 0.4 (Duncan, 2000).

For the Ballina Bypass, dry ordinary Portland cement (OPC) powder was used in general except in sections with an organic content in excess of 10%, where a 75% OPC/25% blast furnace slag mix was used.

An advantage of the DSM equipment is its relatively light weight and stability; it was supported on a 1 m thick sand construction platform. The DSM method also eliminated the generation of soil spoil, which was an important environmental consideration for this site because the soil becomes acidic when exposed to air and any spoil would have needed to be transported and treated with lime prior to reuse or disposal.

9.2.2 Geological and geotechnical models

The geology of the lowlands comprises extensive and often thick deposits of Quaternary alluvium overlying rock. The Quaternary geological and depositional model for the Ballina area is described by Bishop (2004).

Sea levels rapidly rose approximately 150 m between 17,000 and 5000 years ago (Late Pleistocene to Holocene). Unconformable deposition occurred on the Pleistocene horizons and comprised a fining up sequence of lower level gravels and sandy clays to higher level dark gray shelly clays (Pimlico clay). As sea levels rose, estuarine deposition involved infilling of remnant Pleistocene paleochannels up to 40-m depth.

Quantitative x-ray diffraction studies show that the Pimlico clay is composed mainly of clay-size particles that are smectite minerals (75%) with minor kaolinite. As a result of this mineralogy and recent deposition, the Pimlico clay is highly compressible and has low strength. It generally has an organic content of approximately 5%, locally increasing to 11% at one particular location (ECC SA). The groundwater is saline with a salt concentration approximately half that of seawater. The clay has a significant amount of sulfides in it. When the sulfides are exposed to air, they oxide to sulfates and form acidic ground conditions above the water table. Below the water table, the pH is typically approximately 8.

A typical profile of soil parameters with depth is shown in Fig. 9.3, in which C_c is the compression index, C_s is the recompression index, C_α is the creep index, c_h is the coefficient of consolidation in the horizontal direction, and OCR is the overconsolidation ratio. Error bars depicting liquid and plastic limits are plotted with moisture content. Although the data show that the liquid limit is less than the moisture content, which would indicate that the clay is sensitive, the liquid limit data are incorrect because the standard procedure involves drying the sample prior to reconstitution and this alters the behavior of the clay. The sensitivity of the Pimlico clay is typically in the range of 3–5.

9.2.3 Design criteria

The following design criteria were required for the Ballina Bypass:

- A post-construction settlement limit of 50 mm in 40 years within the 6-m approach slab of the bridge abutment. This is referred to as Zone 1 and is shown in Fig. 9.4.
- The zone behind the approach slab (i.e., Zone 2) is required to satisfy 0.5% maximum change in grade in any direction to tie in with the untreated low embankment (i.e., Zone 3).

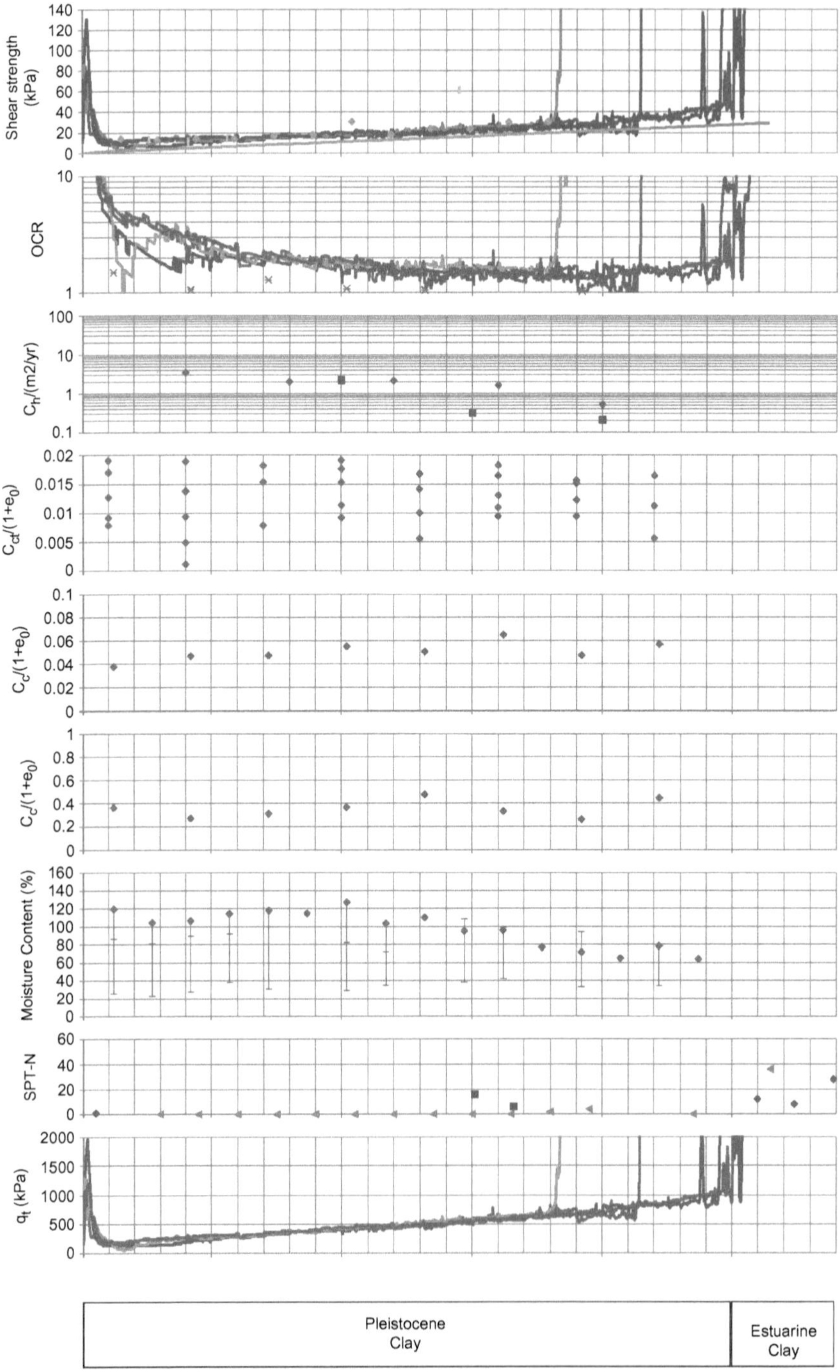

Figure 9.3 Geotechnical model.

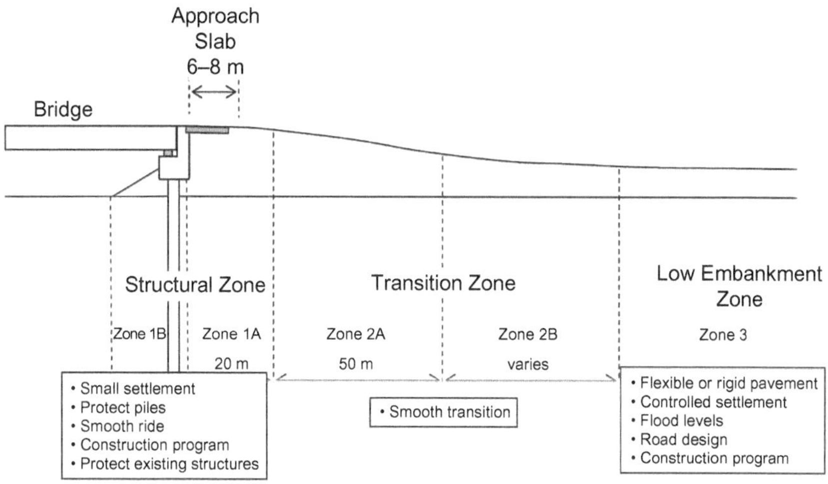

Figure 9.4 Conceptual long section along bridge approach embankments.

- The untreated low embankments were allowed to deform up to 200 mm prior to intervention to reconstruct the pavement back to level.
- The structural design of the new bridge abutment piles allowed for a lateral deformation of 20 mm over 100 years.

9.2.4 Design

Settlement was controlled using DSM columns under the embankment crest while stability was controlled using DSM panels and grids were installed under batter slopes. Design of the DSM was performed generally in accordance with the procedures described in SGF4:95E (Swedish Geotechnical Society, 1997). These procedures consider the DSM reinforced soil can be treated as a composite material.

With reference to Fig. 9.4, DSM columns in the structural zone were designed to the full depth of the soft soil in order to reduce settlement and consequent impacts on the piles due to the lateral soil movement induced by vertical settlement. The area replacement ratio was calculated based on a requirement to limit the stress applied on a column to 75% of its design ultimate strength (C_{col}).

DSM columns in the transition zone were designed to penetrate only to a certain depth into soft clay in order to create a transition of progressively increasing settlement away from the bridge abutment. The partially penetrating DSM columns were assumed to punch into the soft clay, leading to more settlement than would be calculated using the composite material approach. Therefore, three-dimensional analyses using the computer software PLAXIS 3D were performed to assess the optimum length of DSM columns to limit punching into the soft clay.

The consolidation behavior of DSM-treated soil was estimated assuming a DSM column acts as a vertical drain. It has been assumed that creep starts after 90% of consolidation has occurred, with creep indices equivalent to 4% of the coefficient of compression of the composite material.

Stability of embankments with untreated ground, stability berms, and structural geofabric was assessed using a limit equilibrium method to assess whether panels of DSM columns beneath the batters were required. Batter panels were used to support batters, whereas grids of columns were used at the ends of embankments to control three-dimensional loading.

The adopted DSM design parameters are presented in Table 9.1. A summary of the DSM design at various locations is provided in Table 9.2. The acronyms are not explained in this chapter. Excerpts from plan and long section design drawings are shown in Figs. 9.5 and 9.6.

Table 9.1 Adopted design parameters

Parameter	Value
Column diameter	0.8 m
Column Spacing	Varies
Bulk unit weight	20 kN/m^3
Column shear strength, C_{col}	C_{col} = 150 kPa under crest; C_{col} = 100 kPa under batter
Constrained modulus, M_{col}	M_{col} = 30 MPa
Poisson's ratio	0.3

Table 9.2 Summary of design

Location	Embankment height (m)	Depth of soft soil (m)	Area replacement ratio (%)	Width of batter panels (m)
ECS and P2T	5–5.6	9.5–12	30	10
ECC SA	5.2	7.4–8	30	10
ECC NA	5.1–5.7	5.5	30	10
CFRB SA	2.6–3.7	11.5–14	25	8–10
Diversion 2	2–3.8	5.5–9.1	16–30	4–8
North Sandy Flat	4.5–5.8	Up to 10	25–30	9–10

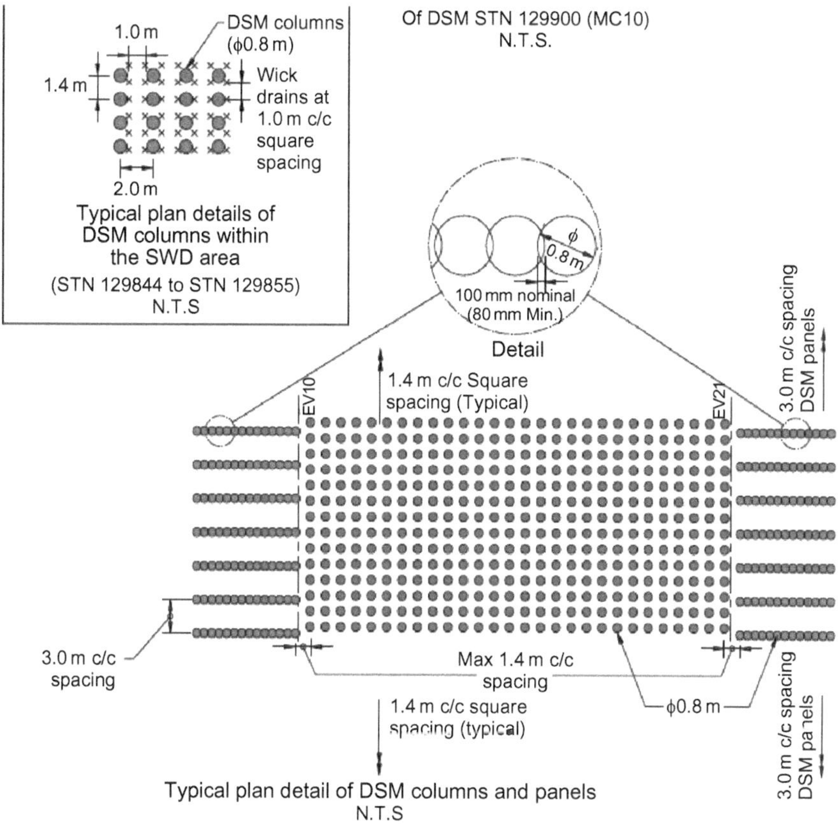

Figure 9.5 Typical plan layout of DSM columns.

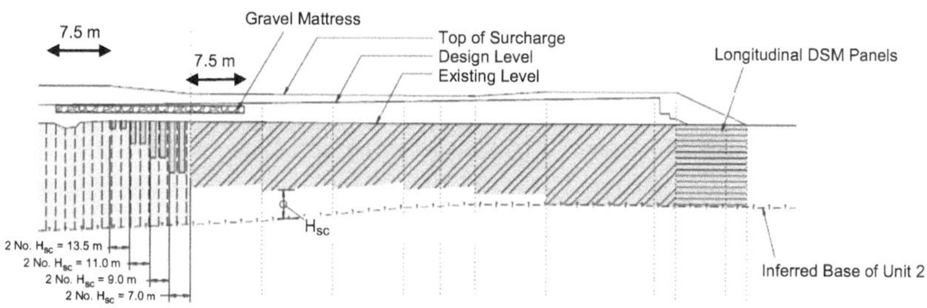

Figure 9.6 Typical long section for structural and transition zones.

9.2.5 Construction specification

The construction specification outlined the process required to construct the columns and the acceptance criteria on the product. The process was as follows:

1. Perform laboratory mix trials to estimate the binder type and quantity prior to mobilizing to site. The tests also provided a trend for strength gain with time.
2. Perform field trials to confirm binder types and quantity along with mixing parameters that optimize construction.
3. Perform production tests to confirm quality has been achieved. Production tests were often performed at 7 days and were extrapolated to 28 days using the trend line developed in the laboratory trials. Production testing was initially specified as three columns with pull out resistance tests (PORT) and three columns sampled by triple tube coring and the samples subjected to unconfined compressive strength (UCS) testing.

The acceptance criteria for the project were that

1. the strength of the columns had to exceed a target minimum strength criteria of 150 kPa beneath the crest of the embankment and 100 kPa in the interlocking panels beneath the batters; but
2. to allow for variations inevitably associated with DSM, the acceptance criteria for DSM allowed for 10% of the test results falling below the target minimum strength criteria provided these test results were equal to or greater than 75% of the target strength.

The definition of column strength as an average column strength or the minimum column strength along the entire length of the column was left vague. Acceptance was at the sole discretion of the geotechnical manager for the project. If acceptance did not occur, it triggered a nonconformance report for further column testing or replacement of columns at the contractor's cost.

In practice, the effect of the acceptance criteria was to increase the mean column strength required in construction above the design column strength to account for the inherent variability in the strength of DSM columns. Reinstallation of nonconforming columns was a cost and time risk to the contractor, so the constructed mean column strength was usually made much higher than implied by the philosophy of the acceptance criteria to minimize rework.

PORTs are described in SGF4:95E (Swedish Geotechnical Society, 1997). In summary, a vane is installed below the columns prior to construction and then pulled out through the columns some time later. The pullout force is corrected for friction on the wire rope and converted using an empirical factor, *N*, which was taken to have a value of 10. The UCS tests were performed measuring axial displacement as well as force to obtain both UCS data and a measure of stiffness for comparison with the design assumptions.

9.3 CONSTRUCTION

Because DSM was a relatively new soil improvement system in Australia, and due to the importance of the soil improvement aspects of the Ballina Bypass, it was originally intended to carry out a pre-construction on-site verification of the DSM process to enable all parties to familiarize themselves with the technique and the finished product and at the same time establish a suitable QA system and implement possible adaptations if deemed necessary to suit local conditions. However, due to restrictions on approvals, the pre-works verification tests could not be executed. With the benefit of hindsight, performance of early works designed to refine the mixing process would have provided significant benefits to the project.

9.3.1 Phase 1: Diversion 2

Work commenced on the southern section of a diversion road acting as a temporary highway. The diversion road was required to be constructed rapidly over the soft soils in the early stages of the project in order to switch traffic off the existing highway to enable construction of the new motorway. At this location, the depth of soft soil was generally in the range of 2–6 m. The depth of soft soil in the northern section of the diversion road increased to 10 m near the location of a creek. A concrete box culvert was to span the creek. The ground was waterlogged and difficult to access prior to construction.

Samples for laboratory testing were obtained from shallow test pits due to the difficult access. The results of the laboratory tests indicated that a Portland cement binder at 140 kg/m^3 was able to achieve the design strength.

The formal process of field trials required by the specification was waived for the diversion road due to a lack of time in the construction program. Initial production testing indicated that a binder content of 140 kg/m^3 was insufficient to achieve the design strength, and the binder content was increased to 160 kg/m^3. The rest of the columns were installed using a binder content of 160 kg/m^3. Excavations exposing the top of the columns and testing showed that the design diameter and strength was achieved in most locations.

However, in the northern section of the road, when the working platform was excavated to allow construction of the box culvert, the exposed columns had diameters of approximately 0.6 m instead of the target 0.8 m. A photo of the undersized columns is provided in Fig. 9.7. The columns shown in this figure were part of a panel and should have had a 100-mm overlap.

Figure 9.7 Undersized DSM columns.

Despite the nonconforming columns being identified, a decision was made by the Alliance to construct the road and the concrete box culvert due to the need to open the diversion to allow construction of the main works to meet the construction program.

The culvert settlements are shown in Fig. 9.8. Up to 450 mm of settlement was measured, and at this stage the culvert units and the 0.5-m-thick base slab had severely cracked. Back-analysis of embankment settlement adjacent to the culvert suggested that the DSM columns provided minimal improvement to the soft ground.

Once the highway had been opened and the diversion was not required, the culvert was removed, timber piles were installed to remediate the area, and a second culvert was constructed.

The reasons for the nonperformance of the northern section of the embankment formed the subject of many discussions between parties under difficult conditions. With the benefit of hindsight, there was consensus that the early time pressures and logistical site restrictions did not enable appropriate testing to be carried out to identify varying different site conditions, nor was there time available to rectify matters once identified.

The problems in the northern section were attributed to a combination of the DSM installation and the excavation/construction procedure for the

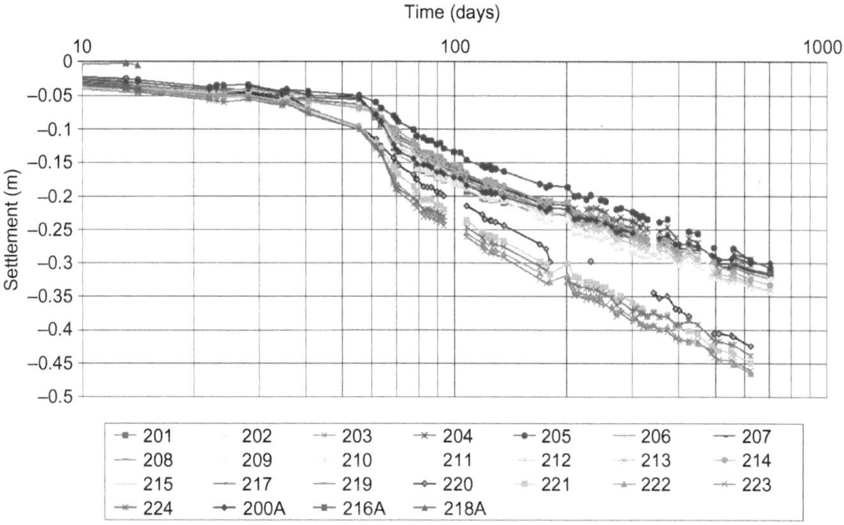

Figure 9.8 Settlement of culvert.

culvert construction. Some of the contributing factors for DSM installation were the following:

- Site conditions: The soils that required mixing ranged from stiff clays at the toe of the columns to very soft clays.
- Sections of trapped vegetation below the working platform.
- Plant and equipment: The plant used in the northern section had lower power, as well as lower torque capacity, compared to the unit used in the south. The same mixing tool was used in both sections.
- Time restraints prevented any verification tests from being carried out in the northern section.

During the initial excavation for the culvert, it was identified that a significant percentage of the columns had diameters on the order of 30% smaller than designed. The cause of this was attributed to the combination of the stiff and soft clays and trapped vegetation, which when treated with a standard DSM tool resulted in the tool becoming clogged with vegetation and stiff clays, resulting in undersized columns being constructed.

A design check was carried out taking into account the undersize columns. It indicated that despite the undersized columns, the settlement criteria could be met, providing the undersized columns had a minimum, although unverified, strength.

Excavation for the culvert was carried out by conventional excavator standing at the top of the cut and working back as the excavation proceeded.

Figure 9.9 Cracking/movements during excavation for culvert.

The excavation in the very soft clays was cut at an angle on the order of 30°–45°, with the excavated material stockpiled at the top of the cut next to the excavator before loading onto trucks. Studying photographs (Fig. 9.9) of the excavation at a later date revealed significant signs of progressive cracking at the top of the excavation and movement of the culvert excavation slope and toe under the surcharge load and of the excavator and the stockpiled spoil. It is believed that movement of the excavation slope may have sheared the DSM columns, contributing to excessive settlement.

Based on the experience gained, the following changes were implemented for the next phase of works:

- The mixing plant was upgraded by increasing the oil flow to the mixing heads.
- The mixing tools were redesigned and provided with blades to cut through vegetation and avoid the risk of clogging by adjusting the blade configuration.
- Slope angles, excavation, and stockpiling procedures were drawn up for future works to avoid damage to DSM columns.

9.3.2 Phase 2: North Sandy Flat

The next phase of work was performed along the new motorway alignment adjacent to the diversion road in the general vicinity of the same creek line. This area was called North Sandy Flat (NSF).

Tests on soil samples showed that the chemical composition and organic content were similar to those of the majority of the site. Field trials were performed approximately 50 m north of the creek. The depth of soft soil in this area was approximately 7 m. The trials demonstrated that conforming columns could be constructed using a binder content of 160 kg/m^3.

The initial sections of DSM treatment of NSF were carried out with the equipment and tools used in Diversion 2 prior to identifying the conditions in Diversion 2 described previously, and making modifications to the tools and equipment as well as excavation procedures.

Back toward the creek, the depth of soft soil was 10 m. Near the creek, the PORT and UCS tests provided conflicting information. PORT tests suggested that the design column strength was being achieved. In contrast, drilling was either unable to get any recovery or the recovery that was achieved was very soft. A series of cone penetrometer tests was performed to provide an independent measure of strength, and these tests found that the column strengths were low in general. Excavations of the columns found that many of the columns had been constructed with a hard outer annulus up to 200 mm thick and a soft inner core.

A decision was made to preload the area with up to 5 m of fill as a proof test. Settlements measured with time are shown in Fig. 9.10. The predicted settlement in this area was approximately 0.13 m. Measured settlement

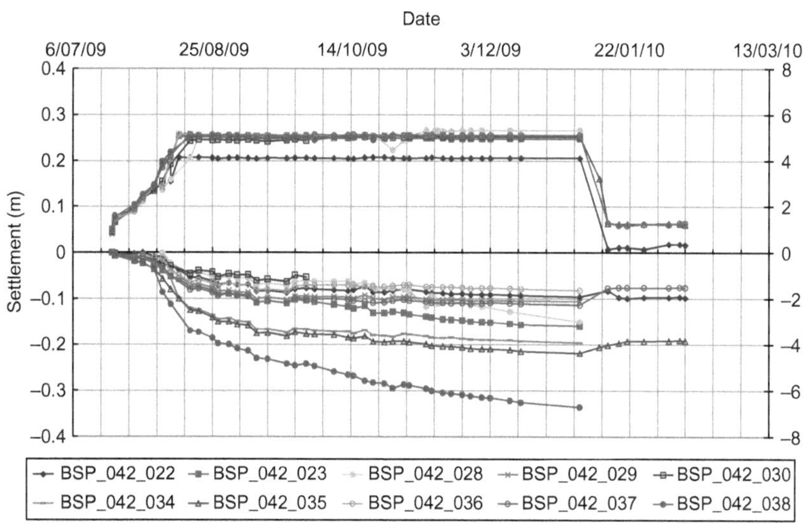

Figure 9.10 Settlement at North Sandy Flat.

exceeded predicted settlement near the creek but was similar to predictions to the north of the creek.

These mixing problems were addressed with the plant modifications, with all subsequent excavations carried out in accordance with the revised excavation procedures.

9.3.3 Phase 3: Majority of project—Successful refinement of process

Experience gained in the first two phases of the project was used to develop procedures that resulted in successful construction for the majority of the project. The refined procedures are presented in this section.

The purpose of laboratory mixing trials is to obtain a reasonably accurate estimate of binder type and content at each location prior to field trials. Previously, near surface samples had been collected, but the moisture content of these samples was not the maximum moisture content with depth. The water content was maximum between 5 and 10 m below ground level. Consequently, the water:cement ratio adopted in the laboratory tests was not representative of the soil profile; therefore, the required binder content was not accurate. The procedure was modified to take soil samples from tubes sampled with depth.

Phases 1 and 2 encountered most difficulties near a creek where the soft soil was deepest. However, the preproduction field trials had not been performed at what appeared to be the worst ground conditions. The process of field trials was modified such that trials were performed at three or four locations within a given site in order to increase the likelihood that the field trials would encounter the worst ground conditions. Nine columns were installed at each location, with three different binder contents and a variety of mixing parameters. After the more extensive field trials were implemented, production testing progressed with few nonconformance reports and no uncertainty in the test results. Although the connection between extensive field trials and production test performance is circumstantial, we believe that extensive field trials allowed the process to be optimized and are a key component of the construction process.

The scope of production testing was increased from three columns per location to approximately 2% of constructed columns, with a minimum of one test performed for each day of production. PORT was the main production test. The main purpose of increasing the quantity of testing was to check each day's production to obtain an early indication of column strength. Action could then be taken if low-strength columns were detected.

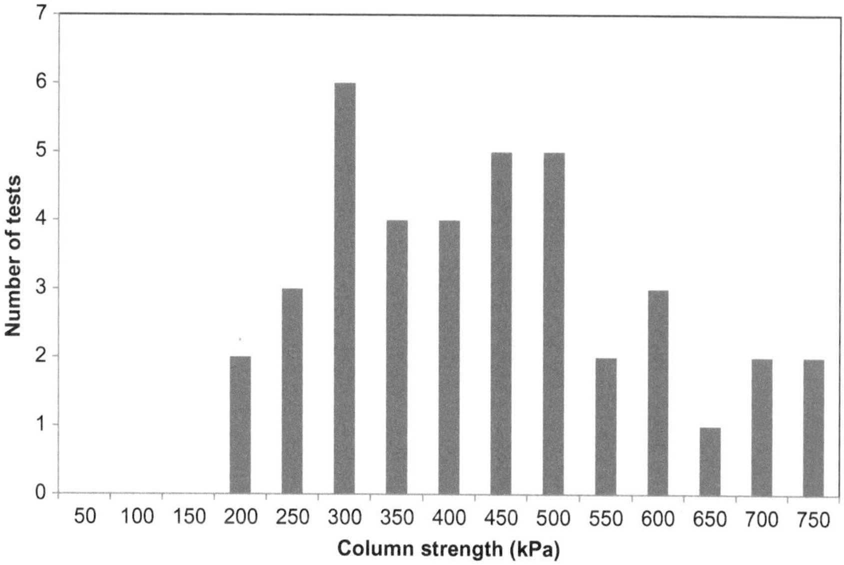

Figure 9.11 PORT test data.

It turned out that few low-strength columns were detected, but the strategy of checking each day's production is considered sound.

A distribution of average column strengths measured by PORT is provided in Fig. 9.11. The data show that a wide range of column strengths occurred but that all of the measurements were greater than the design column strength of 150 kPa.

A summary of average column strengths, standard deviations, and coefficients of variance (COV) for all locations across the project is provided in Table 9.3. The data show that the mean strengths were always greater than the design strength, and the COV ranged from 0.37 to 0.57. High COVs

Table 9.3 Summary of DSM PORT test data interpreted using $N=10$

Location	ECS	P2T	ECC SA	ECC NA	CFPB	NSF
Embankment height (m)	6.0	6.0	5.0	5.5	4.5	5.5
Area replacement ratio (%)	0.3	0.3	0.3	0.3	0.26	0.3
Column spacing (m)	1.3	1.3	1.3	1.3	1.4	1.3
Binder content (kg/m^3)	160	200	200	180	180	160
Mean strength (kPa)	412.1	430.3	331.5	465.4	435.5	327.6
Standard deviation (kPa)	153.4	191.1	188.5	171.6	179.4	182
COV	0.37	0.44	0.57	0.37	0.41	0.56

COV, coefficient of variance.

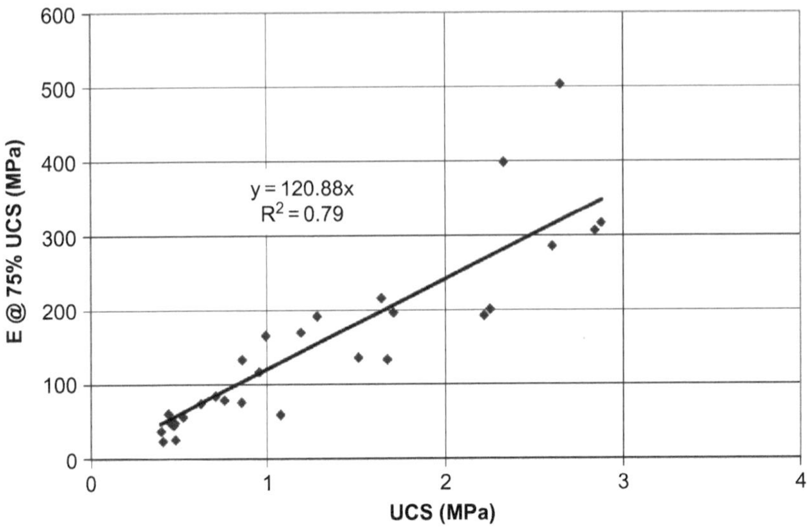

Figure 9.12 Stiffness versus UCS test data.

corresponded to locations with low mean column strength. The magnitudes of the COV are consistent with values reported by Navin and Filz (2005).

A typical plot showing column stiffness at 75% of UCS with UCS is shown in Fig. 9.12. Stiffness was interpreted at 75% of UCS because the columns were designed such that the operational load was 75% of column capacity. The data show that stiffness-to-UCS ratio was approximately 120, which corresponds to a stiffness-to-column strength ratio of 240. The design assumption was that the stiffness-to-column strength ratio would be 200. The data show that the design assumptions were met during construction.

Settlement is plotted with time in Fig. 9.13 for various embankments across the project. Settlement for the majority of the embankments was 30–50% of predicted settlement. Exceptions to this were the Phase 2 works at NSF and an area with high organic content at ECC SA where measured settlement exceeded prediction. Areas of higher than anticipated settlement corresponded with areas having lower mean column strengths and high COV of the columns. An empirical comparison of mean column strength and COV with performance suggests that good performance was achieved when the mean column strength (interpreted using $N=10$) was greater than 2.7 times the design column strength (150 kPa) with a COV of approximately 0.4. A log-normal probability density function (PDF) and a cumulative distribution function (CDF) for these mean and standard deviations are shown in Fig. 9.14. The data show that successful construction required no column strengths less than the design column strength. This observation

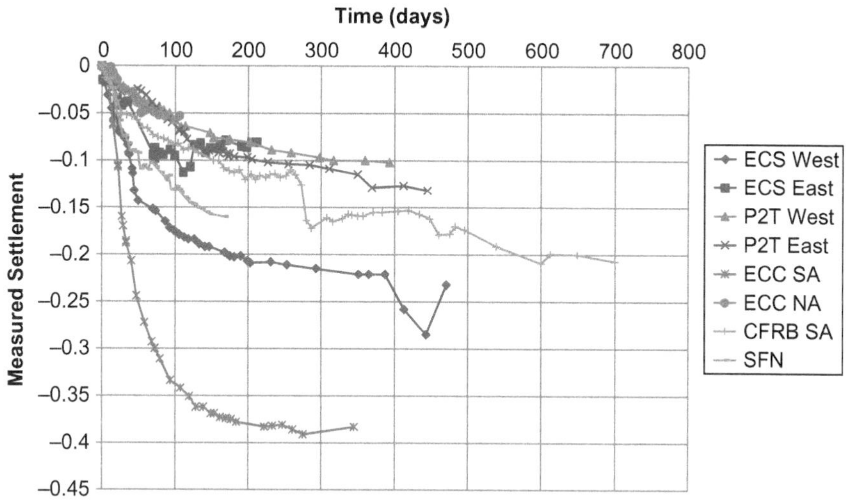

Figure 9.13 Settlement with time.

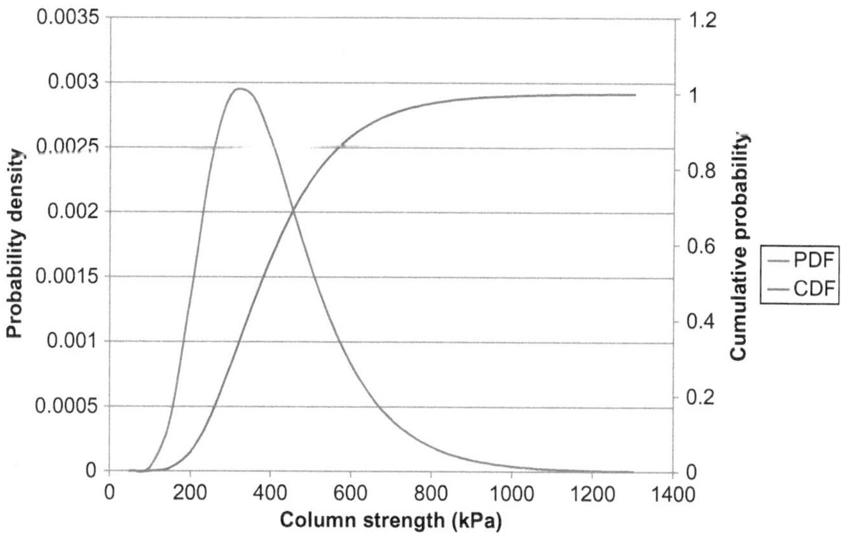

Figure 9.14 PDF and CDF for acceptable production test data.

contrasts with the acceptance criteria in the specification that allowed 10% of columns to have strengths less than the design strength providing these tests had strengths greater than 75% of the design column strength. Construction could probably have been optimized targeting lower mean strengths, but the time and effort required to consistently achieve an optimized result would have outweighed cost benefits from optimizing binder contents. The PDF and CDF shown in Fig. 9.14 could potentially be adopted in future specifications as acceptance criteria.

9.3.4 Effects of organic soils: ECC SA

ECC SA was located in a remnant mangrove area, and the soil had a high organic content of 11%. Field trials performed in this area found that a binder content of 200 kg/m^3 was required to achieve the requirements of the specification comprising a 25% slag:75% Portland cement mix. Production testing found that some of the columns did not achieve the design strength, and additional columns were installed at discrete locations. The tests also demonstrated that many of the columns only just met the acceptance strength criteria. Settlement during construction (Fig. 9.13) was much greater than expected. In these areas, the soft ground was only approximately 6 m deep, and consolidation settlements occurred rapidly and before bridge piles and pavements had to be installed. Minor increases in post-construction creep settlement were anticipated at this location.

This experience confirmed that organic soils can be difficult to stabilize (Kitazume and Terashi, 2013). Further trials using different binder quantities and types could have been beneficial here.

9.3.5 Success of the transition zones

Figure 9.15 shows the settlement with distance from the bridge abutment during construction in the area of partially penetrating columns in the transition zone shown in Fig. 9.6. The data show that progressive settlement

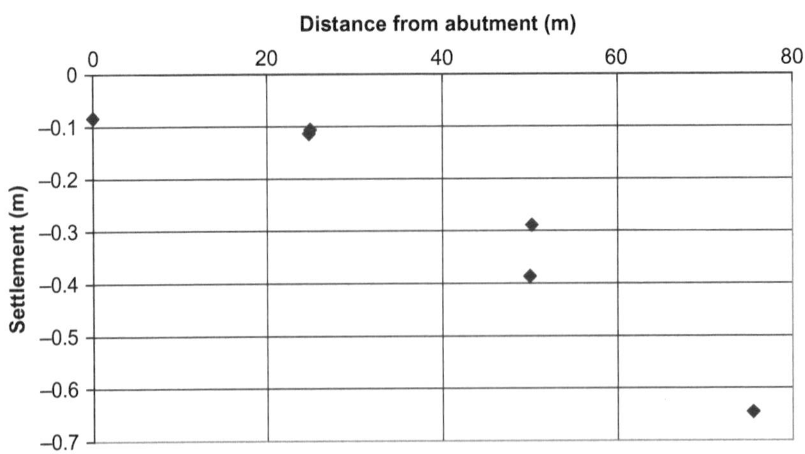

Figure 9.15 Settlement during construction with distance from the abutment.

occurred during construction as intended in design. The DSM treated soil appeared to act as a block of improved ground overlying progressively increasing thickness of soft soil.

These data highlight one of the strengths of the deep mixing method. The relatively large-diameter columns installed at close spacings can be considered to act as a block in transition zones. More rigid types of ground treatment such as concrete-injected columns having smaller diameters and installed at greater spacings are more likely to punch into the underlying soft clay and not act as a block of improved soil.

9.3.6 Embankment stability

No signs of embankment instability were observed at any location during construction. The panels and grids of DSM columns below the batters worked very effectively.

A large proportion of the installed columns were located in the panels and grids. Based on this experience, it may be possible to reduce the dimensions of the panels and grids to achieve cost savings in future projects.

9.4 CONCLUSION

The Ballina Bypass was opened 6 months early, providing benefit to the community. Use of DSM to manage time constraints on the critical path for bridge construction was a key enabler of the opportunity to open the project early. DSM also provided a cost-effective method for controlling settlement, differential settlement, and embankment stability for this project.

Some key lessons were learned during this project, including the following:

1. Soils for laboratory tests should be sampled throughout the depth profile and not just near the ground surface in order to obtain an accurate water: cement ratio.
2. Field trials are key to developing binder and mixing parameters for each location. The field trials should cover several locations within a given site in order to increase the likelihood of encountering the worst ground conditions.
3. The process of DSM construction needs to be monitored and refined to suit the site conditions:
   a. Remnant vegetation
   b. Tool configuration
   c. Ground conditions
   d. Rig capacity
   e. Moisture content and variations

4. Care should be taken when determining temporary load conditions during construction and excavation to avoid damaging or overloading columns to prevent slope failures impacting performance.
5. The quantity of production testing increased from a minimum of six columns per site (approximately 0.2%) to up to 2% of columns depending on the variability of site conditions.
6. The effect of the project acceptance criteria was to increase the mean strength far above the design strength due to the COV of the columns. Nonconformances meant replacing columns, so it was in the interest of the contractor to make stronger columns. Later settlement performance suggested that acceptance criteria were on the conservative side in general, but it was not clear what a more optimized set of criteria could be.
7. Settlement near bridge abutments was generally very good and well within allowable settlement with isolated exceptions that did not impact on outcomes.
8. Settlement in transitions was as expected. DSM is ideal for constructing transitions.
9. There were no problems with stability or lateral displacements of completed road embankments and abutments.

NOTATION

c_h = coefficient of consolidation valid for horizontal drainage
C_c = compression index
C_r = recompression index
C_α = creep index
CR = compression ratio = $C_c/(1+e_0)=\varepsilon_2/\log 2$
CRR = recompression ratio = $C_c/(1+e_0)$
OCR = overconsolidation ratio

ACKNOWEDGMENTS

The authors acknowledge the contribution from the Roads and Maritime Authority of New South Wales and the Ballina Bypass Alliance.

REFERENCES

Al-Naqshabandy, M., Larsson, S., 2013. Effect of uncertainties of improved soil shear strength on the reliability of embankments. J. Geotech. Geoenviron. Eng. 139 (4), 619–632.

Bishop, D.T., 2004. A proposed geological model and geotechnical properties of a NSW estuarine valley: a case study. In: ANZ Conference, February.

Duncan, J.M., 2000. Factors of safety and reliability in geotechnical engineering. J. Geotech. Geoenviron. Eng. 126 (4), 307–316.

EuroSoilStab, 1997. CT97-0351 EuroSoilStab—development of design and construction methods to stabilise soft organic soils. In: Design Guide for Soft Ground Stabilisation. European Industry Publication, Project No. BE96-3177.

Filz, G.M., Navin, M.P., 2010. A practical method to account for strength variability of deep-mixed ground. GeoFlorida, 2426–2433.

Kitazume, M., Terashi, M., 2013. The Deep Mixing Method. Taylor & Francis, New York.

Kitazume, M., Okano, K., Miyajima, S., 2001. Centrifuge model tests on failure envelope of column type deep mixing method improved ground—closure. Soils Found. 41 (4), 108pp.

Lee, F., Lee, Y., Chew, S., Yong, K., 2005. Strength and modulus of marine clay–cement mixes. J. Geotech. Geoenviron. Eng. 131 (2), 178–186.

Madhyannapu, R.S., Puppala, A.J., Nazarian, S., Yuan, D., 2010. Quality assessment and quality control of deep soil mixing construction for stabilizing expansive subsoils. J. Geotech. Geoenviron. Eng. 136, 119–128.

Navin, M.P., Filz, G.M., 2005. Statistical analysis of strength data from ground improved with DMM columns. In: Deep Mixing '05: Int. Conf. Deep Mixing Best Practice and Recent Advances. SGI, Stockholm.

Swedish Geotechnical Society, 1997, February. SGF Report 4:95E: Lime and Lime–Cement Columns—Guide for Project Planning, Construction and Inspection. Swedish Geotechnical Society, Göteborg, Sweden.

CHAPTER 10

Soil Modification by Admixtures

Concepts and Field Applications

Anand J. Puppala, Aravind Pedarla, Tejo Bheemasetti
Department of Civil Engineering, The University of Texas at Arlington, Arlington, Texas

10.1 INTRODUCTION

Problematic soils are those that are weak in strength and will undergo large volume change deformations due to moisture content changes from seasonal fluctuations. Soft soils experience large settlements when they are loaded, and these naturally result in excessive settlements of the structures built over them. In some scenarios, swell and shrinkage of expansive soil can cause significant damage to structures. The magnitude or extent of swell–shrink movements depends on the clayey soil type and environmental conditions, including climatic conditions. Volumetric movements deteriorate the subgrade material that, in due course, leads to the development of cracks and permanent damage. The United States alone spends billions of dollars annually on the repair and maintenance of afflicted buildings, roads, and other structures built on problematic soils, including soft and expansive soils (Nelson and Miller, 1992; Puppala et al., 1997, 2007, 2013).

Enhancement of soil strength and stiffness properties, and mitigation of volume change behavior with additive inclusion, was implemented as early as the 1930s (Chen, 1988). The increase in population and subsequent urbanization has often resulted in the construction of highways on soft and problematic subsoils. Stabilization approaches include soil treatment alternatives such as chemical or organic additives, prewetting, soil replacement, and compaction control, moisture control, surcharge loading, and thermal methods (Hausmann, 1990; Nelson and Miller, 1992; Puppala et al., 1997, 2011). Many of these methods are discussed in the current manual or elsewhere; however, this chapter focuses on chemical additive-based soil treatment methods.

The additive stabilization technique is widely used in the construction of roads, airports, embankments, and canal linings by mixing additives with clayey soils to improve their workability, strength, stiffness, swelling

Chemical, Electrokinetic, Thermal, and Bioengineering Methods
http://dx.doi.org/10.1016/B978-0-08-100191-2.00010-1

characteristics, and bearing capacity of native soils (Puppala et al., 2007, 2009; Pedarla et al., 2011). This chapter provides a comprehensive review of these stabilization methods for modifying soils and their properties.

10.2 SOIL STABILIZATION BY ADDITIVES

Modifications to the soil properties via physicochemical reactions in them prove to be more effective in sustaining the improvements over a long term. The chemical stabilization techniques often enhance shear strength properties and lower the swelling and shrinkage characteristics of soils, thus offering a better foundation base to pavements and other infrastructure (Puppala et al., 1997; Puppala and Musenda, 2000). Use of common additives such as cement, lime, and fly ash has successfully enhanced soil properties to requisite levels. Such enhancements provide stable foundation support to pavement infrastructure, thereby reducing potential maintenance costs.

The following sections detail various stabilization techniques commonly adopted to improve soil properties.

10.2.1 Mix design

Many departments of transportation follow the stabilizer selection criteria based on the plasticity index (PI) value because it has a direct impact on the selection of the additives. After selection of the stabilizer, the optimum dosage of stabilizer needs to be determined using the standard selection criteria discussed later. However, soils with a high content of the mineral montmorillonite do not stabilize effectively with the recommended dosages of lime and cement additives based on PI value. The optimum moisture content (OMC) and maximum dry unit weight (MDUW) need to be determined before further admixture characterization in the laboratory. Figure 10.1 presents a typical stabilization selection process that is followed by the Texas Department of Transportation.

Following stabilizer selection based on PI value, optimum stabilizer dosage required for effective stabilization of subgrade soil needs to be achieved. Determination of the optimum lime stabilizer content in a soil can be achieved based on the pH, unconfined compressive strength (UCS), indirect tensile strength, resilient modulus, and other soil parameters.

10.2.2 Lime stabilization

The application of lime as a stabilizer can be traced back more than 5000 years ago (Schoute, 1999). Lime was used in the construction of several

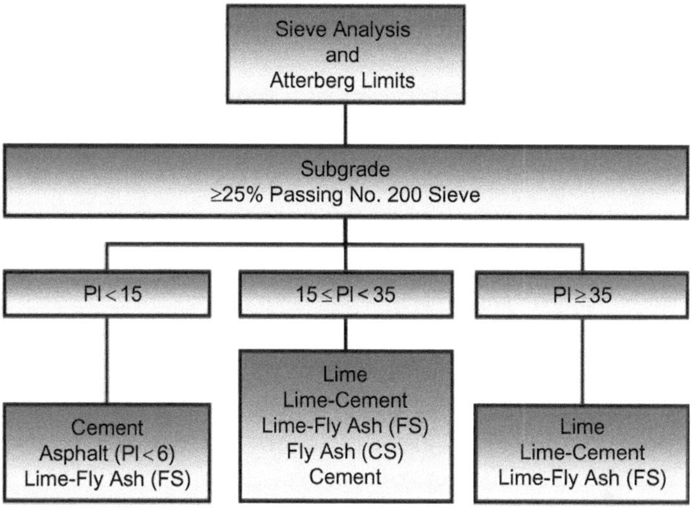

Figure 10.1 Typical stabilizer mix design chart used by the Texas Department of Transportation.

historical places, such as the Appian Way in Rome and the pyramids of Shersi in Tibet (Winterkorn and Pamukcu, 1991). These structures were built by using compacted mixtures of clay and lime (Winterkorn and Pamukcu, 1991).

Typically, there are three types of lime additives, based on their chemical compositions. Quicklime is chemically calcium oxide (CaO), and hydrated lime is calcium hydroxide ($Ca(OH)_2$). The third type, which is used less frequently in soil stabilization, is calcium carbonate ($CaCO_3$). The chemical reactions of these three types are as follows (Sherwood, 1995):

$$CaCO_3 + \text{heat} \rightarrow CaO + CO_2 \tag{10.1}$$

$$CaO + H_2O \rightarrow Ca(OH)_2 + \text{heat} \tag{10.2}$$

$$Ca(OH)_2 + CO_2 \rightarrow CaCO_3 + H_2O \tag{10.3}$$

The basic feature of this treatment is the release of calcium, which participates in ion exchange reactions to pozzolanic reactions. For lime stabilization of soils, pozzolanic reactions depend on the reactive siliceous and alumina provided by the soil. The following are the pozzolanic reactions that occur during stabilization. Calcium hydroxide is a product of the dehydration reaction and a stabilizer itself. $Ca(OH)_2$, when mixed with water, dissociates, thereby increasing the electrolytic concentration and the pH of the soil. The calcium hydroxide dissociation is explained by the following reaction (Schoute, 1999).

$$Ca(OH)_2 = Ca^{2+} + 2[OH]^- \quad (10.4)$$

The released calcium ions also participate in the cation exchange reactions of soils, which results in the diffused double layers around soil particles. This further results in improved workability, reduced plasticity, and certain amounts of shear strength gain through fabric changes in lime-treated soils (Little, 1995; Puppala and Musenda, 2000).

The second phase of reactions takes place between Si and Al ions present in clay, and Ca ions present in lime, resulting in the formation of cementitious products, including calcium silicate hydrate (C-S-H) and calcium aluminate hydrate (C-A-H). These reactions are known as pozzolanic reactions, which occur in a high pH environment and contribute to strength gain with time (Nelson and Miller, 1992; Little, 1995). The following important processes in soils are due to the previously mentioned reactions:

$$Ca^{2+} + 2[OH]^- + SiO_2 \rightarrow C - S - H\,(gel) \quad (10.5)$$

$$Ca^{2+} + 2[OH]^- + Al_2O_3 \rightarrow C - A - H\,(gel) \quad (10.6)$$

This is followed by flocculation and agglomeration of clay particles, causing a change in clay texture and reducing the percentage of fine particles (Chen, 1988; Little, 1995). The addition of lime to a soil results in an increase in the plasticity limit and a decrease in the PI value. This leads to further reduction in swell and shrinkage strains, an increase in shear strength, and a decrease in the compressibility and permeability properties of the soil (Broms and Boman, 1979; Little, 1987; Puppala et al., 1997). Often, the design dosage of lime-treated soils depends on more than one soil property. For example, in the case of expansive soils, lime dosages are dependent on soil pH conditions and volume change properties. For soft soils, they rely on pH and UCS properties. The following describes a laboratory mix design methodology for lime stabilization.

Determination of optimum stabilizer dosage

To determine the optimum amount of stabilizer that is necessary for effective stabilization, the pH-based Eades–Grim test is adopted. A pH value of 12.4 is targeted for designing the soil stabilization. The pH values of the soil–lime mixtures are determined on soils mixed with varying lime dosages, and the pH results are plotted against lime dosages, as in Fig. 10.2. For

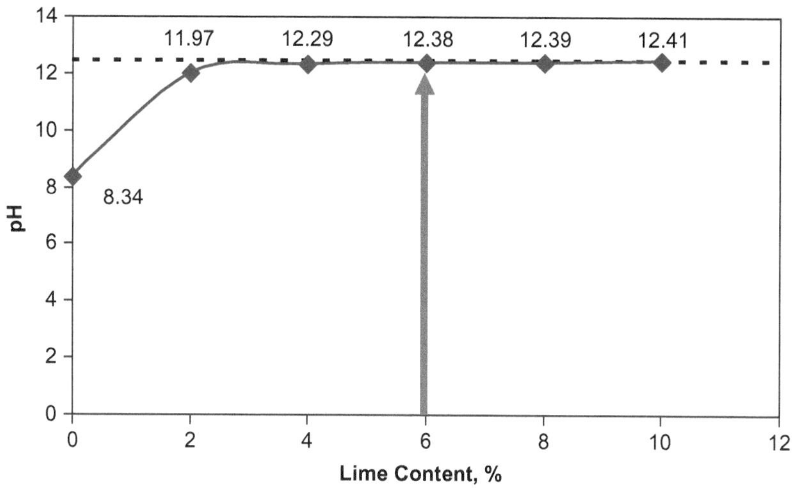

Figure 10.2 Variation in pH with lime content for El Paso clay.

example, based on test results, the minimum amount of lime necessary to stabilize the El Paso clay is 6%.

The second criterion for optimal soil–lime mix is established based on the unconfined compressive strength requirements. The Texas Department of Transportation's Tex-121-E method suggests an unconfined compressive strength value of 50 psi as adequate for soils treated with lime. The minimum number of specimens to complete this test is three per each lime content.

Lime-treated soils are designed and mixed at the OMCs and MDUW conditions for various lime treatment conditions. Soil samples are prepared and tested for UCS properties. The mixtures that provide or meet the UCS requirements will be considered for field applications.

Lime stabilizer application for subgrade treatment

Lime stabilization is commonly adopted for stabilizing unstable subgrades along roadways. Laboratory soil mix design is performed based on plasticity property tests, gradation results, pH conditions, UCS, and, in some scenarios, resilient modulus tests performed on both control and lime-treated soils. The optimum dosage of lime additive once determined is recorded for field application. Often, lime is applied in the form of slurry to minimize stabilizer loss and enhance the efficiency of mixing it with soil. Figure 10.3(a) details the lime application on the subgrade soil. Figures 10.3(b)–10.3(d) detail the mixing, compaction, and construction of roadway subgrade.

Figure 10.3 (a) Slurry application. (b) Mixing with subgrade material. (c) Final compaction of treated section. (d) Finished surface at I-35 W.

The finished surface shown in Fig. 10.3(d) will be evaluated for design specifications using nondestructive techniques. Occasionally, the engineer or quality team requests depth checks of the lime-stabilized layers, which determines whether the lime thickness was done as per the design.

Lime stabilizer application for slope treatment

Surficial slope failures are a common problem in the North Texas region, where moderate-to-high expansive soils experience desiccation cracks. Lime stabilization has been successfully adopted to prevent surficial failures caused by unstable soils (Dronamraju, 2008). Initial laboratory characterization studies need to be conducted to determine the optimum stabilizer.

In the following illustrations, the surficial in situ soil from the slope was cut and then placed on a level pad. The soil in the level pad area was pulverized, moistened, and kept ready for mixing with admixtures. The mixing needed to be initiated as soon as the spreading pass of lime was completed to

Figure 10.4 (a) Placing of fibers. (b) Mixing of fibers and lime-treated soil. (c) Watering the test section. (d) Final compaction.

prevent slurry runoff, as shown in Fig. 10.4(a). Lime and soil were mixed thoroughly on the level pad with the help of a rotary mixer.

The soil treated with admixtures was then transported and placed back in the embankment immediately after mixing. The field construction steps are presented in Fig. 10.4(b). The treated soil was compacted, as shown in Fig. 10.4(c). Adequate water was sprinkled or sprayed on the compacted soil sections during and after compaction to have proper in-place moisture to facilitate the curing process of the lime–soil mixture. The test section was then compacted using a sheep foot roller, as shown in Fig. 10.4(d).

Quality control and assessments of the lime-treated subsoil were done to assess whether the treated soils in the field possessed the required soil properties. This can be done in two ways: by collecting the soil mixture or by preparing soil specimens for laboratory pH-to-strength-to-stiffness tests. Note that the field treatments often provide properties that are less than the laboratory-designed properties; hence, the laboratory mix designs should always aim at higher properties than those that are stipulated by the specifications.

10.2.3 Cement stabilization

Cement stabilization has been practiced for years and is regarded as a popular technology resulting in high-strength soil–cement mixtures, thereby enhancing the performance of the infrastructure built above them. Hydration reactions of Portland cement occur due to the production of compounds and gels that increase the soil strength through complex pozzolanic activity (Chen, 1988; Nelson and Miller, 1992). This bonding not only enhances the cohesive nature of the treated materials so they can withstand the higher loading but also prevents them from undergoing swelling and softening from absorption of moisture and from detrimental freezing and thawing (Bugge and Bartelsmeyer, 1961).

The cementing action in granular soils is similar to that of concrete, except that cement paste does not fill the voids in the aggregates. Cement tends to join at the points of contact between particles, which is dominated by the availability of moisture and the presence of finer particles (Bugge and Bartelsmeyer, 1961). The reactions between cement and expansive soil are similar to those that occur in lime treatment (Chen, 1988; Nelson and Miller, 1992). The reactions during cement chemical processing are as follows (Schoute, 1999):

$$\underset{\text{(tricalcium silicate)}}{2\,(3CaO.SiO_2)} + \underset{\text{(water)}}{6H_2O}\ \underset{\text{(tobermorite gel)}}{3CaO.2SiO_2.3H_2O} + \underset{\text{(calcium hydroxide)}}{3Ca\,(OH)_2} \rightarrow$$

$$\underset{\text{(bicalcium silicate)}}{2\,(2CaO.SiO_2)} + \underset{\text{(water)}}{4H_2O}\ \underset{\text{(tobermorite gel)}}{3CaO.2SiO_2.3H_2O}\ \underset{\text{(calcium hydroxide)}}{Ca\,(OH)_2} \rightarrow$$

$$\underset{\text{(tricalcium aluminate)}}{3CaO.Al_2O_3} + \underset{\text{(calcium hydroxide)}}{2H_2O + Ca(OH)_2}\ \underset{\text{(tetracalcium aluminate hydrate)}}{3CaO.Al_2O_3.Ca(OH)_2.12H_2O}$$

Formation of pozzolanic compounds at the end of the chemical reactions leads to improved strength and workability of the soil. There are different grades of cement material available in the commercial market. Type I is general-purpose cement, appropriate for use on reinforced concrete structures. Type II is used for structures in water or soil containing moderate amounts of sulfates or when heat buildup is a concern. Type III cement provides high strength at an early stage, typically less than 1 week. Type IV moderates heat generated by hydration and is specifically used for massive concrete structures such as dams. Type V cement resists chemical attack by soil and water high in sulfates (Wattanasanticharoen, 2000). Selection of the cement grade type is entirely dependent on the type of soil, desired strength, and design criteria.

Optimal soil–cement mixture is typically established based on strength requirements. The minimum number of specimens to complete this design is three tests on cement-treated soils, at each cement dosage.

Cement stabilizer application for subgrade treatment

Road pavement sections with weak soil subgrade can be strengthened using the cement stabilization technique. Initially, the topsoil is loosened by scarifying the top 6–8 in. of soil with a tiller. The scarification is repeated until there are no soil lumps present in the soil. This state of soil will allow an even and homogeneous mixing of the cement stabilizer with the soil. The cement stabilizer is then mixed with the topsoil using the tiller, as shown in Fig. 10.5(a). A minimum of six passes are needed for the initial soil mixing. This is followed by spraying water uniformly over the entire mixed soil.

The whole section can then be compacted by using a sheep foot roller, as shown in Fig. 10.5(b). Numerous passes need to be made by the compactor to ensure even compaction and to attain the required dry density of the treated subgrade. Both the moisture content and the dry unit weight of the compacted material can be measured in the field using in situ moisture content and density measurement devices.

Case study

Soils are often cement-treated in areas where they are noncohesive in nature. Several cities utilize cement treatment for subgrade stabilization, primarily to enhance soil strength and stiffness properties. Note that the increases in soil strength and stiffness values with cement treatment are significantly higher than those with lime treatment. Many agencies prefer using cement treatments

Figure 10.5 Stabilizer mixed with topsoil. (b) Compaction of cement-treated soil. *(Source: Pillappa (2005)).*

for two reasons: They provide greater sustainable strength for longer periods, and there is less chemical leaching in treated soils. Many cities in North Texas are proposing combined lime–cement treatments of their subgrades, which will provide beneficial advantages of both chemical additives.

Quality control/quality assurance studies of cement treatment often require soil sampling, followed by laboratory testing. Another way to assess these is to perform nondestructive field studies, which can provide stiffness properties. All these methods need to be done prior to the curing process because any delays in sampling after soil–cement treatments would be difficult because of the hardened soil–cement mixture.

10.2.4 Fly ash stabilization

Coal ash is one of the by-products generated when coal is burnt in thermal power plants (Ferguson, 1993). It is typically composed of fly ash (flue gas stream); boiler slag, which coats boiler tubes; and bottom ash (sand-size material and boiler slag). The amount of fly ash recovered from coal ash depends on the type of coal, the burners, and the boiler, and it is generally from 65% to 85%. Fly ash is the fine-grained fraction collected from the suspension of the exhaust gases of the combustion chamber, which may be extracted using an electrostatic precipitator. Boiler ash is a relatively coarser and denser material, and it is collected by gravity at the lower level (Nicholson and Kashyap, 1993).

According to ASTM standard C593 (ASTM International, 1994), fly ash is classified as either Class C or Class F, depending on the original source of the coal used in the power generators. Class F fly ash is produced by the burning of bituminous or anthracite coal found in the midwestern, eastern, and southern regions of the United States, whereas Class C fly ash is obtained from the burning of sub-bituminous or lignite coals. The main difference between these two types is the amount of calcium, silica, alumina, and iron content (Punthutaecha et al., 2006). Details of the compositional properties are presented in Table 10.1.

Typically, in Class F fly ash, the calcium content varies from 1% to 12% and is mainly in the form of calcium hydroxide and calcium sulfate, whereas Class C fly ash may have a calcium content as high as 30–40%. Also, the amount of alkalies (combined sodium and potassium) and sulfates is generally higher in Class C fly ash. All these factors tend to make Class C fly ash a better soil stabilizing agent (Nicholson and Kashyap, 1993). However, in sulfate soil environments, Class C fly ash may lead to ettringite formation (Punthutaecha et al., 2006).

Table 10.1 Chemical requirements for fly ash classification

Properties	Fly ash class	
	Class F	*Class C*
Silicon dioxide (SiO_2) + aluminum oxide (Al_2O_3) + iron oxide (Fe_2O_3) (min%)	70.0	50.0
Sulfur trioxide (SO_3) (max%)	5.0	5.0
Moisture content (max%)	3.0	3.0

Fly ash application for subgrade treatment

Construction procedures similar to cement treatment are applied to the treatment section. Figure 10.6(a) presents the watering of the test section after tilling and mixing with fly ash stabilizer. After mixing, the whole section is then compacted by using a sheep foot roller compactor, similar to the cement-treated subgrade compaction, as shown in Fig. 10.6(b).

Most literature applications of fly ash treatments use fly ash as a co-additive. The most common forms of field treatments are lime–fly ash or cement–fly ash. As seen in Fig. 10.1, many agencies use the combined treatment when soils have plasticity properties that are in the medium range. Also, the use of fly ash is considered as a sustainable practice because many countries term fly ash as a waste material. Because of the waste material characterization, required higher dosages of fly ash additive are often equated to moderate amounts of lime or cement additives.

10.2.5 Ground granulated blast furnace slag stabilization

Blast furnace slag is a by-product of iron production and consists of a combination of the siliceous component of the iron ore and the limestone flux

Figure 10.6 (a) Watering the treated section. (b) Compaction of fly ash-treated soil. *(Source: Pillappa (2005)).*

Table 10.2 Range of composition of blast furnace slag utilized in a study by Pillappa (2005)

Chemical constituents (oxide)	Range of composition (% by mass)
Silicon dioxide (SiO_2)	32–40
Aluminum oxide (Al_2O_3)	7–17
Calcium oxide (CaO)	29–42
Magnesium oxide (MgO)	8–19
Sulfur (S)	0.7–2.2
Iron oxide (Fe_2O_3)	0.1–1.5
Manganese oxide (MnO)	0.2–1.0

used for melting iron (Sherwood, 1995) (Table 10.2). It has a chemical composition similar to that of Portland cement. Different kinds of slags are produced during the manufacture of metals from their ores, but the only product that is suitable for use as a cementitious material is ground, granulated iron blast furnace slag (Ozyildirim, 1990). Molten slag in the furnace floats to the top and is used to form a granulated glassy material that possesses hydraulic properties after it is ground to cement fineness. When the granules are ground to Portland cement fineness, it is called ground granulated blast furnace slag (GGBFS). When this material is mixed with cement, the alkalies released by the hydration of the cement are often adequate for the activation of GGBFS. Activation of GGBFS may be achieved by direct mixing with lime water and used for the production of super-sulfated cement by adding sulfates (Olorunsogo and Wainwright, 1998).

The use of slag significantly reduces the permeability of the soil mixture. The amount of reduction depends on the amount of the slag used (Ozyildirim, 1990). The permeability decreases due to reduction in the pore size associated with the production of dense calcium silicate hydrates in the hydration process that takes place during mixing (Ozyildirim, 1990). Decrease in permeability leads to high chemical resistance in aggressive environments. The slag stabilization not only increases the sulfate resistance in soils but also increases the shear strength and decreases the plasticity index, swelling potential, and shrinkage strains (Ozyildirim, 1990).

Similar to fly ash, GGBFS is also predominantly used as a co-additive. However, in fine-ground form, due to high silica content, GGBFS can be used as a sole chemical additive for soil treatments. Test results from a research study conducted on the stabilization of high-sulfate soils showed that 20% of GGBFS stabilizer treatments of natural subsoils have resulted in a significant strength increase after 3 days of curing (Pillappa, 2005). This laboratory study showed that GGBFS can be used as an effective soil stabilizer.

GGBFS applications for subgrade treatment

Construction procedures similar to those for cement and fly ash treatments can be adopted for this material treatment application. After mixing GGBFS with native control soils, as shown in Fig. 10.7, the subgrade is well mixed and then compacted. Due to dry application of this treatment, dust pollution problems occur, and these can be mitigated by hydrating GGBFS and using it as slurry for soil treatments.

10.2.6 Compost stabilization

Compost is another waste material that has been increasingly used in soil amendment applications. Compost was found to be the most efficient in recovering energy from municipal solid waste (MSW) compared to incineration and landfilling (Mitchell, 1997). The total amount of MSW generated by the United States in 2006 was 251 million tons, of which 21 million tons was recycled by composting. According to the U.S. Environmental Protection Agency (EPA), the energy equivalent of saving 21 million tons of MSW is 2.5 billion gallons of gasoline. Also according to the EPA, the number of community composting facilities has increased from 3227 to 3470 between 2000 and 2003 (USEPA). Composting is defined as the degradation of organic material under controlled conditions into useful end products. Composting degrades the organic matter by aerobic and anaerobic degradation processes.

Soil erosion and cracking often contribute to the poor performance of unpaved shoulders (Fig. 10.8). Soil erosion is affected by many factors, such as the characteristics of the soil, geology of the land, vegetation, and weather patterns. Vegetation provides a positive effect on soil erosion by firmly

Figure 10.7 (a) GGBFS application. (b) Mixing of GGBFS with soil. *(Source: Pillappa (2005)).*

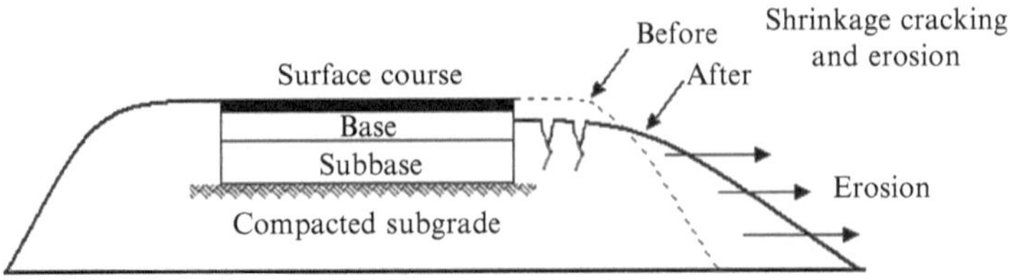

Figure 10.8 Collapsed shoulder section due to erosion. *(Source: Intharasombat et al. (2007)).*

Figure 10.9 Shoulder section at Stephenville, TX. *(Source: Intharasombat et al. (2007)).*

holding the soil together and increasing the permeability of the soil, thus preventing surface runoff.

Shoulder cracks observed on highways eventually result in cracking on travel lanes. An example of this—the cracking on U.S. Highway 108, north of Stephenville, Texas—is presented in Fig. 10.9. The unpaved shoulder subgrades depicted in Fig. 10.8 were amended with compost in a research study conducted at The University of Texas at Arlington. Two types of composts, dairy manure compost and biosolids compost, were utilized for stabilization of the shoulder in this research. Prior to construction, soil from the test plots was collected and evaluated in the laboratory. Composts were acquired from local sources and mixed with the control soil to form different types of compost-manufactured topsoils (CMTs). The performance of CMTs was monitored to address their potential effectiveness in enhancing soil properties and resulting in the reduction of both cracking and erosion.

Following the construction of the CMTs, field instrumentation and monitoring studies were initiated. The results revealed that compost-amended test sections provided better water-holding capacity and strength to the shoulder (Intharasombat et al., 2007). There were fewer erosion problems when biosolids composts, with significant natural fibers, were used for soil amendment. Overall, these studies prove that compost treatment can be considered a sustainable and effective solution for crack prevention in problematic soils. Stabilization of the composts is primarily regarded as a physical amendment process, although certain biosolids composts may provide some chemical reactions in the treatments. More research is needed to address these stabilization methodologies.

10.2.7 Other stabilizers

Bitumen

Utilizing bitumen as a stabilizer was implemented as early as the 5th century BC. Bitumen, composed of 40% heavy hydrocarbons, is a by-product of crude petroleum. When mixed with soil, bitumen forms very thin and strong films, which hold the soil particles together. As a binder, bitumen improves the water-resistant properties of soil and improves the cohesion of naturally noncohesive soils.

The additive proportion of bitumen varies in dosages from 2% to 20%. The compressive strength of the stabilized subgrade increases with the bitumen dosage, up to a certain threshold value. This stabilization method is adopted for sandy or sandy–gravel soils, which lack cohesion. Organic soils and the presence of sulfate in the soils render the stabilization process and prevent the adhesion of bitumen. Bitumen does not impart any color or odor to the stabilized subgrade.

Lime kiln dust and cement kiln dust

Lime kiln dust (LKD) and cement kiln dust (CKD) are by-products of manufacturing and processing of lime and cement, respectively. CKD contains mostly dried raw materials such as limestone, sand, shale, and iron ore. Nearly 4 million tons of CKD is disposed of every year in the United States (Miller and Zaman, 2000). Due to its pozzolanic activator attribute, CKD has been successfully implemented in road soil stabilization (Miller and Azad, 2000). CKD creates a low-ductile asphalt binder, waterproofing, and protection. CKD-treated soils exhibit reduced liquid limit and reduced plasticity indices.

CKD inclusion in soil increases unconfined compressive strength, stiffness, and durability compared to untreated soils (Miller and Azad, 2000). Although LKD and CKD additives are not considered hazardous by environmental regulatory agencies, proper supervision and handling are needed when these additives are used in the field application.

Calcium chloride

Calcium chloride is a by-product of sodium carbonates and has been implemented as a stabilizer in highway construction and maintenance. Calcium chloride ($CaCl_2$) has been used primarily as a dust palliative in roadway maintenance, as well as an accelerator in soil stabilization products. The hygroscopic property and rapid dissolution in water allow for easy application to soil treatment. Calcium chloride exhibits a higher surface tension and lower freezing point than water (Shepard et al., 1991). After the initial application, calcium chloride helps retain moisture and bind the dust particles together. This is critical during compaction because the maximum density can be easily achieved with minimum compaction effort.

Inclusion of calcium chloride in soil improves the engineering properties of the treated soils. With an increase in calcium chloride content, the dry density of the soil increases and optimum moisture content decreases. Calcium chloride has also been implemented as a secondary additive with cement or lime stabilization (Hausmann, 1990).

Sodium chloride

As a stabilizer, sodium chloride acts similar to calcium. Sodium chloride has been adopted as a long-term pavement subgrade. Studies revealed an increase in unconfined compressive strength, California bearing ratio, and indirect tensile strength of treated subgrade. Davoudi and Kabir (2011) showed that the presence of sodium chloride in a soil–lime mixture improves the compressive strength of the mixture due to the increase in pH. Calcium sodium silicate gel improves the cementation process more than calcium silicate gel. Sodium chloride imparts a deleterious effect to the soil mixture when applied in dosages more than 3%. It was found that small dosages of Na^+ surround soil mineral particles with ionic bonds, whereas large dosages (>3%) create a thick double layer, resulting in dispersing soil particles and reducing the overall compressive strength and shear strength (Davoudi and Kabir, 2011).

Bottom ash

Bottom ash is the residue that is fused into heavy particles that drop out of the furnace gas stream (air and combustion gases). Table 10.3 presents the chemical composition properties of bottom ash. These particles are formed at the bottom of the furnace. Bottom ash consists of coarser and heavier particles than fly ash (Torrey, 1978). The color of bottom ash ranges from gray to black. Bottom ash has an angular texture, and it is a relatively well-graded,

Table 10.3 Chemical composition of bottom ash

Chemical composition	%
Silicon dioxide (SiO_2)	21–60
Aluminum oxide (Al_2O_3)	10–37
Calcium oxide (CaO)	0–22
Magnesium oxide (MgO)	0–4
Sulfur trioxide (SO_3)	N/A
Ferric oxide (Fe_2O_3)	5–37

N/A, not applicable.

sand-sized material (Huang and Lovell, 1990). Chesner (1989) found that the moisture content of bottom ash ranges from 20% to 70% of the dry weight of the ash.

The specific gravity of bottom ash ranges from 1.94 to 3.46. Bottom ash is used as a construction material in artificial reefs, roads, liners, and daily covers for landfills, concrete masonry blocks, and lightweight and partial replacement of aggregate.

10.3 CONCLUSION

This chapter provided a summary of soil additive-based stabilization methods that have been commonly adopted. Stabilization methods include lime, cement, fly ash, and GGBFS. Stabilization of problematic soils, using calcium-based additives, involves chemical processes in which pozzolanic reactions take place, inducing maximum bonding-induced soil strength and minimizing the swelling potential. The properties of soil–lime mixtures and soil–cement mixtures depend on the clayey soil type, curing process and period, stabilizer type and dosage, field construction practice, and the amount of chemical additives used in the stabilization. These factors should be investigated in the laboratory before the application of the additive stabilization process in the field. Stabilization by certain additives, including ashes and slags, is a sustainable alternative because it can enhance reuse opportunities. The individual properties of the stabilizers and the field application processes were comprehensively detailed in this chapter.

REFERENCES

ASTM International, 1994. ASTM C593: standard specification for fly ash and other pozzolans for use with lime. In: Annual Book of ASTM Standards, vol. 04.01. ASTM International, West Conshohocken, PA.

Broms, B., Boman, P., 1979. Lime columns—a new foundation method. J. Geotech. Eng. Div., ASCE 4, 539–555.
Bugge, W.A., Bartelsmeyer, R.R., 1961. Soil Stabilization with Portland Cement, Highway Research Board No. 292. National Research Council, Washington, DC, pp. 1–15.
Chen, F.H., 1988. Foundations on Expansive Soils. Elsevier, Oxford.
Chesner, W.H., 1989. Aggregate-related characteristics of municipal solid waste combustion residues. In: Proc. Second Int. Conf. on Municipal Solid Waste Combustor Ash Utilization, Arlington, VA.
Davoudi, M.H., Kabir, E., 2011. Interaction of lime and sodium chloride in low plasticity fine grain soils. J. Appl. Sci. 11 (2), 330–335.
Dronamraju, V., 2008. Studies on Field Stabilization Methods to Prevent Surficial Slope Failures of Earthfill Dams. Dissertation, The University of Texas at Arlington, Texas.
Ferguson, G., 1993. Use of self-cementing fly ashes as a soil stabilization agent. Geotech. Spec. Publ. 36, 1–3.
Hausmann, M.R., 1990. Engineering Principles of Ground Modification. McGraw-Hill, New York.
Huang, W.H., Lovell, C.W., 1990. Bottom Ash as Embankment Material. Geotechnics of Waste Fills—Theory and Practices, ASTM STP 1070. ASTM International, West Conshohocken, PA, pp. 71–85.
Intharasombat, N., Puppala, A.J., Naik, B., Qasim, S.R., Williammee, R., 2007. Experimental investigations to evaluate benefits of using compost amendments to modify expansive soils. Int. J. Environ. Waste Manage. 1 (2/3), 291–307.
Little, D.N., 1987. Fundamentals of the Stabilization of Soil with Lime, Bulletin No. 332. National Lime Association, Arlington, VA.
Little, D.N., 1995. Stabilization of Pavement Subgrades and Base Course with Lime. Kendall Hunt, Dubuque, IA.
Miller, G.A., Azad, S., 2000. Influence of soil type on stabilization with cement kiln dust. Const. Build. Material 14, 89–97.
Miller, G.A., Zaman, M., 2000. Field and laboratory evaluation of cement kiln dust as a soil stabilizer. Transport. Res. Rec. 1714, 25–32.
Mitchell, D., 1997. Compost Utilization by Departments of Transportation in the United States. Department of Environmental Horticulture, Soil and Water Science Department, University of Florida, Gainesville, FL.
Nelson, D.J., Miller, J.D., 1992. Expansive Soils: Problems and Practice in Foundation and Pavement Engineering. Wiley, New York.
Nicholson, P.G., Kashyap, V., 1993. Fly ash stabilization of tropical Hawaiian soils. ASCE Geotech. Spec. Publ. 36, 15–19.
Olorunsogo, F.T., Wainwright, P.J., 1998. Effect of GGBFS particle-size distribution on mortar compressive strength. J. Mater. Civil Eng. 10 (3), 180–187.
Ozyildirim, H.C., 1990. Admixtures and Ground Slag for Concrete, Transportation Research Board No. 365. National Research Council, Washington, DC, pp. 33–43.
Pedarla, A., Puppala, A.J., Chittoori, B., 2011. Influence of mineralogy and plasticity index on the stabilization effectiveness of expansive clays. Transport. Res. Rec. 2212, 91–99.
Pillappa, G.S., 2005. Field and Experimental Studies to Assess the Performance of Stabilized Expansive Clay. Master's Thesis, The University of Texas at Arlington, Arlington, TX.
Punthutaecha, K., Puppala, A.J., Vanapalli, S.J., Inyang, H., 2006. Volume change behaviors of expansive soils stabilized with recycled ashes and fibers. J. Mater. Civil Eng. 18 (2), 295–306.
Puppala, A.J., Musenda, C., 2000. Effect of fiber reinforcement on strength and volume change behavior of two expansive soils. Transport. Res. Rec. 1736, 134–140.
Puppala, A.J., Mohammad, L.N., Allen, A., 1997. Engineering behavior of lime treated Louisiana subgrade soil. Transport. Res. Rec. 1546, 24–31.

Puppala, A.J., Pokala, S., Intharasombat, N., Williammee, R., 2007. Effects of organic matter on physical strength and volume change properties of compost amended expansive clay. J. Geotech. Geoenviron. Eng. 133 (11), 1449–1461.

Puppala, A.J., Manosuthkij, T., Hoyos, L.R., 2009. Better characterization of expansive soils for the design of pavement infrastructure. In: Infrastructure Development on Expansive Soils, INDEX 2009 Proceedings, August 27–29, 2009, pp. 74–89.

Puppala, A.J., Thammanoon, M., Nazarian, S., Hoyos, L.R., 2011. Threshold moisture content and matric suction potentials in expansive clays prior to initiation of cracking in pavements. Can. Geotech. J. 48 (4), 519–531.

Puppala, A.J., Manosuthkij, M., Chittoori, S., 2013. Swell and shrinkage characterizations of unsaturated expansive clays from Texas. Eng. Geol. 164, 187–194.

Schoute, E.J., 1999. Chemical Stabilization of Soft Clay. Master's thesis, Memoirs of the Center of Engineering Geology in The Netherlands 188, pp. 15–39.

Shepard, J.M., Pickett, J., Kienzle, M., 1991. Full-Depth Reclamation with Calcium Chloride. Transportation Research Record 1295. National Academy of Sciences - National Research Council, Washington D.C.

Sherwood, P.T., 1995. Soil Stabilization with Cement and Lime. HMSO, London, pp. 14–55.

Torrey, S., 1978. Coal Ash Utilization: Fly Ash, Bottom Ash and Slag. Noyes, Park Ridge, NJ.

Wattanasanticharoen, E., 2000. Investigate to Evaluate the Performance of Four Selected Stabilizations on Soft Subgrade Soils of Southeast Arlington. Master's ThesisThe University of Texas at, Arlington, Arlington, TX.

Winterkorn, H.F., Pamukcu, S., 1991. Soil stabilization and grouting. In: Fang, H. (Ed.), Foundation Engineering Handbook. Van Nostrand Reinhold, New York, pp. 317–378.

CHAPTER 11

Deep Mixing Method

Equipment and Field of Applications

Nicolas Denies[1], Noël Huybrechts[1,2]
[1]Belgian Building Research Institute, Brussels, Belgium
[2]KU Leuven, Leuven, Belgium

11.1 INTRODUCTION

Ground improvement is a large and important domain in soil mechanics and geotechnical engineering and consists in a wide variety of techniques and methods adapted to a broad range of problems. It cannot be denied that during the past decades the importance of the ground improvement market has increased enormously. New methods, tools, and procedures have been developed and applied in practice. This is especially the case for the deep mixing method. This is not surprising, because it is an outstanding, competitive, and sustainable construction process.

For several decades, the deep mixing method has been used as a ground improvement technique. Although it was first used in the 1950s, the method was not largely employed in the United States until the 1980s. It was introduced in the 1960s in Japan and in the Scandinavian countries. In Europe, initially considered as an alternative to the jet grouting application, the deep mixing method was first used in the late 1980s with the emergence of various execution systems. After the development of various shaft configurations, the market witnessed the emergence of global mass stabilization, trench mixing (in the early 1990s), and the cutter soil mix (CSM) (2003). The historical development of the deep mixing method throughout the world is fully described in Bruce et al. (1998) and Topolnicki (2004). Kitazume and Terashi (2013) and Bruce (2014) respectively concentrate on the historical review of the method in Japan and in the United States.

According to the classification of the ground improvement methods adopted by the ISSMGE TC211 (see Chu et al., 2009), the deep mixing method can be classified as a ground improvement method with grouting-type admixtures. In this method, the weak soil is treated by mixing it with cement, lime, or other binders in situ using a mixing machine. In

Chemical, Electrokinetic, Thermal, and Bioengineering Methods
http://dx.doi.org/10.1016/B978-0-08-100191-2.00011-3

Table 11.1 Deep mixing terminology

CCP: chemical churning pile
CDM: cement deep mixing
CMC: clay mixing consolidation method
DCCM: deep cement continuous method
DCM: deep chemical mixing
DJM: dry jet mixing
DLM: deep lime mixing
DMM: deep mixing method
DSM: deep soil mixing
DeMIC: deep mixing improvement by cement stabilizer
In situ soil mixing
JACSMAN: jet and churning system management
Lime–cement columns
Mixed-in-place piles
RM: rectangular mixing method
Soil–cement columns
SMW: soil mix wall
SWING: spreadable WING method

Source: Porbaha (1998).

response to the growing variety of systems available on the market, Porbaha (1998) proposed a general terminology, as presented in Table 11.1.

Many reviews of the deep mixing method are available, including those by Terashi (2003), Topolnicki (2004), Larsson (2005), Essler and Kitazume (2008), Denies and Van Lysebetten (2012a), and Kitazume and Terashi (2013). Specialty international conferences have been held in Tokyo (1996 and 2002), Stockholm (1999 and 2005), Helsinki (2000), New Orleans (2003 and 2012), and Osaka (2009), with high attendance demonstrating the large worldwide success of this method.

In parallel, the results of national and European research programs have been published in multiple interesting reports (e.g., Coastal Development Institute Tokyo (CDIT), 2002; EuroSoilStab, 2002). The European standard for the execution of deep mixing, "Execution of Special Geotechnical Works—Deep Mixing" (EN 14679), was published in 2005, and in 2013 the U.S. Federal Highway Administration (FHWA) published a design manual titled "Deep Mixing for Embankment and Foundation Support." Currently, in its catalog of technologies, the web database Geotech Tools (Iowa State University, 2010–2014) provides practical guidelines for the design, execution, and control of the deep mixing works.

Most of these research projects and documents have focused on the global stabilization of soft cohesive soils such as silt, clay, and peat. Nevertheless, as illustrated in this chapter, the applicability of the method for structural applications can be no more put in doubt.

11.2 CONSTRUCTION PRINCIPLES AND EQUIPMENT

11.2.1 Introduction and classification

In the deep mixing process, the ground is mechanically (and possibly hydraulically or pneumatically) mixed in place while a binder, generally based on cement or lime, is injected with a specially made (or customized) machine. The deep mixing method can be classified according to its execution process based on the mixing method and the injection mode.

There are two types of installation methods based on the way the binder is injected into the ground (with or without water addition): the wet and the dry mixing methods. In the wet mixing method, which is more frequently applied, a mixture of a binder and water with possibly sand or additives is injected and mixed in place with the soil. Depending on the type of soil and binder, a mortar-like mixture is created that hardens during the hydration process (Essler and Kitazume, 2008).

In the dry soil mixing process, the binder is directly mixed with the soil (generally during the withdrawn phase). The binding agents directly react with the prevailing soil and the contained water and form a soil mortar. Quasthoff (2012) provides a brief state of the art in dry soil mixing and reviews its construction principles, its equipment, and its field of applications. Interested readers can also refer to the proceedings of the International Conference on Dry Mix Methods for Deep Soil Stabilization held in Stockholm in 1999 (Bredenberg et al., 1999).

As previously mentioned, the different types of soil mix systems available on the market can be classified according to the way the mixing is performed into the ground. Such classification has been provided in the past by Bruce et al. (1998), Topolnicki (2004), and Essler and Kitazume (2008). During the short courses of the last International Symposium on Ground Improvement IS-GI 2012 held in Brussels (see Denies and Huybrechts, 2012), Topolnicki (2012) presented an updated classification scheme as illustrated in Fig. 11.1. The different systems are separated according to four levels of classification taking into account (1) the dry or wet mixing; (2) the mechanical, hydraulic, or hybrid method of mixing; (3) the position of the mixing; and (4) the axis of rotation of the mixing tools.

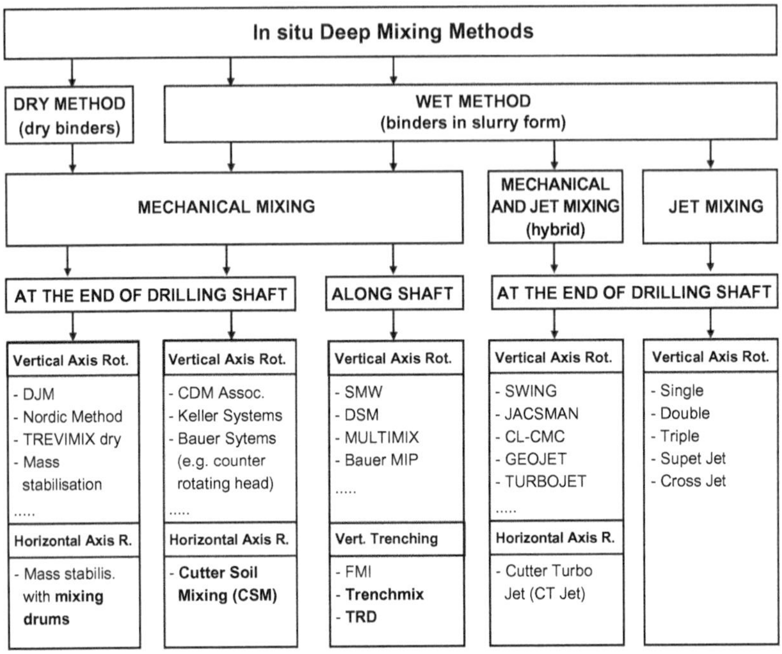

Figure 11.1 Updated classification scheme of soil mixing systems. *(Source: From Topolnicki (2012)).*

Depending on the application, different improvement patterns can be designed with these various systems considering soil–cement columns, rectangular soil mix panels, continuous barriers, or global mass stabilization such as illustrated in Fig. 11.2. Typical improvement patterns of treated soil mass can be found in CDIT (2002), Topolnicki (2004), FHWA (2013), and Kitazume and Terashi (2013).

The deep mixing method can be performed on land or for marine applications. Kitazume and Terashi (2013) describe the equipment available for

Figure 11.2 Typical ground improvement patterns of treated soil (soil–cement columns, soil mix panels, continuous barriers, and global mass stabilization).

Table 11.2 Deep mixing methods and equipment used internationally

Reference	Methods and equipment
Bruce et al. (1999)	DJM, lime–cement columns, Trevimix Dry
CDIT (2002)	CDM, DJM
Topolnicki (2004)	DJM, Nordic method, Trevimix Dry, shallow soil mixing (SSM) method, CDM, DCM, SCC, Hayward Baker–Keller mixing tools, Bauer mixing tools, mix-in-place (MIP), SMW, DSM, Colmix, spread wing (SWING), JACSMAN, Hydramech, Trevimix Wet, Turbojet, Geojet, FMI (cut–mix–injection) machine
Larsson (2005)	CDM, Colmix, CSM, TRD, Hayward Baker–Keller mixing tools, SCC, Geo-Solutions tools, Raito tools, Schnabel DMW (deep mix wall), May Gurney tools, Trevi tools, Rectangular 1 (cutting wheels), Rectangular 2 (box columns), FMI (cut–mix–injection) machine, DJM, Nordic method, SSM method, ALLU mass stabilization mixing tools, SWING, JACSMAN, LDis
Kitazume and Terashi (2013)	CDM, DJM
FHWA (2013)	CDM, DJM, DSM, SMW, Trevimix Wet, Colmix, soil removal technique, SSM, ISS auger method, RAS column method, Rectangular 1 (cutting wheels), Rectangular 2 (box columns), SAM, cementation, single axis tooling, rotomix, CSM method, SWING, JACSMAN, LDis, GeoJetTM, Hydramech, RAS Jet, Turbomix/Turbojet, TRD, Nordic method, Trevimix Dry, MDM (modified deep mixing), dry soil mixing mass, Schnabel DMW (deep mix wall)

near-shore construction, such as port and harbor facilities or man-made island constructions, including the use of deep mixing barges.

Table 11.2 provides a list of references detailing various soil mixing equipment used internationally. Bruce et al. (1999) especially provide a summary of the mixing equipment for the common dry deep mixing methods. Topolnicki (2004) and Larsson (2005) provide a full description of the soil mixing systems used in Europe, Japan, and the United States. Kitazume and Terashi (2013) mainly focus on the application of the cement deep mixing (CDM) and dry jet mixing (DJM) techniques as applied in Japan. FHWA (2013) contains tables detailing deep mixing equipment, tooling data, and treated soil material properties.

The following sections do not provide an exhaustive review of the aforementioned deep mixing systems. Rather, they illustrate the variety of the wet deep mixing equipment available on the market today. Dry methods and shallow mixing methods are not discussed here.

11.2.2 Wet mixing in single or multiple shaft configurations

A large variety of wet mixing systems are available in single shaft configuration. In these soil–cement column systems, the mixing can be mechanically performed at the end of the shaft, such as illustrated in Figs. 11.3 and 11.4, or alternatively along the shaft. In several circumstances, hybrid mixing can be applied with jet assistance, such as in the tubular soil mixing (TSM) system. The TSM technique, as used in Belgium by Smet-Boring, uses both mechanical and hydraulic methods of mixing. Apart from the rotating mixing tool, the soil is cut by the high-pressure injection (up to 500 bar) of the water/binder mixture. As illustrated in Fig. 11.5, an external tube can be used to obtain regular diameter.

To increase production rate, multiple shaft configuration systems have been developed, some of which are equipped with a jet assistance device (Larsson, 2005).

Figure 11.3 Keller wet mixing system in single shaft configuration (available tool diameter for single shaft ranging between 40 and 240 cm). *(Source: From Topolnicki (2012), with courtesy of Keller).*

Figure 11.4 Bauer Single Column Mixing-Double Head, SCM-DH system (available tool diameter for single shaft ranging between 180 and 240 cm). *(Source: After Topolnicki (2012), with courtesy of Bauer).*

Figure 11.5 Smet Tubular Soil Mixing TSM system (column diameter of 38–73 cm). *(Source: After Denies et al. (2012a) and Topolnicki (2012), with courtesy of Smet-Boring nv).*

11.2.3 Wet mixing spreadable systems

Soletanche Bachy has developed the "Springsol" wet soil mixing tool (Guimond-Barrett et al., 2012). As illustrated in Fig. 11.6, this tool is equipped with two mixing blades that spread out under the action of springs. In its folded configuration, the tool diameter is 160 mm, enabling its

Figure 11.6 Soletanche Bachy wet spreadable mixing tool SPRINGSOL. *(Source: From Borel (2012) and Guimond-Barrett et al. (2012), with courtesy of Soletanche Bachy).*

insertion into a temporary casing. By increasing the length of the mixing blades, the column diameter can be adapted (e.g., 40 and 60 cm as illustrated in the top right side of Fig. 11.6). The main advantages of the Springsol tool are the ability to reinforce the ground under an existing railway track or an existing platform (slab and superficial isolated or continuous footings) and the ability to work under low headroom conditions (Borel, 2012). Keller has also developed wet spreadable systems. The first spreadable tool designed by Keller had an external diameter of 300 mm (for its core retractable tool). Keller Foundations has designed the FLAPWINGS system. It consists of a 150-mm core retractable tool that is able to open in order to perform soil mix columns with a diameter of 600 mm (Lambert et al., 2012).

11.2.4 CSM panels

The execution of soil mix rectangular panels can easily be performed using the cutter soil mixing (CSM) system developed by Bauer Maschinen GmbH. As reported in Gerressen and Vohs (2012), the CSM is based on the principle of the trench cutter technique. It is used mainly for the construction of cutoff walls, earth retaining structures, and ground improvement. As it is derived from Bauer Cutter technology, the system extends the applicability of soil mixing to much harder strata. While a self-hardening water/binder mixture is being introduced, soil formations are easily penetrated, broken down, and mixed with the water/binder mixture using the cutter wheels as cutting and mixing tools. The two cutting wheels rotate independently about a horizontal axis. Figure 11.7 illustrates the cutting and mixing tools of the CSM developed by Bauer. Another CSM system that is available on the market has been developed by Soletanche Bachy-TEC (Borel, 2012).

11.2.5 Trench mixing

The principle of the trench mixing method is to produce a soil mix barrier, generally up to a depth of 10 m, in a single continuous pass, which is an advantage particularly in the case of water retaining function (no joints). Figure 11.8 shows the Trenchmix system. For deep and large applications, use of the trench remixing deep (TRD) system can also be envisaged (Burke, 2009).

11.2.6 Advantages and disadvantages of the deep mixing method

The variety of deep mixing equipment used internationally allows the execution of soil mix material in a large range of soil types. Table 11.3 summarizes the main advantages and limitations of the method, as given in Topolnicki (2012). According to the author's experience, the following are additional advantages of the deep mixing method:

Figure 11.7 The cutting/mixing tools of the Bauer CSM system (left) and the QuattroCutter and SideCutter systems (right). *(Source: From Gerressen (2012), with courtesy of Bauer).*

Figure 11.8 Trenchmix tool. *(Source: Courtesy of Soletanche-Bachy/Mastenbroek).*

Table 11.3 Advantages and disadvantages of the deep mixing method

Advantages	Disadvantages
High productivity usually possible, hence economical for large-scale projects	Depth limitations (depending on the method applied)
Can be potentially used in all types of soils and fills (without obstructions)	Not applicable in soils that are very dense, very stiff, or that may have boulders
Column's spacing and patterns highly variable, arrangements tailored to specific needs	Limited or no ability to install inclined columns (depending on the equipment applied)
Engineering properties of treated soil can be closely designed	Uniformity and quality of mixed soil may vary considerably in certain conditions
Causes minimal lateral or vertical stress that could potentially damage adjacent structures	Columns cannot be installed in close proximity to existing structures (except hybrid mixing)
No vibration, medium-low noise	Freeze/thaw degradation may occur
Very low spoil (especially for dry method)	Significant spoil produced with wet method
Can be used for on-land, waterfront, and marine projects	Weight of the equipment may be problematic for weak soils (depending on the method)
Quality of treatment verifiable during construction	Limited ability to treat isolated strata at depth

Source: Topolnicki (2012).

- Use of the existing soil as a construction material.
- Control of the geometry of the soil mix element with depth.
- Contrary to concrete secant pile walls, the execution of the soil mix walls does not suffer from delayed supply (e.g., due to traffic jams) of the fresh concrete
- For the wet mixing method, the amount of spoil return is more limited and more controllable than that for jet grouting or slurry walls.
- Dewatering is not required.

11.2.7 Wet or dry mixing method

The choice of using dry or wet mixing methods is often related to the local historical developments. It can also be linked to the available machines on the local market or to economics. Other aspects may also have to be considered. Topolnicki (2012) compared both processes in Table 11.4.

Table 11.4 Choice of the dry or wet process

Item of concern	Main limitations
Initial water content of the soil to be treated	Cohesive soils with moisture content $w = 60$–200% are best suited for the dry process (lower limit $w > 20\%$, water content below plastic limit is not fully available for hydration).
Quality of mixing	Wet process usually provides better homogeneity of stabilized soil because of easier distribution of slurry across the column area, prehydration of cement, and longer mixing time.
Compressive strength of soil–binder mix	Higher strength is more reliably obtained with the wet process, except for very wet soils.
Ability to penetrate through hard soil layers	Much higher for the wet process due to the "lubrication" effect of the injected slurry and due to higher torque capacity of rigs.
Stratified soils	Wet mixing can provide more uniform strength along the column length due to partial soil exchange/movement in the vertical profile; quality control more difficult for the dry process.
Spoil	Dry mixing creates very little or no spoil.
Use of combined binders and industrial by-products	Quite frequent in dry mixing, slag cement in wet mixing, other binders and by-products very rare.
Air temperature below 0 °C	Dry process is significantly less affected by low temperatures because compressed air is used to transport the binder.
Column reinforcement	Possible in combination with the wet process.

Source: Topolnicki (2012).

11.3 HYDROMECHANICAL CHARACTERIZATION OF THE DEEP SOIL MIX MATERIAL

Several parameters have an influence on the produced deep soil mix material. The quality of the soil mix material depends on the binder type and content, the in situ soil, the execution process, and the curing conditions. The contaminated character of the ground and the exposure conditions of the soil mix elements during their lifetime will have a direct impact on the durability of the soil mix material. Design criteria of permanent soil mix elements may be based on the execution monitoring (of the mixing energy, depth, etc.) and on performance tests, including unconfined compressive strength (UCS), freeze–thaw and wet–dry durability, leachability (in particular cases), porosity, and permeability (if required) tests. Nevertheless, it must be kept in mind that the test procedures directly influence the results of the characterization, possibly resulting in various conclusions.

The topic of mechanical characterization of the soil mix material is beyond the scope of this book. The reader can refer to the following references to obtain information relative to this subject: Bruce et al. (1998), Porbaha et al. (2000), CDIT (2002), Topolnicki (2004), Kitazume (2005), Denies et al. (2012b), Denies and Van Lysebetten (2012a, 2012b), Kitazume and Terashi (2013), FHWA (2013), and, specifically for the dry mixing method, Bruce et al. (1999) and Bredenberg et al. (1999).

11.4 FIELD OF APPLICATIONS AND CASE HISTORIES

Originally, the deep mixing method was developed for ground improvement applications in soft clays and organic soils. Nevertheless, recently, it is increasingly more dedicated to various structural and environmental applications. Table 11.5 summarizes the main applications of the deep mixing method. The following sections illustrate most of them.

11.4.1 Deep mixing method for excavation support: Construction of earth and water retaining structures

In recent years, the deep mixing method has increasingly been used for the construction of earth–water retaining structures. In fact, the soil mix walls represent a more economical alternative to concrete secant pile walls and even, in several cases, king post walls (i.e., soldier pile walls). An historical background of excavation support using soil mix walls is proposed by Rutherford et al. (2005). In this application, the discrete soil mix elements (columns or panels) are placed next to each other, in a secant way. By overlapping the different soil mix elements, a continuous soil mix wall is realized. Steel H- or I-beams are inserted into the fresh soil mix material to resist the shear forces and bending moments. The maximum installation depth of these soil mix walls is on the order of 25 m, but higher installation depths may be observed in the future. The main structural difference between the soil mix walls and the traditional secant piles is the constitutive soil mix material instead of concrete.

Table 11.5 Summary of the deep mixing applications

- Soil reinforcement and foundations
  - As an alternative to classical foundation
  - For underpinning with the help of the spreadable systems
  - For the realization of the foundation of linear structures such as railway tracks and pipelines with the help of the trench mixing systems
- Earth/water retaining walls\
  - For excavation
  - As shaft structure
  - As pit for (micro)tunneling activities
- Cutoff walls, floodwalls, and reinforcement of land levee and embankment
- Slope stabilization and landslide mitigation
- In situ remediation
  - For Permeable Reactive Barriers (PRB) walls
  - For containment walls
  - For soil treatment by stabilization/solidification
- Global mass stabilization
  - For the total shallow treatment of an area (e.g., for industrial installations)
- Liquefaction mitigation
  - With the construction of soil mix caissons
  - With specific arrangement of isolated soil mix elements and soil mix walls
- Land reclamation
  - Particularly in the case of near-shore construction (port and harbor facilities and man-made islands)

In Belgium, the deep mixing method is mainly used for the construction of soil mix walls for earth and water retaining structures. Figure 11.9 illustrates different cases of excavation performed in Belgium with the help of anchored or shored CSM panels or soil–cement columns. The following two case histories illustrate the possibilities of this field of application.

Pinto et al. (2012) report a case history in Portugal for which the CSM technique was applied for the construction of two shafts with temporary earth and water retaining functions. The shafts, approximately 18 m deep and 15 m in diameter, were built to allow the installation of a water supply pipe under a river using microtunneling technology. The CSM panels were built with an overall depth of 24 m and reinforced with steel beams (type IPE 300).

The CSM panels were designed to transmit horizontally the earth and water pressures to these vertical beams. The beams were braced at the top by a capping beam (made of reinforced concrete) and by three lower levels of steel ring beams, as illustrated in Fig. 11.10. For the design, a

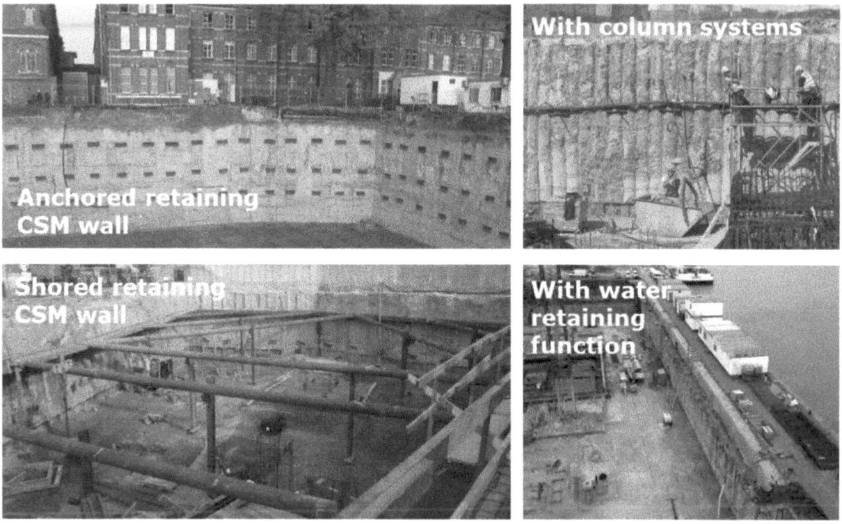

Figure 11.9 Different types of soil mix wall with a retaining function.

two-dimensional finite Element modeling (FEM) axisymmetric model was adopted using Plaxis software.

Pinto et al. (2012) also reported on the execution of an earth–water retaining and foundation structure (with permanent function) realized for the construction of a pumping station. The depth of the excavation was 12 m. CSM panels with an overall depth of approximately 24 m were performed and reinforced with vertical steel beams (type IPE 300) in order to resist the structural loads, earth and water pressures, and to limit the deformations. Buttress panels were also built in order to increase the overall stiffness of the earth retaining structure, as illustrated in Fig. 11.11. The CSM panels were designed to be integrated with the final foundation and earth retaining solution and to limit the water inflow into the excavation. For the design, two-dimensional FEM analyses were again carried out using Plaxis software.

This second case history provides an excellent introduction for the following field of application of the method: the use of the deep mixing method as an alternative to the traditional foundation solutions.

11.4.2 Deep mixing method for foundation applications

Although the deep mixing method has been used for several decades as a ground improvement method, there is a growing trend: the use of soil mix elements with a bearing function. This practice is illustrated with a case

Figure 11.10 Inside view of the shaft after excavation. *(Source: After Pinto et al. (2012)).*

history reporting the use of CSM panels as the foundation of an industrial building in Portugal.

Peixoto et al. (2012) describe an application of the CSM technology for the foundation of an industrial building at Fréjus, France. The main concern during the design was the minimization of the total and differential settlements of the building structure. The loads were transmitted to the

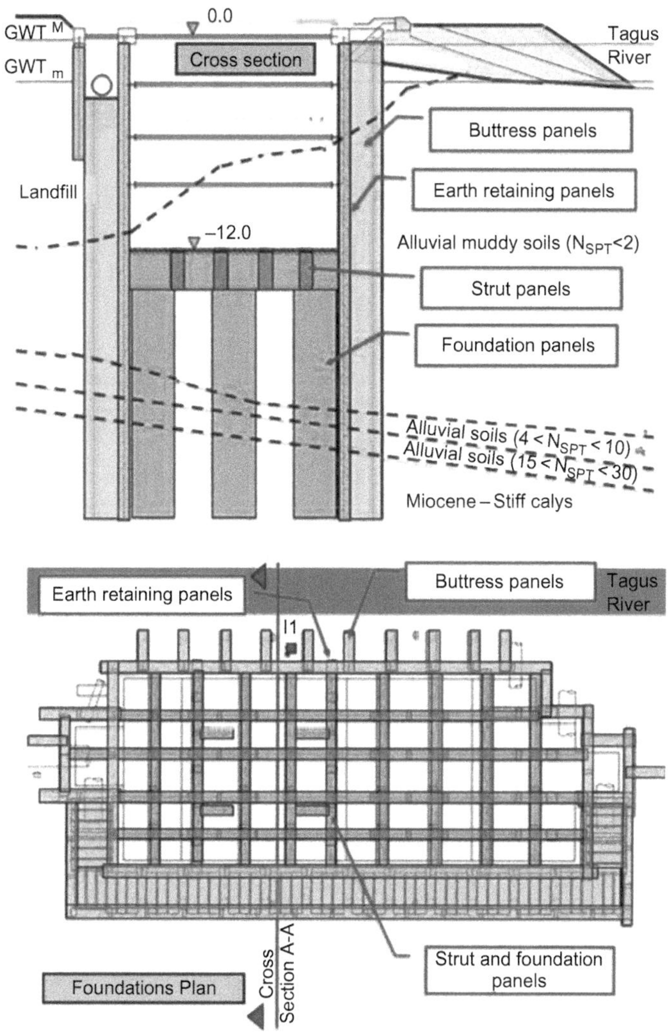

Figure 11.11 Cross section (top) and plan of the adopted solution (bottom). *(Source: After Pinto et al. (2012)).*

marl–sandstone substrate, detected at a depth varying between 3 and 7 m. As illustrated in Fig. 11.12a, enlargement of the top of the panels was also executed to ensure an efficient transfer of the load to the CSM panels. The panel caps were filled by the overflow mixture resulting from the execution of the CSM panels. Figure 11.12b presents the distribution of the CSM panels. A load transfer layer made of granular material and with a thickness of 0.60 m was finally constructed, on which the concrete bottom slab of the building

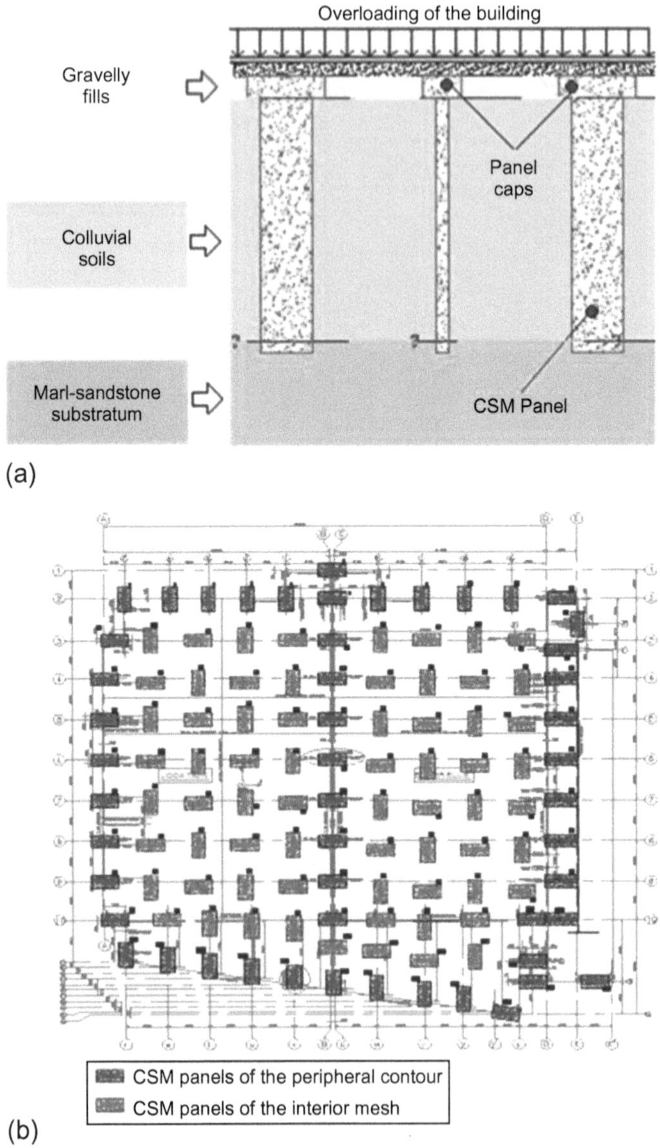

Figure 11.12 (a) Cross section of the performed solution and (b) plan view of the CSM panel distribution. *(Source: From Peixoto et al. (2012)).*

was placed. The analysis of the solution in terms of long-term settlements was carried out using the FEM Plaxis program.

This case study is a reflection of a current growing trend: the comparison during the design phase between at least two alternative solutions—the design of the foundation with the help of a classical pile (or pile raft) solution and the possibility to use soil mix elements covered by a load transfer platform made of granular material (possibly) reinforced with geosynthetics. Although there are many standards and guidelines available for the design of pile foundation, the use of numerical modeling to assess the safety of the combined solution (soil mix elements and load transfer platform) is usually required because of the lack of knowledge of the fundamental mechanisms governing the behavior of this solution.

When considering the long-term structural function of these soil mix elements, the durability of the soil mix material has to be discussed. As reported by Denies et al. (2015b), durability of the soil mix material is a complex topic because it relates to aspects of the evolution and/or the degradation of the hydromechanical characteristics of the soil mix material with time (strength, stiffness, permeability, pH, etc.).

However, there is also the issue of the durability of the soil mix material executed in contaminated grounds or in soils containing compounds that can have a negative effect on the development of its characteristics (chlorides, sulfates, hydrocarbons, heavy metals, etc.). The durability of the soil mix material will also have an impact on the (rate of) corrosion of steel beams integrated into the fresh soil mix material during execution. In the soil mixing process, the contaminants are directly mixed with the injected binder and with the ground. Hence, they will be integrated into the soil mix matrix. As a result, the potential impact of these compounds is more important for the soil mix elements than for cast-in-place or precast concrete elements.

As reported by Denies et al. (2015b), two antagonistic phenomena play a role in the durability of the soil mix material. On the one hand, there is a long-term increase in its strength with time (Terashi, 2002; Topolnicki, 2004; Ganne et al., 2010; Bellato et al., 2012; Filz et al., 2012). On the other hand, there is a progressive degradation of the material observed with time due to several factors (outward diffusion of the cations Ca^{2+}, carbonation process, and freeze–thaw and wet–dry cycles). In addition to these aging effects, designers and soil mix contractors have to consider the contaminated character of the construction site before the start of the works. Some contaminants can be harmful either for the soil mix material (binding and hardening processes) or for the steel beams installed in it (corrosion).

For temporary structural soil mix elements, the presence of the contaminants leads to question the efficiency of the binding and the hardening of the soil mix material. A preliminary study allows the identification and determination of the concentrations of the contaminants that may have a potential deleterious impact on the binding and hardening processes. In function of these concentrations, a preliminary laboratory test campaign may be performed to verify the efficiency of these processes in the presence of these contaminants and to determine the influence of the type and the content of cement.

For permanent structural applications, it is very important that the soil mix material continues to fulfill its function in the long term (arching effect to distribute the earth and water pressures between the steel beams, long-term permeability and bearing functions, etc.). In addition, and if relevant, the risk of corrosion has to be considered.

The short- and long-term terminology as well as the temporary or permanent character of the construction should be defined in the job specifications or in the future standards. For example, a soil mix wall used only during the time of excavation to support the earth pressure would ensure a temporary retaining function. On the other hand, soil mix elements ensuring a bearing function during the lifetime of a definitive construction (building, bridges, etc.) would be characterized as soil mix elements with a permanent bearing function.

11.4.3 Deep mixing method for support of land levees and floodwalls

Hurricane Katrina passed southeast of New Orleans on August 29, 2005. The storm caused more than 50 breaches in drainage canal levees and also in navigational canal levees, and it precipitated one of the worst engineering disasters in the history of the United States. In response, construction and reconstruction of levees were planned, in several cases using the deep mixing technique.

The LPV-111 project consisted of the raising of an existing 8.5-km levee, which rested on a foundation of soft organic clay. LPV-111 is part of the New Orleans East Back Levee, which is an essential component of the New Orleans Hurricane Protection System. The deep mixing method was selected to stabilize and support the burden of the new levee, as illustrated in Fig. 11.13. That project resulted in the production of more than 1,300,000 m^3 of soil mix material. A preliminary laboratory program (bench-scale test) and a field test program (validation tests) were conducted to estimate the appropriate binder type and dosage and equipment configuration capable to efficiently meet the technical requirements of the project. Leoni and Bertero (2012) give a general overview of the project and present the results of the bench-scale and validation test programs with a discussion

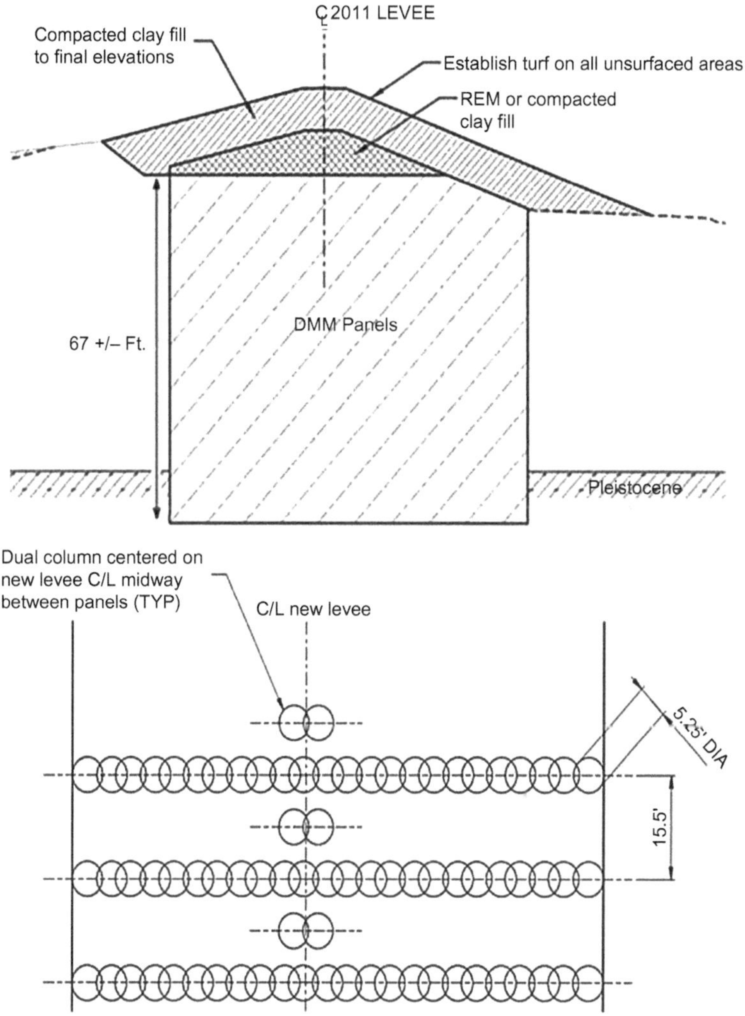

Figure 11.13 Typical design of deep mixing stabilization at LPV-111 (cross section and plan view). *(Source: After Leoni and Bertero (2012)).*

on the UCS design consideration based on the testing of 5000 core samples. At LPV-111, the contractor used two different deep mixing systems: the Trevi Turbo Mix, single and double, and the Contrivance Innovation Cement Mixing Columns.

In New Orleans, dry deep mixing was also used to improve the stability of a section of the land levee and floodwall along the Orleans Avenue Canal.

McGuire et al. (2012) studied the stability of this construction by using the finite difference method (using the FLAC program) and limit equilibrium analysis (using Spencer's method).

In these applications, the soil mix panels not only play the role of support for the embankment but also reinforce the global stability of the levee. For the design aspects related to the deep mixing for land levees and floodwalls, readers can refer to the keynote lecture of Filz et al. (2012) and to the design manual of the FHWA (2013).

11.4.4 Deep mixing method for slope and landslide stabilization

This field of application is similar to the previous one. The stabilization of slope can be realized with the use of soil mix elements installed through its failure surface.

Pinto et al. (2012) describe the widening of an existing road platform (indicated by IP4 in Fig. 11.14) near the city of Amarante in Portugal. CSM panels were executed to stabilize a slope made of heterogeneous landfills that were used for the construction of the existing road platform. A cross section of the geological conditions and the adopted solution are illustrated in Fig. 11.14. CSM panels also serve as foundation for the reinforced fill (maximum height is approximately 20 m) and the motorway traffic with the aid of a load transfer platform. For the design, two-dimensional analyses were carried out using Slide and Plaxis software in order to evaluate the overall stability for static and seismic loads.

Before beginning deep mixing works, it is imperative to identify the potential failure surface of the slope to be stabilized. Moreover, as explained in McGuire et al. (2012), multiple modes of failure must be considered in the design.

Gaib et al. (2012) describe the use of CSM panels for the slope stabilization of the "Fountain Slide" in British Colombia. This slide has been active for decades, and it is part of a massive postglacial earthflow known as the Tunnel Earthflow. The estimated landslide volume is approximately 750,000 m^3. Twenty barrettes, approximately 8 m long and each composed of three individual CSM panels, were constructed with an orientation perpendicular to the slide. If in situ monitoring of the displacement shows a decrease in the movement rate, no global mitigation has been obtained with regard to the important volume of moving soil.

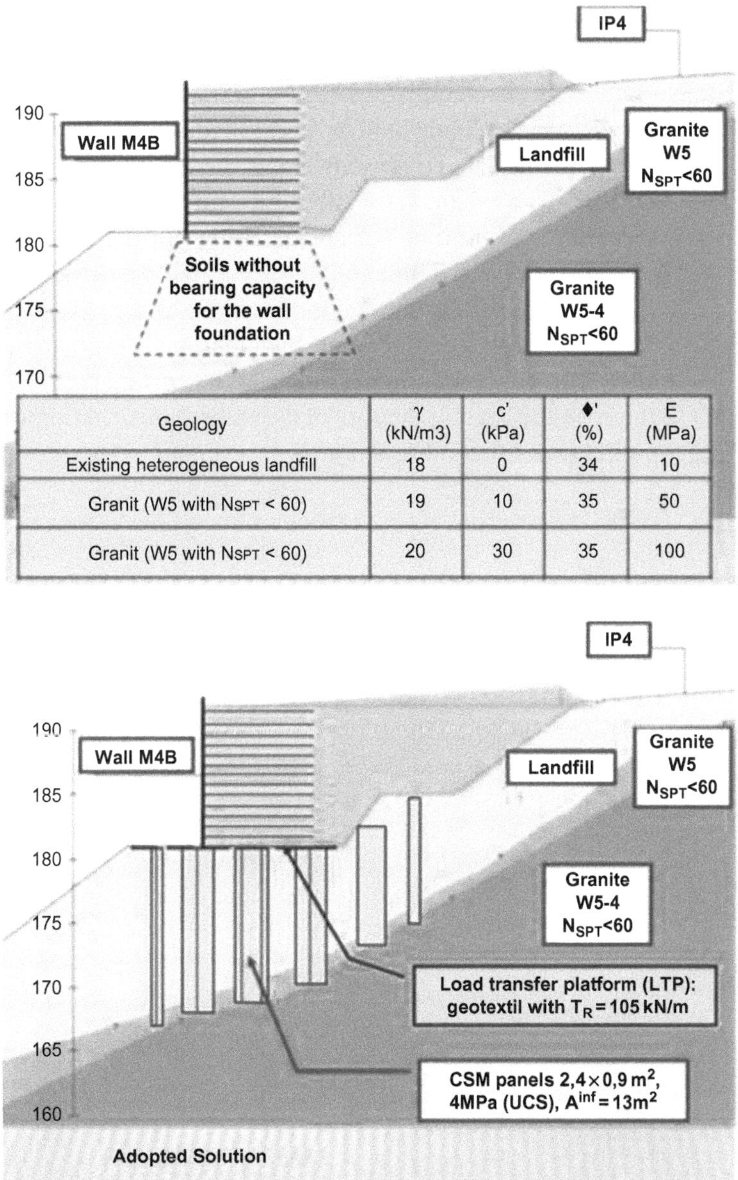

Geology	γ (kN/m3)	c' (kPa)	♦' (%)	E (MPa)
Existing heterogeneous landfill	18	0	34	10
Granit (W5 with N_{SPT} < 60)	19	10	35	50
Granit (W5 with N_{SPT} < 60)	20	30	35	100

Figure 11.14 Cross section of the geological conditions and adopted solution. *(Source: From Pinto et al. (2012)).*

11.4.5 Deep mixing method for cutoff walls

Soil mix walls or trenches can be used as seepage barriers to limit the flow of water (including or not contaminants). Figure 11.15 illustrates the installation of a cutoff wall in a dike body built in Aigle, Switzerland. The role of the dike is to protect industrial installations along the Rhone River against flooding. The core of the dike was realized in a continuous pass with the use of the Trenchmix method of Soletanche Bachy. In this case, the soil mix material has mainly a cutoff function. Its permeability is the key parameter of the design. Nevertheless, the soil mix material has to have sufficient strength to resist internal erosion due to the hydraulic gradient. It is also possible to combine different functions: the soil mix elements are then designed and built to improve the bearing capacity of the embankment, to improve its stability, and to ensure its cutoff function.

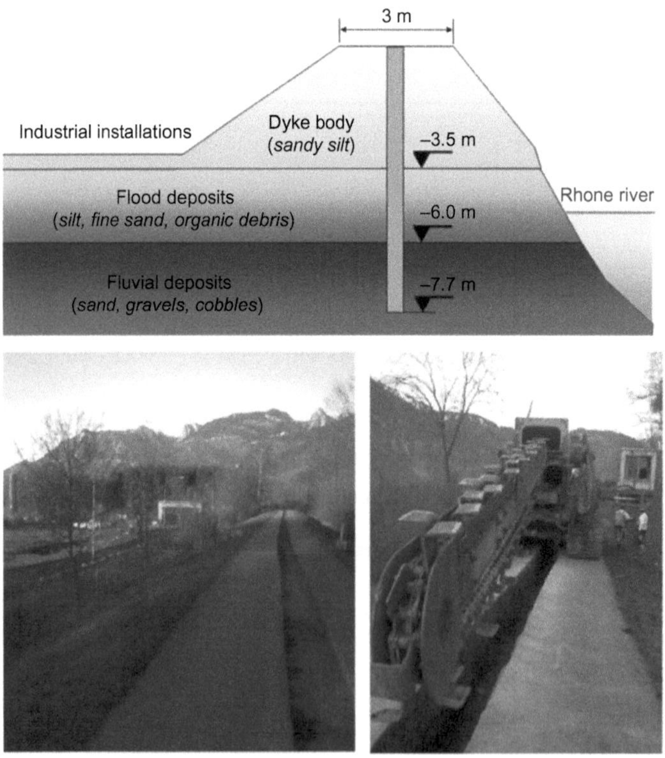

Figure 11.15 Reinforcement of a dike body with the help of the Trenchmix method. *(Source: From Borel (2012), with courtesy of Soletanche Bachy).*

11.4.6 Soil mixing remediation technology

As reported in Al-Tabbaa and Evans (2003), soil mixing was introduced in approximately 1995 in the United Kingdom for geoenvironmental applications, such as the containment of contaminants in and the remediation of brownfields. Since the beginning of the STARNET project in 2001, this field of application has undergone a major evolution.

The STARNET project (http://www-starnet.eng.cam.ac.uk) was established in May 2001 to promote the development of research work on and implementation of stabilization/solidification (S/S) treatment and remediation practices for the numerous brownfield areas present in the United Kingdom. Many references are available on the topic of soil mix remediation technology as reported in the seven "states of practice" reports of the STARNET project (Perera et al., 2005). As a result of the evolution of this application, the deep mixing is currently dedicated to

- the construction of cutoff containment walls;
- the construction of permeable reactive barriers (also called PRB walls); and
- the S/S treatments of the ground.

These are three applications for which permeability and leachability of the soil mix elements are essential parameters but strength often plays a secondary role.

As described in Al-Tabbaa et al. (2012), the permeable reactive barriers are installed in the ground to intersect the flow of contaminated groundwater. Reactive material placed in the barrier removes the contaminants by one or more processes, including sorption, precipitation, oxidation, biodegradation, and encapsulation. In contrast with PRB walls, cutoff containment walls are low-permeability walls designed to isolate a contaminated area from the surrounding area. In addition to these remediation techniques, deep mixing can also be used for soil treatment by S/S. This includes the physical encapsulation and chemical fixation of contaminants in place through a range of processes, including sorption, precipitation, lattice incorporation, complexation, and encapsulation, as reported in Al-Tabbaa et al. (2012). A current growing trend is the use of shallow soil mixing techniques, such as ALLU mass stabilization (ALLU, 2010), for the S/S process.

Soil mix remediation is a cost-effective method with numerous technical alternatives and environmental advantages, including the applicability to sites of any size and to multiple contaminants. Water and soil contaminations consisting of heavy metals and/or organic contamination can be treated.

Combined with recent innovations in deep mixing equipment, this led in October 2008 to the start of the Soil Mix Remediation Technology (SmiRT) project (http://www-g.eng.cam.ac.uk/smirt). Within the framework of this project, a large-scale field trial was initiated on a contaminated site. Various deep mixing systems were used for this experimental campaign. The project studied the influence of several parameters, such as the type of binder as well as the installation variables, including speed of rotation, speed of penetration and withdrawal, and the number of mixing cycles. The setup of the field trials, the different deep mixing systems used for the remediation, and some test results are summarized by Al-Tabbaa et al. (2014), who also present some aspects of field observations, monitoring, in situ testing, coring, and laboratory testing on core samples.

Considering the industrial past of numerous countries throughout the world, soil mixing remediation technology certainly has a bright future.

11.4.7 Deep mixing for liquefaction mitigation

The deep mixing method can also be used to prevent soil liquefaction and post-liquefaction damage. Here, two case histories are presented to describe the growing use of the deep mixing technique.

The first case history, reported by Benhamou and Mathieu (2012), concerns the construction of two buildings on very soft alluvia in Martinique (France). These buildings are located in an area particularly exposed to seismic risks. For the construction of the two buildings, a new type of permanent foundation based on a Geomix caisson (as illustrated in Fig. 11.16) was chosen. The Geomix caissons are built to mitigate the risks of liquefaction damage. The arrangement (36 × 40 m) consists of a grid performed with the use of Geomix trenches. The Geomix technique is based on the hydrofraise technology combined with the CSM principle. Due to the strong inertia and the geometry of the caisson arrangement, the displacements of the Geomix panels are limited during earthquakes. Additional shear stress of the soil and horizontal forces from the structure are concentrated on Geomix bands, and liquefaction of the encased ground is avoided. This treatment also resists external post-liquefaction soil flow. Finally, the Geomix foundations reduce the settlements of the structure.

Another case history concerns the topic of liquefaction susceptibility restrained with the deep mixing method. Yamashita et al. (2013) discuss the measurements performed underneath a piled raft completed with grid-form deep cement mixing walls to reduce the risks of structural damage potentially caused by liquefaction. The structure is a 12-story office building.

Figure 11.16 Arrangement of the Geomix caissons made of CSM panels. *(Source: After Benhamou and Mathieu (2012), with courtesy of Soletanche Bachy).*

Figure 11.17 presents a schematic view of the building and foundation. Figure 11.18 illustrates the layout of piles and grid-form deep cement mixing walls.

The load distribution between the piles, the soil mix walls, and the surrounding soil was monitored for a period of 3 years. After the end of construction, settlements of 20 mm were recorded, as illustrated in Fig. 11.19. As shown in Fig. 11.20, 70% of the load was taken by the piles, 14% by the soil mix walls, and 15% by the soil. The measurements also show that the magnitude 9.0 Tohoku earthquake (March 11, 2011) had almost no effect on the settlements and on the load distribution.

The use of the deep mixing technique combined with typical foundation solutions is being increasingly used in seismic regions such as Japan. To determine the best-suited arrangement (between piles and soil mix elements), new methods of design are currently being developed, as explained in Matsui et al. (2013). The optimum arrangement is one with the lowest volume of treated soil that satisfies the safety limits as determined in the job specifications and standards.

Matsui et al. (2013) present the case of an hybrid application of soil–cement columns used in combination with soil mix walls to reinforce the

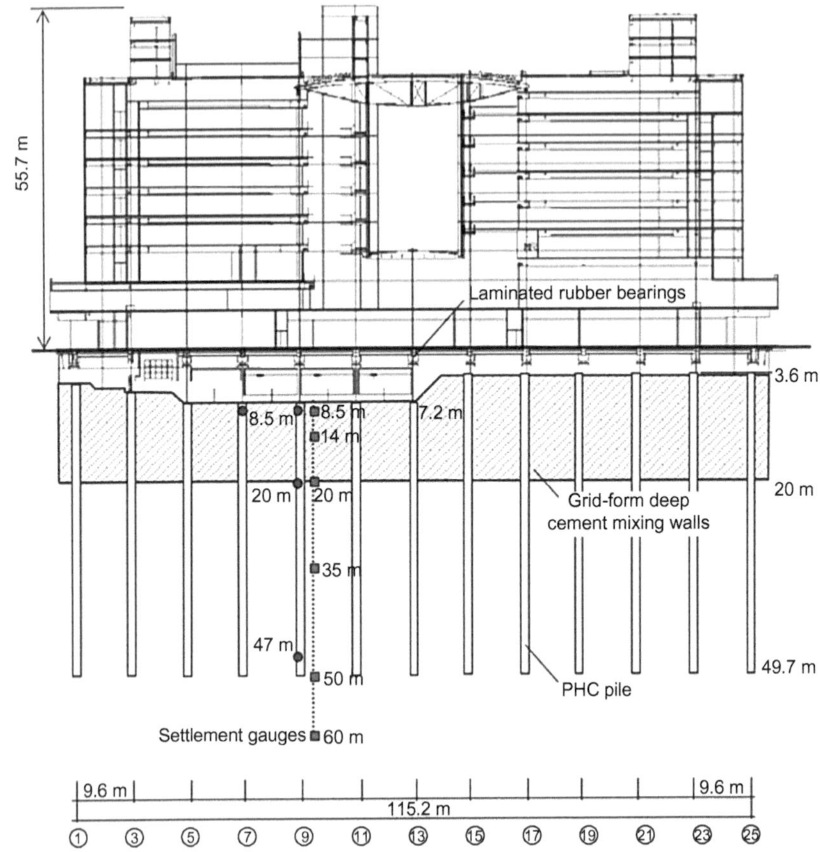

Figure 11.17 Schematic view of the building and foundation. *(Source: From Yamashita et al. (2013)).*

bearing capacity of the soil under an embankment. The basic concept of this method is to install soil mix walls into the ground directly under the slopes of the embankment in order to support the loads of the embankment and to mitigate the lateral movement of the ground. Soil–cement columns are placed both inside and outside the soil mix walls to restrict the vertical and horizontal deformations caused by the loads of the embankment. The authors propose a method to determine the optimum arrangement of the columns and the walls installed under the embankment.

11.5 CASE HISTORY OF HOPMARKT AALST (BELGIUM)

This section presents in detail a case history concerning the construction of a retaining wall in the Hopmarkt square in Aalst, Belgium.

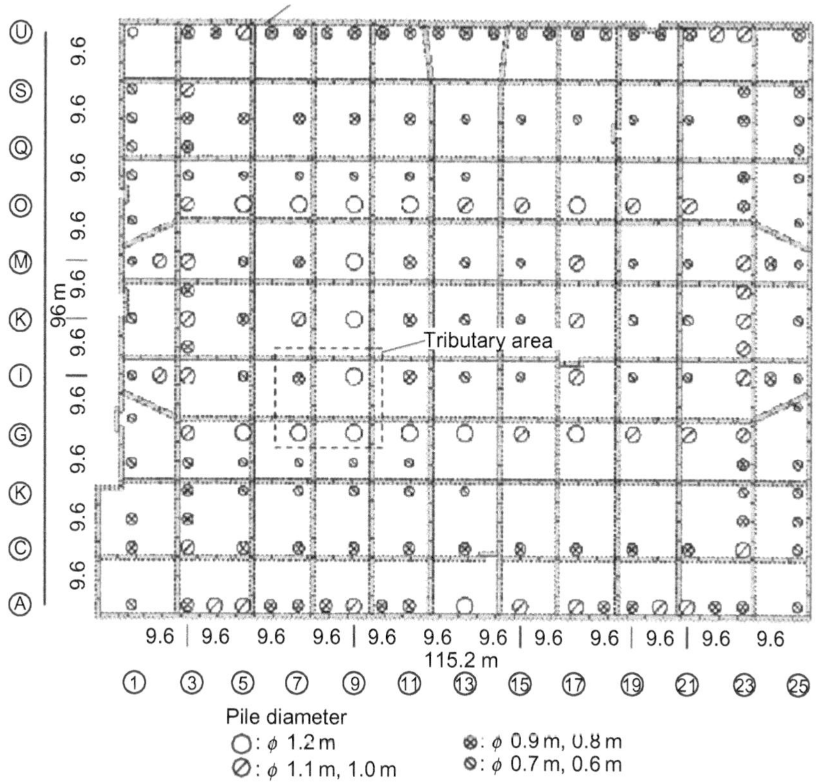

Figure 11.18 Layout of piles and grid-form deep cement mixing walls. *(Source: From Yamashita et al. (2013)).*

Within the framework of the Short Courses of the International Symposium on Ground Improvement held in Brussels in 2012, Eric Leemans from the firm Soetaert nv gave the details of the construction of a composite retaining wall in downtown Aalst. The excavation was dedicated to the construction of a three-story car park below ground level (12-m depth). Figure 11.21 shows the construction site. Dewatering of the area was not allowed because of settlement risks for the neighboring structures. The top soil layers presented a large amount of peat and soft loamy clay. Horizontal permeability inferior to 10^{-8} m/s was required in the project specifications, and the lateral displacement of the retaining walls was limited to 6 cm. For the design solution, a combination of techniques was envisaged with the realization of a composite retaining wall.

First, a CSM wall was constructed to a depth of 21 m into a clayey layer with low permeability. Then sheet piles were sunk into the fresh soil mix material to a depth of 15 m. Figure 11.22 illustrates the cross section of

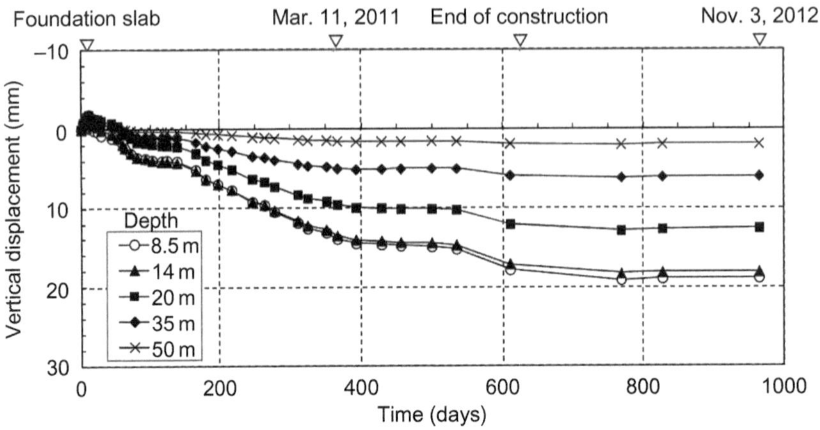

Figure 11.19 Measured vertical ground displacements below raft. *(Source: From Yamashita et al. (2013)).*

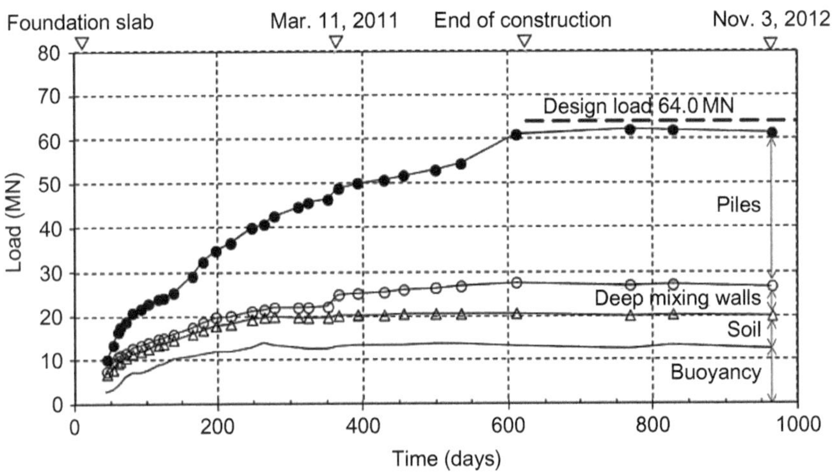

Figure 11.20 Time-dependent load sharing between raft and piles. *(Source: From Yamashita et al. (2013)).*

the CSM and sheet pile wall. Figure 11.23 shows the results of the cone penetration tests performed on the site. In this case, the CSM wall fulfilled a double mitigation function. The sheet pile wall could be installed without significant vibrations, avoiding the risk of damage to neighboring buildings. The CSM wall also had a temporary water retaining function. The dewatering of the building pit could therefore be carried out without significantly affecting the water level outside the building pit, which reduced the risk of

Figure 11.21 Excavation of the Hopmarkt square in Aalst (Belgium). *(Source: From Leemans (2012)).*

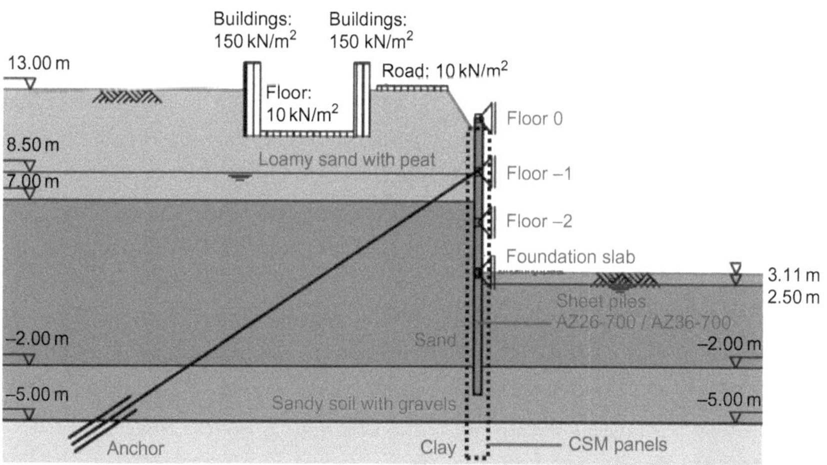

Figure 11.22 Cross section of the excavation. *(Source: After Leemans (2012)).*

settlement of the neighboring buildings. The sheet piles also have a double role. During and after construction, they provide stability to the excavation and they play the role of a permanent watertight barrier. Specific measures were therefore taken to ensure the waterproof qualities of the wall. The interlock of the double AZ sheet piles was welded before they were inserted

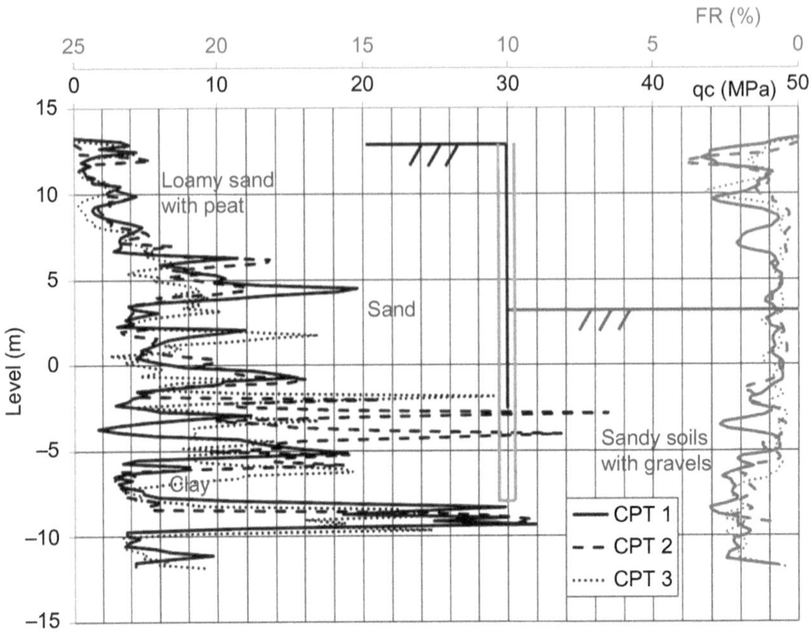

Figure 11.23 CPT results for the site of Hopmarkt Aalst (Belgium).

into the fresh soil mix material. In a similar manner, the anchorage lock and shoe were also welded before placing the sheet piles. The anchor lock was finally injected with expanding polyurethane foam. During excavation, the remaining interlocks between sheet piles were welded as well. Figure 11.24 illustrates the installation of the sheet piles into the fresh soil mix material, and Fig. 11.25 shows the progressive excavation and anchoring of the composite retaining wall.

The soil mix material of the CSM wall was investigated within the framework of the Belgian Building Research Institute (BBRI) Soil Mix Project (2009–2013). Extra CSM panels were executed on the same site with similar execution parameters and slurry properties. After a few days of hardening, these CSM panels were excavated and transported to the laboratory facilities of the BBRI. As described in Table 11.6, the soil mix material was investigated using typical tests performed on core samples and also large-scale tests conducted on large soil mix elements. The first extra CSM panel was cored and cut to obtain core samples and large soil mix blocks in order to perform large-scale UCS tests, such as described in Vervoort et al. (2012). The large-scale UCS tests were conducted to investigate the scale effect and the influence of the unmixed soft soil inclusions included in

Figure 11.24 Installation of the sheet piles into the fresh soil mix material. *(Source: After Leemans (2012)).*

Figure 11.25 Progressive excavation and anchoring of the composite retaining wall. *(Source: After Leemans (2012)).*

the soil mix matrix. The other two extra panels were used for the realization of four large-scale bending tests on half CSM panels. The procedure and the results of these tests were published in Denies et al. (2014, 2015a). These bending tests were performed within the framework of a large experimental

Table 11.6 Mechanical characterization of the soil mix material executed in Aalst (BBRI Soil Mix Project, 2009–2013)

Tests on core samples performed in the laboratory facilities of the BBRI	
UCS according to EN 12390-3 (2009)	7.31 MPa (average value of 31 tests performed on core samples with diameter and height of 105 mm)
Modulus of elasticity, E,[a] according to NBN B15-203 (1990)	8.40 GPa (average value of 4 tests performed on core samples with a diameter of 105 mm and a height of 210 mm)
Tensile splitting strength according to EN 12390-6 (2010)	1.31 MPa (average value of 7 tests performed on core samples with diameter and height of 105 mm)
Volume percentage of unmixed soft soil inclusions into the soil mix material, according to the test procedure of Denies et al. (2012c)	2.6% (obtained on 25 core samples)
Porosity according to NBN B 15-215 (1989)	47.2% (obtained on 4 core samples)
Coefficient of permeability according to DIN 18130-1 (1998)	$<8 \times 10^{-11}$ m/s (obtained on 4 core samples)
Results of two large-scale UCS tests performed on large soil mix samples at KU Leuven	
UCS	5.2 and 4.1 MPa (respectively obtained on rectangular blocks with dimensions $54 \times 50 \times 119$ cm and $54 \times 79 \times 119$ cm)
E_{tg} (tangent)[b]	6.0 and 6.0 GPa
E_{sec} (secant)[b]	6.7 and 6.8 GPa

[a] E is determined in a tangent way varying the applied load between 10% and 30% of the estimated UCS. The deformations of the sample are measured along three axes using DEMEC mechanical strain gages.
[b] Measurement performed with four linear vertical displacement transducer (LVDT) devices with measurement base of approximately one-fourth of the height of the block, installed around the center of each vertical side. The tangent modulus corresponds to the local slope of the stress–strain curve at 50% of the peak strength. The secant modulus is the slope of a straight line joining the origin with the point on the curve at 50% of the peak strength.

campaign of 17 large-scale bending tests conducted to assess the real contribution of the soil mix material to the bending resistance of the soil mix wall.

11.6 CONCLUSION

This chapter introduced the reader to the different construction principles and available equipment for the performance of the deep mixing method.

The chapter also discussed the applications of this technique. Used for ground improvement for more than 50 years, the deep mixing method is being increasingly applied as alternative to traditional foundation and excavation systems or for environmental purposes.

The chapter illustrated the huge field of applications of the method (as reported in Table 11.5). Today, when designers are searching for the best solution to an engineering issue, the deep mixing method is increasingly being considered in the design process. Nevertheless, it is necessary to analyze the problem considering the applicability of the method, the possible uses of the deep mixing equipment, and the produced soil mix material. Figure 11.26 presents these first considerations for determining if the deep mixing method may be an alternative for a particular construction project.

For temporary constructions, the mechanical characteristics of the soil mix material are generally well mastered. The strength and stiffness of the material can be assessed, as well as its adherence with steel. Its permeability can also be determined in the laboratory or with the use of in situ permeability tests. Nevertheless, the durability of the soil mix material with a permanent function remains a controversial topic. As previously mentioned, design criteria of permanent soil mix elements may be based on the execution monitoring (of the mixing energy, depth, etc.) and on performance tests including UCS, freeze thaw and wet–dry durability, leachability

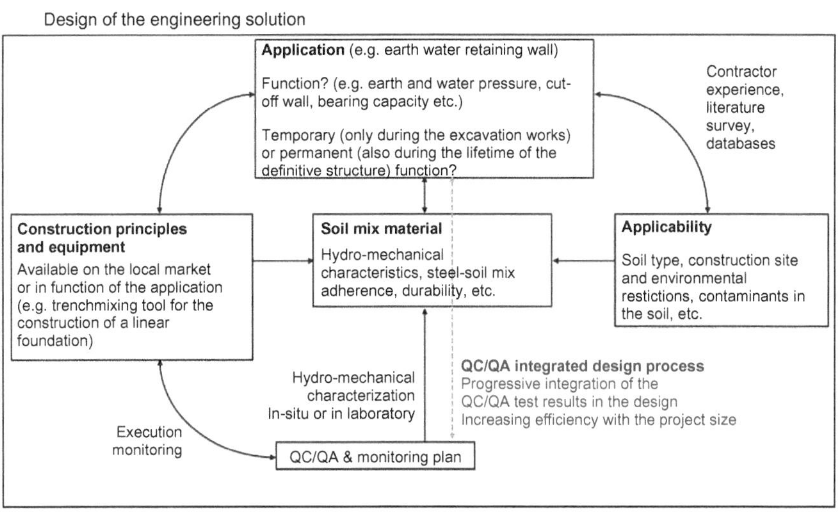

Figure 11.26 First considerations of design for a deep mixing project.

(in particular cases), porosity, and permeability (if required) tests. However, the situation is more complex when contaminants are present in the ground.

Quality control (including execution monitoring) was not highlighted in this chapter. Nevertheless, it remains an essential stage of the construction process. The reader can refer to the following references to obtain information relative to this subject: Maswoswe (2001), EuroSoilStab (2002), Porbaha (2002), Larsson (2005), EN 14679 (2005), Terashi and Kitazume (2011), Denies et al. (2012c), Leoni and Bertero (2012), Filz et al. (2012), FHWA (2013), and Geotech Tools (Iowa State University, 2010–2014).

Deep mixing technology is characterized by continuous innovation with regard to equipment. Considerable experience has been acquired by deep mixing contractors and design engineers, the result of which is the publication of several standards and codes. The European Standard EN 14679 (2005), for example, was elaborated under the umbrella of CEN TC 288 "Execution of Special Geotechnical Works." Nevertheless, although progress has been made in standardization, in practice it seems that there remains a need to develop guidelines for pragmatic aspects, especially regarding quality assurance/quality control (QA/QC) procedures and design. Moreover, the design process of some particular soil mix structures, such as retaining walls, remains vague in most countries. Flexible QA programs that comply with the variable character of the soil mix material should be elaborated in parallel with specific design requirements adapted to the function of the soil mix elements. The publication of the design manual of the FHWA (2013), which concentrates not only on the execution of the method but also on the design aspects of deep mixing applied for embankment and foundation support, is in agreement with that philosophy. For earth–water retaining walls and cutoff walls, new design rules will soon be published jointly by BBRI and SBRCURnet.

In this handbook, the design approach and the QC/QA activities are directly related to the field of application of soil mix walls, to the project restrictions, and to the category of the structure (according to the principles of Eurocode 7—EN 1997-1, 2004). In the future, such guidelines should be developed for each type of application of the deep mixing method.

REFERENCES

ALLU, 2010. Mass Stabilisation Manual. ALLU Finland Oy, Orimattila, Finland.

Al-Tabbaa, A., Evans, C., 2003. Deep soil mixing in the UK: geoenvironmental research and recent applications. Land Contam. Reclam. 11, 1–14.

Al-Tabbaa, A., Liska, M., McGall, R., Critchlow, C., 2012. Soil Mix Technology for Integrated Remediation and Ground Improvement: Field Trials. In: Denies, N., Huybrechts, N. (Eds.), In: Proc. Int. Symp. ISSMGE–TC211: Recent Research, Advances and Execution Aspects of Ground Improvement Works, May 31–June 1, 2012, Brussels, vol. 3, pp. 13–21.

Al-Tabbaa, A., O'Connor, D., Abunada, Z., 2014. Field Trials for Deep Mixing in Land Remediation: Execution and Early Age Monitoring, QC and Lessons Learnt. In: DFI-EFFC Int. Conf. on Piling and Deep Foundations, May 21–23, 2014, Stockholm, pp. 621–630.

BBRI Soil Mix Project, 2009–2013. IWT 080736 soil mix project: SOIL MIX in Constructieve En Permanente Toepassingen–Karakterisatie Van Het Materiaal En Ontwikkeling Van Nieuwe Mechanische Wetmatigheden. Available at, http://www.iwt.be.

Bellato, D., Dalle Coste, A., Gerressen, F.-W., Simonini, P., 2012. Long-Term Performance of CSM Walls in Slightly Overconsolidated Clays. In: Denies, N., Huybrechts, N. (Eds.), In: Int. Symp. ISSMGE–TC211, May 31–June 1, 2012, Brussels, vol. 3, pp. 23–32.

Benhamou, L., Mathieu, F., 2012. Geomix Caissons Against Liquefaction. In: Denies, N., Huybrechts, N. (Eds.), In: Int. Symp. ISSMGE–TC211, May 31–June 1, 2012, Brussels, vol. 3, pp. 33–40.

Borel, S., 2012. Soil Mixing Equipment. In: Latest Advances in Deep Mixing: Presentation of the Short Courses of the International Symposium on Ground Improvement IS-GI Brussels, May 30, 2012.

Bredenberg, H., Holm, G., Broms, B.B. (Eds.), 1999. Int. Conf. Dry Mix Methods for Deep Soil Stabilization, October 13–15, Stockholm. Balkema, Leiden, The Netherlands.

Bruce, D.A., 2014. Deep Mixing in the United States: Milestones in Evolution. In: DFI-EFFC Int. Conf. on Piling and Deep Foundations, May 21–23, 2014, Stockholm, pp. 611–620.

Bruce, D.A., Bruce, M.E.C., DiMillio, A.F., 1998. Deep Mixing Method: A global perspective. In: Geo-Congress 1998: Soil Improvement for Big Digs, ASCE Geotechnical Special Publication No. 81, October 18–21, 1998, Boston, MA, pp. 1–26.

Bruce, D.A., Bruce, M.E.C., DiMillio, A.F., 1999. Dry Mix Methods: A Brief Overview of International Practice. In: Int. Conf. on Dry Mix Methods for Deep Soil Stabilization, October 13–15, Stockholm, Sweden. Balkema, Leiden, The Netherlands, pp. 15–25.

Burke, G., 2009. TRD Soil Mixing at Herbert Hoover Dike. DFI Magazine, Spring.

Chu, J., Varaksin, S., Klotz, U., Mengé, P., 2009. Construction Processes. In: Hamza, M., et al. (Ed.), In: 17th Int. Conf. on Soil Mechanics and Geotechnical Engineering, October 5–9, 2009, Alexandria, Egypt., vol. 4. IOS Press, Amsterdam, pp. 3006–3135.

Coastal Development Institute Tokyo, 2002. The Deep Mixing Method—Principle, Design and Construction. Balkema, Leiden, The Netherlands.

Denies, N., Huybrechts, N. (Eds.), 2012. Proc. Int. Symp. ISSMGE–TC211: Recent research, advances and execution aspects of ground improvement works, May 31–June 1, 2012, Brussels. Five volumes available at, http://www.tc211.be.

Denies, N., Van Lysebetten, G., 2012a. General Report—Session 4: Soil mixing 2—Deep mixing. In: Denies, N., Huybrechts, N. (Eds.), In: Int. Symp. ISSMGE–TC211, May 31–June 1, 2012, Brussels, vol. 1, pp. 87–124.

Denies, N., Van Lysebetten, G., 2012b. Summary of the Short Courses of the IS-GI 2012: Latest Advances in Deep Mixing. In: Denies, N., Huybrechts, N. (Eds.), In: Int. Symp. ISSMGE–TC211, May 31–June 1, 2012, Brussels, vol. 5, pp. 73–123.

Denies, N., Huybrechts, N., De Cock, F., Lameire, B., Maertens, J., Vervoort, A., 2012a. Soil Mix Walls as Retaining Structures—Belgian Practice. In: Denies, N., Huybrechts, N. (Eds.), In: Int. Symp. ISSMGE–TC211, May 31–June 1, 2012, Brussels, vol. 3, pp. 83–97.

Denies, N., Huybrechts, N., De Cock, F., Lameire, B., Vervoort, A., Van Lysebetten, G., et al., 2012b. Soil Mix Walls as Retaining Structures: Mechanical Characterization. In: Denies, N., Huybrechts, N. (Eds.), In: Int. Symp. ISSMGE–TC211, May 31–June 1, 2012, Brussels, vol. 3, pp. 99–115.

Denies, N., Huybrechts, N., De Cock, F., Lameire, B., Vervoort, A., Maertens, J., 2012c. Mechanical Characterization of Deep Soil Mix Material—Procedure Description. In: Denies, N., Huybrechts, N. (Eds.), In: Int. Symp. ISSMGE–TC211, May 31–June 1, 2012, Brussels, vol. 3, pp. 117–126.

Denies, N., Van Lysebetten, G., Huybrechts, N., De Cock, F., Lameire, B., Maertens, J., et al., 2014. Real-Scale Tests on Soil Mix Elements. In: DFI-EFFC Int. Conf. Piling and Deep Foundations, May 21–23, 2014, Stockholm, pp. 647–656.

Denies, N., Huybrechts, N., De Cock, F., Lameire, B., Maertens, J., Vervoort, A., 2015a. Large-Scale Bending Tests on Soil Mix Elements. In: Int. Foundations Congress and Equipment Expo 2015, ADSC, DFI, ASCE G-I Institute and PDCA, San Antonio, TX, March 17–21, 2015.

Denies, N., Huybrechts, N., De Cock, F., Lameire, B., Maertens, J., Vervoort, A., et al., 2015b. Thoughts on the Durability of the Soil Mix Material. In: 16th Eur. Conf. on Soil Mechanics and Geotechnical Engineering, September 13–17, 2015, Edinburgh.

EN 14679, 2005. Execution of Special Geotechnical Works—Deep Mixing. European Standard.

EN 1997-1, 2004. Eurocode 7: Geotechnical design—Part 1: General rules.

Essler, R., Kitazume, M., 2008. Application of Ground Improvement: Deep Mixing. ISSMGE TC211 website:http://www.tc211.be.

EuroSoilStab, 2002. Development of Design and Construction Methods to Stabilise Soft Organic Soils: Design Guide—Soft Soil Stabilisation, EC Project BE 96-3177. .

FHWA-HRT-13-046, 2013. Federal Highway Administration Design Manual: Deep Mixing for Embankment and Foundation Support. U.S. Department of Transportation, Federal Highway Administration, Washington, DC.

Filz, G., Adams, T., Navin, M., Templeton, A.E., 2012. Design of Deep Mixing for Support of Levees and Floodwalls: Keynote Lecture. In: 4th Int. Conf. on Grouting and Deep Mixing, ASCE Geotechnical Special Publication No. 228, February 15–18, 2012, New Orleans, vol. 1, pp. 89–133.

Gaib, S., Wilson, B., Lapointe, E., 2012. Design, Construction and Monitoring of a Test Section for the Stabilization of an Active Slide Area Utilizing Soil Mixed Shear Keys Installed Using Cutter Soil Mixing. In: Denies, N., Huybrechts, N. (Eds.), In: Int. Symp. ISSMGE–TC211, May 31–June 1, 2012, Brussels, vol. 3, pp. 147–157.

Ganne, P., Huybrechts, N., De Cock, F., Lameire, B., Maertens, J., 2010. Soil Mix Walls as Retaining Structures—Critical Analysis of the Material Design Parameters. In: Geotechnical Challenges in Megacities: Int. Geotech. Conf., Moscow, vol. 3, pp. 991–998.

Gerressen, F.-W., 2012. Overview CSM Equipment. In: Latest Advances in Deep Mixing: Presentation of the Short Courses of the Int. Symp. on Ground Improvement IS-GI Brussels, May 30, 2012.

Gerressen, F.-W., Vohs, T., 2012. CSM—Cutter Soil Mixing: Worldwide Experiences of a Young Soil Mixing Method in Challenging Soil Conditions. In: Denies, N., Huybrechts, N. (Eds.), In: Int. Symp. ISSMGE–TC211, May 31–June 1, 2012, Brussels, vol. 3, pp. 159–168.

Guimond-Barrett, A., Mosser, J.-F., Calon, N., Reiffsteck, P., Pantet, A., Le Kouby, A., 2012. Deep Mixing for Reinforcement of Railway Platforms with a Spreadable Tool. In: Denies, N., Huybrechts, N. (Eds.), In: Int. Symp. ISSMGE–TC211, May 31–June 1, 2012, Brussels, vol. 3, pp. 169–178.

Iowa State University, 2010–2014. Geotech Tools: Geo-construction Information and Technology Selection Guidance for Geotechnical, Structural, and Pavement Engineers. Iowa State University, Ames, IA. Available at, http://geotechtools.org.

Kitazume, M., 2005. State of Practice Report—Field and Laboratory Investigations, Properties of Binders and Stabilized Soil. In: Int. Conf. on Deep Mixing, Stockholm, vol. 2, pp. 660–684.

Kitazume, M., Terashi, M., 2013. The Deep Mixing Method. CRC Press, Boca Raton, FL.

Lambert, S., Rocher-Lacoste, F., Le Kouby, A., 2012. Soil–Cement Columns: An Alternative Soil Improvement Method. In: Denies, N., Huybrechts, N. (Eds.), In: Int. Symp. ISSMGE–TC211, May 31–June 1, 2012, Brussels, vol. 3, pp. 179–188.

Larsson, S.M., 2005. State of Practice Report: Execution, Monitoring and Quality Control. In: Int. Conf. on Deep Mixing, Stockholm, vol. 2, pp. 732–785.

Leemans, E., 2012. Mixing is the Future—Case Study. In: Latest Advances in Deep Mixing: Presentation of the Short Courses of the Int. Symp. on Ground Improvement IS-GI Brussels, May 30, 2012.

Leoni, F.M., Bertero, A., 2012. Soil Mixing in Highly Organic Materials: The Experience of LPV111, New Orleans, Louisiana (USA). In: Denies, N., Huybrechts, N. (Eds.), In: Int. Symp. ISSMGE–TC211, May 31–June 1, 2012, Brussels, vol. 3, pp. 189–198.

Maswoswe, J.J.G., 2001. QA/QC for CA/T Deep Soil–Cement. In: Foundations and Ground Improvement, ASCE Geotechnical Special Publication No. 113, June 9–13, 2001, Blacksburg, VA, pp. 610–624.

Matsui, H., Ishii, H., Horikoshi, K., 2013. Hybrid Application of Deep Mixing Columns Combined with walls as Soft Ground Improvement Method Under Embankment. In: Delage, P., Desrues, J., Frank, R., Puech, A., Schlosser, F. (Eds.), In: 18th Int. Conf. on Soil Mechanics and Geotechnical Engineering, Presses des Ponts, September 2–6, Paris, vol. 3, pp. 2545–2548.

McGuire, M., Templeton, E., Filz, G., 2012. Stability Analyses of a Floodwall with Deep-Mixed Ground Improvement at Orleans Avenue Canal, New Orleans. In: Denies, N., Huybrechts, N. (Eds.), In: Int. Symp. ISSMGE–TC211, May 31–June 1, 2012, Brussels, vol. 3, pp. 199–209.

Peixoto, A., Sousa, E., Gomes, P., 2012. Solutions for Soil Foundation Improvement of an Industrial Building Using Cutter Soil Mixing Technology at Fréjus, France. In: Denies, N., Huybrechts, N. (Eds.), Int. Symp. ISSMGE–TC211, May 31–June 1, 2012, Brussels , vol. 3, 243–250.

Perera, A.S.R., Al-Tabbaa, A., Reid, J.M., Johnson, D., 2005. State of Practice Report: UK Stabilisation/Solidification Treatment and Remediation. Part V: Long-Term Performance and Environmental Impact. In: Int. Conf. on Stabilisation/Solidification Treatment and Remediation, CambridgeAvailable at, http://www-starnet.eng.cam.ac.uk.

Pinto, A., Tomásio, R., Pita, X., Godinho, P., Peixoto, A., 2012. Ground Improvement Solutions Using CSM Technology. In: Denies, N., Huybrechts, N. (Eds.), In: Int. Symp. ISSMGE–TC211, May 31–June 1, 2012, Brussels, vol. 3, pp. 271–284.

Porbaha, A., 1998. State of the art in deep mixing technology. Part I: Basic concepts and overview. ICE–Ground Improvement 2 (2), 81–92.

Porbaha, A., 2002. State of the art in quality assessment of deep mixing technology. ICE–Ground Improvement 6 (3), 95–120.

Porbaha, A., Shibuya, S., Kishida, T., 2000. State of the art in deep mixing technology. Part III: Geomaterial characterization. ICE–Ground Improvement 4 (3), 91–110.

Quasthoff, P., 2012. State of the Art in "Dry Soil Mixing"—Basics and Case Study. In: Denies, N., Huybrechts, N. (Eds.), In: Int. Symp. ISSMGE–TC211, May 31–June 1, 2012, Brussels, vol. 3. pp. 285–297.

Rutherford, C., Biscontin, G., Briaud, J.-L., 2005. Design Manual for Excavation Support Using Deep Mixing Technology. Texas A&M University, College Station, TX.
Terashi, M., 2002. Long-Term Strength Gain vs. Deterioration of Soils Treated by Lime and Cement. In: Tokyo Workshop 2002 on Deep Mixing, pp. 39–57.
Terashi, M., 2003. The State of Practice in Deep Mixing Methods. In: 3rd Int. Specialty Conf. on Grouting and Ground Treatment, ASCE Geotechnical Special Publication No. 120, February 10–12, New Orleans, vol. 1, pp. 25–49.
Terashi, M., Kitazume, M., 2011. QA/QC for deep-mixing ground: current practice and future research needs. ICE–Ground Improvement 164 (GI3), 161–177.
Topolnicki, M., 2004. In Situ Soil Mixing. In: Moseley, M.P., Kirsch, K. (Eds.), Ground Improvement, Second ed. Taylor & Francis, New York.
Topolnicki, M., 2012. General Overview. In: Latest Advances in Deep Mixing: Presentation of the Short Courses of the Int. Symp. on Ground Improvement IS-GI Brussels, May 30, 2012.
Vervoort, A., Tavallali, A., Van Lysebetten, G., Maertens, J., Denies, N., Huybrechts, N., et al., 2012. Mechanical Characterization of Large Scale Soil Mix Samples and the Analysis of the Influence of Soil Inclusions. In: Denies, N., Huybrechts, N. (Eds.), In: Int. Symp. ISSMGE–TC211, May 31–June 1, 2012, Brussels, vol. 3, pp. 127–135.
Yamashita, K., Wakai, S., Hamada, J., 2013. Large-Scale Piled Raft with Grid-Form Deep Mixing Walls on Soft Ground. In: Delage, P., Desrues, J., Frank, R., Puech, A., Schlosser, F. (Eds.), In: 18th Int. Conf. on Soil Mechanics and Geotechnical Engineering, Presses des Ponts, September 2–6, Paris, vol. 3, pp. 2637–2640.

CHAPTER 12

Bridge Abutment Made of Cement-Mixed Gravel Backfill

Fumio Tatsuoka[1], Masaru Tateyama[2], Hifumi Aoki[3], Kenji Watanabe[2]
[1]Tokyo University of Science, Chiba, Japan
[2]Railway Technical Research Institute, Tokyo, Japan
[3]Japan Railway Construction Agency, Yokohama-shi, Japan

12.1 INTRODUCTION

This chapter reports the construction of a bridge abutment consisting of the backfill of geogrid-reinforced cement-mixed gravely soil and a thin reinforced concrete (RC) facing structure supporting a bridge girder for a new bullet train line in Kyushu (the southernmost major island of Japan), which was completed in 2003. This new type of bridge abutment (Fig. 12.1) is unique in that the backfill supports laterally the facing, rather than exerting static and dynamic earth pressure on the back face of the facing during static and seismic loading conditions.

The background for the development of this new type of bridge abutment is as follows. A large number of conventional-type railway bridge abutments (Fig. 12.2(a)) were seriously damaged with a large relative settlement between the abutment and the backfill by a number of previous major earthquakes in Japan, including the 1995 Hyogo-ken Nambu earthquake (the so-called Great Kobe Earthquake), as described by Tatsuoka et al. (1997c,d). Such a relative settlement could endanger safe train operation even when it is small (e.g., 10 cm). To prevent this type of damage, in 1978, the Japan National Railway (currently several Japan Railway Companies) introduced a new type of backfill as shown in Fig. 12.2(b)—that is, a triangle-shaped backfill zone immediately behind the abutment, called an "approach block," is constructed with well-compacted, well-graded gravely soil. However, a number of this type of bridge abutments did not perform satisfactorily during several major earthquakes that occurred subsequently. Figure 12.3 shows a typical example.

Most recently, the approach blocks for a number of bridge abutments for new bullet train lines were constructed by using a well-graded gravely soil mixed with cement to decrease as much as possible the settlement of the

Chemical, Electrokinetic, Thermal, and Bioengineering Methods
http://dx.doi.org/10.1016/B978-0-08-100191-2.00012-5

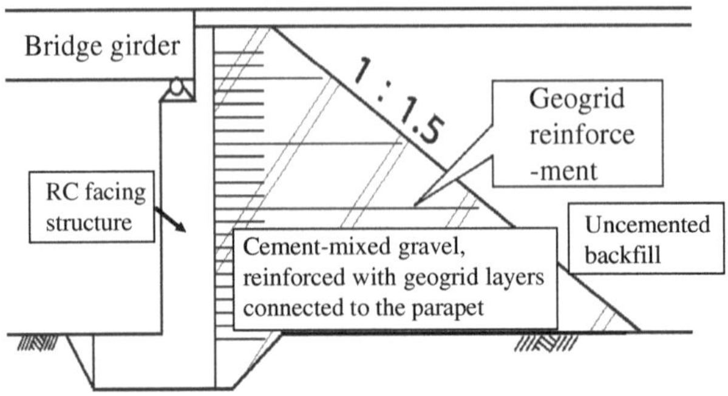

Figure 12.1 A new type of bridge abutment using backfill of cement-mixed gravel.

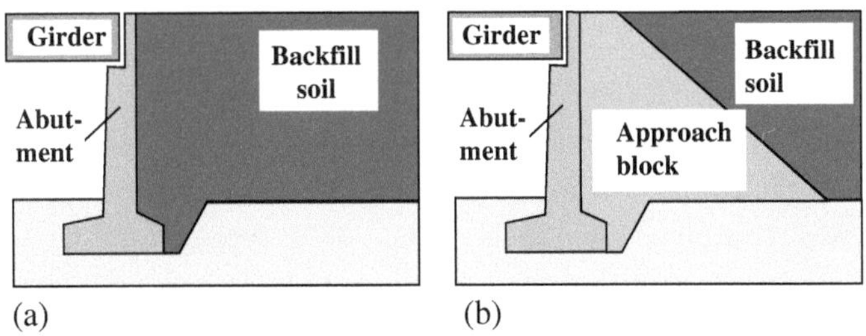

Figure 12.2 Typical conventional bridge abutment types for railway: (a) most conventional type (Model 1 in the shaking table tests described later in this chapter); (b) a new type of structure introduced in 1978 (Model 2 in the shaking table test). *(Source: Watanabe et al. (2002)).*

Figure 12.3 A settlement of the backfill relative to an abutment, Arikawa bridge, Tsugaru-Kaikyo Line, East Japan Railway, Hokkaido Nansei-oki Earthquake, July 12, 1993. *(Source: Courtesy of Railway Technical Research Institute, Japan).*

backfill relative to the RC abutment. However, because the approach block is not connected to the RC abutment, the backfill is not designed to directly support the RC abutment; thus, the dimensions of the RC abutment are not reduced substantially by using the backfill comprising cement-mixed gravely soil.

In view of this situation, the Railway Technical Research Institute, Japan, the University of Tokyo, and the Japan Railway Construction Public Corporation started a long-term joint research project in 1997 with the goal to develop of new seismic-resistant and cost-effective types of bridge abutment. Eventually, the following two types were proposed and studied:

1. The backfill consists of a zone of geogrid-reinforced, cement-mixed, well-graded gravely soil, which should be well compacted, immediately behind the full-height rigid facing structure supporting a bridge girder (Fig. 12.1).
2. The backfill is geogrid-reinforced, well-compacted, well-graded gravely soil that supports a bridge girder (Fig. 12.4). The backfill is preloaded and prestressed by using tie rods, the top ends of which are fixed to the top reaction block placed on the crest of backfill by using a special connection device (the ratchet system; Shinoda et al., 2003b; Uchimura et al., 2003).

With both newly proposed types of bridge abutment, the geogrid layers that are embedded in the backfill to prevent separation into parts are connected to the back of the facing. The first purpose of this connection is to restrain the settlement of the approach block relative to the parapet. As discussed later, the connection is also essential to maintain a high integrity of the abutment structure, particularly under severe seismic loading conditions. By this connection, the steel-reinforced concrete facing structure becomes a continuous beam that is supported at a large number of supports with a small spacing, and therefore it is substantially less massive than the conventional-type RC bridge abutment. Moreover, because the facing is

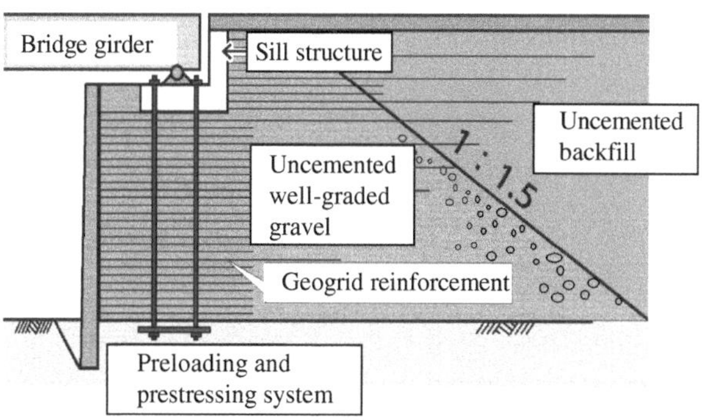

Figure 12.4 A new type bridge abutment using preloaded and prestressed geogrid-reinforced backfill. *(Source: Tatsuoka et al. (1997b,c); Nakarai et al. (2002); Shinoda et al. (2003a,b); Uchimura et al. (2003)).*

constructed after the backfill is completed by a staged construction procedure and the stability of the complete facing is controlled by the bearing capacity of ground immediately below the facing to a much lesser extent than for the conventional-type bridge abutments, it is usually unnecessary to support the facing with piles unless the supporting ground is soft clay. For these reasons, this new type of bridge abutment is considerably much more cost-effective than the conventional-type ones.

In the following sections, first the results from a series of shaking table tests on small-scale models performed to develop the first new type (Fig. 12.1) are summarized. Then, the results from a series of drained triaxial compression (TC) tests on cement-mixed gravely soil (CMG) that were carried out to determine the mixing proportion and the compaction details of CMG are presented. Finally, the design and construction of the first prototype structure and results from field full-scale loading tests are reported.

12.2 SHAKING TABLE TESTS

12.2.1 General

A series of model shaking table tests were performed to evaluate the seismic stability of the new type of abutment (Fig. 12.1) in comparison with that of the conventional-type ones (Fig. 12.2) and to ensure that the new type of abutment could behave satisfactorily even during very high-intensity seismic load (the so-called Level 2 design seismic load, the highest seismic load level to be accounted for in the Japanese seismic design codes and standards) (Watanabe et al., 2002).

12.2.2 Test procedures

Models of bridge abutment

The following five types of abutments were investigated (Table 12.1):

Model 1: The most conventional type with uncemented backfill of ordinary soil type (Fig. 12.2(a))

Model 2: One of the conventional types with an approach block of well-compacted, uncemented, well-graded gravely soil (Fig. 12.2(b))

Model 3: The latest conventional type with an approach block of well-compacted, cemented, well-graded gravely soil that is not reinforced with geogrid reinforcement layers (so the backfill and the facing are not connected to each other) (Fig. 12.2(b))

Models 4 and 5: New types with an approach block of well-compacted, cemented, well-graded gravely soil that is reinforced with geogrid

Table 12.1 Abutment models tested

Model name	Backfill	Reinforcement	Input motion	Width of footing for RC facing (mm)
Model 1 (conventional)	No approach block (Toyoura sand, $D_r = 75\%$)	No	Sinusoidal wave and Kobe wave	390
Model 2 (conventional)	With approach block of dry gravel ($\gamma_d = 1.9$ g/cm^3)	No	Sinusoidal wave	390
Model 3 (conventional)	With approach block of cement-mixed gravel	No (backfill and facing unconnected)	Sinusoidal wave	390
Model 4 (new type)	With approach block of cement-mixed soil	Yes (backfill and facing connected)	Kobe wave followed by sinusoidal wave	290
Model 5 (new type)	With approach block of cement-mixed soil	Yes (backfill and facing connected)	Kobe wave followed by sinusoidal wave	200

Source: Watanabe et al. (2002).

reinforcement layers connected to the back of the facing (Fig. 12.1), having different widths of the footing for the parapet

The facing of these models was made of aluminum. The height was 620 mm, and the footing base had a width of 390 mm (Models 1–3), 290 mm (Model 4), and 200 mm (Model 5). Figure 12.5 shows the details of Model 4. The facing model supported a relatively heavy model bridge girder with a weight of 200 kg through a hinge support on the top of the facing (i.e., the fixed support). Thus, full dynamic lateral load of the girder was activated to the facing. The unreinforced backfill of Model 1 was made by pluviating through air-dried fine sand (Toyoura sand) from a sand hopper at a constant falling height to have a relative density D_r equal to 75%. The approach block of Model 2 was made by compacting a well-graded gravely soil (U_c =10.7; D_{50}=1.1 mm; D_{max} =5.0 mm; fines content=5.2%; water content=5%) to a dry density of 1.9 g/cm^3. Despite the fact that the actual approach block of this abutment type (and also of the new types) is made of cement-mixed, well-graded gravel, the model approach block of Model 3 was made of a cement-mixed loam having an unconfined compressive strength of 200 kN/m^2 considering the model similitude.

Models 4 and 5, which simulated the new structural types of bridge abutment, had a triangle-shaped approach block made of cement-mixed loam (the same as that of Model 3) that was reinforced with geogrid reinforcement layers connected to the back face of the facing. The facing directly supported the model bridge girder at its top. The footing base of the facing was either

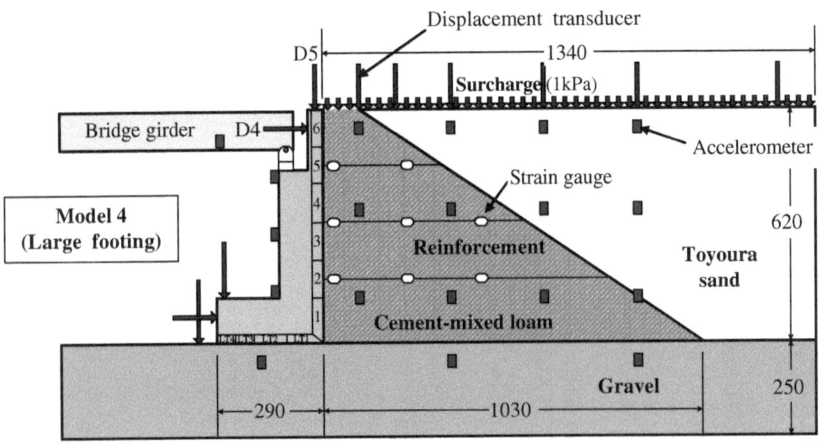

Figure 12.5 Cross sections of reinforced abutment with cement-treated backfill (Model 4). *(Source: Watanabe et al. (2002)).*

relatively wide (Model 4) or relatively narrow (Model 5). The model geogrid was a grid made of 0.2- mm-thick and 3-mm-wide phosphor-bronze strips that were soldered to each other with a center-to-center spacing of 50 and 100 cm in the transversal and axial directions. The tensile force in the model reinforcement was measured by using electric-resistant strain gauges attached to the central strip at three levels. The stiffness of this model reinforcement was larger than that of actual polymer geogrids when based on the model similitude.

The model ground, supporting the model walls, was made by compacting an air-dried gravely soil (U_c=12.1; D_{50}=10.0 mm; D_{max}=32.0 mm; fines content=2.0%) to a dry density of 1.9 g/cm^3. The dynamic response of the abutment and backfill was measured with a number of displacement transducers and accelerometers (Fig. 12.5). The dynamic earth pressure and the dynamic subsoil reaction acting on, respectively, the back face of the facing and the bottom of the base footing of the facing were measured with a number of two-component load cells measuring normal and shear forces (Fig. 12.5).

Dynamic loading at the shaking table

Models 1–3 were subjected to a series of uniform sinusoidal waves, each comprising 50 waves and lasting 10 s at a frequency of 5 Hz. The amplitude of horizontal acceleration at the shaking table, a_{max}, was increased step by step from the initial value of 50 gal with an increment of 50 gal until the displacements of the abutment became considerably large. Models 1, 4, and 5 were subjected to the time history of horizontal ground acceleration recorded at the Kobe Marine Meteorological Observation Station during the 1995 Hyogo-ken Nambu earthquake. The predominant frequency of this input motion had been adjusted to 5 Hz to account for the model similitude. The table maximum amplitude, a_{max}, was increased step by step in increments of 100 gal from 100 to 1400 gal. In the tests on Models 4 and 5, the models did not reach ultimate failure even when the a_{max} value reached 1400 gal. Therefore, subsequently sinusoidal waves were applied to these models stepwise increasing the a_{max} value from 100 gal at increments of 100 gal until the models exhibited ultimate failure.

12.2.3 Test results and discussion

Models 1–3

Figure 12.6 shows the failed models, observed after the respective tests. Figures 12.7(a) and 12.7(b) show the residual displacements at the top of

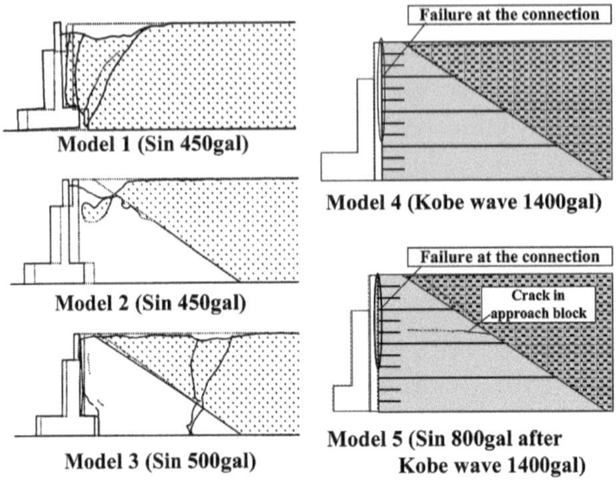

Figure 12.6 Deformations of models after shaking table tests. *(Source: Watanabe et al. (2002)).*

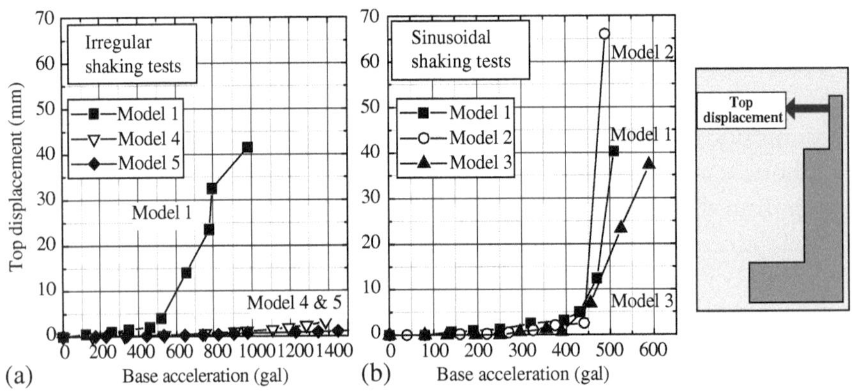

Figure 12.7 Residual displacements at the top of model abutment: (a) irregular (Kobe wave) and (b) sinusoidal table motion. *(Source: Watanabe et al. (2002)).*

the facing plotted against the maximum table acceleration, a_{max}, for all the models. The following trends in behavior are seen from these figures:

1. The deformation of Model 1 when the a_{max} value of sinusoidal motion became 450 gal was very large, showing ultimate failure with a fully developed single major failure plane in the backfill.
2. Model 2 also exhibited brittle failure when the a_{max} value of sinusoidal motion became 450 gal, showing very large deformation of the approach block of gravel, particularly at the upper part.

3. The deformation, particularly the settlement at the crest of the approach block, of Model 3 when the a_{max} value of sinusoidal motion became 450 gal was much smaller than that of Models 1 and 2. However, when the a_{max} value reached 500 gal, the facing started separating from the approach block, exhibiting a high dynamic response because there was no connection between them. Several major cracks developed in the approach block of Model 3, resulting in a loss of structural integrity.

These results clearly indicate that when subjected to Level II seismic load, the seismic stability of these three conventional types of bridge abutment is insufficient. The test results also suggest that the seismic stability of abutment can be increased by taking the following three measures:

1. By constructing the approach backfill using a stiffer material such as cement-mixed gravel soil, the backfill would exhibit substantially smaller settlements immediately behind the parapet.
2. By arranging horizontal reinforcement layers in the cement-mixed soil backfill, the development of major cracks in the zones where the tensile stress exceeds the tensile strength of cement-mixed soil could be effectively prevented.
3. By connecting the reinforcement layers to the back of the facing that directly supports a bridge girder, relative settlements between them could be efficiently prevented while ensuring high integrity of the whole abutment structure.

Models 4 and 5 (simulating the proposed new types of bridge abutment)

Eleven layers of grid reinforcements were arranged horizontally inside the approach block of cement-mixed soil with the ends connected to the back face of the facing (Fig. 12.5). Figures 12.8(a) and 12.8(b) show the relationships between the maximum and residual displacements recorded at the top of the facing and the maximum table acceleration, a_{max}, when subjected to irregular waves and subsequent sinusoidal input waves. In these figures, the residual displacement at the top of the facing of Model 1 is also presented as the reference. The following trends in behavior may be seen from Figs. 12.6–12.8:

1. Models 4 and 5 were substantially more stable than Model 1 (i.e., the most conventional-type abutment).
2. With Model 4, the tensile rupture of the connection between the reinforcement and the facing started when the a_{max} value of irregular waves reached 1400 gal. This phenomenon was also noted by a sudden change in the reading of tensile strain at the respective reinforcement layer. With

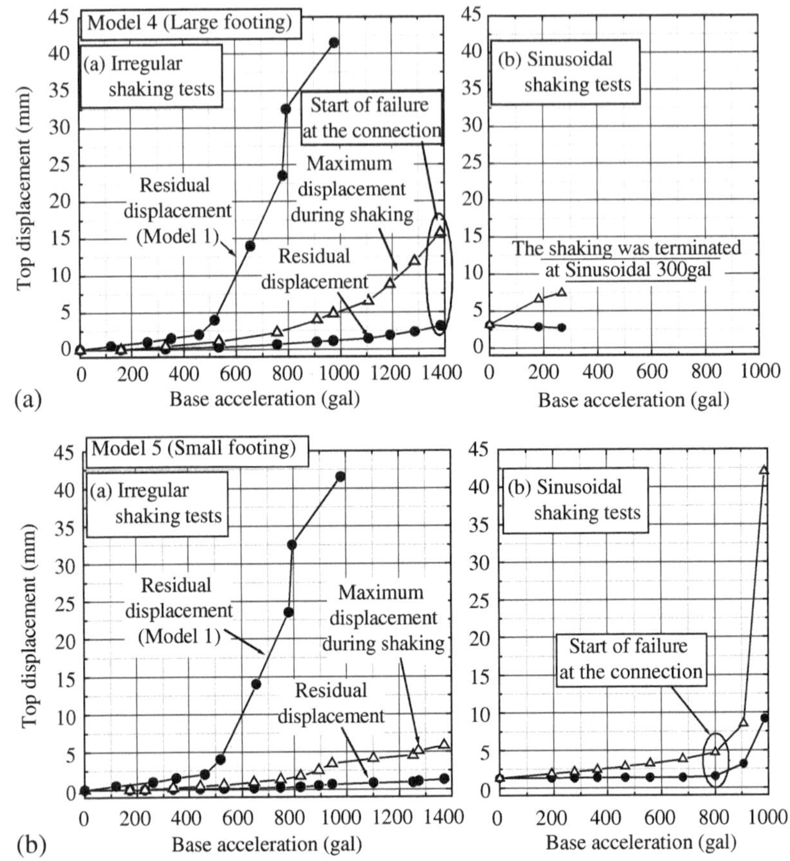

Figure 12.8 Displacement at the top of model abutment of (a) Model 4 and (b) Model 5, compared to that of Model 1. *(Source: Watanabe et al. (2002)).*

Model 5, it occurred when the a_{max} value of sinusoidal wave reached 800 gal, after having been subjected to a series of irregular motions.

3. The connection failure started earlier with Model 4 (with a wider footing) than with Model 5 (with a narrower footing). This unexpected trend in behavior was likely due to larger relative vertical displacements at the interface between the approach block and the facing with Model 4 because of a larger footing width (Fig. 12.9). The fact that Model 5 (with a narrower footing) was more dynamically stable than Model 4 (with a wider footing) can be attributed to the fact that the backfill is more stable than the facing, particularly under dynamic loading conditions.

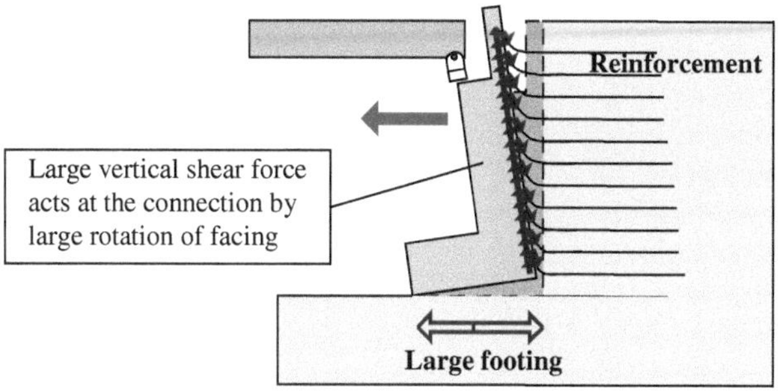

Figure 12.9 Schematic diagram illustrating shear force acting at the connection of reinforcement, Model 4. *(Source: Watanabe et al. (2002)).*

These model test results indicate the essential importance of sufficiently high connection strength between the reinforcement and the parapet for a high seismic stability of this type of abutment structure. With a facing of Model 5 bridge abutment, the connection strength between the reinforcement layers of a polymer geogrid and a full-height rigid facing was confirmed by performing full-scale lateral loading tests as described later in this chapter.

Dynamic disturbing force and resistance

Figure 12.10 shows the time histories of the following physical quantities when Model 4 was subjected to an irregular wave with $a_{max}=539$ gal (see Fig. 12.5 for the measuring points):

1. The dynamic component of the earth pressure acting on the back of the facing
2. The dynamic component of the reinforcement tensile force activated at a place adjacent to the back of the facing
3. The dynamic component of the vertical normal force acting at the toe and heel parts of the footing base of the facing (note that the normal and shear forces acting on the footing base were measured separately at four parts)
4. The dynamic component of the lateral displacement at the top of the facing
5. The input horizontal acceleration at the shaking table

The following trends in behavior may be seen from Figure 12.10:

1. At the dynamically active earth pressure state (i.e., when the dynamic component of the lateral displacement at the top of the facing was directing outward, denoted as A in Fig. 12.10), the increasing resisting

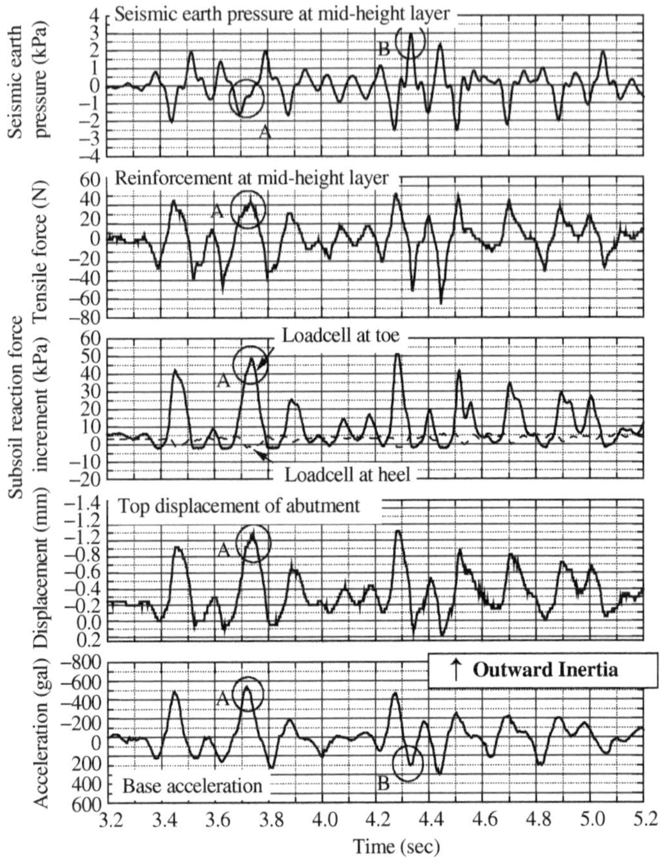

Figure 12.10 Typical time histories of several measured physical quantities, Model 4 (irregular shaking, $a_{max} = 539$ gal). *(Source: Watanabe et al. (2002)).*

components were (1) the reacting vertical normal force near the toe of the footing base and (2) the tensile force in the reinforcement. On the other hand, until this state, the vertical contact force near the heel of the footing had largely decreased, indicating an overturning displacement mode of the facing.

2. At the dynamically active earth pressure state, the dynamic component of the earth pressure activated at the back face of the facing a minimum value, showing that the facing was dynamically less stable than the backfill. On the other hand, the maximum earth pressure was attained at the dynamically passive earth pressure state (denoted as B in Fig. 12.10). Therefore, with this type of bridge abutment, sufficiently high

connection strength between the facing and the reinforcement is essential for a high seismic stability at the dynamically active earth pressure state. These trends of dynamic earth pressure are opposite to those assumed in the conventional seismic design of conventional-type retaining walls, in which the dynamic component of the active earth pressure becomes the maximum, which destabilizes the retaining wall.

Effects of the width of footing

Figure 12.11 shows the relationships between the maximum resistant moment acting at the footing base of the facing defined about the heel of the footing base and the rotation angle of the facing at each shaking stage using irregular waves for Models 4 and 5. Figure 12.12 shows the corresponding relationship between the base acceleration and the reinforcement tensile force that activate adjacent to the back of the facing (see Fig. 12.5). It may be seen from these figures that the resistance moment acting at the footing base was larger with Model 4 (having a wider footing) than with Model 5 (having a narrower footing), whereas the opposite was true with the reinforcement tensile force at the connection. That is, with Model 5, the major resisting force was the tensile force at the connection between the facing and the reinforcement. The maximum reinforcement tensile force consistently increased with an increase in the base acceleration, whereas the resisting moment acting at the footing base started decreasing with an increase in the base acceleration when the rotating angle of the footing base reached a certain value, associated with the start of failure in the ground supporting the footing. Due to such different resisting mechanisms, Model 5 was more

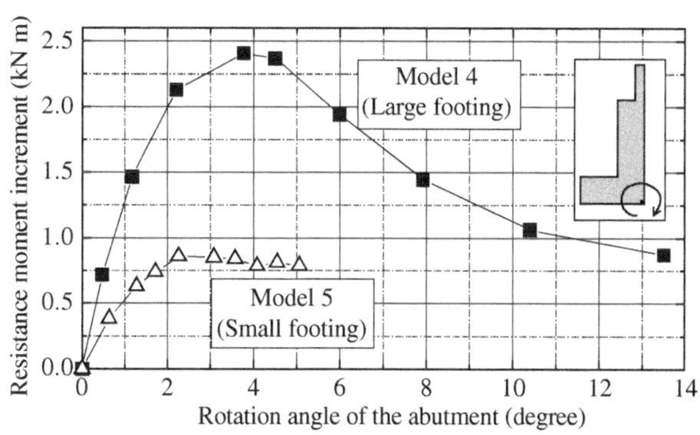

Figure 12.11 Relationships between rotation angle of parapet and resistance moment about the heel of the base of the parapet resulting from the reacting vertical normal force of subsoil. *(Source: Watanabe et al. (2002)).*

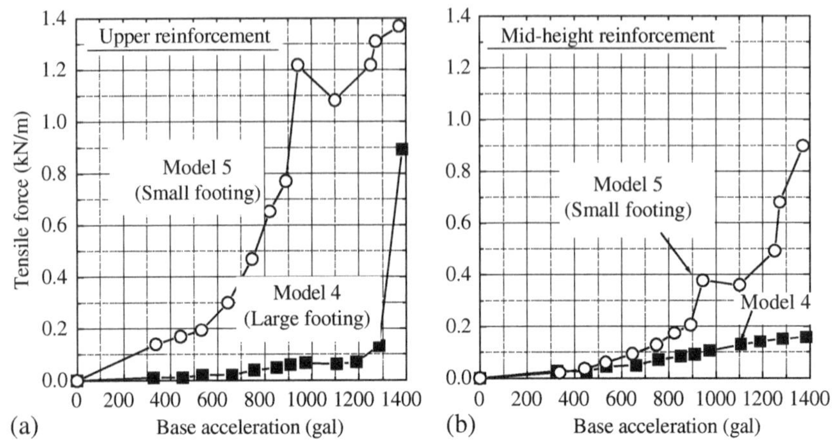

Figure 12.12 Relationships between base acceleration and tensile force in (a) upper reinforcement and (b) mid-height reinforcement, Models 4 and 5. *(Source: Watanabe et al. (2002)).*

stable than Model 4. With Model 5, larger connection force resulted in larger tensile force in the reinforcement with larger load transmitted from the facing to the cement-mixed soil approach block, which resulted in the development of a horizontal crack in the approach block (Fig. 12.6). This type of crack did not develop in Model 4.

The test results presented here suggest that by making sufficiently high connection strength at the reinforcement layers and the facing with sufficiently high strength of the backfill material of the approach block, the size of the base of the footing for the facing of prototype structures can be made rather small as in Model 5, which makes this type of abutment considerably cost-effective.

12.2.4 Summary of the model shaking table tests

The following conclusions can be derived from the test results of the model shaking table tests presented previously:

1. The seismic stability of several conventional types of railway bridge abutments is not sufficiently high.
2. To increase the seismic stability of bridge abutments, it is efficient and cost-effective to construct an approach block using CMG that is reinforced with geogrid layers connected to the back face of a full-height rigid facing supporting a bridge girder.

3. A sufficiently high strength of the connection between the reinforcement layers and the facing is essential not only to restrain the settlement of the backfill relative to the facing but also to ensure a high dynamic stability of the facing, which is less stable than the approach block.

12.3 STRENGTH OF CEMENT-MIXED GRAVEL

12.3.1 Background

It is necessary to evaluate the deformation and displacements of the concerned structure in addition to the safety factor against the ultimate failure of the structure both at working loads and under severe seismic loading conditions. To this end, the strength and deformation characteristics of CMG were evaluated by performing a series of consolidated drained triaxial compression (CDTC) tests as described here. At the same time, the relevant mixing proportion of components as well as the relevant construction method were sought to achieve as high a strength per unit cost of cement-mixed gravel as possible. A high cost-effectiveness is essential for the construction of this type of structure to become one of the standard construction methods for railway as well as highway structures.

CMG has been used to construct large dams (i.e., the roller compacted dam concrete; Hansen and Reinhardt, 1990; Schrader, 1996). However, only a limited amount of research has been performed on the use of CMG in ordinary but critical civil engineering structures requiring a high stability and allowing limited deformation, such as railway and highway bridge abutments. In the current study, a series of CDTC tests were performed on cement-mixed specimens of two types of well-graded quarry gravely soils (Fig. 12.13), which are typical of those used in the field. The first type was a crushed sandstone from a quarry (called Chiba gravel), which has often been used to study the strength and deformation properties of gravel at the University of Tokyo (e.g., Jiang et al., 1999). The other was a crushed gabbro from another quarry (called Kyushu gravel), which was used as the backfill of a prototype of the new type of bridge abutment explained later in this chapter. The effects of compacted dry density, molding water content, and gravel type on the strength and deformation characteristics of cement-mixed gravel were mainly investigated. The effects of grading characteristics were investigated to a limited extent. In addition, the strength characteristics of cement-mixed gravel were compared with those of uncemented original gravel. The effects of molding water content and grading characteristics of gravel evaluated by CDTC tests using

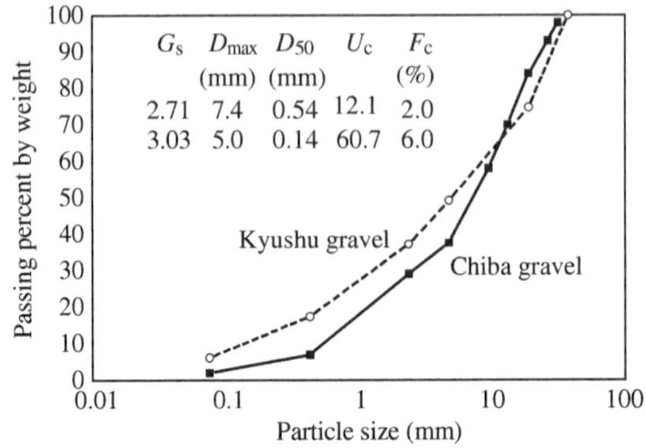

Figure 12.13 Grading curves of two types of gravels tested by large CDTC tests. *(Source: Watanabe et al. (2003a)).*

smaller specimens (7.5 cm in diameter and 15 cm high) of model Chiba gravel are reported in Lohani et al. (2004) and Kongsukprasert et al. (2005). The effects of stress condition during curing on the strength and deformation characteristics evaluated by CDTC tests using model Chiba gravel are reported in Kongsukprasert et al. (2001) and Kongsukprasert and Tatsuoka (2003).

12.3.2 Testing apparatus and preparation of specimens

Gravel type and specimen preparation

Figure 12.14(a) shows the compaction curves of cement-mixed and uncemented specimens of the original Chiba gravel and the one obtained by removing particles larger than 10 mm (i.e., model Chiba gravel) obtained by the standard Proctor compaction tests. Figure 12.14(b) shows similar results for the original Kyushu gravel by the standard and modified Proctor compaction tests. The sizes of mold and compaction energy levels used in these compaction tests are listed in these figures. It may be seen that the respective compaction curve of cement-mixed gravel is nearly the same as the corresponding one of uncemented gravel for these three types of gravel.

Table 12.2 lists five types of specimens prepared for the CDTC tests. The TC specimens were 20 cm in diameter and 40 cm high with cement-mixed Chiba gravel and 30 cm in diameter and 60 cm high with uncemented Chiba gravel. To investigate the effect of grading characteristics on the strength and deformation characteristics of cement-mixed gravel, TC tests were also performed on cement-mixed specimens of model Chiba gravel. The specimens were 20 cm in diameter and 40 cm high. The TC specimens

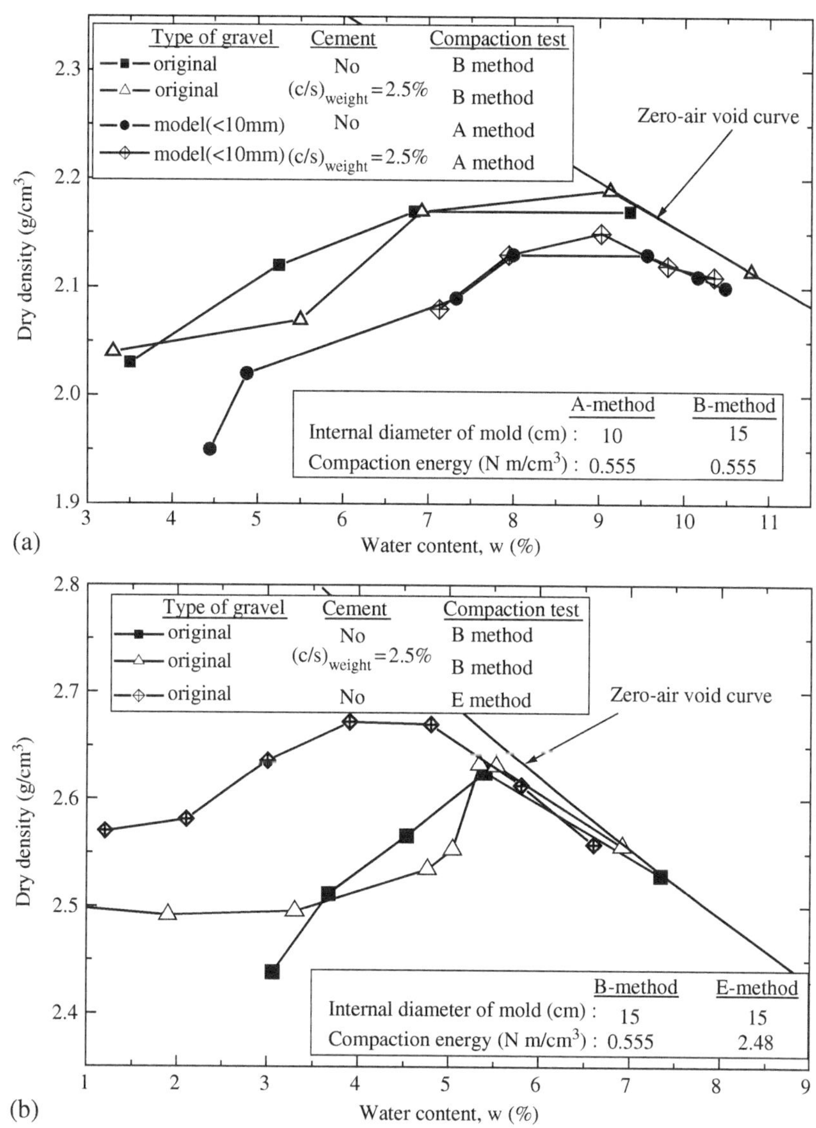

Figure 12.14 Results of compaction tests: (a) Chiba gravel and (b) Kyushu gravel. *(Source: Watanabe et al. (2003a)).*

of cement-mixed Kyushu gravel were 15 cm in diameter and 30 cm high; these dimensions were used because the strength of cement-mixed Kyushu gravel specimens with a diameter of 20 cm exceeded the capacity of the loading system. The specimens of Kyushu gravel without cement-mixing were 20 cm in diameter and 40 cm high. Throughout the study, a fixed

Table 12.2 Specimen types tested

Specimen type	Dry density (g/cm^3)	Confining pressure (kPa)	Water content (%)	Cement/gravel ratio by weight (W_{cement}/W_{gravel} %)	Cement/gravel ratio by volume (V_{cement}/V_{gravel} %)	Size of specimen (cm)
Chiba (cement-mixed)	1.8, 2.0, 2.2	0, 20, 50	3.5, 5.0, 8.7, 10.5	2.5	6.78	20 × 40
Model Chiba ($D < 10$ mm) (cement-mixed)	2.0	20	8.7	2.5	6.78	20 × 40
Chiba (without cement)	1.8, 2.0, 2.2	0, 20, 80	4.0	—	—	30 × 60
Kyushu (cement-mixed)	2.41, 2.56, 2.64	0, 20, 50	4.35	2.5	7.58	15 × 30
Kyushu (without cement)	2.41, 2.56, 2.64	0, 20, 50	4.35			20 × 40

Source: Watanabe et al. (2003a).

cement/gravel ratio by weight (W_{cement}/W_{gravel}) equal to 2.5% was used as a typical value employed in the field.

The cement-mixed specimens were prepared by manual vertical compaction in approximately 6 or 12 sublayers, with a thickness of each sublayer of 50 mm for all the specimens. Three ratios of compacted dry density to the respective maximum dry density attained by using the standard Proctor compaction energy of 0.555 N m/cm^3 (for the compaction curves of-cement-mixed gravel presented in Fig. 12.14) were prepared. These were approximately 80, 90, and 99% for Chiba gravel and 91%, 97%, and 100% for Kyushu gravel. The molding water content was basically 5% (dry of optimum) and 8.7% (optimum) with Chiba gravel, whereas it was 4.35% (optimum for compaction by E-method (the modified Proctor) and slightly dry of optimum for compaction by B method (the standard Proctor)) with Kyushu gravel. To investigate the effect of molding water content on the strength characteristics in more detail, the molding water content was varied between 3.5% and 12%, keeping the dry density constant (2.0 g/cm^3) with model Chiba gravel. The molding water content was 4% with uncemented Chiba gravel and 4.35% with uncemented Kyushu gravel. The compacted cement-mixed specimens were cured for 7 days under atmospheric pressure at constant water content before setting in the triaxial apparatus.

Triaxial test apparatus

All CDTC tests were performed using a large triaxial testing apparatus at the Railway Technical Research Institute (Fig. 12.15). The apparatus is able to control the axial displacement to an accuracy of less than 1 μm (Kohata et al., 1999). The axial load was measured with a load cell set inside the triaxial cell. The axial compression of specimen was measured externally with a pair of proximeters set at the specimen cap and locally with a pair of local deformation transducers (Goto et al., 1991) that were set at the side face of the specimens. The lateral deformations of the specimens were measured by using three pairs of proximeters arranged at three elevations (5/6, 3/6, and 1/6 of the specimen height from the bottom), which were free from the effects of membrane penetration in these CDTC tests in which the effective confining pressure was kept constant.

The specimens were initially isotropically compressed to the prescribed confining pressure, σ_c', and then sheared in drained TC at a constant axial strain rate of 0.01%/min under constant σ_c' until the axial strain became 15%. Three axial unload–reload cycles with a small axial strain amplitude were applied at various deviator stresses during otherwise monotonic

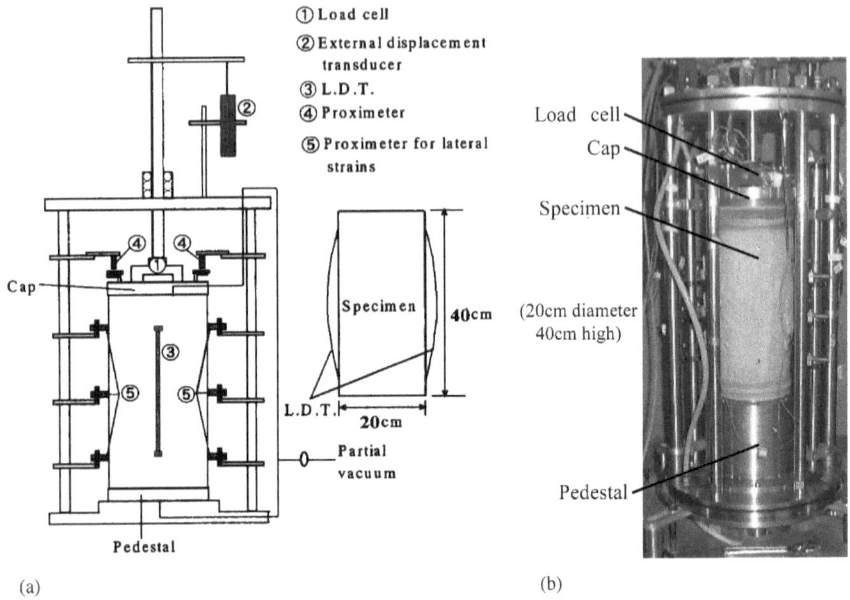

Figure 12.15 Large triaxial apparatus: (a) schematic diagram and (b) picture of the system during a TC test on cement-mixed Chiba gravel. *(Source: Watanabe et al. (2003a)).*

loading to evaluate the equivalent elastic vertical Young's modulus (E_{eq}) and Poisson's ratio (ν_{vh}).

12.3.3 Test results and discussions

Strength characteristic at different compacted dry densities

Figures 12.16(a)–12.16(c) show typical results from CDTC tests on cement-mixed Chiba gravel specimens compacted to different dry densities, ρ_d, at a water content of 5%. It may be seen that, with the same cement content by weight $W_{cement}/W_{gravel}=2.5\%$, the peak strength increases significantly with an increase in the compacted dry density, ρ_d. On the other hand, the effects of ρ_d on the residual strength, defined to be the deviator stress at an axial strain of 15%, are insignificant. Note, however, that the effects of ρ_d on the residual strength are not generally negligible, which is discussed later. Figures 12.17 and 12.18 summarize the peak and residual strengths plotted against the ρ_d value of solid material (gravel and cement). The following trends in behavior may be seen from these figures:

1. With both types of cement-mixed gravel, the peak strength increases with an increase in ρ_d at a rate that is much higher than with the uncemented ones. Therefore, the difference in the peak strength between the

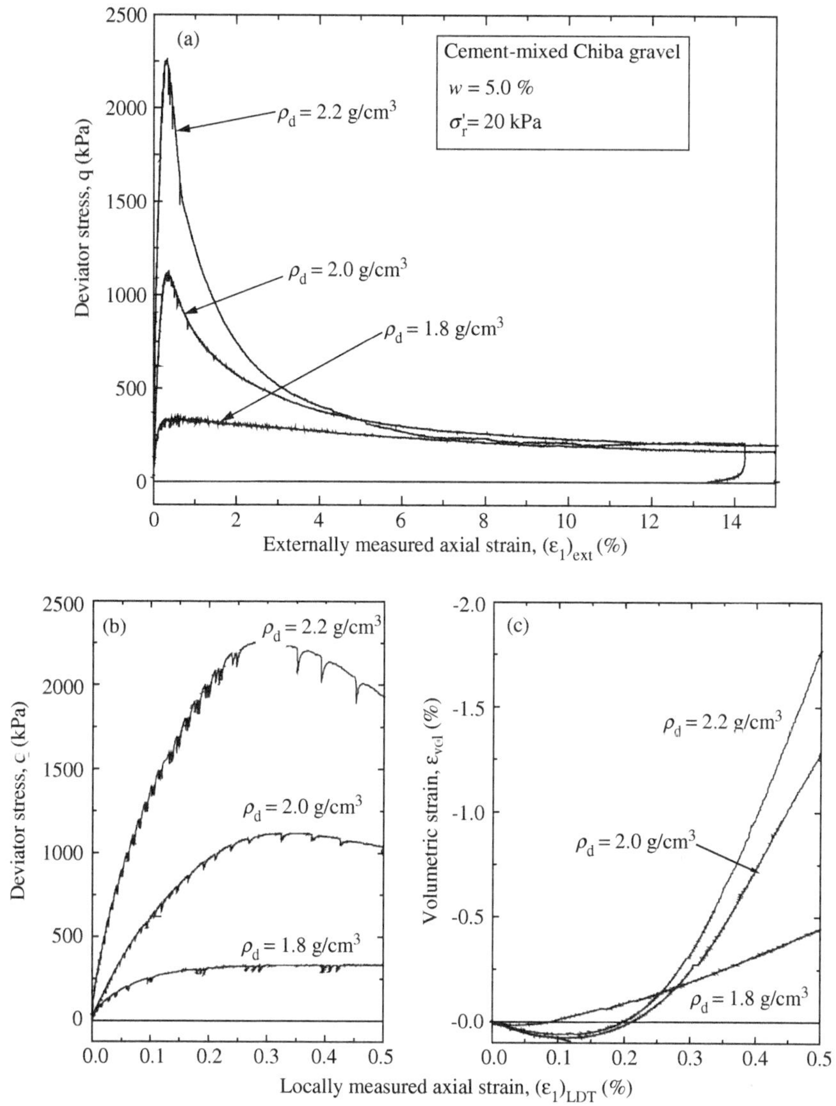

Figure 12.16 (a) Overall deviator stress–axial strain (external) relations; (b) deviator–axial strain (local) relations at small strains; and (c) volumetric and axial strain relations at small strains, cement-mixed Chiba gravel compacted different dry densities. *(Source: Watanabe et al. (2003a)).*

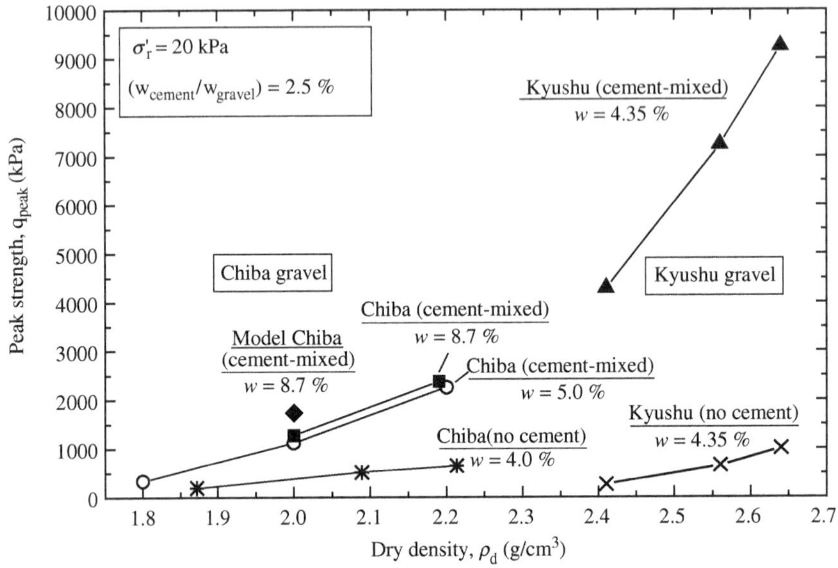

Figure 12.17 Relationships between peak strength and compacted dry density of cement-mixed gravels and uncemented gravels. *(Source: Watanabe et al. (2003a)).*

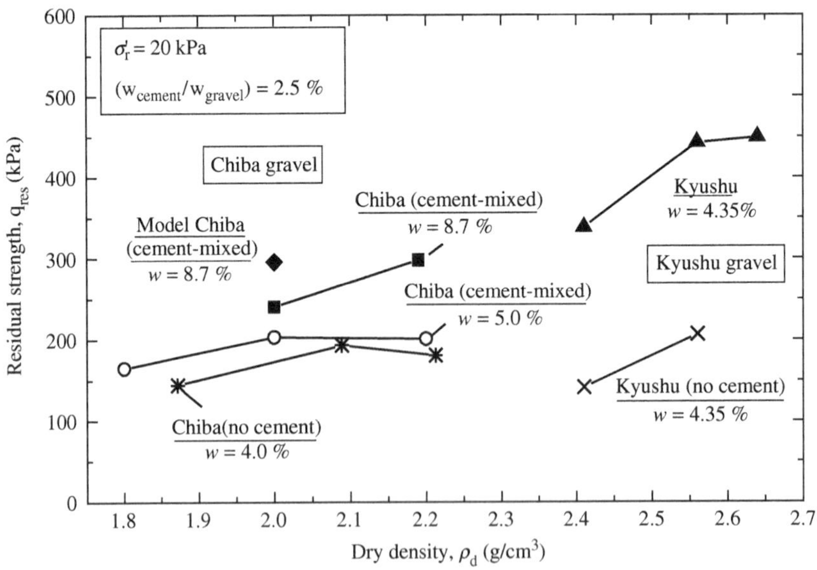

Figure 12.18 Relationships between residual strength and compacted dry density of cement-mixed gravels and uncemented gravels. *(Source: Watanabe et al. (2003a)).*

cement-mixed and uncemented specimens increases with an increase in ρ_d, particularly with Kyushu gravel.

2. When compacted by using nearly the same energy and with the same cement/gravel ratio by weight (W_{cement}/W_{gravel}), the peak strength of cement-mixed Kyushu gravel is significantly higher than that of cement-mixed Chiba gravel. This feature is discussed in detail later.
3. The effects of ρ_d on the residual strength are much smaller than those on the peak strength. Careful observations of the specimens of cement-mixed gravel after the respective TC tests indicated that the residual strength was controlled by frictional resistance along a shear band(s) and not by bonding at interparticle contacts along the shear plane.
4. For the same ρ_d value, the peak strength of the cement-mixed specimen of the model Chiba gravel, not including particles larger than 10 mm and having a smaller coefficient of uniformity, is noticeably higher than that of cement-mixed original Chiba gravel. On the other hand, it is shown in Lohani et al. (2004) and Kongsukprasert et al. (2005) that when compacted by using the same energy, the strengths of the cement-mixed model Chiba gravel and the original type of Chiba gravel with the same cement/gravel ratio by weight are nearly the same. These trends in behavior are consistent with each other because for the same compaction energy, the ρ_d value is higher for the original Chiba gravel than for the model Chiba gravel.

The strength of cement-mixed Kyushu gravel is considerably higher than that for cement-mixed Chiba gravel for the same cement/gravel content by weight and for the same compaction energy. This trend in behavior could be attributed to several factors. First, a considerably higher ρ_d value for the same compacted void ratio (as shown in Fig. 12.19) of Kyushu gravel can be attributed to a specific gravity of soil particles ($G_s=3.03$) that is much higher than 2.71 for Chiba gravel. For the same cement/gravel ratio by weight, $W_{cement}/W_{gravel}=2.5\%$, the cement content ($W_{cement}$/specimen volume V) is proportional to ρ_d. Therefore, for the $q_{peak}-\rho_d$ relation for the same type of gravel plotted in Fig. 12.17, two variables—compacted dry density ρ_d $(=(W_{gravel}+W_{cement})/V)$ and cement content per volume ($=W_{cement}/V$)—change simultaneously. On the other hand, it is likely that the specific gravity of soil particles itself has no direct link to the packing conditions of particles, and therefore the peak strength of cement-mixed gravel is better linked to volume parameters than to mass parameters—that is, the void ratio $(=V/(V_{gravel}+V_{cement})-1.0)$ rather than the ρ_d value and V_{cement}/V rather

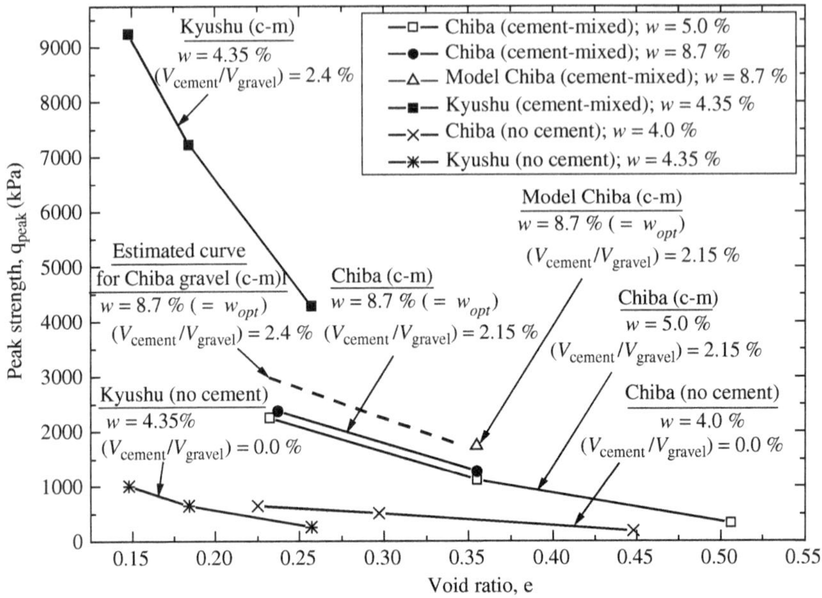

Figure 12.19 Relationships between peak strength and compacted void ratio of cement-mixed gravels and uncemented gravels. *(Source: Watanabe et al. (2003a)).*

than W_{cement}/W_{gravel}. When the peak strengths of cement-mixed Chiba and Kyushu gravels are compared for the same void ratio (as shown in Fig. 12.19), the following two factors should also be considered:

1. For the same W_{cement}/W_{gravel} (=2.5%) and the same void ratio of solid part, the cement/gravel ratio in volume (V_{cement}/V_{gravel}) is different—2.4% with Kyushu gravel and 2.15% with Chiba gravel.
2. The ratios of the molding water content (4.35% with cement-mixed Kyushu gravel and 8.7% with cement-mixed Chiba gravel) to the respective optimum water content are different. It is shown in Lohani et al. (2004) and Kongsukprasert et al. (2005) that for the same type of cement-mixed gravel, even at the same compacted dry density, the peak strength becomes highest when compacted at the optimum water content (w_{opt}). Although it is subtle, this trend in behavior can be seen by comparing the peak strengths of cement-mixed Chiba gravel specimens compacted at water contents of 5% (dry of optimum) and 8.7% (at optimum) in Figs. 12.17 and 12.19.

Figure 12.19 also presents the strength curve for V_{cement}/V_{gravel}=2.4% of cement-mixed Chiba gravel and the molding water content equal to the optimum, which was estimated based on the results of a study on the effects of cement content and water content on the peak strength (Lohani et al., 2004; Kongsukprasert et al., 2005). Compared at the same V_{cement}/V_{gravel} (=2.4%) and the same water content ratio $w/w_{opt}=1.0$, the difference in the peak strength between the two types of cement-mixed gravel at the same void ratio becomes noticeably smaller than when compared at the same W_{cement}/W_{gravel} (=2.5%) and at the different molding water content ratios, w/w_{opt}. However, the peak strength of cement-mixed Kyushu gravel is still noticeably higher than that of the cement-mixed Chiba gravel in Fig. 12.19. Another factor or factors are necessary to explain this difference. The grading characteristics may be one of these factors.

Figure 12.20 shows the relationships between the peak strengths of cement-mixed gravel and corresponding uncemented original gravel measured at the same confining pressure, σ_c' = either 0 or 20 kPa or 50 kPa. The peak strength of uncemented Chiba gravel at $\sigma_c'=50$ kPa was estimated from those extrapolated from the $q_{peak}-\sigma_c'$ relation obtained by CDTC

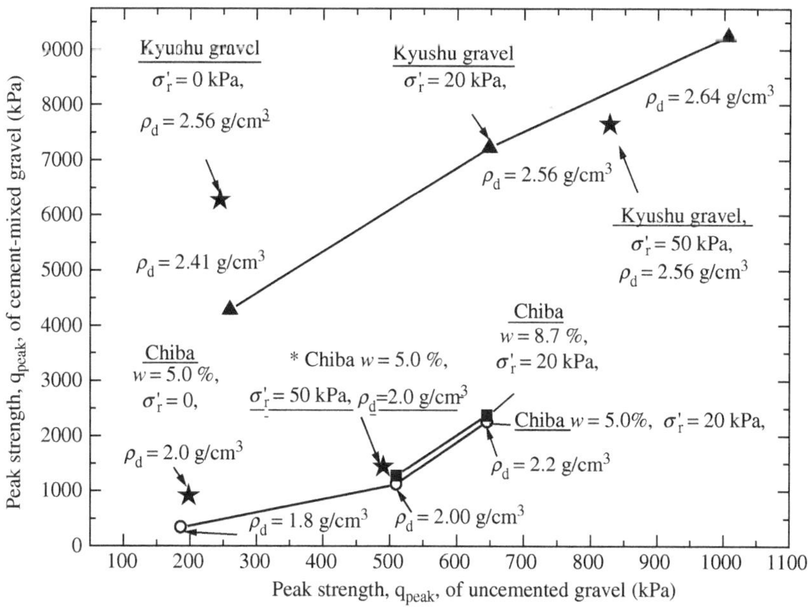

Figure 12.20 Relationships between peak strengths of cement-mixed gravel and uncemented gravel; the water contents shown in this figure are for cement-mixed specimens. *(Source: Watanabe et al. (2003a)).*

tests performed at other similar σ_c' values. The following trends in behavior may be seen:

1. The peak strengths of cemented and uncemented specimens compacted to the same dry density are well correlated to each other. This fact indicates the importance of increasing as much as possible the compacted dry density when the backfill is constructed by using cement-mixed gravel, in the same way with unbound original gravel.
2. The relationship between the peak strengths of cemented and uncemented specimens is not unique, being affected by several other factors, including cement content, molding water content, gravel type, and confining pressure. The ratio of the peak strengths of cement-mixed gravel and uncemented gravel increases with a decrease in σ_c'.

Figure 12.21 compares the residual strength in the same way as Fig. 12.19. Also in this figure, the estimated strength curve for $V_{cement}/V_{gravel}=2.4\%$ of cement-mixed Chiba gravel compacted at the optimum water content is presented. Although they are much less pronounced, the trends in behavior that can be seen with the peak strength in Fig. 12.19

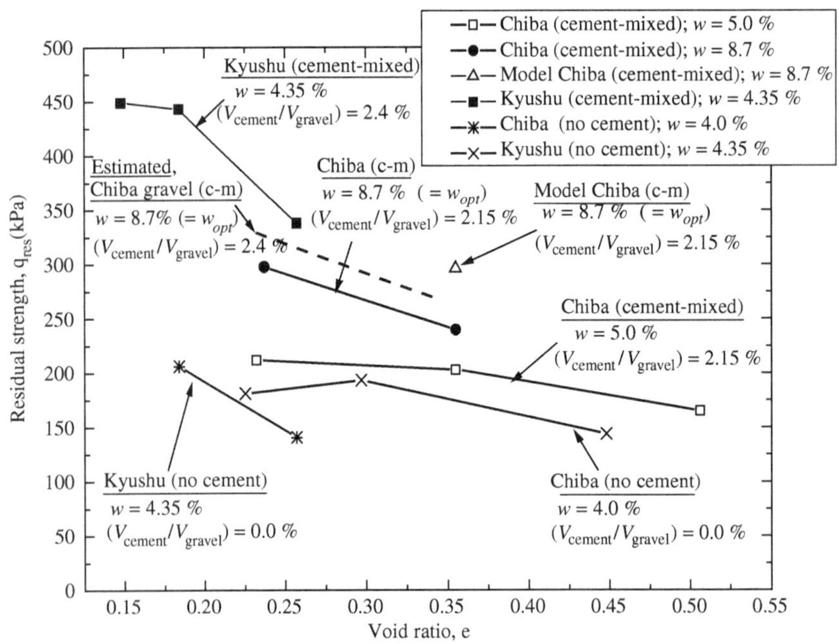

Figure 12.21 Relationships between residual strength and void ratio of cement-mixed gravels and uncemented gravels. *(Source: Watanabe et al. (2003a)).*

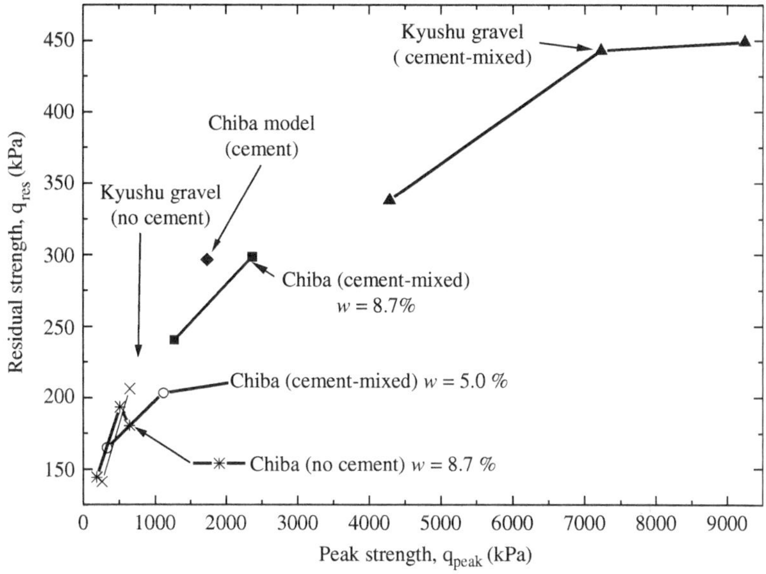

Figure 12.22 Relationships between residual strength and peak strengths of cement-mixed gravel and uncemented gravels. *(Source: Watanabe et al. (2003a)).*

can also be seen in Figure 12.21, indicating that the peak and residual strengths are linked to each other. Figure 12.22 shows the relationships between the corresponding peak and residual strengths of uncemented and cement-mixed specimens of Chiba and Kyushu gravels. It can be seen that the residual strength becomes larger with an increase in the peak strength in a similar way irrespective of gravel type and whether the specimen is uncemented or cement-mixed. That is, the residual strength of cement-mixed gravel is noticeably higher than that of the corresponding uncemented gravel. It seems that the residual strength of cement-mixed gravel is controlled by not only the friction at interparticle contacts but also restraining from free rotation of soil particles by bonding at interparticle contacts; that is, the roughness along a shear plane increases with an increase in the interparticle bond strength.

Effects of molding water content

Figure 12.23 shows the relationships between the peak strength and the molding water content for cement-mixed Chiba gravel. It may be seen that the peak strength became maximum near the optimum water content ($w_{opt}=8.7\%$). This trend in behavior is largely due to the largest compacted

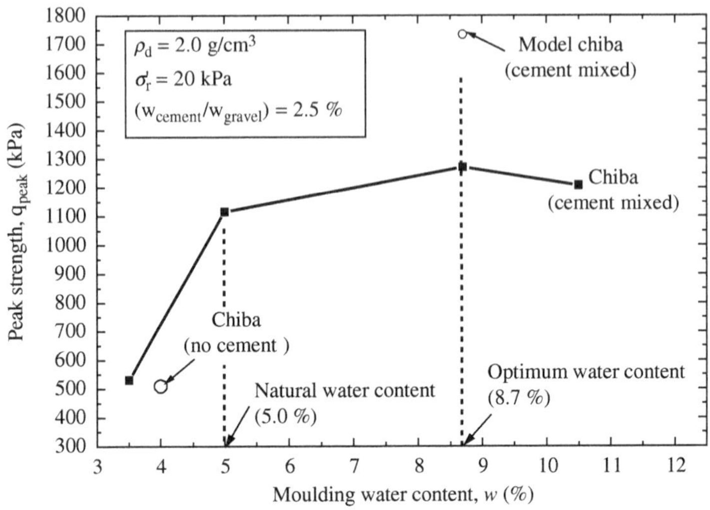

Figure 12.23 Relationships between peak strength and molding water content. *(Source: Watanabe et al. (2003a)).*

dry density (i.e., the smallest compacted void ratio). The other influencing factors include (1) the largest cement content per volume of specimen associated with the maximum compacted dry density (as discussed previously) and (2) the optimum cement paste conditions in terms of the volume of paste and the strength of paste (as discussed later; Lohani et al., 2003, 2004; Kongsukprasert et al., 2005). With respect to the second factor, the strength of cement paste decreases with an increase in the molding water content. It is likely that the decrease in the strength of cement-mixed gravel with an increase in the water content larger than w_{opt} is due to this factor. On the other hand, the amount of cement paste deceases with a decrease in the molding water content. In fact, the peak strength when compacted at $w = 3.5\%$ is very low, close to the peak strength of uncemented gravel. It appears that this very low strength was due to an insufficient amount of cement paste to be distributed uniformly at interparticle contacts. It is shown in Lohani et al. (2004) and Kongsukprasert et al. (2005) that the same trends in behavior were also observed in the TC tests on small specimens of cement-mixed model Chiba gravel. These results indicate that the water content should be controlled as strictly as possible, preferably at the optimum water content, when compacting a given type of cement-mixed gravel in actual construction projects.

Mohr–Coulomb strength parameters

Figure 12.24 shows the peak and residual strengths plotted against the effective confining pressure, σ_c', for the cement-mixed gravel specimens with $W_{cement}/W_{gravel}=2.5\%$ and the corresponding uncemented gravel specimens. The molding water content was equal to the optimum water content for Kyushu gravel, whereas it was 5.0% for cemented Chiba gravel and 4.0% for uncemented Chiba gravel. Table 12.3 shows the angle of internal friction and cohesion intersect of the respective types of specimens obtained from

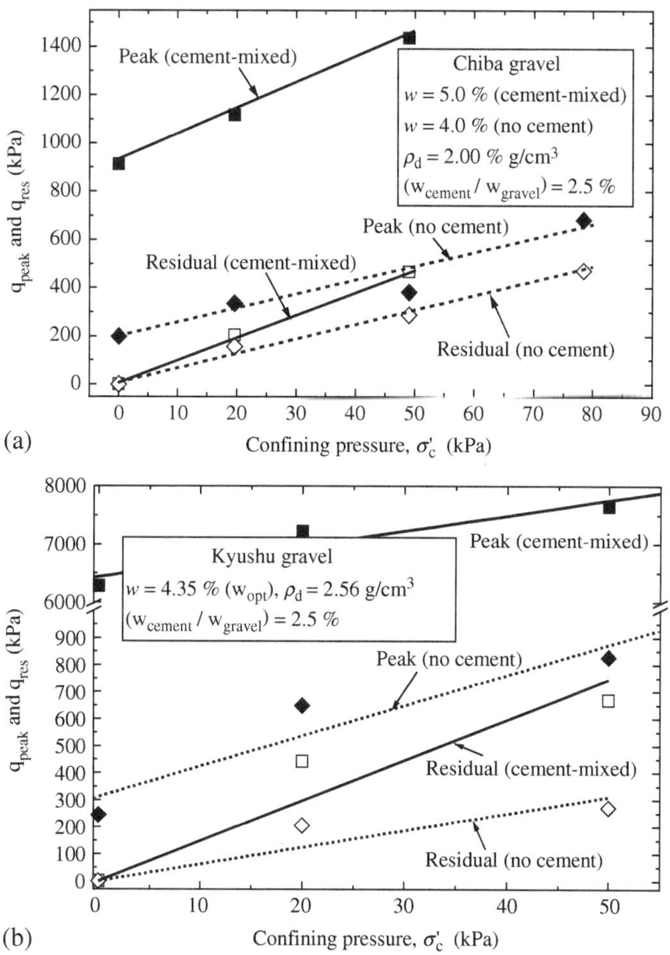

Figure 12.24 Relationships between peak and residual strengths and confining pressure: cemented and uncemented specimens of (a) Chiba gravel and (b) Kyushu gravel. *(Source: Watanabe et al. (2003a)).*

Table 12.3 Internal friction angles and cohesion intersects of cement-mixed and uncemented gravels

	Peak		Residual	
Gravel type	φ_{peak} (deg.)	c_{peak} (kPa)	φ_{res} (deg.)	c_{res} (kPa)
Chiba gravel (cement-mixed)	57.4	132	55.9	0
Chiba gravel (no cement)	48.9	38.4	48.9	0
Kyushu gravel (cement-mixed)	68.4	614	61.8	0
Kyushu gravel (no cement)	58.1	44.7	49.2	0

Source: Watanabe et al. (2003a).

the failure envelopes presented in Fig. 12.24. The residual strength in the unconfined compression tests on cemented and uncemented specimens was equal to zero (i.e., $c_{res}=0$).

The following trends in behavior may be seen from Fig. 12.24 and Table 12.3:

1. Not only the cohesion intersect, c_{peak}, but also the angle of internal friction, φ_{peak}, increase considerably by cement mixing. This trend in behavior is in contrast to the case with cement-mixed soft clay, for which the φ_{peak} value largely decreases while the c_{peak} value largely increases by cement mixing (as reported by Tatsuoka et al., 1997a; Sugai et al., 2000; Sugai and Tatsuoka, 2003). It is likely that the high peak strength of cement-mixed gravel results from restraint to not only sliding but also rotation of soil particles by bonding at interparticle contacts. Note that the effects of the increase in c_{peak} on the shear strength are significant at low σ_c' values.
2. The φ_{peak} values of both cement-mixed and uncemented specimens of Kyushu gravel are considerably higher than the respective values of cement-mixed and uncemented specimens of Chiba gravel. The increase in c_{peak} by cement mixing is much larger with Kyushu gravel than with Chiba gravel. It is not well understood how this feature can be linked to the gravel type, such as fines content, particle shape, grading characteristics, and the mother rock type.

The use of these peak strength parameters in the limit equilibrium-type stability analysis may result in an overestimation of safety factor because of possible major effects of progressive failure (Tatsuoka et al., 2000). The use of values between the peak and residual strengths would be relevant. Further study is necessary in this respect.

12.3.4 Summary of CDTC test results

The following conclusions can be derived from the results of the CDTC tests on large specimens of cement-mixed gravel presented previously:

1. The peak strength of compacted cement-mixed gravel could become significantly higher than that of uncemented gravel. The difference increases with an increase in the compacted dry density. This result indicates the significant importance of compaction control when constructing the backfill of cement-mixed gravel as with the backfill of uncemented gravel.
2. For different types of gravely soil having different specific gravities of particles, the strength of cement-mixed gravel should be considered as a function of compacted void ratio rather than dry density, as well as cement/gravel ratio by volume rather than cement/gravel content by weight. The gravel type is another important influencing factor.
3. The strength of cement-mixed gravel becomes maximum when compacted at the optimum water content, which could be due not only to the maximum compacted dry density (or the minimum compacted void ratio) and the associated maximum cement content per volume of specimen but also to the optimum cement paste conditions in terms of volume and strength. In fact, the strength when compacted significantly drier than optimum could be very low, perhaps due to an insufficient distribution of cement paste at interparticle contacts.
4. Not only the cohesion intercept but also the angle of internal friction could significantly increase by cement mixing, unlike cement-mixed soft clay. The contribution of cohesion intercept to the shear strength becomes particularly large at low effective confining pressures.
5. Despite the fact that the increase is much smaller than that of the peak friction angle by cement mixing, the friction angle at the residual state could noticeably increase by cement mixing. With cement-mixed gravel, the residual strength is well correlated to the peak strength.

12.4 CONSTRUCTION OF A PROTOTYPE OF A NEW TYPE BRIDGE ABUTMENT

12.4.1 Introduction

Based on the results from the shaking table tests on bridge abutment models and the laboratory stress–strain tests on cement-mixed gravely soil described in the preceding sections, in 2002, the Japan Railway Construction Public

Cooperation started the design of a new type of bridge abutment to be constructed at Takada, Kyushu, for a new bullet train line, aiming at high cost-effectiveness (Fig. 12.25) (Aoki et al., 2002). Figure 12.26(a) shows the outline of the bridge abutment, and Fig. 12.26(b) shows a view of the completed abutment. The construction started in November 2002 and was completed 2 months later. To confirm whether the vertical bearing capacity and the lateral stability of the facing structure (RC facing) of the abutment are sufficiently high, which is the most crucial feature of the new structural type of bridge abutment, full-scale field loading tests were performed in February 2003. The results were very satisfactory, as discussed next.

12.4.2 Design

For the new bullet train line, the design train speed is 260 km/h and the standard design train load (live) is P-16. That is, 16 tonf or 157 kN is applied to one axle that has two wheels. Each pair of axles with a minimum center-to-center distance of 2.2 m supports one bogie, whereas two bogies with a center-to-center distance of 15 m support one 20-m-long coach. The abutment supports a simple beam T-shaped RC girder with a length of 15 m with a fixed end at the top of the facing. It was decided that the design compressive strength of cement-mixed backfill is 2 MPa and the design rupture strength of geogrid is 30 kN/m. The soil layers having blow counts (*N* values) by the standard penetration tests equal to or greater than 50 were considered as the supporting ground of the

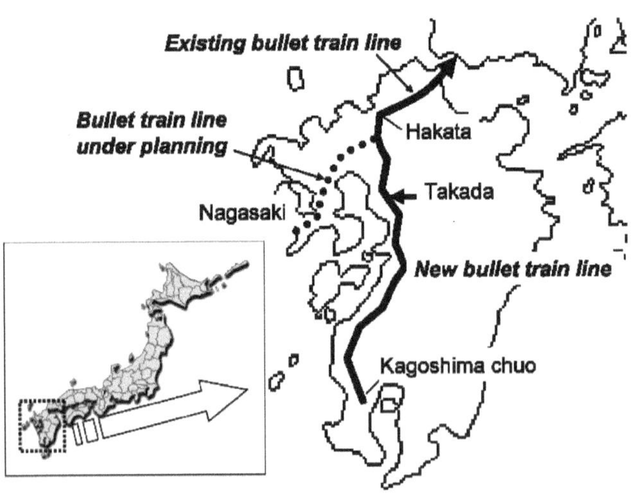

Figure 12.25 Location of the new type bridge abutment using cement-mixed gravel at Takada, Kyushu. *(Source: Aoki et al. (2002)).*

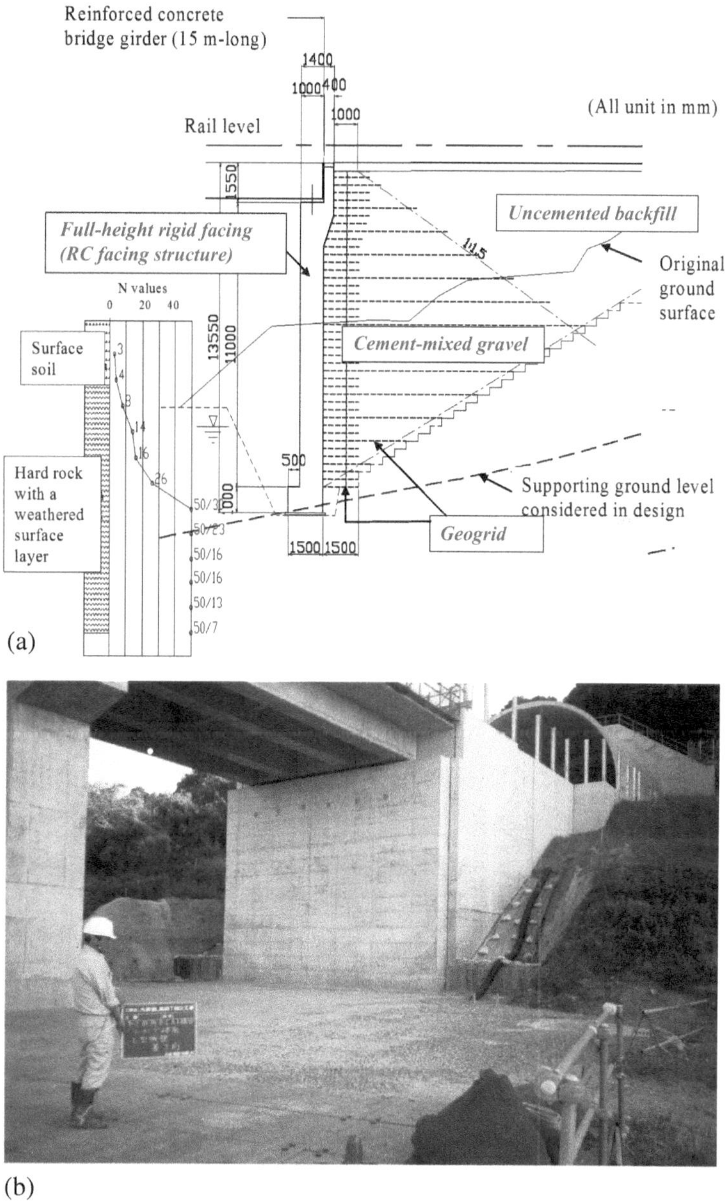

Figure 12.26 (a) New type bridge abutment using cement-mixed gravel constructed at Takada and (b) view of the completed new type of bridge abutment, mid-2003. *(Source: Part a from Aoki et al. (2002)).*

RC facing, whereas the soil layers having N values equal to or greater than 9 were considered as the supporting ground of the approach block (i.e., backfill of CMG).

Figure 12.27(a) shows the conventional-type bridge abutment that would have been constructed if the new structural type (Fig. 12.26) had not been adopted. The RC conventional-type bridge abutment was designed as a cantilever structure resisting against the static earth pressure exerted by the uncemented backfill as well as the static load of the RC abutment and bridge girder and the live load under static loading conditions

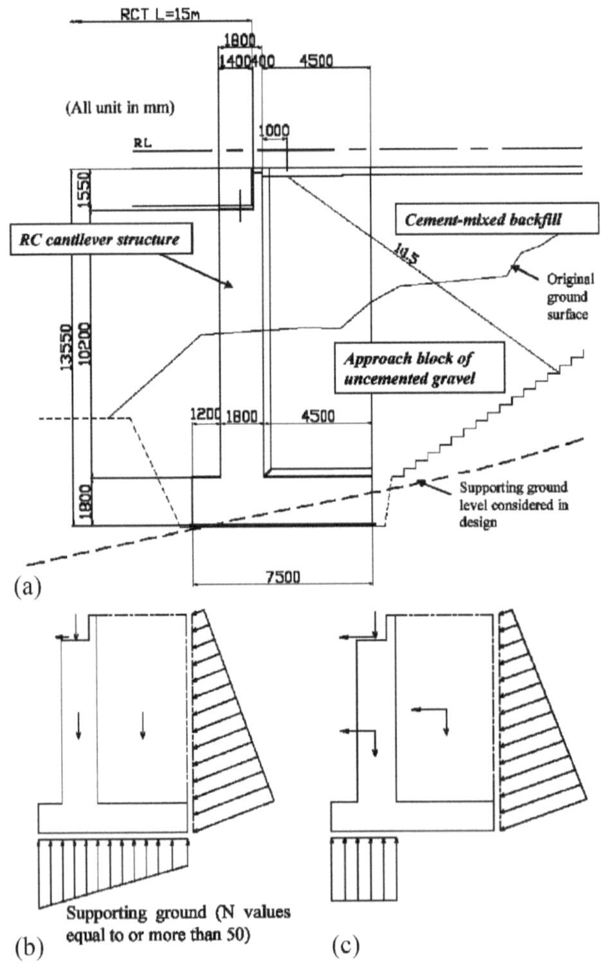

Figure 12.27 (a) Conventional-type bridge abutment and (b) static and (c) seismic loading conditions for conventional-type bridge abutment. *(Source: Aoki et al. (2002)).*

(Fig. 12.27(b)). For seismic design, the dynamic component of earth pressure and the inertia of the RC abutment and bridge girder were also taken into account (Fig. 12.27(c)).

With the new type of abutment, the RC facing and the backfill of CMG were designed as separate structural components (Figs. 12.28(a) and 12.28(b)). The RC facing was designed to support the static and dynamic loads of the facing and those from the bridge girder with the live load. The static and dynamic earth pressures from the backfill were not taken into account considering that the lateral outward movement of the facing should be supported by the cement-mixed backfill under seismic loading conditions. Therefore, the tensile strength of the geogrid layers that connect the facing and the backfill should be sufficiently high to withstand the tensile load activated by the outward lateral movement of the facing during seismic loading conditions. As shown in Fig. 12.28(c), the distribution of tensile force at the connections was obtained by modeling the connections by a series of spring. The backfill of CMG was designed as a gravity-type earth retaining wall resisting the lateral load from the facing and the earth pressure from the unbound soil backfill in back of the cement-mixed gravel backfill. Several long layers of geogrid were

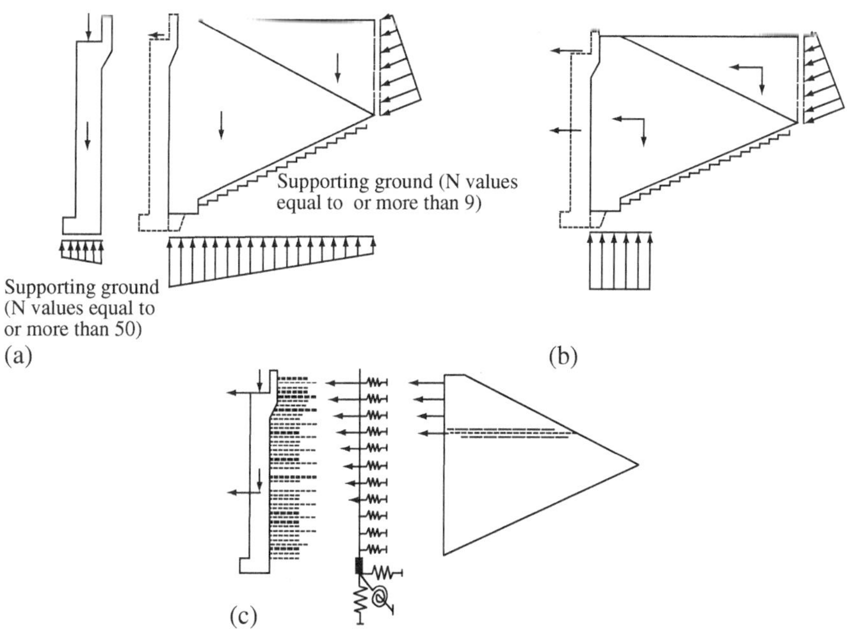

Figure 12.28 (a) Static and (b) seismic loading conditions and (c) seismic conditions at the connection for new type bridge abutment. *(Source: Aoki et al. (2002)).*

arranged at a vertical spacing of 0.9 m, prepared for possible development of tension cracks during severe seismic loading conditions. The strength of the long layers was determined to resist the seismic tensile force acting at the vertical plane at the back end of the short layers, ignoring the tensile strength of the backfill. The tensile force was obtained by assuming the zone in front of this vertical plane to act as a monolith.

The construction cost of the new type bridge abutment was estimated to be 87% of that of the conventional type for good ground conditions at the site. If the ground condition were poorer and the conventional-type bridge abutment should have been supported by a pile foundation (as in many cases), as the new type does not need a pile foundation, the difference in the construction cost could become much higher than in this case.

12.4.3 Construction

The abutment was constructed using the staged construction method (Fig. 12.29). That is, the backfill was constructed before constructing the RC facing. A well-graded crushed gravely soil of gabbro from a quarry (classified as M-40; typically, $D_{max}=37.5$ mm, $D_{50}=5.4$ mm, $U_c=61$, fines content = 6%, and specific gravity = 3.03) was used as the backfill. The stress–strain properties of cement-mixed specimens of this type of gravel were described in the preceding section. The maximum compacted dry density, $(\rho_d)_{max}$, was 2.60 g/cm^3 at the optimum water content, $w_{opt}=4.9\%$, by the compaction tests using a mold with an inner diameter

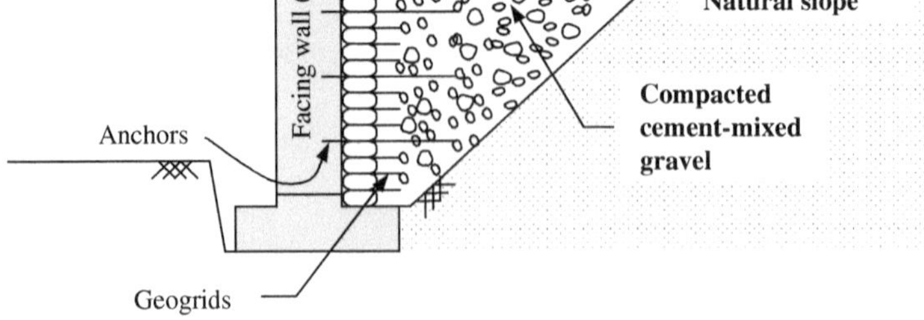

Figure 12.29 Staged construction method for new type bridge abutments with backfill of cement-mixed gravel. *(Source: Aoki et al., 2003; Watanabe et al. (2003b)).*

of 15 cm and an inner height of 12.5 cm with an energy level of the modified Proctor (2.48 N m/cm^3) (the *Eb* method; the Japanese Industrial Standards A1210; Watanabe et al., 2003b). The lift of the compacted soil layer was made as small as 15 cm to obtain a high compacted dry density. The specified compaction energy was achieved by 10 times passing of a small vibratory compaction plant with a weight of approximately 1 tonf for the first 24 lifts (Fig. 12.30(a)) and 6 times passing of a vibratory plant with a weight of approximately 4 tonf for the 25th through 81st lifts (Fig. 12.30b). A cement-mixed ratio by weight equal to 4% was employed. This ratio, which was higher than the value used in the laboratory stress–strain tests, was employed by taking into account a possibly high degree of inhomogeneity of cement mixing in the field.

The field compaction was made with a lift of 15 cm so as to obtain a degree of compaction more than 95%. Figures 12.31(a) and 12.31(b) show the frequency distributions of compacted dry density (by means of RI) and water content measured during the construction. It may be seen that the results were satisfactory. The compaction of the backfill in the field was done with the use of bags filled with unbound gravel placed at the shoulder of each soil layer (Fig. 12.32(a)).

As the bags were wrapped around with polymer geogrid layers, the wall face of the completed backfill was covered with geogrid sheets. The RC parapet was constructed by casting-in-place fresh concrete into a space between an external concrete form supported by steel bars extending from the inside of the backfill and the wall face of the backfill (Fig. 12.32(a)). Steel reinforcement had been arranged in the space. Therefore, the facing had been firmly connected to the geogrid sheets and gravel bags when the facing was completed. Using this staged construction method, the following potential

(a) (b)

Figure 12.30 Compaction of the backfill.

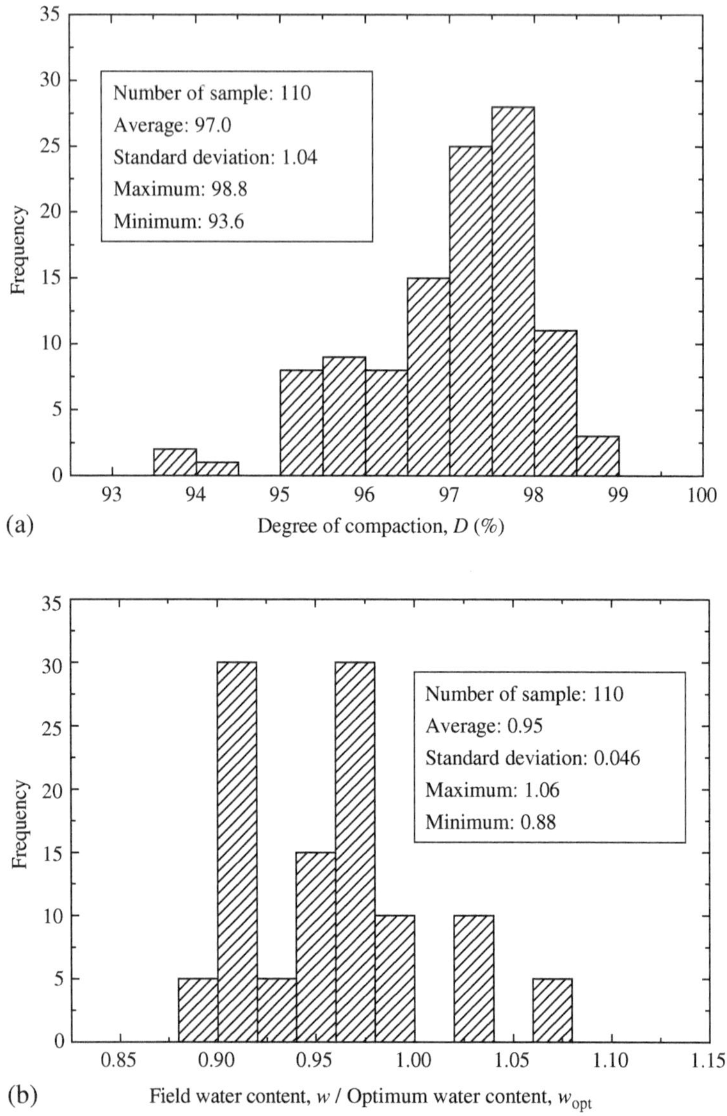

Figure 12.31 (a) Degree of compaction and (b) water content measured during the construction of backfill. *(Source: Watanabe et al. (2003b)).*

problems when the RC facing is constructed prior to the construction of backfill can be avoided:

1. If the facing is first constructed, a very large earth pressure may be exerted on the facing during compaction of backfill. To avoid this problem, the zone of backfill adjacent to the back face of the facing may not be well compacted, forming a structurally weak zone.

(a) (b)

Figure 12.32 (a) Placing gravel-filled bags at the shoulder of each soil layer; steel reinforcement seen in this picture is for the facing that was constructed after the backfill was completed. (b) Bridge abutment with backfill of cement-mixed gravel during full-scale loading test. *(Source: Part a from Watanabe et al. (2003b)).*

2. If the reinforcement layers are connected to the back face of the facing during the construction of the backfill, the connection could be damaged in case the settlement of the backfill relative to the facing during the construction of the backfill becomes too large.
3. To keep the alignment of the front face of the facing as constructed, the facing should be supported with a very stiff propping during the construction of the backfill. Otherwise, the footing of the facing should be made wider while supported with a pile foundation.

Figure 12.32(b) shows the bridge abutment with the completed facing during the full-scale loading test described next.

12.4.4 Full-scale loading tests

To confirm whether the new type of abutment is sufficiently stable under severe seismic loading conditions as considered in the design, full-scale vertical and lateral loading tests were performed on the completed abutment (Aoki et al., 2003) (Figs. 12.33(a) and 12.33(b)). First, vertical load simulating the dead weight of a bridge girder was applied to the top of the RC facing by means of three hydraulic jacks arranged at the shoulder of the facing, which is the fixed end of a bridge girder, and PC steel rods (tendons) anchored into the supporting ground. Then, lateral load was applied to the facing 1.3 m below the shoulder by using six hydraulic jacks arranged at the adjacent pier. Two piers, numbers 5 and 6, were fixed to each other to react as a unit against the lateral load (Fig. 12.34) because it was anticipated that the abutment would be much stronger against lateral loading than would a single pier. It was confirmed that despite relatively small dimensions of the footing for the facing, the

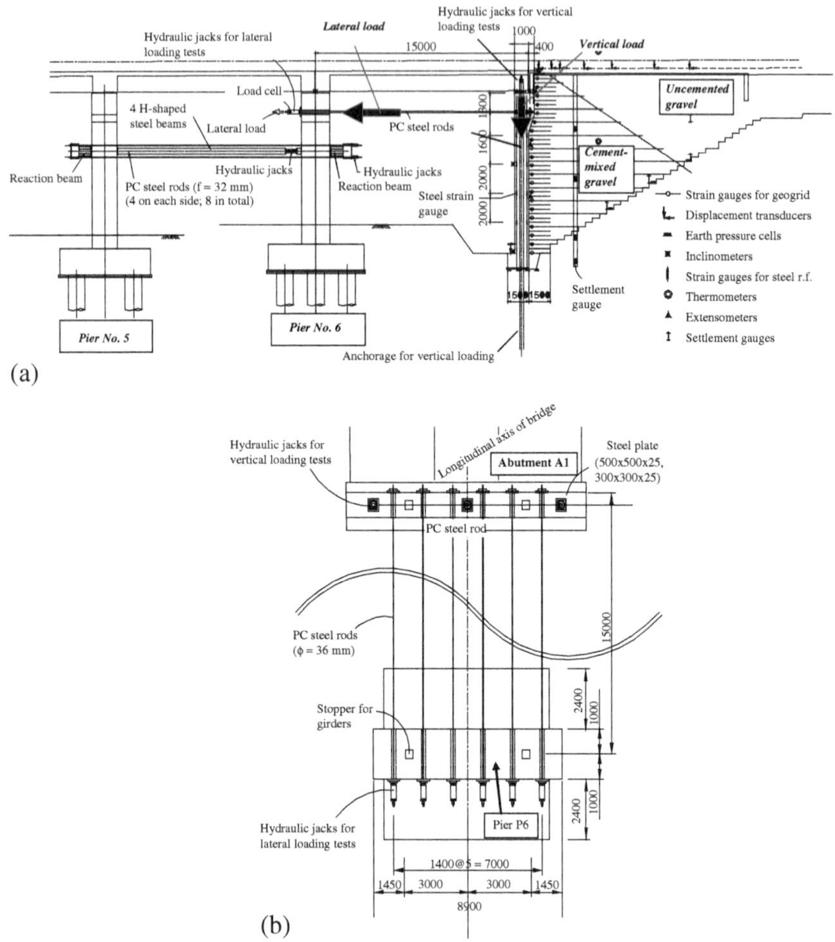

Figure 12.33 (a) Profile and (b) plan of field full-scale loading test of the bridge abutment with backfill of cement-mixed gravel. *(Source: Aoki et al. (2003)).*

vertical bearing capacity of the footing is sufficiently large, as anticipated. Here, the results from the lateral loading tests are reported.

Figure 12.35 shows the time history of lateral load. Figure 12.36(a) shows the relationship between the lateral load and the outward lateral displacement at the top of the facing. Figure 12.36(b) shows the locations of lateral load application and measurement of several quantities. The lateral displacement at 1.3 m below the shoulder of the facing at the maximum load (4 MN) was as small as 15.6 mm. A lateral load of 4 MN is only slightly lower than the design seismic lateral load, equal to 4.171 MN, for a design lateral seismic

Figure 12.34 Overview of field full-scale loading test of bridge abutment with backfill of cement-mixed gravel. *(Source: Aoki et al. (2003)).*

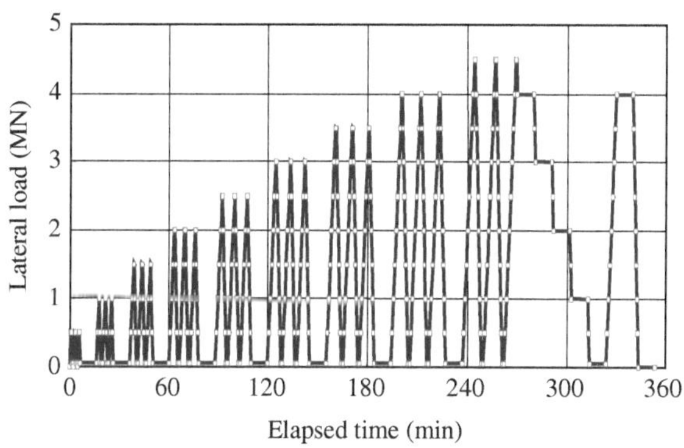

Figure 12.35 Time history of lateral load applied to bridge abutment with backfill of cement-mixed gravel. *(Source: Aoki et al. (2003)).*

coefficient as high as 0.89, corresponding to the design maximum ground horizontal acceleration equal to 871 gal (i.e., the so-called Level II design seismic load). The residual lateral displacement was 9.3 mm, compared with the maximum value of 15.1 mm, which means that the behavior was rather recoverable, showing that the load-displacement state at a horizontal load of 4 MN was far from ultimate failure.

Figure 12.37 shows the vertical distributions of lateral displacement of the abutment and the adjacent two piers. The lateral displacement of the two piers was approximately two times as high as that of the abutment, which means that the lateral stiffness of the abutment is approximately four times as high as that of a single pier. This result clearly indicates that the lateral stability of this new type bridge abutment is much more than sufficient.

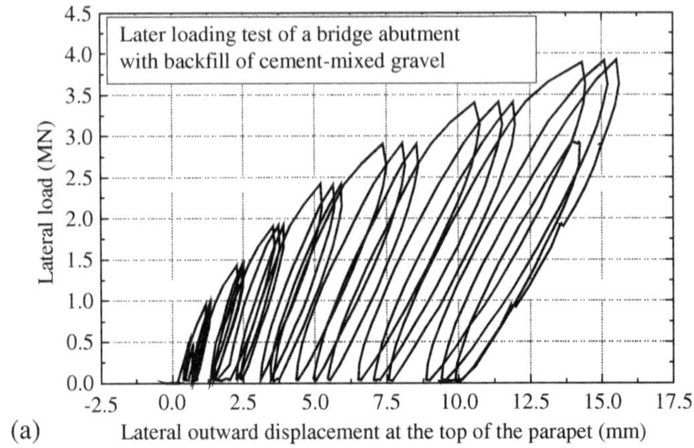

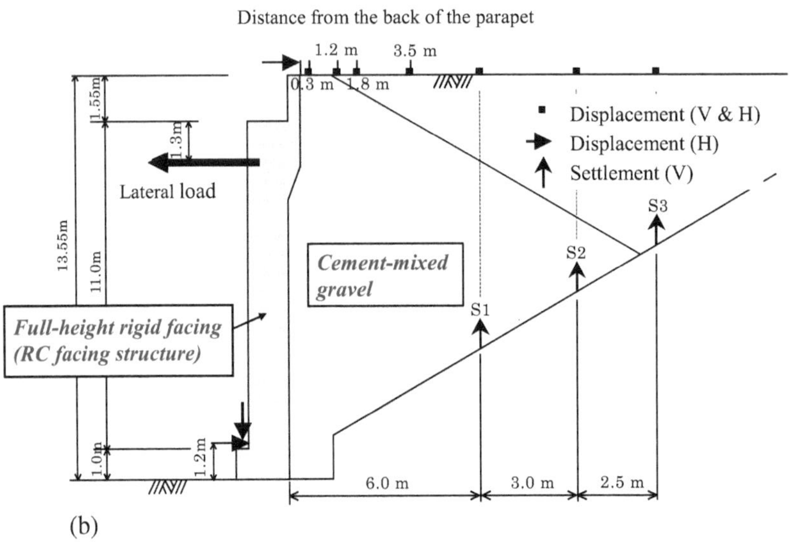

Figure 12.36 (a) Relationship between lateral load and lateral outward displacement at the top of the facing and (b) locations of measurement, bridge abutment with backfill of cement-mixed gravel. *(Source: Aoki et al. (2003)).*

Figures 12.38(a) and 12.38(b) show the distributions of lateral displacement and settlement on the crest of the backfill from the back face of the facing (see Fig. 12.36(b)). Figure 12.38(c) shows the distribution of settlement at the base of the approach block and the distance from the back face of the facing. The following trends in behavior may be seen from these figures:

1. There was no distinct separation between the back of the facing and the approach block, indicating that the geogrid did not show any large deformation at the connection.

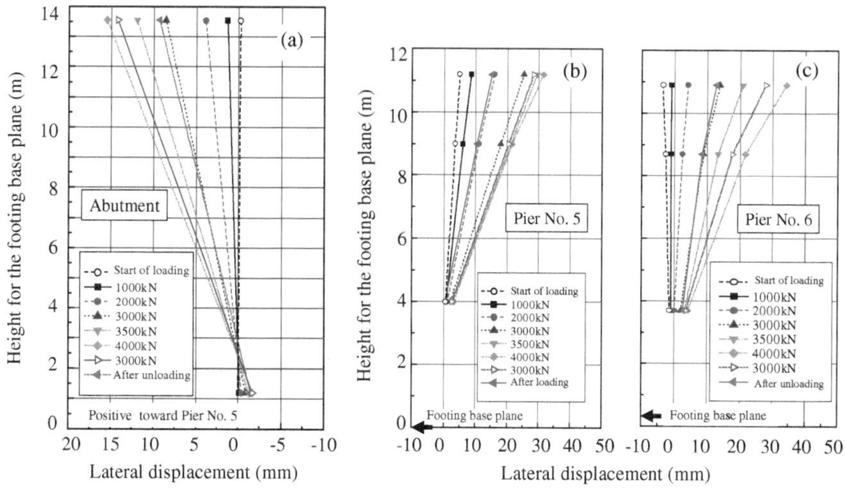

Figure 12.37 Lateral displacements of (a) abutment, (b) pier No. 5, and (c) pier No. 6 from lateral loading tests, bridge abutment with backfill of cement-mixed gravel. *(Source: Aoki et al. (2003)).*

2. The lateral displacement at the crest of the backfill was very uniform for a range of distances from 0 to approximately 5.5 m, indicating that the approach block behaved like a monolith without developing major vertical cracks inside. A major crack with a width of approximately 10 mm was found at a distance of 7.25 m from the back face of the facing at the end of loading (see Fig. 12.40).
3. The crest of the backfill exhibited noticeable settlements, which were larger at places closer to the facing. This trend in behavior was due mainly to a rigid overturning rotational displacement of the approach block caused by lateral loading.
4. The back zone of the approach block exhibited considerable heaving at the base associated with lateral loading (Fig. 12.38(c)). Figure 12.39(a) shows the changes in the earth pressure associated with lateral loading measured at the base of the approach block (see Fig. 12.39(b) for the locations of the earth pressure cells). This result also indicates a rigid overturning rotational displacement of the approach block.

Figure 12.40 shows the overall displacement of the abutment at the maximum lateral load. It may be seen that the parapet and the approach block of cement-mixed gravel behaved like a monolith exhibiting a small overturning rotational displacement, showing no sign of a separation between them, without development of major cracks inside the approach block.

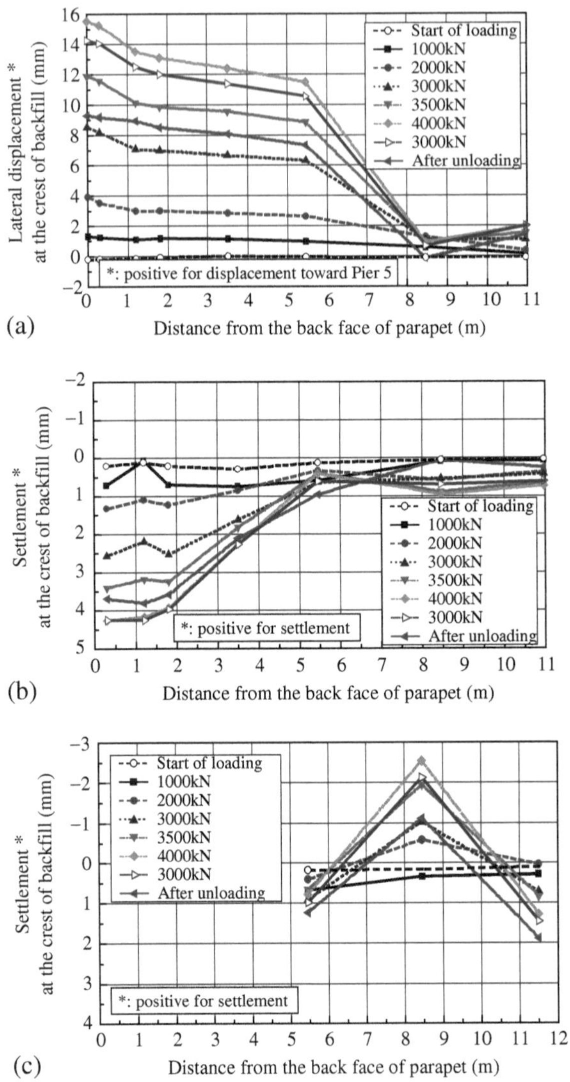

Figure 12.38 (a) Lateral displacements, (b) settlement at the crest, and (c) settlement at the base of backfill of cement-mixed gravel crest. *(Source: Aoki et al. (2003)).*

12.4.5 Summary of results from field full-scale loading tests

The following conclusions can be derived from the test results presented previously:

1. The new type bridge abutment has a lateral stiffness that is higher by a factor of approximately four compared to that of a single pier that was

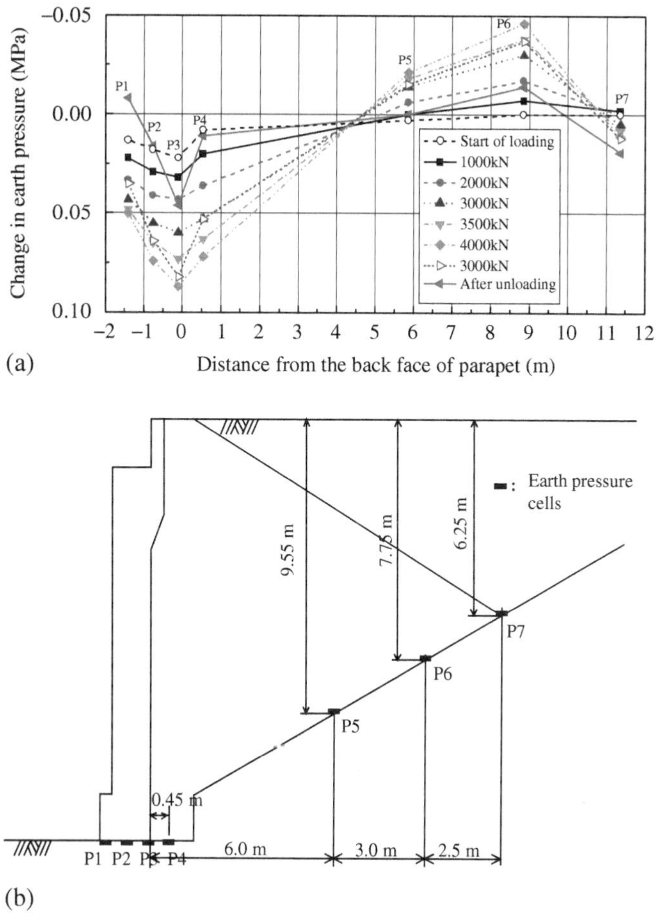

Figure 12.39 (a) Earth pressure at the base of backfill of cement-mixed gravel and (b) locations of earth pressure cells. *(Source: Aoki et al. (2003)).*

designed and constructed following the current seismic-resistant design code against severe design seismic load (the so-called Level II).

2. The maximum and residual outward lateral displacements at the top of the facing when subjected to a very high lateral load simulating severe design seismic load were as small as 15.6 and 9.3 mm, indicating a very high seismic stability of the new type bridge abutment.
3. Against very large lateral load, the facing and backfill behaved like a monolith exhibiting a small overturning rotational displacement, showing no sign of a separation between them, without development of major cracks inside the approach block.

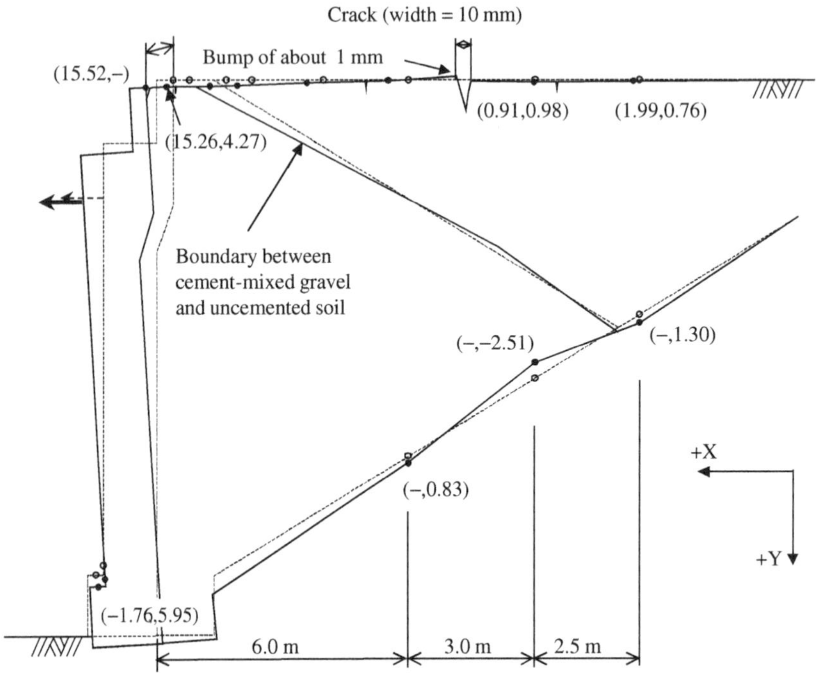

Figure 12.40 Overall residual deformation observed at the maximum lateral load. *(Source: Aoki et al. (2003)).*

12.5 CONCLUSION

Based on the results from the shaking table tests on scaled models in the laboratory and the CDTC tests on cement-mixed gravely soil, a new structural type of bridge abutment was proposed. Subsequently, a prototype bridge abutment was designed and constructed. The advantages of the new type bridge abutment include a high cost-efficiency and a very high seismic stability without showing any serious bump immediately behind the back of the front RC structure (i.e., facing) supporting a bridge girder (or girders). The Japan Railway Construction Public Agency specified "design and construction standards for bridge abutments having an approach block of cement-mixed backfill" in March 2004 to construct bridge abutments of this new type at other locations.

It is now possible to construct permanent civil engineering structures that require a high ultimate stability allowing relatively small displacements by constructing backfill of cement-mixed soil. This chapter presented such a case history. In this case history, a proper understanding of the stress–strain properties (i.e., the strength and deformation characteristics) of cement-mixed soil was essential. To this end, systematic and careful laboratory stress–strain tests (mostly triaxial compression tests) were performed. The test results revealed that the stress–strain properties of different types of cement-mixed soil have common as well as specific features. The results from the laboratory stress–strain tests played a major role in the design.

NOTATION

a_{max} = amplitude of horizontal acceleration at the shaking table
c/g = cement-to-gravel ratio by weight
c_{peak} = cohesion intercept for peak strength
c_{res} = cohesion intercept for residual strength
D_{50} = mean diameter of particle
e = void ratio
q = deviator stress ($= \sigma'_v - \sigma'_h$)
q_{peak} = peak strength (compressive strength)
q_{res} = residual strength
w = water content
ε_v = axial strain
ϕ_{peak} = angle of internal friction for peak strength
ϕ_{res} = angle of friction for residual strength
ρ_d = dry density

ACKNOWLEDGMENTS

The authors express sincere thanks to their past and present colleagues for their help in preparing this chapter, particularly Prof. J. Koseki of University of Tokyo, and Dr. T. N. Lohani, Dr. L. Kongsukprasert, and Dr. T. Uchimura for the report of the new type of bridge abutment.

REFERENCES

Aoki, H., Yonezawa, T., Hashimoto, J., Tateyama, M., Yamada, T., Kohara, A., et al., 2002. Comparative design of bridge abutment with backfill of cement-treated soil and conventional type abutment. In: Proc. 37th Japan Natl. Conf. on Geotech. Eng. Japanese Geotechnical Society, Osaka, Japan, pp. 805–806 (in Japanese).

Aoki, H., Yonezawa, T., Watanabe, O., Tateyama, M., Tatsuoka, F., 2003. Results from field full-scale loading tests on a bridge abutment with backfill of geogrid-reinforced cement-mixed gravel. Geosynth. Eng. J., Japanese Chapter Int. Geosynth. Soc. 18, 237–242 (in Japanese).

Goto, S., Tatsuoka, F., Shibuya, S., Kim, Y.-S., Sato, T., 1991. A simple gauge for local small strain measurements in the laboratory. Soils Found. 31, 169–180.

Hansen, K.D., Reinhardt, W.G., 1990. Roller-Compacted Concrete Dams. McGraw-Hill, New York.

Jiang, G.L., Kohata, Y., Tateyama, M., Tatsuoka, F., 1999. Small strain deformation characteristic at low pressure of dense gravel. In: Jamiolkowski, M. et al., (Eds.), Proc. 2nd Int. Symp. Pre-Failure Deformation Characteristic of Geomaterials, IS Torino '99. Balkema, Rotterdam, The Netherlands, pp. 291–296.

Kohata, Y., Jiang, G.L., Murata, O., Tatsuoka, F., 1999. Elastic-properties-based modelling of non-linear deformation characteristics of gravels. In: Jamiolkowski, M. et al., (Eds.), Proc. 2nd Int. Symp. Pre-Failure Deformation Characteristic of Geomaterials, IS Torino '99. Balkema, Rotterdam, The Netherlands, pp. 533–539.

Kongsukprasert, L., Tatsuoka, F., 2003. Viscous effects coupled with ageing effects on the stress–strain behaviour of cement-mixed granular materials and a model simulation. In: Di Benedetto, H. et al., (Eds.), Proc. 3rd Int. Symp. Deformation Characteristics of Geomaterials, IS Lyon 03. Balkema, Rotterdam, The Netherlands, pp. 569–577.

Kongsukprasert, L., Kuwano, R., Tatsuoka, F., 2001. Effects of ageing with shear stress on the stress–strain behavior of cement-mixed sand. In: Tatsuoka, F. et al., (Eds.), Advanced Laboratory Stress–Strain Testing of Geomaterials. Balkema, Rotterdam, The Netherlands, pp. 251–258.

Kongsukprasert, L., Tatsuoka, F., Tateyama, M., 2005. Several factors affecting the strength and deformation characteristics of cement-mixed gravel. Soils Found. 45 (3), 107–124.

Lohani, T.N., Kongsukprasent, L., Watanabe, K., Tatsuoka, F., 2003. Strength and deformation characteristics of cement-mixed gravel for engineering use. In: Di Benedetto, H. et al., (Eds.), Proc. 3rd Int. Symp. Deformation Characteristics of Geomaterials, IS Lyon 03. Balkema, Rotterdam, The Netherlands, pp. 637–643.

Lohani, T.N., Kongsukprasert, L., Watanabe, K., Tatsuoka, F., 2004. Strength and deformation properties of compacted cement-mixed gravel evaluated by triaxial compression tests. Soils Found. 44 (5), 95–108.

Nakarai, K., Uchimura, T., Tatsuoka, F., Shinoda, M., Watanabe, K., Tateyama, M., 2002. Seismic stability of geosynthetic-reinforced soil bridge abutment. In: Proc. 7th Int. Conf. on Geosynthetics, IGS, Nice, vol. 1, pp. 249–252.

Schrader, E.K., 1996. Compaction of roller compacted concrete. SP96-6, ACI Special Publication 77–101.

Shinoda, M., Uchimura, T., Tatsuoka, F., 2003a. Improving the dynamic performance of preloaded and prestressed mechanically reinforced backfill by using a ratchet connection. Soils Found. 43 (2), 33–54.

Shinoda, M., Uchimura, T., Tatsuoka, F., 2003b. Increasing the stiffness of mechanically reinforced backfill by preloading and prestressing. Soils Found. 43, 75–92.

Sugai, M., Tatsuoka, F., 2003. Ageing and loading rate effects on the stress–strain behaviour of a cement-mixed soft clay. In: Di Benedetto, H. et al., (Eds.), Proc. 3rd Int. Symp. Deformation Characteristics of Geomaterials, IS Lyon 03. Balkema, Rotterdam, The Netherlands, pp. 627–635.

Sugai, M., Tatsuoka, F., Kuwabara, M., Sugo, K., 2000. Strength and deformation characteristics of cement-mixed soft clay. In: Nakase, A., Tsuchida, T. (Eds.), In: Coastal Geotechnical Engineering in Practice, vol. 1. Balkema, Rotterdam, The Netherlands, pp. 521–527.

Tatsuoka, F., Uchida, K., Imai, K., Ouchi, T., Kohata, Y., 1997a. properties of cement-treated soils in trans-tokyo bay highway project. Ground Improve. 1, 37–58.

Tatsuoka, F., Uchimura, T., Tateyama, M., 1997b. Preloaded and prestressed reinforced soil. Soils Found. 37 (3), 79–94.

Tatsuoka, F., Tateyama, M., Uchimura, T., Koseki, J., 1997c. Geosynthetic-reinforced soil retaining walls as important permanent structures: 1996–1997 Mercer Lecture. Geosynth. Int. 4 (2), 81–136.

Tatsuoka, F., Koseki, J., Tateyama, M., 1997d. Performance of reinforced soil structures during the 1995 Hyogo-ken Nanbu earthquake: special lecture. In: Int. Symp. Earth Reinforcement, IS Kyushu '96, vol. 2. Balkema, Rotterdam, The Netherlands, pp. 973–1008.

Tatsuoka, F., Uchida, K., Ouchi, T., 2000. Geotechnical engineering aspects of Trans-Tokyo Bay Highway project. In: Proc. GeoEng2000, Melbourne, vol. 1, pp. 996–1024, (invited lectures).

Uchimura, T., Tateyama, M., Koga, T., Tatsuoka, F., 2003. Performance of a preloaded–prestressed geogrid-reinforced soil pier for a railway bridge. Soils Found. 43 (6), 33–50.

Watanabe, K., Tateyama, M., Yonezawa, T., Aoki, H., Tatsuoka, F., Koseki, J., 2002. Shaking table tests on a new type bridge abutment with geogrid-reinforced cement treated backfill. In: Proc. 7th Int. Conf. Geosynthetics, IGS, Nice, vol. 1, pp. 119–122.

Watanabe, K., Tateyama, M., Jiang, G., Tatsuoka, F., Lohani, T.N., 2003a. Strength characteristics of cement-mixed gravel evaluated by large triaxial compression tests. In: Di Benedetto, H. et al., (Eds.), Proc. 3rd Int. Symp. Deformation Characteristics of Geomaterials, IS Lyon 03. Balkema, Rotterdam, The Netherlands, pp. 683–693.

Watanabe, O., Nasu, Y., Aoki, H., Yonezawa, T., Tateyama, M., Tatsuoka, F., 2003b. Construction control of approach block of cement-mixed gravel backfill. In: Proc. 58th Annu. Conf. JSCE, Tokushima, vol. 3. pp. 1091–1092 (in Japanese).

PART TWO

Electrokinetic, Thermal, and Explosion-Based Techniques

CHAPTER 13

The Use of Electrokinetic Geosynthetics to Improve Soft Soils

Stephanie Glendinning[1], Colin J.F.P. Jones[1], John Lamont-Black[2]
[1]Newcastle University, Newcastle upon Tyne, UK
[2]Electrokinetic Ltd., Newcastle upon Tyne, UK

13.1 INTRODUCTION

The main applications of geosynthetics are primarily related to the civil engineering and environmental industries and are well established as providing filtration, separation, reinforcement, and drainage and acting as barriers. In use conventional geosynthetic materials have a *passive* role; for example, barriers stop the passage of liquids, reinforcement provides tensile resistance, and drains provide a passage for water. New applications for geosynthetics can be identified if the geosynthetic can provide an *active* role, initiating chemical or physical change to the soil matrix in which it is installed as well as providing the established functions. This can be achieved by creating electrically conducting geosynthetics and combining an electrical function with established functions.

Designing with these *active* geosynthetics involves an understanding of electrokinetic phenomena and consideration of soil properties not usually considered in design. This chapter identifies these and illustrates their relevance as well as provides details of case histories of the use of electrically conductive geosynthetics.

13.2 ELECTROKINETIC PHENOMENA IN SOILS

Electrokinetic phenomena, as they occur within soils, may be defined in terms of five categories (Fig. 13.1). Each is related to the movement of charged particles in a fluid medium under the action of an applied potential difference, induces a potential difference, or causes the movement of fluid within a matrix of charged particles that induces a potential difference within the system. The five categories are defined as follows:

Chemical, Electrokinetic, Thermal, and Bioengineering Methods
http://dx.doi.org/10.1016/B978-0-08-100191-2.00013-7

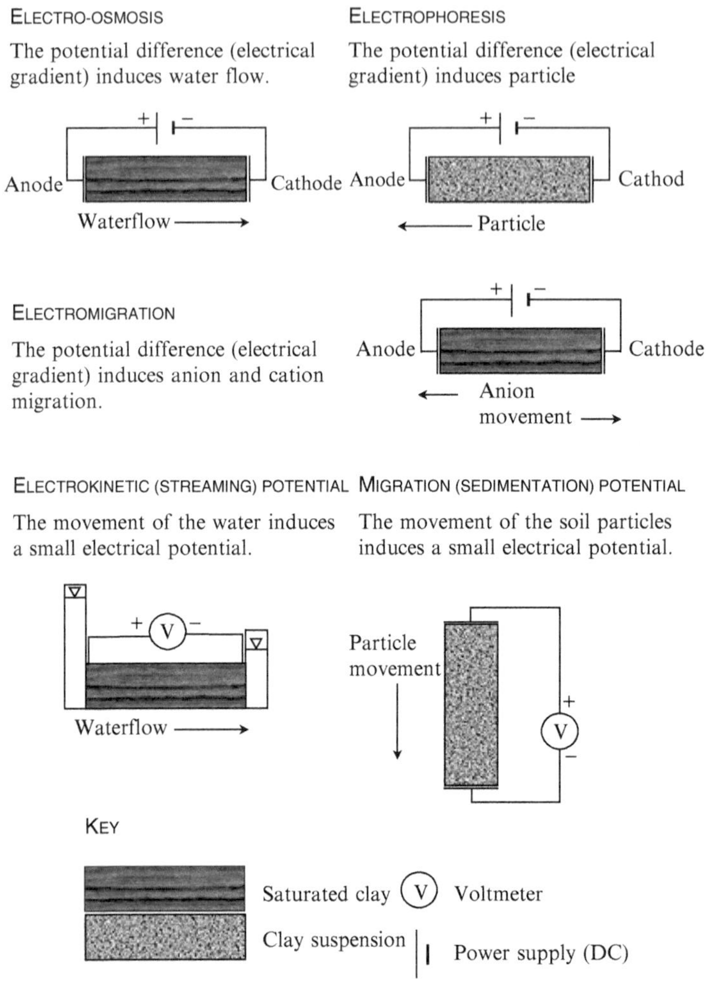

Figure 13.1 Electrokinetic phenomena. *(Source: After Mitchell (1991, 1993)).*

Streaming potential—An electrical potential difference induced by fluid flow in a charged soil particle matrix.

The application of a hydraulic gradient in a clay soil causes the double-layer charges to be displaced in the direction of flow. This displacement results in a deficit of positive charges in the region where the water flow enters the soil element and an increasing excess of positive charges as the soil element is passed through. The overall effect of this charge displacement is the generation of an electrical potential difference over the soil element, which is proportional to the hydraulic flow rate. Streaming potentials of several tens of millivolts have been measured in clays (Mitchell, 1991, 1993).

Migration/sedimentation potential—An electrical potential difference induced by movement of charged soil particles in a fluid medium.

The movement of charged particles, such as clay, during sedimentation of clay particles in a water suspension generates an electrical potential difference caused by the viscous drag of the solution that retards the movement of the diffuse double-layer cations relative to the charged particles.

Electromigration/ion migration—An applied electrical potential difference induces ion migration within the fluid phase of a charged soil particle matrix.

Ions are attracted to the electrode of opposite sign and repelled from the electrode of like sign. The principal difference between ion migration and electrophoresis is that ion migration can take place within the pore fluid of a fixed soil matrix, whereas electrophoresis is the passage of discrete (usually clay) particles within a fluid/particle suspension.

Electrophoresis—An applied electrical potential difference induces movement of charged soil particles within a fluid medium.

When a DC electric field is applied across a colloidal suspension, charged particles are attracted electrostatically to one of the electrodes and repelled from the other. Negatively charged clay particles move toward the anode (Fig. 13.1). Electrophoretic mobilities reported by Lageman et al. (1989) and van Olphen (1977) are in the ranges 1×10^{-10} to 3×10^{-10} and 1×10^{-8} to 3×10^{-8} m^2/Vs, respectively. However, Esrig (1968) reported that the process is inconsequential for most naturally occurring soils.

Electro-osmosis—An applied electrical potential difference induces fluid flow in a charged soil particle matrix.

When a DC electrical potential difference is applied across a wet soil mass, ion migration occurs. As the ions migrate, they drag with them their water of hydration and also exert a viscous drag on the water around them. The flow is initiated in a diffuse double layer, where there are significantly more cations than anions due to the proximity of the negatively charged clay surface. Figure 13.2 shows the development of the electro-osmotic flow with time induced in the double layer.

Electro-osmosis is the electrokinetic phenomenon most applicable to the improvement of soft soils in civil engineering applications.

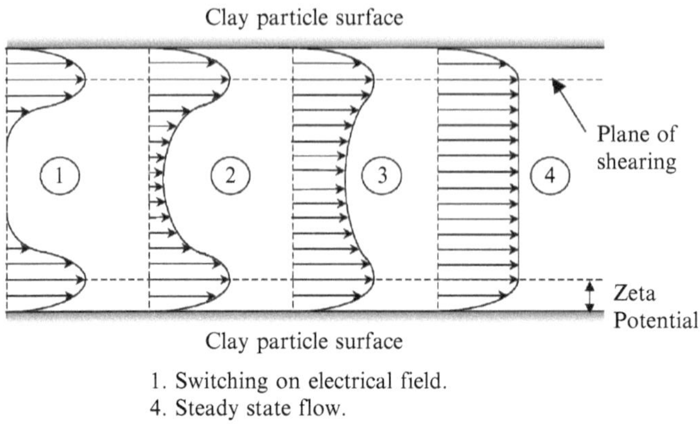

Figure 13.2 Development of electro-osmotic flow velocity. *(Source: After Pamukcu (1996)).*

13.3 ELECTRO-OSMOSIS

Several theories have been developed to explain electro-osmotic transport of water in clay soils. Of these, the Helmholtz–Smoluchowski theory is one of the earliest and the most widely used (Mitchell, 1993).

It is based on an electrical condenser analogy that assumes that the soil capillaries have charges of one sign on or near the surface of the wall and countercharges concentrated in a layer in the liquid a small distance from the wall (Fig. 13.3). The mobile shell of counterions is assumed to drag water through the capillary by plug flow, resulting in a high-velocity gradient between the two plates of the condenser. The balance between the electrical force causing water movement and friction between the liquid and the wall controls the rate of water flow.

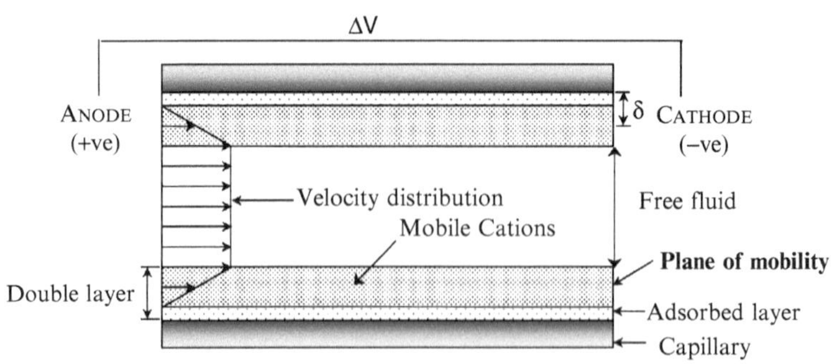

Figure 13.3 Helmholtz–Smoluchowski model. *(Source: After Mitchell (1993)).*

If v is the flow velocity and δ the distance between the wall and the center of the plane of mobile charges, then the velocity gradient between the wall and the center of the positive charges is v/δ; thus, the drag force per unit area is $\eta\, \mathrm{d}v/\mathrm{d}x = \eta v/\delta$, where η is the viscosity. The force per unit area from the electric field is $\sigma \Delta E/\Delta L$, where σ is the surface charge density, and $\Delta E/\Delta L$ is the electrical potential gradient. At equilibrium,

$$\eta \frac{v}{\delta} = \sigma \frac{\Delta E}{\Delta L} \tag{13.1}$$

or

$$\sigma\delta = \eta v \frac{\Delta L}{\Delta E} \tag{13.2}$$

From electrostatics, the potential gradient ζ across a condenser is given by

$$\zeta = \frac{\sigma\delta}{D} \tag{13.3}$$

where D is the relative permittivity or dielectric constant of the pore fluid. Substitution of σ and δ into Eq. (13.3) gives

$$v = \left(\frac{\zeta\, D}{\eta}\right) \frac{\Delta E}{\Delta L} \tag{13.4}$$

The potential ζ is the zeta potential, which is not equal to the surface potential of the double layer. For a single capillary of area a, the flow rate is

$$q_a = va = \frac{\zeta\, D \Delta E}{\eta\ \Delta L} a \tag{13.5}$$

and for a group of N capillaries within total cross-sectional area A normal to the flow direction,

$$q_A = Nq_a = \frac{\zeta\, D \Delta E}{\eta\ \Delta L} Na \tag{13.6}$$

If the porosity is n, then the cross-sectional area of the voids is nA, which must equal Na. Thus,

$$q_A = \frac{\zeta\, D}{\eta} n \frac{\Delta E}{\Delta L} A \tag{13.7}$$

By analogy with Darcy's law, this may be rewritten as

$$q_A = k_e i_e A \tag{13.8}$$

where

$$i_e = \frac{\Delta V}{\Delta L} \text{ (electrical potential gradient (V/m))} \tag{13.9}$$

$$k_e = \frac{\zeta D}{\eta} n \text{ (coefficient of electro-osmotic permeability)} \tag{13.10}$$

This is the equation that describes the flow of water under an electrical potential gradient.

13.3.1 Electro-osmotic permeability

Unlike hydraulic permeability, k_e is relatively independent of pore size (Table 13.1). Casagrande (1952) suggested that for most practical applications, a value for $k_e = 5 \times 10^{-5}$ cm^2/s V can be accepted. Therefore, it can be seen that electro-osmosis can be effective for water movement in fine-grained soils compared to water flow under hydraulic gradients (Mitchell, 1993). This is demonstrated in Fig. 13.4, which shows the hydraulic permeability of a range of typically encountered soils, compared to electro-osmotic permeability for the same soils.

13.3.2 Electro-osmotic efficiency

The efficiency of electro-osmosis relates to the quantity of water moved per unit charge passed (Grey and Mitchell, 1967). According to Mitchell (1993), this factor may vary by several orders of magnitude depending on the soil type.

The efficiency of electro-osmosis depends on the electrical conductivity of the soil mass. The electrical conductivity is in turn controlled by the water content, the cation exchange capacity, and the free electrolyte concentration of the pore fluid, as indicated by Grey and Mitchell (1967) and validated by Lockheart (1983). The mineralogy of the soil also has a major impact on the soil conductivity.

13.3.3 Energy requirements

The quantity of water moved per unit charge passed (l/h/A or mol/F) is a measure of the viability of using electro-osmosis on a particular site. If this quantity is expressed as k_i, then

$$q = k_i I \tag{13.11}$$

where k_i can vary over a wide range. The power consumption is defined by

$$P = \Delta E I = \frac{\Delta E q h}{k_i} \tag{13.12}$$

where ΔE is in volts and I in amps.

Table 13.1 Coefficients of electro-osmotic permeability and other parameters

No.	Material	Water content (%)	$k_e \times 10^{-5}$ (cm^2/s V)	Conductivity σ (S/m)	Approximate k_h (cm/s)	k_i water per unit charge (m^3/s/A)	Reference
1	London clay	52.3	5.8	N/A	10^{-8}	N/A	Casagrande (1952)
2	Boston blue clay	50.8	5.1	N/A	10^{-8}	N/A	Casagrande (1952)
3	Kaolin	67.7	5.7	N/A	10^{-7}	N/A	Casagrande (1952)
4	Clayey silt	31.7	5.0	N/A	10^{-6}	N/A	Casagrande (1952)
5	Rock flour	27.2	4.5	N/A	10^{-7}	N/A	Casagrande (1952)
6	Na-Montmorillonite	170	2.0	N/A	10^{-9}	N/A	Casagrande (1952)
7	Na-Montmorillonite	2000	12.0	N/A	10^{-8}	N/A	Casagrande (1952)
8	Mica powder	49.7	6.9	N/A	10^{-5}	N/A	Casagrande (1952)
9	Fine sand	26.0	4.1	N/A	10^{-4}	N/A	Casagrande (1952)
10	Quartz powder	23.5	4.3	N/A	10^{-4}	N/A	Casagrande (1952)
11	Ås quick clay	31.0	2.0	N/A	2.0×10^{-8}	N/A	
12	Bootlegger Cove clay	30.0	2.4–5.0	0.02	2.0×10^{-8}	1×10^{-7}	Long and George (1967)

Continued

Table 13.1 Coefficients of electro-osmotic permeability and other parameters—cont'd

No.	Material	Water content (%)	$k_e \times 10^{-5}$ (cm^2/s V)	Conductivity σ (S/m)	Approximate k_h (cm/s)	k_i water per unit charge (m^3/s/A)	Reference
13	Silty clay, West Branch Dam	32.0	3.0–6.0	0.25	1.2×10^{-8}–6.5×10^{-8}	9.6×10^{-9}–2×10^{-8}	Fetzer (1967)
14	Clayey silt, Little Pic River, Ontario	26.0	1.5		2×10^{-5}		Casagrande et al. (1961)
15	Decomposed granite, silty clay	N/A	30	0.17	10^{-4}–10^{-5}	1.8×10^{-7}	Chappell and Burton (1975)
16a	Silty clay	23	5.0	0.08–0.12	0.5–8×10^{-5}	6.3–4.2×10^{-9}	Chen and Murdoch (1997)
16b	Silty clay	10	9.0	0.02	1×1^{-5}	4.5×10^{-8}	Chen and Murdoch (1997)
17	Marine silty clay	37	6–9	0.024–0.033	10^{-7}	1.8–3.8×10^{-7}	Eggestad and Føyn (1983)

N/A, not applicable.
Source: After Mitchell (1993) and Wrigley (1999).

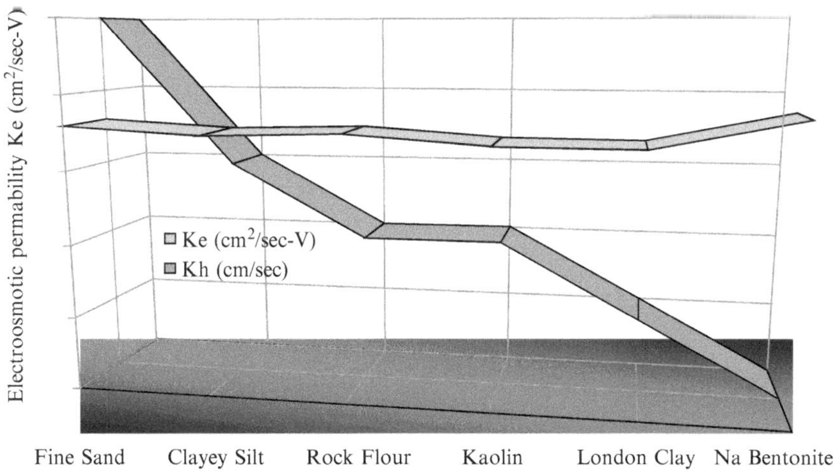

Figure 13.4 Electro-osmotic permeability versus hydraulic permeability. *(Source: Courtesy Electrokinetic Ltd.).*

The power consumption per unit volume of flow is

$$\frac{P}{qh} = \frac{\Delta E}{k_i} \times 10^{-3}\text{kWh} \tag{13.13}$$

The influence of the various soil parameters on the efficiency of electro-osmosis is shown in Fig. 13.5.

13.3.4 Relationship between k_e and k_i

The electro-osmotic flow rate is given by

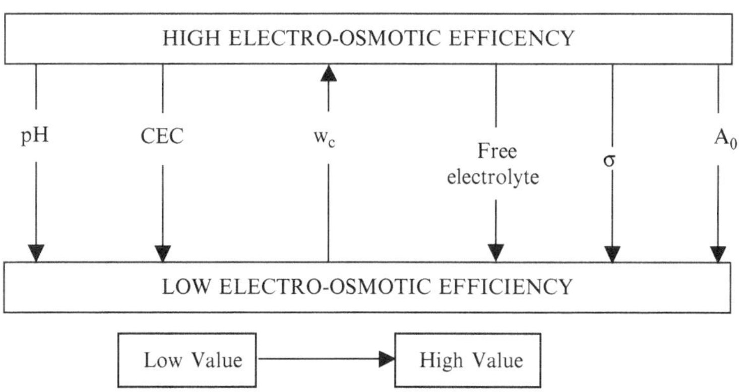

Figure 13.5 Influence of soil variables on electro-osmotic efficiency. *(Source: After Nettleton (1996)).*

$$q_h = k_i I = k_e \frac{\Delta E}{\Delta L} A \tag{13.14}$$

However, $\Delta E/I$ is resistance and (ΔL/resistance*A) is specific conductivity (σ); hence,

$$k_i = \frac{c k_e}{\sigma} \tag{13.15}$$

where k_e is the electro-osmotic permeability (cm/s/V/cm); k_i is the electro-osmotic efficiency (gall/A/h); σ is the specific conductivity (S/m); and C is the constant (1.0 using stated units).

This equation is useful in that k_e varies within narrow limits as shown in Table 13.1. Hence, the electro-osmotic efficiency measured by k_i is dependent on the electrical specific conductivity of the soil (σ).

13.3.5 Applied voltage

Electro-osmosis may be carried out by applying constant current or constant voltage. During treatment, the resistivity of the soil changes due to electrochemical changes and desiccation. When a constant voltage is applied, the magnitude of the current decreases corresponding to the increase in the resistance of the soil mass being treated. Similarly, if a constant current is applied, the corresponding applied voltage will increase. Hamir (1997) carried out a series of basic tests to assess the difference between constant voltage and constant current tests. The findings of his investigation were that, in terms of the properties of the treated soil, no distinction between the two methods could be found.

Polarity reversal (the reversal of electrodes or current direction) has a marked effect on the effectiveness of dewatering in producing more uniform conditions. In addition, current intermittence has been shown to increase the efficiency of electro-osmotic dewatering. The principles of the improvement are quantitatively interpreted in terms of the depolarization of the double layer associated with current intermittence. It has been shown that electro-osmotic strengthening of marine sediments is enhanced by the use of current intermittence, with optimum conditions being obtained with intermittence intervals as 2 min on and 1 min off (Mohamedelhassan and Shang, 2001). Another benefit is the reduction in anode corrosion (Micic et al., 2001).

For certain applications, it should be remembered that the generation of negative pore water pressure is the principal operator in the strengthening process, and negative pore pressure is proportional to the applied voltage.

13.3.6 Pore water pressures

Esrig (1968) presented a solution for the development of pore water pressures under an applied electrical potential field. The solution is based on the following assumptions:

- The soil is homogenous and in a fully saturated state.
- The physical and physiochemical properties of the soil are uniform throughout and are constant with time.
- Electrophoresis of fine-grained soil particles does not occur.
- There is proportionality between the electrically induced velocity of water flow (v) through the soil and the voltage gradient (V). The proportionality factor is the coefficient of electro-osmotic permeability—that is, the Helmholtz–Smoluchowski equation is applicable: $V = k_e \partial V / \partial x$.
- All applied voltage is effective in moving water.
- The electric field produced throughout the soil mass is constant with time.
- No reactions occur at the electrodes (e.g., electrolysis).
- Fluid flows due to an electric field and due to a hydraulic gradient may be superimposed to find the total fluid flow.

Some of these assumptions are inherently incorrect:

- A natural material is very rarely, if ever, homogenous.
- The resistivity of the soil undergoing electro-osmosis has been shown to vary, both with position and time, due to both desiccation and electrochemical changes.
- Separation of the electrodes from the soil may also occur due to gas formation caused by electrolysis.

Hence, the physical and physiochemical properties of the soil vary with time, as does the electric field. Electrolysis occurs during all electro-osmotic treatments.

The assumption that fluid flow caused by an electric field may be superimposed on the flow caused by a hydraulic gradient is important. Grey and Mitchell (1967) demonstrated the validity of the theory indirectly through the measurement of streaming potentials resulting from hydraulic flows. Wan and Mitchell (1976) showed that the superposition of pore water pressures generated by surcharge loading and electro-osmosis was valid. Accepting the validity of superposition of electrical and hydraulically driven flows through an incompressible soil mass and limiting consideration to one-dimension (1-D) flow gives

$$\frac{\partial V}{\partial x} + \frac{\partial V_h}{\partial x} = 0 \tag{13.16}$$

where v is the velocity of flow due to the electrical potential gradient ($\partial V/\partial x$) (i.e., the Helmholtz–Smoluchowski equation), and V_h is the velocity of flow due to the hydraulic gradient (i.e., Darcy's equation).

Differentiation of Eq. (13.16) with respect to x and inserting Helmholtz–Smoluchowski and Darcy's equations gives

$$k_e \frac{\partial^2 V}{\partial x^2} + \frac{k}{\gamma_w} \frac{\partial^2 u}{\partial x^2} = 0 \tag{13.17}$$

Rearranging gives

$$\frac{\partial^2 u}{\partial x^2} + \frac{k_e \gamma_w}{k} \frac{\partial^2 V}{\partial x^2} = 0 \tag{13.18}$$

and substitution of the variable ξ, where

$$\xi = \frac{k_e \gamma_w}{k} V + u \tag{13.19}$$

gives

$$\frac{\partial^2 \xi}{\partial x^2} = 0 \tag{13.20}$$

which is Laplace's equation in a 1-D system. Integrating once gives

$$\frac{\partial \xi}{\partial x} = C_1 \tag{13.21}$$

and integrating again gives

$$\xi = xC_1 + C_2 \tag{13.22}$$

where C_1 and C_2 are constants of integration, which are dependent on the boundary conditions. The possible drainage conditions that may exist at the electrodes, which govern these boundary conditions, are discussed in the following section.

13.3.7 Electrode drainage conditions and pore water pressures

The drainage conditions applied to either the anode or the cathode have one of the following two conditions (Fig. 13.6):

Open: This condition exists when the electrode is open to the atmosphere, such that no excess pore pressure can exist at the electrode—that is, $u = 0$. This condition will occur where an electrically conductive geosynthetic (EKG) electrode is utilized with an external filter in place, or where a more conventional metallic-type electrode is used together with

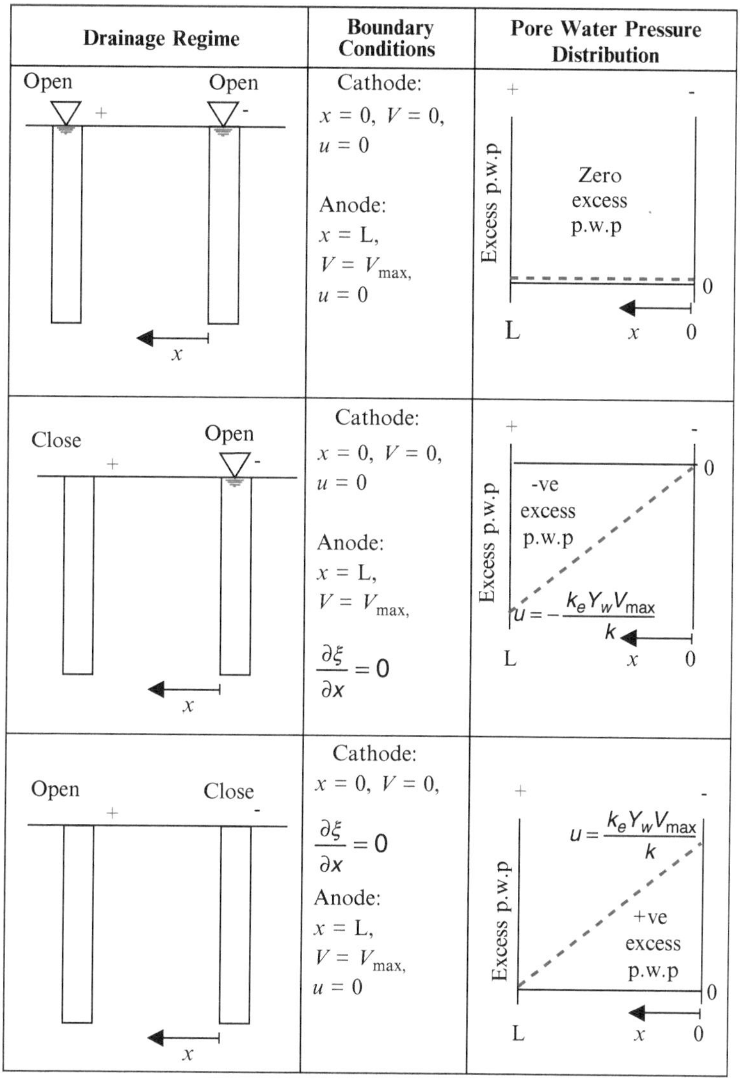

Figure 13.6 Pore water pressures for different electrode drainage configurations. *(Source: After Esrig (1968)).*

an associated drainage path—for example, sand drain or hollow perforated electrode. An open electrode may also be pumped, recharged, or simply allowed to overflow.

Closed: This condition exists when the electrode is sealed such that no passage of gas or fluid can take place along the length of the electrode.

From Fig. 13.6, it can be seen that the following configurations are used for consolidation/dewatering in practice:

- Open anode not recharged–open cathode with overflow
- Closed anode–open cathode with overflow
- Closed anode–open cathode pumped

13.3.8 Electrolysis effects associated with electro-osmosis

During electro-osmotic treatment of a soil mass, the whole system acts as an electrochemical cell in which cations migrate to the cathode and anions migrate to the anode. Reduction reactions take place at the cathode, with the principal reaction being the reduction of water into hydrogen gas. At the anode, there is the possibility of two distinct oxidation reactions—the oxidation of the anode material itself into its oxide or the oxidation of H_2O to yield O_2. Hence, the anode will degrade, with the rate of degradation being dependent on the material from which the anode is manufactured. It is the degradation of the anode that has caused the most problems with the application of electro-osmotic treatment on an industrial scale. Corrosion of the anode is accompanied by a buildup of corrosion products, which degrade the electrical efficiency of the process. The process is also influenced by the production of gas at the electrodes, which must be vented for the process to continue. A schematic of the movement of current and the electrolytic reactions that may occur during electro-osmotic treatment are given in Fig. 13.7.

$$\begin{array}{ll}
\text{(i) } 2H_2O + O_2 + 4e^- \rightarrow 4OH^- \text{ or } 2H^+ + 2e \rightarrow H_2 & \\
\text{(ii) Iron anode:} & Fe + H_2O \rightarrow FeO + 2H^+ + 2e^- \\
\text{or} & 2H_2O \rightarrow O_2 + 4H^+ + 4e^- \\
\text{Copper anode:} & Cu + H_2O \rightarrow CuO + 2H^+ + 2e^- \\
\text{or} & 2H_2O \rightarrow O_2 + 4H^+ + 4e^- \\
\text{Carbon anode:} & C + 2H_2O \rightarrow CO_2 + 4H^+ + 4e^-
\end{array}$$

The liberation of hydrogen ions at the anode causes a reduction in pH, and an increase in hydroxide ions at the cathode causes an increase in pH. If these ions are not removed or neutralized from the anode and cathode, electromigration/ion migration will occur with hydrogen ions moving toward the cathode (acid front) and hydroxide ions moving toward the anode (base front). The advance of acid and base fronts is governed by ion migration as well as electro-osmotic flow, diffusion, and the buffering capacity of the soil medium. Due to a higher ratio of charge to ionic size, H^+ ions move more rapidly than OH^- ions during electromigration. In addition, electro-osmotic

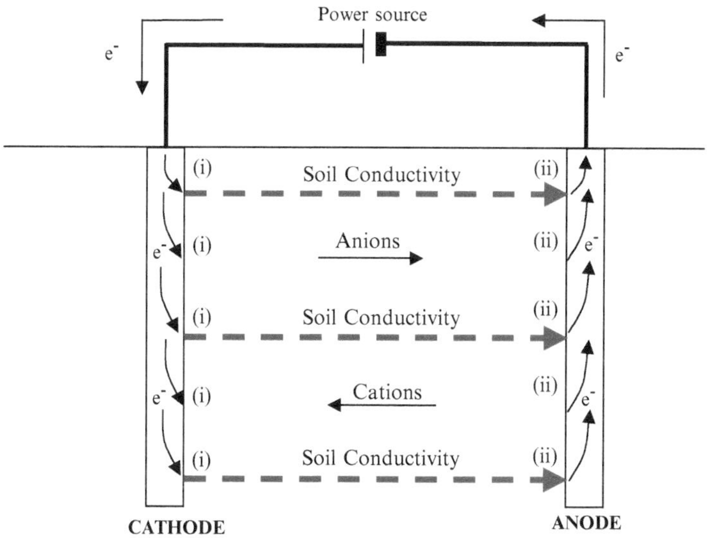

Figure 13.7 Schematic representation illustrating the movement of current through an electrochemical cell set up by electrokinetic phenomena imposed upon a soil mass. *(Source: After Eastwood (1997)).*

flow occurs in the same direction as H^+ migration, but it opposes the direction of OH^- migration. As a consequence, the acid front from the anode moves faster than the base front from the cathode so that acidic conditions prevail throughout most of the soil mass.

13.3.9 Evolution of active geosynthetics

The problems relating to the degradation of the electrodes can be overcome by forming them of materials not susceptible to corrosion and also providing inbuilt drainage facilities for the gases produced and the water collecting at the cathode. This can be achieved by using electrically conducting geosynthetic materials to form the electrodes. The resultant materials differ from conventional geosynthetics in that they are *active* in the sense that they cause change in soils rather than acting in a *passive* manner as is the case with conventional materials (Jones et al., 2005).

Jones et al. (1996) introduced the concept of EKG materials, defining them as a range of geosynthetics that, in addition to providing filtration, drainage, and reinforcement, can be enhanced by electrokinetic techniques for the transport of water and chemical species within fine-grained low-permeability soils, which are otherwise difficult or impossible to deal with. In addition, transivity, sorption, wicking, and hydrophobic tendencies may also be incorporated in the geosynthetic to enhance other properties. The EKG can take the form of a single material that is electrically conductive or a composite material, in which at least one element is electrically conductive. They can be of the same basic form as present-day filter, drainage, separator, and reinforcement materials, but they offer sufficient electrical conduction to allow the application of electrokinetic techniques.

Jones et al. (1996) undertook a series of laboratory studies to evaluate the use of conductive geotextiles as electrodes in electro-osmotic consolidation and reinforced soil. The types of geosynthetics used included needle-punched geotextiles with a 1-mm-diameter copper wire inserted into the geotextile to make it electrically conductive, conductive fiber (carbon) needle-punched material, and modified polyester-reinforcing tape. The latter was made electrically conductive by the addition of a metal stringer aligned parallel to the polyester-reinforcing elements. The results of the tests were favorable and indicated that the EKG behaved as well as a conventional copper electrode.

In the reinforced soil tests, the EKG electrode was used as an anode, with the cathode formed from a needle-punched EKG. The results of pullout tests showed an increase in reinforcement bond of up to 211% and increases in shear strength of up to 200% compared to the values obtained when the geosynthetics were not electrically conductive (Hamir, 1997). Nettleton et al. (1998) continued the work presented by Jones et al. (1996) and suggested that a band drain-type electrode would be the most suitable configuration to fulfill all of the electrode requirements associated with consolidation, bioremediation, and moisture control in embankments.

13.3.10 Development of soil acceptability criteria for electrokinetic treatment

Acceptability criteria for soils have been developed based on standard and nonstandard soil mechanics tests. The basis for advancing these criteria is to permit a relatively rapid assessment of the suitability of a soil for treatment by electro-osmosis based on standard soil mechanics laboratory tests. This does not mean that electro-osmosis-specific tests are not required in order to predict the performance of a particular site installation, but it permits an informed decision to be made at an early stage as to whether to proceed with the more advanced, and expensive, specialized testing.

General classification tests

The liquid and plastic limits of a soil define the range within which a cohesive soil behaves in a plastic state (Fig. 13.8).

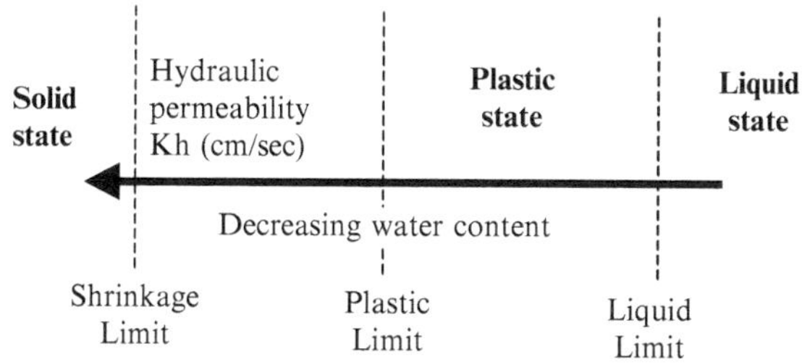

Figure 13.8 Consistency of cohesive soil with variation in water content. *(Source: After Das (1997)).*

Table 13.2 Atterberg limits and associated parameters for different clay minerals

Clay mineral type	Liquid limit (%)	Plastic limit (%)	Activity	Shrinkage limit (%)
Montmorillonite	100–900	50–100	8.5–1.5	
Nontronite	37–72	19–27		
Illite	60–120	35–60	0.5–1.0	15–17
Kaolinite	30–110	25–40	0.5	25–29
Hydrated halloysite	50–70	47–60		
Dehydrated halloysite	33–55	30–45		
Attapulgite	44–47	36–40		
Chlorite	200–250	130–140		

Source: After Lambe and Whitman (1969) and Mitchell (1993).

The value of the Atterberg limits of a cohesive soil depends on several factors, including the quantity and type of clay mineral and type of absorbed cation. Typical values for the Atterberg limits for different clay mineral types are given in Table 13.2. It can be seen that the different clay mineralogies have a direct bearing on the values of the Atterberg limits. It has also been shown that the electrical conductivity of a soil is influenced, to some extent, by the surface charge density (A_0) and is dependent on the clay mineralogy. Hence, the relationship between the Atterberg limits and the electrical conductivity of the soil can be linked to soil mineralogy. In addition to the mineralogy, the electrical conductivity of the soil in situ is partially governed by the electrical conductivity of the pore fluid and the water content (Fig. 13.9).

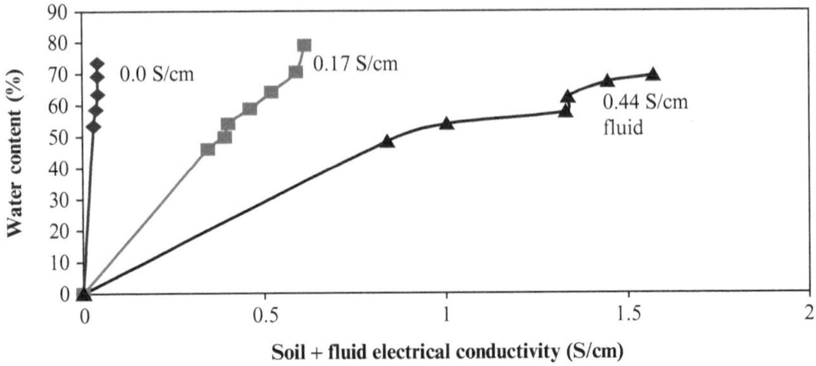

Figure 13.9 Electrical conductivity of grade E kaolin with different conductivity pore fluids. *(Source: After Pugh (2002)).*

Figure 13.9 demonstrates that the overall conductivity of the fluid soil mix is significant, as is the water content of the soil. The chemistry of the pore water, which is also reflected in the exchangeable cations present within the clay minerals, influences the range of results for the Atterberg limits presented in Table 13.2. The monovalent cations (e.g., Na^{+} and K^{+}) give higher values of liquid and plastic limits, whereas the presence of divalent and trivalent cations (e.g., Mg^{2+}, Fe^{2+}, and Al^{3+}) gives lower values. If the soil is tested repetitively, the exchangeable cations can be flushed from the clay, and a change in the Atterberg limits takes place. Thus, for samples that are to be tested with a view to utilizing the results for prediction of the viability of electro-osmosis, it is important that the tests are undertaken in accordance with BS 1377: Part 2 (British Standards Institution (BSI), 1990c) and not repetitively wetted and dried.

The validity of the correlation proposed for the acceptability of soils based on the Atterberg limits and conductivity is demonstrated in Fig. 13.10. The results presented in Fig. 13.10 relate to a review of published electro-osmotic case studies in which both the plasticity indices and electrical conductivities have been given (Casagrande, 1952; Casagrande et al., 1961; Hamir, 1997; Pugh, 1999).

The delineation of acceptable electrical conductivities (σ) and, hence, plasticity indices (PI) is based on the limits proposed by Casagrande (1983) that an acceptable and economic range for the electrical conductivity is 0.05–0.005 S/m. This range gives an associated acceptable range of PI in the range of 5–30%. This criterion is adopted for the assessment of soils.

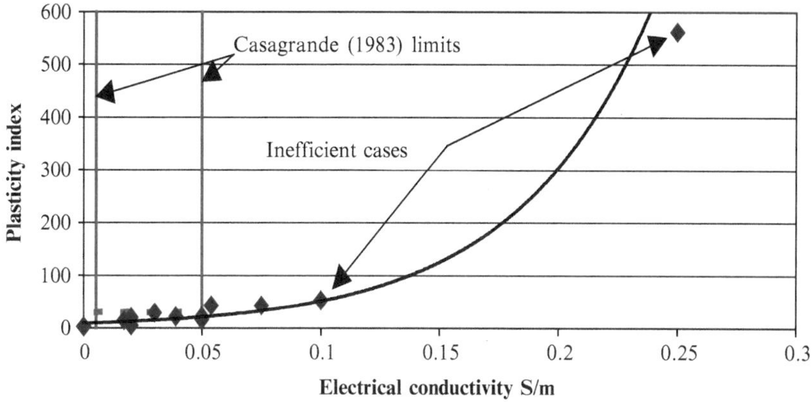

Figure 13.10 Conductivity against plasticity index for a range of natural soils. *(Source: after Pugh (2002)).*

Chemical and electrochemical tests

The standard chemical and electrochemical tests may be carried out in accordance with BS 1377: Part 3 (BSI, 1990d), with the soil samples being taken and prepared in accordance with BS 1377: Part 1 (BSI, 1990b).

Organic content: The presence of organic matter in a soil can have a significant effect on the cation exchange capacity, especially in high pH conditions, and therefore have an effect on the electro-osmotic efficiency of the soil to treatment. Therefore, if the soil may be described as an organic clay or silt, it is recommended that specific electro-osmotic testing be carried out to ascertain the soil's suitability.

Electrical resistivity: The electrical resistivity (ρ) of the soil may be determined in accordance with BS 1377: Part 3:1990 §10 (BSI, 1990d). The disc electrode method is the most appropriate. Electrical resistivity may be related to conductivity (σ) by

$$\sigma = \frac{1}{\rho} \tag{13.23}$$

The range of acceptable and economic values of electrical conductivities (σ) has been proposed by Casagrande (1983), who stated that values within the range of 0.05–0.005 S/m have been found to be acceptable. Values in excess of this range do not indicate that the soil is not susceptible to treatment by electro-osmosis but, rather, that the electro-osmosis installation will draw a high current and may not be economic.

Compressibility and permeability tests: The standard compressibility and permeability tests may be carried out in accordance with BS 1377: Part 5 (BSI, 1990e). The falling head permeability test, which is the most appropriate for fine-grained soils, is not covered by current British or ASTM standards, and the method given in Head (1982) is recommended.

One-dimensional consolidation parameters: Electro-osmotic consolidation is normally two-dimensional in that the electric field and the flow of water/development of pore water pressures are in the same direction, whereas the surface settlements induced are in an orthogonal direction. BS 1377: Part 5 (BSI, 1990e) recommends that the soil specimen should be orientated such that the soil will be loaded in the same direction relative to the stratum as the applied stress in situ. Electro-osmotic consolidation via vertical drains/electrodes is equivalent to consolidation using surcharging and vertical sand drains, in the sense that the applied pressure and drainage occur in two dimensions; therefore, the coefficients of radial drainage are appropriate (Johnson, 1970; Das, 1997).

Shear strength–total stress: A review of the available literature for the range of c_u for soils that have been successfully treated indicates a range of 0.7–81 kPa (Wan and Mitchell, 1976; Mitchell and Wan, 1977; Casagrande et al., 1981; Lo et al., 1991; Milligan, 1994; Abiera et al., 1999). Comparison of the liquidity index criteria with the undrained shear strength criteria indicates that preferable soils (i.e., liquidity index >0.6) will have a c_u of less than 20 kPa, whereas soils that may still be acceptable (i.e., liquidity index >0.2) will have a c_u of less than 55 kPa.

Shear strength–effective stress: Table 13.3 presents typical values for the range of ϕ' for silts and clays that are amenable to treatment by

Table 13.3 Typical ϕ' values for compacted clays

Soil type	Unified classification system designation	ϕ' (deg)
Silty clays	SM	34
Silts and clayey silts	ML	32
Clays of low plasticity	CL	28
Clayey silts, elastic silts	MH	25
Clays of high plasticity	CH	18

Source: After Carter and Bentley (1991).

electro-osmosis. Electro-osmosis may be expected to increase the value of ϕ' for treated soils as Pugh (1999) and Casagrande (1983) have demonstrated for London clay and fine quartz sand, respectively, in that the electro-osmotic process causes a shift in the particle size distribution to a coarser grain size.

13.3.11 Specialist electro-osmotic testing

The suitability of a material for electrokinetic treatment is best determined using an electro-osmotic cell. The cell used with EKG materials was originally developed by Banerjee and Vitayasupakorn (1984) and later modified by Hamir (1997). The cell is shown schematically in Fig. 13.11 and consists of a Perspex cylinder with a fixed base plate and an internal movable piston whose movement may be monitored by means of a displacement transducer (linear variable differential transformer). Provision is made for the location of disc-type electrodes both on the piston and on the fixed base plate by means of cable glands that permit the passage of an electrical cable into the cell, without the loss of pressure. In addition, the cell incorporates side ports through which pore water pressure and voltage gradient may be measured if required by means of a hypodermic needle tipped with porous ceramic. The chamber behind the moveable piston may be pressurized to apply a consolidation pressure to the soil sample. Backpressure may also be applied to the soil sample through tubing that passes through the piston and the base plate; this tubing also acts as a drain for any excess pore water pressure. An alternative to the electro-osmotic cell is the use of an electro-osmosis box described by Nettleton (1996) and Adali (1999).

The usefulness and relevance of the various tests, which can be undertaken to assess the suitability of materials suitable for electrokinetic treatment, are shown in Table 13.4.

13.4 REINFORCED SOIL RETAINING WALL—CASE HISTORY

13.4.1 Introduction

The modern concept of earth reinforcement and soil structures was postulated by Casagrande, who idealized the problem in the form of a weak soil reinforced by high-strength membranes laid horizontally in layers. Polymeric and grid reinforcements were developed in the 1970s. These provided enhanced soil/reinforcement interaction and in the case of polymeric materials permit the use of lower quality (cheaper) and waste materials as backfill (Jones, 1990, 1996).

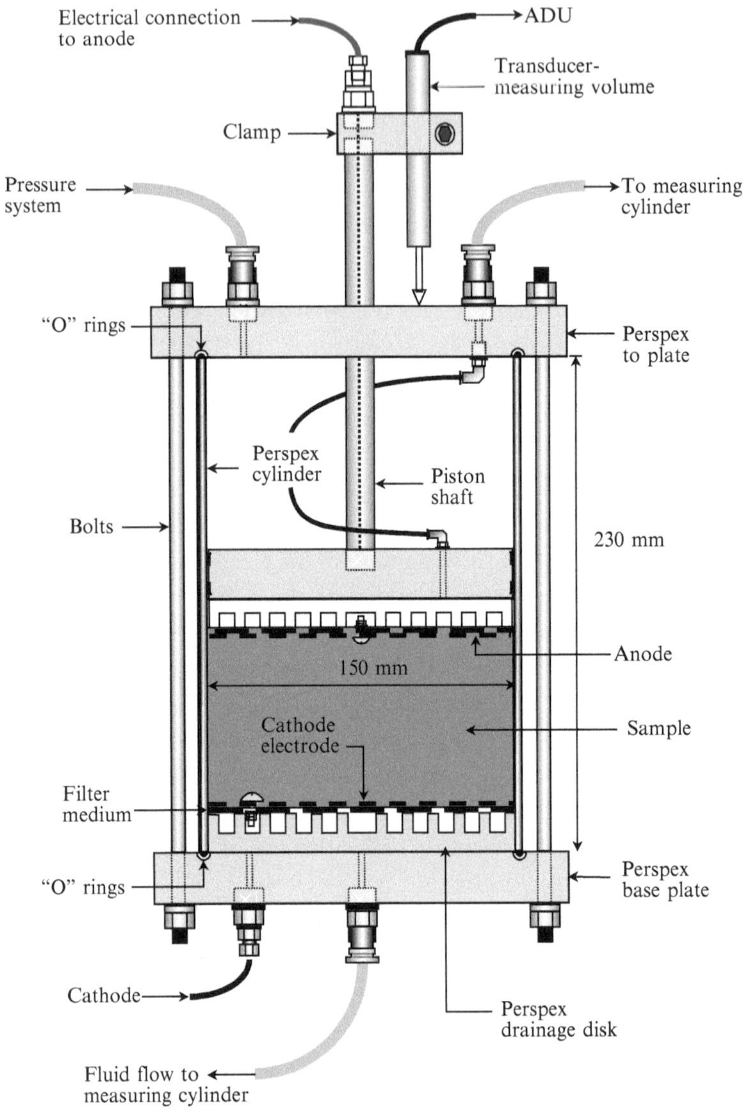

Figure 13.11 Electro-osmotic cell. *(Source: After Hamir (1997)).*

The acute lack of conventional frictional fill in some areas of the world has led to the use of cohesive soils in major reinforced soil structures in these countries. However, experimentation with the use of cohesive and waste fills concluded that the excess pore water pressures generated in the fill during construction can create high horizontal pressures, inhibited the development of effective stress, and so reduced the bond between soil and

Table 13.4 Usefulness of soil tests for assessing acceptability for electro-osmosis

Test	Usefulness	Acceptability range
Atterberg limits	✓✓✓	5–30% PI
Water content	✓✓✓	0.6–1.0 LI
PSD–sieve/sedimentation	✓✓✓✓	See BSI (1986)
Particle density	✓	Not applicable
Organic content	✓✓	Up to ORGANIC
1D consolidation parameters	✓✓✓	$m_v = 0.3$–$1.5\ MN/m^2$
Disc electrode	✓✓✓	0.05–0.005 S/m
Hydraulic permeability	✓✓✓	$<10^{-8}$ m/s
Undrained shear strength	✓✓	<55 kPa
Drained shear strength	✓	$\phi' < 30°$
Electro-osmosis cell	✓✓✓✓	Not applicable
Electro-osmosis box	✓✓✓✓	Not applicable

Note: ✓✓✓✓, excellent; ✓✓✓, good; ✓✓, reasonable; ✓, poor.
Source: After Pugh (2002).

reinforcement. These problems are alleviated close to the face where drainage can occur and with increasing time. The solution has been to include a drainage layer alongside the reinforcement. However, acceptability of the use of cohesive fill is still limited by its hydraulic permeability and its initial water content, thus severely restricting the range of materials utilized in practice. Most codes of practice, including BS 8006 (BSI, 1995), do not permit the use of purely cohesive soil in the construction of reinforced soil structures for permanent works, with the reasons for its exclusion being low strength, high moisture content, high creep, and low bond strength between the soil and the reinforcement. These problems can be overcome by the use of EKG.

13.4.2 EKG reinforced soil wall

The aim of the wall was to demonstrate, by means of a full-scale trial, that electrokinetic phenomena could be applied through the use of EKGs to construct a reinforced soil wall 43 m long and 4.8 m high, using an extremely wet overconsolidated cohesive fill, which under normal circumstances could not be built. The trial demonstrated the synergy between electrokinetic phenomena and reinforced cohesive soil through the use of EKG.

The properties of the fill used to construct the wall are presented in Table 13.5. The undrained shear strength of remolded samples was determined in accordance with BS 1377: Part 7 (BSI, 1990a). Peak and residual

Table 13.5 Soil parameters for fill material for the wall

Parameter	Cohesive fill		
LL (%)	60		
PL (%)	35		
PSD (D_{10}, D_{50}) (mm)	<0.002, 0.03		
G_s	2.61		
Water content (%)	32	35	41
C_u (kPa) (remolded)	89	49	24
Peak ϕ', c' (remolded)	21°, 7.6 kPa	23°, 1.3 kPa	22°, 6.6 kPa
Residual ϕ', c' (remolded)	20°, 5.5 kPa	18°, 7.6 kPa	19°, 6.3 kPa
Peak ϕ', c' (undisturbed)	23°, 10.5 kPa		
Residual ϕ', c' (undisturbed)	12°, 7.8 kPa		
σ (S/m)	0.6[a]		
Electro-osmotic cell (%) improvement	61%		

[a]This value was obtained using the water available on site.

shear strength parameters were determined using a shear box test in accordance with BS 1377: Part 8 (BSI, 1990a). Conductivity was determined using the disc electrode method (BS 1377: Part 3). All remolded samples were prepared using a consolidometer. The results show that there was some variability between the properties, even of the laboratory-prepared samples, and that there was some structure to the in situ soil. Both these factors need to be borne in mind when considering the results of the field experiment.

The data relating to electro-osmotic cell improvement presented in Table 13.5 were obtained using the electro-osmotic cell developed by Hamir (1997), with the percentage improvement being taken as the increase in water removed from the sample in the electro-osmotic cell above that removed in a control cell (no voltage) (Fig. 13.11).

13.4.3 Design philosophy

The design of reinforced soil structures is usually based on design codes that do not permit the use of cohesive fill, and it was not possible to analyze the structure for stability in the short-term using established procedures. Therefore, cohesive reinforced soil design methods were developed. These relied on the geometry of the reinforcement layout being known before the analysis was undertaken. Hence, the stability analysis of the wall was carried out in the long term to ascertain the reinforcement layout (using critical state shear strength parameters for the soil). This layout was then checked for short-term stability. This was achieved using four different analytical methods: critical height, Coulomb, discrete theory, and composite theory. All were used to determine the minimum required undrained shear strength for the clay fill to maintain short-term stability.

The water content required to achieve this strength was derived from the water content–undrained shear strength curve for the fill material ascertained from laboratory testing. The difference between the as-placed water content and the water content corresponding to the required strength was used to calculate the volume of water that needed to be removed from each lift of clay fill during construction.

The treatment time required to remove this volume of water was then determined for an array of electrode configurations, based on a linear voltage gradient and fixed soil parameters. This was followed by a more complex analysis using finite difference and resistance path techniques to establish a more realistic voltage gradient. Variations in the value of k_e, using both an empirical model and a graphical interpretation of the actual variation of k_e (measured in the laboratory), were also considered. The results of these analyses yielded estimated treatment times and estimated power demands drawn by the installation.

Brief descriptions of the results of the short- and long-term designs of the wall are provided next. However, this case history concentrates on the electro-osmotic design. Further details of both methodologies may be found in Pugh (2002).

13.4.4 Long- and short-term design

The wall was designed for long-term stability using effective stress parameters for the laminated clay, established from laboratory testing, using an established reinforced soil wall design package Winwall 6.14 (Netlon Ltd., 1998). A parametric study was conducted for a 4.8-m-high, vertically faced wall to assess what variation took place in the reinforcement layout with changes in the effective stress parameters. It was found that a slightly conservative design could be achieved with a fill shear strength of $\phi' = 15$, $c' = 0$, using secondary reinforcement placed at 600-mm spacing between the main reinforcement. The bottom three layers of polymeric geogrid reinforcement were formed using 80 kN/m (80RE) material, the top layers required 55 kN/m (55RE) material, and the secondary reinforcement was 20 kN/m (SS20) material. Two reinforced soil end blocks constructed using good-quality cohesionless fill supported the ends of the wall. The layout of the wall and the supporting end blocks are shown in

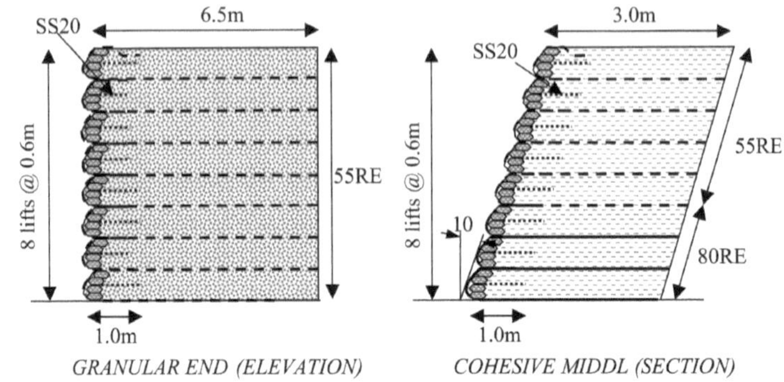

Figure 13.12 Long-term designs produced for the reinforced wall. *(Source: After Pugh (2002)).*

Fig. 13.12. The stability of the wall in the short term was based on the development of cohesion in the clay fill by electro-osmosis.

Four different methodologies were developed for the short-term undrained analysis of the reinforced clay wall:

Critical height: Based on the analysis method proposed by Terzaghi and Peck (1967) for calculating critical vertical cut heights (H_c) in cohesive soil of bulk unit weight (γ) and undrained shear strength (c_u). This method did not consider the contribution of the reinforcement.

Coulomb: A more sophisticated undrained analysis based on a continuation of the work based on Coulomb, 1776. The analysis assumed a failure through the reinforced slope at an inclination of $45° + \phi'/2$. For comparison, a failure plane inclined at 45° (i.e., $\phi' = 0 = \phi_u$) was also analyzed.

Discrete: Considered the undrained shear strength of the clay required to resist the pullout of discrete reinforcing elements and included the influence of the reinforcement.

Composite: Considered the undrained shear strength of the clay-reinforcement composite system.

The results from the different short-term analytical methods are presented in Table 13.6. The significance of these results to the design of the wall was considered to be the following:

- An undrained shear strength of the cohesive fill on the order of 6 kPa would be stable, but the strains required to achieve equilibrium could be excessive.
- An undrained shear strength of the cohesive fill in excess of 10 kPa would be stable as a composite system with both the reinforcement and the

Table 13.6 Minimum undrained shear strength required for short term determined using different analysis methods

	Analysis method	Inclination of failure plane to vertical $\theta = 45°$	$\theta = 45° + \phi'/2$
No contribution from reinforcement	Critical height	$c_u = 21.6$ kPa	
	Coulomb	$c_u = 21.6$ kPa	$c_u = 20.9$ kPa
Contribution from reinforcement	Discrete	$c_u = 9.6$ kPa	$c_u = 8.1$ kPa
	Composite	$c_u = 6.2$ kPa	$c_u = 6.0$ kPa

shear strength of the clay being utilized to maintain the stability of the system.

- An undrained shear strength of the cohesive fill in excess of 22 kPa would be sufficiently high, such that the system would be stable with little, if any, load being taken by the reinforcement.

In conclusion, it was considered that if the undrained shear strength of the cohesive fill could be increased to 10–20 kPa, then the wall would remain stable in the short term and allow construction to be completed to the design height.

13.4.5 Electro-osmotic design

The long- and short-term designs of the wall established that an undrained shear strength (c_u) of between 10 and 20 kPa was required from the clay to ensure the stability of the wall.

The purpose of the electro-osmotic design was to establish the following variables:

- The voltage and current to be drawn
- The length of treatment time required to improve the shear strength of the clay fill to a maximum of 20 kPa

To assess these variables, a design method was developed based on the quantity of water that needed to be removed from the soil to achieve the desired increase in undrained shear strength.

13.4.6 Preliminary electro-osmotic design

Using the soil parameters obtained from the laboratory testing presented in Table 13.5, a relationship was established relating the undrained shear strength of the clay to the water content for remolded samples

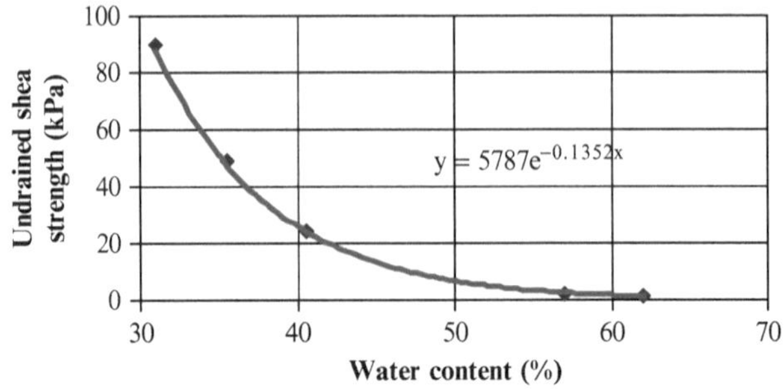

Figure 13.13 Relationship between c_u and water content for remolded fill.

(Fig. 13.13). The use of remolded samples was justified because the clay for the wall was remolded before being placed.

Assuming that the clay was placed in a very fluid state with an undrained shear strength of approximately 1–1.5 kPa with an associated water content on the order of 75–65%, as shown in Fig. 13.15, and knowing that the required shear strength of 20 kPa is associated with a water content of 42%, it was possible to establish that the required reduction in water through electro-osmosis was approximately 33–23%.

The volume of soil to be treated in each 600-mm lift of the 24-m-long, 3-m-wide walls was 43.2 m^3. For a 23% and 33% reduction in water content from 65% to 42% and from 75% to 42%, the volume of water that needed to be removed was 9.7 and 12.7 m^3, respectively. If the value of k_e is assumed to be that suggested by Casagrande (1952), $k_e = 5 \times 10^{-5}$ cm/s/V/cm, and V/L is established by simply dividing the applied voltage by the distance between the anode and cathode assuming point electrodes, then a preliminary treatment time of between 3.7 and 9.0 days is obtained for each 600-mm lift of clay (Table 13.7). It is also worth noting that if 9.7 and 12.7 m^3 of water are removed from the soil mass, then the change in volume associated with the

Table 13.7 Results of simplistic electro-osmosis analysis

Horizontal electrode spacing (m)	Assumed voltage gradient at 30 V (V/cm)	Assumed k_e (cm^2/s V)	Water content reduction required (%)	Treatment time (days)
0.4	0.83	5×10^{-5}	23–33	3.7–4.9
0.8	0.60	5×10^{-5}	23–33	5.2–6.8
1.2	0.45	5×10^{-5}	23–33	6.9–9.0
Whole wall	0.63	5×10^{-5}	23–33	4.9–6.5

removal of this volume of water would cause a surface settlement of approximately 130–175 mm over the whole surface area.

From Table 13.7, it can be seen that by varying the electrode spacing and, hence, the voltage gradient, the theoretical treatment time could be altered. The treatment times calculated in this manner are simplifications because they do not take into account the desiccation of the soil with time, nor do they take into account electrochemical changes that take place within the soil mass during electro-osmosis treatment. As a result, the times calculated in this way were considered as lower bound values.

13.4.7 Advanced electro-osmotic design

The preliminary design can be enhanced by refining the input parameters:

- Electro-osmotic permeability (k_e): The value of k_e applicable to the soil in question at the relevant voltage gradient was established from laboratory testing, and the variation with time may be taken into account.
- The voltage gradient (V/L) can be established more realistically using Laplace's equation and a finite difference analysis and taking into account the geometry of the electrode layout.

Refinement of the electro-osmotic permeability (k_e)

Mitchell (1993) states that the value of the parameter k_e is generally in the range of 1×10^{-5} to 1×10^{-4} cm^2/s V (cm/s per V/m), and Casagrande (1952) states the range as being 2×10^{-5} to 5×10^{-5} cm^2/s V. However, as treatment progresses and electrochemical reactions take place, desiccation of the soil occurs due to the removal of water by the electro-osmosis process. As a result, the quantity of water moved per unit of voltage decreases. The significance of this variation in k_e is that initially the flow of water achieved by electro-osmosis increases to a maximum within the first 12 h of treatment followed by a rapid decrease in the volume of water moved per unit time, followed in turn by a lower steady-state flow.

To model these phenomena, for practical application to the wall, a constitutive model for the variation of k_e with time was developed under different voltage gradients (Fig. 13.14). The calculation of treatment times based on the graphical interpretation presented in Fig. 13.14 was undertaken by digitizing the curve and using a spreadsheet to calculate the volume of water that flows in a time increment of 0.1 day. In this way, when the cumulative flow volume is equal to the volume of water required to be removed, the corresponding cumulative treatment time can be established.

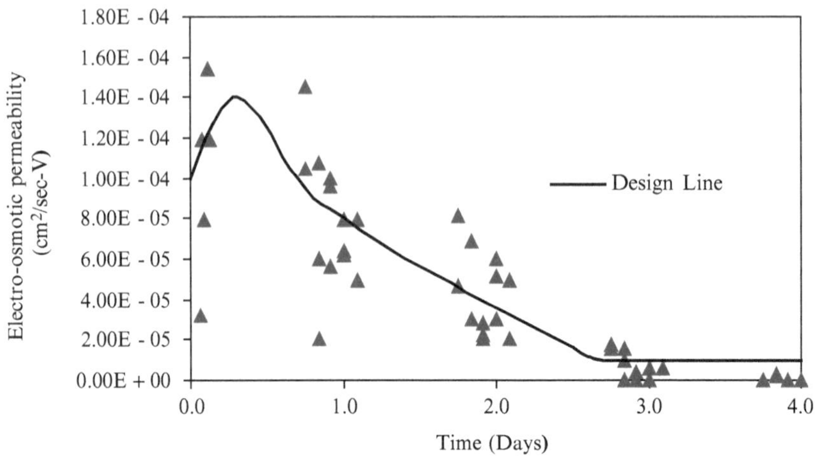

Figure 13.14 Graphical interpretation of k_e against time for design purposes.

Refinement of the voltage gradient parameter (V/L)

It was assumed in the simplistic design that the voltage gradient could be obtained by dividing the applied voltage by the spacing between anodes and cathodes. This is a simplification of what occurs in reality. The true voltage distribution obtained by the application of a potential difference by point electrodes is given by Laplace's equation (Eq. (13.26)) (Stroud, 1990; Young and Freedman, 1996). Laplace's equation was used in a conventional spreadsheet program to provide a more realistic distribution of the potential electrical field as demonstrated by Williams et al. (1993).

In practice, the voltage distribution will change with time as a result of the variation of the resistance of different zones of the soil due to desiccation, electrochemical changes within the soil mass, and closure of the electrode spacing due to settlement of the fill. This has been observed both in the field and in the laboratory by several researchers (Mitchell and Wan, 1977; Lo et al., 1991). However, to model the complexity that this continual variation of resistance with time would introduce was not considered viable for design purposes. The treatment times predicted by the simplistic and refined analyses are shown in Table 13.8.

The relationship between the water content and the undrained shear strength (c_u), together with the variation of electro-osmotic permeability (k_e) obtained from the laboratory testing, provides a means to predict the variation of the undrained shear strength as treatment progressed. Figure 13.15 shows the variation of c_u against time for the voltage gradients

Table 13.8 Summary of estimated treatment times using different voltage gradients and k_e variations for 23–33% reductions in water content

Electrode spacing (m)	Treatment time at 30 V for 23–33% w_c reduction (days)			
	Simplistic linear voltage variation		Finite difference voltage variation	
	Assumed k_e	*Measured* k_e	*Assumed* k_e	*Measured* k_e
0.4	3.7–4.9	3.1–9.0	6.5–8.5	16.0–26.0
0.8	5.2–6.8	9.5–17.5	9.4–12.3	30.7–45.2
1.2	6.9–9.0	18.2–28.8	12.4–16.3	45.8–50+

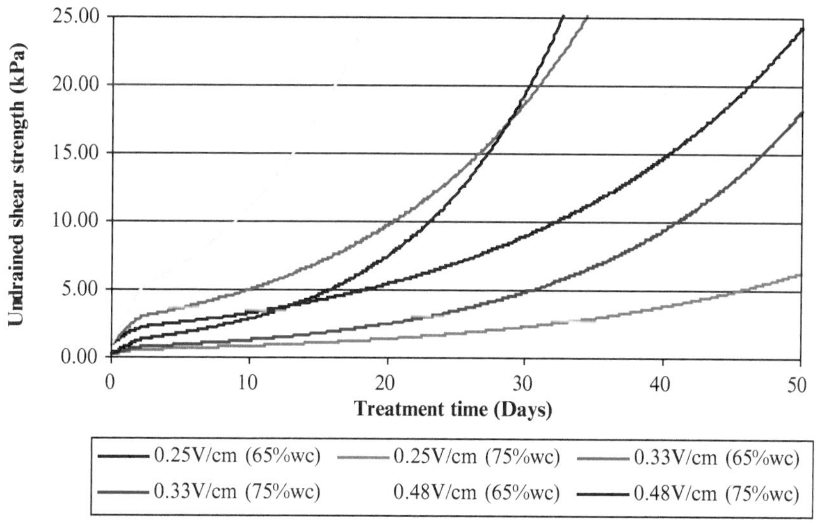

Figure 13.15 Variation of c_u against time at different voltage gradients and initial w_c.

established from the finite difference analysis and for the two assumed initial water contents.

Figure 13.15 shows that the curves, predicting the undrained shear strength, contain a kink at a treatment time of approximately 2.7 days. This kink relates to the change in the electro-osmotic permeability to a constant value, as shown in Fig. 13.13(b). Knowledge of the variation of c_u with treatment time is useful because strength is a parameter that can be measured rapidly in the field by means of a shear vane. In turn, this made it possible to confirm that the electro-osmotic treatment was progressing as anticipated.

Inspection of Table 13.8 reveals that the treatment times predicted using the linear voltage variation are the shortest due to the simplifications used to obtain the voltage gradient. They were thus considered as a lower bound solution for the treatment time. It is important to note that scale effects were not considered in the extrapolation of the laboratory results to the field prediction.

13.4.8 Construction of the wall

The wall was constructed using a wraparound design, utilizing sandbags for temporary stability of the front face. The ends of the "cohesive fill" trial section were retained using conventional reinforced soil blocks, constructed before starting the cohesive section. A small additional cohesive trial section was constructed at one end of the wall contemporaneously with the main trial. No electricity was supplied to this zone so that it would act as a control. This area was retained on one side using gabions.

The main trial section was subdivided into three sections, each having a horizontal electrode spacing of 1.2, 0.8, and 0.4 m, respectively. Geosynthetic drains were placed midway between the electrodes to provide a drainage path for the excess pore water pressure. The reason for different electrode spacing was to achieve different electric field intensities; thus, a variation in ΔV could be achieved using a single power source. The electrical potential applied across the electrodes was 30 V DC. This gave initial voltage gradients of 0.45, 0.6, and 0.83 V/cm based on the anode/cathode spacing. Current intermittence was not used, but the current was turned off during the night and on weekends. It was observed that this resulted in an increase in the dewatering rate when the current was reapplied in the morning.

The wall was constructed using a staged construction technique, such that a single lift of clay fill was constructed and dewatered *vertically* by electro-osmosis applied via horizontally placed pairs of electrodes and drains. The bottom electrode acted as a *closed* anode and the upper electrode as an *open* cathode. In this way, negative pore pressures were developed next to the anode, resulting in drainage. Once one lift had been successfully treated, the next lift was constructed and the process repeated until the full height of the wall was achieved—a total of eight lifts. The construction and dewatering processes is shown in Fig. 13.16.

The electrodes used were EKG consisting of a geonet construction manufactured using a counter-rotating die process. The clay used for the construction was slurrified using a 360° excavator within a lake located at the front of the trial area. Slurrification was achieved by excavating a hole within the clay in the lake and adding water. The mixture was worked with the bucket of the excavator for approximately 1 h until its consistency was that of fluid slurry. When placed within the wall structure, the slurry fill was

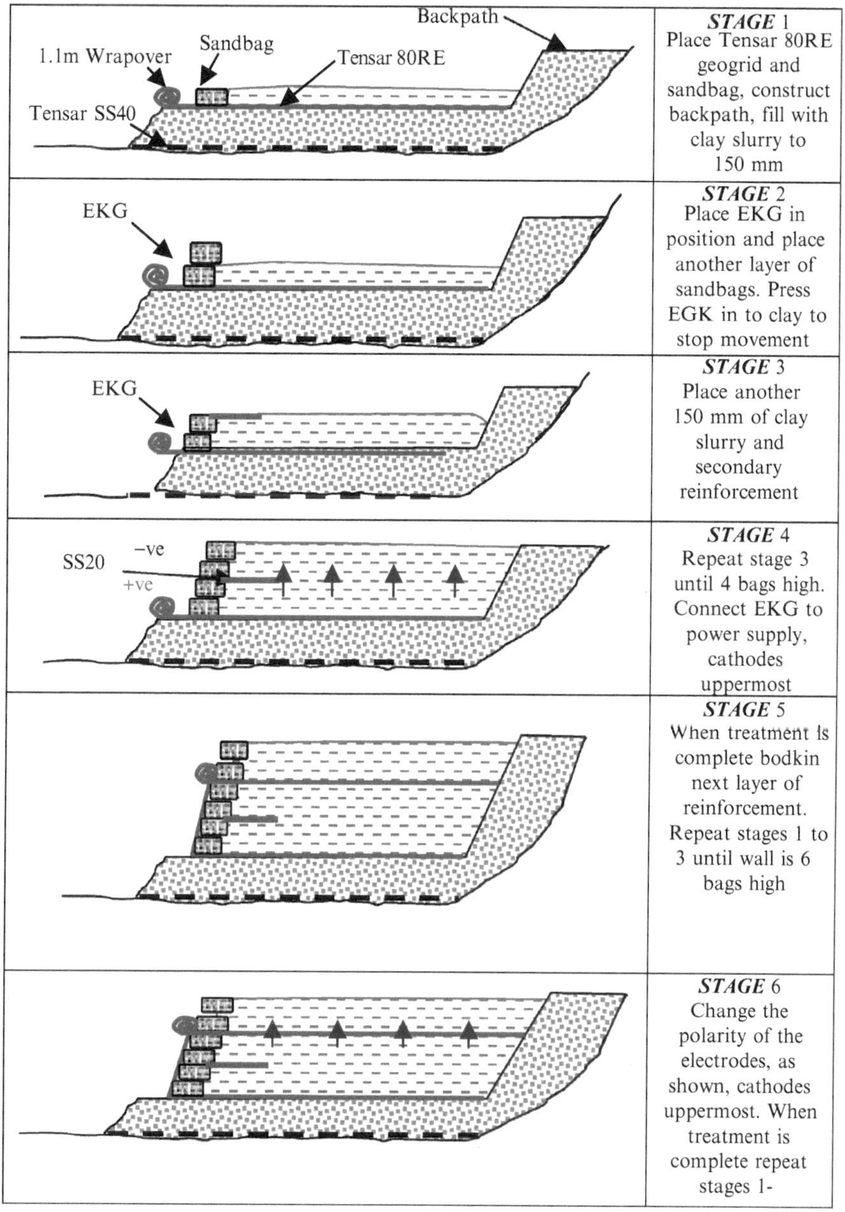

Figure 13.16 Construction sequence for cohesive wall.

self-leveling. Laboratory tests on the slurry gave a water content of approximately 75% (approximately liquid limit + 20%), which corresponded to a c_u of approximately 1–1.5 kPa (Fig. 13.13). During construction of the first lift, the moisture content during placing was slightly lower at approximately

50%, with a corresponding c_u of 5 kPa; this was due to a lack of mixing and inexperience of the construction technique.

13.4.9 Monitoring and analysis of the wall

During construction of the wall, several aspects of the construction were monitored, including fuel consumption, electrical power (voltage and current), undrained shear strength (c_u) of the cohesive fill, and movement of the front face. Pore water pressures and surface settlements were not measured due to the practical difficulties posed by the construction sequence and the corrosion of metallic instrumentation. The results of current and shear strength measurements are presented next.

Electrical current

The quantity of current drawn at the prescribed voltage (30 V) was measured using the analog dials located on the transformer/rectifier. The frequency of readings was varied to reflect the rate of change of current drawn. That is, during the initial powering up of a lift for treatment, readings were taken every 15 min. As the rate of change of current declined, the time interval between readings was increased until one reading was taken every hour during the working shift. During initial powering up, the electrodes were connected in groups of five—that is, five anodes and five cathodes.; in this way, it was possible to record the initial current drawn by the different electrode spacing configurations and also allowed any discrepancies in electrical resistance to be identified to a set of five electrode pairs for further investigation. It also enabled a quantification of the difference in resistance of the different electrode spacing used in the trial.

The electrical currents measured in the field could not be compared directly with those obtained from the laboratory testing due to the difference in volume of soil being treated. (Conductivity is related to the properties of the soil, the electrical contact between the soil and the electrode, the properties of the electrode, and the shape of the electric field. Current, and indeed conductance, is *additionally* related to the distance between the electrodes and the area of electrode–soil contact.) To allow a more direct comparison of the results, the current drawn was converted into an overall electrical conductivity (σ) of the system:

$$\sigma = \frac{1}{R}\frac{L}{A} \tag{13.24}$$

where

$$R = \frac{V}{I} = \frac{30V}{I} \tag{13.25}$$

The results of the electrical conductivity for the wall are presented in Fig. 13.17 for the different lifts of the wall, together with the design line obtained from laboratory testing.

It was apparent that the actual electrical conductivity measured in the field was approximately 10 times less than that obtained from the laboratory electro-osmosis cell. This result is logical when the configuration of the two different situations is considered. The electro-osmosis cell used plate electrodes and hence the electrical field established was 1-D, and theoretically the voltage gradient that occurs within the cell was uniform. In the field construction, the voltage was applied to the wall by means of EKG linear strips, which may be considered as point electrodes, thus generating an essentially 2-D electrical field, with the effect being more pronounced at greater spacings between electrodes of the same polarity.

In addition, Fig. 13.17 shows that the electrical conductivity of the fill in the wall underwent a reduction with time on the order of 95% from an initial value of approximately 0.026–0.0013 S/m. This compares with the values suggested in the design method.

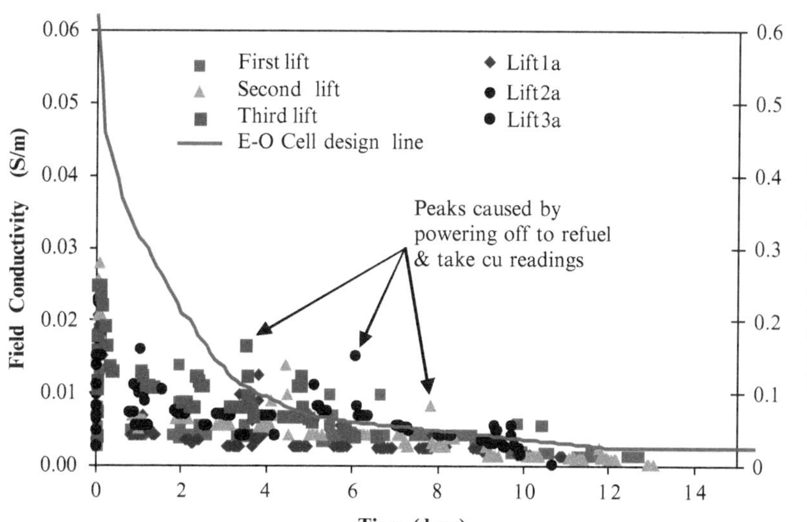

Figure 13.17 Variation of electrical conductivity, laboratory and field. *(Source: After Pugh (2002)).*

Undrained shear strength

The undrained shear strength of the cohesive fill undergoing electro-osmotic treatment was measured using a Pilson hand shear vane with a $1^1/_4$-in. (31-mm) vane. The measurements of undrained shear strength were taken at two depths, 0.25 and 0.5 m, in each lift to distinguish the variation in shear strength in the fill between the anode and cathode positions. To reduce errors, five readings were taken at each location and averaged at each depth, at approximately the same locations along the wall. The undrained shear strength was measured every 2 m along the full 24-m length of the electro-osmosis zone and at three different locations in the control zone. In addition, samples were taken of the soil from the shear vane test locations for laboratory testing to establish their water content.

The interpretation of the field measurements of the undrained shear strength obtained from the hand shear vane and corrected for conversion to field shear strength was undertaken by superimposing the theoretically calculated shear strength based on an initial water content and voltage gradient calculated by the finite element analysis. Due to the large variation in the results obtained from the hand shear vane, even within the zones of the same electrode spacing, a zonal average was calculated for each of the electrode spacings used (i.e., 0.4, 0.8, and 1.2 m) and the control zone. This allowed easier interpretation of the results by eliminating the large degree of scatter that was present in the unrefined field results. These results are plotted in Figs. 13.18 and 13.19 for the two test depths of 0.25 and 0.5 m. The variation between the 0.25- and the 05-m depth readings illustrates the difference in the strength changes with proximity to the anodes or cathodes.

Inspection of Figs. 13.18 and 13.19 reveals that the initial undrained shear strength (c_u) demonstrated a large degree of scatter but generally was in the range of ≈3–15 kPa. The higher values were attributable to lumps of harder clay in the slurry due to ineffective mixing. The design line for an initial water content of 65% at a voltage gradient of 0.48 V/cm shows a relatively good correlation with the field measurements of shear strength. Nearly all the undrained shear strength results measured in the field fell within realistic ranges defined by the theoretical curves. The field measurements of zonal averages show an obvious improvement with increasing treatment time.

The results obtained at a test depth of 0.25 m are generally lower than those obtained at a test depth of 0.5 m. Consideration of the electrode depths, with the anode at 0.45-m and cathode at 0.15-m depth, respectively,

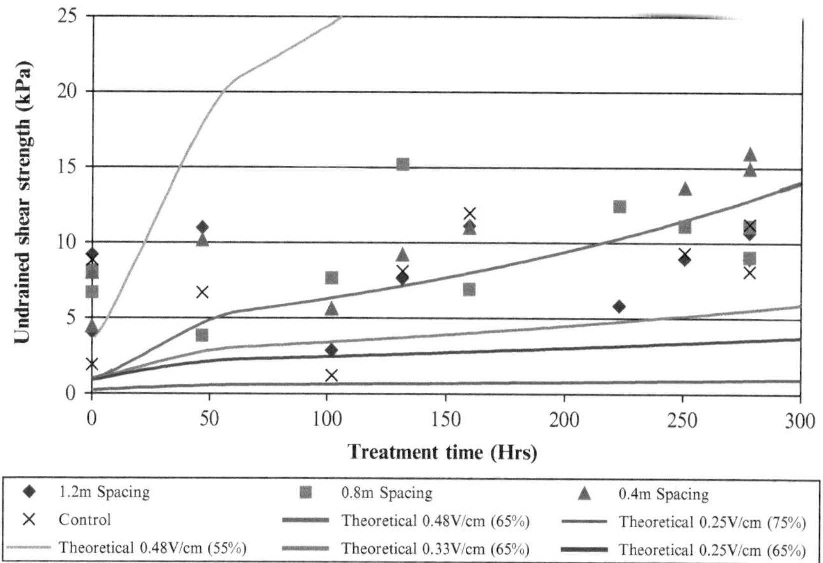

Figure 13.18 Theoretical and zonal results of c_u against treatment time, 0.25-m test depth. *(Source: After Pugh (2002)).*

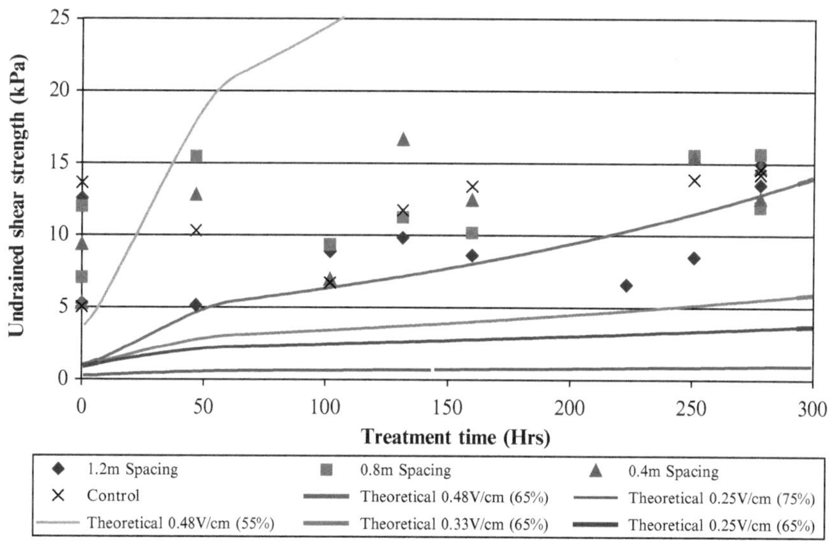

Figure 13.19 Theoretical and zonal average results of c_u against time, 0.5-m test depth. *(Source: after Pugh (2002)).*

explains these results. The results at a depth of 0.5 m are located immediately below the anode, whereas the results obtained at a depth of 0.25 m are located closer to the cathode.

The control zone also showed an improvement in shear strength with time. This was caused by self-weight consolidation, which was aided by the inclusion of drainage paths through the electrodes and filter elements. However, due to the reduced nature of the improvement that took place within the control zone, the continued construction of the zone became increasingly more difficult as the height of the construction increased.

The initial undrained shear strength assumed for the theoretical analysis appeared to be critical in the theoretical prediction. This was attributable to two factors:

- The relationship between c_u and w_c is not linear but, rather, approximately exponential, as shown in Fig. 13.13; hence, a greater increase in c_u occurs for a smaller reduction in w_c at lower water contents.
- The variation of the electro-osmotic permeability with time is not linear, as demonstrated in Fig. 13.14, with a significant decrease in k_e taking place after a period of approximately 2 days (48 h). This is reflected in Fig. 13.15 by the change of slope in the predicted value of c_u that occurred after a treatment time of approximately 50 h.

The combination of these two factors exaggerates the effect of electrokinetic treatment on soils with lower water contents but also *minimizes* the effect of treatment on soils with an initial high water content as shown by the curve for 75% initial water content in Fig. 13.13.

The electrode spacing, and hence the voltage gradient, also had a significant effect on the treatment process, as demonstrated by the increased improvement of the 0.4-m electrode spacing zone over the other zones. The theoretical analysis also predicted this, as shown in Fig. 13.14 by the curves of 0.48, 0.33, and 0.25 V/cm corresponding to electrode spacing of 0.4, 0.8, and 1.2 m, respectively, at an initial water content of 65%.

13.4.10 Treatment time

During construction, treatment time could only be measured indirectly by means of the shear strength and by means of the variation of current against generator time. Figure 13.15 shows that the conductivity of the fill in the structure had reached its residual value after a period of 10–12 days of generator operation. After this time, the efficiency of the installation would be extremely low, with the majority of the voltage being dropped across the high-resistance zone adjacent to the anode.

13.4.11 Practical applications

The trial demonstrated the successful use of EKG in the construction of a reinforced wall using a cohesive "slurry" fill. Although it is recognized that the properties of the fill used were extreme in the sense that they fell well outside the bounds of what would normally be regarded as fill material, some important principles have been established. It is possible to construct reinforced soil structures using cohesive fill and EKG because drainage is not dependent on hydraulic permeability. Thus, construction may be permitted using very poorly draining material and/or material with a high water content without the risk of generating very high pore water pressures.

With increasing environmental pressures requiring reuse of construction materials on site, this may be of significant benefit. There may no longer be the need to import high-quality fill (and dispose of poor-quality fill) for construction of, for instance, embankments, bunds, or abutments. Repairs to clay slopes, particularly where access is problematic or where high-quality fills are in short supply, may be possible using material found on site.

Additional applications include improvement of waste materials, particularly in cases in which water contents are very high, such as dredgings and tailings.

13.4.12 Conclusions

On the basis of the results presented herein, it may be concluded that the proposed electro-osmotic design method is a valuable predictive tool for estimating the change in undrained shear strength with time during an electro-osmotic treatment process.

The accurate input of the initial soil and treatment parameters used in the analysis is critical to its correct function. It is suggested that a sensitivity study be undertaken using the analysis method and an envelope of shear strength/treatment times established for a realistic range of conditions that may exist in the field and that appropriate laboratory testing be used to establish the variation of k_e to be used in the analysis.

The laboratory method for predicting current drawn produces an overestimation by a factor of 10 of that which occurs in practice; this is due to the differences in the shape of the respective electrical fields. It would appear sensible to use this to calculate extreme upper bounds of likely power consumption during any full-scale application.

Although not discussed in detail, the design approach adopted for the initial design of the electrode spacings would appear sensible and offer a pragmatic approach that could be adopted using skills and software available in most design offices.

Lastly, the applications of this technology are numerous, particularly with increasing pressures on the reuse of waste and sustainable construction. "R-EKG wall" could be the next generation of earth structures.

13.5 CIVIL AND ENVIRONMENTAL APPLICATIONS OF ACTIVE GEOSYNTHETICS

It has been established that conductive geosynthetics, acting as electrodes, can be used to effect the movement of contaminants through soil to the electrodes and then adsorb them. Using electrophoresis, it is possible to dewater industrial wastes, which are currently untreatable and can only be disposed of in tailing lagoons.

Conductive geosynthetics can be used to consolidate soil or to reduce the volume of industrial wastes by electro-osmosis, thus significantly lowering the cost of disposal. The use of conductive reinforcement can permit the use of fine, very wet material as fill for reinforced structures. Initial studies have shown that the use of electrically conductive band drains could prevent liquefaction of susceptible soils during earthquakes.

Other applications can be seen in different fields, such as sport and horticulture. The use of conductive geosynthetics can resolve some of the problems inherent in many large sports stadiums, such as the detrimental effect of shade on growing surfaces. An electrically conductive geosynthetic laid as a continuous porous membrane at root level provides oxygen directly to the root system as well as offering a method to control drainage, aeration, and ball bounce (Lamont-Black et al., 2003). A working model of this technology was exhibited in the Science Museum in London during the 2002 Football World Cup.

Three basic applications can be identified for active geosynthetics:

1. Accelerated settlement of solids from liquids (electrophoresis)
2. Dewatering to reduce volume alone (electro-osmosis)
3. Consolidation to improve strength by consolidation (electro-osmosis)

The application areas of active geosynthetics can be further identified by consideration of the electrical conductivity and the water content of the material to be treated (Fig. 13.20).

The application of EKGs for the construction of reinforced soil consolidation of weak soil and sports applications has been reported by Pugh et al. (2000), Lamont-Black (2001), Lamont-Black et al. (2003).

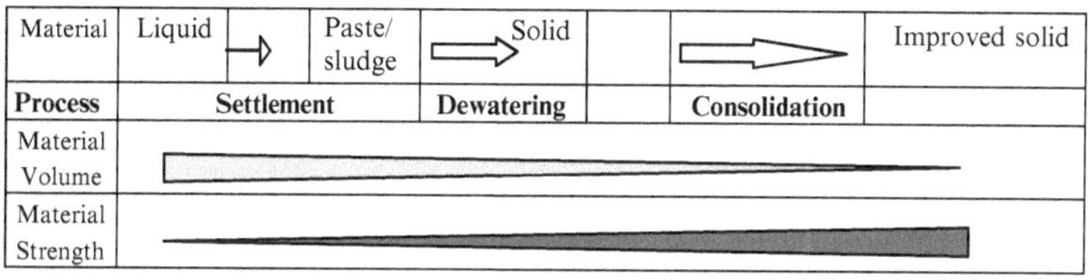

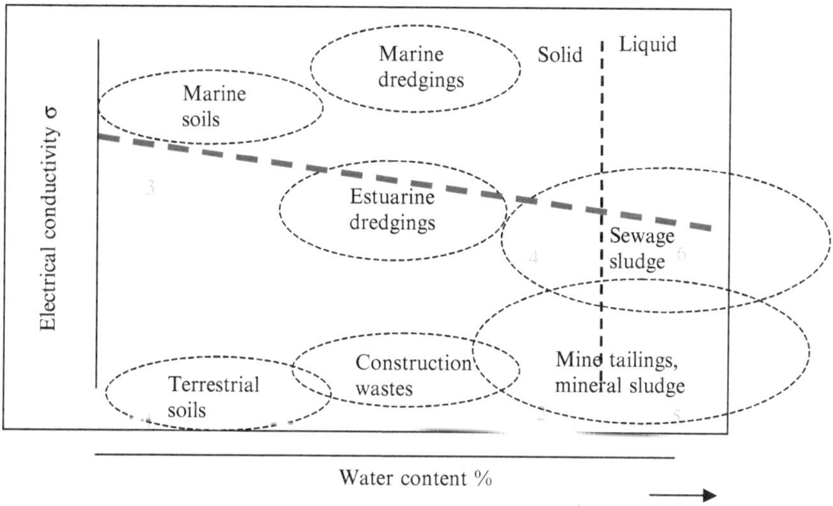

Figure 13.20 Schematic plot of electrical conductivity versus water content. The red dashed line represents an estimated viable treatment limit; the black dashed line represents the transition from solid (digable) to liquid (pumpable).

The results presented demonstrate that EKG can lead to improvements and forms of construction beyond those that can be achieved by conventional methods. The use of active geosynthetics to stabilize mine tailings has been reported by Fourie et al. (2002).

13.5.1 Novel applications of active geosynthetics

A number of novel applications for active geosynthetics have been proposed, including those discussed here.

13.5.2 Trenching and excavation

It would be possible to combine EKG dewatering with conventional well pointing technology in low-permeability soils to achieve a more rapid draw-down of the phreatic surface than that currently possible (Fig. 13.21). An

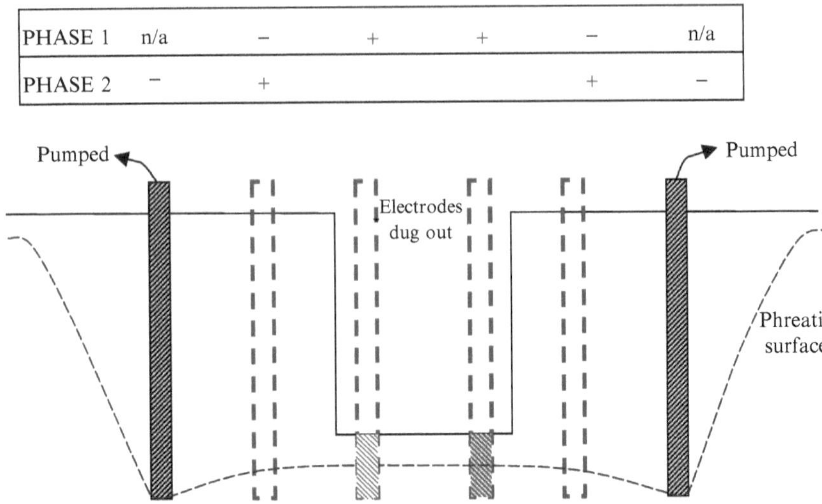

Figure 13.21 Multiphase drawdown of the phreatic surface using EKG for excavation and trenching.

additional benefit of electro-osmotic dewatering is that the soil properties may be improved, which will make excavation easier and more stable.

Tunneling and pipe jacking

Earth Pressure Balance Machine (EPBM) or mining methods could be enhanced using EKG technology to improve the soil conditions in the vicinity of the tunnel or to reduce post[construction settlements associated with the tunnel. If access to the surface was also available, the EKG could be installed at the surface to assist the process. In addition, electrokinetics could be used in very soft soils to reduce the possibility of chimney formation (Fig. 13.22).

Piling

Electrokinetic phenomena can assist with piling in two principal ways. If EKGs are installed in close proximity to the piling area and the steel piles are made into the cathodes, the result would be to locally soften the soil in the vicinity of the piles and facilitate driving. Once the desired piling depth has been achieved, the electrokinetic process can be shut down and the soil will then gradually regain its equilibrium and the soil properties will return to their former state around the piles (Fig. 13.23). Preferentially, the polarity could be reversed, making the piles anodic and therefore improving the soil properties in the vicinity of the piles. The latter process has been successfully employed in Canada (Milligan, 1994).

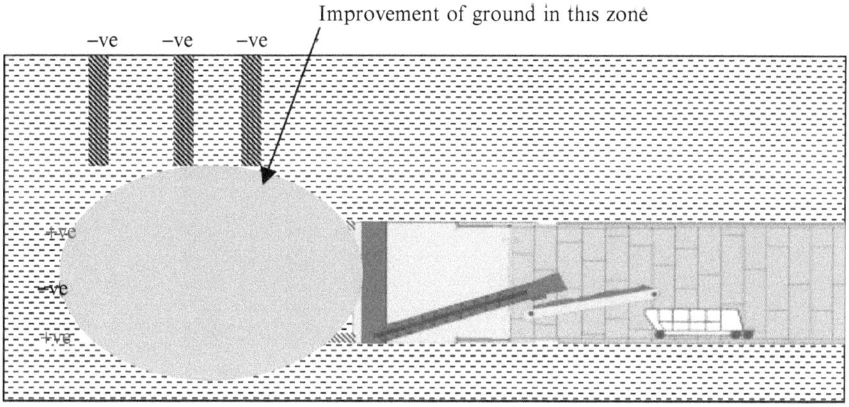

Figure 13.22 Application of EKG to tunneling.

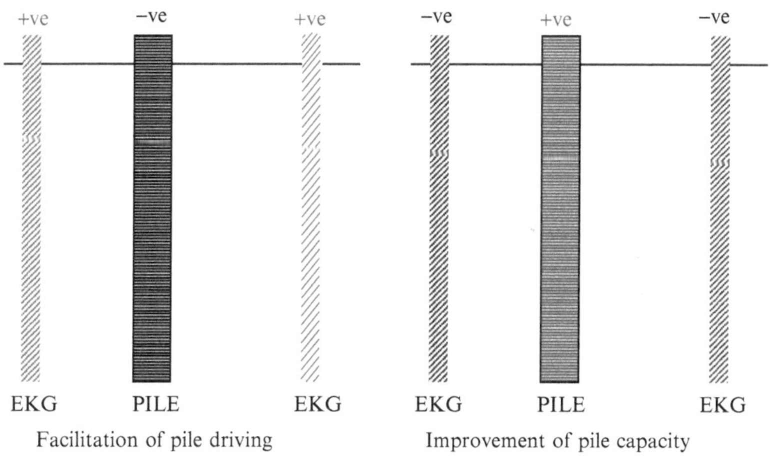

Figure 13.23 EKG applications to piling.

Enhanced lime migration

EKG also has a potential application in improving existing lime pile techniques. In this case, the EKG could be used as a conductive element to establish electro-osmotic flow to induce the migration of calcium ions through the soil, thereby increasing the zone of influence of the pile by causing lime migration to a greater radius than occurs by conventional diffusion (Fig. 13.24).

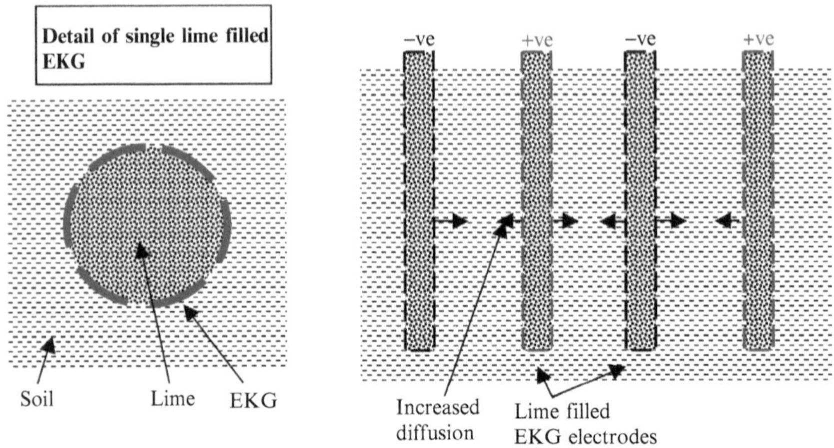

Figure 13.24 Enhancement of lime piles through the use of EKG electrodes.

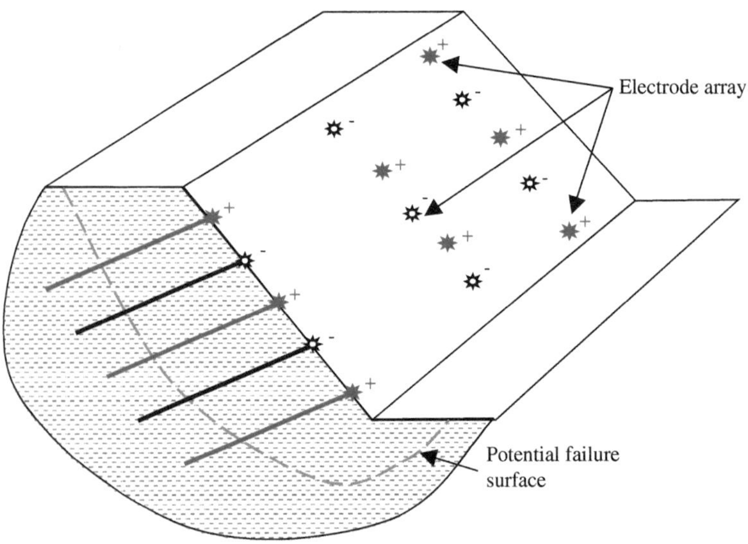

Figure 13.25 EKG used for stabilization of shallow foundations.

Slope stability

The stability of slopes may be improved through the use of EKG to achieve a more rapid dewatering of the soil and to act as drains to reduce pore water pressures within an unstable cohesive slope (Fig. 13.25). The application of a potential difference generates negative pore water pressures at the anode, an increase in the soil strength, and an increase in bond between the EKG nail or anchor and the soil resulting in an overall increase in the stability of the slope.

Shrinkage and swelling prevention of shallow foundations and pipelines

EKG technology may be able to offer a remedy to the problem associated with low-rise structures built with shallow foundations, which are constructed on soils prone to shrinkage and swelling. In this case, the EKG would enable moisture control of the susceptible strata and either add or remove water as necessary to prevent the volume change of the founding stratum.

13.6 CONCLUSION

Active geosynthetics introduce a new chapter in ground engineering technology. They have already shown advantage in being able to make use of poor soil as construction materials and provide enhanced dewatering capabilities. The latter has major implications in the treatment of sludges and tailings, and it is in these areas where most interest is being shown in their use.

Recent studies have shown that soils suitable for treatment by electrokinetic geosynthetics can readily be identified using conventional soil mechanics tests, although additional soil testing to determine the specific electrokinetics is required to determine treatment times and power consumption.

NOTATION

A_0 = surface charge density per unit pore volume, C/m^5
C_v = coefficient of consolidation, m^2/s^1
C_{vr} = coefficient of consolidation for radial flow, m^2/s^1
C_1, C_1 = constants of integration, variable
D = dielectric constant, dimensionless
D_{10}, D_{50} = largest particle size in smallest 10/50% of particles by mass, m
E = voltage difference, V
electrical energy, Wh
E_D = dilatometer modulus, N/m^2
E_u = undrained Young's modulus, N/m^2
E^{ϕ}_{red} = reduction voltage, V
G = shear modulus, N/m^2
G_s = specific gravity, dimensionless
I = electrical current, A
K = Boltzman constant, J/K^1
double-layer thickness, m
K_a = active earth pressure coefficient, dimensionless
K_p = passive earth pressure coefficient, dimensionless
L_{ii} = conductivity, variable

N = Avogadro's number, molecules/mol^{1}
number, dimensionless
P = electrical power, W
P_R = reinforcement load, N
P_s = shear load, N
P_v = vertical load, N
Q = volume of fluid, m^3
R = ratio anions/cations, dimensionless
resistance, Ω
T = temperature, K
tension in reinforcement, N
T_r = time factor for radial flow, dimensionless
T_v = time factor, dimensionless
U = average degree of consolidation, dimensionless
V = valence, dimensionless
voltage, V
X_{ii} = driving gradient, variable
c = efficiency factor, dimensionless
C_c = effective cohesion, N/m^2
c_u = undrained shear strength, N/m^2
e = electronic charge, C
voids ratio, dimensionless
i_e = electrical potential gradient, V/m^1
k = hydraulic permeability, m/s^1
k_e = electro-osmotic permeability, m^2/s^1/V^1
k_i = electro-osmotic efficiency, m^3/s^1/A^1
m_v = coefficient of volume compressibility, m^2/N^1
n = porosity, dimensionless
number of electrode pairs, dimensionless
n_0 = electrolyte concentration, ions/m^3
p' = stress invariant, N/m^2
q = stress invariant, N/m^2
unconfined compressive strength, N/m^2
q_a = flow through area "a," m^3/s^1
t = time, s
u = pore water pressure, N/m^2
v = velocity in free pore fluid, m/s^1
specific volume, dimensionless

v_e = velocity of flow induced by electrical potential gradient, m/s^1
v_h = velocity of flow induced by hydraulic gradient, m/s^1
w = water content, dimensionless
w_k = critical water content, dimensionless
Z = depth, m
A = reinforcement adhesion factor, dimensionless
α_b = bond coefficient, dimensionless
α_{ds} = direct sliding coefficient, dimensionless
γ = bulk unit weight of soil/fluid, N/m^3
$\overline{\gamma}$ = mean activity coefficient in double layer, dimensionless
γ_w = bulk unit weight of water, N/m^3
δ = double layer thickness, m
settlement, m
angle of friction soil/reinforcement, degrees (°)
ε_a = axial strain, dimensionless
ζ = zeta potential, V
potential across a condenser, V
η = viscosity, N s/m^2
θ = angle, degrees (°)
μ_g = geological settlement correction factor, dimensionless
v = Poisson's ratio, dimensionless
ξ = electro-osmotic equation variable, kg/m^2
π = pi, dimensionless
ρ = resistivity, Ωm
ρ_c = consolidation settlement, m
ρ_{oed} = oedometric settlement, m
ρ_w = density of water, kg/m^3
σ = surface charge density, C/m^2
electrical conductivity, S/m^1
σ'_b = effective bearing stress, N/m^2
σ'_n = effective normal stress, N/m^2
σ_1 = major principal stress, N/m^2
σ_2 = intermediate principal stress, N/m^2
σ_3 = minor principal stress, N/m^2
τ = shear stress, N/m^2
ϕ = angle of shearing resistance, degrees (°)
Ψ_0 = surface potential, V
Ω = resistance, Ω

REFERENCES

Abiera, H.O., Miura, N., Bergado, D.T., Nomura, T., 1999. Effects of using electro-conductive PVD in the consolidation of reconstituted Ariake clay. Geotech. Eng. J. Southeast Asian Geotech. Soc. 30 (2), 67–83.

Adali, E., 1999. Study of Electro-Osmosis in Multi-Layered Soils. M.Sc. dissertation, Geotechnical Group, University of Newcastle upon Tyne, UK.

Banerjee, S., Vitayasupakorn, V., 1984. Oedometer tests. J. Geotech. Eng., ASCE 110 (8), 1007–1023.

British Standards Institution, 1986. BS 8004: Code of Practice for Foundations. British Standards Institution, Hemel Hempstead, UK.

British Standards Institution, 1990a. British Standard Methods of Test for Soils for Civil Engineering Purposes. British Standards Institution, London.

British Standards Institution, 1990b. British Standard Methods of Test for Soils for Civil Engineering Purposes: Part 1. General Requirements and Sample Preparation, BS 1377. British Standards Institution, London.

British Standards Institution, 1990c. British Standard Methods of Test for Soils for Civil Engineering Purposes: Part 2. Classification Tests, BS 1377. British Standards Institution, London.

British Standards Institution, 1990d. British Standard Methods of Test for Soils for Civil Engineering Purposes: Part 3. Chemical and Electro-Chemical Tests, BS 1377. British Standards Institution, London.

British Standards Institution, 1990e. British Standard Methods of Test for Soils for Civil Engineering Purposes: Part 5. Compressibility, Permeability and Durability Tests, BS 1377. British Standards Institution, London.

British Standards Institution, 1995. BS 8006: Code of Practice for Strengthened/Reinforced Soil and Other Fills. British Standards Institution, London.

Carter, M., Bentley, S.P., 1991. Correlations of Soil Properties. Pentech, London.

Casagrande, L., 1952. Electro-osmotic stabilisation of soils. J. Boston Soc. Civil Eng., ASCE 39, 51–83.

Casagrande, L., 1983. Stabilization of soils by means of electro-osmosis. J. Boston Soc. Civil Eng., ASCE 69 (2), 255–302.

Casagrande, L., Loughney, R., Matich, M.A.J., 1961. Electro-osmotic stabilization of a high slope in loose saturated silt. In: Proc. 5th Int. Conf. on Soil Mechanics and Foundation Engineering, Paris, vol. 2, pp. 555–561.

Casagrande, L., Wade, N., Wakely, M., Loughney, R., 1981. Electro-osmosis projects, British Columbia, Canada. In: Proc. 10th Eur. Conf. on Soil Mechanics and Foundation Engineering, Rotterdam, pp. 607–610.

Chappell, B.A., Burton, P.L., 1975. Electro-osmosis applied to unstable embankment. J. Geotech. Eng. Div., ASCE 101 (GT8), 733–740.

Chen, J.L., Murdoch, L., 1997. Effects of electro-osmosis on natural soils: field test. J. Geotech. Geoenviron. Eng., ASCE 125 (12), 1090–1098.

Coulomb, C.A., 1776. Essai sur une application des régeles des maximus et minimus a quelque problémes de statique re'latif à l'architecture. In: Memoirs Divers Savants 7. Académie Sciences, Paris.

Das, B.M., 1997. Advanced Soil Mechanics, second ed. Taylor & Frances, London.

Eastwood, B.J., 1997. A Fundamental Study of the Electrochemical Failure Mechanism of a Novel Impressed Current Cathodic Protection System. PhD thesis, Department of Chemistry, University of Newcastle upon Tyne, UK.

Eggestad, Å., Føyn, T., 1983. Electro-osmotic improvement of a soft sensitive clay. In: Proc. 8th Eur. Conf. on Soil Mechanics and Foundation Engineering, Helsinki, vol. 2, pp. 597–602.

Esrig, M.I., 1968. Pore pressures, consolidation and electrokinetics. J. Soil Mech. Div., ASCE 94 (SM4), 899–921.
Fetzer, C.A., 1967. Electro-osmotic stabilisation of West Branch Dam. J. Soil Mech. Found. Div., ASCE 93 (SM4), 85–106.
Fourie, A.B., Johns, D., Jones, C.J.F.P., 2002. In-situ dewatering of mine tailings using electrokinetic geosynthetics. In: Proc. 6th Int. Conf. on Geosynthetics, Nice.
Grey, D.H., Mitchell, J.K., 1967. Fundamental aspects of electro-osmosis in soils. J. Soil Mech. Found. Div., ASCE 93 (SM6), 209–236 (Closure discussion in 95(SM3), 875–879, 1969).
Hamir, R., 1997. Some Aspects and Applications of Electrically Conductive Geosynthetic Materials. PhD thesis, University of Newcastle upon Tyne, UK.
Head, K.H., 1982. Manual of Soil Laboratory Testing: Vol. 2. Permeability, Shear Strength and Compressibility Tests. Pentech, London.
Johnson, S.J., 1970. Precompression for improving foundation soils. J. Soil Mech. Found. Div., ASCE 96 (SM1), 145–175.
Jones, C.J.F.P., 1990. Construction influences on the performance of reinforced soil structures: state-of-the-art review. In: Proc. Int. Conf. on Reinforced Soil, Glasgow, pp. 97–116.
Jones, C.J.F.P., 1996. Earth Reinforcement and Soil Structures. Telford, London.
Jones, C.J.F.P., Fakher, A., Hamir, R., Nettleton, I.M., 1996. Geosynthetic materials with improved reinforcement capabilities. In: Proc. Int. Symp. on Earth Reinforcement, Fukuoka, Kyushu, Japan, November 12–14, vol. 2, pp. 865–883.
Jones, C.J.F.P., Lamont-Black, J., Glendinning, S., Pugh, R.C., 2005. New applications for smart geosynthetics. In: Rathje, E.M. (Ed.), Geo-Frontiers 2005, ASCE Geotechnical Special Publications 130–142 & GRI-18, Austin, TX.
Lageman, R., Pool, W., Seffinga, G.A., 1989. Electro-reclamation: state of the art. In: NATO/CCMS, Third International Conference on Demonstration of Remedial Action Technologies for Contaminated Land and Groundwater, November 6–9, Montreal, Canada, pp. 115–136.
Lambe, T.W., Whitman, R.V., 1969. Soil Mechanics. Wiley, New York.
Lamont-Black, J., 2001. EKG: The next generation of geo synthetics. Ground Eng. 34 (10), 22–23.
Lamont-Black, J., Glendinning, S., Jones, C.J.F.P., Huntley, D.T., 2003. Electrokinetic processes for the management of natural sports turf. In: Proc. Natl. Turfgrass Foundation Annual Conference, Harrowgate, UK, p. 4.
Lo, K.Y., Ho, K.S., Inculet, I.I., 1991. Field test of electro-osmotic strengthening of soft sensitive clays. Can. Geotech. J. 28, 74–83.
Lockheart, N.C., 1983. Electro-osmotic dewatering of clays: II. Influence of salt, acid and flocculants. Colloid. Surface. 6, 239–251.
Long, E., George, W., 1967. Turnagain slide stabilization, Anchorage, Alaska. J. Soil Mech. Found. Div., ASCE 93 (SM4), 611–627.
Micic, S., Shang, J.Q., Lo, K.Y., Lee, N.Y., Lee, S.W., 2001. Electrokinetic strengthening of a marine sediment using intermittent current. Can. Geotech. J. 38 (2), 287–302.
Milligan, V., 1994. First application of electro-osmosis to improve friction pile capacity—Three decades later. In: Proc. 13th Int. Conf. on Soil Mechanics and Foundations Engineering, January 5–10, New Delhi, India., vol. 5. Balkema, Rotterdam, The Netherlands, pp. 1–5.
Mitchell, J.K., 1991. Conduction phenomena: from theory to geotechnical practice. Géotechnique 41 (3), 299–340.
Mitchell, J.K., 1993. Fundamentals of Soil Behaviour, second ed. Wiley, New York.
Mitchell, J.K., Wan, T.Y., 1977. Electro-osmotic consolidation—its effects on soft soils. In: Proc. 9th Int. Conf. on Soil Mechanics and Foundation Engineering, Tokyo, Japan., vol. 1. Balkema, Rotterdam, The Netherlands, pp. 219–224.

Mohamedelhassan, E., Shang, J.Q., 2001. Effects of electro materials and current intermittence in electro-osmosis. Ground Improv. 5 (3), 3–11.
Netlon Ltd, 1998. Winwall Version 6.14, Reinforced Soil Design Package. Tensar International, Blackburn, UK.
Nettleton, I.M., 1996. Electro-Bioremediation. Internal report, Geotechnical Group, University of Newcastle upon Tyne, Newcastle upon Tyne, UK.
Nettleton, I.M., Jones, C.J.F.P., Clarke, B.G., Hamir, R., 1998. Electrokinetic geosynthetics and their applications. In: Proc. 6th Int. Conf. on Geosynthetics, Atlanta, GA, vol. 2., Intl. Geosynthetics Society, Georgia, pp. 871–876.
Pamukcu, S., 1996. Electro-chemical technologies for in-situ restoration of contaminated subsurface soils. J. Geotech. Eng. available at, http://www.ejge.com/1997/Ppr9704/Abs9704.htm.
Pugh, R.C., 1999. Mid-Term Review Report, EPSRC Research Report. University of Newcastle upon Tyne, UK.
Pugh, R.C., 2002. The Application of Electrokinetic Geosynthetic Materials to Uses in the Construction Industry. Ph.D. thesis, University of Newcastle upon Tyne, UK.
Pugh, R.C., Clarke, B.G., Jones, C.J.F.P., 2000. An electro-osmosis consolidation trial using electrokinetic geosynthetics. In: Proc. 4th Int. Conf. on Ground Improvement Geosystems, Helsinki, June 7–9, vol. 2, Finnish Geotechnical Society, Finland, pp. 533–540.
Stroud, K.A., 1990. Further Engineering Mathematics—Programmes and Problems, second ed. Macmillan, London.
Terzaghi, K., Peck, R., 1967. Soil Mechanics in Engineering Practice, second ed. Wiley, New York.
van Olphen, H., 1977. An Introduction to Clay Colloid Chemistry, second ed. Wiley–Interscience, New York.
Wan, T.Y., Mitchell, J.K., 1976. Electro-osmotic consolidation of soils. J. Geotech. Eng. Div., ASCE 102 (GT5), 473–491.
Williams, B.P., Smyrell, A.G., Lewis, P.J., 1993. Flownet diagrams—the use of finite differences and a spreadsheet to determine potential heads. Ground Eng. 26 (6), 32–38.
Wrigley, N.E., 1999. The practical use of electro-osmosis. Personal communication, October.
Young, H.D., Freedman, R.A., 1996. University Physics, ninth ed. Addison-Wesley, Reading, MA.

CHAPTER 14

Electro-Osmotic Stabilization

Amnart Rittirong, Julie Shang
Department of Civil and Environmental Engineering, The University of Western Ontario, London, Ontario, Canada

14.1 THEORETICAL BACKGROUND

Electro-osmosis is a technique used for the consolidation and strengthening of soft, saturated clayey soils. As shown in Fig. 14.1 (Shang, 1998), when a direct current (DC) voltage is applied to soil via electrode poles, the soil pore water will be attracted toward the direction of the negative terminal (cathode) due to the interaction of the electric field, the ions in the pore water, and the soil particles. If drainage is provided at the cathode and prohibited at the anode, consolidation will be induced by electro-osmosis, resulting in lower soil water content, higher shear strength, and lower compressibility. In addition, electrochemical reactions associated with an electro-osmotic process alter the physical and chemical properties of the soil and lead to a further increase in shear strength (Mitchell, 1993).

Casagrande (1941, 1949, 1959) first applied the technique of electro-osmosis to strengthen and stabilize soft silty clays in the mid-1930s. Since then, successful field tests have been reported that used electro-osmosis to strengthen silty clays and soft sensitive clays, to stabilize earth slopes, and to reinforce steel piles installed in clayey soils (e.g., Bjerrum et al., 1967; Casagrande, 1983; Lo et al., 1991). Electro-osmotic consolidation has been considered for projects requiring a rapid improvement in the properties of soft clayey soils.

When an open cathode and sealed anode condition is present, a negative pore water pressure in soil is generated upon the application of a DC electrical field. In one dimension, the pore pressure generated by electro-osmosis, $u_{eo}(x, t \rightarrow \infty)$ (kPa), at a distance x (m) to the cathode is given by (Esrig, 1968)

$$u_{eo}(x) = -\frac{k_e}{k_h}\gamma_w U(x) \tag{14.1}$$

where k_e (m^2/sV) is the electro-osmotic permeability, k_h (m/s) is the hydraulic conductivity, γ_w (9.81 kN/m^3) is the unit weight of water, and $U(x)$ (V) is the electrical potential at distance x to the cathode.

http://dx.doi.org/10.1016/B978-0-08-100191-2.00014-9

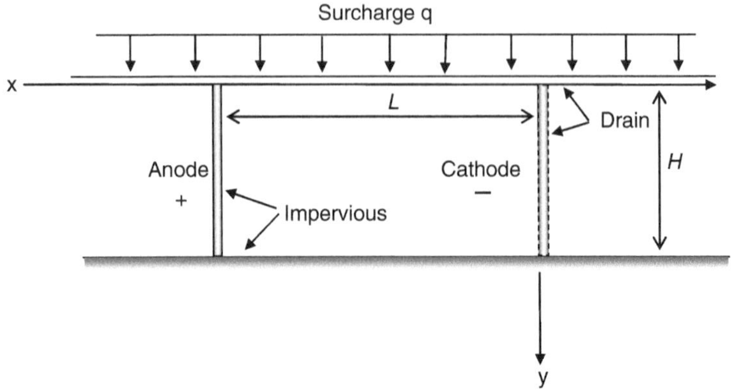

Figure 14.1 Schematic of electro-osmotic consolidation. *(Source: Modified from Shang (1998)).*

Equation (14.1) states that the pore pressure induced by electro-osmosis is negative and proportional to the electrical potential (i.e., it has a maximum magnitude at the anode and zero at the cathode). The negative pore pressure results in an increase in the effective stress in the soil, leading to consolidation, as described in the conventional consolidation theory. Knowing the pore pressure generated by electro-osmosis, the time rate of electro-osmotic consolidation can be estimated by conventional consolidation theory.

The electro-osmotic permeability, k_e, governs the water flow in a soil mass under an electrical gradient in a similar way as the hydraulic conductivity governs the flow in soil under a hydraulic gradient. When both the anode and the cathode are open to drainage and the hydraulic gradient is set to zero, k_e can be determined by measuring the flow velocity across a soil plug using an empirical relation (Mitchell, 1993):

$$q_e = k_e E \tag{14.2}$$

where q_e is the water flow vector due to an electrical gradient (m/s), and E is the electric field intensity vector, defined as

$$E = -\nabla U \tag{14.3}$$

The power consumption per cubic meter of soil mass per hour is calculated from

$$p = \kappa E^2 \tag{14.4}$$

where p is the unit power consumption (kW/m^3), and κ is the electrical conductivity of the soil (1/Ω-m).

Table 14.1 Design parameters and common soil properties in electro-osmotic consolidation

Parameter	Unit	Typical range
k_h, hydraulic conductivity	m/s	10^{-10}–10^{-8}
k_e, electro-osmotic permeability	m^2/sV	~10^{-9}
κ, electrical conductivity of soil	S/m (Ω-m)	0.01–0.5
E, electric field intensity	V/m	20–100
c_v, coefficient of consolidation	m^2/s	10^{-8}–10^{-7}
P, hourly power consumption	kWh/m^3	0.01–1

Equation (14.4) indicates that the power consumption of electro-osmotic treatment increases with the soil electrical conductivity and applied electric field. Table 14.1 summarizes the typical ranges of soil and electrical properties that are suitable and have been used for electro-osmotic consolidation.

A two-dimensional electro-osmotic consolidation model was developed by Shang (1998) that can take the effects of both preloading and electro-osmotic consolidation into account. The most predominant electrochemical effects during an electro-osmotic process include the development of a pH gradient, the generation of gases, and heating. The pH of soil water will increase rapidly to as high as 11 or 12 at the cathode and decrease to almost 2 at the anode. Consequently, metallic anodes will corrode at the anode. Oxygen gas is generated at the anode and hydrogen gas at the cathode due to hydrolytic reactions. The electrical current also generates heating. The seriousness of these effects is directly related to the applied voltage and current. It is usually desirable to minimize heating effects to reduce power consumption. It has been found that applying polarity reversal and intermittent (pulse) current can reduce pH gradient and corrosion and increase the electro-osmotic permeability of the soil, thus improving the efficiency of electro-osmotic treatment (Shang et al., 1996).

The evaluation of electro-osmotic consolidation on a specific soil can be conducted according to the procedure outlined here.

14.1.1 Determination of parameters

In addition to conventional soil properties such as the grain size, preconsolidation pressure, shear strength, water content, hydraulic conductivity, k_v, and coefficient of consolidation c_v, the parameters required for a treatability analysis include the electro-osmotic permeability, k_e; electrical conductivity, κ; intensity of electric field, E; and power consumption, p. All these parameters can be determined from laboratory tests prior to field application

(Shang and Mohamedelhassan, 2001). Table 14.1 lists the typical ranges of the major parameters for soils that are suitable for electro-osmotic treatment.

14.1.2 Electrical operation system in field applications

The electrical operation system can be designed based on the parameters obtained from laboratory tests and from the geotechnical investigation of the site. Typically, the electrode poles consist of metallic rods or pipes installed vertically into the ground. Sometimes, prefabricated vertical drains may be installed at the cathode as the drainage channel. The depth of the electrode insertion should be equal to the thickness of the soil layer to be treated. The upper portion of the electrodes in contact with the ground surface crust or top drainage layer should be electrically insulated using a dielectric coating to avoid short circuiting due to the presence of surface water (Lo et al., 1991).

The material, layout, and spacing of electrodes and the applied voltage are of utmost importance to a successful field application. Among the most commonly used conducting metallic materials, the best results were reported using electrodes made of iron and copper rather than aluminum (Sprute and Kelsh, 1980; Mohamedelhassan and Shang, 2001). Electrodes made of carbon-coated steel rods and graphite have been used in laboratory studies to prevent electrode corrosion (Lockhart and Stickland, 1984).

The typical spacing between anodes and cathodes reported in the literature ranged from 1 to 3 m (Casagrande, 1983; Lo et al., 1991). In general, an approximately uniform electric field gives the best results (Casagrande, 1983). To achieve an approximately uniform electric field, the spacing between electrode rods of the same polarity should be much less than spacing of the opposite polarity.

Power supply capacity can be estimated based on the soil's electrical conductivity and electrode layout. It has been found that a more dramatic voltage drop takes place at the soil–electrode contacts at a higher applied voltage, which made the treatment less efficient (Casagrande, 1983; Shang et al., 1996). It was also observed that the voltage drop at the soil–electrode interface is affected by the electrode materials (Mohamedelhassan and Shang, 2001). Therefore, a lower voltage applied across smaller anode–cathode spacing is desirable to generate the required electric field, and special attention should be paid to the electrode materials and configurations. However, the cost of electrodes and installation should also be considered. The final design will be based on a balance of the cost of electrodes and electrode installation as well as the treatment efficiency. For additional information, see Arman (1978),

Broms (1979), Mitchell (1981, 1993), U.S. Navy (1983), Van Impe (1989), Hausmann (1990), and Micic et al. (2003a,b).

Electrokinetic stabilization is a hybrid between electro-osmosis and chemical grouting. The infusion of certain stabilization chemicals into silty and sandy soils is made more efficient by the application of an electrical potential difference to the soil mass. The procedure is more effective in silty soils that are otherwise difficult to grout ordinarily. Information on this technique can be found in Broms (1979) and Mitchell (1981). Electrokinetic-assisted chemical stabilization has also been applied to offshore calcareous soils (silts and sands) for stabilization of petroleum platforms (Mohamedelhassan and Shang, 2003; Shang et al., 2004a,b).

In this chapter, several well-documented field tests using electro-osmosis to stabilize soft clay soils are reviewed. These cases have been established as the classic examples of the applications of electro-osmotic stabilization.

14.2 CASE HISTORIES

14.2.1 Stabilization of an excavation in Norwegian quick clay (Bjerrum et al., 1967)

The first well-documented case trial was carried out by Bjerrum et al. (1967) on a site located in Ås, 30 km south of Oslo, Oslofjord, Norway. The soil on the site is a quick clay with sensitivity of approximately 100. The soil properties are summarized in Table 14.2. Although the site is located on a flat slope (1:8), the slope stability analysis indicated that the factor of safety was only 0.77, attributed to the nature of the quick clays. The construction of a sewage treatment plant required an excavation to a depth of 4.5 m. However, a factor of safety 1.3 against bottom heave failure would allow the excavation to only a depth of 2.3 m. Removing the topsoil surrounding the excavation was not possible because of a railway nearby. Therefore, electro-osmotic stabilization was selected to stabilize the excavation.

The electrode installation and the instrumentation on the site are shown in Fig. 14.2. Reinforcing steel bars, 19 mm in diameter and 10 m long, were used as electrodes. Anode rows alternated with cathode rows covered the area of 200 m^2. The embedded depth of electrodes was 9.6 m with 2-m spacing between anode–cathode rows and 0.6–0.65 m apart in the same row. It was recognized that during electro-osmotic treatment, the shear strength increases at the anode and decreases at the cathode in the early stage of the treatment. Therefore, to prevent the slope failure, the electrode rows were installed in the direction of the maximum inclination of the slope. The

Table 14.2 Geotechnical properties of soft clay

Parameter	Unit	Typical range
w, natural water content	%	31
w_p, plastic limit	%	14
w_L, liquid limit	%	19
I_p, plastic index	%	5
Clay content	%	37
Activity index	–	0.14
Salt concentration in pore water	g/l	0.9
Content of organic matter	%	0.9
G_s, specific gravity	–	2.75
γ, bulk unit weight	kN/m^3	19.0
s_u, average undrained shear	kPa	8.83
Strength	–	0.14
s_u/p, shear strength ratio	–	~100
S_t, sensitivity	–	0.40
C_o, compression index	m^2/s	
c_v, coefficient of consolidation	cm/s	0.5×10^{-7}
k, hydraulic conductivity		2×10^{-8}

Source: Modified from Bjerrum et al. (1967).

stability of the slope was improved by 10% by unloading the upper soil in the vicinity of the railway during the treatment.

The electrode installation was finished on April 13, 1964. Then the power was applied on April 14. The settlement rate was approximately 8 mm per day. The flow rate of water on days 4 and 7 was 1.7 and 1.5 m^3 per day, respectively. The flow rate decreased with time. On day 51, because of short circuit between rows 1 and 2, row 1 was disconnected. The polarity was reversed to investigate the effect of polarity reversal between days 51 and 58. The settlement rate decreased from 4.5 mm per day just before the short circuit to 2.4 mm per day. Due to a decrease in the settlement rate, the polarity was reversed back to the original arrangement at day 58. During the first 18 days, the current was set to be 210 A. After day 18, the current was increased to 350 A. Attempts were made to maintain this level of current, but the electric resistance of soil surrounding anodes increased. On day 8, the potential drop at the anode was 10 V from the applied voltage of 40 V. This indicated that only 75% of the applied voltage was effective. On day 80, only 50% of the applied voltage was effective. The treatment lasted for 120 days. The measured settlements are shown in

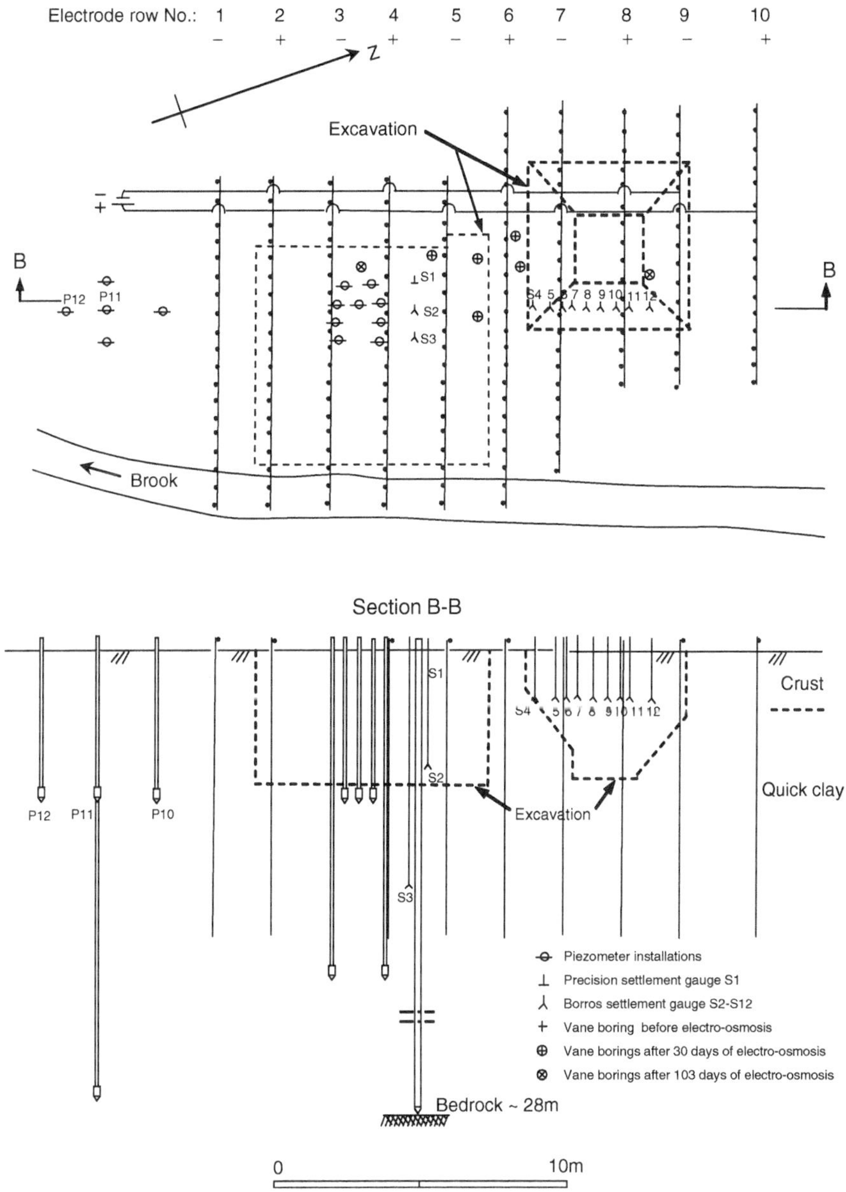

Figure 14.2 Plan and cross section of the electrode installation. *(Source: Modified from Bjerrum et al. (1967)).*

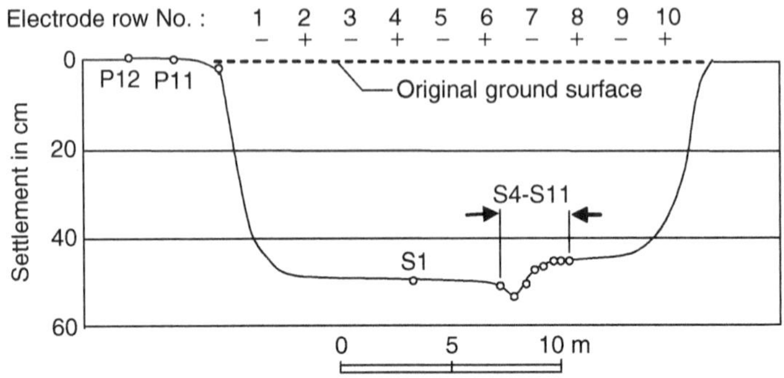

Figure 14.3 Final settlement profile. *(Source: Modified from Bjerrum et al. (1967)).*

Fig. 14.3. No settlement occurred outside the treated area. A long crack was found along the western boundary. The change in water content in soft clay was back-calculated from the consolidation of soft clay.

The total energy consumption was 30,000 kWh or 17 kWh/m^3 of clay. After treatment, cathodes were removed easily, whereas developed adhesion prevented removing the anodes. The total corrosion of steel bar anodes was 740 kg from the original weight of 1990 kg; in other words, 37% of steel anodes were corroded, or 2.25 kg/m^3 of iron was used for stabilizing the clay. As the water moves from the anodes to the cathodes, the negative pore pressure occurs in the vicinity of the anodes and excess (positive) pore pressure occurs in the vicinity of the cathode. With the hydraulic conductivity of 2×10^{-8} cm/s, and applied voltage of 40 V, theoretically a difference of pore pressure would be 32 atm. However, in reality, the pressure at the cathode would not exceed the overburden pressure. In addition, the suction pressure at the anode would not exceed 1 atm because of gas generated by the electro-osmotic process.

It was noted that when the polarity was reversed on day 58, the negative pore pressure suddenly changed to positive pressures, and vice versa. The pore pressure at the cathode and the anode remained positive and negative, respectively, until the power was shut down. After 103 days of treatment, the vane shear strength increased from approximately 8 to 15–20 kPa down to a depth of 6–8 m. No significant change in shear strength was found below this depth. During excavation, many soil samples at a depth of 4.5 m were taken. The results of measurements are illustrated in Fig. 14.4. The variation of undrained shear strength is shown in Fig. 14.4(a). The shear strength before treatment was 8.8 kPa. After the treatment, the shear strength increased to 108 kPa at the anode. The shear strength was

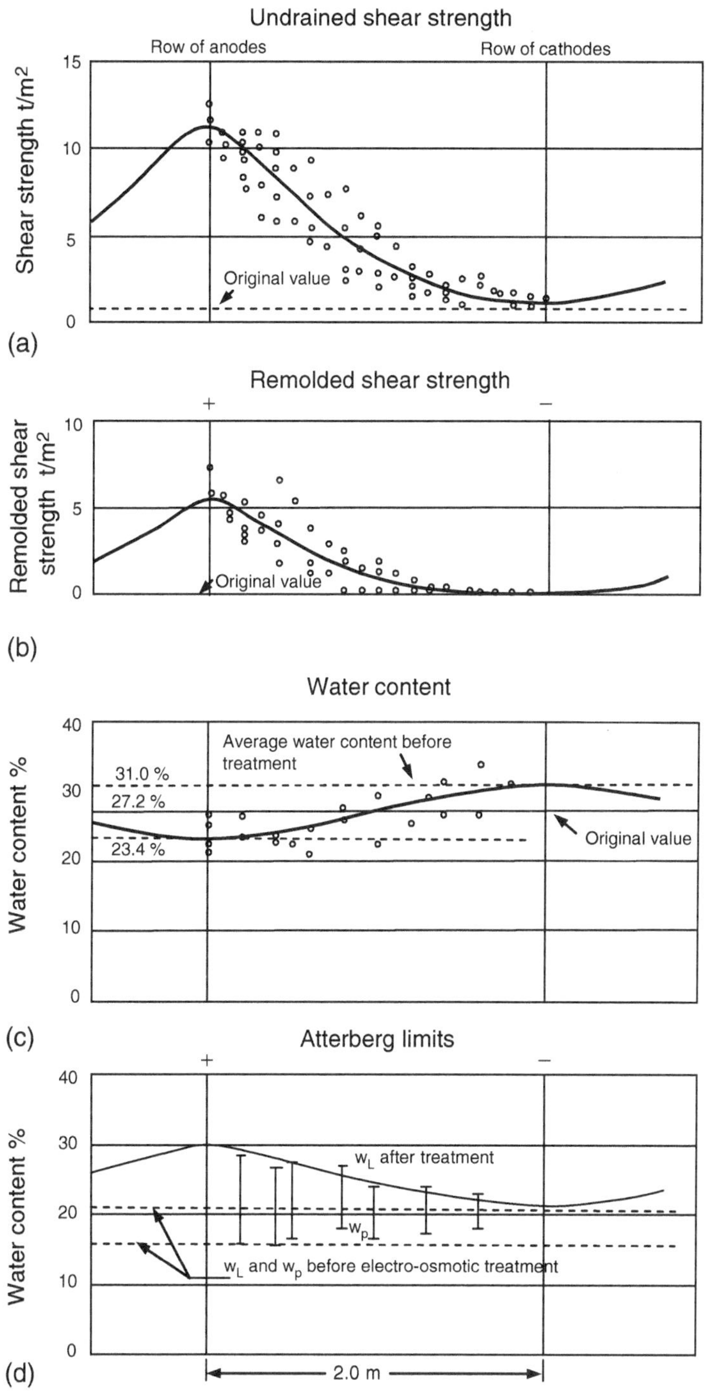

Figure 14.4 Results of vane boring and sampling at the bottom of the excavation. *(Source: Modified from Bjerrum et al. (1967)).*

39 kPa halfway between the electrodes. There was no increase in shear strength at the cathode. The average shear strength was 37 kPa. The remolded shear strength is shown in Fig. 14.4(b). The sensitivity reduced from exceeding 100 to a value of approximately 2–4. The average initial water content was 31.0%. After the stabilization, the water content was reduced 7% at the anode. The average reduction of water content was 3.8%, approximately the same as the designed value of 3.4%. The liquid limits and plastic limits are shown in Fig. 14.4(c). At the anode, the liquid limit increased up to 30% from the original value of 21%. The plastic limit was insignificantly changed. An increase in the liquid limit was attributed to acidity at the anode (i.e., pH=5.8–6.3). Where pH of clay at the cathode was 7.7, the increase in the liquid limit was negligible in this area. It was believed that the acid conditions led to the formation of proton-saturated clay, which is unstable and changes to an aluminum-saturated clay. The original untreated clay was considered a sodium–calcium clay of illitic–chloritic character, and its conversion to an aluminum clay would regularly lead to a higher plasticity. In addition, iron ions dissolved from the anode led to the formation of iron-saturated clay yielding the same effects on plasticity. The increase in the liquid limit was an important factor in the increase in shear strength. That is, an increase in the liquid limit reduces the liquidity index. The soil becomes stiffer and has higher shear strength. Further study indicated that the preconsolidation pressure increased. The original preconsolidation at a depth of 4.5 m was 78 kPa, whereas the preconsolidation pressure halfway between the electrodes was increased to 343 kPa.

14.2.2 Electro-osmotic stabilization of West Branch Dam (Fetzer, 1967)

West Branch Dam is a compacted earth dam located on Mahoning River in northeastern Ohio. The dam is 3000 m long and 24 m high. The concrete outlet conduit is located at the middle of the dam. The conduit is 3.5 × 7.0 m. Each joint was connected by a concrete collar. The project site is covered with glacial deposits with 25-m-thick gray clay. The clay layer is interspersed by gray silt and silty sand with the bedrock of sandstone at a depth beyond 32 m. Table 14.3 summarizes the soil properties.

The construction of the dam commenced in May 1963. The fill at the central part was rapidly constructed. While the fill was raised from El 999 ft (El 305 m) to El 1007 ft (El 307 m), the separation of the conduit joint was detected at the dam axis due to settlement. The fill placement was ceased. Cracks occurred in the central body of the dam with a maximum

Table 14.3 Geotechnical properties of soft silty clay

Parameter	Unit	Typical range
w, natural water content	%	21–40
w_L, liquid limit	%	32–40
I_p, plastic index	%	9–28
s_u, undrained shear strength	kPa	21–54
γ_d, dry unit weight	kN/m^3	14.5–15.1

Source: Modified from Fetzer (1967).

opening of approximately 2.5 cm. A severe dip of the conduit, the maximum opening in the joint was 22 cm. Three joints near the axis were broken. On November 14, the fill of 4 m was removed to stop the settlement and separation of the joints.

Soil strength and pore water pressure were investigated. Piezometers were installed in the clay layer underlying the dam and in silty sand underlying the clay layer. It was found that 1 year after the construction stopped, the excess pore pressure underneath the dam was still not completely dissipated because of the low-hydraulic conductivity of the clay layer. The undrained shear strength of the soil was as low as 0.2 tsf (19 kPa), and the factor of safety of the slope was merely 1.0. It would have been impossible to raise the dam to the design height because the safety factor would have been 0.62. Remediation measures, such as berms and sand drains, were considered. However, preliminary analyses indicated that the berms would be excessively wide and block the entrance and outlet channels. The 30-cm-diameter sand drains with spacing 1.5 m were considered as well. However, installing sand drains would cause temporary instability of the foundation. Eventually, electro-osmotic stabilization was selected as the best approach to stabilize the dam.

The feasibility of electro-osmotic stabilization was investigated by L. Casagrande of Harvard University. The electro-osmotic conductivity, k_e, ranged from 3.0×10^{-5} to 6.2×10^{-5} cm/s computed on the basis of 1 V/cm. Casagrande concluded that the dam foundation could be stabilized by electro-osmotic process. The electrodes layout consisted of three strips of anodes and cathodes over a 230 × 300 m area. An eight-row strip at 6-m spacing was installed along the dam axis, and a six-row strip at 6-m spacing was installed at the outside edge of each berm. A total of 660 anodes and 320 cathodes were installed. The anodes were made of 6.5-cm-diameter double-extra-strength black steel pipes with a plug in the tip. Steel railroad rails were used in the vicinity of the conduit. Each cathode consisted of an electrode and a pumping

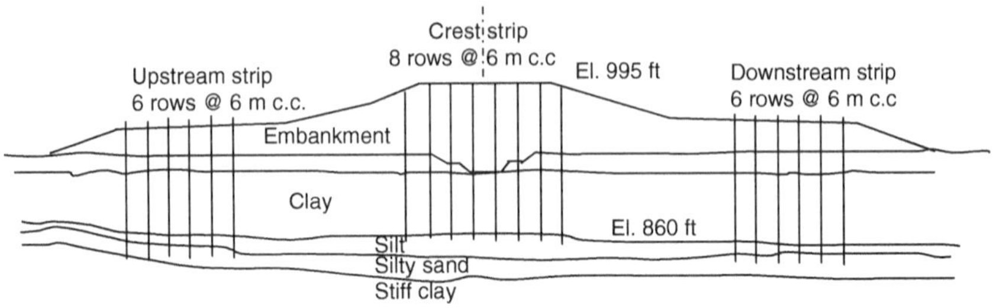

Figure 14.5 Section through electrodes. *(Source: Modified from Fetzer (1967)).*

system placed in a 35.5-cm diameter. The cathode electrodes consisted of a 5-cm-diameter steel pipe. The schematic section of the installation is shown in Fig. 14.5. The anodes and cathodes were installed through 40-cm-diameter drilled holes. To reduce the artesian pressure in the lower sand layer, the cathodes on the berms were installed into the sand layer. Before the dam was impounded, the upstream cathodes were sealed. The downstream cathodes were left opened to relieve artesian pressure in the lower silty sand.

Initially, the power generators were installed on the upstream berm. Unfortunately, the vibrations from the generators significantly increased the excess pore pressure underneath the dam. Therefore, they were moved downstream. Two 300-kW, ten 200-kW, and two 90-kW generators were used in the operation with capacities of 11,000 A at 150 V, 2500 A at 140 V, and 950 A at 60 V, respectively. Because the power was insufficient to supply 150 V to all electrode strips as recommended by Casagrande, the treatment was implemented in several steps. On August 10, 1965, power was applied to the crest strip near the outlet conduit to relieve pore pressure in this area. On September 18, continuous power with 50–70 V was applied to the downstream strip. After the pore pressure underneath the downstream berm decreased, the power was moved to the crest and upstream strips. The treatment at the crest strip was completed in May 1966. The embankment was raised to the designed crest (El 1012 ft or El 308 m) in June–August 1966.

The changes in piezometric levels at El 890 ft (El 271 m) are shown in Fig. 14.6. Significant reduction in piezometric levels took place during the electro-osmotic process. On the upstream berm, the piezometric levels within the strip reduced as much as 6.0 m in a month when 100 V was applied. On the crest strip, the piezometric level dropped with an average of 2.0 m per month. The voltage in the vicinity of piezometer 3-DC ranged

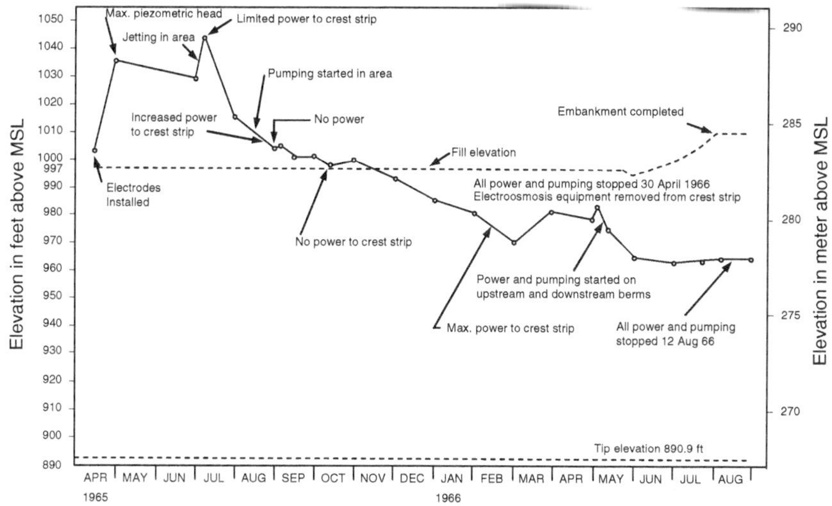

Figure 14.6 Piezometric levels. *(Source: Modified from Fetzer (1967)).*

from 40 to 56 V. On February 13, 1966, the voltage was increased to 150 V, inducing the rise in the piezometric level in the vicinity of the cathodes in March 1966. It was found that the pore pressures were very low and very high in the vicinity of the anodes and cathodes, respectively. From April 30 to May 9, 1966, all power and pumping were stopped; the recovery of piezometric levels was relatively low. During the period from June 1 to August 5, 1966, the embankment was raised from El 995 ft (El 303) to El 1012 ft (El 308). The pore pressure increased slightly.

After removing the 4-m-thick fill in November 1964, the average rate of settlement was 20 mm per month. During the electro-osmotic stabilization, the average settlement rate was 43 mm per month, and then it decreased to 18 mm per month. This indicated that the treatment accelerated the consolidation of foundation. After the construction was completed, the stability analysis was performed. A factor of safety 1.16 was achieved from the total stress analysis. For effective stress analysis, assuming the friction angle of 18°, the factor of safety was computed as 1.56.

14.2.3 Slope stabilization by electro-osmosis (Wade, 1976)

The Kootenay Canal hydroelectric project is situated on the Kootenay River in southern British Columbia, Canada. The project consists of a gated canal intake structure near Corra Linn Dam, a 4.8-km partially lined canal, a forebay, and four steel surface penstocks heading to the powerhouse with four 125-MW units. The net head at full operation is approximately 75 m. The deposit thickness varied from 16 m in the forebay area to 33 m in the penstock area. This deposit consisted of silt and clayey silt with interbeds of fine sand with a thickness up to 2 m. Generally, the water table was at around mid-depth of the soil layer, with perched water in some locations. The most problematic soil was a layer of loose, sensitive silt just above the bedrock throughout the forebay–penstock area. This silt was flat bedded, with the thickness varying from 2 to 12 m. Properties of the sensitive silt are summarized in Table 14.4. In unconsolidated undrained triaxial tests, some samples were so soft that they slumped by themselves. The slope stability analysis of the penstock area indicated that the slope had to be excavated to 3.5:1 (horizontal:vertical) to attain a unit factor of safety. Therefore, it was decided to apply electro-osmosis to increase the shear strength and to reduce the soil water content.

The sequence of electrode installation is shown in Fig. 14.7. The layout of the treatment consisted of five double rows of electrodes installed along the penstock slope and two double rows on the forebay slope. Each double row consisted of a single row of anodes and a single row of cathodes at 3-m spacing. The anode rows were installed closer to the surface slope in order to induce seepage force toward the slope. Five-centimeter-diameter steel pipe was used for the electrodes. A 30.5-cm casing was installed at the cathode with water jets. An eductor pipe (pumping pipe) and a perforated pipe were placed in the casing. The hole was filled with sand. The groundwater was pumped out from the cathode. The electro-osmotic treatment was carried out in steps prior to excavation. The electrodes were installed at the first

Table 14.4 Geotechnical properties of sensitive silt

Parameter	Unit	Typical range
w, natural moisture content	%	30–40
w_L, liquid limit	%	26–35
w_p, plastic limit	%	NP–28
Clay content	%	0–27
γ_d, dry unit weight	kN/m^3	12.8–15.2
Coefficient of permeability	cm/s	10.6–10.7
Compression index	—	0.16–0.33
ϕ', effective angle of internal friction	degrees	27–32
c', effective cohesion	kPa	0
s_u/p_c, shear strength ratio	—	0.29
Calcium carbonate content	%	10–13

Source: Modified from Wade (1976).

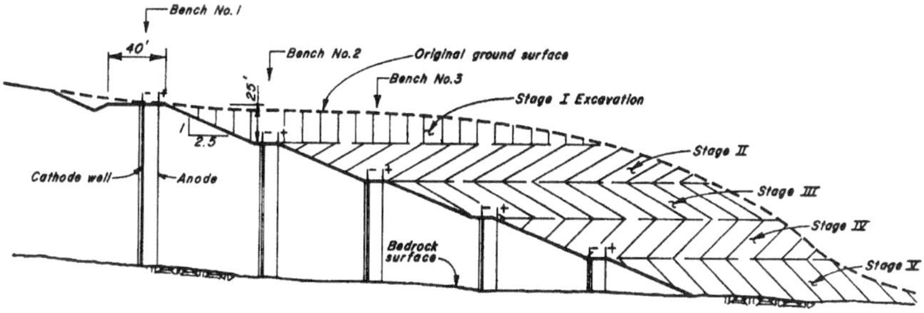

Figure 14.7 Sequence of excavation. *(Source: Wade (1976)).*

bench, and the power was applied for 7 days. Then the soil was excavated to the level of the second bench to install the electrodes and pumping wells. The power was applied for another 7 days before the excavation proceeded to the next level. The sequence was continued to the bottom of the slope.

The treatment was carried out over a 9-month period, from September 1972 to the end of May 1973. The volume of silt treated was approximately 0.4 million m^3 with the power consumption of 2.67×10^6 kWh. Significant drops in piezometric level were observed in all piezometers within 3 days after the voltage was applied. After 2 weeks of the treatment, the water table was within 3 m above the bedrock, and it has remained at that level ever since. The discharge rate of extracted water was 270 l/min in the first few weeks of treatment to approximately 45 l/min before the power was shut down. The average discharge was 70 l/min with the total volume of water of 27,000 m^3. No significant slope movement was registered by markers installed on the ground surface. To assess the efficiency of the electro-osmotic treatment, borehole samples were taken and two deep trenches were excavated. The trenches did not deform during 3 or 4 weeks after the excavation. Blasting rock in the powerhouse area caused vibration in silt. During the blasting operation, the ground motion was measured using a portable seismograph. The highest particle velocity registered was in the range of 5 cm/s, with the maximum value approximately 9 cm/s and a peak acceleration of 2.8 *g*. No movement of the trench walls was observed during this period, indicating that the electro-osmotic treatment played an important role in improving the soil shear strength.

Because of operational and equipment problems, the applied voltage varied from approximately 95 to 175 V during the treatment. The average voltage was 120 V over 9 months. The voltage variations are illustrated in Fig. 14.8. A significant voltage drop was found at the electrodes, caused by the high electrical resistance due to excessive drying and base exchange

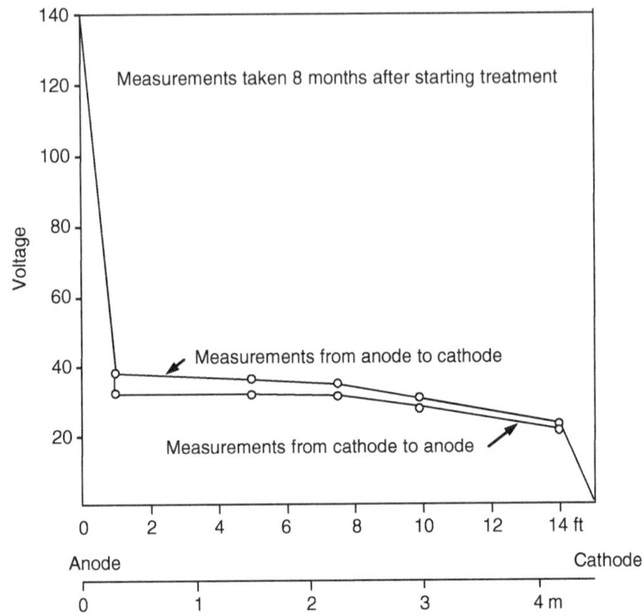

Figure 14.8 Voltage variation between an anode and a cathode. *(Source: Wade (1976)).*

effects. After the slope was cut and covered by rock fill, seepage of groundwater was measured at the toe of the slope. The maximum flow was approximately 23 l/min occurring during spring. Consolidated-undrained (CU) triaxial tests were carried out on specimens taken from the test trenches. The results showed that the residual friction angle of the silt increased from 27–32° to 35° after treatment.

14.2.4 Electro-osmosis applied to an unstable embankment (Chappell and Burton, 1975)

A dock with a capacity to handle a 400,000-ton deadweight vessel was constructed in Singapore. The dock is located partly offshore behind a large cellular cofferdam. A 60-m-long and 8-m-high embankment was constructed on the shore area, which was filled by pouring silt dredged from the seabed into the water. Instability developed on the embankment during dewatering of the cofferdam, including large longitudinal cracks that appeared on the slope. Soil properties of subsoil are shown in Table 14.5. The instability was mainly caused by the silt slurry with low permeability

Table 14.5 Geotechnical properties of soft soil

Material	Undrained cohesion, C_u (kPa)	Undrained friction angle, Φ_u (deg)	Water content, w (%)	Saturated unit weight (kN/m^3)
Fill				
Before treatment	15.8	19	40	17.9
After treatment	48.0	14	36	18.4
Soft blue clay				
Before treatment	4.8	—	72	16.5
After treatment	13.8	14	35	17.0
Gray/white clay				
Before treatment	28.7	3	39	18.6
After treatment	38.3	7	20	19.5

Source: Modified from Chappell and Burton (1975).

(10^{-4}–10^{-5} cm/s). To stabilize the embankment, sheet piles were installed through the crest to reduce seepage and through the toe to reduce the movement of the embankment, as shown in Fig. 14.9. However, both approaches could not stop the movement. Therefore, it was decided to implement electro-osmotic stabilization.

The power consumption was a main issue in the feasibility analysis of the electro-osmotic treatment. It was assumed that the power consumption for the electro-osmotic treatment was relatively high. However, one should recognize that the power consumption is related to the soil electrical conductivity. In Southeast Asia, the salt content in silts and clays is typically in the range

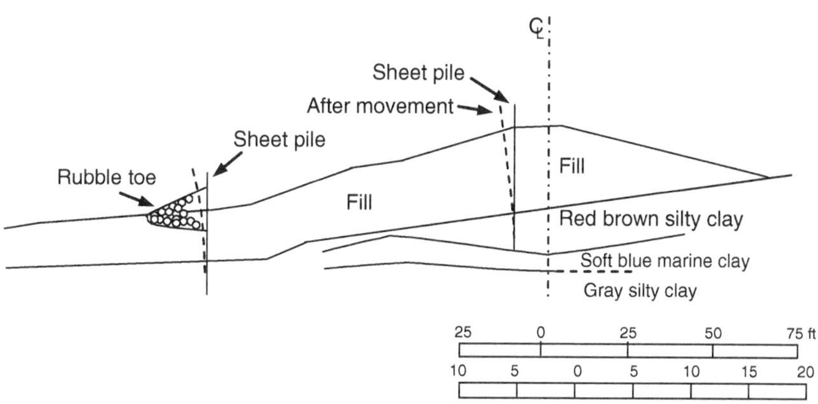

Figure 14.9 Soil profile on Section A-A. *(Source: Chappell and Burton (1975)).*

of 3–5%, compared to no more than 1% as reported in some cases in Europe and North America. Therefore, the power consumption for soil in this region could be much lower due to higher electrical conductivity of soil.

Before treating the embankment, a trial field test was performed for the purpose of preliminary study. It was estimated that a minimum dewatering rate of 140 l/day was sufficient to dry out the silt and remove water seeping into the soil. Electrodes were spaced 3 m apart. The depth of stabilization was 5 m (16.5 ft) to include the whole embankment and part of the foundation. The applied voltage was 90 V. The electro-osmotic permeability, k_e, was measured as 0.5×10^{-4} cm/sV. An approximate flow of 180 l/day was obtained. In the field test, 10-cm-diameter holes were drilled to a depth of 5 m for the 2.5-cm-diameter anodes and cathodes made of reinforcing bars. The spacing between the electrodes was 3 m. A portable welding generator was used as a power supply to supply a voltage of 40 V and a current of 25–30 A. The treatment was operated for 24 h, and water extraction rate was 550 l/day. Because the flow rate was approximately four times higher than the design value, the distance between the anode and the cathode was increased.

After the efficiency of the electro-osmotic treatment was proved by the trial field test, the electrode layout adopted a pattern of four to six anodes in a semicircle surrounding a central cathode at a 12-m radius. The layout of the electrode arrangement is shown in Fig. 14.10. Four electrode groups were installed. Because of the limitation of the power supply, only two groups could be operated simultaneously. These pairs were alternately operated at 1-day intervals in the early stages. When the embankment movement was slower, the interval was changed to 3 days. It was found that hydrogen gas emitted at the cathode brought the extracted water to the surface, even though there was no sand drain or plastic pipe. After 9 days of the electro-osmotic treatment, the movement of the embankment reduced from up to 1 m/day to less than 1 cm/day. As tabulated in Table 14.5, the shear strength of treated soil more than doubled, and the water content significantly decreased. It was found that the soil surrounding the anode was irreversibly hardened due to electrochemical process. The power consumption was only 0.5 kWh/m^3 of soil, which was relatively low.

14.2.5 First application of electro-osmosis to improve friction pile capacity (Soderman and Milligan, 1961; Milligan, 1995)

Big Pic River Bridge is on the Trans-Canada Highway going around the north shore of Lake Superior. This bridge consists of three-span steel truss,

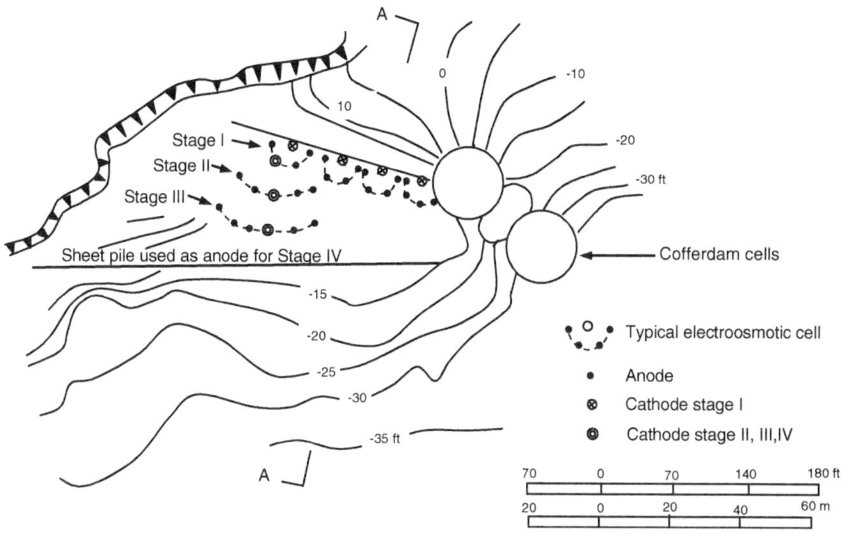

Figure 14.10 Layout of electrode arrangement. *(Source: Chappell and Burton (1975)).*

more than 180 m in length. Valleys in this area are overburdened by very thick silts and clays. The load of the bridge is supported by friction pile foundations. With artesian pressure in the soil, the shaft friction became insufficient to achieve the design load. Soil properties of varved clay are summarized in Table 14.6. The top layer was compact fluvial silty sand overlying 18-m-thick medium to stiff varved silty clay. The varved clay consisted

Table 14.6 Geotechnical properties of varved clay

Parameter	Unit	Typical range
Thickness of bands		
Light band (clayey silt)	mm	12
Dark band (silty clay)	mm	25
w, natural moisture content		
Light band	%	22–25
Dark band	%	40–60
w_L, liquid limit		
Light band	%	25–30
Dark band	%	40–60
w_p, plastic limit		
Light band	%	20–25
Dark band	%	20–25
S_u, undrained shear strength	kN/m^2	25–50

of 2-cm-thick dark gray, brittle clay laminae and 1-cm-thick light gray, clayey silt laminae. The standard penetration resistance (*N* value) decreased with depth from 20 to 50 m. Artesian pressure was observed at a depth of 50 m. At a depth of 80 m, the pressure head rose up to 6 m above ground surface, and the *N* value became zero.

A friction pile foundation was selected because the thickness of the low-strength and highly compressible deposit was excessive. Before the embedded length was decided, test piles with different lengths were tested. Steel H-section piles, 300 × 300 mm, were driven to depths from 16.5 to 50.5 m, as shown in Fig. 14.11. With the artesian pressure, the ultimate load decreased with increasing pile length. The planned design load was 350 kN per pile. The maximum ultimate load obtained from the 16.5-m-long pile was 360 kN, yielding the allowable load of only a half of the design load. The piles were tested up to 400 days after driving. No significant increase in the ultimate capacity was observed despite the fact that 90% of excess pore pressure induced by driving the piles dissipated within 3 days.

To achieve the design load, electro-osmotic stabilization was applied. A trial test was conducted using two 16.5-m-long test piles. The piles were treated as anodes. The power was supplied by using an electric welding machine. With a voltage of 115 V, the pile with an original ultimate load

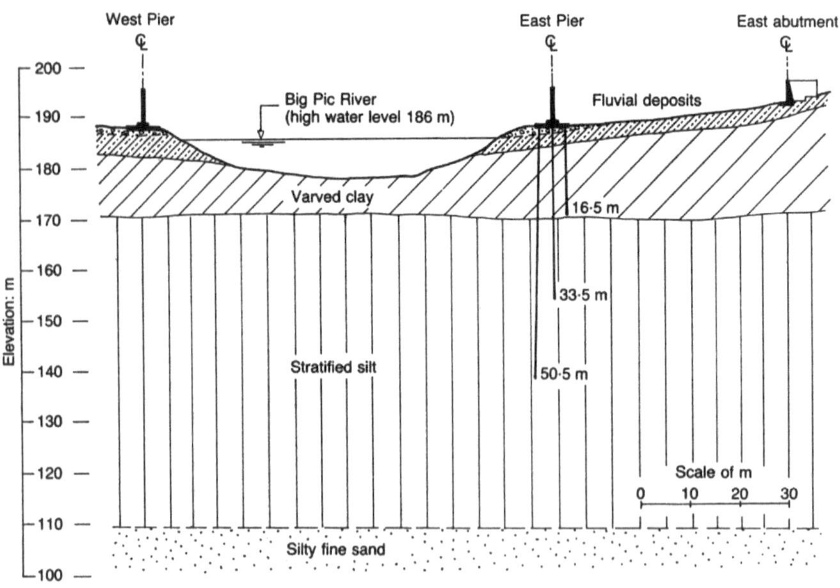

Figure 14.11 Soil profile at Big Pic River. *(Source: Milligan (1995)).*

of 260 kN had increased up to 500 kN after 3 h. Based on these results, the pile group consisting of 16.5-m-long pile with electro-osmotic treatment was selected. The design load was changed to 135 kN per pile. The electrode arrangement is shown in Fig. 14.12. The H-piles were used as anodes. The average distance between the anodes and the cathodes was 7 m. The perimeter cathodes were 21 m long. The arrangement at the West Pier was similar to that at the East Pier. Three diesel generators with an output of 70–120 V and 600–1000 A per unit were used. For the cathode, a steel pipe was put into a plastic pipe to prevent clogging by calcium carbonate. The water was discharged through the plastic pipe without any pumping. At first, it was intended to treat the pile group as a whole, but the power was insufficient. Therefore, the exterior piles were disconnected. The treatment period was 1060 h or 44 days.

The test results of Pile E-16 in the East Pier are shown in Fig. 14.13. The ultimate load increased from 300 kN prior to treatment to more than 600 kN after 34 days of treatment. Bored soil samples adjacent to the piles. No obvious trend in the increase of shear strength and decrease in water content on bored soil samples adjacent to the piles. Piezometers were installed adjacent to the piles and the cathodes. The tips of piezometers were located

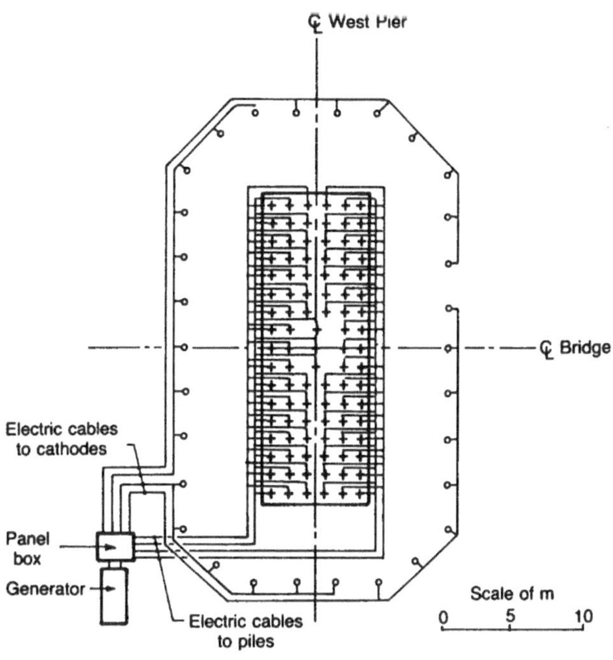

Figure 14.12 Electrode layout. (Source: *Milligan (1995)*).

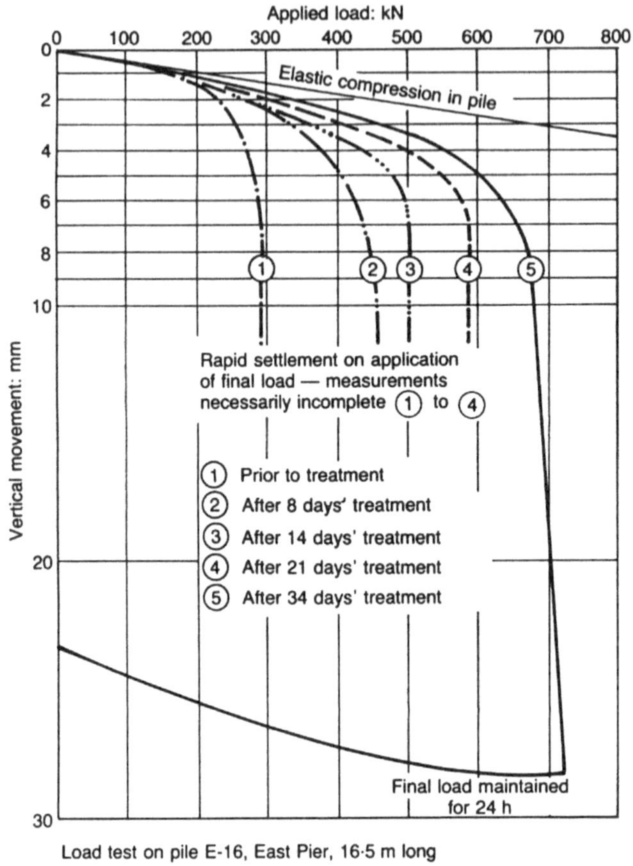

Figure 14.13 Pile load tests during electro-osmotic treatment (1959). *(Source: Milligan (1995)).*

at a depth of 50 m, where the artesian pressure was detected. The piezometric level at the piezometer adjacent to the piles decreased to 9 m below the normal water level. However, this piezometric level was completely recovered 90 days after terminating treatment. The piezometric level at the cathode was 2 m above the normal water level. Similarly, the piezometric level was recovered 100 days after stopping treatment. The settlement was measured at the pile cap. During treatment, the settlement was 40 mm. The long-term performance of the piles was monitored over a period of three decades (i.e., from 1960 to 1992). Pile load tests were carried out on adjacent Pile G-5. The test results are illustrated in Fig. 14.14. No reduction in ultimate capacity over 31 years was detected, and the settlement was within acceptable limits.

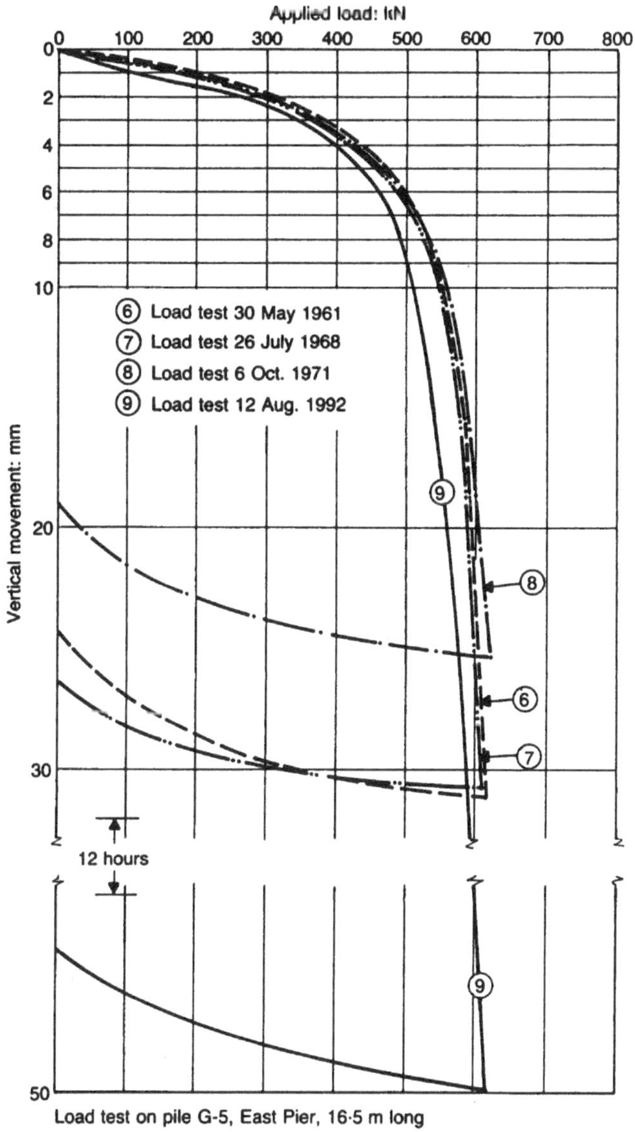

Figure 14.14 Pile load tests after the electro-osmotic treatment (1961–1992). *(Source: Milligan (1995)).*

14.2.6 Field test of electro-osmotic strengthening of soft sensitive clay (Lo and Ho, 1991; Lo et al., 1991)

A field test was carried out in July and August 1989 to assess the effectiveness of the electro-osmotic stabilization on a soft sensitive clay in Ottawa Valley, Canada. The Champlain sea clay in this region is well-known for its high sensitivity (Quigley, 1980). Many landslides have taken place in this area. The site was located 21 km southeast of Ottawa, at the Gloucester test fill site at Canada Forces Station in which a test embankment was constructed in 1967. Soil properties are tabulated in Table 14.7. The in situ vane shear tests and moisture content tests were performed before treatment. The average moisture content was 80%. The liquidity index was obviously over unity. The remolded shear strength was virtually zero, pronouncing very high sensitivity. The soft soil was treated from depths of 1.5 to 5.5 m, where the vane shear strength was lower than 20 kPa.

The detail of electrodes is shown in Fig. 14.15. A 60.3-mm-diameter copper pipe was used as the electrode. The electrode was designed so that water could flow through the electrode itself. Therefore, the pipe was perforated with cone-shaped steel shoe to ease penetration. The top 1.22-m crust was considered relatively conductive. To prevent short circuiting, the electrode was insulated at the top part, and to avoid short circuiting due to rainfall and flood, the top 0.3 m was also insulated. No hole was drilled on the insulated portion. The electrode was pushed into the soil by a drilling rig. During installation of the electrodes, remolded clay became slurry and filled the electrodes. Electrodes were cleaned by flushing water.

Table 14.7 Geotechnical properties of soft clay

Parameter	**Unit**	**Typical range**
w, water content (%)	%	65–90
w_L, liquid limit (%)	%	43–47
w_p, plastic limit (%)	%	21–25
S_t, sensitivity	—	
Above 3.5-m depth		20
Below 3.5-m depth		~100
S_u, field vane strength	kPa	17.6–19.7
Preconsolidation pressure	kPa	
At 2.5-m depth		48.5
At 4.5-m depth		53
Cation concentration in pore water	mE/100 g soil	
Sodium		4.87
Potassium		2.46
Calcium		0.50
Magnesium		0.38
Cation-exchange capacity	mE/100 g soil	18–20
Salinity	g/l	1.0–1.5
pH		8.4–8.8

Source: Modified from Lo and Ho (1991).

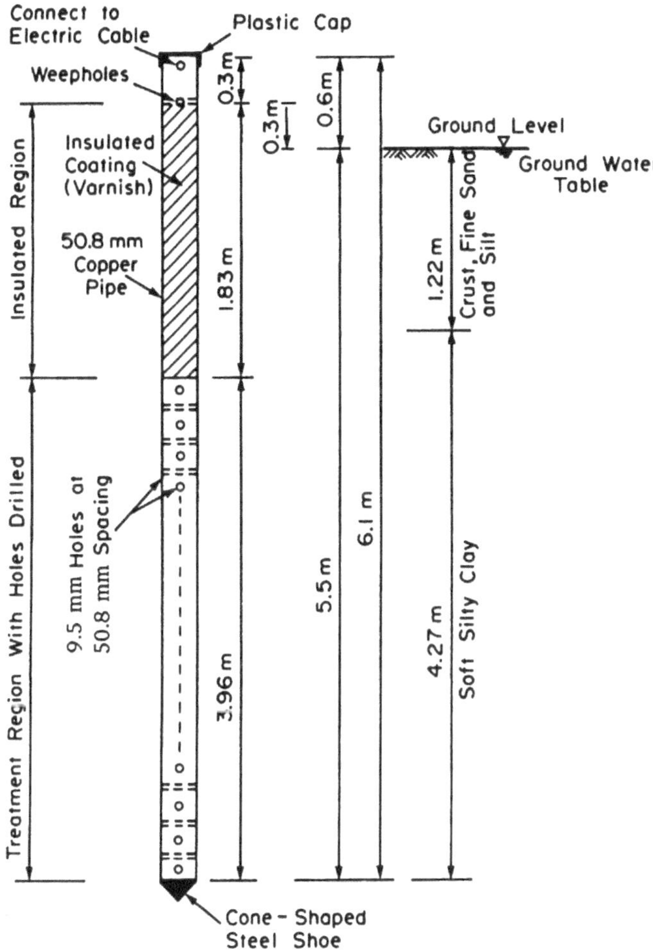

Figure 14.15 Electrode for electro-osmotic treatment. *(Source: Lo et al. (1991)).*

Instruments were installed to measure settlement, vane shear strength, and voltage distribution during treatment.

The layout of the electrode installation is shown in Fig. 14.16. Nine electrodes were installed. Initially, row A and row C were anodic (positive), whereas row B was cathodic (negative). The polarity was reversed midway during the treatment period. The installation was completed in 2 weeks. The treatment was started on July 24, 1989, and completed on August 25, 1989, lasting 32 days. The initial polarity lasted for 17 days, and then the reversed polarity was maintained for 15 days. Initially, a voltage of 25 V was applied. Due to an increase in electric resistance of soil, the current subsequently

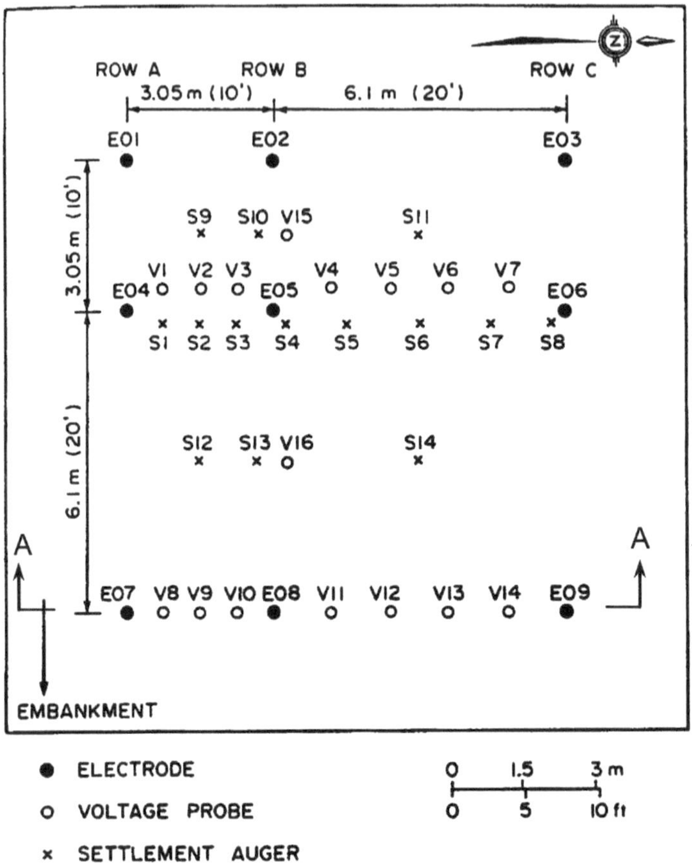

Figure 14.16 Electrode layout and instrumentation. *(Source: Lo et al. (1991)).*

decreased. The voltage was adjusted periodically to maintain the current of 40 A. Approximately 50 min after treatment, water started to flow from the cathodes. Hydrogen gas bubbles were also observed. Before treatment, the anode was filled with water, but no water was found during the treatment.

The records of ground settlement are shown in Fig. 14.17. The maximum settlement on day 18 was 62 mm. At the cathode, a net heave before polarity reversal was 18 mm recorded at settlement gauge S4. The heave reduced to settlement after polarity reversal, which made the settlement relatively uniform. Because the polarity was reversed on day 18, no significant settlement appeared. At the end of treatment, the settlement was 38–68 mm, with an average settlement of 51 mm. The variations of shear strength at different locations are shown in Figs. 14.18 and 14.19. The shear strength

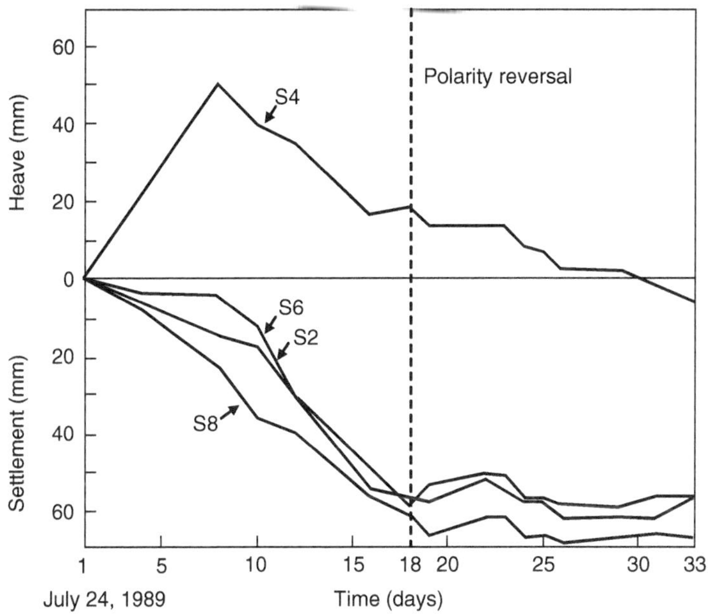

Figure 14.17 Measured settlement with time. *(Source: Lo et al. (1991)).*

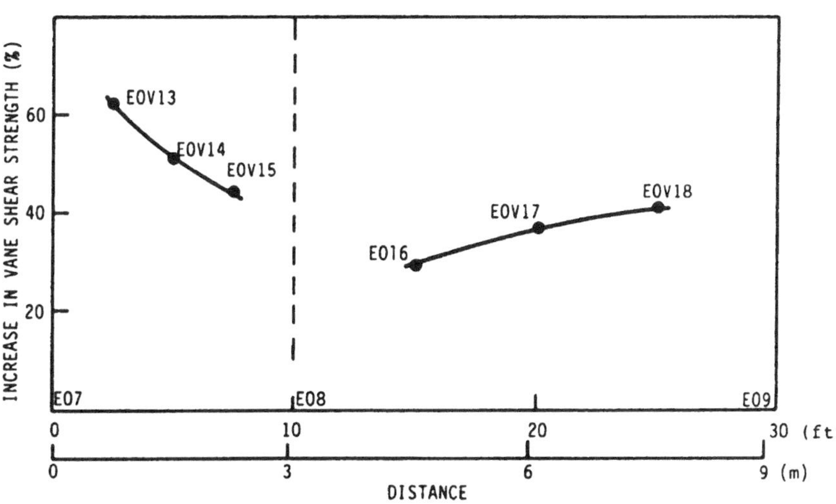

Figure 14.18 Variation of average shear strength between electrodes along Section A-A. *(Source: Lo et al. (1991)).*

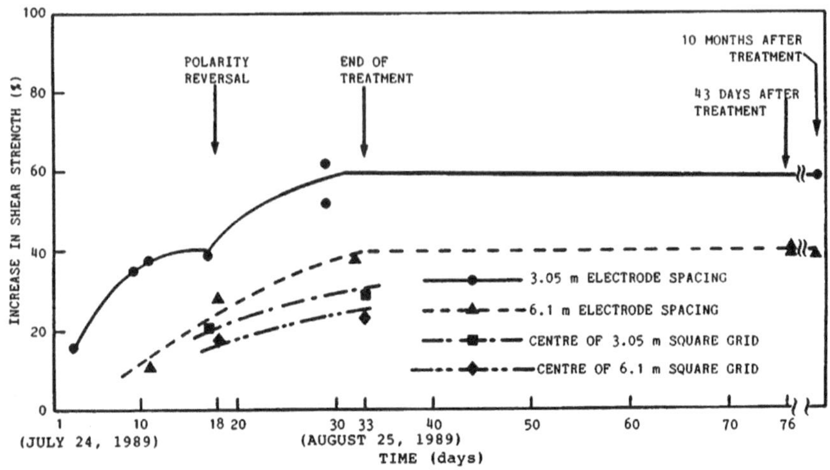

Figure 14.19 Percentage increase in average shear strength. *(Source: Lo et al. (1991)).*

increased significantly within 32 days. Halfway between a pair of electrodes of 3.05-m spacing, the average shear strength was 50%, and it was 36% halfway between electrodes of 6.1-m spacing. Although previous applications reported that the center of the square grid was inactive for electro-osmotic treatment, this field test demonstrated that with the improved electrode design, the shear strength in the inactive zone can be efficiently improved. It was found that at the centers of 3.05-m and 6.1-m square grids, the average shear strength increase of the soil after treatment was 24% and 23%, respectively. A rapid increase in shear strength occurred in the 3.05-m electrode spacing, with a 16% increase on day 3. Bjerrum et al. (1967) reported that the treatment of a soil mass was nonuniform and that there was no shear strength increase in the vicinity of the cathode. However, as shown in Fig. 14.18, the polarity reversal improved uniformity of shear strength increase. When the polarity was reversed, shear strength increased again, as shown in Fig. 14.19.

To investigate long-term effects, vane shear tests were performed 43 days and 10 months after treatment. No reduction in shear strength was observed. The shear strength improvement tended to be permanent. The shear strength profiles in this field test were different from those reported by Bjerrum et al. (1967), who embedded the electrode down to 9.6 m, but the shear strength of soil below a depth of 6 m was not improved. The gas at the lower part could not be released to the atmosphere. The accumulation of gas diminished the efficiency of the treatment. However, with perforated electrodes, this field test resulted in an increase in shear strength throughout the depths of electrodes.

According to the record of voltage variation, the voltage drop in the vicinity of the anode was high, whereas the voltage gradually dropped at halfway between electrodes. This indicates that the power consumption in the vicinity of an anode is relatively high. The total power consumption was 2136 kWh, approximately 1% of the total cost of the treatment project. It is worth noting that no pumping of extracted water is required. This is advantageous in terms of energy saving. Considering the increase in shear strength and the energy consumption, this project was economical and efficient.

To investigate the change in geotechnical properties, two boreholes were drilled and 127-mm-diameter Osterberg tube samples were recovered after treatment down to a depth of 6 m. The boreholes (EOS1 and EOS2) were halfway between electrodes of 3.05- and 6.1-m spacing. The treatment was stopped on August 25, 1989. The samples were recovered in mid-October or approximately 1½ months after stopping treatment. The decrease in water content is quite uniform with depth. The water content decreases approximately 10%. The change in sensitivity was found between depths of 3.5 and 5 m. It decreased from 100 to approximately 60. The change in shear behavior of clay is illustrated in Fig. 14.20. CU triaxial tests with pore pressure measurement were performed on soil samples between 3 and 4 depths. The effective stress failure envelopes significantly expand due

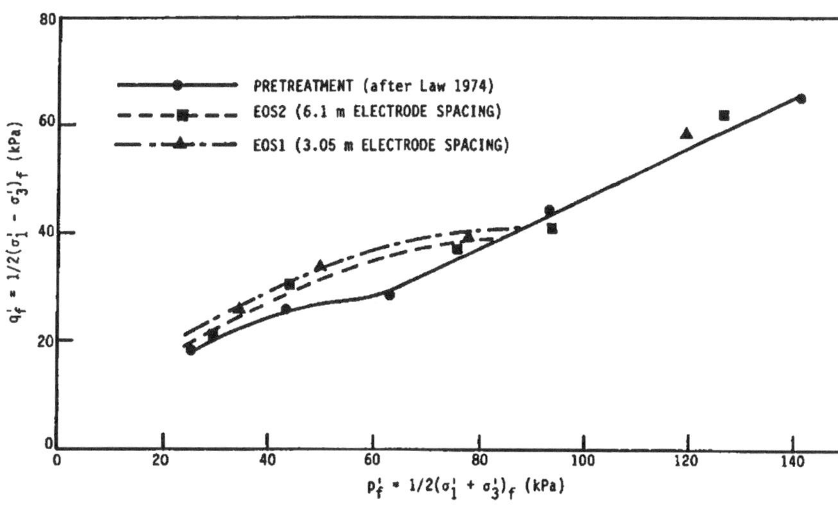

Figure 14.20 Variation of strength envelope before and after electro-osmotic treatment. *(Source: Lo and Ho (1991)).*

to an increase in preconsolidation pressure. At the higher mean stress, the failure envelopes of both pre- and post-treatment samples are merged into a straight line. One-dimensional consolidation tests were conducted on the samples at depths of 2.5 and 4.5 m. The results of one-dimensional consolidation tests are shown in Fig. 14.21. After treatment, the void ratio was substantially reduced with an increase in preconsolidation pressure. At 2.5-m depth, the preconsolidation pressure increased up to 50% and 30% for EOS1 and EOS2, respectively. At 4.5-m depth, the preconsolidation pressure increased up to 85% and 70%, respectively. It was concluded that the improved shear strength by electro-osmotic process is permanent due to the increase in preconsolidation pressure.

Some effects on physical and chemical properties were also investigated. It was found that the liquid limits increased to an average of 50%, whereas the plastic limit was not significantly changed. Accordingly, the plastic index increased approximately 8%. The salinity of soil after treatment increased from

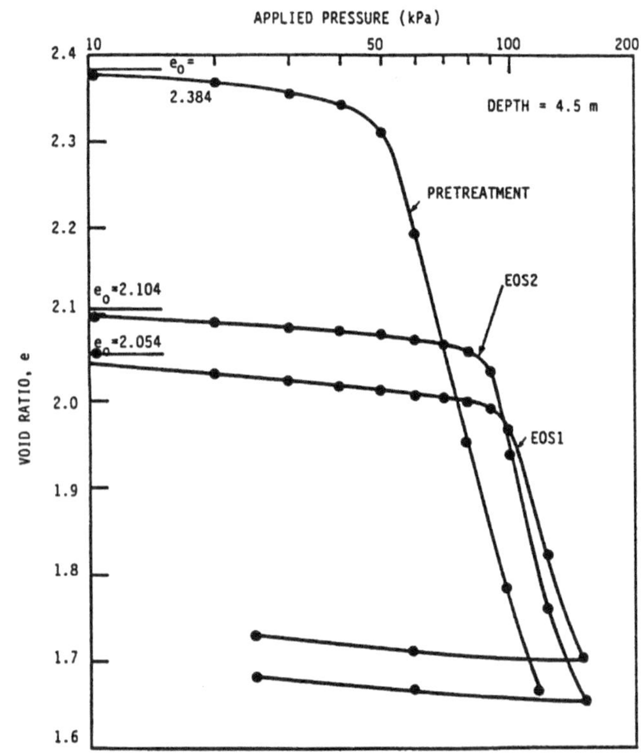

Figure 14.21 Typical consolidation curves before and after treatment. *(Source: Lo and Ho (1991)).*

an initial value of 1.3–2.18 g/l for an approximately 70% increase. The pH of water expelled from the cathode was tested immediately in the field. The pH rapidly increased 1 day after starting treatment. The value increased with increasing applied potential. After polarity reversal, the pH dropped to 7.5 and then gradually increased to 10.5 before treatment was stopped.

14.3 CONCLUSION

The case histories demonstrate that the electro-osmotic treatment is a vital stabilization technique that is cost-effective for soft silty clay and soft clayey silt. This technique can be adopted to many geotechnical engineering applications on stabilization and strengthening a dam, embankment, foundation, and slope. The electro-osmotic process induces change in physical and chemical properties of soil. The treatment extracts water from soil and induces negative pore pressure, which results in consolidation settlement and a decrease in water content. The electrochemical reaction increases the liquid limit, produces cementation bonds, and decreases sensitivity. The efficiency of treatment can be improved by proper design of electrodes and by electrical polarity scheme. With proper electrode design, the soil is strengthened throughout the entire depth of electrodes.

The most significant increase in shear strength occurs in the vicinity of anodes. With polarity reversal, an increase in shear strength is more uniform across the electrodes. The improvement in shear strength is permanent because of a significant increase in preconsolidation pressure and change in soil inherent properties. With further development, electro-osmotic stabilization should make a significant contribution to geotechnical engineering.

REFERENCES

Arman, A., 1978. Current practices in the treatment of soft foundations. In: Soil Improvement, History, Capabilities, and Outlook. Geotechnical Engineering Division, ASCE, Reston, VA, pp. 30–51.
Bjerrum, L., Moum, J., Eide, O., 1967. Application of electro-osmosis to a foundation problem in a Norwegian Quick Clay. Geotechnique 17, 214–235.
Broms, B.B., 1979. Problems and solutions to construction in soft clay. In: Proc. Sixth Asian Regional Conf. on Soil Mechanics and Foundation Engineering, Singapore, vol. 2. Guest Lecture, pp. 3–38.
Casagrande, L., 1941. Zur Frage der Entwässerung feinkörniger Böden (On the Problem of Drainage of Fine Soils). Deutsche Wasserwirtschaft 11.
Casagrande, L., 1949. Electro-osmosis in soils. Geotechnique 1, 159–177.
Casagrande, L., 1959. Review of Past and Current Work on Electro-Osmotic Stabilization of Soils, Harvard Soil Mechanics Series No. 45. Harvard University Press, Cambridge, MA.
Casagrande, L., 1983. Stabilization of soils by means of electro-osmosis—state of the art. J. Boston Soc. Civil Eng., ASCE 69 (2), 255–302.
Chappell, B.A., Burton, P.L., 1975. Electro-osmosis applied to unstable embankment. J. Geotech. Eng. Div., ASCE 101 (8), 733–740.
Esrig, M.L., 1968. Pore pressure, consolidation, and electrokinetics. J. Soil Mech. Found. Div., ASCE 94 (4), 899–921.
Fetzer, C.A., 1967. Electro-osmotic stabilization of West Branch Dam. J. Soil Mech. Found. Div., ASCE 93 (4), 85–106.
Hausmann, M.R., 1990. Engineering Principles of Ground Modification. McGraw-Hill, New York.
Lo, K.Y., Ho, K.S., 1991. The effects of electroosmotic field treatment on the soil properties of a soft sensitive clay. Can. Geotech. J. 28, 763–770.
Lo, K.Y., Ho, K.S., Inculet, I.I., 1991. Field test of electro-osmotic strengthening of soft sensitive clay. Can. Geotech. J. 28, 74–83.
Lockhart, N.C., Stickland, R.E., 1984. Dewatering coal washery tailings ponds by electro-osmosis. Powder Tech. 40, 215–221.
Micic, S., Shang, J.Q., Lo, K.Y., 2003a. Electrocementation of a marine clay induced by electrokinetics. Int. J. Offshore Polar Eng. 13 (4), 308–315.
Micic, S., Shang, J.Q., Lo, K.Y., 2003b. Load carrying capacity enhancement of skirted foundation element by electrokinetics. Int. J. Offshore Polar Eng. 13 (3), 182–189.
Milligan, V., 1995. First application of electro-osmosis to improve friction pile capacity—three decades later. Geotech. Eng. 113 (2), 112–116.
Mitchell, J.K., 1981. Soil Improvement—state-of-the-art report, Session 12. In: Proc. 10th Int. Conf. on Soil Mechanics and Foundation Engineering, Stockholm, vol. 4, pp. 506–565.
Mitchell, J.K., 1993. Fundamentals of Soil Behavior, second ed. Wiley, New York.
Mohamedelhassan, E., Shang, J.Q., 2001. Effects of electrode materials and current intermittence in electroosmosis. Ground Improv. 5 (1), 3–11.
Mohamedelhassan, E., Shang, J.Q., 2003. Electrokinetics generated pore fluid and ionic transport in an offshore calcareous soil. Can. Geotech. J. 40, 1185–1199.
Quigley, R.M., 1980. Geology, mineralogy and geochemistry of Canadian soft soils: a geotechnical perspective. Can. Geotech. J. 20, 288–298.
Shang, J.Q., 1998. Electroosmosis-enhanced preloading consolidation via vertical drains. Can. Geotech. J. 35, 491–499.
Shang, J.Q., Mohamedelhassan, E., 2001. Electrokinetic Dewatering of Eneabba West Mine Tailings: An Experimental Study. ASCE Special Publication No. 112: Soft Ground Technology, ASCE, Reston, VA, pp. 346–357.

Shang, J.Q., Lo, K.Y., Huang, K.M., 1996. On influencing factors in electro-osmotic consolidation. J. Geotech. Eng. 27 (2), 23–26.
Shang, J.Q., Mohamedelhassan, E., Ismail, M., 2004a. Electrochemical cementation of offshore calcareous soil. Can. Geotech. J. 41 (5), 877–893.
Shang, J.Q., Mohamedelhassan, M., Ismail, M.A., Randolph, M.F., 2004b. Electrochemical cementation of calcareous sand for offshore foundations. In: Proc. 14th Int. Offshore and Polar Engineering Conference, Toulon, France.
Soderman, L.G., Milligan, V., 1961. Capacity of friction piles in varved clay increased by electro-osmosis. In: Proc. 5th Int. Conf. on Soil Mechanics and Foundation Engineering, Paris, vol. 2, pp. 143–147.
Sprute, R.H., Kelsh, R.H., 1980. Dewatering fine-particle suspensions with direct current. In: Proc. Int. Symp. of Fine Particle Process, Las Vegas, NV, vol. 2, pp. 1828–1844.
U.S. Navy, 1983. Soil Dynamics, Deep Stabilization, and Special Geotechnical Construction, Design Manual 7.3. Naval Facilities Engineering Command, Washington, DC.
Van Impe, W.F., 1989. Soil Improvement Techniques and Their Evolution. Balkema, Rotterdam, The Netherlands.
Wade, M.H., 1976. Slope stability by electro-osmosis. In: Proc. 29th Canadian Geotechnical Conference, Vancouver, pp. 44–66, Section 10.

CHAPTER 15

Electrokinetic Improvement of Soft Clay Using Electrical Vertical Drains

Eng Choy Lee
Infra Tech Pty Ltd., Burwood, NSW, Australia

15.1 INTRODUCTION

Preloading is a common and cost-competitive technique to consolidate soft clay. Surcharge load such as granular fill is placed on the soil surface to increase the confining stress in the soft clay, which results in the development of excess pore water pressure. The soil is consolidated when the soil pore water is drained from the soil.

Prefabricated vertical drains (PVDs) are commonly used in combination with the preloading technique. PVDs shorten the drainage path in the soil and, hence, the time for primary consolidation. The principles governing consolidation by PVDs are documented extensively in the literature, and design approaches of consolidation with PVDs are available for different boundary conditions.

The principal drawbacks of this ground improvement methodology of preloading with PVD include the following:

- Considerable rest time is required to achieve the required degree of consolidation. This typically lasts between 6 months and 1 year. This puts considerable pressure on project scheduling because lengthy rest times need to be allocated.
- In some locations, fill materials for preloading may not be easily available or available at low cost.

The previous factors provide initiative to seek out alternative ground improvement methods that will address the these issues. Research programs have been developed to assess the effectiveness of electrokinetic (EK) treatment of soft clay.

Innovative applications of geosynthetics in ground treatment of soft clay have been progressed by incorporating the electro-osmotic phenomena with the traditional functions of geosynthetic materials. This entails the

Chemical, Electrokinetic, Thermal, and Bioengineering Methods
http://dx.doi.org/10.1016/B978-0-08-100191-2.00015-0

development of a new range of geosynthetic materials that are electrically conductive. The emergence of electrically conductive polymers since the early 2000s has enabled the development of electrically conductive PVDs in which copper electrodes are sandwiched between the electrically conductive polymer cores of PVDs. Such electrically conductive PVDs, or electric vertical drains, serve the usual function of conducting water hydraulically from the soil to the top and bottom drainage layers and also act as electrodes that can be connected to direct current (DC) sources to cause electro-osmotic (EO) flow of water from anode to cathode (Fig. 15.1). Various researchers have published literature on the phenomenon of electro-osmosis in soils (e.g., Mitchell, 1993; Shang, 1998; Jones et al., 2002).

These electrically conductive geosynthetics can be applied in accelerating the consolidation of soft clay while also increasing the shear strength of the soft soil. This innovative product takes advantage of the effectiveness and simplicity of PVDs to provide vertical drainage, and it incorporates electrical conductivity along the core to facilitate electric current flow into the soil, enabling the electro-osmotic ground treatment process.

EO is a process in which the flow of water between soil particles is induced under an applied electrical potential. EO-based soil improvement is suitable for fine-grained soils (clay) that possess a net surface negative charge. It must have equal charges of cations and anions to maintain electrical neutrality of the clay soil (Mitchell, 1993). Under an applied electrical potential, cations are attracted to the anode while anions are attracted to the cathode. As ions move, they drag along water of hydration and the surrounding free water by viscous force. This will induce a flow of pore water

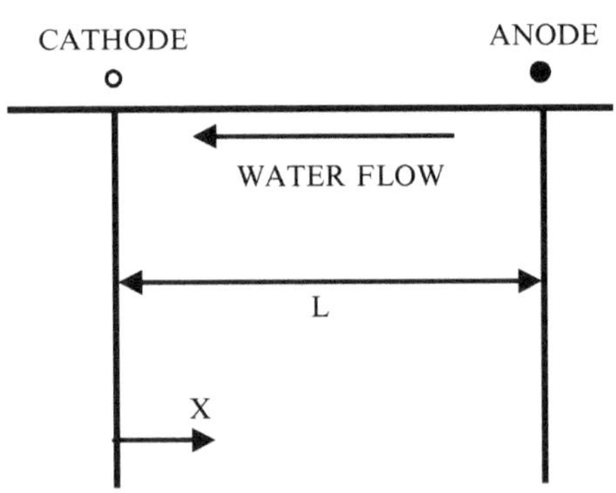

Figure 15.1 Geometry of electrode system for one-dimensional flow (without surcharge).

toward the cathode. Hence, this will initiate the consolidation and strengthening of clay.

These electro-osmotic vertical drains (EVDs) enhance the shear strength and reduce the compressibility of the soft clay. A distinct benefit to the end user is the significantly shortened period of treatment to achieve the required shear strength. Furthermore, with the EVD system, the need for additional surcharge fill material is negated. This translates into cost savings for the end user.

15.2 BACKGROUND OF ELECTROKINETIC STABILIZATION

The electrokinetic stabilization technique has been used to stabilize pile foundations, slopes, excavations, embankments, and dams constructed in soft clays and silts. The treatment utilizes principles of electro-osmosis and involves applying a direct voltage across electrodes embedded in the soil, generating electrical potential in the soil. The electrical potential induces the movement of soil pore water from the anode (positive pole) to the cathode (negative pole). Typically, the soil water is permitted to drain at the cathode and is prevented from entering the anode, which leads to soil consolidation at the anode. This electro-osmotic consolidation process rapidly decreases the soil water content and rapidly increases the soil shear strength. Treatment will be more efficient if the pore water surrounding the cathode is assisted with continuous removal.

The electrodes used in past electrokinetic treatment applications were made of steel, aluminum, and copper. These metallic electrodes corrode rapidly at the anode during treatment, thus diminishing treatment efficiency as time progresses. Advancements in the development of electrically conductive polymers led to the development of electrokinetic geosynthetics called EVDs. The EVD combines the drainage function of the geosynthetics and the electrokinetic function of electrically conductive polymers. The structure of the EVD is similar to that of the PVD (Fig. 15.2). It comprises the following:

- A core structure that is made from conductive polymer encapsulating a thin film of copper foil; the copper foil is added to enhance the conductivity of the core structure and enhance its function as an electrode.
- A filter jacket composed of nonwoven geotextile.

The electrokinetic consolidation technique combines the established technology of PVDs and electro-osmotic treatment of groundwater. The concept is driven by the fact that electro-osmotic conductivity of groundwater is of a

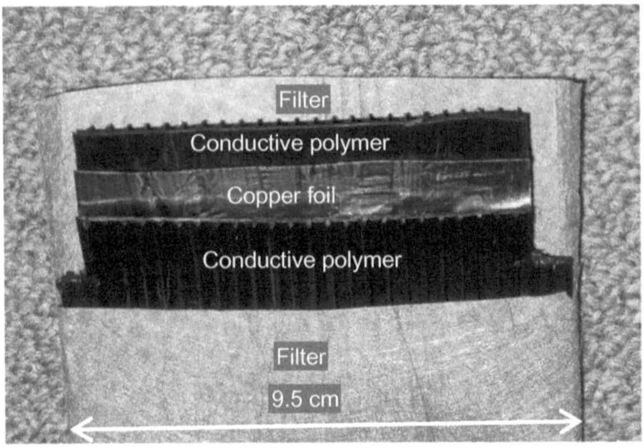

Figure 15.2 Structure of EVD.

much higher order than the hydraulic conductivity. Other significant advantages of the electrokinetic consolidation technique include the following:

- The rate of shear strength increase is much quicker than that of conventional preloading with PVD.
- The requirement for large quantities of surcharge fill is eliminated. Only minimal preload material is required as a working platform.
- The treatment duration is significantly reduced.

Previous studies demonstrated applications of EVDs, such as soil consolidation in a land reclamation, stabilization of a reinforced soil wall, stabilization of mine tailing deposits, separation of solids in sewage sludge, and wastewater treatment. Researchers used EVDs developed from PVDs in a land reclamation project. They reported that the soil shear strength increased from 25–30 kPa to 40–60 kPa within 13 days of treatment and that electrical resistance of the EVD–soil system substantially increased during treatment. However, they did not indicate the source of the increase (Chew et al, 2004).

The effect of EVDs is twofold, namely electro-osmotic consolidation and electrochemical effects generated by the application of a DC current. A case study was undertaken to investigate these two important aspects involved in soil improvement using EVDs; it studied the following (Lee et al. 2010):

1. Changes in the electrical resistance of the soil–EVD interface during electrokinetic treatment
2. The rate of electro-osmotic flow
3. Changes in shear strength of the soil due to the electrokinetic treatment

15.3 PRINCIPLES OF ELECTROKINETIC CONSOLIDATION

15.3.1 General

Electrokinetic treatment applies the principle of electro-osmosis in combination with PVDs to improve the engineering properties of soft clay and, in the process, accelerate the consolidation and strengthening of soft clay. Among the many processes occurring during EO treatment is the reduction of pore water in the soil. This reduction in soil moisture indirectly improves the shear strength and compressibility characteristic of the soil. This type of soil improvement is potentially more effective in reducing the water content that conventional PVD, which relies on hydraulic gradient created by external loading.

In electro-osmotic ground treatment, the hydraulic conductivity or flow of water under an electric gradient potential is a function of the coefficient of electro-osmotic permeability, k_e (m^2/sV); the electric potential gradient, i_e (V/m); and the cross-sectional area of flow, A (m^2). The relationship is similar to Darcy's law and may be written as follows:

$$Q_A = k_e\, i_e\, A \tag{15.1}$$

where Q_A is the discharge capacity (m^3/s). The electro-osmotic groundwater flow for clay soils is on the order of 10^{-9} m^2/sV.

15.3.2 Effective stress and rate of consolidation

The equilibrium pore pressure for electro-osmotic consolidation is governed by Eq. (15.2):

$$\Delta u = -\frac{k_e}{k_h}\gamma_w V \tag{15.2}$$

where Δu is the change in pore water pressure, k_e is the coefficient of electro-osmotic permeability, k_h is the coefficient of hydraulic permeability, γ_w is the density of water, and V is the applied voltage (in volts).

Combined with applied external loading ($\Delta\sigma$), the change in effective stress, $\Delta\sigma'$, is governed by Eq. (15.3):

$$\Delta\sigma' = \Delta\sigma - \Delta u$$

or

$$\Delta\sigma' = \Delta\sigma + \frac{k_e}{k_h}\gamma_w V \tag{15.3}$$

The rate of consolidation is influenced by the coefficient of horizontal consolidation, c_h, of the soil and hence its hydraulic permeability, but not by its electro-osmotic permeability, k_e. For the one-dimensional case of parallel electrodes,

$$T_h = \frac{c_h t}{L^2} \tag{15.4}$$

where L is the distance between electrodes.

Shang (1998) showed that a "pole" exists at 0.637 L from the cathode at which the rate of consolidation is exactly the same for surcharge loading and for electro-osmotic consolidation. It is thus convenient to use the pole as the reference line at which stresses and pore pressures are determined.

Shang (1998) solved the two-dimensional electro-osmotic consolidation involving vertical drains and preloading and presented relationships between T_h and the degree of consolidation U at the pole, as shown in Fig. 15.3.

15.3.3 Voltage loss and pore pressures

The variation of voltage due to electro-osmotic consolidation is shown in Fig. 15.4. The voltage drop with depth is due to the resistance of the wires

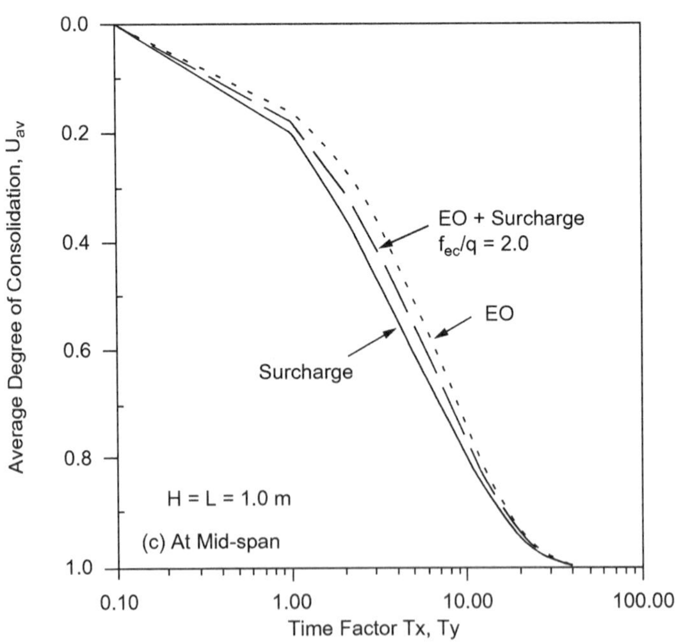

Figure 15.3 Degree of consolidation versus time at the pole. *(Source: After Shang (1998)).*

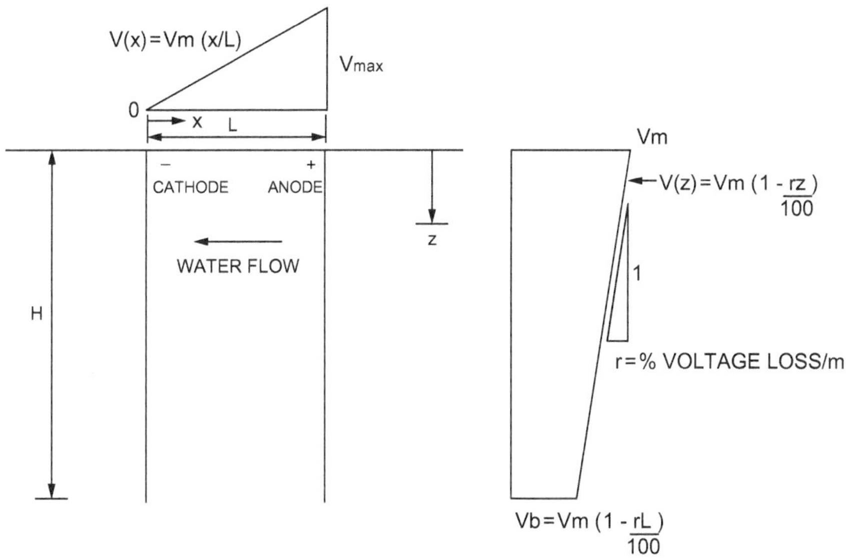

Figure 15.4 Variation of voltage between cathode and anode and with depth.

and the electrodes. Ignoring the small losses at the interface between electrodes and soil, the average pore water pressure due to electro-osmosis between the anode and the cathode at mid-depth of the clay layer and at the pole is thus represented by Eq. (15.5) (Mohamad, 2008):

$$u = -0.637\frac{k_e}{k_h}\gamma_w V\left(1 - r\frac{H}{200}\right)U_{EO} + \sigma_q \qquad (15.5)$$

where r is the voltage gradient, H is the depth of treatment or thickness of clay layer, V_m is the voltage, U_{EO} is the average degree of consolidation, and σ_q is the vertical stress at mid-depth due to the surcharge.

15.3.4 Gain in shear strength

The increase in the soil shear strength after electrokinetic treatment is attributed to (1) the reduction of soil water content (*change in effective stress*) and (2) soil hardening by *electrochemical reactions*. The reduction in soil water content leads to a change in effective stress. Hence, the gain in undrained shear strength of the treated soil will be a function of the increase of effective stress within the soil. Mohamad (1992) showed that the following relationship that exists between undrained shear strength, the overconsolidation ratio (OCR), and effective overburden stress is applicable to soft clays:

$$\frac{S_u}{\sigma'_v} = 0.259OCR^{0.78} \quad (15.6)$$

where S_u is the undrained shear strength, and σ'_v is the effective overburden stress.

The effects of electrochemical reactions are primarily reflected as an increase in Atterberg limits and the development of soil cementation. Previous studies observed that the soil liquid limit, w_L, and plastic limit, w_P, increase after EK treatment. The change in soil water content, w, and soil Atterberg limit is represented by the soil liquidity index, $I_L = (w - w_P)/(w_L - w_P)$. Previous studies illustrated that the undrained shear strength of the reconstituted soil is a logarithmic function of the liquidity index. The soil shear strength increases with decreasing liquidity index. Therefore, the increase in the soil Atterberg limits increases the soil shear strength.

Esrig and Gemeinhardt (1967) report an increase in the liquid limit and plastic limit for the electrokinetic treatment in an illitic clay. The liquid limit and plastic limit are linearly increased with increasing pH, especially in the vicinity of the cathode. The strength of the clay increases more than an order of magnitude compared to the clay with the same water content. Bjerrum et al. (1967) reported that the electrokinetic treatment increases the soil liquid limit near the anode and that either a decrease in the soil water content or an increase in soil liquid limit can decrease the liquidity index of the soil. Furthermore, the preconsolidation pressure of the soil after the electrokinetic treatment is 100% higher than that of the soil without electrokinetic treatment at corresponding water content.

Cementation after electrokinetic treatment is reported in the literature. Cundy and Hopkinson (2005), Mohamedelhassan et al. (2005), and Rittirong et al. (2007) observed the deposition of cements after electrokinetic treatment. Ozkan et al. (1999) experimented with electrokinetic treatment to stabilize kaolinite clay. Soil shear strength increased by 500–600% in the anode region within 15 days of treatment. This could be a result of the soil cementation developed by electrokinetic treatment.

15.3.5 Generation of oxygen and hydrogen gases in electrokinetic treatment

Applying an electrical potential across electrodes generates electrochemical reactions such as oxidation and reduction at the anode and the cathode, respectively:

$$\text{At anode, } 2H_2O\,(l) - 4e^- \rightarrow O_2(g) + 4H^+\,(aq) \quad (15.7)$$

$$\text{At cathode, } 4H_2O\,(l) + 4e^- \rightarrow 2H_2(g) + 4\,OH^-(aq) \tag{15.8}$$

The reactions produce oxygen and hydrogen gases at the anode and the cathode, respectively. The accumulation of gas at the electrode–soil interface increases the electrical resistance of the system and hence decreases the treatment efficiency. Consideration must thus be taken for a drop of voltage gradient at the electrode–soil interface.

15.3.6 Effects of electrode materials

The efficiency of electrokinetic treatment is affected by the electrode material used. Lockhart (1983) used steel, copper, and carbon electrodes to study the electro-osmotic dewatering of kaolinite. The author reported that the copper electrode proved best for dewatering of Cu-kaolinite. Segall and Bruell (1992) reported that the flow rate generated by iron electrodes was twice the flow rate generated by graphite electrodes. Casagrande (1983) suggested a reduction factor of 0.6–0.9 to account for the voltage drop at the soil–electrode interfaces based on field applications. Mohamedelhassan and Shang (2001) reported that the carbon electrodes cause significant voltage drops at the anode. The voltage drop at the electrode reduces the voltage gradient in the soil and hence decreases the efficiency of the treatment.

Different types of electrokinetic geosynthetics, EVDs, that function as both electrodes and filters for drainage are used in the electrokinetic treatment of soils. Hamir et al. (2001) investigated the following four types of electrokinetic geosynthetics material:

Type 1: A needle-punched geosynthetic material containing a copper wire stinger in which copper wires are woven into a conventional needle-punched polypropylene geotextile.

Type 2: A needle-punched geosynthetic material incorporating stainless-steel fibers. Polymeric and stainless-steel fibers are mixed to form needle-punched electrically conductive geotextile.

Type 3: A composite polypropylene and carbon fiber nonwoven sheet. A needle-punched carbon fiber felt is encapsulated in nonwoven polypropylene sheets and is used to transmit electrical current; the polypropylene sheet is used as a filter for drainage.

Type 4: Geocomposite strip reinforcement with a copper wire stinger. Polyester fiber is encapsulated in polyethylene. Copper wires are woven into the polyethylene casing.

The effectiveness of Types 1–3 is comparable to that of copper electrodes for the purpose of electro-osmotic consolidation, whereas Type 4 is used for the investigation of soil reinforcement. The bond strength increases significantly after electrokinetic treatment.

A major problem with conventional metallic electrodes such as copper and steel is the corrosion of the electrode during treatment. The current design of EVD uses metal coated with conductive polymer made by mixing black carbon with conventional polymers. Chew et al. (2004) used EVDs developed from a PVD for a land reclamation project. The EVD was designed to be electrically conductive, with a copper foil encapsulated in a conductive polymer core and then wrapped in a nonwoven textile as a filter. The electrokinetic treatment uses EVD to accelerate soil consolidation. Glendinning et al. (2005) used steel meshes coated with conductive polymer to stabilize a 4.8-m-high reinforced soil wall. Clay slurry was used as fill and was consolidated by the electrokinetic treatment using electrokinetic geosynthetics.

15.4 CASE STUDY 1

15.4.1 Site of field trial and geotechnical conditions

The project site is located in Dengkil, Selangor, near Kuala Lumpur International Airport in Malaysia. The site is part of an approach embankment to a bridge structure, with embankment heights ranging from 2.5 to 5 m. The total width of the embankment, measured at toe level, is approximately 40 m.

The site is underlain by soft to very soft alluvial silty clay of up to 10-m thickness. Beneath the soft silty clay lies medium-dense silty sand. Some decayed wood was encountered at various depths. Groundwater table was encountered at approximately 1 m below existing ground level (EGL). The initial properties of the soil are as shown in Table 15.1.

The preinstallation peak shear strength as measured by geonor vane shear apparatus (push-in) ranged from 8 kPa at 1 m below EGL to 18 kPa at 10 m below EGL, with undrained shear strength increasing at a rate of approximately 0.75 kPa/m. An exceptionally high shear strength of 23 kPa was recorded at a depth of 7 m; this was thought to be due to the presence of fibrous and decayed wood material. Figure 15.5 shows the variations of undrained shear strength with depth.

Table 15.1 Initial soil properties

Liquid limit (%)	Plastic limit (%)	Plasticity index (%)	Moisture content (%)
154	79	75	132

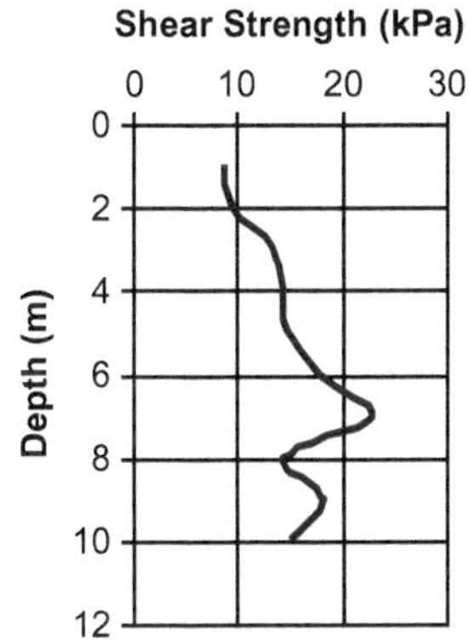

Figure 15.5 In situ shear strength (preinstallation of EVD).

15.4.2 Setup of field trial

The test plot size is 20 × 20 m, with a vertical drain (EVD) spacing of 1.5 m. A total of 196 points were installed in 14 rows, as illustrated in Fig. 15.6.

Figure 15.7 illustrates the equipment used for the installation of the EVD. A 0.5-m-thick sand drainage blanket was provided to enable the discharge of groundwater to the ditches on the edges. The electrodes at the top of each EVD were connected to an electric power generator supplying DC current.

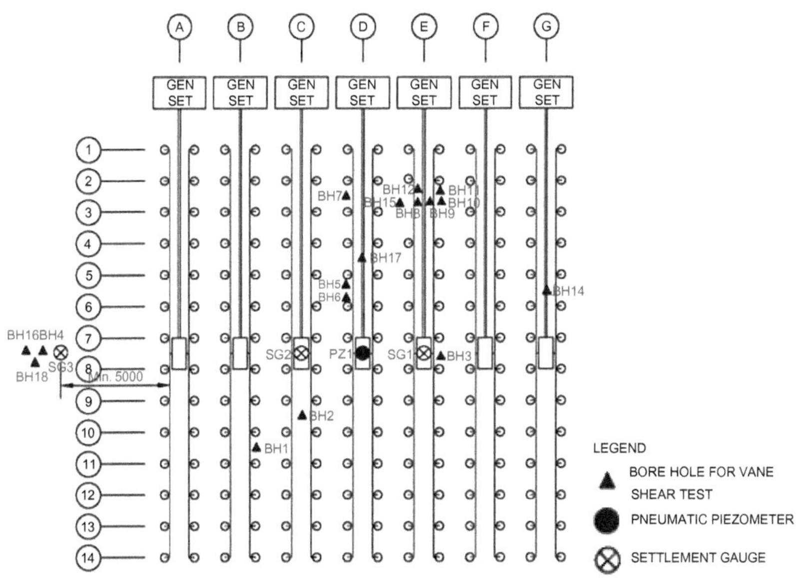

Figure 15.6 Layout of charging pattern, instrumentation, and testing.

Figure 15.7 Installation of EVD.

The electric current was applied for approximately 14 days on a 12-h cycle with reversals in the polarity of the current from the day 10 onward.

Upon completion of the charging, the embankment fill was placed to a height of 2.5 m. Figure 15.8 shows a view of the test plot.

15.4.3 Instrumentation and testing

The instrumentation scheme within the test plot and the control section includes the following:

- Settlement plates (3 numbers)
- Pneumatic piezometer (1 number)

Figure 15.8 A view of the test plot.

These instrumentations were monitored on a regular basis. The positions of these instruments are indicated in Fig. 15.6. Settlement plates SG 1 and SG 2 were installed within the test plot, whereas settlement plate SG 3 was installed approximately 5 m outside the boundary of the test plot. SG 3 was utilized as a control because it was located within the conventional PVD ground treatment method. The piezometer, PZ 1, was installed at a depth of 8 m below existing ground level.

The shear strength profile of the test plot was also measured on a regular basis following commencement of the trial. This was carried out using the penetration field vane (geonor vane) at 1-m-depth intervals.

At the end of the charging period and following the placement of embankment fill, further measurements were taken, include the following:

- Penetration vane shear test to measure the in situ shear strength of the soil (peak).
- Monitoring of the settlement gauges and piezometer.
- Measurement of the moisture content and density profile with depth by obtaining undisturbed samples. The determination of the final consolidation properties of the soil shall be obtained by carrying out consolidation tests on these samples.

15.4.4 Control section

A control section was also set up for comparison purposes. This control section was located on a nearby plot with similar ground conditions and similar embankment height. The conventional ground improvement method using PVDs (spaced at 1.5 m c/c) and surcharge of approximately 3-m height was applied to the control section. For this control section, the instrumentation includes one settlement plate and field vane shear testing.

15.4.5 Charging pattern and charging efficiency

Seven generators were employed on site to supply direct electrical current to the electrodes, with each generator supplying current to two rows of electrodes. The average voltage supplied was approximately 80 V. The rows of electrodes were alternately set as positively (anode) and negatively (cathode) charged. Figure 15.6 also shows the pattern of charging.

The efficiency of the charging achieved was approximately 50%. This means that up to 50% of the electrodes were unable to transmit the required electrical charge to the ground. This was attributed to some faulty EVDs in which the copper element was damaged during initial trial manufacture or during installation.

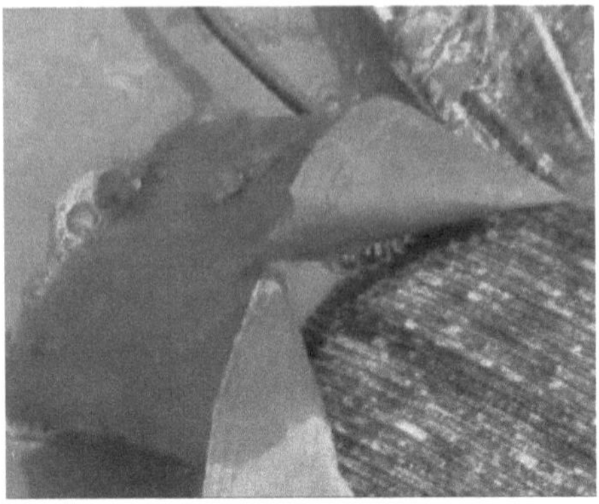

Figure 15.9 Groundwater drawn up the cathode.

15.4.6 Test results and observations

EVD treatment

The most distinct observation of the behavior of the EVD during charging was the constant discharge of groundwater through the cathodes, as shown in Fig. 15.9. This was accompanied by the discharge of gas bubbles as well. This occurred approximately 20 min after commencement of charging. It indicated that the electrodes were successfully transferring the electrical potential to the soft ground and the EO treatment. There was some slight depression in the area immediately surrounding the anodes. This indicated that consolidation of the soft soil was occurring surrounding the anode, as discussed previously.

The settlement gauge readings from the instrumentation monitoring program are shown in Fig. 15.10. Rapid settlement was recorded by the settlement plates, especially SG 1 and SG 2, upon placement of embankment fill material. An immediate settlement of up to 1.0 m was recorded at SG 1. This immediate settlement suggests a brittle collapse of a cemented soil structure due to the imposition of the embankment load. The phenomenon is in line with the chemical precipitation at the anode, as indicated in Section 15.2. The settlement plate SG 2 is somewhat nearer to the boundary with the PVD treatment and hence recorded a lower settlement. SG 3 is located within the control area, and the settlement recorded approximates the computed PVD treatment. However, there is a small EVD boundary effect.

Figure 15.11 shows the piezometer readings. There was a slight drop in the pore water pressure during the charging period, consistent with the consolidating soil. This was reflected in an increase in undrained shear strength, as shown in Fig. 15.12.

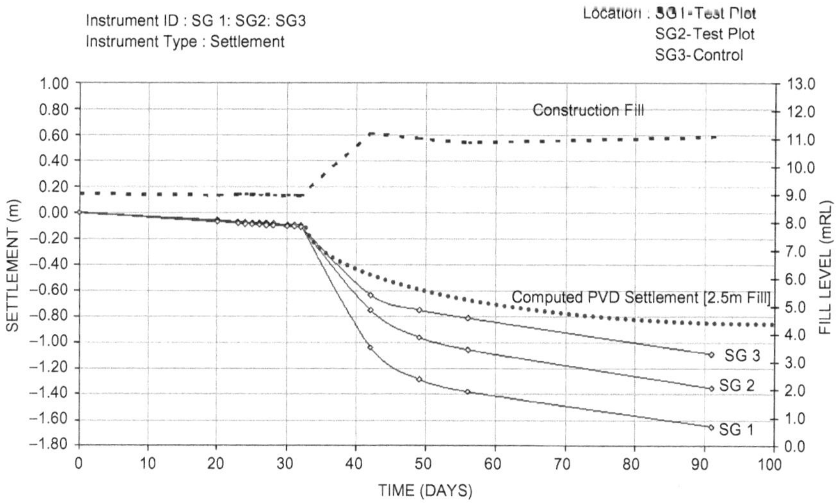

Figure 15.10 Settlement gauge readings.

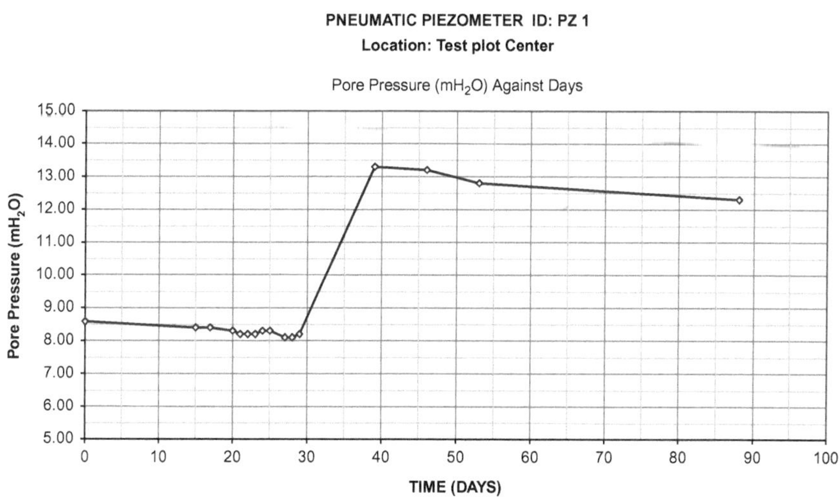

Figure 15.11 Piezometer readings.

Figure 15.12 shows the increases in undrained shear strength as measured in the vane shear tests. Initially, there was an increase in shear strength, particularly within the top 5 m. This increase became more distinct at 72 days, when a significant increase in the undrained shear strength is seen. This increase may be attributed to the electro-osmosis treatment of the soil as well as the PVD effect of the EVD during the placement of the embankment

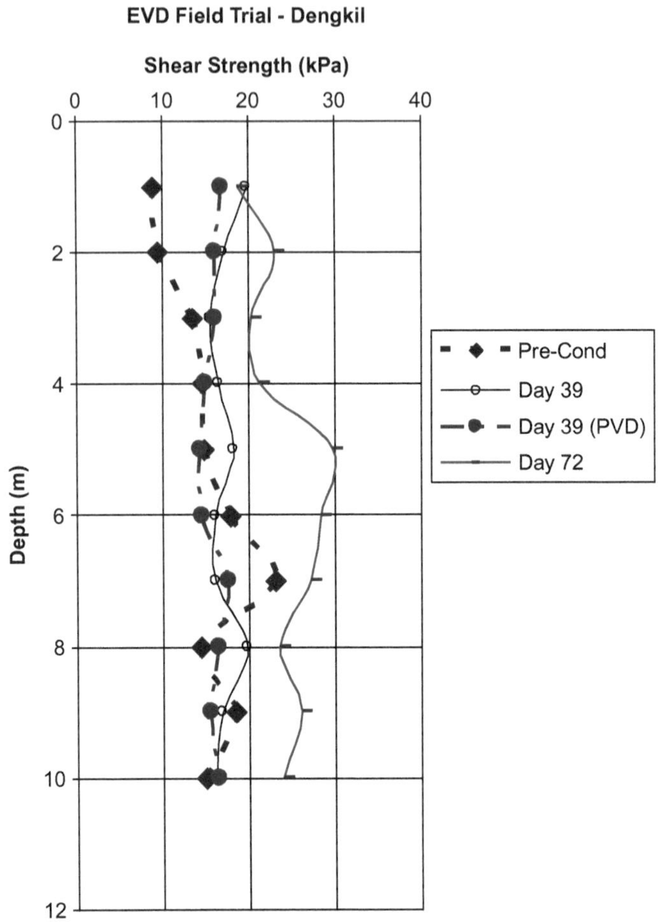

Figure 15.12 Vane shear test results.

load. The percentage of shear strength increase exceeded 100% of the initial values. The final undrained shear strength recorded was 20 kPa at 1-m depth, increasing at a rate of approximately 0.5 kPa/m.

PVD treatment (control section)

Figure 15.13 shows the vane shear strength measurements in the control area (i.e., with PVD treatment). It shows that the undrained shear strength remained constant at approximately 14 kPa, up to a depth of approximately 5 m.

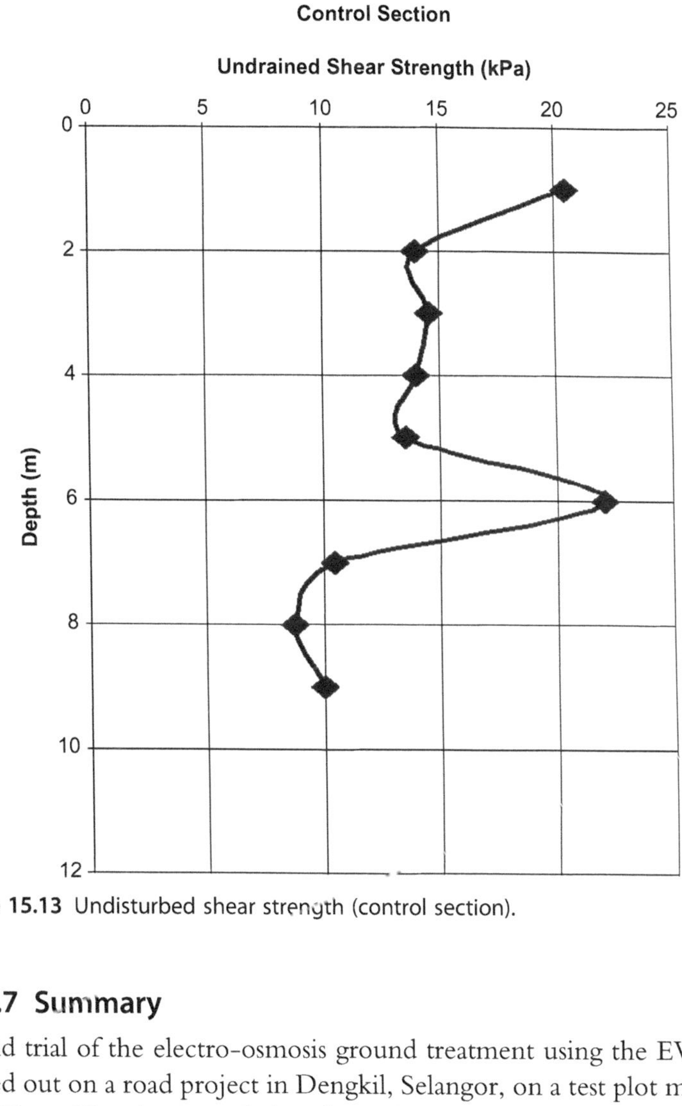

Figure 15.13 Undisturbed shear strength (control section).

15.4.7 Summary

A field trial of the electro-osmosis ground treatment using the EVD were carried out on a road project in Dengkil, Selangor, on a test plot measuring 20 × 20 m. The site is underlain by soft silty clays with undrained shear strength ranging from approximately 8 to 18.5 kPa. The drains (EVD) were charged on a 12-h cycle for approximately 14 days. Instrumentation in the form of settlement plates and pneumatic piezometer was installed. Vane shear tests were carried out at regular intervals. Results of the instrumentation suggest the formation of cementation due to EO treatment, and this relatively weak bond was broken by the placement of the embankment load. The undrained shear strength appeared to have increased to values ranging from 20 to 30 kPa, indicating a significant increase of more than 100% of the initial shear strength values.

These results suggested that the EVD can be used for foundation treatment of soft soils, especially when a high embankment is designed to be placed over such soft soil profiles. Traditionally, foundations for high embankment over soft soils are treated with the use of stone columns, piled rafts, or similar systems. The EVD will be a suitable alternative method to these traditional ground treatment methods.

15.4.8 Discussion

The electrokinetic treatment of the soil was performed prior to the placement of fill for the road embankment. Because no surcharge fill was placed on the test plot during charging of the electrodes to collapse the voids into stable structure, the drainage of pore water could have resulted in a fragile macro structure. The subsequent placement of embankment fill increases the excess pore water pressures within the treated ground and also results in large strains within the treated zone.

The test plot was placed within the footprint of the embankment; hence, no ground treatment was performed outside the toe to the embankment. These factor resulted in loss of shear strength and insufficient toe support to the embankment, which led to slope instability of the embankment at its design height.

15.5 CASE STUDY 2

A ground improvement project using EVD for electrokinetic treatment was carried out. Observations and results in the field test were compared with the laboratory test. The investigation focused on an increase in electrical resistance, the observation of electro-osmotic discharge, and an improvement in soil undrained shear strength.

15.5.1 Site and ground conditions

The upgrade of an access road to an oil terminal in Kuching, Sarawak, Malaysia required widening the road from an existing road width of 8 m to a new road width of 16 m, constituting a four-lane single carriageway. This required that the existing carriageway be widened by 4 m on either size. The proposed road structure consisted of 1-m-high embankment fill with an approximately 0.5-m-thick pavement. The soil on the site comprised very soft to soft, light to dark gray, clayey silt with traces of fine sand up to 15-m depth. The void ratio and the compression index were approximately 0.8 and 0.5, respectively. The soil was classified as high plasticity silt, with silt content of 58%, clay content of 39%, and sand content of 3%. The groundwater table existed at near surface levels.

Figure 15.14 shows the vane shear strength, c_u, soil water content, plastic limit, and liquid limit of the soft clay. The undisturbed vane shear strength of the original soil ranged from 5 kPa at 2-m depth to 13 kPa at 6-m depth. The water content was 60% at a depth of 2 m, which was higher than the liquid limit. The water content increased to 77% at 5.5-m depth and 70% at 7-m depth, which were slightly less than the liquid limits. Because the soil shear strength was too low for the road construction, the shear strength of the soil had to be increased to greater than 20 kPa.

15.5.2 Test setup

The treatment area measured approximately 560 × 4 m on each side of the existing road. Site preparation works included placement of a 1-m-thick sand fill above the soft soils. This served both as a drainage blanket and as a working platform. To prevent the ingress of incoming tidal waters into the site, a dike was built along the external boundaries of the treatment.

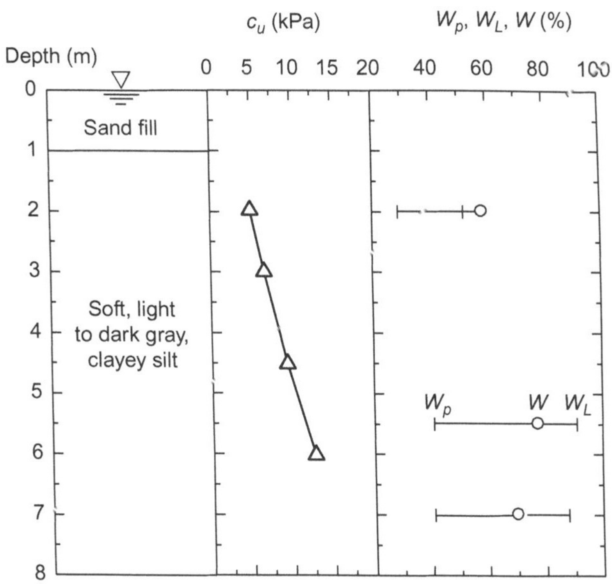

Figure 15.14 Soil profile and properties of the original soil.

It was decided to install EVDs to a depth of approximately 6 m below the original ground surface because of budget constraints. The schematic diagram and the photograph of the test setup are shown in Figs. 15.15(a) and 15.15(b), respectively. Three rows of EVDs were installed in the treated area, with distances between the rows of 1.2 m, and the spacing between EVDs in the same row was 1.0 m. Electric cables were connected to the copper foil inside the EVDs. The treatment area was subdivided into sections of 28 × 4 m to fit individual power units. The electrification period was 20 h daily for 5 days in each section (28 × 4 m). Five power supplies were used and circulated for the treatment in different sections. The total treatment period for the entire area was 14 days.

The discharged water and gas bubbles were observed at the cathode as shown in Fig. 15.16 within 30 min after the application of the electric potential. The discharged water was black, and it turned brown within a day after discharge. The water may contain iron ions that oxidized when exposed to the atmosphere. It was observed that the electro-osmotic flow stopped on the first day of the voltage application. Therefore, it was necessary to induce the electro-osmotic flow using polarity reversal on the second day. No electro-osmotic flow was found on days 3–5 of the treatment, although the polarity reversal was applied daily. This could also be due to the short treatment depth. As a result, the electrification was terminated at the end of day 5.

15.5.3 Current and resistance

Typical electrical current and applied voltage are shown in Fig. 15.17(a). The DC voltage between the anode and cathode rows was approximately 10 V initially and gradually increased to maintain the electrical current. The specific electrical resistance (R/H) was estimated to be the average resistance of the EVD and soil between each electrode pair. As shown in Fig. 15.17(b), the total electrical resistance, R_{total}, was initially 12 Ω/m and increases suddenly after 73 h of treatment to 76 Ω/m.

The electric field distribution on a repetitive area is shown in Fig. 15.18. The average electric field intensity on the EVD–soil interface ranged from 29 V/m at $\Delta V = 10$ V to 87 V/m at $\Delta V = 30$ V. As a result, the efficiency factor ranged from 50% to 78% for the applied voltage. The efficiency factor was used to estimate the resistance of EVD. Figure 15.5(b) shows the specific electrical resistance of EVD during treatment. The electrical resistance was dominated by the resistance of the soil and EVD–soil interface during

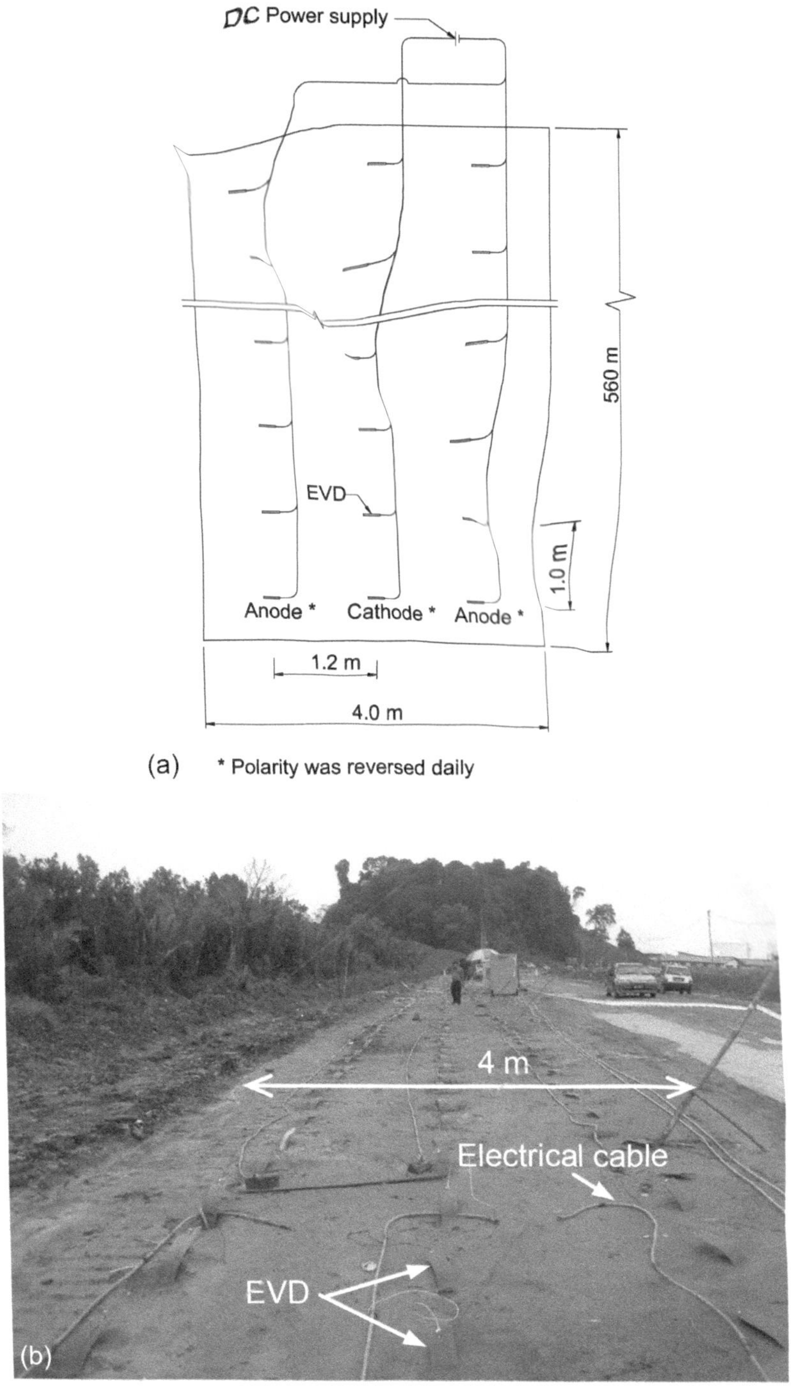

Figure 15.15 Setup of field test: (a) plan view and (b) photograph of the setup.

Figure 15.16 Discharge of groundwater from an EVD cathode.

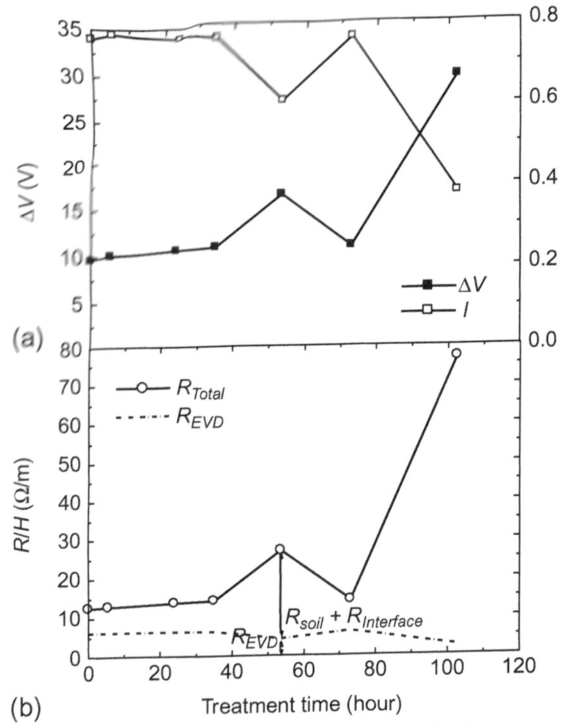

Figure 15.17 Electrical resistance measured during the field test.

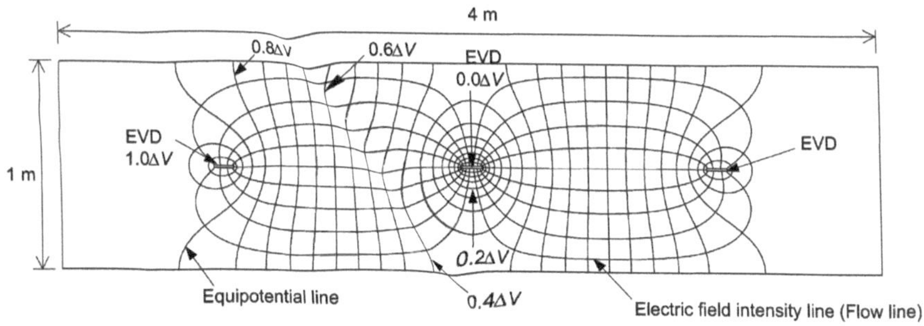

Figure 15.18 Electric field distribution in a repetitive area.

treatment. The increase in total electrical resistance was probably due to gas generated at the EVD–soil interface.

Energy consumption is a major part of treatment cost. The energy consumption in the treatment using EVD reported by Chew et al. (2004) and in this study was 1.8 and 0.7 kWh/m^3, respectively. The energy consumption of the electrokinetic treatment using the EVD electrodes was at the lower bound for the electrokinetic treatments.

15.5.4 Increase in vane shear strength

Field vane shear tests were carried out at five locations over the entire treatment area after the completion of the electrokinetic treatment. Figure 15.19 shows the shear strength profile of treated soil compared to the original shear strength profile. The original undrained shear strength of the soil increased linearly with depth from 5 to 13 kPa from depths of 2 to 6 m. The shear strength of the treated soil ranged from 15 to 50 kPa. The average undrained shear strengths ranged from 22 to 39 kPa after only 5 days of treatment. Compared to the original shear strength, the soil undrained shear strength increased noticeably within the short treatment period.

This improvement indicated that the electrokinetic treatment using EVDs can improve the soil effectively. It was surprising, however, that the strength increased significantly although the electro-osmosis stopped within 2 days of treatment. Electro-osmosis alone may not account for this distinct increase in shear strength.

15.5.5 Discussion

The increase in the soil undrained shear strength in the electrokinetic treatment is derived from three factors:

1. An increase in the soil density due to consolidation induced by the surcharge pressure

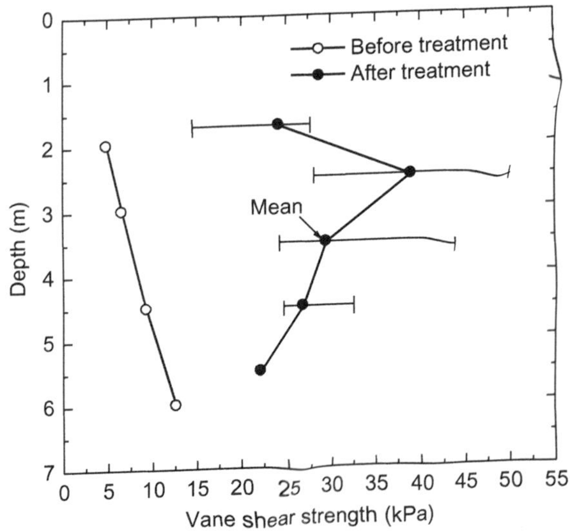

Figure 15.19 Vane shear strength of soil after electrokinetic treatment.

2. An increase in soil density due to the electro-osmotic consolidation
3. Soil cementation and an increase in Atterberg limits due to electrochemical reactions

Note that the surcharge pressure from the 1-m-thick gravel blanket is relatively low compared with the significant increase in the undrained shear strength. Therefore, the increase in the shear strength in this field test can be attributed to the second and third factors primarily.

The observations in the field test are similar to the results in the laboratory test. The gas is generated in EVD, which results in a substantial increase in electrical resistance. The electro-osmotic flow occurs in a short period. The laboratory test shows that the electro-osmotic flow occurs only on the first day of treatment. Likewise, the electro-osmotic flow in the field test is observed for only approximately 2 days after the start of treatment, whereas it should be noted that the electro-osmotic consolidation is extended by polarity reversal. Therefore, the distinct improvement of the soil undrained shear strength appears to be a combination of the results of electro-osmotic and electrochemical effects. The former generates consolidation, whereas the latter generates soil structural changes, which will cause increases in soil Atterberg limits and cementation. This study shows that the electrokinetic treatment using EVDs improves the soil shear strength efficiently. Although the effects of the electro-osmosis in the improvement of soil shear strength are reported extensively in the literature, the effects of electrochemical reactions have not been investigated in detail. This study suggests that the improvement in soil shear strength due to the electrochemical effect is significant compared to the improvement due to electro-osmotic consolidation.

15.6 CONCLUSION

The performance of EVDs was investigated via case study and a road project on clay soils. It was observed that voltage loss at the EVD surface decreased the efficiency of the electrokinetic treatment. An increase in applied voltage can compensate the voltage loss. It was observed that applying a low voltage can minimize this time-dependent voltage loss. It was found that the electro-osmotic flow stops soon after the start of the electrokinetic treatment in the field test—that is, after 2 days in the field test in which the electro-osmotic flow was extended using polarity reversal.

The field test showed that the soil shear strength increased with electrokinetic treatment due to electro-osmotic consolidation and electrochemical effects. The former reduced soil water content, whereas the latter increased soil liquid and plastic limits and developed cementation bonds. The vane shear strength increased from between 5 and 16 kPa in original soil to an average between 22 and 39 kPa after 5 days of treatment in the field test. This soil shear strength increase was noticeable within a short treatment period, indicating that the electrokinetic treatment using EVDs was efficient.

NOTATION

Q = electro-osmotic flow rate
k_e = coefficient of electro-osmotic permeability
k_h = coefficient of hydraulic permeability
E = electric field intensity
A = cross-sectional area normal to the direction of the flow
L = electrode spacing
Δu = change in pore pressure
$\Delta \sigma'$ = change in effective stress
$\Delta \sigma$ = change in total stress
V = applied voltage (in volts)
ΔV = change in applied voltage

V_m = voltage
c_h = coefficient of consolidation with horizontal drainage
T_h = time factor, dimensionless
U = degree of consolidation
r = voltage gradient
I = current in amperes
H = depth of treatment or thickness of clay layer
U_{EO} = average degree of consolidation
σ_q = vertical stress at mid-depth due to the surcharge
S_u = undrained shear strength
σ'_v = effective overburden stress
C_u = undrained shear strength
R/H = specific electrical resistance
R_{total} = total electrical resistance
γ_w = unit weight of the water
k = hydraulic conductivity
k_e = electro osmotic permeability
x = distance from the cathode to the anode
e^- = negative ion
I_L = liquidity index
w_L = liquid limit
w_P = plastic limit

REFERENCES

Bjerrum, L., Moum, J., Eide, O., 1967. Application of electro-osmosis to a foundation problem in a Norwegian quick clay. Géotechnique 17, 214–235.

Casagrande, L., 1983. Stabilization of soils by means of electro-osmosis – State-of-the-art. J. Boston Soc. Civil Engineers, ASCE 69 (2), 255–302.

Chew, S.H., Karunaratne, G.P., Kuma, V.M., Lim, L.H., Toh, M.L., Hee, A.M., 2004. A field trial for soft clay consolidation using electric vertical drains. Geotext. Geomembr. 22, 17–35.

Cundy, A.B., Hopkinson, L., 2005. Electrokinetic iron pan generation in unconsolidated sediments: implications for contaminated land remediation and soil engineering. Appl. Geochem. 20 (5), 841–848.

Esrig, M.I., Gemeinhardt, J.P., 1967. Electrokinetic stabilization of an illitic clay. J. Soil Mech. Found. Div. 93 (SM3), 109–128.

Glendinning, S., Jones, C.J.F.P., Pugh, R.C., 2005. Reinforced soil using cohesive fill and electrokinetic geosynthetics. Int. J. Geomech. 5 (2), 138–146.

Hamir, R.B., Jones, C.J.F.P., Clarke, B.G., 2001. Electrically conductive geosynthetics for consolidation and reinforced soil. Geotext. Geomembr. 19 (8), 455–482.

Hansbo, S., 1981. Consolidation of fine-grained soils by prefabricated drains. In: Proc. 10th Int. Conf. on Soil Mechanics and Foundation Engineering, vol. 3. Balkema, Rotterdam, The Netherlands, pp. 677–682.

Jones, C.F.P., Glendinning, S., Shim, G.S.C., 2002. Soil consolidation using electrically conductive geosynthetics. In: Delmas, P., Gourd, J., Girard, H. (Eds.), Geosynthetics–7th Int. Conf. on Geosynthetics. Swets & Zeitlinger, Lisse, Holland.

Lee, E.C., Douglas, R.S., Saberi, F.I., Razak, A.S.A., 2010. Electrokinetic consolidation of soft clay using the electric vertical drains (EVD). In: Proc. 17th Southeast Asia Geotechnical Conf., May. Taipei, Taiwan, pp. 10–13.

Lee, E.C., Neoh, C.A., Douglas, R.S., 2007. Performance of electrically conductive vertical (wick) drains in the treatment of soft clay. In: Proc. Transportation Conf. Geosynthetics 2007, Washington, DC, pp. 217–223.

Lockhart, N.C., 1983. Electroosmotic dewatering of clays: III. Influence of clay type, exchangeable cations, and electrode materials. Colloids Surface. 6, 253–269.

Mitchell, J.K., 1993. Fundamentals of Soil Behavior, second ed. Wiley, New York.

Mohamad, R., 1992. Engineering the north–south expressway on soft ground. In: Geotropika '92. Johor Bahru, Malaysia.

Mohamad, R., 2008. Precompression of Soft Clay by Electro-Osmosis using Electrically Conductive Prefabricated Vertical Drains. Technical Note 002-08, Straits Consulting Engineers, Malaysia.

Mohamedelhassan, E., Shang, J.Q., 2001. Effects of electrode materials and current intermittence in electro-osmosis. Ground Improve. 5, 3–11.

Mohamedelhassan, E., Shang, J.Q., Ismail, M.A., Randolph, M.F., 2005. Electrochemical cementation of calcareous sand for offshore foundations. Int. J. Offshore Polar Eng. 15, 71–79.

Ozkan, S., Gale, R.J., Seals, R.K., 1999. Electrokinetic stabilization of kaolinite by injection of Al and PO_4^{3-} ions. Ground Improve. 3, 135–144.

Rittirong, A., Shang, J.Q., Ismail, M.A., Randolph, M.F., 2007. Effect of electric field intensity on electro-cementation of caissons in calcareous sand. Int. J. Offshore Polar Eng. 17, 74–79.

Segall, B.A., Bruell, C.J., 1992. Electro-osmotic contaminant removal process. J. Environ. Eng. 118, 84–100.

Shang, J.Q., 1998. Electro-osmosis enhanced preloading consolidation via vertical drains. Can. Geotech. J. 35 (3), 491–499.

CHAPTER 16

Experimental and Numerical Investigations of the Behavior of a Heat Exchanger Pile

Lyesse Laloui, Mathieu Nuth, Laurent Vulliet
Ecole Polytechnique Fédérale de Lausanne, EPFL, Switzerland
Soil Mechanics Laboratory, Lausanne, Switzerland

16.1 INTRODUCTION

Managing energy resources is becoming one of the crucial issues of the 21st century. Following the global trend of population increase and improvement of living standards, primary energy consumption also continues to increase. Although no short-term exhaustion of energy resources is expected, the environmental consequences of an overuse of natural energetic resources could well become problematical. Among the resulting problems, the greenhouse effect, linked to the undeniable increase of CO_2, probably remains the most acute.

This chapter deals with the development of a new sustainable technology for the intermittent storage of energy in soils. The goal of our research work is to contribute, by a geotechnical approach, to obtaining reliable, environmentally and resource-friendly heating and cooling of residential, office, and commercial buildings at low additional cost. The idea behind energetic geostructures is to take advantage of the thermal storage capacity of the ground as an energy storage system by using the foundation of a building (e.g., piles or retaining walls). The key factor in the sustainability of such system is the use of the building elements, which are already needed for structural reasons.

From an energetic standpoint, the main advantages of this new technology are the following:

- Up to 50% CO_2 emission reduction for new buildings
- Reduction in the use of primary energy and increase in the use of renewable energy by intermittent heat and cold storage in soils as an innovative new technology
- Reduction of transportation of energy, including a reduction of trucking traffic and accident risk as an indirect benefit

Chemical, Electrokinetic, Thermal, and Bioengineering Methods
http://dx.doi.org/10.1016/B978-0-08-100191-2.00016-2

From a geotechnical standpoint, heating of foundations may also have an important advantage in the improvement of the soil characteristics (Cekerevac and Laloui, 2004). This may result in a reduction in foundation cost. Another potentially significant practical and technological aspect concerns our results showing that the thermal pretreatment of clays may have a very positive effect on their resilience under cyclic loading, which results in higher resistance of the buildings against earthquakes (Laloui et al., 2005). In any case, freezing of the piles is to be avoided by continuous monitoring and control systems to prevent thaw-induced defects.

This simple yet rational technology has already been used in many installations in Switzerland (e.g., buildings Pago in Grabs, 600 kW; Kino in Buchs, 70 kW; or the Malerva apartment complex in Sargans, 70 kW) and worldwide (>300 installations in Europe). Figure 16.1 shows two examples of buildings using heat exchanger piles: the 200-m-high Frankfurt Main Tower constructed in 1999 is founded on a $30 \times 50\ m^2$ base supported by 112 drilled piles 1.5 m in diameter and 30 m in length, and Dock E at the Zurich airport makes use of 315 energy piles. Currently, and despite the large number of existing installations, the main obstacle in the industrial development of this technology is the lack of reliable knowledge concerning the thermal effects on the structural behavior of the foundations. In particular, no analytical, physical, or numerical tools are available to consider the complex interactions between thermal storage and the mechanical behavior of geostructures.

Figure 16.1 Examples of buildings using heat exchanger piles: (left) the Frankfurt Main Tower and (right) Dock E at the Zurich Airport.

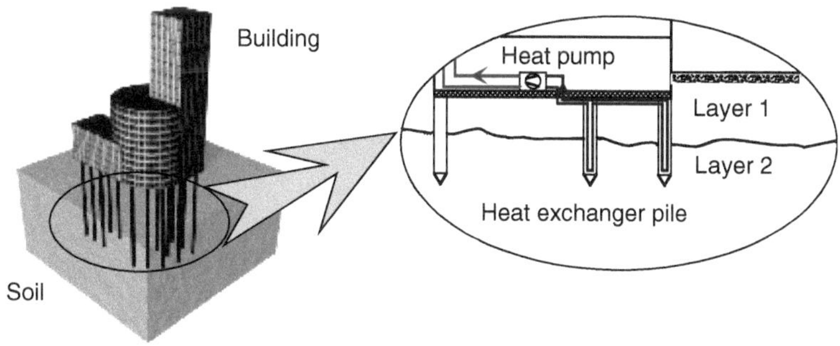

Figure 16.2 Schematic representation of a heat exchanger pile system.

The parallel combination of several heat exchanger piles, hydraulically connected and linked to a heat pump, permits the extraction of warmth from the ground to satisfy the need for heat in winter and to expel excess heat resulting from air conditioning in summer (Fig. 16.2). With this geothermal use of geostructures, buildings can be cooled at minimal cost and cheaply heated with a heat pump, utilizing the available geothermal energy in the ground and the natural thermal properties of the concrete.

In this chapter, the behavior of a typical energetic geostructure—that is, a heat exchanger pile—is investigated through experimental and numerical approaches.

An in situ test of a heat exchanger pile under actual conditions was carried out at the Swiss Federal Institute of Technology in Lausanne (EPFL) (Laloui et al., 2003). Some of the significant results of this study have been used to analyze the capabilities of a thermo-hydro-mechanical finite element model to simulate the thermal soil–structure interactions. This contributes to the use of numerical approaches for the evaluation of the optimal design for the soil–energy storage system, the determination of typical design parameters (soil type, concrete surface, etc.) and the development of a standard calculation method, as well as the verification criteria, for design purposes.

16.2 EXPERIMENTAL IN SITU TEST ON A HEAT EXCHANGER PILE

A four-story building under construction at the Swiss Federal Institute of Technology in Lausanne (Switzerland) was chosen for an in situ test of a heat exchanger pile. The building is 100 m in length and 30 m in width (Fig. 16.3), and it is founded on 97 piles approximately 25 m in length.

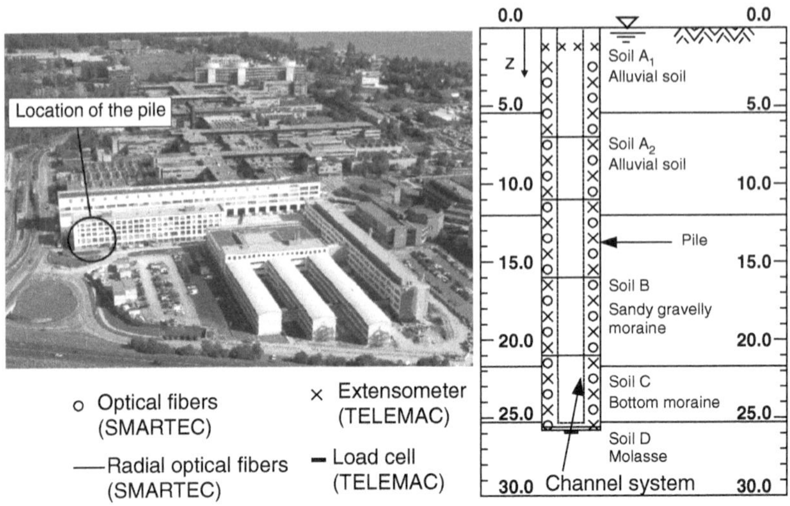

Figure 16.3 View of the test site in Lausanne (Switzerland), the soil profile, and the instrumentation of the tested pile.

The tested pile was located at the side of the building. The drilled pile diameter was 88 cm and the length was 25.8 m. The schematic soil stratigraphy profile is presented in Fig. 16.3 and described in Table 16.1. This soil profile was obtained on the basis of various geotechnical investigations and two static pile tests. The values of the ultimate shaft friction (q_s) and the tip capacity (q_p) are given in Table 16.1. The groundwater table in this zone is very close to the ground surface.

Polyethylene (PE) tubes were installed vertically in the reinforcing structure with a U-shaped configuration to permit the passage of the heat-carrying fluid (Fig. 16.4). The instrumentation chosen for the measurement of strain, temperature, and load in this in situ test was made up of 58 gauges placed as indicated in Fig. 16.3:

- Twenty-eight vibrating-wire extensometers to measure vertical strain and temperature
- Twenty-four fiber-optic extensometers (1 m long) to measure vertical strain (Inaudi et al., 2000)
- Five fiber-optic extensometers (2 m long) to measure radial strains at five depths
- One load cell to measure load at the bottom of the pile
- Four extensometers located at the head of the pile to measure vertical strain from which axial load at the pile top can be determined

Table 16.1 Material parameters used in thermo-hydro-mechanical finite element modeling

Soil layer	Unit weight, ρ (kg/m^3)	Water content, w (%)	Porosity, n	Permeability, K^* (m/s)	Ultimate shaft friction, q_s (kPa)	Tip capacity, q_p (kPa)	Resistance, c_{us} (kPa)	Bulk modulus, K (MPa)	Shear modulus, G (MPa)	Friction angle, Φ (°)	Cohesion, c (kPa)	Thermal conductivity, Λ (W/m/°C)	Heat capacity, ρ_c (J/m^3.°C)	Solid dilation coefficient, β's (°C^{-1})
A1	2000	30	0.1	2×10^{-6}	0	—	15–20	122	113	30	5	1.8	2.4×10^6	10^{-5}
A2	1950	30	0.1	7×10^{-7}	0	—	15–20	122	113	27	3	1.8	2.4×10^6	10^{-5}
B	2000	20–40	0.35	1×10^{-6}	30	—	20–30	59	1,000	23	6	1.8	2.4×10^6	10^{-4}
C	2200	14–20	0.3	1×10^{-6}	165	—	70–150	83	1,400	27	20	1.8	2.4×10^6	10^{-4}
D	2550	—	—	—	300	11,000	$R_c = 12$ MPa	620–3,100	550–2,800	—	—	1.1	2.0×10^6	10^{-6}
Pile	2500	—	—	—	—	—	-	17,381	14,313	—	—	2.1	2.0×10^6	10^{-5}

Note: Fluid dilation coefficient, $\beta_f' = 10^{-5}$ °C^{-1}; R_c, unconfined compressive strength.
Source: Adapted from De Cérenville (1997).

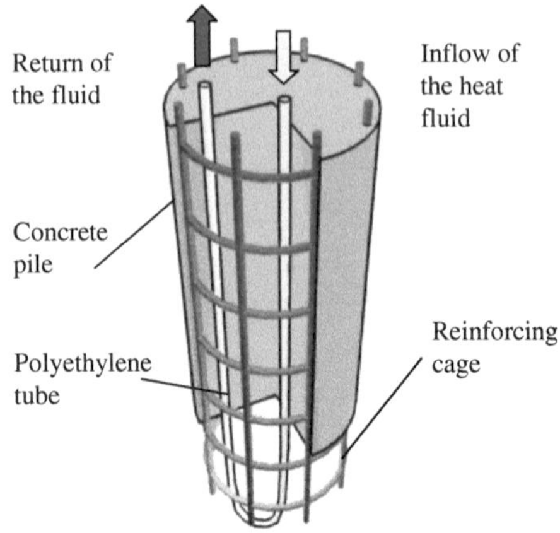

Figure 16.4 PE tubes in the pile.

16.2.1 Additional tests

Several additional tests were run to characterize the materials, including triaxial tests on samples from the A, B, and C soil layers and ultrasonic transmission and integrity tests in the pile.

Ultrasonic transmission tests

The elastic modulus of the concrete in the pile, $E_{concrete}$, was found based on laboratory tests, as well as on cross-hole ultrasonic transmission tests, having determined a unit weight of approximately 2500 kg/m^3. $E_{concrete}$ values for two different assumptions for the Poisson's ratio ($\upsilon=0$ and 0.16) are between 26 and 32 GPa. Introducing the percentage of reinforcing steel φ and E_{steel}, the elastic modulus of steel, the following relationship gives the total estimated elastic modulus for the pile:

$$E_{pile} = E_{concrete}\left[1+\varphi\frac{E_{steel}}{E_{concrete}}\right]$$

The elastic modulus of the pile, E_{pile}, was finally estimated to be 29,200 MPa.

Integrity test

Pile integrity and pile geometry were checked by using the reflected wave method carried out with the PIT integrity device. No anomalies were discovered, with the exception of a slight increase in pile section with depth (Fig. 16.5). For the remaining analyses, the real diameter at various depths will be taken into account, and not the nominal one. Thus, the section

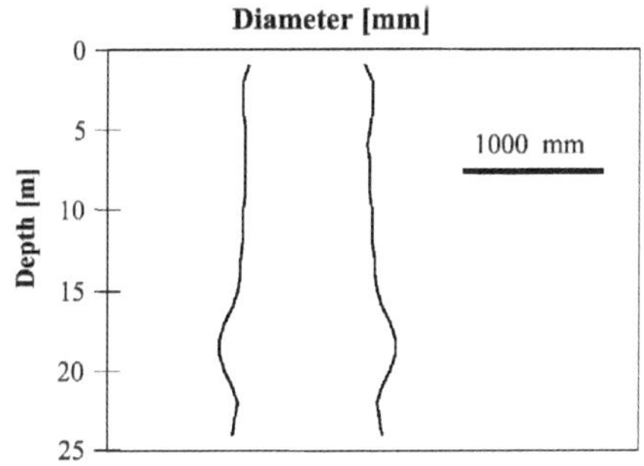

Figure 16.5 PIT integrity test results: diameter of the pile versus depth.

varied between 7200 and 10,800 cm^2. However, for the numerical simulations, an average value of 9000 cm^2 was adopted.

Readers interested in more details of this in situ test may refer to Laloui et al. (2003).

16.2.2 Loading history

The test pile was subjected to two types of loading: mechanical and thermal. The mechanical load was simply applied by the dead weight of the building under construction; the thermal load was applied by a heating device controlling the water temperature in the PE tubes installed in the pile. The two types of loads were applied separately and alternately in order to decouple the thermal and mechanical effects. Thus, at the end of the construction of each story, a thermal loading cycle was applied to the pile (for the complete loading history, see Fig. 16.6). Without taking into account the measurements made during the

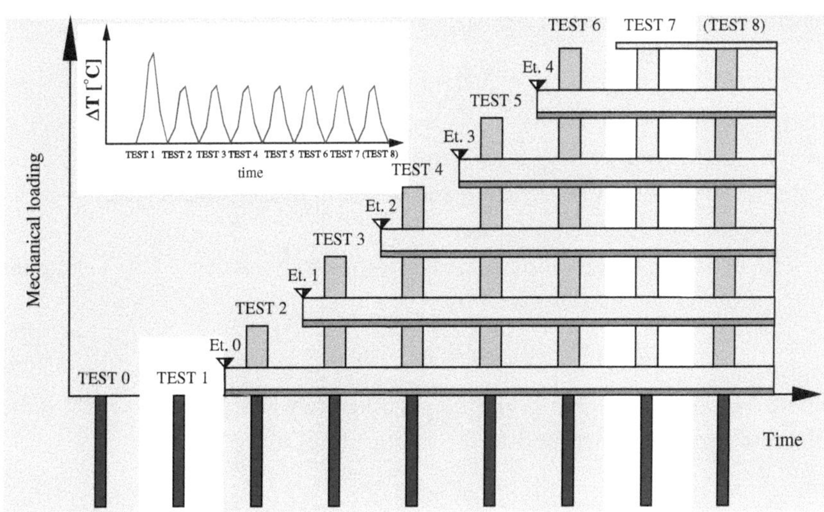

Figure 16.6 Thermo-mechanical loading history.

concreting of the pile (Test 0), seven tests were carried out. The maximum mechanical load on the pile head was approximately 1300 kN.

Test 1 differs from the others with regard to the free displacement boundary condition of the pile head. For the other tests, the pile was blocked in its movement by the applied load due to the building stories under construction. In addition, the thermal loads were on the order of $\Delta T = 21$ °C for Test 1, whereas they were 15 °C for the others (values usually encountered in heat exchanger piles). The measurements for Test 8 were made at the natural temperature of the ground.

16.3 THERMO-HYDRO-MECHANICAL MODEL

To simulate the behavior of the heat exchanger pile and the surrounding soil, a numerical model coupling the mechanical behavior to the thermal and hydraulic phenomena is needed. Here, a thermo-hydro-mechanical (THM) model for saturated porous media is used. This Biot-type model is derived using homogenization theory (Modaressi et al., 1991a; Laloui, 1993). Fully coupled equations govern the evolution of pore water pressure, solid displacement, and heat flow under mechanical, hydraulic, and thermal loading.

16.3.1 Mathematical formulation

The field equations of the THM model are

Momentum conservation

$$\text{div}\ \sigma + \rho_{sat}\ \mathbf{g} = 0 \tag{16.1}$$

where σ is the total (Cauchy) stress tensor with tensile stresses taken as positive; ρ_{sat} is the total average mass density; $\rho_{sat} = n\rho_f + (1-n)\rho_s$, where ρ_s and ρ_f are the mass density of the solid and the fluid phases, respectively; n is the porosity; and $\boldsymbol{g}$ the gravity vector.

The behavior of the solid matrix is assumed to be governed by Terzaghi's concept of effective stress given by

$$\boldsymbol{\sigma} = \boldsymbol{\sigma}' - p\delta \tag{16.2}$$

where δ is the Kroenecker's symbol, p is the pore water pressure, and σ' is the effective stress tensor related to the elastic strain ε^e by

$$\boldsymbol{\sigma}' = \boldsymbol{D} : \varepsilon^e \tag{16.3}$$

where $\boldsymbol{D}$ is the elastic constitutive tensor depending on the Young's modulus, E, and Poisson's ratio, ν. Thus, the momentum conservation equation takes the form

$$\mathbf{div}\ (\boldsymbol{D}:\varepsilon^{e}) - \mathbf{grad}\ p + \rho_{sat}\ \boldsymbol{g} = 0 \tag{16.4}$$

Mass conservation

The continuity equation reads

$$\frac{\partial_t p}{Q} - \frac{\partial_t T}{Q'} + \text{div}\ \partial_t \boldsymbol{u}_{rf} + \text{div}\ \partial_t \boldsymbol{u}_s = 0 \tag{16.5}$$

where T is the temperature, $\boldsymbol{u}_{rf}$ is the relative fluid velocity, and $\boldsymbol{u}_s$ is the solid displacement. Assuming linear compressibility of phases, Q and Q' are given by

$$Q = \frac{1}{[n\beta_f + (1-n)\beta_s]};\quad Q' = \frac{1}{[n\beta'_f + (1-n)\beta'_s]} \tag{16.6}$$

where β_α is the compressibility of the phase α (α = s for solid or f for fluid) at constant temperature T, β_α is the coefficient of thermal (volumetric) expansion of the corresponding phase at constant pore water pressure. The temperature is assumed to remain the same in both phases. Darcy's law is used to describe the relative fluid velocity

$$\partial_t \boldsymbol{u}_{rf} = -\boldsymbol{K}^{*}\mathbf{grad}\ (p + \rho_f\, \boldsymbol{g}\boldsymbol{x}) \tag{16.7}$$

where $\boldsymbol{K}^{\star}$ is the kinematic permeability tensor, and $\boldsymbol{x}$ is the position vector of the material point.

Energy conservation

With c, c_f, and Λ being the total equivalent specific heat, the specific fluid heat, and the coefficient of the equivalent-medium thermal conductivity, respectively, and assuming Fourier's linear law, energy conservation is given by

$$\begin{aligned}&\rho_c \partial_t T + \rho_f c_f \partial_t \boldsymbol{u}_{rf}\,\mathbf{grad}\ \ T + \rho c T\ \text{div}\ \partial_t \boldsymbol{u}_s + \rho_f c_f T\ \text{div}\ \partial_t \boldsymbol{u}_{rf}\\ &- \left[\frac{\beta'_f}{\beta_f}\ \text{div}\ (n\partial_t \boldsymbol{u}_s + \partial_t \boldsymbol{u}_{rf}) + (1-n)\partial_T \sigma' : d_t \varepsilon_s\right] T - \text{div}[\Lambda \text{grad}\ \ T] = 0\end{aligned} \tag{16.8}$$

In the case of heat exchanger piles, the heat convection as well as the heat induced by the solid displacement may be neglected. The energy equation is then decoupled and takes the following simple form:

$$\rho c \partial_t T - \text{div}[\Lambda\ \mathbf{grad}\ \ T] = 0 \tag{16.9}$$

As may be seen, the fundamental laws governing static equilibrium (Eq. (16.4)), fluid flow (Eq. (16.5)), and heat flow (Eq. (16.9)) are coupled throughout the dependent variables solid displacement vector, pore water pressure, and temperature in the medium.

Constitutive behavior of the solid matrix: Drucker–Prager thermo-elastoplastic model

The behavior of clays and other fine-grained soils (usual host media for deep geostructures) is strongly affected by temperature variations and exhibits thermo-plastic behavior (Cekerevac and Laloui, 2004). To describe such complex behavior, several advanced thermo-mechanical models are available (Laloui, 2001; Laloui and Cekerevac, 2003). In this work, a simple thermo-elastoplastic model is used for the simulation of the behavior of the concrete pile as well as that of the surrounding soil.

In incremental form, the total strain, $\dot{\varepsilon}$, is divided into two parts:

$$\dot{\varepsilon} = \dot{\varepsilon}^{\mathrm{Te}} + \dot{\varepsilon}^{\mathrm{p}} \tag{16.10}$$

where $\dot{\varepsilon}^{\mathrm{Te}}$ is the linear thermo-elastic strain, and $\dot{\varepsilon}^{\mathrm{p}}$ is the plastic strain.

Thermo-elastic part

The thermo-elastic strain rate is expressed as

$$\dot{\varepsilon}_{v}^{\mathrm{Te}} = \frac{\dot{p}'}{K} + \beta_{\mathrm{s}}'\dot{T} \tag{16.11}$$

$$\dot{\varepsilon}_{\mathrm{d}}^{e} = \frac{\dot{q}}{G} \tag{16.12}$$

where $\dot{\varepsilon}_{v}^{\mathrm{Te}}$ is the volumetric strain rate and $\dot{\varepsilon}_{\mathrm{d}}$ the deviatoric strain rate; $\dot{q}$ is the deviatoric effective stress; $'$ is the effective mean pressure; K and G are the bulk and shear moduli, respectively; $\dot{T}$ is the temperature increment; and β_{s}' is the thermal expansion coefficient of the solid skeleton.

Plastic part

The plastic strain rate is given by

$$\dot{\varepsilon}^{\mathrm{p}} = \lambda.\boldsymbol{\Psi} \tag{16.13}$$

where λ is the plastic multiplier, and $\boldsymbol{\psi}$ is the direction of the strain increment derived from the plastic potential. An associated model with the Drucker–Prager yield limit is adopted here:

$$f = \frac{2\sin\phi}{\sqrt{3}(3-\sin\phi)}\sigma_{\mathrm{I}} + \sigma_{\mathrm{II}}^{\mathrm{d}} - \frac{6c\sin\phi}{\sqrt{3}(3-\sin\phi)} \tag{16.14}$$

where σ_{I} and $\sigma_{\mathrm{II}}^{\mathrm{d}}$ are the first invariant of the effective stress tensor and the second invariant of the deviator of the effective stress tensor, respectively; ϕ is the internal friction angle; and c is the cohesion.

16.3.2 Numerical integration

Eqs. (16.4) (equilibrium), (16.5) (flow), and (16.9) (heat), together with the constitutive thermo-elastoplastic law, constitute the coupled system of equations to be solved. To be able to resolve this formulation for complex geometries and boundary conditions, a finite element approach is adopted. The weak form of the governing equations was obtained using a variational formulation (Laloui, 1993).

Initial/boundary conditions

Let Ω be the domain occupied by the solid skeleton and Γ its boundary. The problem is to find $\boldsymbol{u}_s(\boldsymbol{x}, t)$, $p(\boldsymbol{x}, t)$, and $T(\boldsymbol{x}, t)$, where $\boldsymbol{x}$ is the position vector and t the time such that Eqs. (16.4), (16.5), and (16.9) are satisfied on Ω, and such that

$$\boldsymbol{u}_s = \boldsymbol{u}* \text{ on } \Gamma_u, \quad \sigma \boldsymbol{n} = \boldsymbol{t} \text{ on } \Gamma_\sigma$$

$$p = p* \text{ on } \Gamma_{\mathrm{p}}, \quad \partial_{\mathrm{t}} \boldsymbol{u}_{\mathrm{rf}} \boldsymbol{n} = \phi \text{ on } \Gamma_\phi$$

$$T = T* \text{ on } \Gamma_T, \quad q_T \boldsymbol{n} = \Theta \text{ on } \Gamma_{\phi_t}$$

together with the initial conditions:

$$\boldsymbol{u}_{\mathrm{s}}(\boldsymbol{x}, 0) = \boldsymbol{u}_{\mathrm{s}0}(\boldsymbol{x}), \quad p(\boldsymbol{x}, 0) = p_0(\boldsymbol{x}), \quad T(\boldsymbol{x}, 0) = T_0(\boldsymbol{x})$$

where q_T is the heat flow given by Fourier's law. Γ_u, Γ_σ, Γ_ρ, Γ_ϕ, Γ_T, and $\Gamma_{\phi T}$ are the parts of the boundary on which the solid displacements, the total stress vector, the pore water pressure, the fluid flux, the temperature, and the thermal flux, respectively, are prescribed.

Weak formulation and finite element discretization

Let V, Q, and F be the spaces of admissible virtual displacements, pressures, and temperatures, respectively:

$$V = \{\mathbf{v}/\mathbf{v} \text{ smooth on } \Omega/\mathbf{v} = 0 \text{ on } \Gamma_u\}$$

$$Q = \{q_{\mathrm{v}}/q_{\mathrm{v}} \text{ smooth on } \Omega/q_v = 0 \text{ on } \Gamma_p\}$$

$$F = \{\theta/\theta \text{ smooth on } \Omega/\theta = 0 \text{ on } \Gamma_T\}$$

Thus, Eqs. (16.4), (16.5), and (16.9) of the initial/boundary value (THM) problem, after applying the Green–Gauss theorem with the boundary conditions previously described, become

$$\forall v \in V,\ \ (\mathbf{div}\{D:\varepsilon(\boldsymbol{u}_s)\},\boldsymbol{v})_\Omega - (\beta_s' T,\ \mathrm{div}(\boldsymbol{v}))_\Omega - (p,\ \mathrm{div}(\boldsymbol{v}))_\Omega \\ = (\rho_{\mathrm{sat}}(\boldsymbol{u}_s,\ T)\boldsymbol{g},\boldsymbol{v})_\Omega + \langle \boldsymbol{t},\ \boldsymbol{v}\rangle_{\Gamma\sigma} \tag{16.15}$$

$$\forall q \in Q,\ \ (\mathrm{div}(\partial_t \boldsymbol{u}_s), q_v)_\Omega + (\boldsymbol{K}^*\ \mathbf{grad}\ p,\ \mathbf{grad}\ q_v)_\Omega + (\partial_t p/Q,\ q_v)_\Omega \\ -(\partial_t T/Q',\ q_v)_\Omega = \left(\boldsymbol{K}^*\ \mathbf{grad}\ \left(\rho_f g\cdot x\right),\ \ \mathbf{grad}\ q_v\right)_\Omega - \langle \phi,\ q_v\rangle_{\Gamma\phi} \tag{16.16}$$

$$\forall \theta \in F,\ \ (\rho c \partial_t T,\theta)_\Omega + (\Lambda\ \mathbf{grad}\ T,\ \mathbf{grad}\ \theta)_\Omega + \langle \Theta,\ \theta\rangle_{\Gamma_{\phi T}} \tag{16.17}$$

where $(,)_\Omega$ denotes the integration over the domain, and $<,>\Gamma$ denotes the integration on its boundary Γ.

With the semidiscretization in space by means of isoparametric finite elements, displacements, pore water pressures, and temperatures (respectively $\boldsymbol{u}_s$, p, and T) defined in spaces V, Q, and F can be respectively approximated by $\boldsymbol{u}_{\mathrm{sh}}$, p_{h}, and T_{h}, and expressed as follows on V_{h}, Q_{h}, and F_{h}, subspaces of V, Q, and F:

$$\boldsymbol{u}_{\mathrm{sh}} = \sum_{\mathrm{I}} \left(\sum_{j=1}^{N} u_{\mathrm{sI}}^{j}(t)\, W_{\mathrm{SI}} \mathbf{e}_j \right) \\ p_{\mathrm{h}} = \sum_{\mathrm{I}} p_{\mathrm{I}}(t)\, W_{\mathrm{fI}} \\ T_{\mathrm{h}} = \sum_{\mathrm{I}} T_{\mathrm{I}}(t)\, W_{\mathrm{I}} \tag{16.18}$$

where N is the dimension of V_{h}, and $\{\mathbf{e}_i\}_i{=}_{1,\mathrm{N}}$ is a basis of V_{h}. W_{I}, W_{fI}, and W_{SI} are the interpolation functions on node I of temperature, pore water pressure, and displacement, respectively, considered here to be identical. Hence, the semidiscretized system derived from Eqs. (16.15)–(16.17) can be written in matricial form:

$$\forall j, \forall J, \Sigma_{\mathrm{I}}\left(\Sigma_{\mathrm{I}} K_{\mathrm{IJ},ij} u_{\mathrm{sI}}^{i} - L_{\mathrm{IJ},j} p_{\mathrm{I}} - L_{\mathrm{IJ},j}' T_{\mathrm{I}} = F_{\mathrm{SJ},j} \right. \\ \forall \mathrm{J}, \Sigma_{\mathrm{I}}\left(\Sigma_i L_{\mathrm{JI},i} \partial_t u_{\mathrm{sI}}^{i} + H_{\mathrm{IJ}p\mathrm{I}} + M_{\mathrm{IJ}} \partial_t p_{\mathrm{I}} - M_{T\mathrm{IJ}} \partial_t T_{\mathrm{I}}\right) = F_{\mathrm{PJ}} \\ \left(\forall \mathrm{J}, \Sigma_{\mathrm{I}}\left(M_{T\mathrm{IJ}}' \partial_t T_{\mathrm{I}} + H_{T\mathrm{IJ}} T_{\mathrm{I}}\right) = F_{T\mathrm{J}}\right) \tag{16.19}$$

Different solution procedures to solve this system of equations have been developed (Modaressi et al., 1991b; Laloui, 1993) using a thoroughly coupled formulation with global resolution and a prediction–correction method. The three-dimensional formulation of this THM model is implemented in a general-purpose finite element code Gefdyn (Aubry et al., 1985).

16.4 FINITE ELEMENT MODEL

The test pile was statically part of a pile group but the only one to be heated. It will therefore be treated as a single pile for the sake of simplicity. It appears, however, useful to consider this analysis, which is similar to but less complex than the full analysis of the whole group of surrounding piles. Because the considered pile was circular in cross section and installed vertically in the ground, which has a horizontal lithology and horizontal water table, an axi-symmetric geometry was adopted. The mesh takes into account a heat exchanger pile with a radius of 0.5 m (instead of 0.44 m as in the experiment) and a length of 26 m, as well as the five soil layers. The bottom of the mesh is located at a depth of 52 m (Fig. 16.7). The elements (534 in total) are quadratic for displacements and bilinear for pore water pressures and temperatures. The contact between pile shaft and soil was described as perfectly rough. This implies that the finite element nodes of the pile and the soil are constrained by no relative movement. The material behavior of the contact area was simulated by the material behavior of the soil. To overcome the problems that can rise at this interface, a refined finite element mesh was used at that location.

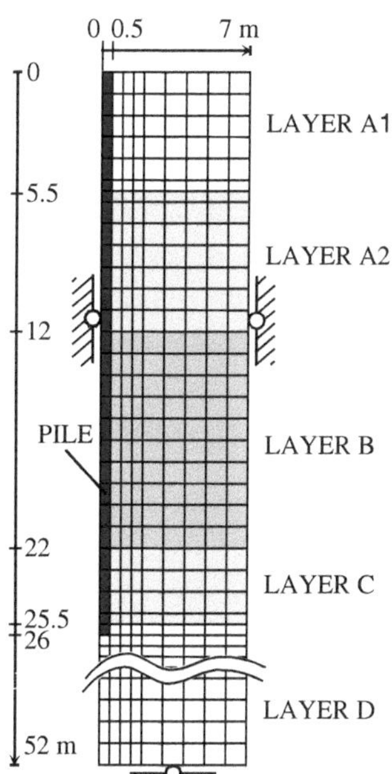

Figure 16.7 Finite element mesh and mechanical boundary conditions.

16.4.1 Material characteristics

The material parameters are summarized in Table 16.1. The soil is represented by five layers obeying the Drucker–Prager thermo-elastoplastic model. The pile itself is modeled with solid elements and behaves as a thermo-elastic material. The mechanical parameters (K, G, ϕ, c) were determined from triaxial tests at three confining pressures for each soil layer (Laloui et al., 1999). The pile is considered impervious, as is the layer D (molassic formation), whereas the other soil layers are considered drained.

The horizontal permeability coefficients were determined from in situ measurements, and isotropic permeability was further assumed. The thermal parameters (Λ, ρc, β'_s) were estimated based on the geotechnical characteristics, whereas the porosity of each layer was determined experimentally.

16.4.2 Initial and boundary conditions

The mechanical boundary conditions are shown in Fig. 16.7. Restrictions were applied to both vertical and horizontal displacements on the base of the mesh and to horizontal displacements on the sides. The initial stresses in the pile, as well as in the soil layers, were calculated assuming saturated bulk unit weights and a coefficient of earth pressure at rest $K_0 = 1$.

The hydraulic boundary conditions consisted in assuming the pile as impervious, allowing the drainage of the soil layers to take place at the top surface as well as on the right-hand side of the mesh. Undrained conditions were applied along the axis of symmetry and for the layer D. The initial pore water pressure corresponds to the hydrostatic profile with a water table located at the top surface.

The thermal boundary conditions consist in allowing the heat to flow through the right-hand side of the mesh as well as through the bottom of the mesh. Constant temperatures were imposed on the top surface of the mesh, and the heat flux was supposed null along the axis of symmetry.

16.5 EXPERIMENTAL AND NUMERICAL RESULTS

In this section, the THM finite element results are presented and compared with the experimental ones. The behavior of the concrete pile is considered first and then that of the soil layers.

16.5.1 Thermal behavior

To analyze the thermal behavior of the pile, Test 1 is considered (Fig. 16.6). In this test, a heating–cooling cycle was imposed (12 days of heating and then 16 days of cooling). Two thermal boundary conditions were applied:

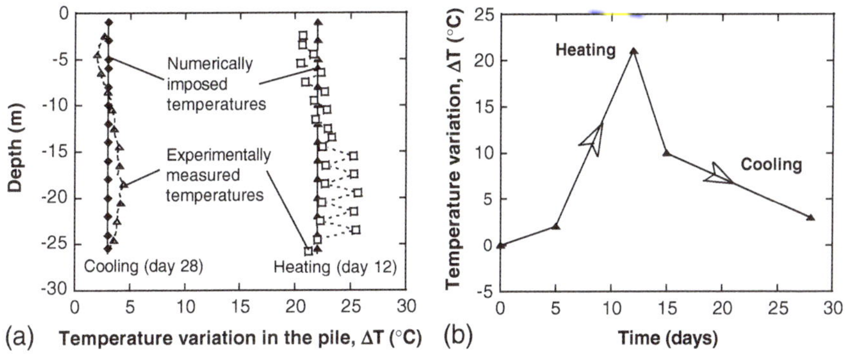

Figure 16.8 Temperature values imposed in the pile.

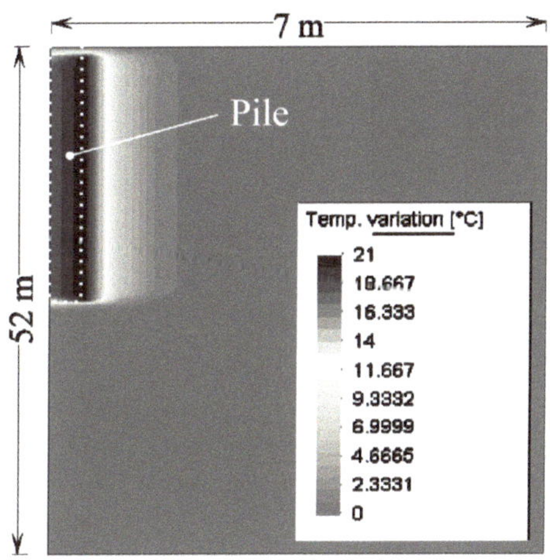

Figure 16.9 Isothermal values at the end of the heating period.

- Constant temperature at the top surface.
- Imposed incremental temperatures in the concrete pile. These values are determined from the measured ones (Fig. 16.8(a)). Time variation of the thermal load is shown in Fig. 16.8(b).

Figure 16.9 shows the temperature field after the heating period. Note that due to the thermal parameters of the molasses layer, the thermal diffusion is mainly horizontal. As a result, the part of the mesh affected by the applied loads is limited to the depth of the pile. Some of the following results will be thus presented only for the upper portion of the mesh.

16.5.2 Thermal effect on the mechanical behavior of the pile

Thermal strains

In the case of Test 1 (Fig. 16.6), the thermal heating–cooling cycle was applied to the pile. No mechanical load was imposed at the pile top, and the pile was free to move upwards. The imposed thermal field generated strains in the pile. Figure 16.10(a) compares the simulated thermo-elastic vertical strains with respect to the measured ones for the heating and cooling periods. Strains are not uniform during the heating period and are influenced by the friction along the pile shaft. In fact, the measurements show different straining according to the type of surrounding soil, and the layer boundaries (A–D) may be identified in Fig. 16.10(a). The model is able to reproduce this effect.

Thermo-elastic linear behavior is observed and computed during the cooling phase. This reversibility means that the displacement of the pile with respect to the soil has not yet reached the threshold where the friction would no longer permit the pile to return to its initial state. The modeled radial strains fit the measured ones well, as may be seen for one of them (at a depth of 16 m) in Fig. 16.10(b). The radial strain behavior shows that lateral contact is still maintained between the pile and the ground after a thermal cycle. Note that the final temperature values are different from the initial ones; this explains why the strains do not return to zero at the end of the heating–cooling cycle.

Pile uplift

Test 1 enabled the measurement of the maximum uplift produced by a temperature increase of 21 °C. Values obtained by precision leveling (extensometers and the optical fibers) compared well with the numerical results (Fig. 16.11).

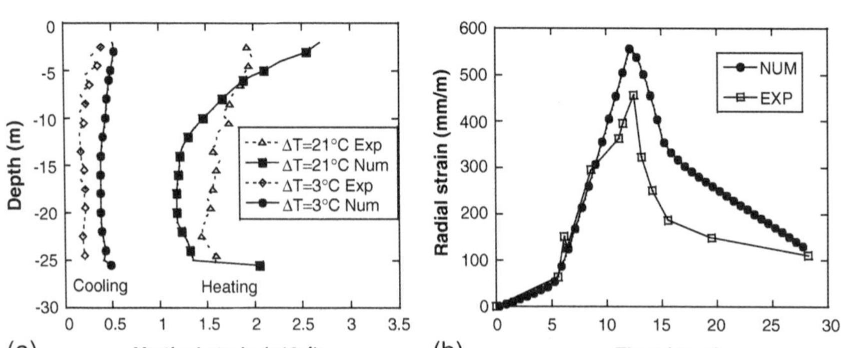

Figure 16.10 Thermal strains during a heating and cooling period.

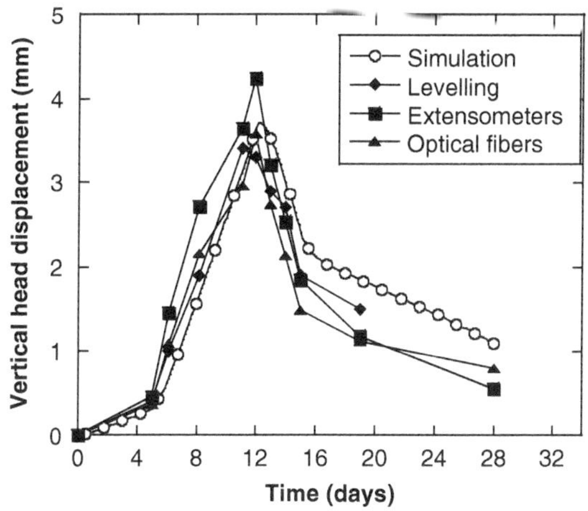

Figure 16.11 Thermal pile uplift during a heating–cooling cycle.

Induced thermal stresses

Under the effect of a temperature increase, a pile with free boundary conditions dilates. In the case in which it is not entirely free to move (e.g., due to side friction or blocked head and toe), part or all of the deformation will be prevented. The constrained strains produce thermal stresses. Figure 16.12(a) shows a comparison between measured and simulated variations in vertical stresses corresponding to Test 1 under a temperature variation of 13.4 °C. Figure 16.12(b) presents the vertical effective stresses during the heating–cooling cycle.

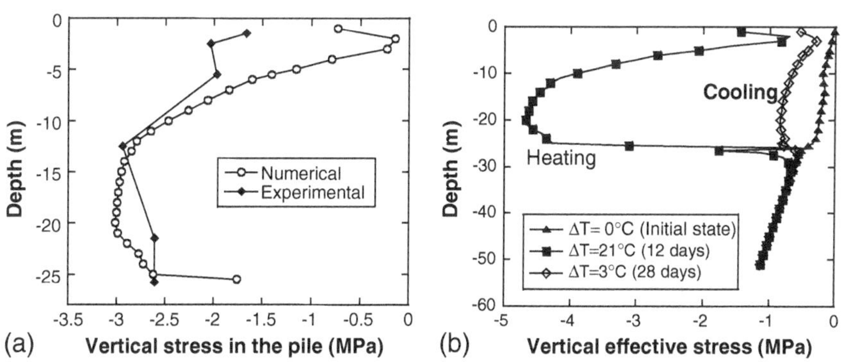

Figure 16.12 Thermal vertical stresses: (a) under a thermal load of 13.4 °C and (b) during a heating–cooling cycle.

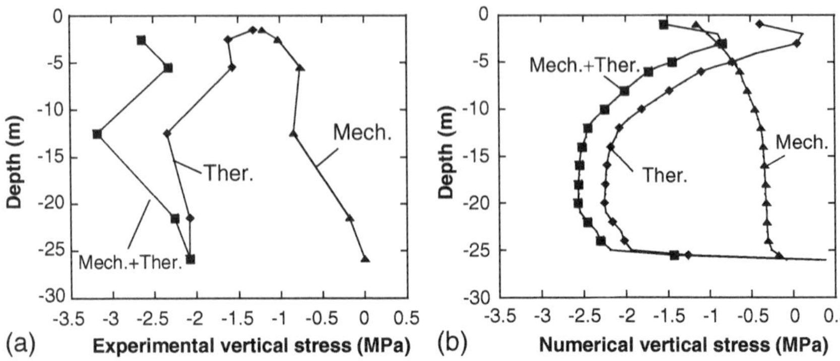

Figure 16.13 Thermo-mechanical vertical stresses in the pile: (a) experimental results and (b) numerical simulations.

16.5.3 Thermo-mechanical behavior of the pile

To analyze the thermo-mechanical behavior of the pile, Test 7 (Fig. 16.6) is now considered. The experimental results clearly show the differences between mechanical and thermal loading (Fig. 16.13(a)). Although the mechanical axial stress is large at the pile top (on the order of 1.3 MPa) and diminishes with depth (the toe carries almost no load), the thermal load is larger and rather uniform. This results in an overstress on the order of 1.2 MPa at the pile head and strongly solicits the toe (2 MPa). An analysis of Tests 5 and 6 shows that a temperature increment of 1 °C results in an additional temperature-induced load on the order of 100 kN. As a consequence, the total axial load in the pile is twice as large as the one due to purely mechanical loading, with a large solicitation of the toe.

The mechanical loading is modeled by applying increments of vertical stress to the top of the pile. The thermal loading is imposed in the pile as discussed previously. The model is able to reproduce the decrease in the mechanical vertical stresses with depth as well as the increase in the thermally induced vertical stresses with depth (Fig. 16.13(b)).

16.5.4 Thermo-hydro-mechanical behavior of the soil

The THM finite element modeling shows the following results:

In the case of Test 1 (thermal load only)

- Heating of the pile produces a global swelling of the top surface of the soil (Fig. 16.14). The calculated displacement at the pile top corresponds well to the measured value (Fig. 16.11).

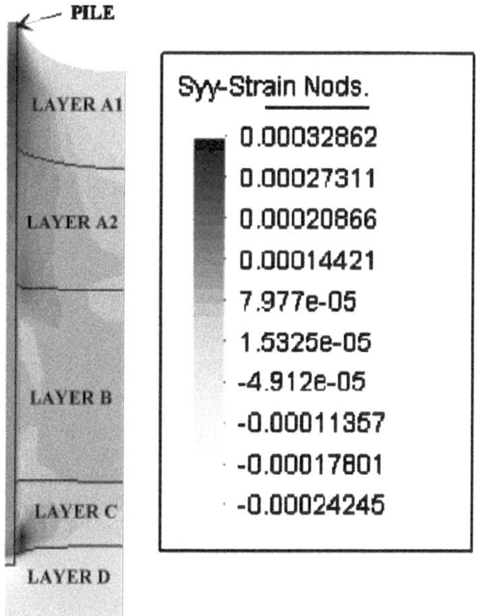

Figure 16.14 Test 1—heat-induced deformations (magnification × 1425.15).

- The induced vertical strains are mainly dilative. However, as was experimentally observed, compressive strains are generated at the bottom of the pile.

In the case of Test 7 (thermo-mechanical load)

- The mechanical load (induced by the weight of the building) on the pile does not produce any strain in the soil at a distance as short as 1 m from the pile shaft.
- The heat effect is more pronounced and induces small strains at 1 m from the pile shaft (Fig. 16.15).
- Due to the limited mechanical effects and the small heat-induced strains, the pore water pressures, as well as the effective stresses, remain almost constant.

16.6 CONCLUSION

Heat exchanger piles offer a good opportunity for the use of sustainable energy for building heating and cooling. Despite the existence of several installations of this type, the understanding of the THM behavior is limited. This chapter presented new experimental work and modeling capabilities.

Strains in the test pile are thermo-elastic in nature. Their intensity depends on the type of surrounding soil. The additional thermal-induced axial load in

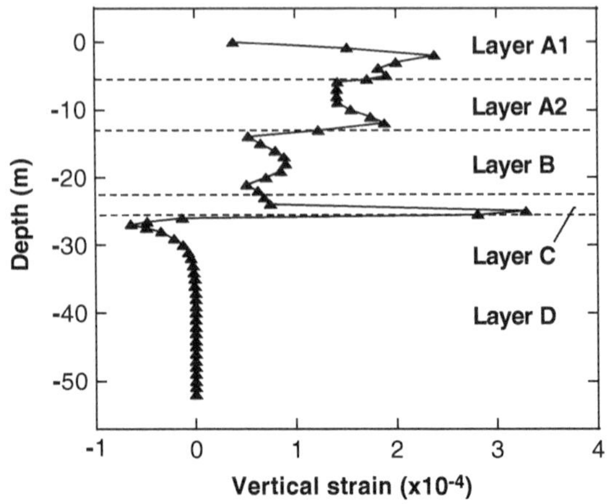

Figure 16.15 Test 7—thermo-mechanical-induced vertical strains in the soil at 1 m from the pile shaft.

the pile may be rather large. At the pile toe, thermal effects produce much larger axial stresses than those (almost zero) created by the dead weight of the building; this is due to the uniform nature of the thermal effects. Finally, the difference in the direction of movement (uplift for thermal and settlement for static) acts such that friction resistance is not affected by temperature and relief of side friction mobilization is seen during heating. Even if the thermal effect propagates more in the soil than does the mechanical load, the induced strains are limited and do not affect the pore water pressure or the effective stress.

A coupled displacement–pore water pressure–temperature finite element analysis was carried out. It was assumed that the concrete and the soil were saturated and governed by a thermo-mechanical law, that the heat flow followed Fourier's law, and that the pore fluid respected Darcy's law.

Applying the principle of effective stress, and determining the heat fields using energy conservation, from the equilibrium and the continuity statements a coupled THM system of equations is obtained. This coupled THM model is capable of modeling the complex behavior of a heat exchanger pile and reproduces the experimental results quite well.

REFERENCES

Aubry, D., Chouvet, D., Modaressi, A., Modaressi, H., 1985. Gefdyn: Logiciel D'Analyse Du Comportement Mécanique Des Sols Par Éléments Finis Avec Prise En Compte Du Couplage Sol-Eau-Air, Report Ecole Centrale Paris.
Cekerevac, C., Laloui, L., 2004. Experimental study of the thermal effects on the mechanical behaviour of a clay. Int. J. Numer. Anal. Meth. Geomech. 28, 209–228.

De Cérenville Géotechnique, 1997. Rapport géotechnique, Extension Quartier Nord Message 96, Ecublens, Switzerland.
Inaudi, D., Laloui, L., Steinmann, G., 2000. Looking below the surface. Concrete Eng. Int. 4(3).
Laloui, L., 1993. Modélisation Du Comportement Thermo-Hydro-Mécanique Des Milieux Poreux Anelastiques. Doctoral thesis, Ecole Centrale Paris.
Laloui, L., 2001. Thermo-mechanical behaviour of soils. RFGC 5, 809–843.
Laloui, L., Cekerevac, C., 2003. Thermo-plasticity of clays: an isotropic yield mechanism. Comput. Geotech. 30 (8), 649–660.
Laloui, L., Moreni, M., Steinmann, G., Vulliet, L., Fromentin, A., Pahud, D., 1999. Test En Conditions Réelles Du Comportement Statique D'Un Pieu Soumis à Des Sollicitations Thermo-Mécaniques. Report of the Swiss Federal Office of Energy, Bern, Switzerland.
Laloui, L., Moreni, M., Vulliet, L., 2003. Comportement d'un pieu bi-fonction, fondation et échangeur de chaleur. Can. Geotech. J. 40 (2), 388–402.
Laloui, L., Cekerevac, C., Vulliet, L., 2005. Non-isothermal modelling of the cyclic mechanical behaviour of MC clay. In: Proc. 11th Int. Conf. of IACMAG, Torino, Italy, pp. 377–384.
Modaressi, H., Forester, E., Laloui, L., 1991a. Stability analyses of coupled thermo-hydro-mechanical schemes for deformable porous media. In: Lewis, R.W., Chin, J. H., Homsy, G.M. (Eds.), In: Numerical Methods in Thermal Problems, vol. 7. Pineridge Press, Swansea, UK, pp. 1073–1083.
Modaressi, H., Laloui, L., Aubry, D., 1991b. Numerical modelling of thermal consolidation. In: Second European Speciality Conference on Numerical Methods in Geotechnical Engineering, Santander, pp. 280–292.

CHAPTER 17

Vacuum Consolidation and Vacuum Consolidation with Heating

Maria Esther Soares Marques[1], Serge Leroueil[2]
[1]Military Institute of Engineering, Rio de Janeiro, Rio de Janeiro, Brazil
[2]Department of Civil and Water Engineering, Laval University, Québec City, Québec, Canada

17.1 INTRODUCTION

In order to study the vacuum consolidation technique and heating applied over a typical eastern Canadian clay deposit, two trial embankments, 13 × 13 m, were executed approximately 30 km north of Montreal at Saint-Roch-de-l'Achigan, Québec (Fig. 17.1). The deposit was subjected to vacuum consolidation (trial embankment A) and vacuum consolidation and heating (trial embankment B), down to approximately 7.3 m depth. The heating was applied for 5 months, and the vacuum consolidation was also carried out for 5 months.

The main advantage of vacuum preloading is that the equivalent stress of a 4.5-m conventional preloading fill can be applied to the clayey deposit in 4–6 months as a function of the horizontal coefficient of consolidation of the soil and the vertical drains spacing. The vacuum consolidation can be applied in a single step, even on very soft soils, because the increase in vertical effective stress is due to the decrease in pore pressure; thus, there is no decrease in the general stability safety factor.

Studies on strain rate and temperature effect on compression, strength parameters, critical state, limit state, and permeability were summarized by Leroueil and Marques (1996), and those effects were discussed by Marques et al. (2004) after tests were carried out on Saint-Roch-de-l'Achigan clay. The effect of temperature on in situ soil behavior was also studied to determine the mechanical behavior of soil and its volume variation caused by temperature changes (Miliziano, 1992), the amount of energy storage (Bergenståhl et al., 1994; Moritz, 1995), and whether there is improvement of foundation characteristics (Edil and Fox, 1994). Studies of temperature effect on barriers for nuclear waste were also carried out by Lingnau et al. (1995), Hueckel and Peano (1987), and others.

Chemical, Electrokinetic, Thermal, and Bioengineering Methods
http://dx.doi.org/10.1016/B978-0-08-100191-2.00017-4

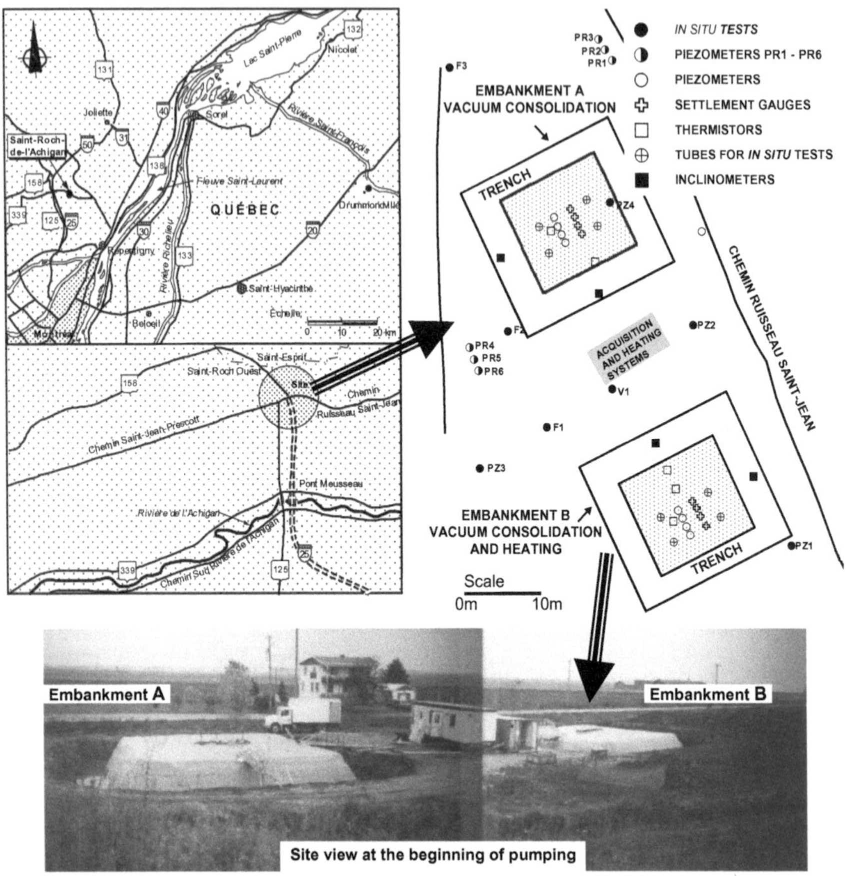

Figure 17.1 Localization and layout of Saint-Roch-de-l'Achigan site.

The purpose of the trial embankments was to improve the soft clay characteristics using both vacuum preloading and vacuum preloading with heating. It was the first time that a vacuum consolidation preloading was installed in Canada and the first time that vacuum consolidation was associated with heating. The idea of in situ heating originated from the fact that permeability, deformations, and strain rate increase with temperature in the normally consolidated range (Marques et al., 2004).

17.2 GENERALITIES ON VACUUM PRELOADING

The vacuum preloading technique was proposed by Kjellman (1952). Nevertheless, until 1980 this technique was almost never used because high-quality membranes and efficient water–air pumps were necessary to apply vacuum. Since 1980, this technique has been successfully used throughout the world (Choa, 1989; Qian et al., 1992; Cognon et al., 1994, Jacob et al., 1994; Magnan, 1994; Shang et al., 1998; Dong et al., 2001; Van Impe et al., 2001; Indraratna et al., 2004; Rujikiatkamjorn et al., 2007).

Figure 17.2 shows a section of the vacuum preloading platform installed at Saint-Roch-de-l'Achigan, Québec. The method consists of applying

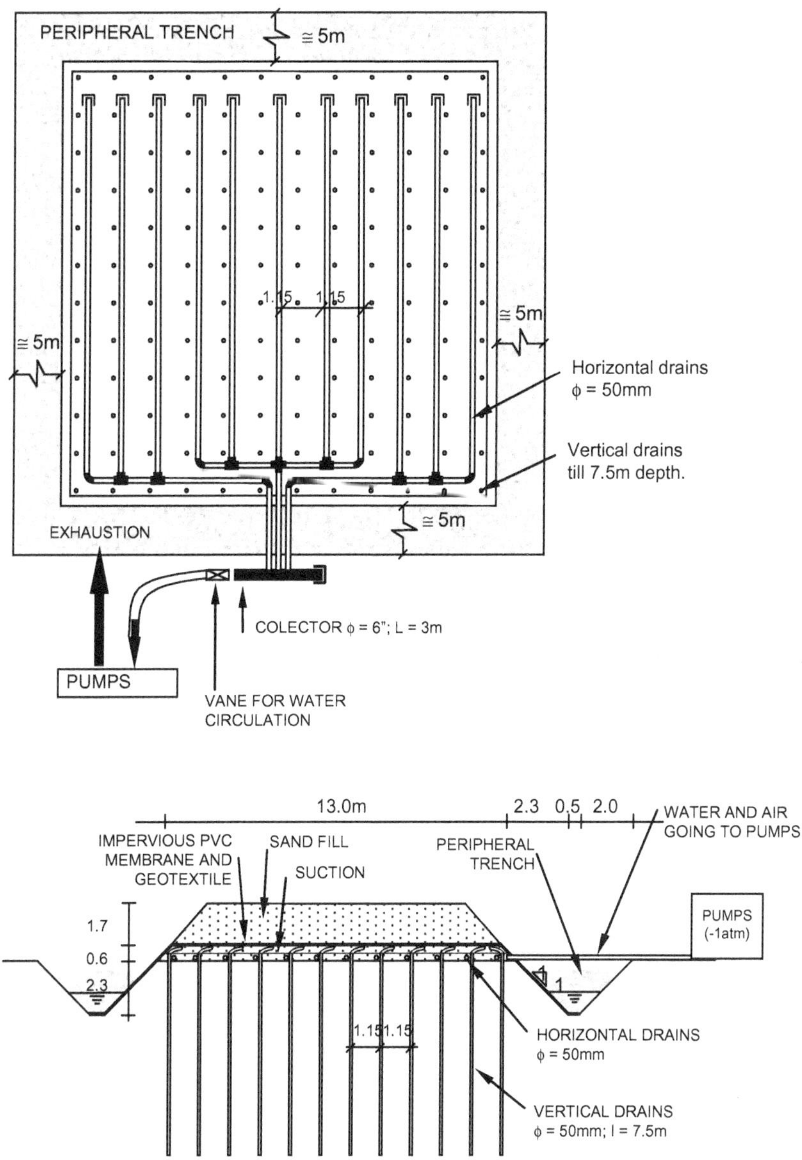

Figure 17.2 Vacuum consolidation preloading at Saint-Roch-de-l'Achigan.

vacuum in a clay deposit by means of a pumping system, in association with grids of vertical and horizontal drains. An impervious PVC membrane is installed at the ground level and goes below the water table in a peripheral trench in order to confine vacuum inside the soil mass. Instrumentation can pass through the membrane, but it must be well sealed to avoid air leakage.

Water and air are pumped inside a pump container in which pressure is close to −1 atm. During vacuum preloading, a pore pressure variation, Δu_v, of 70–75% of the atmospheric pressure (70–75 kPa) can usually be obtained under the membrane. The number of pumps used for vacuum preloading depends on the area to be treated and site conditions, and drain spacing is often chosen to consolidate the clay deposit within a 4- to 6-month period.

Vacuum preloading can also be applied in association with a conventional fill that can be built above the membrane after the air-tightness of the membrane has been checked. Once the designed degree of consolidation has been achieved, pumping is stopped and the fill height can be increased to final grade, without generating primary settlements.

Vacuum preloading has some advantages compared to the conventional technique:

- It is recommended when stability is a major concern. As previously indicated, it generally allows completion of the embankment after one stage of consolidation and can thus reduce construction duration.
- Horizontal displacements are smaller than those for conventional embankments, which can reduce horizontal forces applied on nearby structures (Magnan, 1994).
- Conventional fills are often partially submerged due to the development of settlements, which reduces the total vertical stress and the final effective stress. With vacuum preloading, the stress increase remains constant with time when loading.
- Fill volumes involved are generally smaller because vacuum preloading does not require berms or fill disposal when unloading.

Underwater vacuum consolidation has also been used as an improvement technique or to decrease volume of deposed dredging material. Van Impe et al. (2001) described an unconventional use of vacuum consolidation underwater without the PVC membrane and with horizontal drains installed inside a silty clay in order to decrease dredging storage volumes.

Vacuum preloading can be applied on very soft clay, thus eliminating the need to remove the upper layer material, which is sometimes difficult to accomplish due to environmental problems caused by the disposal of the material.

There are also some limitations when applying vacuum preloading with membrane:

- The membrane must be airtight, and vacuum efficiency decreases in clay deposits with granular layers. Also, in case of high permeable layer at the surface, the membrane must reach greater depths.
- The pump system needs technical maintenance while pumping; it may thus not be cost-effective for small areas.
- In northern countries, pumping systems cannot be used under extremely low temperatures, unless the hydraulic system is protected against frost.
- Vacuum pumping, when using a membrane system, increases the reference water level from the natural water table level to that of the horizontal drains where vacuum is applied. The maximum pore pressure change or the maximum change in effective stress caused by vacuum pumping depends on the relative position of the natural water table and the horizontal drains. If the water table is at a large depth, it is then preferable to install the horizontal drains at the level of the water table; the membrane has to go below the water table level in the peripheral trenches, which can be disadvantageous with regard to stability.

Schematic stress paths under the center of a conventional embankment (C) and a vacuum preloaded platform (V) are shown in Fig. 17.3. The yield curve shown in this figure comes from triaxial tests carried out at 20 °C on samples of the silty clay taken at a depth of 5.5 m at Saint-Roch-de-l'Achigan (Marques et al., 2004).

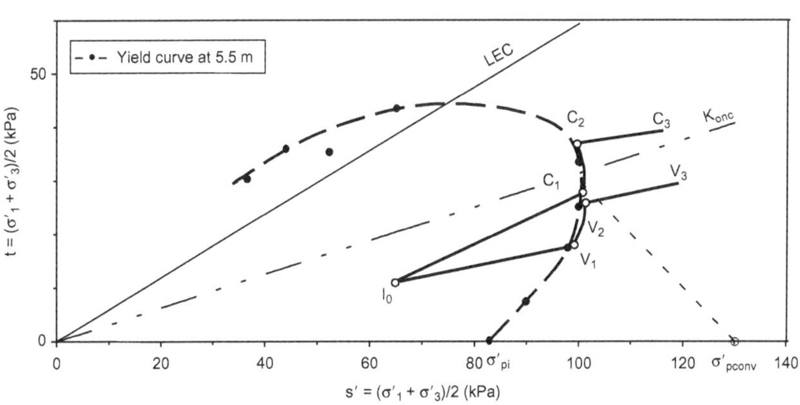

Figure 17.3 Schematic stress paths under conventional embankment and under a platform preloaded by vacuum.

For a conventional fill, the stress path is similar to I_0C_1 in the overconsolidated domain. Yielding is reached at C_1, and then the stress path follows the yield surface up to C_2 at the end of construction, under an almost constant vertical effective stress (Leroueil et al., 1978). During consolidation following construction, the stress path moves in the normally consolidated domain, along a stress path such as C_2C_3, above the K_{onc} line. As a consequence, important settlements and outward horizontal movements occur, typically equal to 20% of the settlement (Tavenas et al., 1979).

During vacuum preloading, effective stresses increase because the pore pressure decreases. Because pore pressure acts isotropically, the stress path tends to be parallel to the isotropic axis. In fact, due to gravity forces and boundary conditions, the stress path is probably not horizontal but, rather, like I_0V_1 in the overconsolidated domain, with yielding at V_1. The stress path then follows the yield surface from V_1 to V_2. Due to the shape of the limit state curves of natural clays (Díaz-Rodriguez et al., 1992), the vertical yield stress in these conditions (V_1) should be slightly smaller than that under a conventional embankment (C_1).

In the normally consolidated domain, the stress path should be similar to V_2V_3, below the K_{onc} line, resulting in horizontal displacements toward the treated area. Such behavior has been observed in particular by Magnan (1994) and Dong et al. (2001).

17.3 SITE CONDITIONS

The site was located at Saint-Roch-de-l'Achigan, close to the construction site of a viaduct over highway 25, as shown in Fig. 17.1. At that time, there were three other viaducts nearby, in different phases of construction, over this highway, with a typical embankment height of 7 m.

Figure 17.4 shows some geotechnical characteristics of Saint-Roch-de-l'Achigan clay deposit. It consists of a 2.5-m-thick and stiff clay crust overlying 10 m of plastic and overconsolidated silty clay, which covers a glacial till. The plasticity index of the silty clay varies between 41 and 46, and its liquidity index decreases from 1.4 at a depth of 3 m to 1.1 at a depth of 9 m. The overconsolidation ratio established on the basis of the conventional oedometer test is typically equal to 2.1. The ratio between isotropic yield stress (σ'_{pi}) and oedometer yield stress (σ'_{pconv}), $\sigma'_{pi}/\sigma'_{pconv}$, is typically equal to 0.65, well inside the range for Champlain sea clays (Díaz-Rodriguez et al., 1992). Based on the previous discussion in reference to Fig. 17.3, the in situ vertical yield stress mobilized with vacuum preloading is expected to be between σ'_{pi} and σ'_{pconv}.

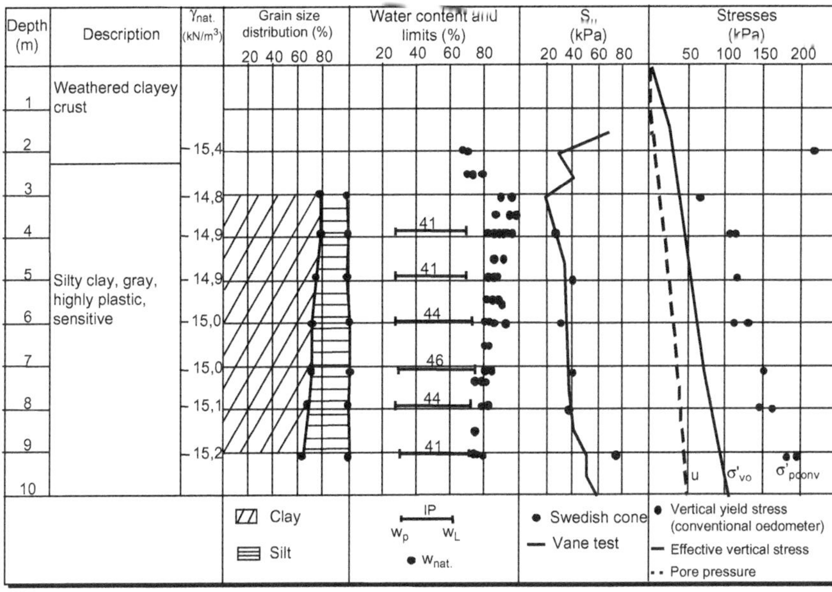

Figure 17.4 Geotechnical characteristics of Saint-Roch-de-l'Achigan clay deposit.

17.4 VACUUM PRELOADING AND HEATING

17.4.1 Site preparation

Prefabricated vertical drains were installed with a spacing of 1.15 m up to a depth of approximately 7.5 m in the silty clay deposit, and horizontal drains were placed in the approximately 60-cm-thick sand layer (Fig. 17.2). These drains were axially flexible and radially rigid PVC circular tubes that were perforated with geotextile around them.

Instrumentation was installed soon after the drains and the peripheral trench was excavated. Although not always required, geotextile was placed under the membrane in order to protect it against puncture. After installation of the membrane, the vacuum system was tested and a 0.7-m-thick sand layer was applied on the membrane.

At embankment B, thin copper tubes (3/8-in. diameter) were installed inside the vertical drains, and hot water was pumped inside these tubes down to 7.3 m depth in a closed hydraulic system, independent of the vacuum pumping system, as shown schematically in Fig. 17.5.

First, all the instrumentation was installed in the trial embankments and the heating was turned on at trial embankment B. At embankments A and B,

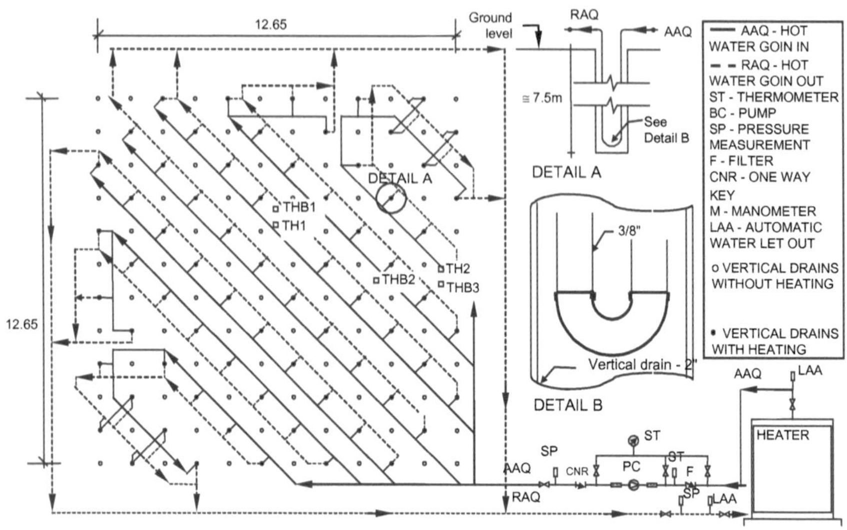

Figure 17.5 Layout of the heating system.

the vacuum pumping was turned on 30 and 58 days after beginning of heating, respectively. The fills were raised 60 days after the beginning of vacuum application so the deposit could achieve the normally consolidated domain.

The experimentation was installed in summer and the vacuum and the heating application went through a Canadian winter. The heating and vacuum system and the instrumentation were specially protected against the low ambient temperatures of the Canadian winter.

17.4.2 Instrumentation

Six Casagrande piezometers were installed outside the preloading area in order to obtain the original pore pressure profile. Under both embankments, the clay deposit was divided into the weathered crust and three sublayers of silty clay, with vibrating wire settlement gauges installed at depths of 3.25, 4.70, 6.15, and 7.60 m, as shown in Fig. 17.6. Also, four vibrating wire piezometers, capable of measuring some negative pore pressures, were installed in the middle of the sublayers (Fig. 17.6).

Three vertical wood thin piles, with six thermistors each (Fig. 17.7), at different depths, were installed inside the clay deposit under embankment B (THB1–THB3 in Fig. 17.5), and two others were installed under embankment A. Thermistors were also installed inside the sandy fill (TH1 and TH2 in Fig. 17.5).

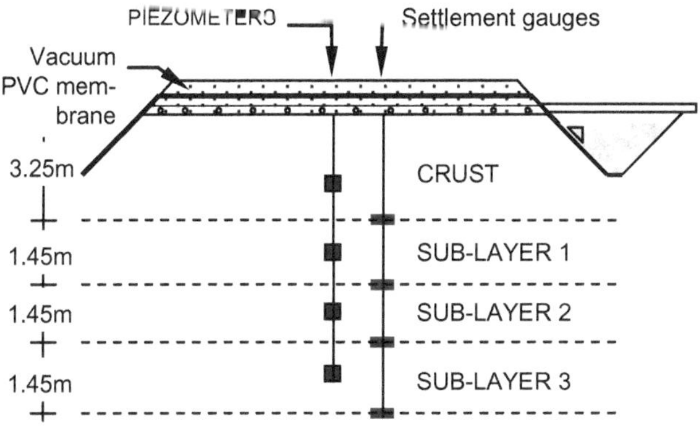

Figure 17.6 Instrumentation profile.

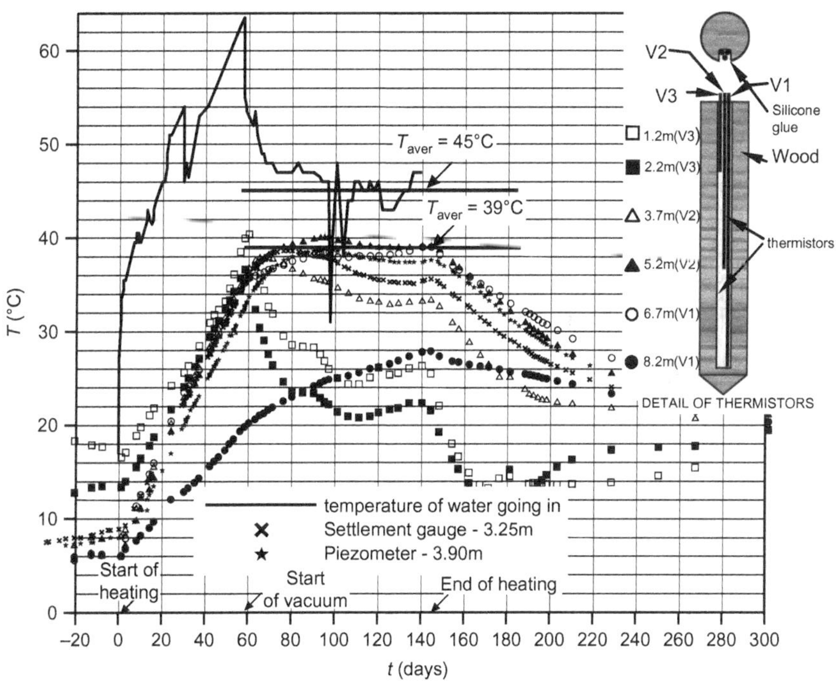

Figure 17.7 Temperature variation below embankment B.

The applied suction was measured inside the sand fill, below the membrane. Two inclinometers were installed just outside each platform, inside the peripheral trench. Surface settlement measurements were also carried out at the center and along the contour of the platform by 16 settlement plates. A vibrating wire piezometer was installed outside the treated area in order to provide regular and automatic measurements of the water table during vacuum preloading.

17.5 OBSERVATIONS AND ANALYSIS

17.5.1 Temperature

Seasonal temperature changes were observed inside the crust, but below the crust the initial temperature of the deposit was constant from 6° to 7 °C. Figure 17.7 shows the variation in temperature with time for the thermistors installed in the middle of embankment B (THB1), at 1.2, 2.2, 3.7, 5.2, 6.7, and 8.2 m. The piezometers and settlement gauges also measured temperature. These measurements were in accordance with those of the thermistors, as shown in Fig. 17.7.

The initial temperature increase rate was 0.5 °C/day. After 58 days of heating, the vacuum pumping was turned on and a decrease in temperature was observed due to the saturation of the crust and the sandy fill with cold water. Because the heating system had to compensate for the influx of cold water, the temperature rate decreased. The heating was carried out for 5 months, and the average soil temperature below the crust, from 4 to 7 m, increased from 7° to 39 °C.

17.5.2 Pore pressures

Figure 17.8 shows the pore water pressure profile obtained with Casagrande piezometers at different times (Fig. 17.8(a)). The water table was at 1.5-m depth, inside the weathered crust, and there was a water flow toward the underlying till layer.

The vibrating wire piezometers were installed in the center of the platform soon after the installation of the vertical drains. Due to the small distance between drains (1.15 m), the pore pressure profile became hydrostatic, in equilibrium with the natural water table level (Fig. 17.8(b)). Hahn et al. (1998) described a vacuum preloading on a Bangkok soft clay, where due to pumping of the sand layer underneath the clay, the pore pressure was lower than the hydrostatic pore pressure. Also in this case, when the vertical drains were installed, the pore pressure shifted to hydrostatic.

The hydrostatic profile presented in Fig. 17.8(b) was used as a reference for subsequent pore pressure measurements. The pore pressure profile remained constant until the excavation of the trench, when a temporary drop of approximately 4 kPa was observed due to stress release. As vacuum

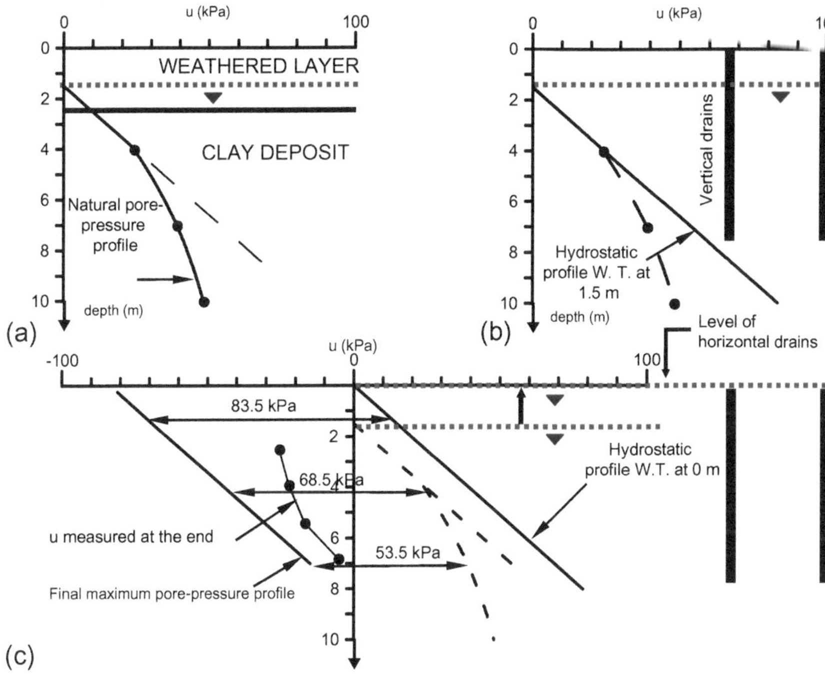

Figure 17.8 Pore water pressure profile: (a) initial conditions, (b) after installation of vertical drains, (c) and during pumping.

application began, the water increased to the level of the horizontal drains (hydrostatic profile in Fig. 17.8(c)).

A pump system efficiency as high as 81% was obtained due to the small dimensions of the site; thus, the vacuum system applied a suction, Δu_v, equal to 81 kPa. Considering this value and a final settlement of approximately 25 cm, the maximum variation of pore pressure should have been −83.5 kPa, as indicated by the final maximum pore pressure profile shown in Fig. 17.8(c).

The maximum increase in vertical effective stress ($\Delta \sigma'_v$) that could be achieved was thus the difference between the final maximum pore pressure profile (Fig. 17.8(c)) and the natural pore pressure profile shown in Figs. 17.8 (a) and 17.8(c). In this case, this increase varies with depth. It is approximately 68.5 kPa at a depth of 4 m, but it decreases to 53.5 kPa at a depth of 7 m. This confirms that it is preferable to have the horizontal drains at the level of the water table. Another way to avoid this is to use the vacuum

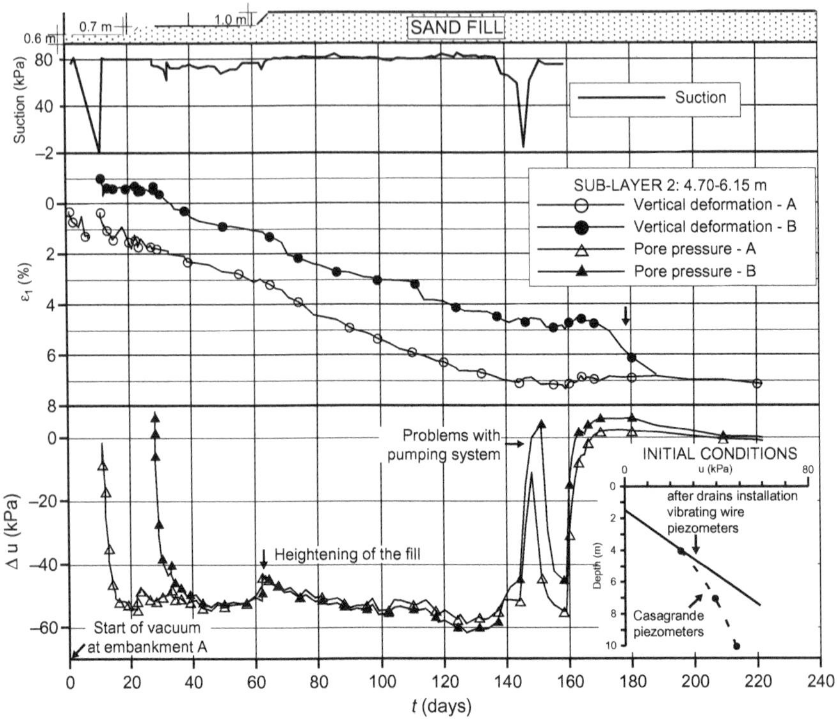

Figure 17.9 Suction, vertical deformation, and pore pressure variation.

technique without membrane, where the vacuum pumps are connected with a line of horizontal pipes connected to each vertical drain, and in this case the vertical drain is sealed (blind drain) until the water table is reached.

Figure 17.9 shows the applied suction and the fill height, as well as the average vertical strain (ε_1) and the pore pressure variation with time in the sublayer 2 (4.7–6.15 m). The ε_1 values were calculated from settlement gauges, and pore pressure changes were calculated from the hydrostatic profile with water table at a depth of 1.5 m (Fig. 17.8(b)).

After 61 days of pumping, the height of the fill was increased by 1 m. A pore pressure increase of approximately 8 kPa was observed, smaller than $(\gamma_h)_{fill}$ but close to the stress increase when the applied load is uniformly distributed at the level of the bottom of the trench (Fig. 17.2). An increase in strain rate was observed on both embankments when the fill was raised. After 159 days, when pumping was stopped, the pore pressure increased, but it did not return immediately to its initial value. There was a time lag due to the lowering of the water table from the level of the horizontal drains to its natural level.

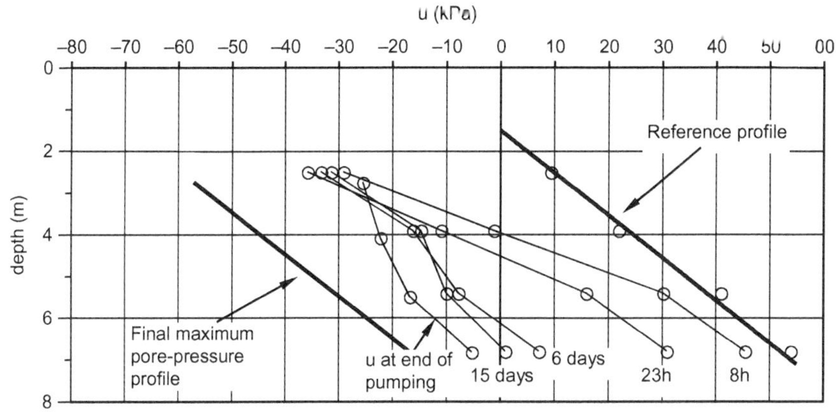

Figure 17.10 Pore pressure variation with depth under the platform subjected to vacuum preloading.

Figure 17.10 shows the pore water pressure as a function of depth at different times after the beginning of pumping. The piezometers installed in the crust presented pore pressure values higher than expected, possibly due to desaturation of the pore pressure transducer subjected to negative pore pressures in excess of approximately 30 kPa. However, the piezometers in sublayers 2 and 3 presented good performance under higher pore pressures.

At the end of pumping, the measured pore pressures in sublayers 2 and 3 were approximately 10 kPa greater than the final maximum pore pressure profile (Fig. 17.8(c)) and were still decreasing, indicating that primary consolidation was not completed.

17.5.3 Vertical displacements

Sixteen settlement plates were also installed over the trial embankments, and the total vertical displacements of the eight vibrating wire settlement gauges were also measured. Figure 17.9 shows the vertical deformation of sublayer 2.

The heating of embankment B was turned on before the vacuum, when the deposit was overconsolidated and there was expansion—a behavior that was also observed in the laboratory. However, after the vacuum consolidation was applied, the settlements of the fills were very similar.

An increase in strain rate was observed on both embankments when the fill was raised. The strain rates were approximately $10^{-9}\,s^{-1}$ on both embankments when pumping was turned off, which is a very high in situ strain rate compared to that of conventional fills.

Due to the limited dimensions of the platform, surface settlements measured on the contour were only 14–33% of those in the center. This boundary effect is quite common with conventional fills due to stress distribution; with vacuum, it is due to the fact that soil inside the treated area is subjected to vacuum, whereas the natural soil outside is not. From the experience gained by Soltraitement, reducing the drain spacing on the periphery of the treated area can attenuate this effect.

17.5.4 Horizontal displacements

Before the excavation of the trench, four inclinometers were installed at the embankment toes and one outside the preloading area. In order to measure in wintertime, antifreeze was mixed with the water inside the inclinometer tubes. However, the acquisition system did not work properly under very low temperatures, below −25 °C (−13 °F).

Horizontal displacements due only to vacuum consolidation are expected to be very small since the stress path due to vacuum preloading is almost parallel to the isotropic axis and the relationship (σ'_v/σ'_h) remains less than K_0. However, when vacuum consolidation is associated with conventional fill, the (σ'_v/σ'_h) could be higher than K_0. Because the fills were not very high at this site, the horizontal displacements were very small even after the end of pumping, when some expansion of the soil mass takes place.

Two inclinometers were installed at the embankment toe, and although measures were not accurate laterally, displacements appeared to be insignificant. Dong et al. (2001) observed that when vacuum consolidation is associated with surcharge preloading, lateral displacements are negative at the upper portion of the soft deposit.

17.5.5 Compression and yield stresses

Deep settlement gauges allowed the determination of the settlement, and thus of the strain, as a function of time for the different sublayers. Also, the vertical effective stresses were calculated at the level of the piezometers, in the middle of the sublayers. These calculations took into account (1) the stress distribution with depth due to platform dimensions, also considering peripheral trench depth; (2) the change in total vertical stress due to settlements; and (3) the average pore pressure at the level of piezometers.

The effective vertical stress versus vertical strain curves were then obtained. Note, however, that due to a defective deep settlement gauge

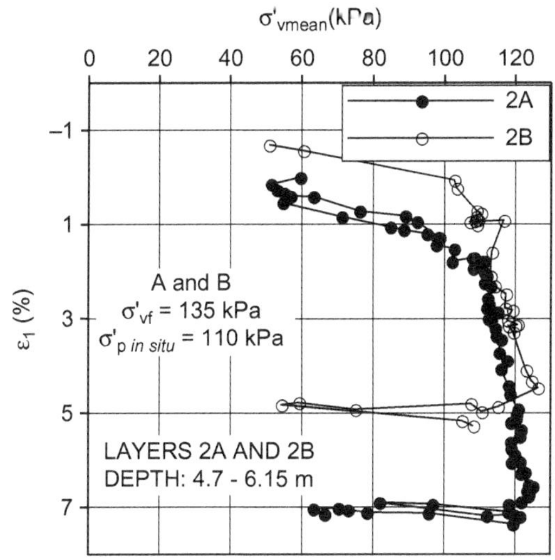

Figure 17.11 Compression curve of sublayers 2A (embankment A) and 2B (embankment B).

at a depth of 4.7 m, it had been assumed that the settlements of sublayers 1 and 2 were equal. This is certainly slightly wrong, but it does not change the general shape of the compression curve.

Figure 17.11 shows the compression curve for sublayer 2. Although due to excavation of the trench and changes in reference pore pressure profiles, the maximum final effective stress ($\sigma'_{vf}=135$ kPa for sublayer 2; Fig. 17.11) was lower than expected, the normally consolidated domain was reached.

Yielding was clearly obtained at a $\sigma'_{p\ in\ situ}$ value equal to 110 kPa; then compression continued, whereas the vertical effective stress was progressively increasing. At the end of pumping, the vertical effective stress was approximately 10 kPa smaller than the maximum effective stress (σ'_{vfmax}) that could be reached with vacuum at the end of primary consolidation.

Vertical yield stresses obtained by different procedures are shown in Fig. 17.12. They were obtained from conventional oedometer tests (σ'_{pconv}), piezocone tests ($\sigma'_{pcone}=(q_T - \sigma_v)/N_{\sigma T}$, with $N_{\sigma T}=3.4$), and in situ compression curves ($\sigma'_{p\ in\ situ}$) in the middle of each sublayer. Yield stress obtained from triaxial isotropic compression tests (σ'_{pi}) is also shown. The in situ yield profile is between the isotropic and oedometer yield profile, as expected for vacuum preloading, as discussed in reference to Fig. 17.3.

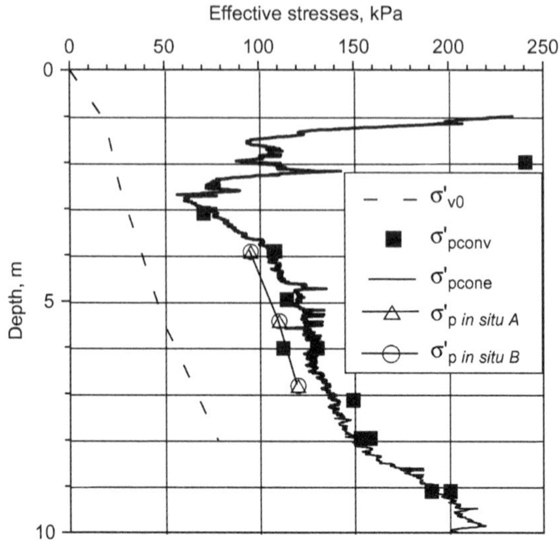

Figure 17.12 Vertical yield stress and in situ vertical effective stress profiles.

17.6 CONCLUSION

The main conclusion from the Saint-Roch-de-l'Achigan test is that vacuum preloading can be used in the geotechnical and climatic conditions that prevail in the province of Québec if the hydraulic system is protected against frost.

Lessons were also learned from this experience, particularly the following:

- In hydrogeologic conditions encountered in eastern Canada, with the water table often at a depth on the order of 1 m, horizontal drains should be installed at the level of the water table in order to optimize the use of the technique, or the vacuum technique without membrane could be used.
- Vibrating wire piezometers were shown to give reliable measurements up to negative pore pressures on the order of 25 kPa. Although the final effective stresses were lower than expected, effective stresses achieved normally consolidated range.

The main objectives of the experimental site were to observe the performance of the vacuum system applied on a sensitive Canadian clay deposit and to study the in situ viscous behavior. Better results under vacuum consolidation at this site could not be obtained due to the high depth of the water table and the initial pore pressure conditions, which decreased the efficiency of the vacuum system. However, the instrumentation used proved to be adequate for the very special conditions on this site.

Heating proved to be a very expensive technique, and the in situ behavior contradicted the laboratory observations (Marques, 2001; Marques et al., 2004) with respect to temperature variation of embankment B. The stress variation due to vacuum application plus fill was only high enough to surpass the preconsolidation pressure, which could be the main reason for such behavior. However, it is possible that if the deposit had been well inside the normally consolidated domain, the heating effect on the behavior would be more important, mainly with respect to vertical deformations.

ACKNOWLEDGMENTS

The authors are grateful to the Ministère des Transports du Québec, which made this study possible. Geopac Soil Treatment, Ménard Soltraitement, Denis Jobin, and Serge Paré are also sincerely acknowledged for their valuable technical support.

REFERENCES

Bergenståhl, L., Gabrielsson, A., Mulabdic, M., 1994. Changes in soft clay caused by increases in temperature. In: Proc.13th Int. Conf. on Soil Mech. and Found. Eng., New Delhi, pp. 1637–1641.

Choa, V., 1989. Drains and vacuum preloading pilot test. In: Proc. 12th Int. Conf. on Soil Mech. and Found. Eng., Rio de Janeiro, pp. 1347–1350.

Cognon, J.M., Juran, I., Thevanayagam, S., 1994. Vacuum consolidation technology—Principles and field experience. In: Vertical and Horizontal Deformations of Foundations and Embankments, Settlement '94, College Station. ASCE, pp. 1237–1248, Geotech. Spec. Publ. No 40.

Díaz-Rodriguez, J.A., Leroueil, S., Alemán, J.D., 1992. Yielding of Mexico City clay and other natural clays. J. Geotech. Eng. Div., ASCE 118 (7), 981–995.

Dong, Z.L., Zhen, X.L., Kan, M.M.K., 2001. Effect of vacuum preloading and road base surcharge on the lateral displacement of surrounding soil in highway ground improvement. In: Lee, C.F. et al., (Eds.), Soft Soil Engineering. Swets & Zeitlinger, London, pp. 459–464.

Edil, T.B., Fox, P.J., 1994. Field test of thermal precompression. In: Vertical and Horizontal Deformations of Foundations and Embankments, Settlement '94, College Station. ASCE, pp. 1274–1286, Geotech. Spec. Publ. No. 40.

Hahn, L.T., Bergado, D.T., Shibuya, S., 1998. Engineering properties of soft Bangkok clay improved using PVD with surcharge and vacuum preloading. In: Yangisawa, E., Moroto, N., Mitachi, T. (Eds.), Problematic Soils. Balkema, Rotterdam, The Netherlands, pp. 681–684.

Hueckel, T., Peano, A., 1987. Some geotechnical aspects of radioactive waste isolation in continental clays. Comput. Geotech. 3, 157–182.

Indraratna, B., Bamunawita, C., Khabbaz, H., 2004. Numerical modelling of vacuum preloading and field applications. Can. Geotech. J. 41, 1098–1110.

Jacob, A., Thevanayagam, S., Kavazajian, E., 1994. Vacuum-assisted consolidation of a hydraulic landfill. In: Vertical and Horizontal Deformations of Foundations and Embankments, Settlement '94, College Station. ASCE, pp. 1249–1261, Geotech. Spec. Publ. No 40.
Kjellman, W., 1952. Consolidation of clay soil by means of atmospheric pressure. In: Proc. Conf. on Soil Stabilization. MIT, Cambridge, pp. 258–263.
Leroueil, S., 1996. Compressibility of clays: fundamental and practical aspects. J. Geotech. Eng. 122 (7), 534–543.
Leroueil, S., Marques, M.E.S., 1996. Importance of strain rate and temperature effects in geotechnical engineering. In: Measuring and Modeling Time-Dependent Soil Behavior. ASCE, pp. 1–60, Geotech. Spec. Publ. No. 61.
Leroueil, S., Tavenas, F., Mieussens, C., Peignaud, M., 1978. Construction pore pressures in clay foundations under embankments: Part II. Generalized behavior. Can. Geotech. J. 15, 54–65.
Lingnau, B.E., Graham, J., Tanaka, N., 1995. Isothermal modeling of sand–bentonite mixtures at elevated temperatures. Can. Geotech. J. 32, 78–88.
Magnan, J.P., 1994. Maîtrise des amplitudes et vitesses de tassement des remblais sur argiles molles: Évolutions récentes. Bull. Liaison des Laboratoires des Ponts et Chaussées 194, 27–40.
Marques, M.E.S., 2001. Laboratory and Field Viscous Behaviour of Clayey Soils (in Portuguese). DSc. thesis, In: COPPE/UFRJ, Rio de Janeiro, Brazil, research performed at Université Laval, Québec.
Marques, M.E.S., Leroueil, S., Almeida, M.S.S., 2004. Viscous behaviour of a Champlain Sea clay from Saint-Roch-de-l'Achigan. Québec. Can. Geotech. J. 41, 25–38.
Miliziano, S., 1992. Effetti della temperatura sul comportamento meccanico delle terre coesive. DSc. thesisConsorzio entre l'Universitá de Rome La Sapienza e de Napoli Federico II, Itália.
Moritz, L., 1995. Geotechnical properties of clay at elevated temperatures. In: Yoshikuni, H., Kusakabe, O. (Eds.), Compression and Consolidation of Clayey Soils. Balkema, Rotterdam, The Netherlands, pp. 267–272.
Qian, J.H., Zhao, W.B., Cheung, Y.K., Lee, P.K.K., 1992. The theory and practice of vacuum preloading. Comput. Geotech. 13, 103–118.
Rujikiatkamjorn, C., Indraratna, B., Chu, J., 2007. Numerical modelling of soft soil stabilized by vertical drains, combining surcharge and vacuum preloading for a storage yard. Can. Geotech. J. 44, 326–342.
Shang, J.Q., Tang, M., Miao, Z., 1998. Vacuum preloading consolidation of reclaimed land: a case study. Can. Geotech. J. 35 (5), 740–749.
Tavenas, F., Mieussens, C., Bourges, F., 1979. Lateral displacements in clay foundations under embankments. Can. Geotech. J. 16 (3), 532–550.
Van Impe, W.F., Van Mieghem, J., Vandycke, S., Van Impe, P.O., 2001. Underwater vacuum consolidation of dredged silt—a case history. In: Lee, C.F. et al., (Eds.), Soft Soil Engineering. Swets & Zeitlinger, London, pp. 499–506.

CHAPTER 18

Use of Explosion in Soil Improvement Projects

Shuwang Yan[1], **Jian Chu**[2]
[1]Geotechnical Research Institute, Tianjin University, Tianjin, China
[2]Professor and James M. Hoover Chair in Geotechnical Engineering, Iowa State University, USA; Nanyang Technological University, Singapore

18.1 REVIEW OF DIFFERENT EXPLOSIVE METHODS

The use of explosion as a soil improvement method is not common. Nevertheless, several explosive methods have been developed and used in practice, including the explosive compaction method, the explosive piling method, and the toe shooting method.

The explosive compaction method uses buried explosives to cause the densification of loose granular soil ground. This method has been used in practice in the United States for approximately 50 years. A number of theoretical and case studies have been published in the past (Hall, 1962; Chadwick et al., 1964; Mitchell, 1970; Charlie et al., 1992; Narin van Court and Mitchell, 1995, 1998; Gandhi et al.,1998; Gohl et al., 2000; Dembicki et al., 1992). When adopting this method, the usual procedure consists of installing the explosive charges through a pipe installed to the required depth, filling the hole after the pipe is withdrawn, and firing the charges. So far, this method is limited to the application of granular soils.

Explosion is also used to form a borehole for the installation of compacted cement or fly ash mixed soil piles. Those piles are normally only 2 or 3 m long and are used to improve the ground of clay fill or loess soil. In using this method, a 22-mm-diameter hole is first formed. The borehole is then filled with a TNT and wood-cutting scraps mixture. The ratio is adjusted depending on the soil type and the required pile diameter. A detonator is placed at the bottom of the explosive mixture. After firing the charges, a cylindrical borehole of roughly 350 mm is formed. The soil surrounding the borehole is also compacted at the same time by the explosion. The compacted zone is approximately 2.5 to 3 times the borehole diameter. Cement or fly ash mixed soil can then be placed inside the borehole and compacted by heavy tamping in layers to form a pile. More detailed construction procedures and some case studies are presented in Zhu et al. (2003). This method is economical and quick.

Chemical, Electrokinetic, Thermal, and Bioengineering Methods
http://dx.doi.org/10.1016/B978-0-08-100191-2.00018-6

However, it may only be effective when the groundwater table is relatively low or when a relatively dry borehole can be formed without support. This kind of site conditions is commonly encountered in the western part of China, where the method originated.

When the soft ground to be improved is 6 m deep or less, it may be an economical option to excavate the soft clay and replace it with sand, gravel, or rock fill. Explosives may be used in this case as a mean to remove the soft clay. One such method is described in *Soil Mechanics for Road Engineers* (Road Research Laboratory, 1952) for construction over peat mires. In this so0called toe shooting method, charges are installed in the peat mires to blast away the peaty soil before competent soils can be replaced. However, this method may not remove the weak soil completely. It may also be used only for a shallow depth. Therefore, the method has been used so far only in muddy and swampy areas where it is difficult to employ heavy construction machinery and to transport the excavated soil. A similar method for removing and replacing seabed mud with rocks for offshore dike construction has also been used in China. A case study was presented in Jin and Shi (1999).

18.2 EXPLOSIVE REPLACEMENT METHOD

18.2.1 Principle and technique

The principle of the explosive replacement method is to remove soft clay and replace it with crushed stones using controlled blasting. The procedure of the method is illustrated in Fig. 18.1. As shown in Fig. 18.1(a), 18.explosive charges are first installed in the soft clay, and then crushed stones are piled up next to it on the side of the road that has been improved. When the charges are detonated, the soft clay is blown out and cavities are formed. At the same time, the crushed stones collapse into the cavities. In this way, the soft soil is replaced with crushed stones in a rapid manner. The soil that is blown into the air will form a liquid and flow away after it falls to the surface. The crushed stones after collapsing form a slope of 1V:3H or 1V:5H, as shown in Fig. 18.1(b). The impact of the explosion also causes an instantaneous reduction in the shear strength of the soil below the level of explosion (Osipov et al., 1984) so that the crushed stones can sink into the soft clay layer. The stones help the soil at the bottom to consolidate, and the clay itself will also gain part of its original strength after explosion (Osipov et al., 1984). More crushed stones are backfilled to form a leveled ground and a steeper slope, as shown in Fig. 18.1(c). The process is then repeated to remove and replace the soil in another section.

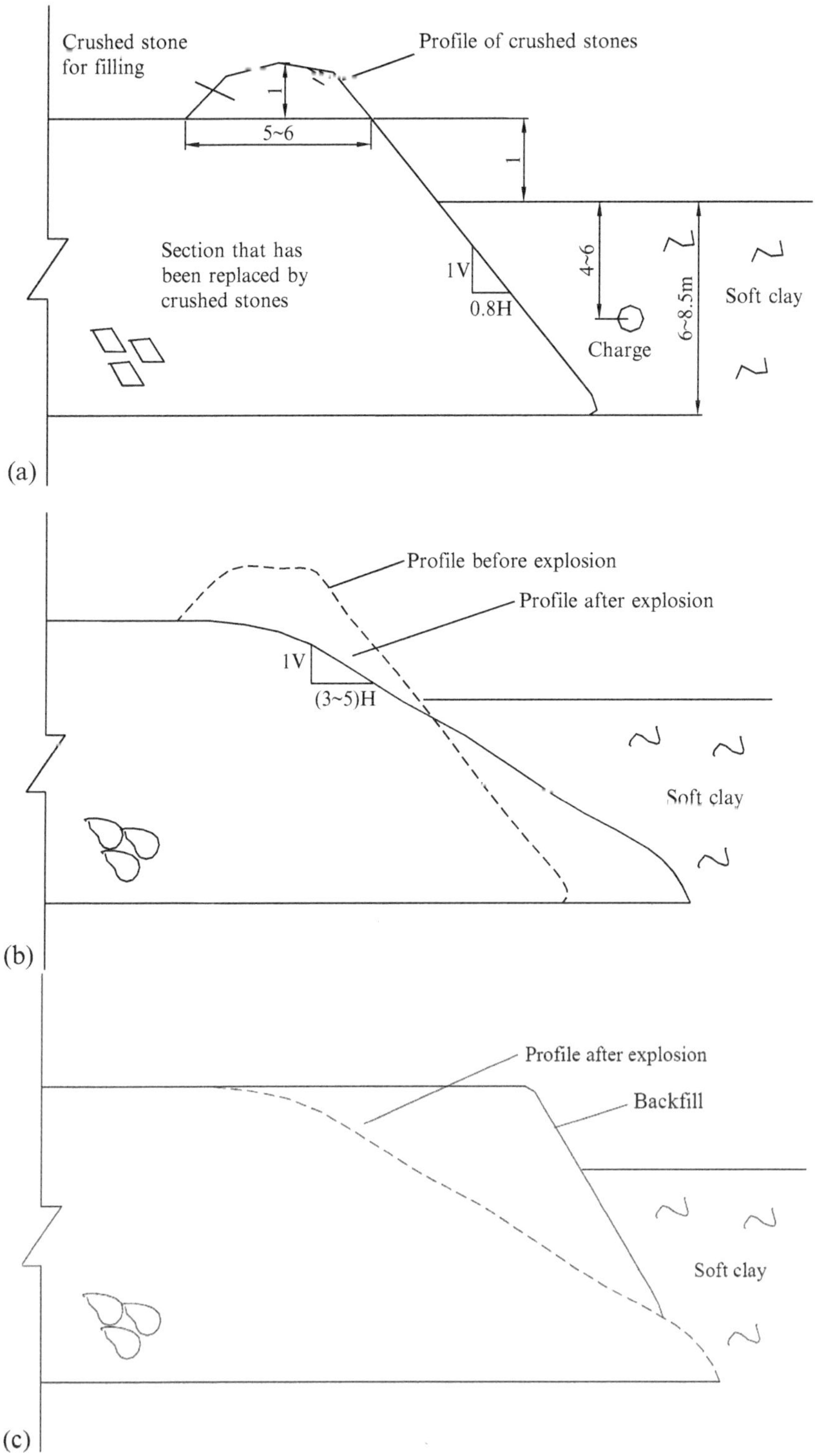

Figure 18.1 Explosive displacement procedure: (a) Before explosion, (b) after explosion, and (c) after backfill.

18.2.2 Analysis

The success of the explosive replacement method depends on the following design parameters: charge placement (in plan and elevation), charge weights, and detonation sequence. The Hopkinson's number (HN) is often used for explosive compaction (Ivanov, 1967; Narin van Court and Mitchell, 1995):

$$\mathrm{HN} = \left(Q^{1/3}\right)/R \tag{18.1}$$

where Q is the charge weight in kilograms, and R is the effective radius in plan in meters. However, for a given value of HN, there are infinite combinations of charge weight with radius. Furthermore, it is difficult to select the suitable values of HN in practice. Based on blasting mechanics (Henrych, 1979), the following equation has been derived by Yan and Chu (2004):

$$R = 0.493 P_k^{\frac{5}{36}} \left(\frac{D}{\rho_e}\right)^{\frac{2}{9}} \left(P_a + \sum_{i=1}^{n} \gamma_i (h_c)_i\right)^{-\frac{1}{4}} Q^{\frac{1}{3}} \tag{18.2}$$

where P_k is a pressure constant in Pa, ρ_e is the density of the explosive in kg/m^3, D is the velocity of the explosive in m/s, P_a is the atmospheric pressure in Pa, γ_i is the unit weight of soil in N/m^3, and h_c is the thickness of the soft clay above a cavity in meters.

For a TNT type of explosive, the following parameters can be chosen: density $\rho_e = 1100$ kg/m^3, velocity $D = 6200$ m/s, and $P_k = 3.0 \times 10^8$ Pa. The unit weight of the soil can be taken as 18,000 N/m^3. Substituting all the parameters into Eq. (18.2) gives

$$R = \frac{10.88 Q^{\frac{1}{3}}}{(98000 + 18000 h_c)^{0.25}} \tag{18.3}$$

Because the purpose of blasting is to remove the soft clay, the thickness of the soft clay above a cavity should be controlled to be small. Using $h_c = 0$ in Eq. (18.3) we have

$$R = 0.62 Q^{\frac{1}{3}} \tag{18.4}$$

Eq. (18.4) is equivalent to Eq. (18.1) when NH = 0.61. For Q = 10–20 kg, we get $R_{vd} = 1.34$–1.68 m from Eq. (18.4). The distance between charges, L, can be estimated as

$$L = \lambda R \tag{18.5}$$

where $\lambda = 1.5$–2.0. When $\lambda = 2$, $L =$ the diameter of the cavity. To allow some overlapping, $\lambda < 2$ should be used.

18.3 CASE STUDY

18.3.1 Soil conditions

Due to rapid economic development, a highway connecting Ganzhou and Dingnan in Jiangxi Province (Fig. 18.2) was constructed from January 2002 to January 2004. The highway is 127 km long and 26.5 m wide. A few sections of highway run through valley zones. The valleys were 20–50 m wide with water table typically at the ground surface. A typical soil profile is shown in Fig. 18.3. The soils were alluvial in nature. The first layer was silty clay 6–8.5 m thick with a 0.5- to 0.6-m-thick vegetation layer on top. The second layer was silty gravel 1.3–2 m thick overlying weathered sandstone. The top 1 to 2 m of the sandstone was highly decomposed. Typical soil properties of the soft clay are shown in Fig. 18.4. The water content of the soil was generally higher than the liquid limit (Fig. 18.4(b)). The undrained shear strength of the silty clay was only approximately 20 kPa (Fig. 18.4(c)). Improving the soft clay layer quickly for highway construction was one of the challenges of this project. Some of the commonly adopted soil improvement methods could not be applied because these

Figure 18.2 Location map.

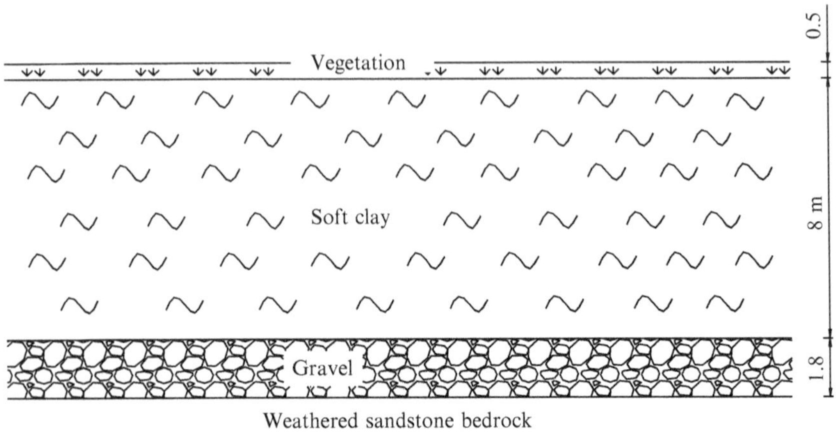

Figure 18.3 Typical soil profile in the valley.

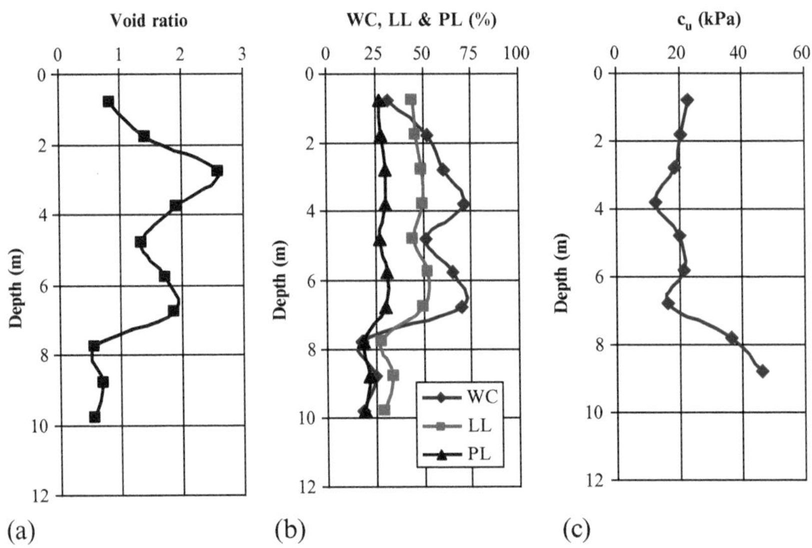

Figure 18.4 Typical soil properties. WC, water content; LL, liquid limit; PL, plastic limit; c_u, undrained shear strength.

methods could not offer an expedient solution to meet the project schedule. For example, surcharge or vacuum preloading together with vertical drains (Yan and Chu, 2003) is normally used for similar projects. However, the time required for consolidation was too long for this project. Deep cement mixing is also sometimes used for highway construction in China (Lin and

Wong, 1999). However, this method was considered too expansive to be used over a long section. Furthermore, the use of this method might not meet the construction schedule either. The explosive replacement method was considered feasible for this project for the following reasons: (1) The construction was in a remote mountainous area, so blasting was permitted; plenty of rocks were generated from the tunneling work for the same project; and the soft clay layer to be replaced was only approximately 6–8.5 m thick. After some experiments, a workable method was established and adopted for this highway project.

18.3.2 Soil improvement work

Explosive charges were placed in the soft soil to be improved as shown in Fig. 18.1(a). A crushed stone pile 5 or 6 m wide and 1 or 2 m high was placed behind the area where dynamite was installed. The charges were installed at a horizontal spacing of 2 m in one row. The embedded depth and the weight of the charge were determined based on the depth of the clay layer as calculated using Eq. (18.3), as shown in Table 18.1. A TNT type of explosive was used. The charges were installed using a 16-ton excavator, as shown in Fig. 18.5. A pipe 21.3 cm in diameter and 12 m long was driven into the soft clay using an 11-kW vibrator. Once the pipe reached the required depth, a cylindrical charge of 19 cm in diameter was placed. For a charge weight of 16–24 kg, the length of the charge would be 50–80 cm. The pipe was filled with water as an overburden pressure to the charge before the pipe was pulled out. Sequential detonation was used to reduce the impact of the explosion. A picture taken at the moment of explosion is shown in Fig. 18.6. The explosion removed most of the soft soil within the top 5 or 6 m in the exploded area and caused the pile of crushed stones to slide into the area where the soft clay was removed by explosion. After collapsing, the crushed stones form a slope of 1V:3H or 1V:5H, as shown in Fig. 18.1(b). More

Table 18.1 Calculation on the weight of each charge

Depth of soil, h (m)	5	6	7	8
Distance between charges, L (m)	2	2	2	2
λ in Eq. (18.15)	1.7	1.6	1.5	1.5
Radius of cavity, R_{vd} (m)	1.18	1.25	1.33	1.33
Thickness of soft clay above cavity, h_c (m)	2.65	3.50	4.33	5.34
Embedded depth for charges, $h_c + R_{vd}$ (m)	3.83	4.75	5.66	6.67
Weight of each charge, Q (kg)	9.5	12.2	15.7	16.8
Weight of each charge used, Q_u (kg)	16	20	24	24

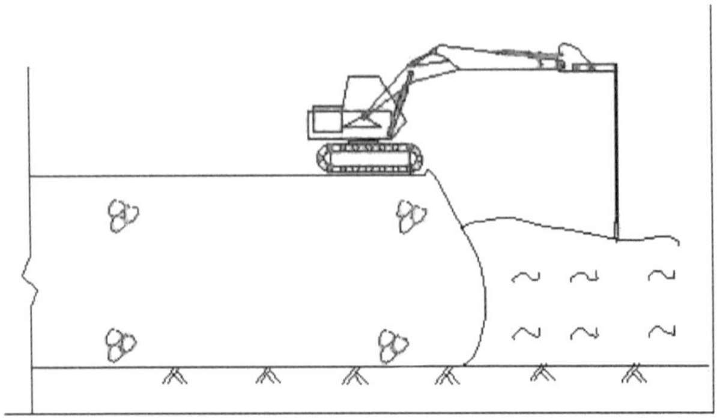

Figure 18.5 Instalation of charges.

Figure 18.6 During explosion.

crushed stones were backfilled to form a leveled ground and a steeper slope, as shown in Fig. 18.1(c). This process was then repeated to remove and to replace the soil in another section. Using this method, the soft clay layer was replaced by crushed stones in a rapid manner. The potential delay to the project was thus avoided.

The source of the crushed stones was the sandstone rock excavated from tunneling construction for the same highway project. The particle size of the stones ranged from 10 to 70 cm. The pile of crushed stones was typically 1 or 2 m high and 5 or 6 m wide (Fig. 18.1(a)). After the explosion, the stones fell into the cavities and formed a slope of 1V:3H to 1V:5H. More stones were backfilled to form a leveled ground and a boundary slope of 1V:0.8H. With every round of explosion, the improvement section could be advanced 6–8 m.

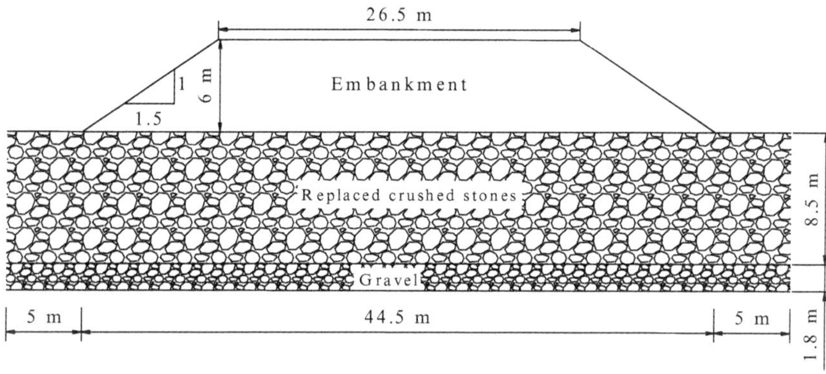

Figure 18.7 A typical cross section of the road.

Table 18.2 Basic properties of the fill used for embankment

Soil type	Liquid limit (%)	Plastic limit (%)	Plasticity index (%)	Fines content, <75 μm (%)	Organic content (%)
Clayey sand (SC)[a]	25.3	12.3	13	18	2.7

[a]According to the Unified Soil Classification System.

An embankment typical 6 m high (Fig. 18.7) was constructed after the soil improvement work. The fill used for the embankment was mainly a clayey sand; its basic properties are given in Table 18.2. The maximum dry density and the optimum water content as obtained from standard Proctor compaction tests were 1.8 Mg/m^3 and 14.5%, respectively. Roller compaction with 0.3–0.5 m for each lift was used. The relative compaction values specified were 95% for the top 0.8 m, 93% for the fill between 0.8 and 2.3 m, and 90% for the fill below 2.3 m, respectively.

18.3.3 Verification

Boreholes were drilled to examine the depth of the crushed stones after soil improvement. One of the borehole logs is shown in Fig. 18.8. The stones were found to be present up to 9 m, in which the top 5 or 6 m was densely packed, whereas the remaining 4 or 5 m was embedded in clay. Below the crushed stone layer were a silty gravel layer and the weathered sandstone layer.

Plate load tests were conducted using a 1.0 × 1.0-m square plate. The load was applied via a hydraulic jack reacted against a steel beam that was

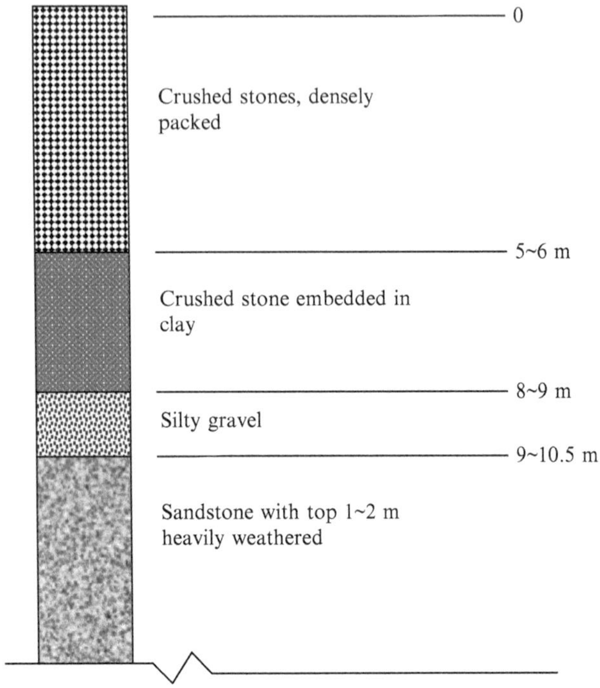

Figure 18.8 Borehole log of the replaced soil layer.

counterbalanced by dead weight. The plate was placed on the ground surface before the 6 m of embankment was built. The results of a typical plate load test are shown in Fig. 18.9. The results indicate that the improved ground had adequate bearing capacity. Using the load versus settlement curve shown in Fig. 18.9, the modulus of subgrade reaction, k_s, which is used for pavement design (American Association of State Highway and Transportation Officials, 1993), can be determined as the secant modulus for a specified point on the curve (Bowles, 1996). The modulus of subgrade reaction determined from the initial linear portion of the curve was 120 MPa (see Fig. 18.9). Note that the plate load test results only reflected the condition of the upper layer of 1.5–2 m deep in the compacted stone layer. The critical area for settlement would be the deeper zone where the stone was mixed with soft clay, which was not significantly stressed by the plate load tests. Therefore, the plate load test results gave an optimistic picture of the load settlement behavior. The settlement of improved foundation soil measured 3 months after the opening of the highway was more than the maximum settlement shown in Fig. 18.9 but less than 30 mm. The total

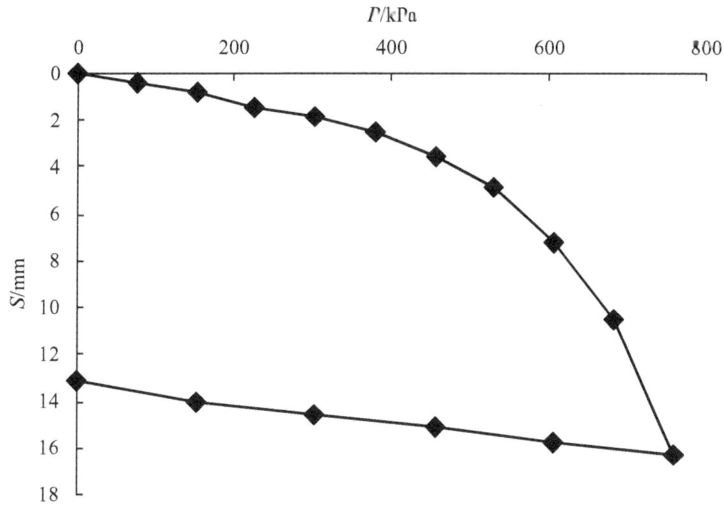

Figure 18.9 Results of plate load test.

settlement of the highway measured at the same time was less than 100 mm. The total allowable settlement as specified by the Ministry of Transport for design of expressways in China was 300 mm.

Ground-penetrating radar (GPR) was used to detect the distribution of the crushed stones in the soft clay. The radar system transmits repetitive, short-time electromagnetic waves into the ground from a broad bandwidth antenna. Some of the waves are reflected when they hit discontinuities in the subsurface, whereas some are absorbed or refracted by the materials to which they come into contact. The reflected waves are picked up by a receiver, and the elapsed time between wave transmission and reception is automatically recorded. This method is explained in detail in Koerner (1984).

The GPR system used in this project adopted a frequency of 100 MHz. This frequency was chosen to suit the depth of the crushed stone layer. GPR tests were conducted along six lines of a total length of 417 m. Two lines were along the longitudinal direction and four lines along the transverse direction of the highway. One scanned profile is shown in Fig. 18.10. The crushed stones in the top 5 m of the soil profile were detected. Soft clay pockets within this layer could also be identified from the image, as indicated by arrows in Fig. 18.10. However, the stones in the deeper layer could not be identified clearly from the image. This may be because the radar wave became much less ineffective when it penetrated the layer of stones embedded in clay.

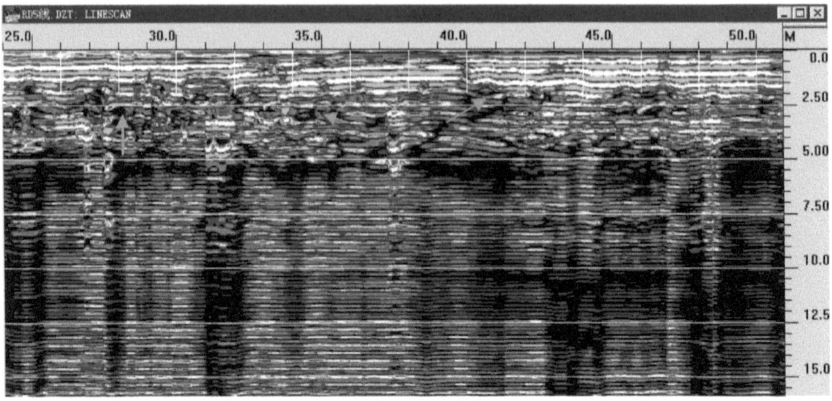

Figure 18.10 An image of ground-probing radar.

18.4 CONCLUSION

A new explosive replacement method was developed and adopted for the soil improvement work of a highway project in Jiangxi, China. The method uses the energy of explosion to remove and replace soft clay with crushed stones. Explosive charges were placed in the soil to be improved according to a certain pattern. Crushed stones were piled up next to the area where charges were installed. The explosion created cavities in the soil and caused the pile of crushed stones to slide into the exploded area. Based on blasting mechanics (Henrych, 1979), a set of equations regarding the weight of the charge and the radius of the explosive cavity were derived to calculate the weight of the charges and the spacing between the charges. The design of the charges was developed using the equations, and the results were found to be satisfactory. A method to install charges was also developed and used in this project. The borehole exploration after the soil improvement revealed that the crushed stones were densely packed in the top 5 or 6 m but were embedded in clay in the bottom 3 or 4 m. The results of the plate load tests indicated that the top layer of the improved ground had sufficient bearing capacity and high modulus of subgrade reaction. The GPR adopted was effective in detecting the layer of the densely packed crushed stones.

REFERENCES

American Association of State Highway and Transportation Officials, 1993. AASHTO Guide for Design of Pavement Structures. American Association of State Highway and Transportation Officials, Washington, DC.

Bowles, J.E., 1996. Foundation Analysis and Design. McGraw-Hill, New York.

Chadwick, P., Cox, A.D., Hopkins, H.G., 1964. Mechanics of deep underground explosions. Proc. R. Soc. 256 (1970), 235–300, London A

Charlie, W.A., Jacobs, P.J., Doehring, D.O., 1992. Blasting-induced liquefaction of an alluvial sand deposit. Geotech. Test. J. 15, 14–23, ASTM.

Dembicki, E., Imiolek, R., Kisielowa, N., 1992. Soil Compaction with Blasting Method. In: Chowdhury, R.N. (Ed.), Geomechanics and Water Engineering in Environmental Management. Balkema, Rotterdam, The Netherlands, pp. 599–622.

Gandhi, S.R., Dey, A.K., Selvam, S., 1998. Densification of pond ash by blasting. J. Geotech. Geoenvironm. Eng. 125 (10), 889–899, ASCE.

Gohl, W.B., Jefferies, M.G., Howie, J.A., Diggle, E.D., 2000. Explosive compaction: design, implementation and effectiveness. Geotechnique 50 (6), 657–665.

Hall, C.E., 1962. Compacting a dam foundation by blasting. J. Soil Mech. Found. Div 88 (3), 33–51, ASCE.

Henrych, J., 1979. The Dynamics of Explosion and Its Use. Elsevier, New York.

Ivanov, P.L., 1967. Compaction of Noncohesive Soil by Explosions, National Technical Information Service Report No. TT70-57221. U.S. Department of Commerce, Springfield, VA.

Jin, L., Shi, F., 1999. Blasting techniques for underwater soft clay improvement. China Harbour Eng. 2, 10–17; 3, 16–27.

Koerner, R.M., 1984. Construction and Geotechnical Methods in Foundation Engineering. McGraw-Hill, New York.

Lin, K.Q., Wong, I.H., 1999. Use of deep cement mixing to reduce settlements at bridge approaches. J. Geotech. Geoenviron. Eng. 125 (4), 309–320, ASCE.

Mitchell, J.K., 1970. In-place treatment of foundation soils. J. Soil Mech. Found. Div. 96 (SM1), 73–110, ASCE.

Narin Van Court, W.A., Mitchell, J.K., 1995. New Insights into Compaction of Loose, Saturated, Cohesionless Soils. In: Soil Improvement for Earthquake Hazard Mitigation. ASCE, Reston, VA, pp. 51–65, ASCE Geotechnical Special Publication No. 49.

Narin Van Court, W.A., Mitchell, J.K., 1998. Investigation of Predictive Methodologies for Explosive Compactions. In: ASCE Specialty Conference on Geotechnical Earthquake Engineering and Soil Dynamics. ASCE, Reston, VA, pp. 639–653, Seattle, WA, ASCE Geotechnical Special Publication No. 75.

Osipov, V.I., Nikolaeva, S.K., Sokolov, V.N., 1984. Microstructural changes associated with thixotropic phenomena in clay soils. Geotechnique 34 (2), 293–303.

Road Research Laboratory, 1952. Soil Mechanics for Road Engineers. HMSO, London.

Yan, S.W., Chu, J., 2004. Soil improvement for a road using the vacuum preloading method. Ground Impro. 157 (4), 259–265.

Zhu, B.H., Liao, H.J., Zhou, L.X., 2003. Soil Improvement and Reinforcement of Building Foundation. Xi'an Jiaotong University Press, Xi'an, China.

PART THREE

Bioengineering

CHAPTER 19

Use of Biogeotechnologies for Soil Improvement

Jian Chu[1,2], Volodymyr Ivanov[2], Jia He[2], Maryam Maeimi[2], Shifan Wu[1]
[1]Iowa State University, USA
[2]Nanyang Technological University, Singapore

19.1 INTRODUCTION

Cement or chemicals have often been used as binders for soil to either increase the shear strength or reduce the water conductivity of soil. However, the use of cement or chemicals for soil improvement is not sustainable in the long term because these require a considerable amount of natural resource (e.g., limestone) and energy to produce. The use of cement or chemicals for soil improvement is expensive and time-consuming. Also, more cost-effective soil improvement methods are needed to allow disaster mitigation measures to be applied to larger areas with a limited budget. For example, liquefaction has long been identified as a major hazard, as seen in the 2011 earthquake disasters in New Zealand and Japan. However, ground treatment for mitigation of liquefaction hazard is normally only carried out for sites where important infrastructures such as sea ports and airports are constructed. This is because the cost involved in ground treatment for the prevention of liquefaction is high. If a much cheaper liquefaction mitigation method can be developed, ground treatment for prevention of liquefaction can thus be applied over a wider range of projects to reduce the impact of the damages associated with liquefaction. There is an urgent need to develop new and sustainable construction materials that can reduce the use of cement or chemicals and reduce the cost for soil improvement.

Using the latest microbial biotechnology, a new type of construction material, biocement, has been developed as an alternative to cement or chemicals (Whiffin, 2004; Ivanov and Chu, 2008). Biocement is made of naturally occurring microorganisms at ambient temperature and thus requires much less energy to produce. It is sustainable because microorganisms are abundant in nature and can be reproduced easily at low cost. The microorganisms that are suitable for making biocement are usually nonpathogenic and environmentally friendly. Furthermore, unlike the use of cement, soils

Chemical, Electrokinetic, Thermal, and Bioengineering Methods
http://dx.doi.org/10.1016/B978-0-08-100191-2.00019-8

can even be treated or improved without disturbing the ground or environment because microorganisms can penetrate and reproduce in soil. Harnessing this natural, unexhausted resource may result in an entirely new approach to geotechnical or environmental engineering problems and bring enormous economic benefit to the construction industry. The application of microbial biotechnology to construction will also simplify some of the existing construction processes. For example, biocement can be in either solid or liquid form. In liquid form, the biogrout has much lower viscosity and can flow like water. Thus, the delivery of biocement into soil is much easier compared with that of cement or chemicals. Furthermore, when cement is used, one usually has to wait for weeks for the full strength to be developed, whereas when biocement is used, the reaction time can be reduced if required.

A number of methods for the use of biotechnologies in construction have been developed in recent years. A summary of the various methods is given in Fig. 19.1. In this chapter, the following specific applications to geotechnical engineering are presented: a liquefaction mitigation method using biogas desaturation, construction of a temporary water storage pond in sandy soil using the biocrust method, the use of biogrouting for the prevention of levee failure due to overtopping and erosion, and road construction using biogrout made of waste. Some of the methods presented here need to be developed further before they can be used in practice. Nevertheless, it is only a matter of time before some of these methods become sound technologies.

19.2 PRINCIPLES OF MICROBIAL GEOTECHNOLOGY

The principle of microbial treatment is to use the microbially induced calcium carbonate precipitation or other approaches to produce bonding and cementation in soil so as to increase the strength and reduce the water conductivity of soil. A number of studies have been carried out in recent years (Mitchell and Santamarina, 2005; Ivanov and Chu, 2008; Van der Ruyt and van der Zon, 2009; Van Paassen et al., 2010). Much of the work remains at the experimental stage. However, the scale of treatment has increased rapidly with time in recent years (Van Paassen et al., 2010).

The microbiological processes induce calcium carbonate crystals, other minerals, or slimes as illustrated by examples shown by Van der Ruyt and van der Zon (2009), Van Paassen et al. (2010), and Chu (2012). Those minerals or slimes act as cementing agencies between sand grains to increase the shear strength of soil and/or to fill in the pores in soil to reduce the water conductivity as illustrated schematically in Fig. 19.2. The two processes to increase

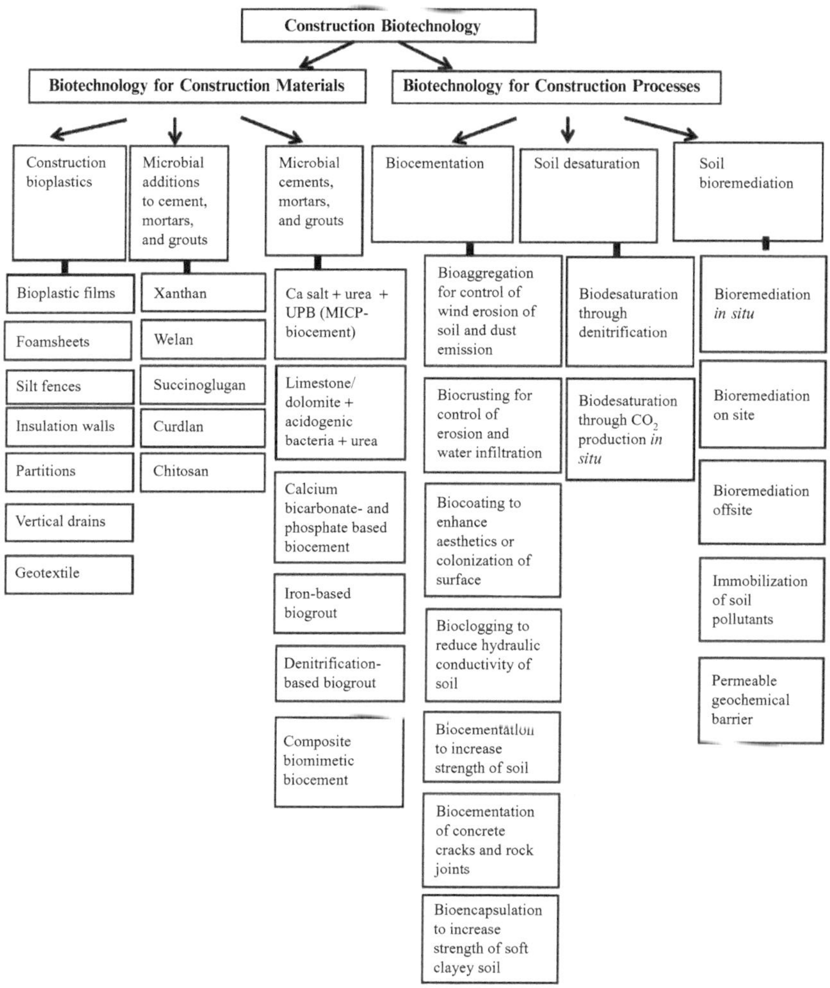

Figure 19.1 Category of biotechnology used for construction. *(Source: After Ivanov et al. (2014)).*

the shear strength and reduce the permeability of soil have been called biocementation and bioclogging, respectively (Ivanov and Chu, 2008). The process to deliver the biocement in situ to achieve biocementation or bioclogging is called biogrouting. The materials used to produce biocementation or bioclogging effect are called biocement. The existing studies so far show that the biocement method is effective at both increasing the shear strength and reducing the permeability of soil. Normally, bonding materials to produce biocementation will also reduce the permeability of soil at the

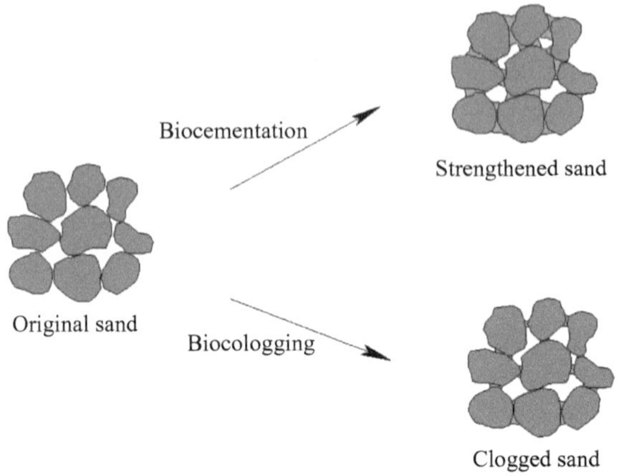

Figure 19.2 Schematic illustration of (a) biocementation and (b) bioclogging processes.

same time due to the reduction in the pore space. However, there are biocements that will only produce biocementation or bioclogging. The viscosity of biogrout is generally as low as that of water. Thus, it is possible to pump the biogrout into the ground without mixing with soil. This will enable the construction process to be simplified.

There are also other processes for microbial treatment, such as the generation of slime for bioclogging. Another microbial process is the generation of biogas, which can also be used beneficially for geotechnical problems, as illustrated in the next section.

The biogrouting methods have the following advantages:

1. Biocement is produced at ambient temperature and thus requires much less energy to produce.
2. Biogrout has much lower viscosity and can flow like water.
3. The bacteria that are used for biocement are usually nonpathogenic and environmentally friendly.

19.3 BIOGAS DESATURATION FOR MITIGATION OF LIQUEFACTION

Soil liquefaction often occurs in sandy soil during earthquakes. It is one of the major causes of earthquake-related disasters. Liquefaction was largely responsible for extensive damage to residential properties in the eastern suburbs and satellite townships of Christchurch, New Zealand, during the 2010 Canterbury earthquake and more extensively again following the Christchurch earthquakes that occurred in early and mid- 2011. The common methods that can be adopted for mitigation of liquefaction include the following broad categories: (1) replacement or physical modification; (2) densification; (3) pore water pressure relief; and (4) foundation reinforcement, as summarized by Chu et al. (2009). However, it is difficult to treat ground for the purpose of mitigation of liquefaction because the scale of the ground to be treated is normally huge, and it is thus too expensive to be implemented. Therefore, there is a need for more cost-effective solutions for liquefaction treatment.

A new approach for the mitigation of liquefaction potential of sand is to introduce gas bubbles into potentially liquefiable, saturated sand. Soil liquefaction is caused by a large generation of pore water pressure. Several studies (e.g., Okamura and Soga, 2006; Yegian et al., 2007) have shown that when saturated sand is made slightly unsaturated (with a degree of saturation between 85 and 95%) by inclusion of gas bubbles, the excess pore water pressure generated in soil under a dynamic load will be greatly reduced.

However, it is difficult to introduce gas into the ground for desaturation and maintain the same amount of degree of saturation for a long time. This is because the gas bubbles tend to dissolve in water or escape from the ground over time. Pumping can be used as illustrated conceptually in Fig. 19.3. However, the distribution of gas bubbles introduced by pumping will not be uniform. Furthermore, the gas pumped into the ground tends to form aggregated gas pockets rather than individual bubbles. As a result, the sizes of the gas bubbles or aggregates are not small enough to be kept in the ground for a long period of time.

One innovative way to overcome the previously mentioned problems is to generate tiny gas bubbles in situ using microorganisms. This method is

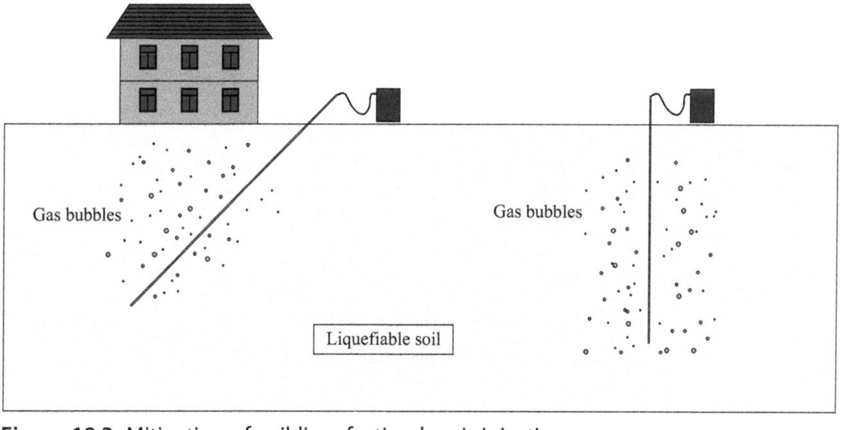

Figure 19.3 Mitigation of soil liquefaction by air injection.

promising because it has the following advantages: (1) It consumes the least energy because the low-viscosity bacteria and nutrient fluid can be delivered easily into sand; (2) the gas generated by bacteria can be distributed more evenly than by other means; and (3) the gas bubbles generated by bacteria are tiny, so the gas bubbles are less prone to escape from the ground. This so-called biogas desaturation method for mitigation of liquefaction has been developed by our research group (He et al., 2013; He and Chu, 2014).

To verify the effectiveness of the biogas method, shaking table tests using a laminate box as shown in Fig. 19.4 were carried out on saturated sand and sand desaturated to a degree of saturation (S_r) of 95, 90, and 80%,

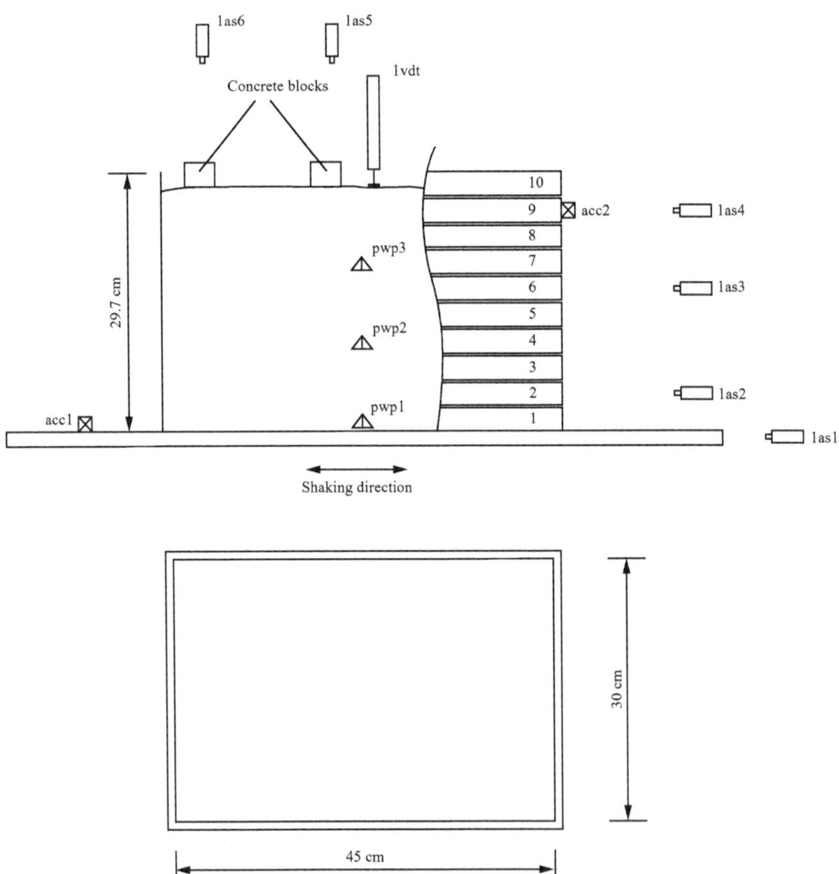

Figure 19.4 Laminar box and instrumentations used for shaking table tests. acc, accelerometer; las, laser displacement transducer; pwp, pore water pressure transducer.

respectively, using biogas. Some of the results are presented in Fig. 19.5, in which the pore water pressures generated during ground shaking under an acceleration of 1.5 m/s^2 are plotted. Figure 19.5(a) shows the pore water pressure for a saturated sample, and Fig. 19.5(b) shows the pore water pressure for a sample with a degree of saturation of 80%. The pore pressures were measured at three different locations as shown in Fig. 19.4. It can be seen from Fig. 19.5 that by merely reducing the degree of saturation using biogas, the amount of pore water pressure generation can be significantly reduced.

Figure 19.6 shows the pore water pressure ratio obtained from each shaking table test plotted against the relative density of the sand for samples with different degrees of saturation S_r. The pore water pressure ratio is defined as

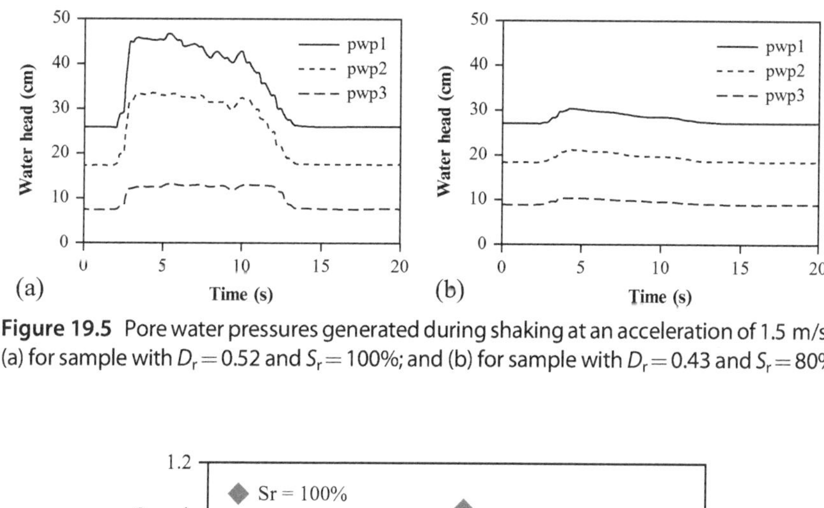

Figure 19.5 Pore water pressures generated during shaking at an acceleration of 1.5 m/s^2: (a) for sample with $D_r = 0.52$ and $S_r = 100\%$; and (b) for sample with $D_r = 0.43$ and $S_r = 80\%$.

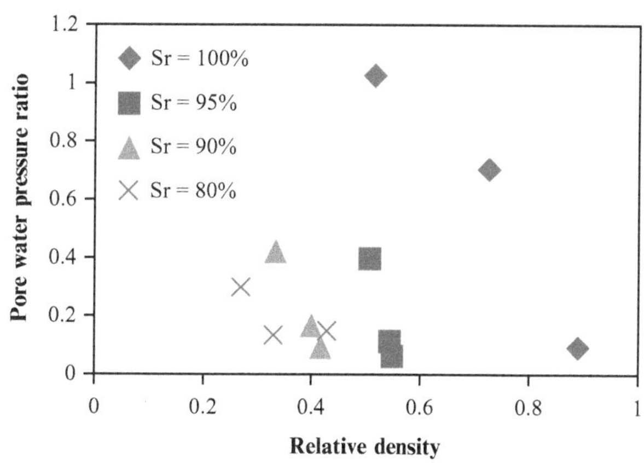

Figure 19.6 Plot of pore water pressure ratios against relative densities under a_{max} 1.5 m/s^2 for samples with different degrees of saturation (with the pore water pressure measured by transducer pwp2, as shown in Fig. 19.4).

the ratio of maximum excess pore water pressure to the initial effective overburden stress. When the pore water pressure ratio is close to 1, liquefaction becomes possible. The normal design requirement is to control the pore pressure ratio to be less than 0.5. The samples with the degree of saturation less than 100% in these tests were created using nitrogen gas produced by denitrifying bacteria. It can be seen from Fig. 19.6 that the pore water pressure ratio for the saturated sand with a relative density of 50% is as high as 1 and thus the sand liquefied under the applied ground acceleration.

One of the conventional ways to prevent liquefaction from occurring is to increase the relative density of sand. This requires the use of energy-intensive soil improvement methods such as dynamic compaction. However, if the degree of saturation of the sand is reduced to 95%, the pore pressure ratio would reduce to 0.4 for sand with the same density (Fig. 19.6). The cost and energy required to reduce the degree of saturation from 100 to 95% is much less. If the degree of saturation is reduced to 90%, the pore water pressure ratio can be controlled to be even less than 0.2, as shown in Fig. 19.6. Therefore, the biogas method can be very effective for liquefaction mitigation. It is also less expensive compared to conventional methods.

A comparison of ground settlement for a fully saturated sand layer and a sand layer treated with biogas with a degree of saturation of 95% is shown in Fig. 19.7. The settlement is expressed as a ratio of the current and the

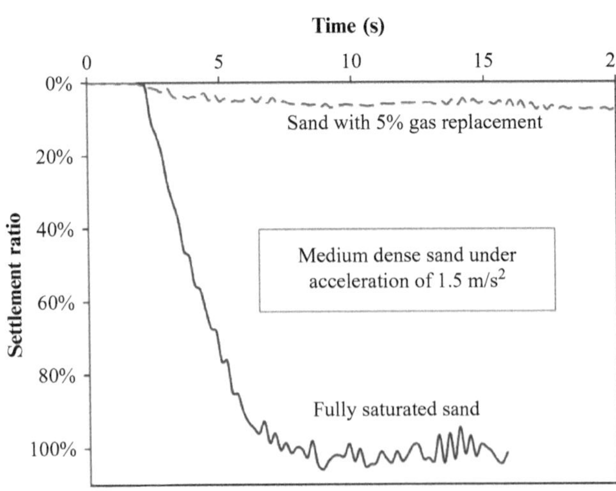

Figure 19.7 Comparison of ground settlement induced by ground shaking under an acceleration of 1.5 m/s^2 for a saturated sand layer and a sand layer with 5% gas replacement.

maximum settlement for fully saturated sand as 100%. It can be seen from Fig. 19.7 that with only 5% of gas replacement, the ground settlement generated under ground shaking with an acceleration of 1.5 m/s^2 can be reduced by more than 90%. Thus, the biogas method is effective in preventing the occurrence of soil liquefaction as well as reducing the damage caused by liquefaction.

19.4 BIOCRUST FOR CONSTRUCTION OF WATER POND

Biocement can also be used to reduce the water conductivity of sand through bioclogging, as illustrated in Fig. 19.1. One of the methods developed by our research group is biocrust. This method uses urease-producing bacteria to precipitate a layer of calcium carbonate on top of sand as shown in Fig. 19.8. This hard crust layer has a permeability of less than 10^{-7} m/s (Chu et al., 2012) and thus can be used as an impervious layer for water storage or for erosion control for beach or sandy riverbank. One model for water storage is shown in Fig. 19.8(b). Because the layer of crust is rather thin, 2 or 3 mm, the amount of biogrout used is small. Thus, this method is much more economical than conventional methods. The details of this method are described in Chu et al. (2012).

However, this thin calcium carbonate layer is brittle and may break under bending. For this reason, it may be suitable only for temporary use. Another method, the so-called bulk bioclogging method, has also been developed (Chu et al., 2012). In this method, bacterial suspension and calcium ions and urea solutions can be sprayed on top of the sand surface three to six times (Chu et al., 2012). An impervious layer of 20–25 mm thick can thus be formed by surface spraying, as shown in Fig. 19.9(a). Using this method, an aquiculture pond model has been made, as shown in Fig. 19.9(b) (Chu et al., 2013). The permeability of this layer is on the order of 10^{-7} m/s.

Figure 19.8 Formation of (a) a thin impervious layer on top of sand using the biogrouting method and (b) a water pond model in sand using this method.

Figure 19.9 Bulk bioclogging method to (a) form an impervious layer and (b) make an aquiculture pond.

The bioclogging effect also increases the shear strength of sand by creating cementation for the sand grains. As a result, the sand in the impervious layer has gained an unconfined compressive strength in the range of 215–930 kPa (Chu et al., 2013).

The two bioclogging methods discussed here have been used for the construction of inexpensive, temporary water storage facilities for various needs. However, large-scale field applications are yet to be carried out.

19.5 BIOGROUTING FOR PROTECTION OF LEVEES

By using the microbially induced calcium carbonate precipitation (MICP) method, the shear strength of soil can be increased through the mechanisms illustrated in Fig. 19.2. When cement or chemicals are used to treat soil, the amount of improvement in the shear strength of soil is dependent on the amount of cement or chemical used. Similarly, when biocement is used, the shear strength of soil is affected by the amount of minerals precipitated. A summary of the unconfined compressive strength (UCS) of biocement-treated sand versus the calcium carbonate (or calcite) content by weight obtained by our research group and other published data is shown in Fig. 19.10. It can be seen that the UCS increases with increasing calcium carbonate content. The highest UCS obtained is 27 MPa at approximately 33% calcite content (Van der Ruyt and van der Zon, 2009). For normal applications, the UCS required is less than 2 MPa. This will require a calcium carbonate content of only 5–10% or 90–180 kg/m^3. To achieve the same UCS for sand using cement grouting, the amount of cement used would be between 250 and 300 kg/m^3. Because the production of biocement can be cheaper, as discussed by Ivanov and Chu (2008), the overall cost for biogrouting can potentially be lower.

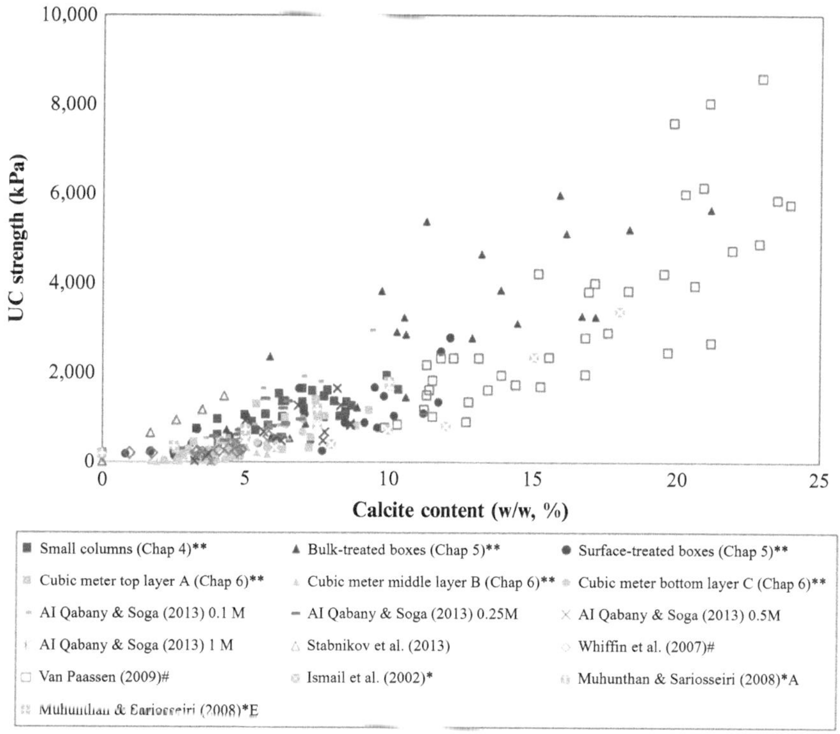

Figure 19.10 The unconfined compression strength (UCS) versus calcium carbonate content relationship for different biocement treated sands. *(Source: After Li (2014)).*

At the same time, the biogrout will reduce the permeability of soil. The permeability versus calcium carbonate content relationships for different sands obtained in different projects by different researchers are shown in Fig. 19.11. It can be seen that the permeability of sand can be reduced from an order of 10^{-4} to 10^{-7} m/s using the normal MICP method.

This biogrout method can be used for levee construction or rehabilitation. Many previous case histories have shown that erosion due to overtopping is one of the major causes of levee failure. A method for levee rehabilitation using biogrout to enhance its ability against erosion under overtopping and seepage through the levee and the base of the levee has been proposed. This so-called bioshotcrete method is illustrated in Fig. 19.12. In this method, bacterial suspension and biogrout consisting of fiber, calcium source, and urea are sprayed on top of the levee surface to form a biocoating. Biogrouting columns can also be installed under the levee to form a cutoff wall, as shown in Fig. 19.12.

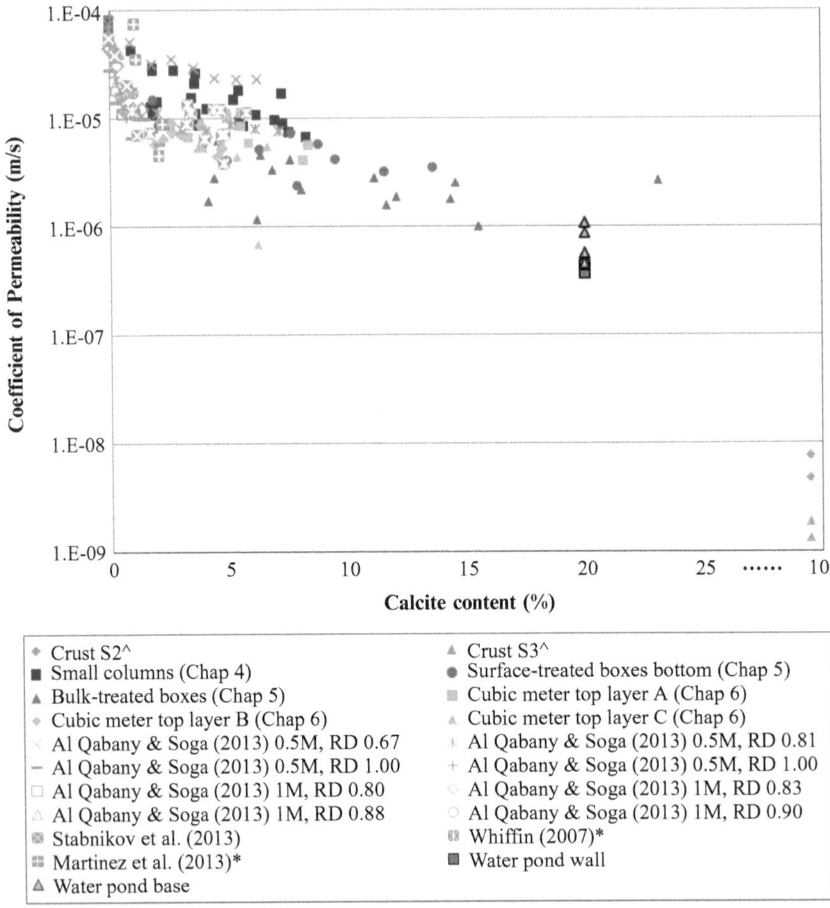

Figure 19.11 The permeability versus calcium carbonate content relationship for different biocement-treated sands. *(Source: After Li (2014)).*

To evaluate the effectiveness of this method, model tests using a hydraulic flume were performed. In conducting these tests, a levee model was made of standard Ottawa sand and then treated using biogrout. The dimensions of the levee model are shown in Fig. 19.13. The biogrouting solution was applied by spraying it six times onto the surface of the levee model. The biogrout included urease-producing bacteria suspension, solution of calcium chloride, and urea. The model was then placed inside a hydraulic flume for testing, as shown in Fig. 19.14. The model did not erode or collapse after it being subjected to an overtopping flow under different flow rates for 30 days. More details on the model tests are given in Naeimi (2014).

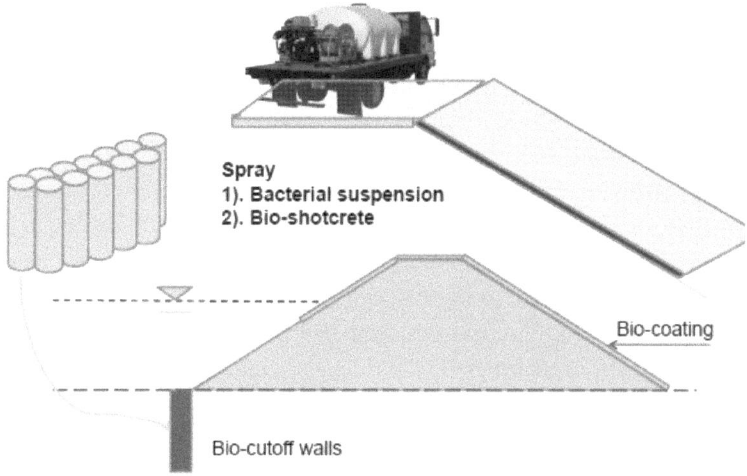

Figure 19.12 Methods for using biogrout for levee protection against erosion due to overtopping and seepage through the base of the levee.

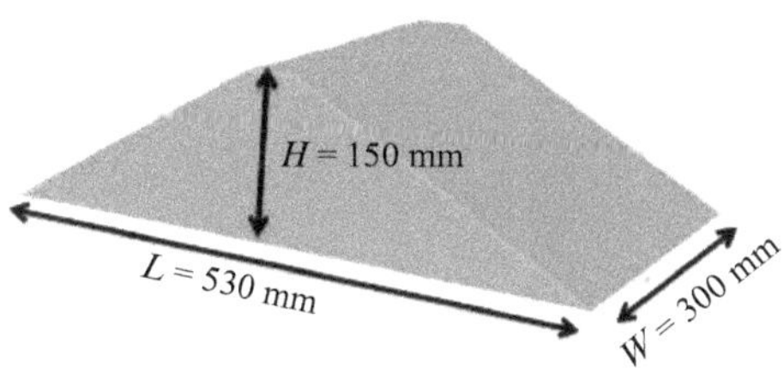

Figure 19.13 Dimensions of the levee model used for the model tests in a hydraulic flume.

Figure 19.14 A levee model undergoing overtopping flow in a hydraulic flume.

After model tests in the hydraulic flume, samples were taken from the model to determine the UCS. The UCS values determined for samples taken from different depths are shown in Fig. 19.15. It can be seen that UCS within the top 30 mm is much higher than that at other positions, indicating the formation of a hard layer on top of the levee. It is this hard layer that protects the levee from erosion and seepage flow through the levee. The calcium carbonate content distribution with depth in the model levee is also shown in Fig. 19.15. It can be seen that there is a good correlation between the UCS strength and the calcium carbonate content. As shown in the figure, there is also precipitation of calcium carbonate in the soil below the 30-mm crust, which contributes to the overall stability of the levee. The permeability of the soil is also reduced with the calcium carbonate content, as shown in Fig. 19.16. The permeability value can be reduced to below 10^{-7} m/s, as shown in Fig. 19.16. Thus, this hard layer also forms an impervious layer to prevent water from flowing through the levee.

A method to inject biocement into the ground to form biogrouted columns can be adopted to install cutoff walls for the base of the levee as shown in Fig. 19.12. The method to install the biogrouted columns is similar to that

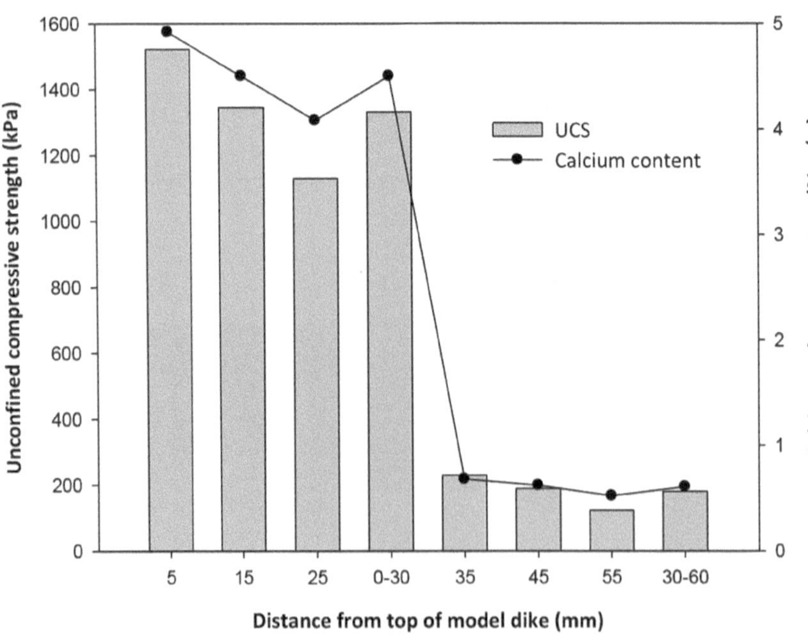

Figure 19.15 Unconfined compressive strength and precipitated calcium carbonate content distribution along the depth of the levee model.

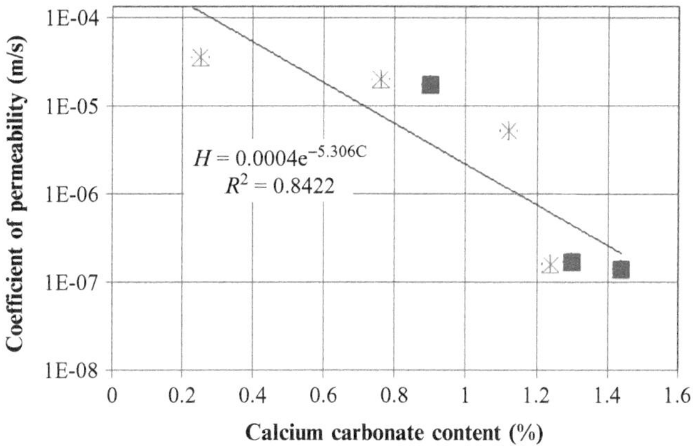

Figure 19.16 Change in permeability of sand with the calcium carbonate content in sand (% w/w).

for compaction grouting. The biogrout is injected into the ground point by point from the bottom to the top along an injection shaft to form a biogrouted column. Note that the viscosity of the biogrout is very low, and the particle size of the biogrout is smaller than the pore size in sand. Thus, the biogrout should be able to permeate into sand to form a biogrouted ball as shown in Fig. 19.17 at one injecting point.

Figure 19.17 Formation of a biogrouted ball through injection of biogrout in sand at one injection point.

The model tests presented previously, such as the one shown in Fig. 19.14, have demonstrated that the biogrouting method is effective in protecting the levee from erosion due to overtopping flow. However, note that the hard layer formed on top of the levee is only 30–40 mm thick, which is relatively thin. The thickness of this hard layer can be increased by changing the type of biogrout used and the number of injections adopted, as demonstrated by Naeimi (2014). To increase the ductility, synthetic fibers should be used together with the biogrout. The use of fiber will also increase the strength of the crust and prevent it from cracking easily. The fiber-reinforced biogrouting method should be viable for use in levees. Because the biogrouted layer is relatively thin, this method will be more cost-effective than any other methods for levees.

19.6 BIOGROUTING FOR ROAD CONSTRUCTION OR REPAIR

Biogrout can be used for road construction or repair in a manner similar to the use of cement grout. To further reduce the cost of biogrout, a method to produce biogrout using mining and agricultural by-product has been developed. In this method, limestone fines, a by-product of limestone mine and corn cobs, or any other agricultural by-product can be used to produce soluble calcium salt using an acidogenic fermentation process. The soluble calcium salt can be used together with biomass to form biogrout for road construction and repair.

Some laboratory tests have been carried out to establish the procedure for this method. A fermenter was used to produce soluble calcium solution with limestone residuals in it, as shown in Fig. 19.18 as the white solution in the cylindrical container. Bacterial suspension and urea were then added to the soluble calcium solution to form a biogrout solution. This solution can be applied directly on top of soil or a crushed stone layer, as shown in Fig. 19.19. Once this layer dries, a biogrouted slab is formed, as shown in Fig. 19.20.

The cost of biotreatment may be lower than that of conventional methods such as the use of cement or chemicals. However, a direct cost comparison may not be made at this stage because the biocement may have to be produced in a factory setting. For example, the fermentation process adopted to produce biogrout introduced previously is very complicated, and its operation may not be well controlled. This has to be done in a factory when a large quantity of biocement is required for construction. Until the production lines are in place, the cost of biocement cannot be evaluated realistically.

Figure 19.18 Production of biogrout using solutions made of limestone fines and corn cobs with the addition of biomass and urea.

Figure 19.19 Applying the biogrout directly on top of the base and subbase of the road.

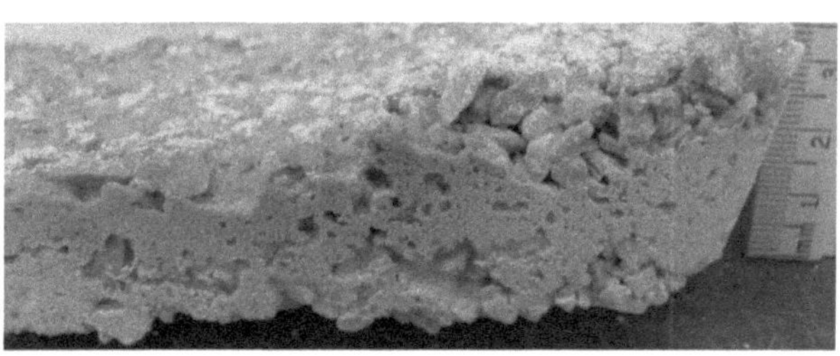

Figure 19.20 Formation of a hard biogrouted slab.

19.7 CONCLUSION

In this chapter, the principles of microbial geotechnologies are discussed. Studies on the application of different microbial methods are presented. These include the biogas desaturation method for mitigation of soil liquefaction, which may become the most cost-effective method for the reduction of liquefaction risks; the biocrust method for the construction of a temporary water storage pond in sandy soil; the biogrouting method for the protection of levees from erosion and seepage failure; and the biogrouting method for road construction and repair using biogrout made of waste materials. The main conclusions are summarized as follows:

1. The biogrouting method has the following major advantages: It is produced at ambient temperature and thus requires much less energy to produce; it has much lower viscosity and can flow like water; and the bacteria that are used for biocement are nonpathogenic and environmentally friendly.
2. Denitrifying bacteria can be used to produce nitrogen gas in situ and enhance the liquefaction resistance of soil through desaturation. This method has the potential to be the most cost-effective method for mitigation of liquefaction hazard.
3. A biogrouting method is proposed for the rehabilitation of levees. In this method, fiber mixed biogrouting is used to treat the surface of levees to form a strong crust to protect the levees from erosion and seepage failure.
4. Biogrouting materials can be produced using mining and agricultural by-products, and they can be used for road construction and repair.

ACKNOWLEDGMENTS

The contributions of other team members Dr. Viktor Stabnikov, Bing Li, and Lu Liu are gratefully acknowledged.

REFERENCES

Chu, J., 2012. General Report on Biogrout and Other Grouting Methods. In: General Report, Proc. ISSMGE-TC 211 Int. Symp. on Ground Improvement, Brussels, May 31–June 1, pp. 177–187.

Chu, J., Varaksin, S., Klotz, U., Mengé, P., 2009. Construction Processes: State-of-the-Art Report. In: Proc. 17th Int. Conf. on Soil Mechanics and Geotechnical Engineering, Alexandria, Egypt, October 5–9, vol. 4, pp. 3006–3135.

Chu, J., Stabnikov, V., Ivanov, V., 2012. Microbially induced calcium carbonate precipitation on surface or in the bulk of soil. Geomicrobiol. J. 29, 544–549.

Chu, J., Ivanov, V., Stabnikov, V., Li, B., 2013. Microbial method for construction of aquaculture pond in sand. Geotechnique 63 (10), 871–875.
He, J., Chu, J., 2014. Undrained responses of microbially desaturated sand under static monotonic loading. J. Geotech. Geoenviron. Eng. 140 (5), 04014003, ASCE.
He, J., Chu, J., Ivanov, V., 2013. Mitigation of liquefaction of saturated sand using biogas. Geotechnique 63 (4), 267–275.
Ivanov, V., Chu, J., 2008. Applications of microorganisms to geotechnical engineering for bioclogging and biocementation of soil in situ. Rev. Environ. Sci. Biotechnol. 7 (2), 139–153.
Ivanov, V., Chu, J., Stabnikov, V., 2014. Basics of Construction Microbial Biotechnology. In: Pacheco-Torgal, F., et al. (Ed.), Biotechnologies and Biomimetics for Civil Engineering. Springer, New York.
Li, B., 2014. Geotechnical Properties of Biocement Treated Soils. PhD thesis, Nanyang Technological University, Singapore.
Mitchell, J.K., Santamarina, J.C., 2005. Biological considerations in geotechnical engineering. J. Geotech. Geoenviron. Eng. 131 (19), 1222–1233, ASCE.
Naeimi, M., 2014. Biocementation of Sand in Geotechnical Engineering. PhD thesis, Nanyang Technological University, Singapore.
Okamura, M., Soga, Y., 2006. Effects of pore fluid compressibility on liquefaction resistance of partially saturated sand. Soils Found. 46 (5), 695–700.
van der Ruyt, M., Van der Zon, W., 2009. Biological in situ reinforcement of sand in near-shore areas. Geotech. Eng. 162, 81–83.
van Paassen, L.A., Ghose, R., van der Linden, T.J.M., van der Star, W.R.L., van Loosdrecht, M.C.M., 2010. Quantifying biomediated ground improvement by ureolysis: large-scale biogrout experiment. J. Geotech. Geoenviron. Eng. 136 (12), 1721–1728, ASCE.
Whiffin, V.S., 2004. Microbial $CaCO_3$ Precipitation for the Production of Biocement. PhD thesis, Murdoch University, Australia.
Yegian, M.K., Eseller-Bayat, E., Alshawabkeh, A., Ali, S., 2007. Induced-partial saturation for liquefaction mitigation: experimental investigation. J. Geotech. Geoenviron. Eng. 133 (4), 372–380, ASCE.

CHAPTER 20

The Role of Native Vegetation in Stabilizing Formation Soil for Transport Corridors

An Australian Experience

Behzad Fatahi[1], Udeshini Pathirage[2], Buddhima Indraratna[2], Muditha Pallewattha[2], Hadi Khabbaz[1]
[1]Centre for Built Infrastructure Research, School of Civil and Environmental Engineering, University of Technology Sydney (UTS), Sydney, Australia
[2]Centre for Geomechanics and Railway Engineering, Faculty of Engineering and Information Sciences, University of Wollongong, NSW, Australia

20.1 INTRODUCTION

Since the mid-1990s, increasing public concern over the appearance of slopes has encouraged the development of a systematic and integrated approach to using native vegetation to stabilize them. Bioengineering applications, including the use of vegetation in civil engineering, began with slope stabilization and erosion control, but vegetation also can influence the soil water balance and hydrogeology of the site. A railway line is an engineering structure that is greatly influenced by hydrological and environmental conditions, and given the lengthy rail networks in coastal areas of Australia, many rail tracks were built on clayey soils that are sensitive to moisture, which is why bioengineering methods for improving the ground are becoming increasingly popular for stabilizing railway corridors. Based on current research and published literature, Southeast Asian countries such as the Philippines, Malaysia, and Hong Kong have used bioengineering ground treatment more than others, so there is greater acceptance of this type of modification, particularly for stabilizing slopes. Landscape treatment and the bioengineering technical guideline (Geotechnical Engineering Office, 2000) concluded that future development will include a wider use of native vegetation and the potential application of alternative measures. This technique can be cost-effective, flexible, and applicable in remote areas, and it adapts to the environment without loss of life or damage to property.

Chemical, Electrokinetic, Thermal, and Bioengineering Methods
http://dx.doi.org/10.1016/B978-0-08-100191-2.00020-4

Tree roots provide three stabilizing functions: reinforcement of the soil, dissipation of excess pore pressure, and establishment of sufficient matric suction to increase the shear strength. Available studies indicate that most attempts to quantify the effects of vegetation have focused on the structural reinforcement provided by the roots, whereas few attempts have been made to relate the changes in soil strength and stiffness to the rate of transpiration. Trees can provide suction up to the wilting point of a soil–root system (approximately 3 MPa), and therefore, ground consolidation associated with transpiration increases the strength and stiffness of the soil. This process may be compared to improving soft soil via prefabricated vertical drains and vacuum preloading. In fact, transpiration is similar to a vacuum in that it results in an increase in the hydraulic gradient and effective stress in soil, which leads to consolidation without positively increasing the initial excess pore pressure. Although transpiration depends on the size of the root zone, which can be shallow or deep depending on the type of vegetation and condition of the soil, the root-based suction (negative pore water pressure) is much greater (>10 times) than vacuum for a prefabricated vertical drain system. The suction induced by transpiration increases the effective stresses, which in turn increases the settlement and stiffness of unsaturated soil. Despite these advantages, geotechnical engineers have been reluctant to adopt vegetation-based stabilization partly because vegetation is perceived to be unpredictable, and therefore nonuniform patterns of soil moisture reduction may develop, and also because of time dependency. It is often assumed that a 5- to 10-year delay is required for the roots to strengthen and dry the soil, and when modeling a vegetated vadose zone, a detailed description of the root water uptake is required, but existing methods only consider a simplified model that is implemented mainly in the flow equation. Although current design standards such as the Uniform Building Code (1997) and Australian Standard AS2870 (1996) provide guidelines for the design and construction of footings and structures on expansive clays, none of them provide any guidelines on how ground desiccation caused by native vegetation should be included. Given the importance of the vadose zone in most geo-environmental projects, there is a need to develop a better understanding of how trees, including root-based suction, influence behavior within this zone.

As mentioned previously, using native vegetation in semi-arid climates and coastal regions of Australia to stabilize railway corridors built over expansive clays and compressive soft soils has become increasingly popular. In these regions, ballast pocket is but one of the existing problems, and using native vegetation can be a solution. Heavy trains or ballast tamping to reshape and level the ballast can create a ballast bowl (or ballast pocket) in the base region where water sits and softens the ground. Because this region is the foundation of the rail track, determining how it will perform under load is absolutely critical. Although maintenance programs pay more attention to drainage, it is often unsuccessful in itself due to site restrictions such as limiting topography, and because the interaction between vegetation and ground is not known, complete and appropriate design procedures for strategically planted vegetation are still unavailable. In fact, the soil moisture profile close to vegetation, the influencing factors, and the zone influenced by vegetation and its properties are not clearly established for design purposes. Previous studies reported that civil engineers are aware that trees and shrubs help deform foundations placed on expansive and collapsible soils, and this effect will be more severe in regions with wet seasons and very dry seasons. As Biddle (1998) and McInnes (1986) reported, engineers have mainly focused on the detrimental effect of vegetation on buildings and pavements, whereas newer field observations have shown that where there are trees beside railway tracks, their localized, undrained failure is minimized. Using native vegetation beside remote railway lines in Australia to stabilize existing railway corridors built over expansive clays and compressive soft soils has become increasingly popular. Properly selected and used, vegetation that includes native trees and shrubs can reduce soil moisture by root water uptake and also increase its shear strength and stiffness by increasing the matric suction and controlling erosion (Fig. 20.1).

This chapter examines the results of a field investigation conducted near an Australian native Black Box tree (*Eucalyptus largiflorens*) located in Miram village in western Victoria. Boreholes were drilled, soil samples were taken, and the moisture content and suction were measured. Moreover, trenches were dug in order to observe the distribution of tree roots under the ground. The soil samples were tested in the laboratory to measure the required mechanical and physical parameters of the soil. The soil moisture content and suction profiles were compared with the predicted values in order to verify the model. Furthermore, the profiles of ground settlement near the tree are presented and interpreted, and the results of preliminary studies on the effects of root growth and osmotic suction on the root water uptake are explained.

20.2 TRANSPIRATION

Vegetation affects the hydrogeological conditions of the soil profile. Trees, shrubs, and grasses deplete moisture from the soil through transpiration. Furthermore, local field studies by Premchitt et al. (1992) revealed that tree

Figure 20.1 Trees planted along railway lines, Coalcliff, NSW, Australia.

canopies reduce rainfall intensity (I_R) below the vegetation canopy. This means that rainfall intensity below the tree canopy (I_T) is less than I_R (e.g., $I_T = 0.45\ I_R$). In most cases, if soil permeability (k) is greater than I_T, the soil in the root zone remains unsaturated. According to White et al. (1997) and Blight (2005), of the different hydrological features, transpiration is the key and most uncertain factor in the water balance model.

Transpiration is defined as water vaporized from the plant. This loss of moisture through pores in the leaf is caused by a difference in vapor pressure between the interior leaf space and ambient air. Hopkins (1999) summarized that more than 90% of the moisture escapes from leaves, whereas only small amounts escape from other parts, such as lenticles (small openings in the bark) of young twigs and branches. Hence, the transpiration rate relates mainly to the specification of the leaves. The outer surface of a leaf is covered with a multilayered waxy deposit called cuticle, which is very hydrophobic and resists the diffusion of water and water vapor from the underlying cells. Leaves also have small pores on the surface called stomata. Hopkins showed that their most important role is to provide a route for the exchange of gasses (principally carbon dioxide, oxygen, and water vapor) between the internal air space and the bulk atmosphere surrounding the leaf. Approximately 90–95% of water loss induced by transpiration stems from stomatal transpiration, whereas the remaining 5–10% stems from cuticular transpiration. Although the cuticle is generally considered to be an impermeable skin, small amounts of moisture can pass through it. When the skin is thin or dry, the stomata close to prevent stomatal transpiration, and thus cuticular transpiration is greater. Because transpiration involves the evaporation of water, it has a significant role in cooling the leaves.

According to Alarcon et al. (2000) and Hopkins (1999), the following three principal methods are commonly used to measure the rate of transpiration: weight loss, gas exchange, and sap flow. Sealing a well-watered potted plant to prevent evaporation through the pot or the surface can illustrate the weight loss method because the plant may be weighed at intervals and any weight loss can be attributed to water transpiring through the shoots. As Kramer (1983) reported, the weight loss method, also known as the lysimeter method, was scaled up for agricultural field studies by constructing large containers filled with soil (perhaps several cubic meters), mounted on weighing devices buried in the ground. In these cases, records of water input (rainfall and irrigation) and evaporation must be kept. The lysimeter method is generally considered to be the most reliable and accurate for field studies, but lysimeters are expensive to construct and are not really considered as portable. The gas exchange method is often used in conjunction with experiments on photosynthesis; this involves sealing a leaf or branch in a transparent chamber with a flowing air system. Transpiration can be estimated as the difference in water content of the air entering and leaving the chamber. The temperature, content of carbon dioxide, and other parameters can also be measured. This method has also been scaled up for field studies by enclosing entire trees or other large plants with a plastic canopy. Gas exchange methods, whether on a small scale in the laboratory or in large-scale field measurements, are usually limited to short-term studies. The act of enclosing the plant may, over the long term, significantly alter the microclimate surrounding the leaves, and therefore the temperature, humidity of the incoming airstream, and air velocity must be carefully monitored and controlled. Alternatively, chambers and measuring systems can be made quite portable, and a number of commercial instruments are now available for field studies.

As Smith and Allen (1996) mentioned, the sap flow method provides direct and continuous measurement of whole plant water use with a long time resolution. With this method, the heat pulse velocity is measured by temperature sensors inserted upstream and downstream of a line heater, and then the heat pulse velocity is related to the rate of sap flow using theoretical calibration factors. The theoretical calibration factors commonly used can be found in Swanson and Whitfield (1981) and Fernandez et al. (2001). As Alarcon et al. (2000) reported, the compensation heat pulse technique developed by Swanson and Whitefield is one of the popular sap flow methods used to measure transpiration. In this method, one set of heat pulse probes is installed per tree, and each set consists of a 1.8-mm-diameter heather needle and two thermocouple probes with the same diameter. Heather and thermocouples are inserted into parallel holes drilled radially into the trunk 300 mm above the surface, and then the heat pulse velocity is measured. The method recommended by Swanson and Whitfield considers the effects of wounding caused by a probe, but after that, the heat pulse velocity is related to the actual sap flow using developed equations such as that by Edwards and Warwick (1984). The sapwood area of the tree stems is determined from an analysis of trunk cores taken by cutting the trunk down at the end of the experiment.

20.3 TREE ROOT WATER UPTAKE

The loss of moisture from the soil may be categorized as (1) water used for metabolism in plant tissues and (2) water transpired to the atmosphere. However, as Radcliffe et al. (1980) suggested, the volume of water required for photosynthesis or metabolism in plant tissues compared to the total water uptake by roots is negligible. Total transpiration can then be assumed to be the same as the water uptake through the root zone.

Clearly, the soil conditions (soil suction, hydraulic conductivity, penetration resistance, etc.), the type of vegetation (root distribution, the relative proportion of active roots and leaf area, etc.), and atmospheric conditions (net solar radiation, temperature, humidity, etc.) affect the rate of root water uptake and hence transpiration. To formulate a comprehensive equation for calculating the rate of root water uptake, any interaction between the previously mentioned features should be taken into account, but because this is a complex problem, some of these features are simplified in this study, as summarized here:

1. Because the potential transpiration rate (potential water discharge rate) was used instead of the difference between the water potential of root and soil, hydraulic conductivity was not directly included in the transpiration rate formulation.
2. Resistance to penetration by the soil influences root distribution, but this was only considered indirectly through the effect of root distribution.

3. In an assessment of the rate of water uptake, active roots have a major role, so their relative proportion was considered in the root distribution effect.
4. Because the potential transpiration rate is related to interaction between the atmosphere and foliage, the leaf area (canopy) was included in potential transpiration. The influences of net solar radiation, air density, temperature, and humidity changes were ignored in the model proposed here.

In brief, the three independent features that were considered in detail in the root water uptake model are soil suction, root distribution, and potential transpiration. It was assumed that the function of the root water uptake rate can be expressed as a product of three separate functions. Consequently, as reported by Indraratna et al. (2006), the rate of tree root water uptake is given by

$$S(x, y, z, t) = G(\beta(t))F\left(T_p(t)\right)f(\psi(t)) \tag{20.1}$$

where $\beta(t)$ is the root length density at time t, $G(\beta(t))$ is the root density factor, $T_p(t)$ is the rate of potential transpiration at time t, $F(T_p(t))$ is the potential transpiration factor, and $f(\psi)$ is the soil suction factor.

For a given transpiration rate, the rate of water uptake from any particular unit volume of wet soil is proportional to the root length density (β). Eastham et al. (1988) suggested that this relationship is nonlinear. Furthermore, according to Raper (1998), the rate of root water uptake must reach a limiting value when the density of a tree root is such that the water uptake approaches its physiological maximum. Therefore, based on agronomical research, an asymptotic relationship can be assumed for root water uptake. Because the hyperbolic tangent function represents a nonlinear asymptotic curve, the following equation is suggested for the root density correction factor $G(\beta)$:

$$G(\beta) = \frac{\tanh\left(k_3\beta(t)\right)}{\int_{V(t)} \tanh\left(k_3\beta(t)\right)dV} \tag{20.2}$$

where k_3 is an experimental coefficient. The denominator is applied to this factor to normalize the function within the root zone, where $\int G(\beta)dV = 1$.

According to Fatahi (2007), it can be assumed that potential transpiration is not distributed uniformly within the root zone because of the root resistance term, and a linear distribution with depth for potential transpiration is a more appropriate distribution. Accordingly, Eq. (20.3), which considers the linear distribution of transpiration with depth, is suggested to take into account the effect of potential transpiration:

$$F(T_P) = \frac{T_P(t)(1 + k_4 z_{\max} - k_4 z)}{\int\limits_{V(t)} G(\beta)(1 + k_4 z_{\max} - k_4 z)\, dV} \tag{20.3}$$

where $G(\beta)$ is given by Eq. (20.2), and k_4 is an experimental coefficient to involve the effect of depth in the transpiration rate. The denominator of Eq. (20.3) was applied to satisfy the potential transpiration mass balance. By comparing Eq. (20.3) and Nimah and Hanks' (1973) model, k_4 can be established by $k_4 \cong -(1 + R_c)/H_{\text{root}}$, where H_{root} is an effective water potential in the root at the soil surface, and R_c is a flow coefficient in the plant root system.

To calculate $f(\psi)$, different approaches have been recommended by various researchers. The equation suggested by Feddes et al. (1978), which is a simple and appropriate formula to determine soil suction effects, was used in this study. Function $f(\psi)$ can be written as

$$\left\{\begin{array}{ll} f(\psi) = 0 & \psi < \psi_{\text{an}} \\ f(\psi) = 1 & \psi_{\text{an}} \leq \psi < \psi_{\text{d}} \\ f(\psi) = \dfrac{\psi_{\text{w}} - \psi}{\psi_{\text{w}} - \psi_{\text{d}}} & \psi_{\text{d}} \leq \psi < \psi_{\text{w}} \\ f(\psi) = 0 & \psi_{\text{w}} \leq \psi \end{array}\right\} \tag{20.4}$$

where ψ_{w} is the soil suction at wilting point—that is, the suction limit at which a particular vegetation is unable to draw moisture from the soil. ψ_{d} is the highest value of ψ and ψ_{an} (soil suction at anaerobiosis point) is the lowest value of ψ at $S = S_{\max}$, where $S_{\max}$ is the maximum rate of root water uptake. An experimental study by Kutilek and Nielsen (1994) also confirmed the same trend given by Feddes et al. (1978).

As a result, according to Eq. (20.1), the rate of root water uptake $S(x, y, z, t)$ can be estimated by multiplying three functions that represent soil suction, $f(\psi)$, the root density, $G(\beta)$, and the potential transpiration, $F(T_p)$. To determine $f(\psi)$, Eq. (20.4), suggested by Feddes et al. (1978) to consider the effects of soil suction, was used in this study. Figure 20.2 shows the flow chart used to calculate the rate of root water uptake as the sink term in Darcy's flow equation. Further information about the development of the root water update model can be found in Fatahi (2007).

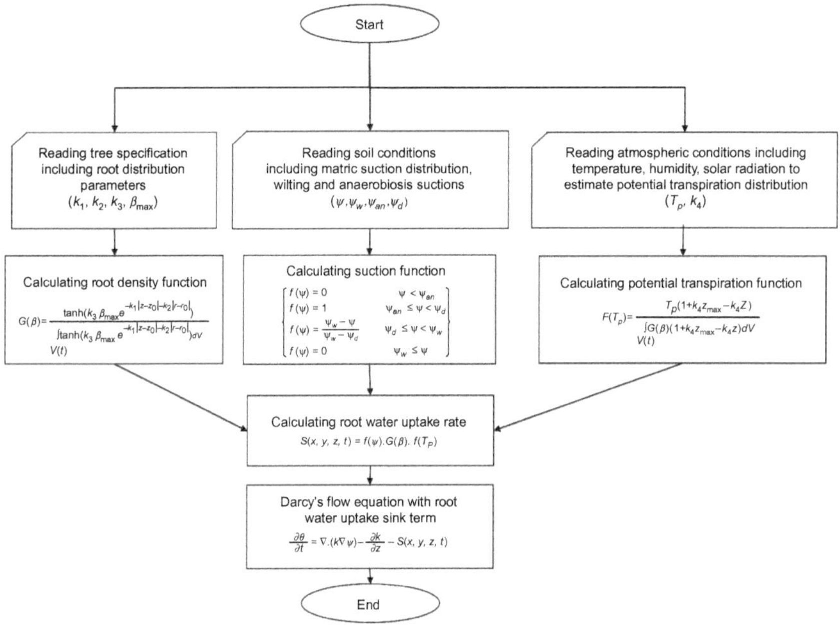

Figure 20.2 Flow chart to calculate root water uptake rate as the sink term in the moisture flow equation.

20.4 CASE STUDY OF AN AUSTRALIAN NATIVE TREE

20.4.1 Site location, geology, and climate

This case history is related to the results of a geotechnical investigation conducted near an 11-m-high *E. largiflorens* tree. The proposed site is at Miram village, which is 15 km away from Kaniva, a city in the state of Victoria, Australia. According to Lawrence (1975), the oldest surfacial unit is Tertiary Plicoene marginal marine Parilla sand, where the cross sections show a tertiary sequence between 100 and 330 m thick. In some areas in Miram, there are lenses of Quarternary Holocene Paludal and Peilian that include lagoon and swamp deposits of silt and clay and lunette deposits of sand, silt, and clay, respectively. Lawrence showed one basement fault in the Horsham area, which was largely interpreted from gravity data.

The climate in Miram is semi-arid with mild winters and long hot summers. Kaniva (141.2427 E, 36.3725 S), Nhill (141.6367 E, –36.3347 S), and St. Arnaud (Tottington) (143.1206 E, –36.7911 S) in western Victoria are the closest weather stations to Miram (141.3000 E, –36.4000 S) for temperature, rainfall, and evaporation readings, respectively. The mean daily maximum temperature ranges from 13.7°C in July to 29.7°C in January.

The mean monthly rainfall ranges from 20.9 mm in January to 47.7 mm in August, with a mean annual rainfall of 415.3 mm. The mean monthly potential evaporation ranges from 30.45 mm in July to 257.9 mm in January (Bureau of Meteorology, 2006). On an annual basis, the potential evaporation (1483.7 mm/year) is more than three times the average annual rainfall (415.3 mm/year).

20.4.2 Geotechnical site investigation and laboratory tests

A mobile drilling rig with a 76-mm-diameter drill was used to drill vertical boreholes based on the rotary-dry method. The cores obtained by push-sampling tubes were waxed immediately after taking. The geotechnical section for the soil found at the site is shown in Fig. 20.3.

The average soil profile can be described as 200 mm of sandy clay topsoil underlain by brown, firm to hard sandy lean clay to approximately 3 m below the surface. Underneath the clay is a medium-dense to dense, multi-colored clayey sand layer with rust-colored partings approximately 3–6 m deep. The groundwater level is below 6 m. Soil changes are gradual, with no distinct layer boundaries evident below the base of the topsoil. The cores were laid in special boxes and photographed after drilling was completed. In order to classify the subsurface layers and determine their physical and mechanical properties, the required tests were conducted according to the available standard methods. Figure 20.4 shows the particle size distribution of soil samples taken from various depths. Figure 20.5 shows the average soil profile used at the site for this case study.

Figure 20.5 shows profiles of the void ratio away from the tree, the initial deformation modulus, preconsolidation pressure, and the overconsolidation

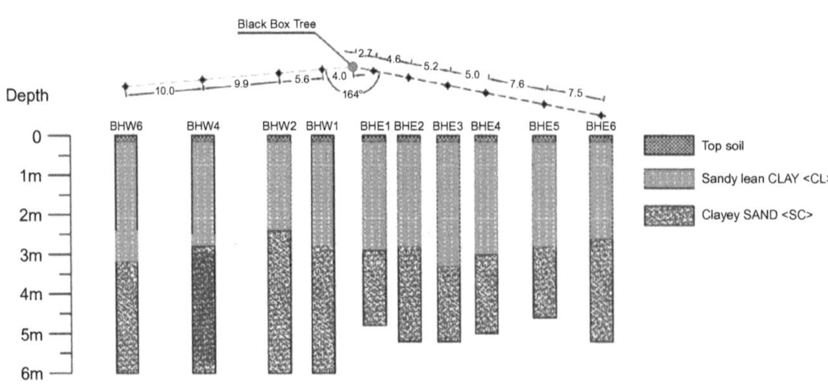

Figure 20.3 Geotechnical section of Miram site (dimensions are in meters).

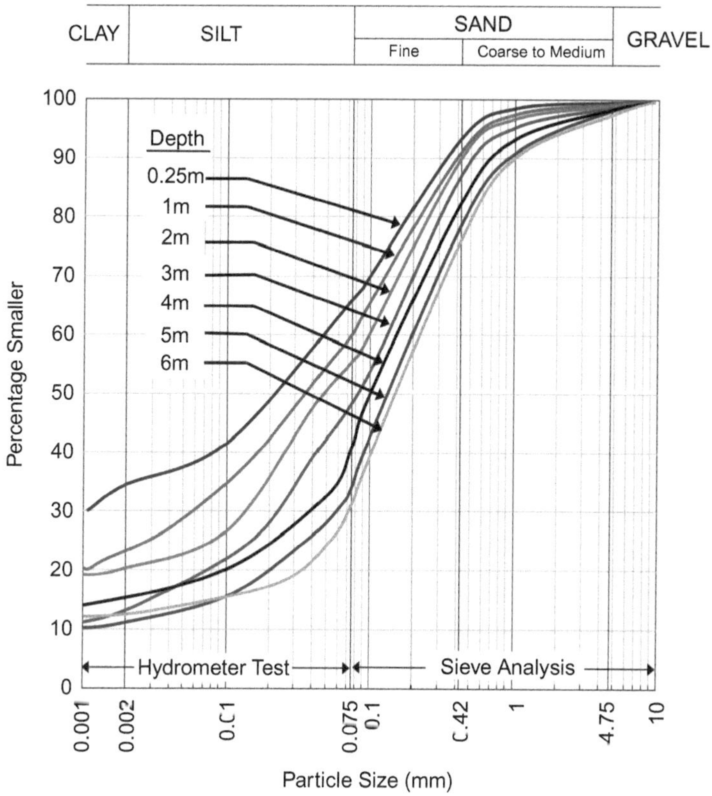

Figure 20.4 Soil particle size distribution.

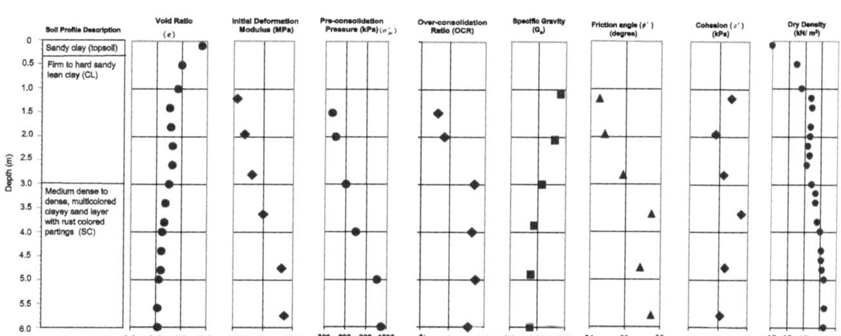

Figure 20.5 Profiles of soil properties at Miram site, VIC, Australia. *(Source: Modified after Fatahi et al. (2014)).*

ratio. In addition, this figure reports the variation of specific soil gravity, effective friction angle and cohesion coefficient, and the dry density.

The saturated hydraulic conductivity of the soil was measured using double-ring permeameter (DRP) on the remolded soil. The DRP is able to detect leakage from the sidewall, and saturation time can be reduced by conducting the test under high hydraulic gradients. The soil was compacted in a standard compaction mold (D = 101.6 mm and H = 116.5 mm) to its field dry density. Because the samples were taken from different depths away from the trunk, the dry densities differed, so permeability tests were conducted at different depths. Table 20.1 shows the laboratory results of the average ground permeability at Miram site.

Pressure plate (ASTM D2325) and buried filter paper (ASTM 5298) tests were used to measure the soil–water characteristic curve (SWCC). Figure 20.6 shows the setup in the laboratory that was used for both methods. Due to the value of air entering the ceramic plates used in the pressure plate apparatus, the suction measurements were limited to 1500 kPa and the full suction range was not covered. The buried filter paper method could measure a wider range of suction, and in this study, the matric suction was measured using Whatman No. 42 filter paper. Because the SWCC of undisturbed samples was obtained through a drying process using a filter paper technique, determining the water retention curve at low values of suction was practically impossible. In this study, therefore, the pressure plate test was used to determine the SWCC of undisturbed samples with low suction, whereas the filter paper method was used to measure the high suction of drier samples. To avoid damaging the soil samples used for the pressure plate test, stainless-steel sampling tubes were cut at 10-mm lengths without removing the sample; the rings were then used as sample retainer rings. Figure 20.7 shows the results of SWCC for the soil samples taken from different layers. The profile of average matric suction measured near a tree in mid-May 2005 is shown in Fig. 20.8.

Table 20.1 Average soil permeability at the Miram site, VIC, Australia

	Depth (m)				
	1	**2**	**3**	**4**	**5**
Permeability (m/s)	6×10^{-10}	1.8×10^{-10}	5.4×10^{-11}	1.6×10^{-11}	5×10^{-12}

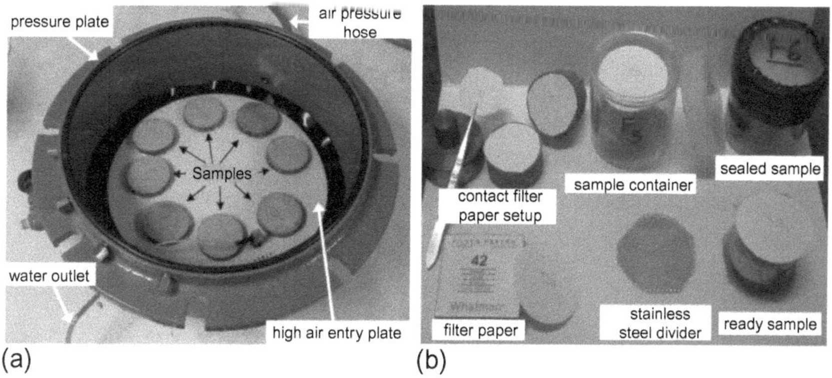

Figure 20.6 Laboratory soil suction measurements: (a) Pressure plate test and (b) filter paper method.

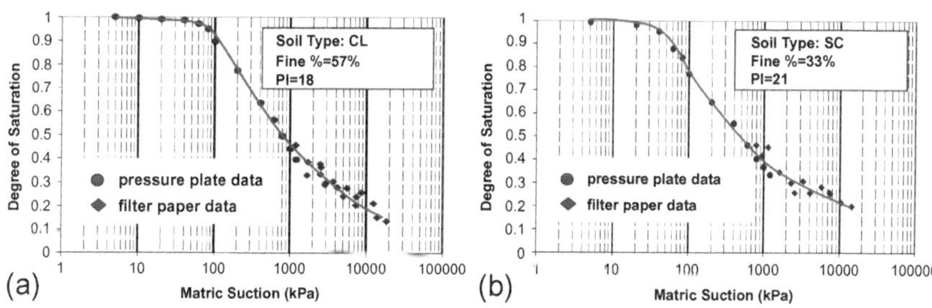

Figure 20.7 Soil water characteristic curves for soil types (a) CL and (b) SC. *(Source: Modified after Fatahi et al. (2014)).*

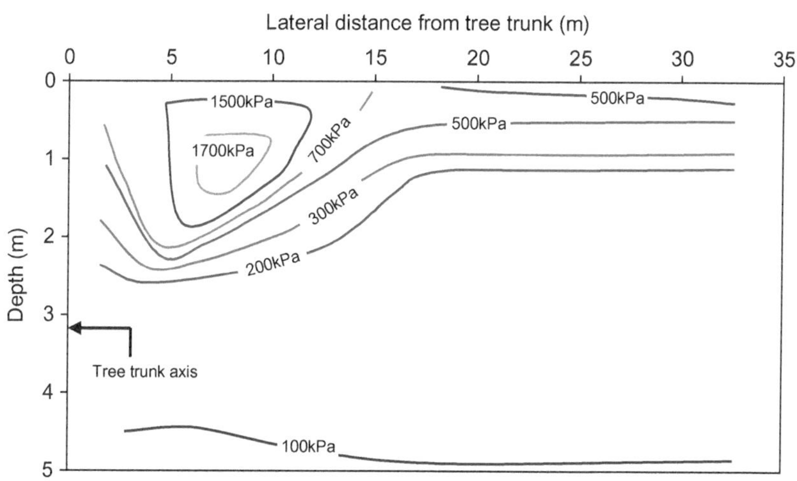

Figure 20.8 Profile of the distribution of average matric suction near eucalyptus largiflorens tree.

20.4.3 Root distribution measurement

Organic soils have solid constituents that consist predominantly of decomposed vegetable materials. The major part of a tree root is normally organic matter, and the soil, including the roots of native vegetation, is known as humus. Indeed, the parent material of organic soil is vegetation. In this study, it was assumed that the root content of the soil matrix was directly proportional to the organic content of soil. Thus,

$$\beta = a \times \Gamma \tag{20.5}$$

where a is an experimental constant, and Γ is the organic content of the soil matrix. By considering the exponential distribution of active roots in the vertical and radial directions, Eq. (20.6) can be obtained:

$$\Gamma(r, z, t) = \Gamma_{max}(t)\, e^{-k_1(t)|z - z_0(t)| - k_2(t)|r - r_0(t)|} \tag{20.6}$$

where Γ_{max} is the maximum percentage of the organic content.

As Scott et al. (1995) noted, fresh organic matter in the soil is liable to be attacked by bacteria when in air. Therefore, the organic matter of soil samples (disturbed and undisturbed) at different depths and lateral distances away from a tree was measured using the method described in ASTM D2974. The organic content was measured by igniting a sample of dried soil placed into a muffle furnace set at 440°C. If the samples contained any visible woody reinforcing roots (non-active roots), they were removed before testing. By fitting the field data to Eq. (20.6), the root distribution factors can be calculated, as given in Table 20.2. Equation (20.7) represents the distribution of organic content near an *E. largiflorens* tree:

$$\Gamma(r, z) = 6.2 e^{-3.5|z - 1.2| - 0.5|r - 8.5|} \tag{20.7}$$

To define the experimental constant a, some active roots were separated from the soil matrix, and the dry weight and length of roots were measured. The relationship between the weight of the dry roots and their length is as presented in Eq. (20.8).

$$W_{root} = 0.16 L_{root} \tag{20.8}$$

where W_{root} is measured in grams, and L_{root} is measured in meters.

Considering the soil dry density equal to $\gamma_d = 17\ \text{kN/m}^3$, the value of a is equal to $a = \gamma_d\ (\text{g/m})/0.16\ (\text{g/m}) = 1.06 \times 10^7\ \text{m/m}^3$. Therefore, the maximum density of a given root length is equal to $\beta_{max} = 659{,}000\ \text{m/m}^3$. This value is the density of the root length for the equivalent active roots with the diameters varying from 0.4 to 1mm.

Table 20.2 Parameters of interaction between tree and ground of a Black Box tree at Miram

Parameter	Measured value	Comments
T_p	80 L/day	Estimated from Slavich et al. (1998) and Jolly and Walker (1996)
ψ_w	1,700 kPa	Estimated from field measurements
ψ_{an}	4.9 kPa	Clayey soil with air content of 0.04 (Feddes et al., 1976)
r_{max}	20 m	Estimated from field observation
z_{max}	3 m	Estimated from field observations
r_0	8.5 m	Radial coordinate of the maximum root density point
z_0	1.2 m	Vertical coordinate of the maximum root density point
β_{fmax}	659,000 m^{-2}	Measured according to organic content
k_1	0.35	Measured according to organic content
k_2	0.55	Measured according to organic content

Two 1-m wide × 35-m-long × 3.5-m-deep trenches were dug by an excavator to study the distribution of tree roots and the dimensions of the root zone. Another trench was excavated between these two trenches to check the relatively homogeneous distribution. The main field observations can be summarized as follows:

- Excavation of three trenches in three different directions indicated that roots were developed 20 m horizontally and 3 m vertically; that is, $r_{max} =$ 20 m and $z_{max} = 3$ m.
- The maximum active density of root length occurred somewhere between 0.5 and 1.5 m deep with a horizontal distance between 7 and 9 m; this is in agreement with the measurements of the organic content (Table 20.2).
- At 23 m away from the trunk, the soil became wetter and stuck to the bucket during excavation. This suggested there was a deficit in the moisture content close to the *E. largiflorens* tree that extended approximately 20–25 m away.
- The maximum reinforcing root density occurred underneath the trunk and decreased with lateral and vertical distances. Figure 20.9 shows field observations of reinforcing roots from the walls of the trench.

Figure 20.9 Reinforcing roots of a Black Box tree observed at Miram.

Field measurements revealed that the minimum moisture content and maximum matric suction of the top 3 m of soil were 9% and 1700 kPa, respectively. Therefore, the soil wilting point suction ψ_w was estimated to be approximately 1700 kPa. The parameters used in this analysis relating to the interaction between the tree and the atmosphere are given in Table 20.2.

20.4.4 Finite element simulation

A two-dimensional finite element analysis was used to predict the distribution of the soil moisture content and matric suction near the tree. The numerical analysis in this case study was based on the basic effective stress theory of unsaturated soils incorporated in the ABAQUS finite element code. The axisymmetric model and specified boundary conditions are shown in Fig. 20.10.

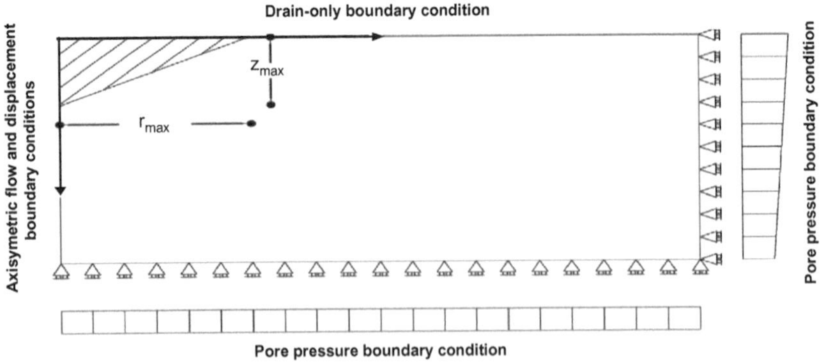

Figure 20.10 Geometry and boundary conditions of the model.

The mesh used in this simulation contains axisymmetric, bilinear strain quadrilateral elements (CAX4P) with four displacements and pore pressure nodes at the corners of each element. The entire FE mesh consists of 3993 nodes and 3840 elements. The flux boundary at the surface was controlled by climatic conditions that included the rainfall and evaporation rates. It was assumed in this study that the influence of other hydrological features such as irrigation, surface runoff, and surface storage was negligible. Evaporation and rainfall rates on a daily basis were applied at the surface using nonuniform distributed flow that represented the magnitude of the flow rate on the surface. Figure 20.11 shows the monthly rainfall rate from 2003 to 2006 applicable to the site. Further meteorological data from the site can be found in Fatahi (2007) and Fatahi et al. (2010).

The model used in this study to estimate the actual evaporation rate as a function of potential evaporation was based on the approach originally proposed by Aydin (1998). It was assumed that any initial evaporation from a saturated soil proceeds at the potential rate until air entry suction is reached at the soil surface. When the air entry suction value is reached, soil evaporation begins to drop; in fact, beyond the value of the air entry suction, evaporation progressively falls and the surface becomes progressively drier until it reaches wilting point suction. When this point is reached, the amount of

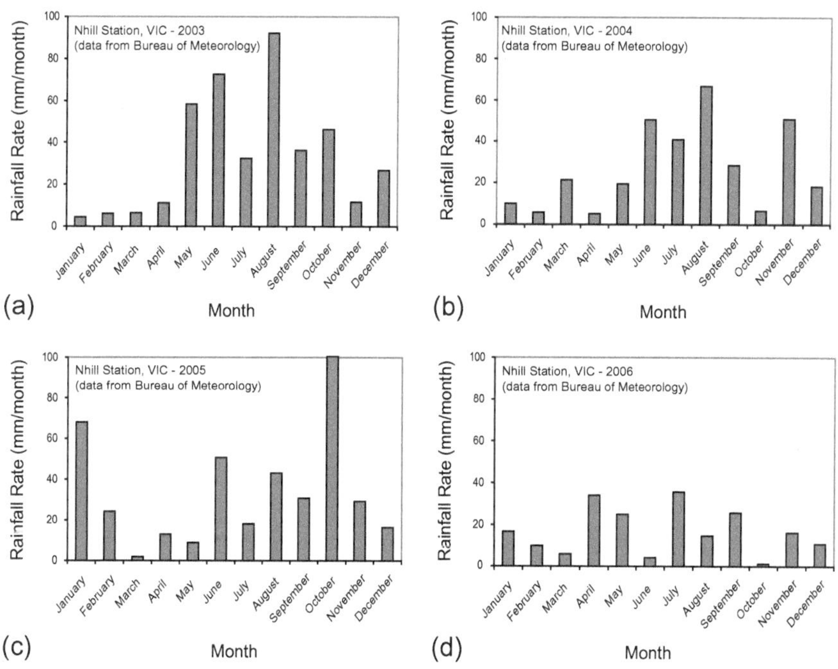

Figure 20.11 Monthly precipitation rate from 2003 through 2006.

evaporation is insignificant. According to Aydin et al. (2005), it can be concluded that the ratio of actual to potential evaporation could be linearly correlated to the log-transformed soil matric suction. Note that although soil water evaporates slowly when it is beyond the wilting point suction, water is still lost through the slow process of vapor diffusion. In this study, it was assumed that water lost by vapor diffusion was negligible.

According to Xu et al. (2006), the evapotranspiration rate (ET_{ref}) is proportional to the pan-evaporation rate (E_p); hence,

$$ET_{ref} = k_p E_p \tag{20.9}$$

where k_p is the pan coefficient, which depends on the crop characteristics.

In general, evapotranspiration can be calculated by the following equation:

$$ET_{ref} = E_p + T_p \tag{20.10}$$

Substituting Eq. (20.9) into Eq. (20.10) results in

$$T_p = E_p \left(k_p - 1\right) \tag{20.11}$$

Therefore, it can be concluded that the potential transpiration rate is proportional to the evaporation rate. For a fully grown tree, k_p can be assumed to be a constant value, and thus

$$\frac{T_P}{T_{P_0}} = \frac{E_p}{E_{P_0}} \tag{20.12}$$

where T_{p0} and E_{p0} and are the values of potential transpiration rate and potential evaporation rate at a reference point, respectively. Hence, it is assumed that any variation in the potential transpiration rate has the same trend as any variation in evaporation. In other words, the effects of seasonal changes including net solar radiation and variations in temperature and humidity are indirectly considered in the transpiration rate.

According to the field measurements, the initial pore water pressure can be assumed as hydrostatic, and the water table was located 15 m below the surface at the time of observation. Because this analysis would last for more than 2 years, the profile of the initial moisture content only had a minor influence on the final predictions of the top layers of soil. The root water uptake model was included in the numerical scheme via Visual Fortran subroutines. The main subroutine included the rate of root water uptake as a moisture flux boundary applied to the four sides of every element within the root zone. In other words, Eq. (20.1), incorporating Eqs. (20.2)–(20.4), was used in the numerical model as boundary flux to determine the rate of root water uptake within the root zone at each increment of time. After a geostatic analysis for controlling and modifying the initial conditions defined for a steady-state equilibrium of ground under geostatic loading, a consolidation analysis for the transient response of partially saturated soil under transpiration was conducted. This stage included a time-dependent analysis from January 1, 2003 until May 15, 2005 at daily intervals with continuous root water uptake.

The basic effective stress concept adopted in this study to analyze unsaturated soil behavior was inspired by Bishop (1959):

$$\sigma'_{ij} = \sigma_{ij} - u_a\delta_{ij} + \chi(u_a - u_w)\delta_{ij} \tag{20.13}$$

where σ'_{ij} is the effective stress of a point on a solid skeleton, σ_{ij} is the total stress in the porous medium at the point, u_a is the pore air pressure, u_w is the pore water pressure, δ_{ij} is Kronecker's delta ($\delta_{ij} = 1$ when $i = j$ and $\delta_{ij} = 0$ when $i \neq j$), and χ is the effective stress parameter attaining a value of unity for saturated soils and zero for dry soils. In unsaturated soil mechanics, the term $(u_a - u_w)$ is usually called the matric suction. The quantity $|\chi u_{w\delta_{ij}} + u_a\delta_{ij} - \chi u_a\delta_{ij}|$ may be considered as an equivalent pore pressure, which is a portion of the effective stress in a soil resulting from fluid pressure in the pores. Khalili and Khabbaz (1998) showed that by plotting the value of χ against a more appropriate parameter such as the ratio of matric suction over the air entry (suction ratio), a unique relationship may be obtained for most of the soils used in this study.

In this study, the coefficient of soil permeability (k) described by Brooks and Corey (1964) was used:

$$k = k_s(e)s_e^{\frac{2+3\lambda}{\lambda}} \tag{20.14}$$

$$S_e = \left[\frac{S_r - (S_r)_{res}}{1 - (S_r)_{res}}\right] \tag{20.15}$$

where $k_s(e)$ is the saturated coefficient of permeability estimated based on the Kozeny–Carman equation, S_e is the effective degree of saturation, S_r is the degree of saturation, $(S_r)_{res}$ is the residual degree of saturation, and λ is the slope of the soil water characteristic curve on a log–log plot (i.e., $\Delta\log S_e/\Delta\log \psi$). Kozeny (1927, 1928) and Carman (1938, 1956) developed the following semi-empirical, semi-theoretical formula for predicting the saturated permeability of porous media:

$$k_s(e) = \frac{\gamma}{C_{k-c}\mu S_0^{\ 2}} \frac{e^3}{1+e} \tag{20.16}$$

where γ is the unit weight of the fluid, μ is viscosity of the fluid, C_{k-c} is the Koezy–Carman empirical coefficient, S_0 is the specific surface area per unit volume of particles, and e is the void ratio of the porous media. The previous permeability equations in conjunction with the soil water characteristic curve were included in the ABAQUS input file.

The material behavior of the medium to highly overconsolidated soil was modeled with an elastoplastic cap model that was used by Reul (2004) and Reul and Randolph (2003). Here, the analysis was governed by the stiffness of the soil rather than its strength, where the deformation would occur well below the peak shear stress. The cap model consists of three yield surface segments: the pressure-dependent, perfectly plastic shear failure surface F_s, the compression cap yield surface F_c, and the transition yield surface. Changes of stress inside the yield surfaces cause elastic deformation, whereas changes of stress on yield surfaces cause plastic deformation. The shear failure surface was almost perfectly plastic, whereas volumetric plastic strains causes hardening of the cap. Plastic flow is defined by a non-associated flow potential G_s of the shear surface, and the associated flow potential G_c of the cap. In the cap plasticity model, the conus yield surface can be defined as follows Reul (2004); Reul and Randolph, (2003):

$$F_s = t - d - \text{p}.\tan\beta' = 0 \tag{20.17}$$

where t (deviatoric stress), d (intersection of the conus yield surface with the t-axis or material cohesion), and β' (slope of the conus yield surface at p–t plane) can be calculated using the equations proposed by Reul (2004) and Reul and Randolph (2003). The conus and transition yield surfaces in the cap plasticity model were defined by the following two equations (Reul, 2004; Reul and Randolph, 2003):

$$F_c = \sqrt{\left(p - \frac{p_b - Rd}{1 + R\tan\beta'}\right)^2 + \left(\frac{Rt}{1 + \alpha - (\alpha/\cos\beta')}\right)^2} - R\left(d + \frac{p_b - Rd}{1 + R\tan\beta'}\tan\beta'\right) = 0 \tag{20.18}$$

$$F_t = \sqrt{\left(p - \frac{p_b - Rd}{1 + R\tan\beta'}\right)^2 + \left[t - \left(1 - \frac{\alpha}{\cos\beta'}\right)(d + p_a\tan\beta')\right]^2} - \alpha\left(d + \frac{p_b - Rd}{1 + R\tan\beta'}\tan\beta'\right) = 0 \tag{20.19}$$

where p_b is the compression yield stress, and α is the shape parameter of the transition yield surface. The plastic potential cap, transition, and conus surfaces were defined by Reul (2004) and Reul and Randolph (2003) as

$$G_c = G_t = \sqrt{\left(p - \frac{p_b - Rd}{1 + R\tan\beta'}\right)^2 + \left(\frac{Rt}{1 + \alpha - (\alpha/\cos\beta')}\right)^2} \quad (20.20)$$

$$G_s = \sqrt{\left(p - \frac{p_b - Rd}{1 + R\tan\beta'}\right)^2 \tan^2\beta' + \left(\frac{t}{1 + \alpha - (\alpha/\cos\beta')}\right)^2} \quad (20.21)$$

The non-associated flow applied onto the failure and transition yield surfaces implied that the material stiffness matrix was not symmetric, and therefore the unsymmetrical solver should be invoked in any analysis. Nevertheless, if the region of the model where non-associated inelastic deformation was occurring is confined, a symmetric approximation to the material stiffness matrix would probably give an acceptable convergence rate. In this case study, the initial deformation modulus (E), the effective angle of friction (θ'), and the effective cohesion coefficient (c') of the soil were derived from the laboratory and field tests, whereas the shape parameter of the yield surface (K) was calculated from the angle of friction, and an estimated value was used for the shape parameter of yield surface (R). A complete description of the cap model can be found in the ABAQUS manual (ABAQUS, 2003).

In this study, the top 6 m of soil strata was divided into four layers, 1.5 m deep. The boundary between the upper clayey soil and the lower sandy soil was at a depth of 3 m, but below 6 m, it was assumed that the properties are identical to those at 4.5–6.0 m deep and do not change. The soil water characteristic curves used in this study are shown in Fig. 20.7. It was proposed that ψ_d (the highest value of soil suction when $S = S_{max}$) was equal to the air entry value suction (ψ_{air}). The material properties and parameters used in the finite element analysis were given in Table 20.2, and the additional measured parameters are given in Table 20.3.

20.4.5 Field measurements in May 2005

Figure 20.12 shows a comparison between the field measurements and the predictions of the numerical model for volumetric moisture content. The numerical results incorporating the developed root water uptake model basically agree with the field measurements, although according to Fig. 20.16, the field measurements for the moisture content reduction are noticeably different from the finite element predictions close to the tree trunk. This

Table 20.3 Parameter values used in the finite element analysis

Parameter	Layer I	Layer II	Layer III	Layer IV	Description
E (MPa)	25	40	56	87	Initial deformation modulus
c' (kPa)	12	10	13.5	10.7	Cohesion
θ' (deg)	25	26.5	31	30.5	Internal friction angle
p'_c (kPa)	400	550	700	950	Preconsolidation pressure
γ_d (kN/m^3)	16.6	17.3	17.8	18.2	Dry density
k_s (m/s)	5.7E-10	1.2E-10	1.8E-11	2.7E-12	Saturated permeability
e_0	0.61	0.52	0.47	0.42	In situ void ratio
β' (deg)	44.53	46.35	51.20	50.70	Slope of the conus yield surface in the p–t plane
d (kPa)	25.32	21.03	27.94	22.19	Intersection of the conus yield surface with t axis
K	0.753	0.741	0.707	0.711	Shape parameters of the conus
α	0.01	0.01	0.01	0.01	Shape parameter of the transition surface
R	0.1	0.1	0.1	0.1	Shape parameter of the cap
ψ_d (kPa)	100	80	20	30	Maximum ψ when $S = S_{max}$
ψ_{air} (kPa)	100	80	20	30	Air entry value

was not surprising because the trunk and its foliage altered the uniform distribution of rainfall, and the shadow under the canopy changed the rate of evaporation due to variations in temperature and humidity. As a consequence, these effects probably caused the disparity between the field data and finite element predictions. Figure 20.13 shows the comparison between the field measurements and model predictions of the soil matric suction in the top 6 m of the soil layer at different lateral distances from the tree trunk.

It is important to note that part of the difference between the field measurements and numerical predictions (Figs. 20.12 and 20.13) is related to the individual effect of roots. In other words, in the numerical analysis, root water uptake as a sink term was considered in the flow equation, but the effect of each root was not considered individually because as the main roots penetrate the soil, there may be a gap between them where water can collect. Because the woody roots are in a denser pattern under the trunk and in close

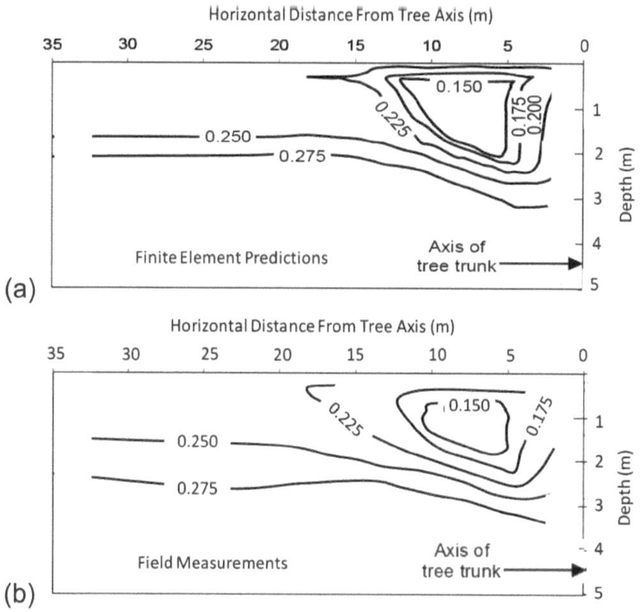

Figure 20.12 Contours of volumetric soil moisture content reduction in the vicinity of the tree: (a) Current numerical analysis results and (b) field measurements in May 2005. *(Source: Modified after Fatahi et al. (2014)).*

proximity to each other, a disparity between the field measurements and predictions in this area seems more likely. Furthermore, the actual field data are expected to be affected by the soil heterogeneity. Figure 20.14 shows the model predictions of the profile of ground settlement distribution as a result of evapotranspiration and precipitation in mid-May 2005.

According to Fig. 20.19, the maximum ground settlement of 200 mm occurs approximately 6 m away from the tree where, as expected, the ground settlement induced by soil consolidation decreased with depth. This ground settlement stemmed from root water uptake and evaporation from the soil surface. As Fig. 20.14 shows, the soil surface settlement caused by evaporation was approximately 110 mm, so the remaining settlement was assumed to be induced by transpiration.

Figure 20.15 illustrates the lateral displacement vectors near the tree, whereas Fig. 20.20 indicates that the maximum lateral displacement occurred away from the trunk but still within the root zone. By applying axisymmetric boundary conditions where lateral displacement at the tree trunk was zero, the maximum lateral displacement occurred 12 m away from the trunk.

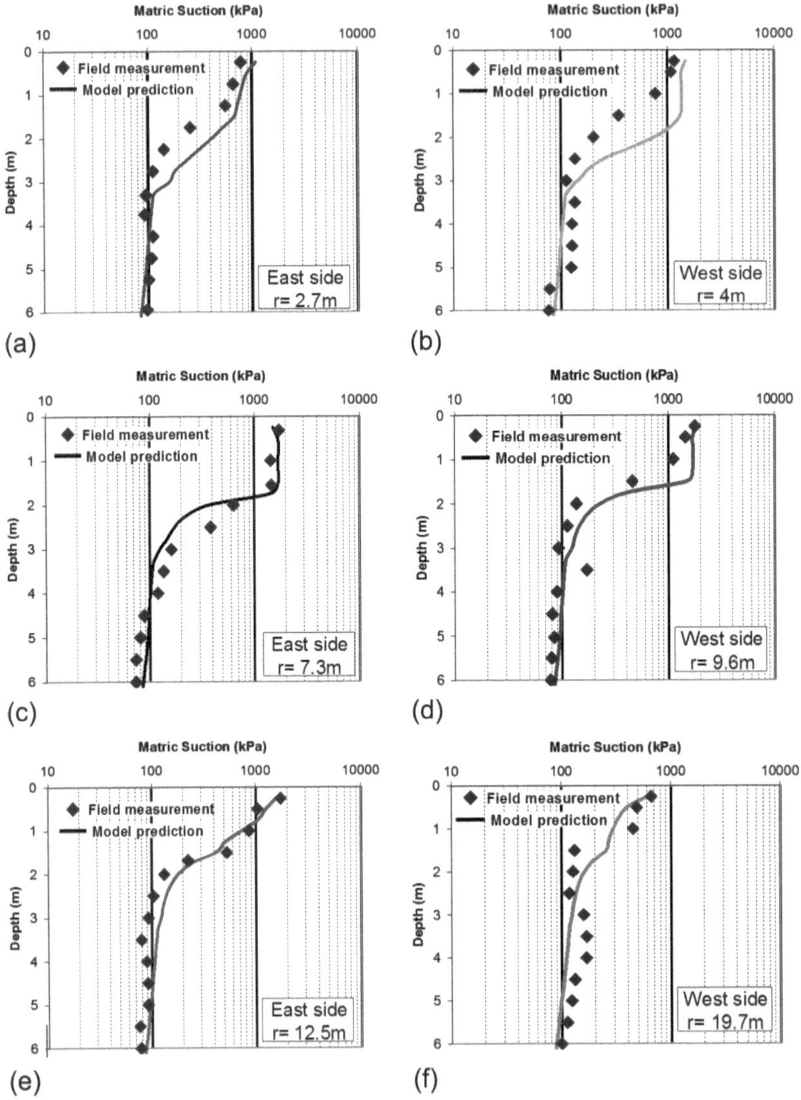

Figure 20.13 Soil matric suction profiles at (a) 2.7 m, (b) 7.3 m, (c) 12.5 m, (d) 17.5 m, (e) 26.1 m, and (f) 33.6 m away from the tree trunk at the eastern side of the trunk in May 2005. *(Source: Modified after Fatahi et al. (2014)).*

Active tree roots drain the pore water in all directions, so as an analogy, the behavior of active roots can be compared with vertical and horizontal prefabricated drains with vacuum preloading where transpiration through the active roots removes water from the soil and thus increases the matric suction and the effective stress. This increased effective stress causes

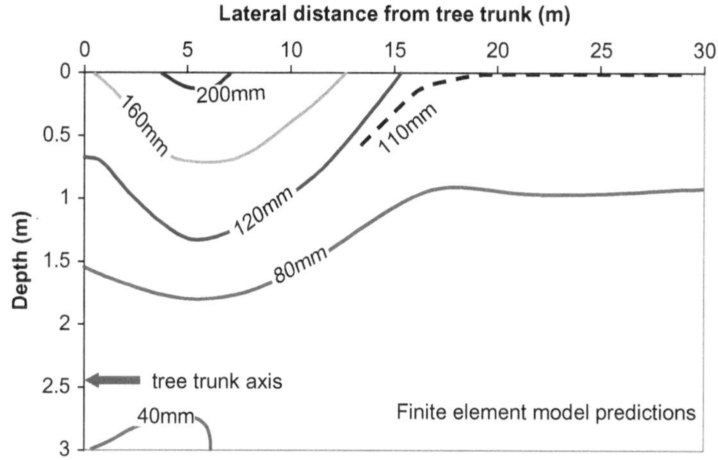

Figure 20.14 Model predictions for the contours of vertical displacement. *(Source: Modified after Fatahi et al. (2014)).*

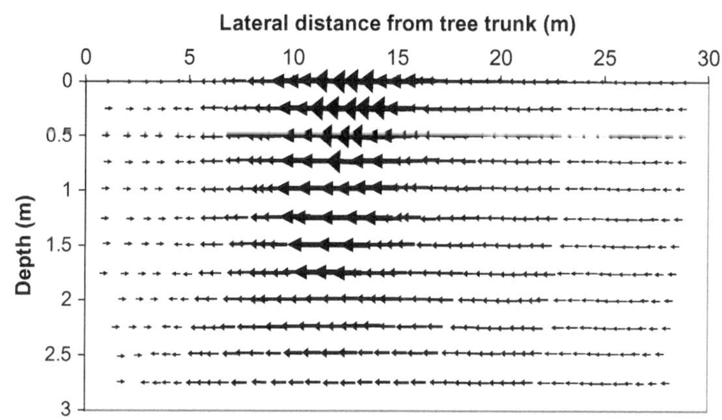

Figure 20.15 Lateral displacement vectors.

consolidation while the active roots reduce the drainage path in the radial and vertical directions, just like the prefabricated vertical drains. Because consolidation time is inversely proportional to the square of the length of drainage path, active roots directly reduce the consolidation time. Indeed, referring to Figs. 20.14 and 20.15, it can be seen that tree roots cause vertical and lateral consolidation with negative pore water pressure, which accelerates consolidation. Note that evaporation from the soil surface is also analogous to a uniformly distributed vacuum pressure on the surface that generates a negative pore water pressure on the soil surface, which then propagates vertically and results in an increase in the effective stress causing soil consolidation. A comparison between vertical and horizontal displacement indicates there was more vertical displacement than horizontal displacement due to evaporation from the soil surface.

20.5 IMPORTANCE OF A ROOT REINFORCEMENT AND SUCTION INTEGRATED SYSTEM

Trees can improve the matric suction in adjacent soil through the process of evapotranspiration, as discussed previously, whereas tree roots can provide mechanical strengthening to the adjacent soil matrix. The effect of root reinforcement in terms of improving the shear strength of soil has been studied under saturated soil conditions (Waldron, 1977; Waldron and Dakessian, 1981; Docker and Hubble, 2001, 2008), whereas soil suction induced by transpiration has been considered separately (Cameron, 2001; Indraratna et al., 2006; Fatahi, 2007; Fatahi et al., 2009). However, it is imperative that these two phenomena should be studied together to obtain more realistic results because the soil surrounding a tree is in a partially saturated condition most of the time and the tree's evapotranspiration influences the root reinforcement capacity (Indraratna et al., 2014).

When a role that a tree plays in ground improvement is considered, understanding its root system is important regardless of whether strengthening is achieved by suction development or mechanical strengthening. The primary roots help to strengthen soil mechanically, whereas the hair roots/active roots mainly contribute to the suction and fibrous effects. According to Indraratna et al. (2014), the total shear strength generation in terms of mechanical strengthening and suction induced by root water uptake is mainly governed by the spatial distribution of the root system.

20.5.1 Root systems

Two main root systems (i.e., the fibrous root system and the tap root system) can be distinguished between trees in the environment, but their spatial distribution indicates that they behave according to the availability and requirement of water, minerals and nutrient absorption, storage of food in root crops, and the synthesis of various essential compounds (i.e., growth regulators). In well-drained soils, roots typically have a fibrous root system that often results from a combination of lateral and oblique roots with no real tap root; in fact, only a small percentage of tree root systems have a real tap root (Kramer, 1995).

Environmental factors such as the texture and structure of soil, its aeration, moisture, temperature, pH, salinity, and the presence of toxic elements such as lead, copper, and aluminum—including competition with other plants and the presence of bacteria-, fungi-, and soil-inhabiting animals—also influence the growth and spatial distribution of tree roots. The general spatial distribution variations of root systems in relation to the soil bulk density, temperature, and rainfall are illustrated in Figs. 20.16 and 20.17.

According to Figs. 20.16 and 20.17, tree root systems are highly sensitive to environmental parameters and tend to complicate the shape of their root systems as engineering parameter in ground improvement. Therefore, in relation to mechanical strengthening in saturated condition, Docker and

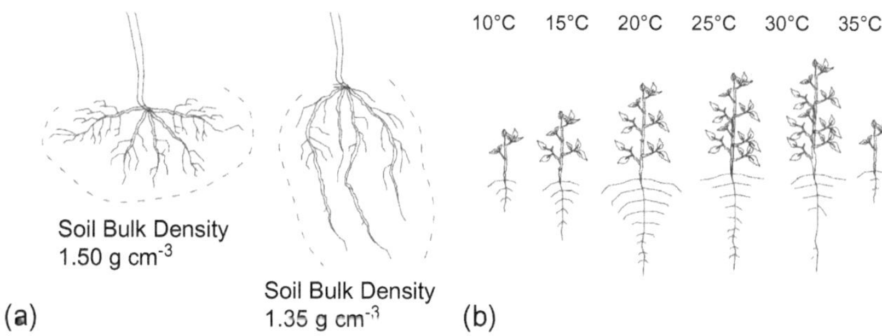

Figure 20.16 (a) Root system of young barley plants grown in the field in soils with different bulk densities. (b) Influence of root zone temperature on root morphology and shoot growth of potato seedlings. *(Source: (a) Modified after Gilmen (1980); (b) modified after Sattelmacher et al. (1990)).*

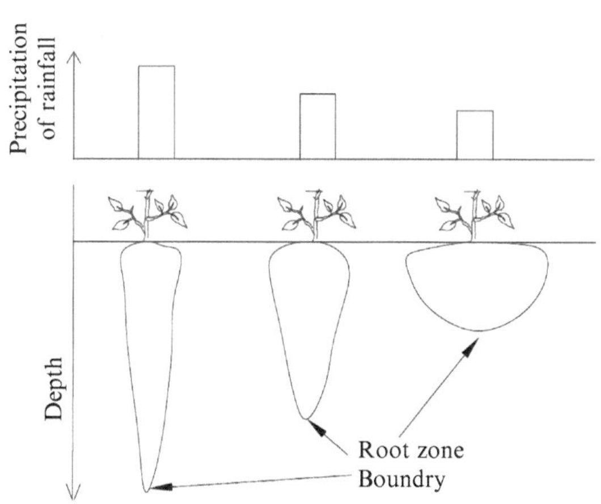

Figure 20.17 Schematic diagram of the effect of the amount of rainfall on the depth of the roots of winter wheat (very fine sandy loam–silty loam in the Great Plains). *(Source: Modified after Sattelmacher et al. (1990)).*

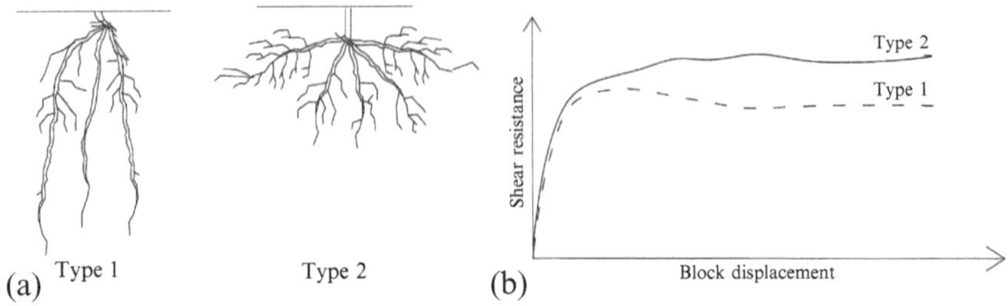

Figure 20.18 (a) The distinct root morphology through an in situ shear test block. (b) Diagrammatic representation of two generally distinct root systems at direct shearing. *(Source: Modified after Docker and Hubble (2009)).*

Hubble (2009) introduced two root system shapes (Fig. 20.18) by considering the shear resistance generated against the block displacement in an in situ shear test. These two root system shapes can be used with root length density (β) as an integrated system that can increase the suction and mechanical strength of soil.

20.5.2 Mechanical strengthening induced by roots

The effect of root reinforcement with regard to stabilizing slopes and controlling erosion has been widely discussed during the past several decades. Waldron (1977) and Wu et al. (1979) developed a simple root model for vertical roots, and Gray and Leiser (1982) modified the model for inclined roots.

Waldron (1977) defined the increment of shear strength ΔS value, and Fig. 20.19 shows the model of a flexible elastic root extending vertically through a horizontal shear zone. Three different values (ΔS) of shear stress related to three failure modes such as stretching, slipping, and breaking were defined, and several experimental coefficients relating to the bond strength

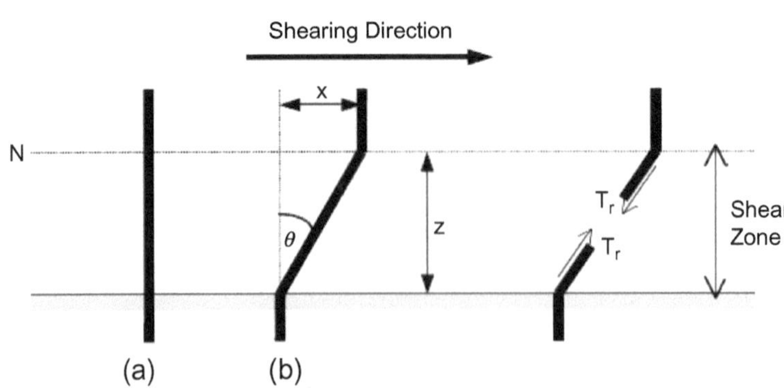

Figure 20.19 Model of flexible elastic root extending vertically through a horizontal shear zone of thickness *Z*: (a) Undisturbed soil; (b) soil above *N* displaced by *x*. *(Source: Modified after Waldron (1977)).*

between root and soil and also tensile strength of roots and other dimensional parameters were introduced by Waldron (1977). According to Waldron (1981), the increment of the shear strength (ΔS) can be directly added to the coulomb equation because there is no change in the friction angle:

$$\tau = c + \Delta S + \sigma_N \tan \phi \tag{20.22}$$

Pollen and Simon (2005) introduced a "RipRoot" model by applying a fiber bundle model to evaluate root reinforcement. In this model, the roots break progressively, unlike the simultaneous breaking in models such as that of Wu et al. (1979). Thomas and Pollen-Bankhead (2010) developed a Rip Root model using Monte Carlo simulation to evaluate the effect of root reinforcement. Most of the previously mentioned predicted values varied by 60–110% from the experimental results (Docker and Hubble, 2001; Fan, 2012).

Docker and Hubble (2009) carried out a large number of in situ direct shear tests for root permeated saturated soil surrounding four species of Australian trees, and they also introduced an emperical equation (Eq. (20.23)) for the shear strength increment values that were mainly based on the root area ratio (RAR):

$$\Delta S = B.(\mathrm{RAR}) + \mathrm{C} \tag{20.23}$$

$$\text{Root area ratio (RAR)} = \frac{\textit{Area of root along the shear plance}}{\textit{Area of the shear plane}} \tag{20.24}$$

In Eq. (20.23), C and B are experimental coefficients related to the different species of trees.

20.5.3 Development of coupled model concept

Although the previously discussed models and experiments were introduced for saturated soil conditions, the soil matrix surrounding the tree was generally in partially saturated conditions, as described in Section 20.4; therefore, the parameters developed for root reinforcement in saturated soil conditions where the tree was considered to be a nonliving object may not be realistic. Because the matric suction changes the bond strength between the main roots and the soil, the tensile strength of the root changes as the root cell structure alters due to potential variations in the root water (Yajun and Cosgrove, 2000), and according to classical botanical understanding, osmosis at the tip of the root binds the soil matrix together (see Fig. 20.20) and it tends to alter the methods of root failure described in Section 20.5.2.

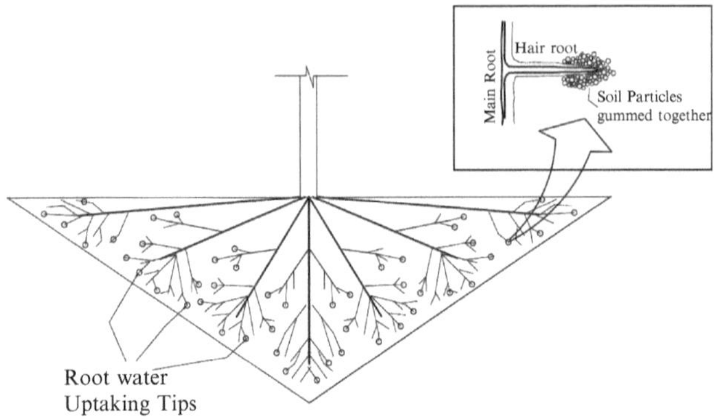

Figure 20.20 Schematic diagram of root system with distribution of water-uptaking active root tips. (Inset) Enlarged root tip.

Figure 20.21 shows a schematic diagram of soil elements with different boundary conditions under railway ballast in a soil system permeated with roots. Conventional unsaturated soil mechanics theories are valid for soil element A, which is directly underneath the railway ballast but farther away from the tree root zone. Therefore, to evaluate the shear strength of soil element A, Eq. (20.25) (Fredlund et al., 2012) can be used:

$$\tau(t) = c' + (\sigma_n - u_a(t))\tan\phi' + (u_a(t) - u_W(t))\left[\tan\phi'\left(\frac{\theta(t) - \theta_r}{\theta_s - \theta_r}\right)\right] \tag{20.25}$$

where u_a, u_w, and θ are the time-dependent pore air pressure, pore water pressure, and moisture content, respectively. These parameters can be evaluated

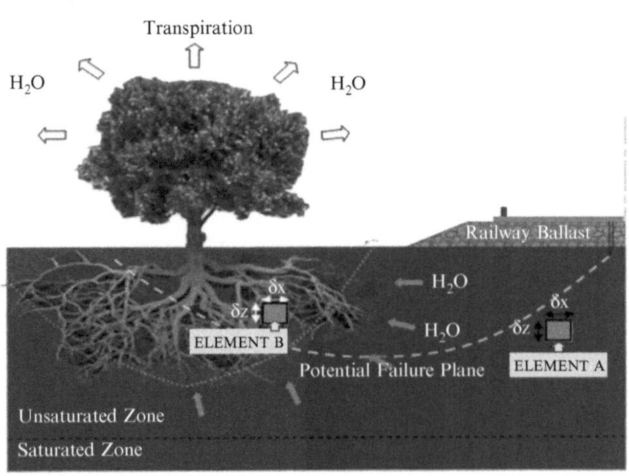

Figure 20.21 Schematic diagram of root water uptake process and root reinforcement.

using the model described in Section 20.4. Soil element B in Figure 20.21 is at the root zone, and therefore its root suction and root reinforcement exert a significant influence. Moreover, the soil water characteristic curve that is related to the surrounding soil properties is not valid for element B due to the high suction generated near the tips of the roots. Furthermore, the stiffness of the root zone increases the shear strength in soil adjacent to element A, and therefore it is identified as the "shear capacity" of the root area zone, which acts as an external stiffener that includes the root reinforcement and suction. Computing the effects from root reinforcement and suction separately and then superimposing, even when all the previously mentioned facts are considered, may not give realistic answers because the suction influences root reinforcement and alters the SWCC of the root zone.

20.6 ION UPTAKE BY ROOTS AND ASSOCIATED OSMOTIC SUCTION VARIATION IN SURROUNDING SOIL

20.6.1 Osmotic suction in soil

Soil suction is considered as the summation of matric suction (ψ_m) and osmotic suction (π). Most studies have been carried out to facilitate the importance of matric suction in unsaturated soils, whereas less attention has been given to the role of osmotic suction and its influence on unsaturated soil properties. Osmotic suction results from the presence of dissolved solutes in pore water. High concentrations of salts in pore water cause massive changes in the physical and volumetric properties within the soil structure (Sreedeep and Singh, 2006). Thus, the magnitude of π is directly proportional to the concentration of salts, which is a function of the water content (w) (Sacchi et al., 2001); this is why it is important to study the effect of π independently of ψ_m.

Because π is therefore important, the uptake of solute by the roots to the surrounding soil has a significant effect; indeed, the uptake of nutrients and other ions from the roots creates a concentration gradient at the root surface and also affects the osmotic suction of the soil. In order to study this effect precisely, understanding the nutrient uptake and demand at the root surface of the tree is obviously vital. These measures have been addressed quantitatively by previous botanical studies, and the well-established solute uptake model developed by Tinker and Nye (2000) is explained next.

20.6.2 Solute transport in the soil near the root surface

According to Tinker and Nye (2000), the rate at which a root absorbs solutes is related to its individual and collective concentration in the soil solution at the boundary, whereas the rate of solute intake is proportional to the root demand coefficient (αa). Figure 20.22 shows the solute transport process near a root surface:

$$I = 2\pi\alpha a C_{La} \tag{20.26}$$

where I is the rate of uptake per unit length, a is the radius of the root, and C_{La} is the concentration in solution at the root surface.

C_{La} depends on the concentration of soil pore solution (C_L), and transpiration also helps carry these solutes near the root surface, along with water due to mass flow. The relationship between the uptake rate of solutes into roots and the concentration of solutes in the external pore water is called an isotherm (Tinker and Nye, 2000), where different types of isotherms are presented for different scenarios. Marschner (1995) introduced the Michaelis–Menten relationship for ions with a minimum concentration (C_{Lmin}), where there is no uptake. The shape of a single concentration pretreatment isotherm is shown in Fig. 20.23, where I_{max} is the highest value of I and K_m is a constant. This relationship is for low and medium concentrations, when the high-affinity uptake is dominant, as given by Eq. (20.27). When the concentration around the root surface is high, the inflow rises gradually to an ill-defined maximum (I'_{max}) due to low affinity, and then it begins to decline again:

$$I = I_{max}\, C_{La}/(K_m + C_{La}) \tag{20.27}$$

The straight-line intersecting model (Fig. 20.24) was introduced for solute uptake by plants in flowing solution culture systems with constant concentrations, where plants are in equilibrium (Tinker and Nye, 2000). The important components of this isotherm are the crossover point (I_{crit}) and the slope of the graph, which can be obtained by Eq. (20.28):

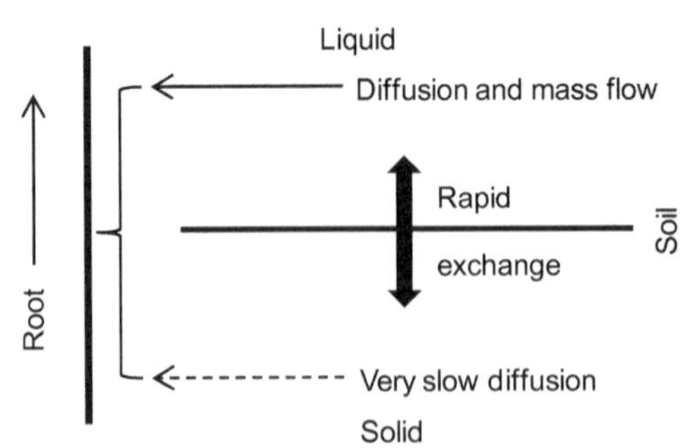

Figure 20.22 Solute transport process near a root surface. *(Source: After Tinker and Nye (2000)).*

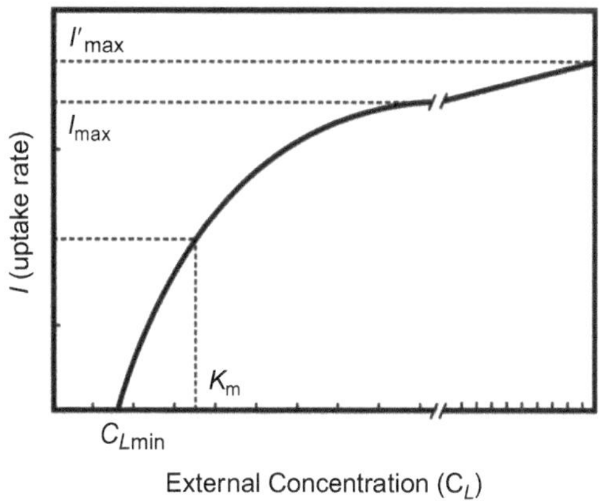

Figure 20.23 Typical Michaelis–Menten relationship. *(Source: After Marschner (1995)).*

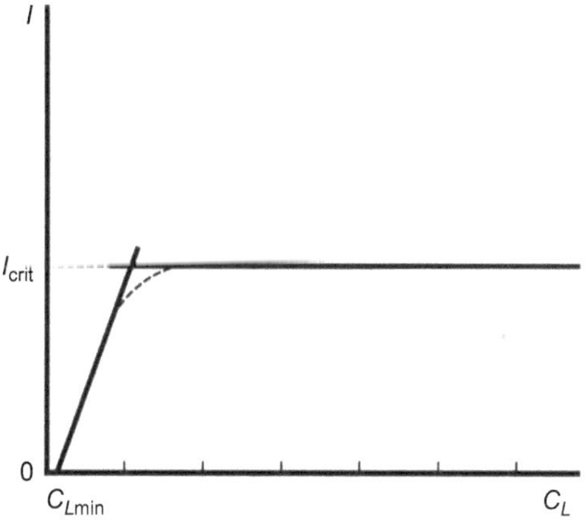

Figure 20.24 Straight-line intersecting model. *(Source: After Tinker and Nye (2000)).*

$$\begin{aligned} I &= 2\pi\alpha a C_{\mathrm{La}} \quad & C_{\mathrm{La}} < C_{\mathrm{Lcrit}} \\ I &= I_{\mathrm{crit}} \quad & C_{\mathrm{La}} > C_{\mathrm{Lcrit}} \end{aligned} \tag{20.28}$$

20.6.3 Coupling osmotic suction into the root water uptake model

The Fatahi (2007) model captured the effect of soil suction, root distribution, and potential transpiration in the root water uptake model, as shown in Eq. (20.1). In a similar manner, Eq. (20.1) can also be modified by incorporating the osmotic suction factor ($f(\pi(t))$) as shown in Eq. (20.29), where S^* is the uptake of solute water:

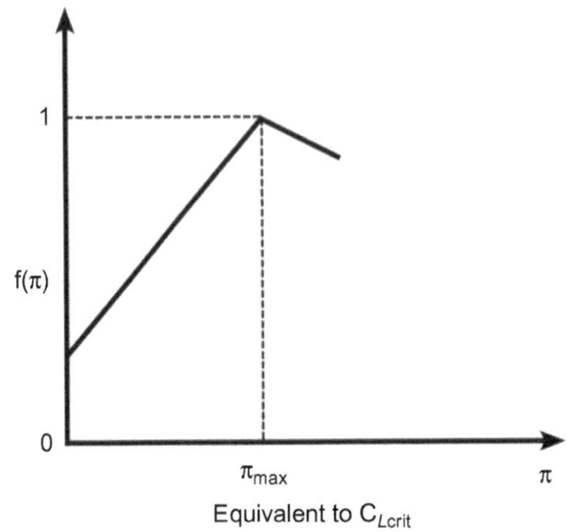

Figure 20.25 Osmotic suction factor (preliminary development).

$$S^*(r, z, t) = G(\beta(t))F\left(T_p(t)\right)f(\psi(t))f(\pi(t)) \quad (20.29)$$

The osmotic suction factor ($f(\pi(t))$) can be addressed in the same way that the soil suction factor ($f(\psi(t))$) was approached in Eq. (20.4) and Fig. 20.25:

$$\begin{aligned} f(\pi) &= f(\pi_0) \quad \pi = 0 \\ f(\pi) &= 1 \quad\quad\ \ \pi = \pi_{max} \end{aligned} \quad (20.30)$$

Note that the slope of the graph must be determined through experiments.

20.6.4 Osmotic suction measurement

Osmotic suction can be obtained by measuring the electrical conductivity because it is directly proportional to the concentration of solute. Abedi-Koupai and Mehdizadek (2007) presented a range of empirical relationships obtained by extracting samples of pore water and then measuring the electrical conductivity to find the osmotic suction values.

20.7 CONCLUSION

In this chapter, a profile of the moisture content and suction around an *E. largiflorens* (Black Box) tree in western Victoria, Australia, was determined by taking disturbed and undisturbed samples of moisture content and soil suction. The properties of the soil, including its physical and mechanical parameters, along with the specifications of a tree and the atmospheric conditions near the tree were measured and used in a validation exercise.

The distribution of the roots of this tree seen through trenches excavated in different directions showed that they had developed 20 m horizontally and 3 m vertically. Field observations showed that a moisture deficit close to the *E. largiflorens* tree had extended approximately 20–25 m away from the trunk.

The model proposed for the rate of root water uptake was included in a numerical analysis using the ABAQUS finite element code to examine the distribution of soil suction and profile of the moisture content near the tree. The results of a numerical prediction of the matric suction and moisture content around this Black Box tree were compared with the field data taken in May 2005. Although there were some uncertainties in some field and laboratory measurements of the soil parameters, including the actual distribution of tree roots and atmospheric parameters, a good agreement was generally obtained between the measured and simulated distribution of soil moisture. A comparison of the results indicated similar trends. It was also shown that a numerical analysis that included the proposed root water uptake model can reasonably predict the region of maximum matric suction away from the axis of the trunk, as measured in the field, whereas ground deformation contours for this case study indicated that maximum settlement after approximately 3 years occurred on the surface and away from the trunk. Although more field data will assist in further verification of the root water uptake model, the proposed model and associated numerical analysis is a promising tool for predicting the matric suction induced by tree roots within a soil matrix. Although it is well-known that high rainfall and infiltration can rapidly reduce suction, because transpiration is a continuous process of discharging water from the soil matrix, rainfall will be quickly taken in through tree roots. Thus, in vegetated ground, the effects of matric suction on the shear strength and deformation of partially saturated soil should be considered. Moreover, the role of matric suction and osmotic suction must be coupled in the water uptake model, and therefore a preliminary model is being developed for this purpose.

ACKNOWLEDGMENTS

The authors appreciate the financial support from Rail-CRC to conduct the field investigations. In addition, the authors acknowledge the support of Elsevier and Taylor & Francis for allowing the use of portions of the authors' previously published journal articles. Furthermore, authors appreciate

Oxford University Press, Elsevier, and Soil Science Society of America for permission to reproduce data from previously published articles as cited and referenced in this book chapter.

REFERENCES

ABAQUS, 2003. User Manual Version 6.5. ABAQUS, Inc, Providence, RI.
Abedi-Koupai, J., Mehdizadek, H., 2007. Estimation of osmotic suction from electrical conductivity and water content measurements in unsaturated soils. Geotech. Testing J. 31 (2), 142–148.
Alarcon, J.J., Domingo, R., Green, S.R., Sanchez-Blanco, M.J., Rodrigez, P., Torrecillas, A., 2000. Sap flow as an indicator of transpiration and the water status of young apricot trees. Plant Soil 227, 77–85.
Australian Standard AS2870, 1996. Residential Slabs and Footings Construction, AS2870-1996.
Aydin, M., 1998. Hydraulic properties and water balance of a clay soil cropped with cotton. Irrigation Sci. 15, 17–23.
Aydin, M., Yang, S.L., Kurt, N., Yano, T., 2005. Test of a simple model for estimating evaporation from bare soils in different environments. Ecol. Modelling 182, 91–105.
Biddle, P.G., 1998. Tree Root Damage to Buildings. Willowmead, Wantage, UK.
Bishop, A.W., 1959. The principle of effective stress. Teknish Ukebadk 39, 859–863.
Blight, G.E., 2005. Desiccation of a clay by grass, bushes and trees. Geotech. Geol. Eng. 23, 697–720.
Brooks, R.H., Corey, A.T., 1964. Hydraulic Properties of Porous Media. Colorado State University, Fort Collins, CO.
Bureau of Meteorology, 2006. Victoria State, Australia
Cameron, D.A., 2001. The extent of soil desiccation near trees in a semi-arid environment. Geotech. Geol. Eng. 19 (3–4), 357–370.
Carman, P.C., 1938. The determination of the specific surface of powders. J. Chem. Indian Trans. 57, 225.
Carman, P.C., 1956. Flow of Gases through Porous Media. Butterworths, London.
Docker, B.B., Hubble, T.T.C., 2001. Strength and Stability of Casuarina Glauca Roots in Relation to Slope Stability. In: 14th Southeast Asian Geotechnical Conf, pp. 745–749, Lisse, The Netherlands.
Docker, B.B., Hubble, T.T.C., 2008. Quantify root-reinforcement of river bank soils by four Australian tree species. Sci. Direct 100, 401–418.
Docker, B.B., Hubble, T.T.C., 2009. Modelling the distribution of enhanced soil shear strength beneath riparian trees of southeastern Australia. Ecol. Eng. 35, 921–924.
Eastham, J., Rose, C.W., Cameron, D.M., Rance, S.J., Talsma, T., 1988. The effect of tree spacing on evaporation from an agroforestry experiment. Agric. Forest Meteorol. 42, 355–368.
Edwards, W.R.N., Warwick, N.W.M., 1984. Transpiration from a kiwifruit vine as estimated by the heat-pulse technique and the Penman–Monteith equation. N. Z. J. Agric. Res. 27, 537–543.
Fan, C.C., 2012. A displacement-based model for estimating the shear resistance of root-permeated soils. Plant Soil 355, 103–119.
Fatahi, B., 2007. Modelling of Influence of Matric Suction Induced by Native Vegetation on Subsoil Improvement. University of Wollongong, Australia, PhD thesis.
Fatahi, B., Khabbaz, H., Indraratna, B., 2009. Parametric studies on bioengineering effects of tree root-based suction on ground behavior. Ecol. Eng. 35 (10), 1415–1426.
Fatahi, B., Khabbaz, H., Indraratna, B., 2010. Bioengineering ground improvement considering root water uptake model. Ecol. Eng. 36 (2), 222–229.

Fatahi, B., Khabbaz, H., Indraratna, B., 2014. Modelling of unsaturated ground behaviour influenced by vegetation transpiration. Geomech. Geoeng. Int. J. 9 (3), 187–207.
Feddes, R.A., Kolinska-Malinka, K., Zaradny, H., 1976. Simulation of field water uptake by plants using a soil water dependent root extraction function. J. Hydrol. 31, 13–26.
Feddes, R.A., Kowalik, P.J., Zaradny, H., 1978. Simulation of Field Water Use and Crop Yield, Simulation Monograph. Pudoc, Wageningen.
Fernandez, J.E., Palomo, M.J., Diaz-Espejo, A., Clothier, B.E., Green, S.R., Giron, I.F., et al., 2001. Heat-pulse measurements of sap flow in olives for automating irrigation: tests, root flow and diagnostics of water stress. Agric. Water Manage. 51, 99–123.
Geotechnical Engineering Office, 2000. Technical Guidelines on Landscape Treatment and Bioengineering for Man-Made Slopes and Retaining Walls. Geotechnical Engineering Office, Government of Hong Kong SAR, Hong Kong.
Gilmen, E.F., 1980. An Illustrated Guide to Pruning. Environmental Horticulture Department, IFAS, University of Florida.
Gray, D.H., Leiser, A.T., 1982. Biotechnical slope protection and erosion control. Van Nostrand Reinhold, New York.
Hopkins, W.G., 1999. Introduction to Plant Physiology. Wiley, New York.
Indraratna, B., Fatahi, B., Khabbaz, H., 2006. Numerical Analysis of Matric Suction Effects of Tree Roots. In: Proc. Inst. Civil Eng. Geotech. Eng., 159, pp. 77–90, 2.
Indraratna, B., Pallewattha, M., Fatahi, B., Pathirage, U., 2014. Improvement of Soil Stability Along Rail Corridors Through Native Vegetation. In: Paper presented at the 7th International Congress on Environmental Geotechnics, Melbourne, Australia.
Jolly, I.D., Walker, G.R., 1996. Is the field water use of *Eucalyptus largiflorens* F. Muell. affected by short-term flooding? Aust. J. Ecol. 21, 173–183.
Khalili, N., Khabbaz, M.H., 1998. A unique relationship for shear strength determination of unsaturated soils. Geotechnique 45 (8), 681–688.
Kozeny, J., 1927. Ueber kapillare leitung des wassers im boden. Wien Akad. Wiss 136 (2a), 271
Kozeny, J., 1928. Berechnung der senkung skurve in regelmassigen breiten gerinnen (Computation of surface curve in uniform broad channels]. Wasserkraft Wasser Wirtschaft 23 (16), 232–234.
Kramer, P.J., 1983. Water Relationship of Plants. Academic Press, New York.
Kramer, P.J., 1995. Water Relation of Plant and Soil. Academic Press, London, Chap. 5.
Kutilek, M., Nielsen, D.R., 1994. Evapotranspiration. In: Soil Hydrology. Catena-Verlag, Reiskirchen, Germany, pp. 195–216.
Lawrence, C.R., 1975. Geology, Hydrodynamics and Hydrochemistry of the Southern Murry Basin. Geological Survey of Victoria Memoir 30.
Marschner, H., 1995. Mineral Nutrition of Higher Plants, second ed. Academic Press, London.
McInnes, D.B., 1986. Drying Effect of Different Verge Planted Tree Species on Urban Roads. In: 13 ARRB and 5th REAAA Conference, pp. 54–66.
Nimah, M.N., Hanks, R.J., 1973. Model for estimating soil water, plant and atmospheric interrelations: I. Description and sensitivity. Proc. Soil Sci. Soc. Am. 37 (4), 522–527.
Pollen, N., Simon, A., 2005. Estimating the mechanical effects of riparian vegetation on stream bank stability using a fiber bundle model. Water Resour. Res. 41 (7), W07025.
Premchitt, J., Lam, T.S.K., Shen, J.M., Lam, H.F., 1992. Rainstorm Runoff on Slopes. Geotechnical Engineering Office, Hong Kong.
Radcliffe, D., Hayden, T., Watson, K., Crowley, P., Phillips, R.E., 1980. Simulation of soil water within the root zone of a corn crop. Agron. J. 72, 19–24.
Raper, G.P., 1998. Agroforestry Water use in Mediterranean Regions of Australia. In: Water and Salinity Issues in Agroforestry No. 2, Publication No: 98/62. Rural Industries Research and Development Corporation, Barton, Australia.
Reul, O., Randolph, M.F., 2003. Piled raft in overconsolidated clay: comparison of in situ measurements and numerical analyses. Geotechnique 53 (3), 301–315.

Reul, O., 2004. Numerical study of the bearing behaviour of piled rafts. Int. J. Geomech. 4 (1), 59–68, ASCE.
Sacchi, E., Michelot, J.-L., Pitsch, H., Lalieux, P., Aranyossy, J.-F., 2001. Extraction of water and solutes from argillaceous rocks for geochemical characterisation: methods, processes, and current understanding. Hydrogeol. J. 9, 17–33.
Sattelmacher, B., Marschner, H., Kuhne, R., 1990. Effects of the temperature of the rooting zone on the growth and development of roots of potato (Solanum-tuberosum). Ann. Bot. (Lond) 65 (1), 27–36.
Scott, E.M., Rattray, E.A.S., Prosser, J.I., Killham, K., Glover, L.A., Lynch, J.M., et al., 1995. A mathematical model for dispersal of bacterial inoculants colonizing the wheat rhizosphere. Soil Biol. Biochem. 27, 1307–1318.
Slavich, P.G., Hatto, T.J., Dawes, W.R., 1998. The Canopy Growth and Transpiration Model of WAVES: Technical Description and Evaluation. CSIRO Land and Water, Melbourne, Australia.
Smith, D.M., Allen, S.J., 1996. Measurement of sap flow in plant stems. J. Exp. Bot. 47 (305), 1833–1844.
Sreedeep, S., Singh, D.N., 2006. Methodology for determination of osmotic suction of soils. Geotech. Geol. Eng. 24, 1469–1479.
Swanson, R., Whitfield, D.W.A., 1981. A numerical analysis of heat pulse velocity theory and practice. J. Exp. Bot. 32, 221–239.
Thomas, R.E., Pollen-Bankhead, N., 2010. Modeling root-reinforcement with a fiber-bundle model and Monte Carlo simulation. Ecol. Eng. 36, 47–61.
Tinker, P.B., Nye, P.H., 2000. Solute Movement in the Rhizosphere. Oxford University Press, New York.
Uniform Building Code, 1997. International Building Code. International Code Council, Washington, DC.
Waldron, L.J., 1977. The shear resistance of root-permeated homogeneous and stratified soil. Soil Sci. Soc. Am. J. 41 (5), 843–849.
Waldron, L.J., Dakessian, S., 1981. Soil reinforcement by roots: calculation of increased soil shear resistance from root properties. Soil Sci. 132, 427–435.
White, I., Melville, M.D., Wilson, B.P., Sammut, J., 1997. Reducing acidic discharges from coastal wetlands in eastern Australia. Wetlands Ecol. Manage. 5, 55–72.
Wu, T.H., McKinnell, W.P., Swanston, D.N., 1979. Strength of tree roots and landslides on Prince of Wales Island, Alaska. Can. Geotech. J. 16, 19–33.
Xu, C.Y., Gong, L., Jiang, T., Chen, D., Singh, V.P., 2006. Analysis of spatial distribution and temporal trend of reference evapotranspiration and pan evaporation in Changjiang (Yangtze River) catchment. J. Hydrol. 327, 81–93.
Yajun, W., Cosgrove, D.J., 2000. Adaptation of roots to low water potentials by changes in cell wall extensibility and cell wall proteins. J. Exp. Bot. 51, 1543–1553.

CHAPTER 21

Environmentally Friendly Slope Stabilization Using a Soil Nail and Root System in Canada

Myint Win Bo[1], Mike Fabius[1], Arul Arulrajah[2], Suksun Horpibulsuk[2,3]
[1]DST Consulting Engineers Inc., Thunder Bay ON, Canada
[2]Swinburne University of Technology, Melbourne, Australia
[3]Suranaree University of Technology, Nakhon Ratchasima, Thailand

21.1 INTRODUCTION

There are generally two main types of slopes in which geotechnical engineers are interested—natural slopes and man-made slopes. Whereas the former is formed due to geological and geomorphological processes in nature, the latter is formed by human activity usually during the process of construction or waste disposal. Man-made slopes may be in the form of either elevated fills or excavated depressions. Major geotechnical issues encountered in the stabilization of both natural and man-made slopes include gradual and often continuing displacements, sudden slope mass failure, and hydraulic upheaval. Because they are designed by professional engineers with an advanced and improved understanding of soil and rock mechanics, geotechnical engineering, and mechanics of slope stability, man-made slopes can be created to stand safely at virtually any required gradient. Similarly, natural slopes can be made safe without compromising the environment or aesthetics. For example, even at locations with severe space constraints, slopes can be designed to achieve any acceptable factor of safety (FOS) by introducing a suitable form of reinforcement or supports.

Even slopes that were adequately stable at one time may require stabilizing as a result of changes to slope geometry, soil or rock strength, loading conditions, or environmental conditions. Such changes, usually unplanned, to many natural and man-made slopes that formed years ago have led to economic losses, aesthetic losses, losses of infrastructure constructed near the slopes, losses to the ecosystem, and loss of lives. This chapter discusses several such case histories of existing natural and man-made slopes and how these have been stabilized in Ontario, Canada.

Chemical, Electrokinetic, Thermal, and Bioengineering Methods
http://dx.doi.org/10.1016/B978-0-08-100191-2.00021-6

21.2 MECHANISMS OF EXISTING SLOPE FAILURES

The following are typical trigger mechanisms of existing slope failures:
1. Change in slope geometry
2. Loss of soil or rock mass strength
3. Change in loading conditions
4. Change in environmental conditions

A brief discussion of these mechanisms follows.

21.2.1 Change in slope geometry

Typical changes to the slope geometry occur due to natural disturbances such as erosion at the toe or to man-made changes such as filling at the crest (often to gain additional useable land), as well as excavation at the toe of the slope. Although the planned filling or excavation is typically implemented with reasonable planning, design, and construction methods, inadequately planned activity can result in unexpected consequences. Examples are slopes initially constructed of loose waste materials and that later affect new development nearby.

21.2.2 Loss of soil mass strength

Many geotechnical engineers have witnessed natural or excavated slopes standing at near vertical conditions. Soil mass strengths, which contribute to stability of slopes under long-term conditions, are characterized using drained strength parameters and pore water effects. The latter can involve either strength reduction (typically saturated pore pressures, for example, as a groundwater table rises) or strength increase (typically matric suction within partially saturated pores).

Most soils encountered in existing slopes are overconsolidated, which in the case of natural slopes is due to their geological history (unloading due to erosion) and in the case of man-made slopes is due to mechanical compaction processes applied during the placement of the fill. Therefore, particularly for cohesive soils, such overconsolidated soils often possess sufficient drained cohesion and drained friction angle to ensure stable slopes. Examples of exceptions to this are active natural slopes such as riverbanks that are continually being eroded and in a state of incipient failure and also loosely placed (often waste) materials placed without compaction.

Groundwater has a significant impact on soil and rock mass strength. A rising groundwater table or groundwater pressure from uphill sources reduces soil strength significantly. For noncohesive soil above the groundwater table, matric suction has a significant strength contribution. However, this component can be easily lost upon entry of more water into the pores, for example, due to infiltration from rainfall, surface water runoff, or a rise in the groundwater level. In many cases, strength from matric suction is increased by the suction generated from tree roots, and therefore dying trees will reduce the soil mass strength.

Loss of strength can also occur as a result of soil disturbance. Vibration disturbance from earthquakes can affect soil strength, for example, weakening certain cohesive soils and causing liquefaction of loose fine granular material. Frost, desiccation, and slippage are other causes of disturbance-induced strength reduction.

21.2.3 Change in loading condition

Changes in loading conditions are not unusual; for example, they can occur due to construction of permanent infrastructure supported on the crest or on the slope, temporary parking of vehicles or equipment on the crest or slope, ponding of water on the crest or slope, entry of water into tension cracks, changes in the groundwater level, and vibration loads (from earthquakes, blasting, or moving equipment). In addition, many natural slopes are filled with trees, which cause additional loads on the slope as they mature. Examples are tree loads due to increases in diameter and height, increasing density of certain trees, and increasing exposure to wind load.

21.2.4 Change in environmental condition

The effects of changes in environmental conditions on natural slopes are typically much more extensive than those of slope geometry, soil strength, or loading conditions. Environmental changes are typically regional instead of location specific, and therefore increase slope and landslide hazards along extensive areas of similar terrain, such as along floodplains and in mountainous regions.

Impacts of environmental changes to natural slopes have been extensively discussed by Bo et al. (2008). Due to global warming, patterns of precipitation, wind, type of vegetation, average temperature, and flooding will change. For natural soil slopes, these changes may result in the loss of certain resisting forces available and incur additional driving forces. Furthermore, landslide-prone terrain may become more susceptible to natural hazards such as earthquakes.

Examples of climate change impacts on slopes include a water infiltration increase (causing loss of soil suction), a reduction in effective stress due to rising groundwater levels, a loss of root reinforcement due to changes in the type of vegetation or dying of vegetation, an increase in seepage forces due to frequent and intense storms, an increase in the magnitude of rapid drawdown conditions, a reduction of resisting forces due to erosion of the slope toe, an increase in seepage velocity due to an increased groundwater temperature, a reduction of soil cohesion due to increased temperature, and vegetation losses as a result of wind changes.

Increasing amounts of precipitation typically have a major destabilizing effect on slopes. Seepage-induced slope failures were studied by Budhu and Gobin (1995). Based on their study, an increase in the hydraulic gradient and an increase in the slope of the seepage line significantly reduce a slope's stability (typically characterized with a FOS). An increase in seepage forces can also increase the potential for piping (internal hydraulic erosion) of the soil within a soil slope (Tomlinson and Vaid, 2000). In addition, an increase in infiltration replaces the air voids in the soils with water, and the significant strength obtained from soil suction is lost.

Increases in precipitation will also result in a greater water surplus in catchment areas, causing greater flows in water courses and extreme flooding. This causes accelerated erosion at the toe of natural slopes in the floodplains of streams or riverbanks. This process can trigger a landslide of a previously inactive natural slope due to reduction of the toe resistance. These fluctuations can also create frequent occurrences of rapid drawdown conditions in front of slopes, a well-known trigger of instability to natural slopes. Surface erosion can also occur, resulting in the removal of vegetation and increase in runoff infiltration.

21.3 METHODS OF STABILIZATION

Various stabilization methods are available to the geotechnical engineer for control of the four previously described failure mechanisms.

21.3.1 Control of geometry

To restore or to reform an acceptable geometry of slope with sufficient stability (often characterized with a FOS), a common approach is to simply flatten the slope. Alternatively, a counter berm can be provided at the toe (Figs. 21.1 and 21.2). However, these options are only possible when space is available and environmental impacts are acceptable. In addition, fill often induces new ground settlements if placed over compressible soil. If a new slope geometry is desired—for example, to create more land through oversteepening of a slope or for an excavation—then various kinds of retaining structures are available. These include a gravity wall, cantilever wall, gabion wall, and sheet pile wall, some of which often incorporate tieback anchors (Figs. 21.3 and 21.4). Soil reinforcement or replacement with stronger materials can also be used to achieve a new steeper geometry.

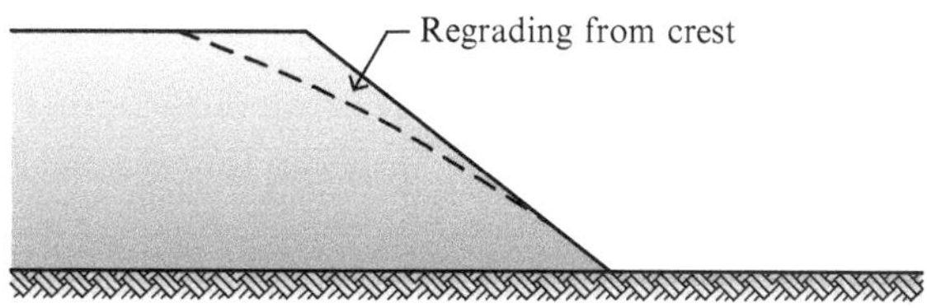

Figure 21.1 Flattening of a slope for stabilization.

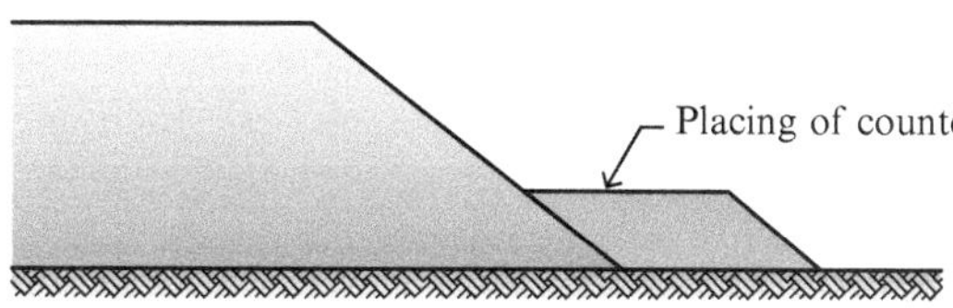

Figure 21.2 Placing of counter berm to stabilize.

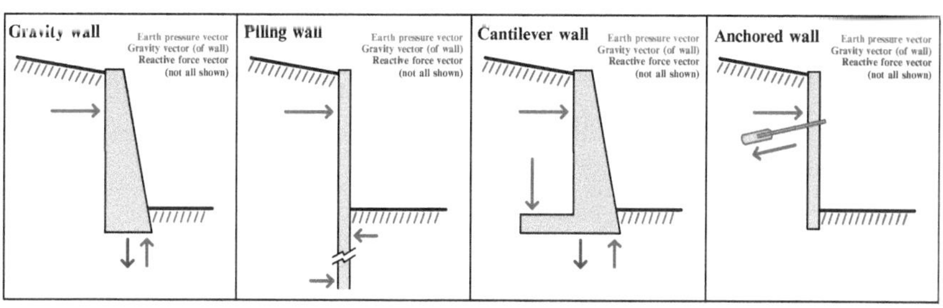

Figure 21.3 Types of retaining structures to stabilize slope by reshaping of geometry.

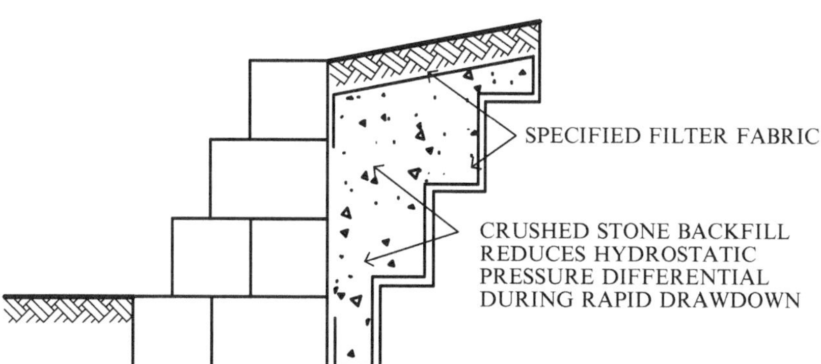

Figure 21.4 Stabilization of slope using gabions.

21.3.2 Controlling the soil mass strength

Loss of soil mass strength due to loss of suction can be controlled by minimizing infiltration of surface water through improving drainage and maintaining vegetation. The latter increases strength through the consumption of moisture from the soil (Fig. 21.5). As tree roots uptake water, beneficial soil suction in the soil is maintained or increases in the near surface soils (Indraratna et al., 2006).

Groundwater drainage is another effective method for increasing the soil mass strength within an existing slope. Slopes in low-permeable soil can be drained with trench drains constructed down the slope at frequent spacings along the slope. For more permeable soils, horizontal drains installed near the toe (typically by drilling) can be very effective at increasing strength due to reduced groundwater pressures.

Alternatively, improvement of soil mass strength is feasible through several well-developed construction processes. These include introducing inclusions (e.g., installation of stone, cement, or lime columns), reinforcing (e.g., with geotexile, soil nails, or piles), and applying electro-osmosis (with an electrical current through the soil) (Figs. 21.6–21.9).

21.3.3 Controlling the loading conditions

Construction of new infrastructure or traffic from heavy vehicles near the crest of a slope (which can contribute to the driving forces) are common loads that can destabilize a slope. New fill placed at the crest or excavation

Figure 21.5 Roots increase soil mass strength through reinforcement and moisture uptake.

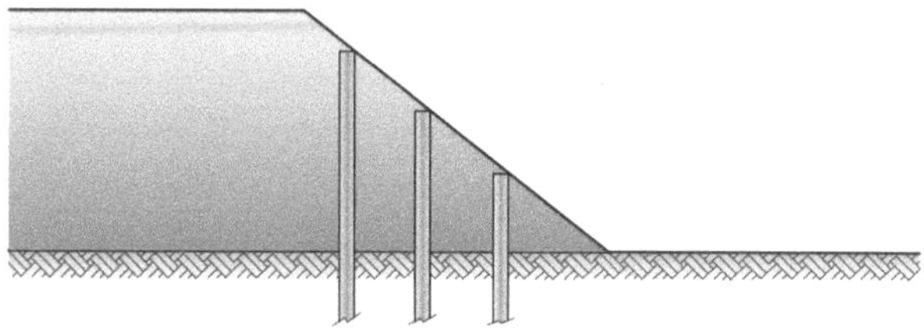

Figure 21.6 Stabilizing a slope using stone or lime columns.

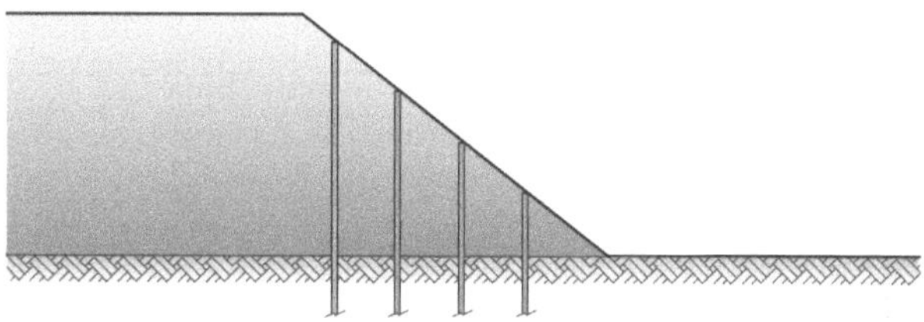

Figure 21.7 Stabilizing a slope using piles.

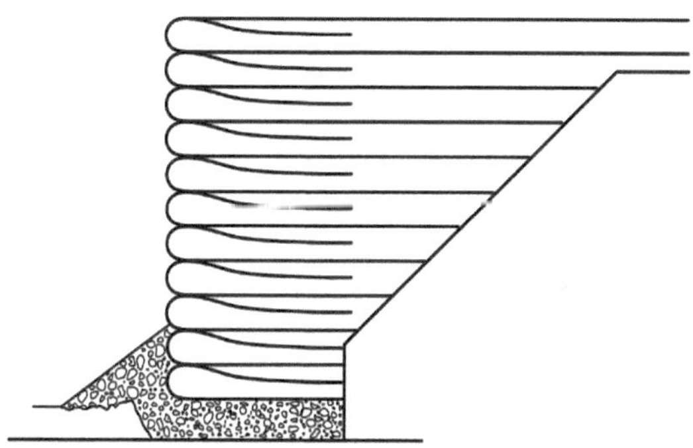

Figure 21.8 Geosynthetic reinforced slope.

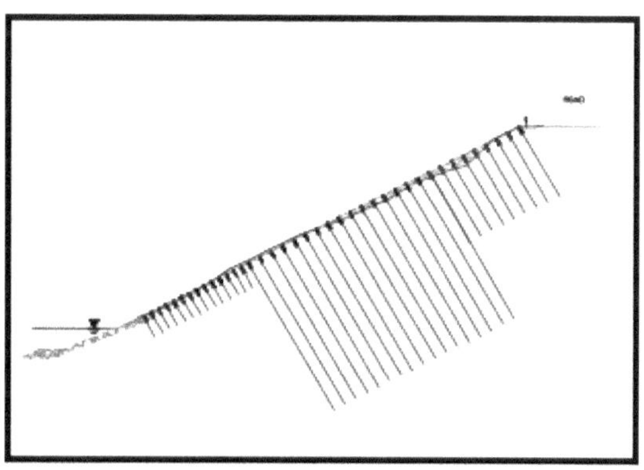

Figure 21.9 Stabilizing a slope using soil nails (Fabius et al. (2008)).

at the toe are also examples that can be addressed. These additional loads can either be prohibited or moved away from the slope.

Alternatively, the slope can be reinforced and strengthened to safely withstand these additional loads. This typically involves constructing works that apply additional resistance from tension, compression, or bending forces, such as with anchors, walls, or micropiles. Reinforcement of a slope with geotextiles or soil nails at strategic locations is also an available option to remedy these situations.

Slope instability caused by increased tree loads and wind loads transferred from the trees can be minimized by controlling the density of trees, as well as trimming and pruning to minimize wind loads and control the growth rate. Selecting certain species of trees that have acceptable growth limits in terms of size and height can also be implemented.

21.3.4 Controlling changes in environmental conditions

Stabilization or improvement of a specific slope impacted by environmental changes can typically be implemented by applying appropriate stabilization methods as discussed previously for controlling geometry, material strength, and loading conditions. Note that some methods, such as mechanically (geosynthetic) stabilized earth, may be unsuitable for an existing slope if it is impractical to reconstruct without undue disturbance. Others, such as soil nails, micropiles, and electro-osmosis, can be applied in situ without reconstruction of the slope.

Environmental changes often require only minor adjustments for minimizing their effects. These can include one or a combination of improving drainage, controlling infiltration, maintaining groundwater levels, or maintaining vegetation.

Whereas addressing environmental changes to an individual slope may be straightforward as described previously, when an entire region of similar slopes in a certain region such as a mountainous area or floodplain is affected, then a different approach to address the new hazards is necessary. Detailed intrusive geotechnical investigations are no longer practical, and often remote sensing together with probabilistic approaches are applied for hazard assessments. Typical solutions involve mapping areas with different risk levels and providing different approaches commensurate with risk. For example, planning can involve establishing a cautionary zone for potentially affected facilities or infrastructure together with periodic monitoring of the regression of an active slope (Fabius et al., 2004). A detailed investigation and stabilization solution are implemented only when the slope crest regresses into the cautionary zone.

21.3.5 Selecting a stabilization method

From the foregoing, it is apparent that some stabilization methods control more than one of the four slope failure mechanisms. For example, drainage can control both soil mass strength and environmental changes, whereas soil nails provide both soil mass strength improvement and resisting forces. Similarly, retaining structures control both geometry and loading conditions. Selecting the optimum solution for a particular slope requires establishing suitable design criteria and then developing a full understanding of the ground conditions, the environmental conditions, the failure mechanisms, and the tools available for stabilization.

With regard to slopes that have yet to be constructed, typically where new development is involved, a wide variety of stabilizations are available to the geotechnical engineer. Existing slopes, however, provide a greater challenge due to the impracticability of reconstructing the slopes and particularly the economic constraints when the instability was unplanned for. These are often slopes destabilized by climate change or rendered inadequate by otherwise previously unplanned land use. Examples are active riverbank slope erosion encroaching on an urban environment, or a side-hill railroad fill destabilized as a result of heavier train loads.

Optimal stabilization selection requires identifying the technology that best fits the design criteria. For example, heavy civil works such as retaining walls are typically not cost-effective for smaller localized treatments, and they often do not meet requirements for controlling site disturbance or addressing sites with difficult access to the slope. Similarly, expenditures can be delayed if the solution can be applied surgically to only those areas of a slope currently requiring treatment, on an as-required basis. In other cases, a drainage solution may not be feasible if impacts from groundwater discharge are predicted. In some cases, a physically flexible system is preferred to minimize live loading and earthquake effects on rigid works. Typically, aesthetic and environmental criteria are increasingly important considerations. These are all examples for which the installation of soil nail systems is preferred.

Next, this chapter focuses on stabilizing existing soil or waste slopes with an inadequate degree of stability and where stabilization with additional fill is not an option because of land or access constraints. Methods that minimize impacts to the slope and to the environment are important. Four case studies of nine slopes are presented—two for railroad fills and two for riverbanks. These vary in length from 30 m to more than 1 km, and they vary in height from 5 to 30 m.

21.4 HISTORY OF SOIL NAILS

Soil nails generally comprise metallic or polymeric members installed in the ground to provide tension and bending resistances to reinforce the soil mass for stabilization of natural or existing slopes. Installation can be either grouted in prebored holes or by insertion using a displacement technique with either static push or dynamic forces (vibration, percussion, and/or rotation). Soil nails are installed either horizontally or at various degrees from the horizontal, depending on the slope geometry. Horizontal nails develop significant tension resistance, whereas nails perpendicular to a failure plane develop significant bending resistance.

Soil nailing originated from rock-bolting, multianchorage systems and from reinforcement soil techniques (Phear et al., 2005). Many retaining walls were built in France using numerous bars as rock anchor in the early 1960s (Bonazzi and Colombet, 1984). Similar techniques of supporting tunnel and underground galleries with reinforced bars and bolts were developed by New Austrian Tunneling method (NATM) in the mid-1960s (Bruce and Jewell, 1986). In 1972, the first reported soil nail wall was constructed in Versailles, France (FHWA, 1993).

By 1988, at least 50,000 m^2 of soil-nailed walls and steep slopes had been constructed in France (Bruce and Jewell, 1986). In the mid-1990s, more than 500 soil-nailed walls were constructed in Germany (Federal Highway Administration, 1997). By 2001, more than 500 soil-nailed walls had been constructed for highway works in United States. In the United Kingdom, soil nailing for reinforcing of slopes and walls was first used in the late 1980s. In Canada, soil nailing was started in the early 1980s as an excavation support technology.

Soil nails were originally applied to near-vertical cut slopes with steel nails grouted immediately after each successive cut level into near-horizontally drilled holes. Traditional design methods assume the nails act entirely in tension, even though slope failure observations have indicated nails experience significant bending resistance at the failure surface. The economics of such a system is typically within 20% of other excavation stabilization systems.

Existing unstable soil slopes, on the other hand, are typically flatter than 45 degrees. In such cases, insertion of nails can often be accomplished without drilling or grouting. In most cases, these nails are installed perpendicular to the slope surface. This method has environmental and economic appeal because installations are quick and implemented in situ without having to reconstruct the existing slopes or embankment.

The soil nail system applied to the case studies herein was originally developed as an attempt to develop a faster, better, and cheaper option for existing unstable slopes. Key development challenges were the need for a rapid insertion method without costly drilling or grouting, finding an alternative to the traditional mesh and shotcrete facing, addressing long-term corrosion, and developing analytical approaches for bending resistance. Insertion was accomplished with the development of percussion technologies and efficient splicing techniques. The facing was achieved through the development of a natural biotechnical facing of shallow vegetation roots together with a nail head widening. A sacrificial approach was applied to long-term corrosion, similar to piles. Finally, suitable analytical approaches were developed, supplemented by advanced modeling techniques.

21.5 STABILIZATION OF NATURAL SLOPES USING ENVIRONMENTALLY FRIENDLY SOIL NAILS IN CANADA

Deep mechanically inserted soil nails supplemented with shallow vegetation have been termed a soil nail and root technology (SNART). This engineered system was developed and patented by DST Consulting Engineers, Inc., in Canada in the 1990s (Tozer and Fabius, 2000). The materials used are steel bar of various diameters, typically less than 60 mm in diameter and with a square or rectangular plate head that is a few hundred millimeters in size. The plate head is oriented along the nails and across the slope, just below the ground surface.

Installation is by means of pushing using static or dynamic driving forces depending on the type of soil and resistance, using lightweight equipment operating on or adjacent to the slope. The shallow soil between the nails is typically reinforced with a vegetation system comprising long root grass as a facing, which usually provides reinforcing as well as maintaining suction in the soil mass at the shallow depth. Biodegradable coconut fibers are frequently used to initially enhance the vegetation growth and protection of topsoil. Length and spacing of nails are engineered to address all potential modes of failure. These include deeper failures below the nails and through the nails, shallower failures in which the soil flows around the nails, and surficial slipping between the nails. The design of the steel materials incorporates a sacrificial allowance for the design life of the system. Only the vegetation is visible once the system is installed. Using this technology, several existing slopes have been stabilized.

The following sections discuss stabilizations of three railway embankment slopes at two locations and six riverbank stabilizations at two locations, all implemented within the past 15 years in Ontario, Canada.

21.5.1 Railway Embankment 1

This remote site is located on a class 1 railroad in northwestern Ontario, Canada. Following a culvert failure and its repair, which involved cutting into the slope of an 11-m-high soil embankment, the track required resurfacing and had problems with insufficient shoulder width. The slopes were as steep as 40° to the horizontal. The subsequent geotechnical investigation concluded that deep internal failures through both embankment soils (dry sands and noncohesive silts) and foundation soils (stiff clay) were unlikely to present long-term problems. However, to achieve stability of the embankment fill, it was necessary to stabilize both fill slopes. The study area is shown in Fig. 21.10.

The diagnosis of the problem was as follows:

- Difficulties maintaining shoulders due to lateral train forces and "sun-kink" forces (from thermal expansion) in an extensive curve with steep slopes
- Very loose soil sides (both) as a result of culvert repair work
- Very loose, steep sides as a result of previous bank widening (through dumping from the top) of shoulders on the outside of curve
- Localized loose embankment zones due to some soil loss into culvert prior to its repair
- Low FOS against shallow slips (parallel to slope) due to steep, incompletely vegetated slopes
- Problems exhibited over approximately 50 m of embankment, centered on the culvert

The SNART system was selected in conjunction with an epoxy-stabilized shoulder to stabilize the 11-m-high embankment slopes. The embankment materials are dry and granular, with sand and gravel on the outside of a silt core. Underlying soil beneath the embankment is stiff clay. Therefore, no deep-seated sliding planes were expected to have occurred in the past. The detailed soil profile was determined by intrusive investigation and laboratory testing.

Following a detailed design process, 3-m-long by 38-mm-diameter nails with 200 × 200-mm shear plate heads were installed at a rate of 6 nails per lineal meter of embankment slope. These were concentrated within the upper half of the slope due to the drier nature of the soils (less soil suction

Figure 21.10 Study area (Railway Embankment 1)

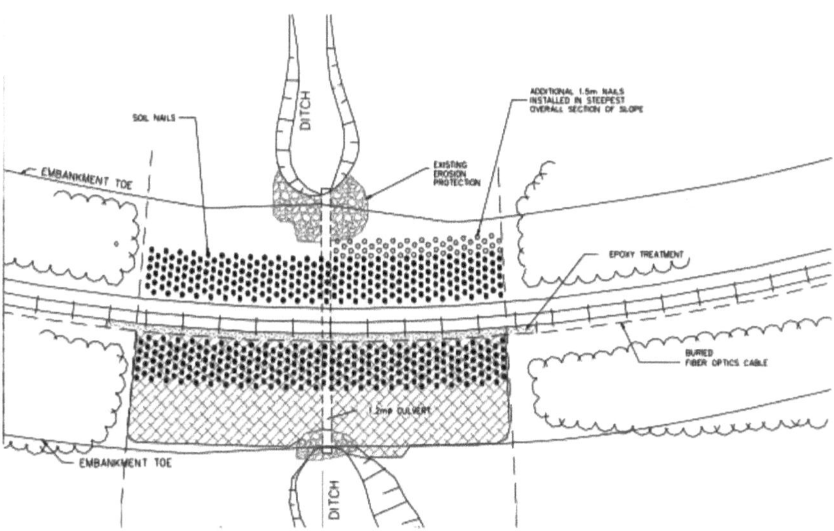

Figure 21.11 Layout of soil nails installed at Railway Embankment 1.

effects) and the increased effects of train vibrations. The design root depth for the facing was 150 mm, and a seed mix together with a biodegradable coconut fiber mat was installed for areas where existing vegetation did not meet this requirement. A total of 560 nails were installed. The pattern of soil nail layout is shown in Fig. 21.11. The details of stabilized

Table 21.1 Details of Railway Embankment 1, fill, foundation soils, and soil nail design

Details of embankment	Description	Selected design	Description
Embankment height	11 m	Size of nail	38 mm
Slope gradient	40° from horizontal	Length of nail	3 m
Length of stabilization	50 m	Nail spacing	1.3 m
Type of fill	Silt core with granular fill	Nail angle	40°
Type of foundation	Stiff clay	Nail specification	400 MPa steel
Design parameters	$C'=0$ for saturated materials, 1 kPa for unsaturated granular fill, 2 kPa for unsaturated silt $\phi'=40°$ (outer granular fill) and 35° (silt core)	Vegetation (grass)	Root depth 150 mm with biodegradable coconut fiber mat
FOS achieved	1.3	No. of nails	560

embankment in terms of descriptions, fill, and foundation soils, as well as soil nail installation, are shown in Table 21.1.

Both slopes at this location have continued to perform adequately.

21.5.2 Railway Embankment 2

This 30-m-long slope instability is located on the south side of a double-track railway embankment on a class 1 railroad near St. Catherines in southern Ontario, Canada, where double tracks cross a 360-wide creek valley on an embankment up to 20 m in height. At the location of the problem, the embankment is a side-hill fill with the fill slope as steep as 43° to the horizontal.

In the unstable area, the embankment was 12–16 m in height. An intrusive geotechnical investigation identified essentially stiff clay fill and foundation materials, with some noncohesive materials. The geotechnical analyses identified potential failure planes, both shallow and deep, and concluded that (1) the south embankment slope was in borderline stability and (2) further movements could be expected to occur. Furthermore, the toe of the south embankment had signs of active erosion from runoff along the gully at the toe and could be contributing to the problem.

Based on the geotechnical investigation for this site, the embankment materials consist of silt and clay fills, some mixed with ash waste, to a depth of 4.6 m below base of rail. These overlie a stiff clay till, which in turn overlies a deep deposit of firm to very stiff clay below 9 m. On the slope, a layer of sandy materials covers the clay fill. Given the general topography, this embankment was identified as a side-hill fill on ground that slopes to the southeast.

To address the instability, SNART was selected. Both shallow and deep slip failure modes were analyzed using various potential slip locations as well as various nail lengths, diameters, and spacing.

Based on the analytical results, 2.5- to 8-m-long, 38-mm-diameter nails with 200 × 200-mm shear plate heads were required at a rate of 8 nails per lineal meter of embankment slope. A total of 275 nails were installed. Following nail installation, a coconut fiber biotechnical facing blanket was installed in order to promote new vegetation growth and assist in controlling erosion of embankment materials. Riprap was applied at the toe to control further gully erosion. A cross section illustrating the site is shown in Fig. 21.12, and soil nail installation is shown in Fig. 21.13. The details of stabilized embankment descriptions, fill, and foundation soil, as well as soil nail installation, are shown in Table 21.2.

The stabilized slope at this location has continued to perform adequately without further problems reported.

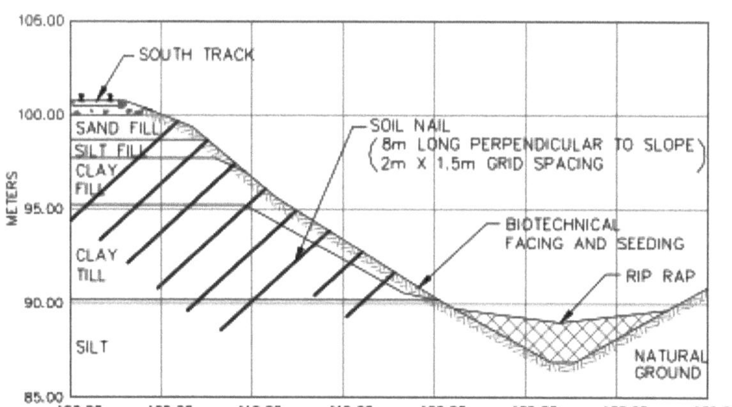

Figure 21.12 Site 2 fill and soil profile together with soil nail pattern.

Figure 21.13 Soil nail installation at Railway Embankment 2.

Table 21.2 Details of Railway Embankment 2, fill, foundation soils, and soil nail design

Details of embankment	Description	Selected design	Description
Embankment height	12–16 m	Size of nail	38 mm
Slope gradient	43°	Length of nail	2.5–8 m
Length of stabilization	30 m	Nail spacing	1.4 m
Type of fill	Silt and clay	Nail angle	Perpendicular to slope
Type of foundation	Stiff clay	Nail specification	400 MPa steel
FOS achieved	1.3	Vegetation (grass)	Root depth 150 mm with biodegradable coconut fiber mat
		No. of nails	275

21.5.3 Kaministiquia riverbank stabilization, Ontario

The Kaministiquia River meanders through deep glaciofluvial and glaciolacustrine deposits on the way to its delta on Lake Superior in Canada. The study area comprises a 2-km stretch of steep 30-m-high riverbank section in Thunder Bay (Fig. 21.14). The slope is partially covered with mature trees. Where past slope failures have occurred, gullies down the slope have formed that are devoid of vegetation. Perched groundwater in some areas seeps from sand layers near the top of the slope. Slope angles vary from

Figure 21.14 Aerial photo view of Kaministiquia River.

27° to 45° to the horizontal, with the steeper sections being generally between gullies or at the scarps of gullies. Slope failures observed in the past have always been shallow and have occurred after rainfall or during spring thawing. Since 1972, the Lakehead Region Conservation Authority (LRCA) had a concern for several houses, a street, and a water main at the crest of the bank (Fabius and Suke, 1990). In 1977, Thunder Bay experienced a major storm and flood. Erosion caused some residents to lose as much as 10 m of soil from their yards (in multiple shallow slides) during that event, which lasted several days. In 1982, a study concluded that the riverbank was regressing at a rate averaging 0.5 m/year. This natural process involved erosion at the toe, resulting in an oversteepened slope that periodically failed and slid into the river. Figure 21.15 shows a failed riverbank.

Subsurface conditions

The soil conditions at the site comprise (1) an upper 3 m of loose sand over (2) a medium plastic firm to stiff clay layer varying in thickness from nil to 2 m along the slope and (3) a deep deposit of nonplastic silt, faintly varved, extending to more than 25 m below river level. The silt is compact with some dense zones. Some of the slopes have perched water tables above and within the upper clay layer. Monitoring results have indicated that the main water table has a

Figure 21.15 The failed river bank at Kaministiquia River.

maximum level at 19 m below the crest at elevation 187.4 m, as measured over the course of 1982. A higher level at 189 m was selected for design, based solely on engineering judgment. A minimum water table at 185 m was selected, 2 m above nearby Lake Superior where the river discharges.

This background research on rainfall and its effects on slope stability was useful for understanding the Kaministiquia River slope behavior. Observations during a 25-year period indicate that the depth of slides has always been shallow, circular, and less than 3 m (Fabius and Suke, 1990). Non-flood-related failures usually occur during spring runoff from snowmelt or during periods of heavy rainfall. With the toe protected from further river erosion, continuing failures are attributed to loss of soil matrix suction as a result of water infiltration from precipitation.

In 1985, the LRCA purchased and demolished several homes at the crest of the riverbank and installed a 1.5-km length of riprap toe protection. Further stabilization was then provided where a 350-m length of the oversteepened riverbank encroached on the right of way for an urban street at the crest. Stabilization involved the installation of 12-m-deep bored soldier piles with concrete lagging, anchored to bored deadman structures behind the street. Rather than retain the remaining 1.15 km of crest, the solution included allowing this portion of the slope to naturally flatten and regress toward predetermined cautionary zones for infrastructure and houses at the crest (Fabius et al., 2004; Fig. 21.16). By 2000, five areas of slope had encroached on the cautionary zones, indicating the need for further stabilization measures (Fabius et al., 2004).

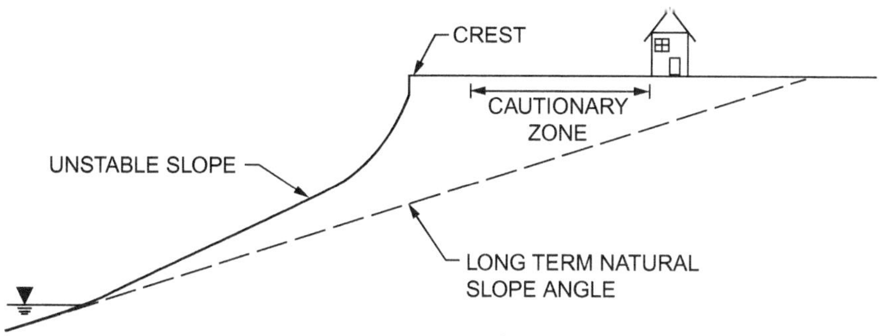

Figure 21.16 The cautionary zone approach.

Stabilization options

As an alternative to the soldier pile wall previously used along a portion of the riverbank, a soil nailing system developed specifically for oversteepened slopes (Tozer and Fabius, 2000) was implemented to stabilize the five new critical areas (toe protection had previously been placed). This soil nail system eliminates the need for drilling/grouting, and it replaces the traditional shotcrete facing with a biotechnical facing. The soil nails in this case provide shear (bending) resistance along potential slip surfaces. As such, the soil nails must be installed (in this case by driving, vibrating, or rotation) to below potential slip surfaces that have an inadequate FOS. For potential slip surfaces near the surface of the slope, additional resistance is provided through vegetation (roots) and/or the installation of special heads near the top of the soil nail. Figure 21.11 shows a typical soil nail design layout. The limiting factor for economical soil nails is the ability to install the nails to sufficient depths without drilling. Based on numerical analyses, soil nails driven to a depth of 11 m along the upper half of the slope increase the FOS to approximately 1.3, equivalent to that of the nearby soldier pile retaining wall already in place. Subsequent field trials confirmed that soil nails could be economically installed on the slope to the required depth without adverse impacts to the existing vegetation. A major attraction of this method is that it can be surgically applied to specific areas where the slope has regressed to within the cautionary zone without treating an entire length of slope as in the case of a retaining wall. This is also a flexible system that can be improved or modified as required, based on performance. Costs were less than half that of other systems. The details of the stabilized riverbank with descriptions, riverbank soils, as well as soil nail installation details are provided in Table 21.3. Figure 21.17 shows the soil nail configuration, and Fig. 21.18 shows nail

Table 21.3 Details of riverbank and soil nail design

Details of riverbank	Description	Selected design	Description
Bank height	30 m	Size of nail	38 mm
Slope gradient	27–45°	Length of nail	4.5–11 m
Length of stabilization	Five areas up to 50 m each	Nail spacing	1.5 m
Type of soils	Sand, stiff clay underlain by nonplastic silt	Nail angle	Perpendicular to slope
Design parameters	$C = 10$ kPa (clay layer) $\phi' = 28°$ (upper sand) to 35° (silt) Suction $\phi^b = 15°$	Vegetation	Root depth 100–350 mm (grass) and 450 mm (trees) with biodegradable coconut fiber mat
FOS achieved	1.3		

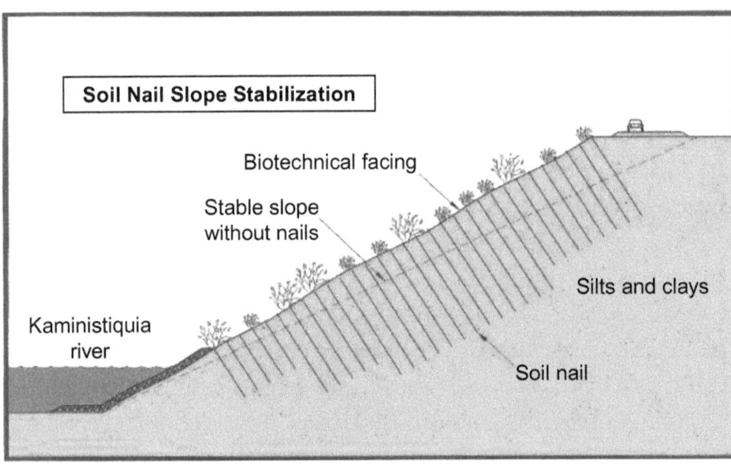

Figure 21.17 Soil nail and root solution for Kaministiquia riverbank slopes.

installation in progress. The steel nails were designed with a sacrificial corrosion allowance for a 100-year design life.

Following stabilization, the five slopes were monitored for nail head movements as the nails developed their resistance. Although these measurements include annual effects of frost action on the soil, they indicate less than 30 mm of downslope movement (Fig. 21.19).

Figure 21.18 Installation of soil nails using slope-mounted equipment.

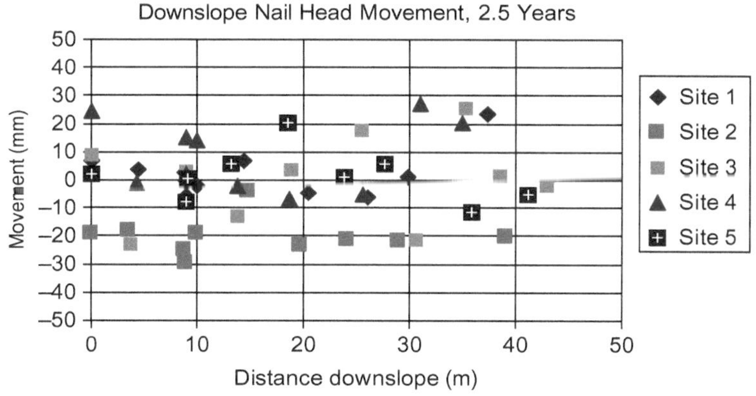

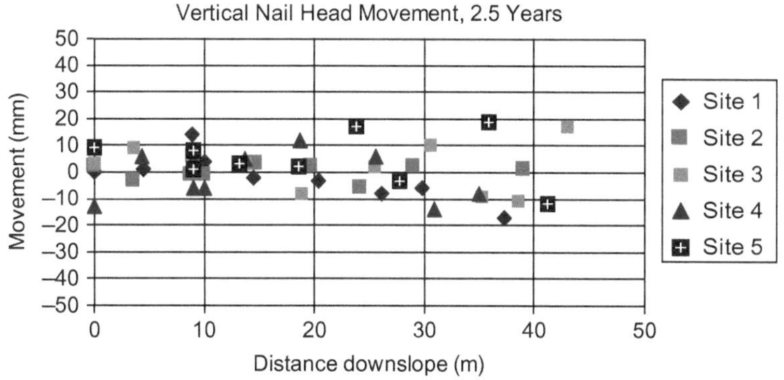

Figure 21.19 Results of 2½ years of horizontal and vertical nail head monitoring, Kaministiquia riverbank.

21.5.4 Garden River slope stabilization

A comprehensive geotechnical investigation was carried out to address the slope instability noted along the north bank of the Echo River immediately south of Echo River Road in 2007. The area of concern extends approximately 1.2 km along Echo River Road easterly from Pioneer Road Bridge toward the Garden River First Nation, Ontario, Canada. The purpose of the investigation was to assess the subsurface soil and groundwater conditions, to determine the causes of the slope instability, and to provide geotechnical recommendations for rehabilitation of the distressed slope. The instability of the slope was visually identified by the developed tension cracks of widths ranging from approximately 10 to 100 mm and lengths extending up to 1200 m easterly from the bridge. It was also noted that the water level of Echo River had lowered by 1 m during the past few years, and tension cracking had recently worsened. Feasible slope stabilization work options along with their cost-effectiveness were developed.

The generalized stratigraphy of the site, based on the conditions at boreholes, consists of fill soils overlying soft to firm clay layer. Fill soil layers were found at approximately 2.2 m. The thickness of the soft to firm clay layer was not known because boreholes were terminated at 10.7 m, where similar clay soil was encountered. The height of the riverbank is approximately 5 m. The geotechnical engineering parameters shown in Table 21.4 were adopted as design parameters for each formation of the soils.

Based on the information collected from the site investigations and laboratory testing, potential failures of the embankment were analyzed, including deep circular failures through the soils and shallow slides through the riverbank soils.

SLOPE-W (Geostudio 2004, version 6.22) slope stability analysis software was used for the analysis and determined the geometry of slip, slip mass,

Table 21.4 Geotechnical parameters

Soil	Unit weight (kN/m^3)	Undrained shear strength (kPa)	Drained internal friction angle (deg.)
Sand and gravel (fill)	20	N/A	30
Black organic soil (fill)	12	50	29
Gray silt (fill)	19	50	29
Clay (fill)	18	40–80 (40)	28
Soft clay	18	20–55 (35*)	24 (14*)

Note: Values in parentheses are residual strength or critical state value. Design values selected are indicated with an asterisk.
N/A, not applicable.

and driving forces. Resulting driving forces and slip mass from the stability analyses were used for stability analyses with soil nails for various spacing, sizes, and length. Various soil nail configurations were carried out through trial and error until riverbank stability was achieved with a sufficient FOS of 1.3. In the analysis, a reduced dimension of soil nail was used as an input data considering a sacrificial corrosion allowance for 100 years of life under the prevailing soil and groundwater conditions as established through soil testing.

Stress-deformation analyses were also carried out using PLAXIS 2-D finite element modeling software to determine horizontal displacement, stresses on the nails, and stability of the riverbank. Both undrained and drained conditions with suitable live loads were analyzed, and the stability condition under short- and long term conditions was determined.

Under the existing condition, the riverbank had a FOS of approximately unity, which is an unacceptably low margin of safety given the road that it supports. This confirmed the need for stabilization.

Several options were considered for the remedial design, which included regrading of the riverbank but that would unacceptably encroach on either the road at the crest or the river at the toe. Geogrid reinforced riverbank stabilization was also considered, but this required rebuilding of the riverbank. This option would require additional engineered fill to be imported as well as a long construction duration for the installation and compaction process, negatively affecting access along the road at its crest. In addition, significant energy and natural soils are required to transport the material and to construct the embankment, which is environmentally unfriendly. Furthermore, handling of fill materials on the riverbank during construction could increase siltation in the river, causing a significant impact on the ecosystem within the river. Moreover, reconstruction of the riverbank would require removal of existing vegetation.

Therefore, SNART was selected, given that reconstruction and importing of additional fill material would not be required and also would minimize the environmental impacts, including siltation and noise pollution due to heavy construction.

Soil nailing with a suitable spacing and size and length of soil nail deeper than the slip depth provided a sufficient FOS of greater than 1.3. Table 21.5 shows size, length, and spacing of soil nails for the riverbank under two gradients of slope.

Figures 21.20 and 21.21 show stress deformation analyses using PLAXIS version 8 (see also Tables 21.6 and 21.7). Figure 21.22 shows soil nail installation in progress.

Table 21.5 Soil nail dimensions analysis for stability of embankment at Echo riverbank

Drainage condition	Slope gradient	Nail size (mm)	Nail length (m)	Nail spacing (m)	FOS
Undrained	1.5:1	35	4.5	1.2 × 1.2	1.3
Drained	1.5:1	35	4.5	1.2 × 1.2	1.3
Undrained	1.5:1	35	4.5	1.5 × 1.5	1.3
Drained	1.5:1	35	4.5	1.5 × 1.5	1.3

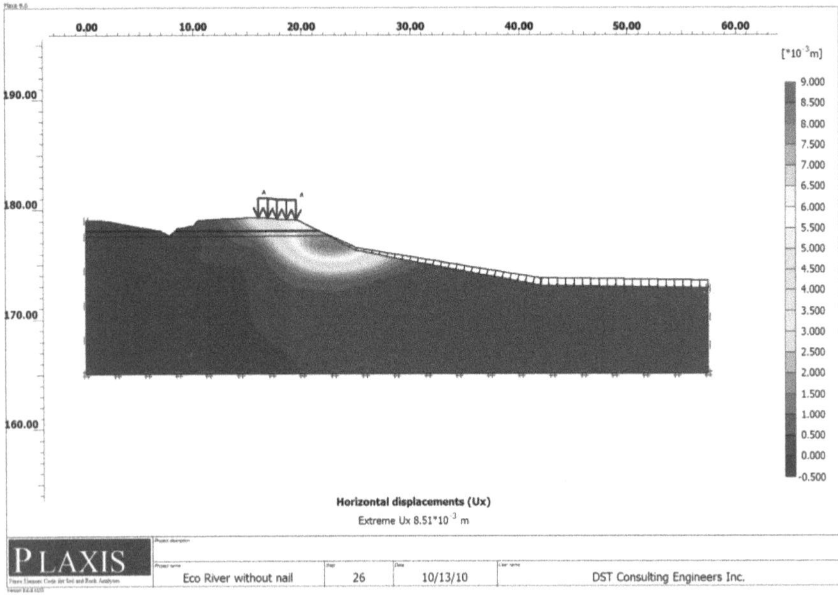

Figure 21.20 Slope stability analysis before stabilization using finite element modeling.

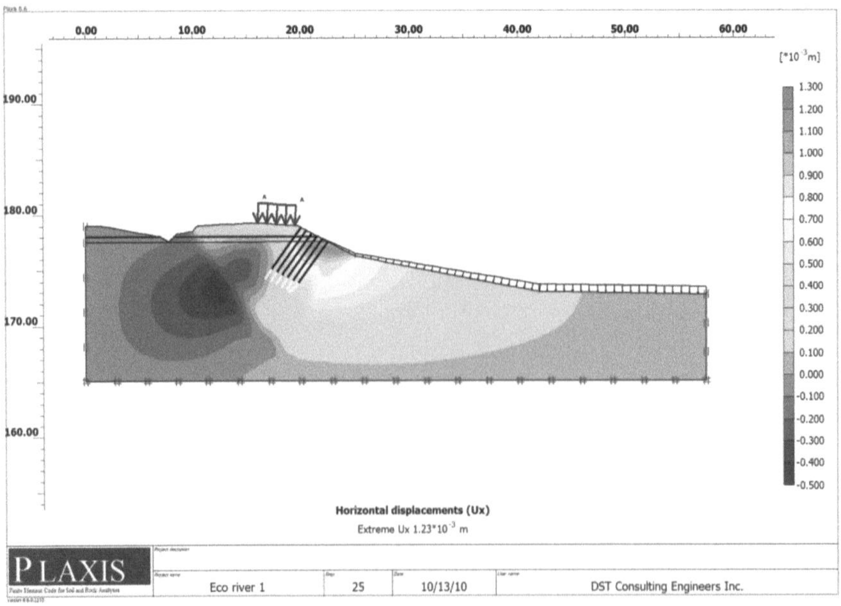

Figure 21.21 Slope stability analysis with soil nail stabilization using finite element modeling.

Figure 21.22 Soil nailing in progress.

Table 21.6 Stress–deformation analysis from PLAXIS

Drainage condition	Analysis type	Size (mm)	Length (m)	Spacing (m)	Horizontal displacement (m)	FOS
Undrained	Without soil nail	—	—	—	0.0085	1.0
Drained	Without soil nail	—	—	—	0.0032	1.0
Undrained	With soil nail	35	4.5	1.5 × 1.5	0.0012	1.3
Drained	With soil nail	35	4.5	1.5 × 1.5	0.0008	1.3

Table 21.7 Soil nail reinforcement treatment sections

Section		Soil nail			
Chainage	Length of riverbank (m)	Diameter (mm)	Effective length (m)	Grade	Grid spacing (m)
Section from 0+050 to 0+080	30	35	4.5	400	1.5 × 1.5
Section from 0+080 to 0+180	100	35	4.5	400	1.2 × 1.2
Section from 0+180 to 0+550	370	35	4.5	400	1.5 × 1.5
Total (m)	500				

21.6 CONCLUSION

This chapter focuses on case studies of slope stabilization works in Canada. Techniques using SNART systems have proved to be both technically and economically effective as well as environmental friendly for stabilization of existing natural and man-made soil slopes. Environmentally friendly techniques in slope stabilization are of particular interest to site owners, geotechnical practitioners, and researchers alike, considering the current interest in preserving the environment while also addressing negative impacts on slopes through climate change. The performance of these environmentally friendly slope stabilization methods is demonstrated by several case studies successfully applying SNART.

REFERENCES

Bo, M.W., Fabius, M., Fabius, K., 2008. Impact of Global Warming on Stability of Natural Slopes. In: Paper presented at the 4th Canadian Conference on Geohazards, Université Laval, May 20–24, 2008, Québec, Canada.

Bonazzi, D., Colombet, G., 1984. Réajustement et entretien des ouvrages de talus. In: Proc. Colloque Renforcement en Place des Sols et des Roches. ENPC Press, Paris.

Bruce, D.A., Jewell, R.A., 1986. Soil nailing: application and practice—Part 1. Ground Eng., 10–15, November.

Budhu, M., Gobin, R., 1995. Seepage-induced slope failures on sandbars in Grand Canyon. J. Geotech. Eng. 121 (8), 601–609, ASCE.

Fabius, M., Suke, S.M., 1990. The Investigation and Stabilization of a High River Bank in Thunder Bay. In: 43rd Canadian Geotechnical Conference, Quebec, Canada, pp. 69–75.

Fabius, K.A., Fabius, M., Vanapalli, S.K., Garga, V.K., 2004. Managing Landslide Hazards with the Cautionary Zone Approach. In: Paper presented at the 9th International Symposium on Landslides, Rio De Janiero, Brazil.

Fabius, M., Bo, M.W., Villegas, B., 2008. Stabilization of a 30 M High Riverbank in Canada with Nails, Plates and Roots. In: Paper presented at the 6th International Conference on Case Histories in Geotechnical Engineering, Arlington, VA.

FHWA, 1993. "French National Research Project Clouterre 1991-Recomandations Clouterre 1991", (English Translation) Soil Nailing Recommendations, Publication FHWA-SA-93-026. Federal Highway Administration, Washington, D.C..

Highway Administration, Federal, 1997. International Scanning Tour for Geotechnology September–October 1992, Soil Nailing Summary Report. http://ntl.bts.gov/DOCS/scangeo.html.

Indraratna, B., Fatahi, B., Khabbaz, H., 2006. Numerical analysis of matric suction effects of tree roots. Proc. Inst. Civil Eng. 159 (GE2), 77–90.

Phear, A., Dew, C., Ozsoy, B., Wharmby, N.J., Judge, J., Barley, A.D., 2005. Soil Nailing—Best Practice Guidance. CIRIA, London.

Tomlinson, S.S., Vaid, Y.P., 2000. Seepage forces and confining pressure effects on piping erosion. Can. Geotech. J. 37, 1–13.

Tozer, S.E., Fabius, M., 2000. Steep Nailed Embankment Technology: 2 Case Studies. In: Proc. 51st Annual Highway Geology Symposium, Seattle.

CHAPTER 22

Clay Soil in Suburban Environments

Movement and Stabilization through Vegetation

Stacey Vorwerk, Don Cameron, Gunnar Keppel
School of Natural and Built Environments, University of South Australia, Adelaide, South Australia

22.1 INTRODUCTION

Expansive clays are found worldwide in arid and semi-arid regions (Al-Rawas et al., 2006). Settlement or heave of expansive clay occurs with water insertion or extraction (Navarro et al., 2009). As a result, infrastructure built on or in these clays is known to fail structurally and/or deform unacceptably, resulting in a combined annual cost of $30 billion to the United States and China in rectification works (Li et al., 2014).

Ground heave or settlement of expansive clay soil depends on several factors: soil mineralogy and composition, the level of water in the soil, and the presence of vegetation (species dependent) on a site (Kalantari, 2012). Currently, a widely accepted geotechnical testing regime to characterize expansive soil (Fityus et al., 2005) is not available, and debate continues over current methods of assessment.

Tree roots are thought to intensify clay shrinkage through soil moisture extraction (Cameron, 2001; Richards et al., 1983). In the United Kingdom, 15–20% of disclosed insurance claims, in a normal nondrought year, involve cases in which trees are present (Deakin, 2005). On the other hand, roots stabilize slopes through soil reinforcement and increase of the shear strength through development of matric suction with soil drying (Fatahi, 2007). There is evidence of (stabilizing) improvement to soil engineering properties on sloped embankments (Stokes et al., 2009; Hubble et al., 2013) and under rail tracks (Potter and Cameron, 2005; Fatahi, 2007) occurring in the presence of vegetation. However, the movement of soil moisture by the roots of different tree species within a range of soil and climate environments (O'Malley and Cameron, 2005; Standards Association of Australia, 2011) is not known. Therefore, it is imperative that moisture, ground movement,

Chemical, Electrokinetic, Thermal, and Bioengineering Methods
http://dx.doi.org/10.1016/B978-0-08-100191-2.00022-8

and the reinforcement effect of specific tree species root architecture (De Baets et al., 2008) on expansive clay soils in built environments are identified and quantified.

In this chapter, the impact of vegetation growing in expansive soils in both suburban and rail corridor environments is reviewed, with focus on (1) characterization of expansive soils through relevant tests to predict ground movement, (2) tree impacts on soil moisture and drying settlement, and (3) case studies of stabilization of and damage to buildings and infrastructure on expansive clays, with or without trees. A review of information in these areas may provide a platform for developing improved design guidelines for suburban infrastructure on clay sites with trees.

22.2 CHARACTERIZING EXPANSIVE SOILS

Expansive soils cover approximately 20% of Australia's surface (Richards et al., 1983). These soils have been described as "calamitous soils" (Chen et al., 2007) due to the unpredictable volumetric change with changing water content (Gromko, 1974), and expansive soils are often poorly managed after construction. It is therefore essential to adequately characterize these soils using standardized tests to achieve efficient management. Existing methods used by geotechnical engineers to characterize expansive soils are discussed in Section 22.2.1. Section 22.2.2 details continuing research on development of testing methods to provide clarity in classifying expansive clays in a simple systematic manner.

22.2.1 Quantifying properties of expansive clay soils

The swelling characteristics of expansive clays vary with changes in testing method and sample preparation (Mielenz and King, 1955). The methods discussed here are used to understand expansive clays and cohesive soils, as part of general geotechnical practices. An effective, globally relevant testing regime to classify expansive soils does not exist (Fityus et al., 2005).

In Australia, swelling and movement potential of clay soils may be assessed indirectly by simple index testing (i.e., Atterberg limits), or by more direct measurements of soil reactivity (the linear strain per unit change in the logarithm of soil suction), through tests such as the shrink–swell test (Fityus et al., 2005). Measurement of the amount and pressure of water in soils is also important. Finally, knowledge of soil permeability and potential changes with increasing soil wetness aids in the understanding of soil moisture redistribution over time.

Reactivity

In Australia, local knowledge of soil reactivity is usually combined with Atterberg limits to predict soil behavior (Mitchell, 1993). In South Australia, Mitchell and Avalle (1984) proposed correlations between reactivity (termed instability index in their paper) and plasticity index for the dominant soil types found in Adelaide, South Australia, as shown in Fig. 22.1. The overall correlation is not strong, and the trends indicated the dependence of soil type on the correlation.

Similar research by Cameron (1989) over a wider range of Australian soils also could not uncover a simple universal correlation between soil reactivity and either Atterberg limits or linear shrinkage. Covar and Lytton (2001) delved deeper into the plasticity chart (plasticity index plotted against liquid limit) to propose that position on the chart inferred basic soil mineralogy. Thereafter, they correlated position on the chart with a volume change parameter (coefficient of linear extensibility or COLE) based on data from an extensive database. The approach requires liquid limit, plastic index, and the percentage of grains passing a 63-μm sieve (approximate fines content of 75 μm). Numerous reports documenting observations of damage due to expansive clays are available from all over the world (e.g., Li et al., 2014; Puppala et al., 2014). However, predictive models to forecast ground movements are rare (Erguler and Ulusay, 2003).

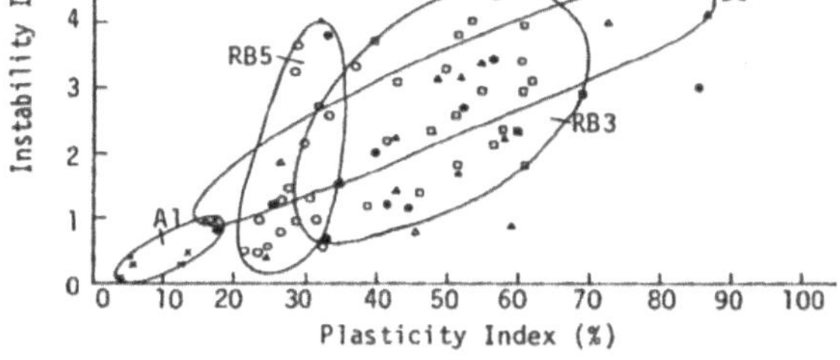

Figure 22.1 Correlation between reactivity (instability index) and plasticity index. *(Source: Mitchell and Avalle, 1984; first published by @Engineers Australia, 1984. Reprinted with permission).*

Currently, soil swelling is described through Atterberg limits, COLE, cation exchange capacity, and the saturated moisture content test.

Kariuki and van der Meer (2004) studied a range of different soils (47 samples) and found that, like Covar and Lytton (2001), soil properties that reflected clay mineralogy were most useful to correlate with volume change. They were able to predict values of potential volume change derived from the COLE test, a form of reactivity testing common in the United States, which is performed on soil clods. They found that combinations of different index tests, chemical tests, and cation exchange capacity were best for predicting potential volume change.

Shrink–swell soil reactivity

The shrink–swell testing procedure used in Australia is detailed in the Australian Standards (Standards Association of Australia, 1997). Both shrinkage and swelling test components are (usually) carried out on intact field soil samples (Fityus et al., 2004). The shrinkage component involves slowly air drying an unrestrained cylindrical core of soil and then oven drying it to find the ultimate axial shrinkage strain (Fityus et al., 2004). The swell component is determined by an oedometer test, wherein the undisturbed sample is placed under a vertical pressure of 25 kPa and then flooded with distilled water. The soil becomes saturated, and soil swelling eventually ceases. By measuring the height of the fully swollen soil, the swelling strain for the laterally restrained sample is able to be calculated.

The shrink–swell index (I_{ss}) represents the rate of change of axial strain with soil suction change for a laterally unconstrained soil specimen (Fityus et al., 2004). An empirical formulation is used for calculation of I_{ss} by first applying a correction to the swelling strain for the exaggeration of swell due to confinement, summing the corrected swell strain and the shrinkage strain together, and dividing the sum by an assumed change in the logarithm of total soil suction (discussed in the next section) from near the shrinkage limit to the quasi-saturated condition. The suction change represents the range of total suction over which soil volume change responds relatively linearly. Ground movements can be calculated if the change in I_{ss} with soil profile is known, along with soil suction changes.

Water stored in soil and soil suction

Knowledge of water stored in soil and the negative soil water potential, or suction, developed in partially saturated soil at various times of the year aids

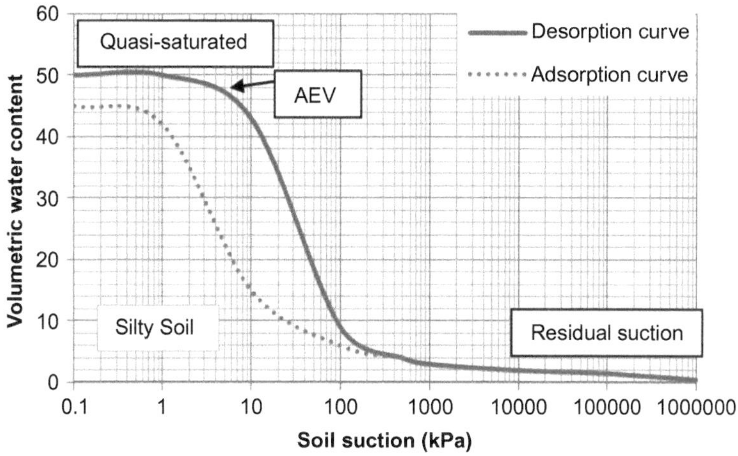

Figure 22.2 Typical drying/desorption and wetting/absorption curve. *(Source: Selected data from Fredlund et al., 1994).*

decision-making in civil engineering and construction. Relevant terms associated with soil water storage are as follows:

Total soil suction: The sum of matric suction and solute/osmotic suction (Krahn and Fredlund, 1972; Nelson and Miller, 1992). The total suction pressure is the opposite (or negative) pressure applied by the external gas on the soil water resulting in a pool of pore water in equilibrium through a semipermeable membrane with the soil water (Krahn and Fredlund, 1972).

Solute/osmotic suction: The negative pressure (gauge) required to create equilibrium between a pool of pure water, semipermeable membrane barrier, and a solution pool identical in composition with the soil water (Krahn and Fredlund, 1972).

Matric suction: Matric suction is the tenacity of the soil matrix to hold onto water (Hillel, 1998) in the absence of solute in the water. Krahn and Fredlund (1972) described matric suction as the negative pressure (gauge) compared to gas pressure (external) on water in the soil, where a solution, equal in composition to the soil water, reaches equilibrium through a porous permeable membrane wall with the soil water.

Soil water characteristic curve (SWCC): Matric suction and water content are uniquely related for each soil and may be plotted as an SWCC (Fig. 22.2; Fredlund and Xing, 1994). Soil density and structure influences this relationship (Nelson and Miller, 1992). The use of the SWCC began initially in the agricultural industry and has been adopted by geotechnical engineers (Fredlund and Xing, 1994; Nimmo and Landa, 2005). It is a useful tool to assist in modeling surface deformation of pavements due to volumetric change of expansive clays (Vanapalli and Adem, 2013).

Soil permeability

Soil permeability, or the coefficient of hydraulic conductivity, may be measured with the falling head permeability test (Fortin, 2014). Clays have low permeability, and their permeability is affected by the initial soil state prior to wetting (density, soil suction, and soil structure). Expansive clay soils generally have very low permeability. The decrease of permeability with initial soil suction can be predicted by using the soil–water characteristic curve (Fredlund et al., 1994). A typical example is illustrated in Fig. 22.3.

22.2.2 Research trends and needs

Considering the economic and infrastructure impacts of clay soils, research into modeling and characterizing their properties and behavior is important. Areas of research that are currently expanding and those that are of vital importance to facilitate better management of clay soils in the future are discussed in the next few sections.

Shrinkage and swelling strain

Volumetric prediction of shrinkage and swelling of clays is currently impossible without knowing the shrinkage and swelling strain. Puppala et al. (2014) attempted to provide correlations between shrinkage and/or swelling strain with change in gravimetric moisture content, Δw, to predict movements in expansive clays. Five Texan soils were extensively tested over three compaction moisture contents. Load was not applied to the samples as they

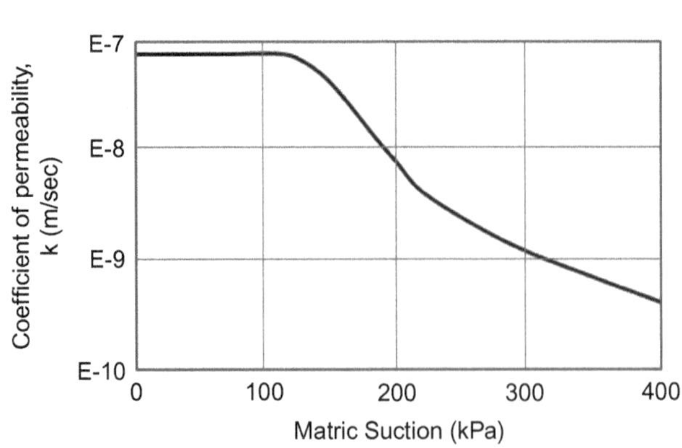

Figure 22.3 Typical soil permeability curve.

were allowed to change volume, and volumetric strains were recorded using digital cameras. Simple correlations were found that were reasonably applicable for all test results—for example, for axial swell and shrinkage strains, ε_{a-sw} and ε_{a-sh}, respectively:

$$\varepsilon_{a-sw} = 0.36\Delta w \text{ and } \varepsilon_{a-sh} = 0.23\Delta w$$

The shrinkage strain correlations were not as strong as those for swelling strain. Volumetric strains correlated more strongly than axial strain with Δw; both shrinkage and swelling coefficients were similar, being 0.70 and 0.66, respectively. Interestingly, the axial swelling strain was approximately half that of the volumetric swelling strain, as assumed in the derivation of the shrink–swell index.

Clay mineralogy

The minerals within clay interact in a variety of ways, hence affecting the volume change ability of the clay. Mineralogy of soils in Australia is not included generally in geotechnical classification. Kalonite, illite, and smectite are the three main groups of clay minerals (Chabrillat and Goetz, 1999). Particle thickness, diameter, and specific surface area (SSA) of these clay minerals are outlined in Table 22.1 (Chen, 1988; Yitagesu et al., 2011). Montmorillonite, a member of the smectite expansive clay group, has the most dramatic shrink–swell capacity due to its particle size (Chen, 1988) and moisture extraction (due to dry weather) and holding ability (Rogers et al., 1993). High quantities of montmorillonite expansive clay minerals are found in soils that exhibit deep vertical cracking (Rogers et al., 1993). These deep vertical cracks in the soil contract with the addition of moisture. Soils exhibiting these behaviors are referred to as vertisols. Vertisols have high cation exchange capacities along with low infiltration rates and poor drainage ability (Yerima et al., 2005).

Table 22.1 Particle thickness, diameter, and SSA of clay minerals

	Morphological characteristics		
Clay mineral	**Particle thickness ($\times 10^{-6}$ m)**	**Particle diameter ($\times 10^{-6}$ m)**	**SSA (m^2/g)**
Kaolinite	0.5–2	0.5–4	10–20
Illite	0.003–0.1	0.5–10	65–180
Montmorillonite (smectite)	<9.5 A	0.05–10	50–840

SSA, specific surface area.
Source: Data from Yitagesu et al. (2011).

Clay mineralogy is most readily identified with the X-ray diffraction (XRD) technique. XRD identifies soil mineralogy and quantification of crystalline materials (expansive clays) through identification of diffraction patterns (CSIRO, 2012). The diffraction patterns of minerals within the soil sample are produced by intensity of reflected radiation, through an initial fixed wavelength X-ray (CSIRO, 2012). Each individual reflection of the reflected radiation from the soil sample results from individual diffractive wave contributions of all parts of the soil sample (Rhodes, 2000).

Note, however, that expansive clays are composed of a mixture of clay minerals and quartz, with the quartz representing the sand and silt fractions of the soil. The quartz component is important because it is non-expansive.

XRD has been shown to be useful in determining the effectiveness of lime-treated soils through identification of unconsumed lime (Al-Mukhtar et al., 2010).

Clay soil stabilization

Clay soil stability can be increased physically, chemically, or biologically. Physical improvement is through compaction. The addition of certain chemicals is well-known to stabilize highly plastic soils (e.g., lime). Vegetation and microorganisms may also affect soil stability, but the amount of stabilization caused by the roots or microbes of specific species within a clay profile is not readily quantifiable. Plant systems may also assist in prevention of erosion of vertisols. The key research needs in the areas of biological and chemical stabilization of clay soils are discussed in the following sections.

Biological stabilization

Biological mechanisms affect the aggregation of soil (Watts et al., 2005). Microbes (Rillig et al., 2010) and soil protein (Wu et al., 2012) may play an important part in soil stabilization within a suburban environment. Wu et al. demonstrated significantly positive correlations between soil proteins and soil water-stable aggregates, carbohydrates, and microbes (mycorrhization) found around roots of *Citrus unshiu*. Soil water repellency has been shown to be induced by biotic activity (Spohn and Rillig, 2012). However, the impact of biotic activity on volumetric changes at depth in a clay soil profile is unknown.

The use of microbes has been reported by many researchers (e.g., Samudre et al., 2014) to facilitate an increase in compressive strength of concrete. Investigation of treating concrete cracking damage using microbes is underway (Samudre et al., 2014). Fungi or bacterial colonies surround tree roots (Dorioz et al., 1993), so it is possible that bacterial mechanisms in soils and/or bacteria working in collaboration with specific plant roots could help stabilize soil, while also protecting infrastructure, in ways not currently understood.

Plants are often located on slopes. Where the initial natural native vegetation was removed, slope failures have often been reported subsequently (Gray, 1995). The following indirect analyses and test methods have been employed to evaluate the strength of the soil–tree root system in studies of soil slope stability (Abe and Ziemer, 1991):

- Statistical correlations between landslide frequencies and forest conditions
- Slope stability analysis incorporating roots and either root tensile strength or pull-out resistance
- Evaluation of the shear strength of the soil with roots

Vetiver grass (*Veitveria zizanioides*) has been used to complement engineering design to enhance slope stability and control erosion (e.g., Hengchaovanich, 1996). This particular grass species has a large and dense root system offering shear strength in the range of 6–10 kPa per kilogram of root and a tensile strength of 75 MPa. A numerical study on vetiver grass by Jotisankasa et al. (2014) suggested that slope stability could be improved with vetiver grass on moderate soil slopes, but it could also lead to reduction of factor of safety in steeper applications (60°) over weathered rock. Their modeling assumed greater permeability due to the root system of the grass, thereby allowing moisture penetration into the body of the slope.

In suburban environments, vegetation is commonly thought to be the cause of deformation of infrastructure due to deep ground movements in the soil profile where moisture is extracted from the ground by the tree roots. Again, data basing of information about vegetation growing in Australian expansive clay suburban environments and its impact on infrastructure would provide important information. It would be useful to record the following as a minimum:

- Growth of plants over time
- Soil fertility required to support various vegetation species
- The soil profile and the reactivity of soil layers

At a research level, it would be ideal to obtain information on the transpiration rates and root architecture of specific vegetation species; ground movement away from and near the vegetation; and, where applicable, the reinforcement effect of the root system in the soil

Chemical stabilization

Soil can be treated chemically to help prevent erosion, provide soil stability (Tingle et al. 2007; Muttuvel, 2008) and act as barriers to deter penetration of soils by roots (trifluarin herbicide; Wilson and Lester, 2002). Addition of chemicals, such as lime, cement, or gypsum, to stabilize expansive clay subgrade materials for geotechnical purposes has been practiced for decades (Tayabji et al., 1982), although deep stabilization is difficult in heavy clays, even with high-pressure injection of slurries. The slurry can penetrate through cracks and fissures but cannot penetrate much of the intact soil.

Nontraditional stabilization additives, such as ionic stabilizers, enzymes, and salts (calcium or magnesium chloride), may be applicable for fine-grained soils, according to Tingle et al. (2007). These additives can vary in constituents; hence, long-term stabilization is currently difficult to predict (Muttuvel, 2008) and so further research is needed on their durability and the interaction of various plants with the available cations from clays to either destabilize or stabilize soils.

Prediction of clay soil movement or stability

Numerous reports documenting observations of damage due to expansive clays are available from throughout the world (e.g., Li et al., 2014; Puppala et al., 2014). However, predictive models to forecast ground movements are rare (Erguler and Ulusay, 2003). In Australia, soil reactivity tests are combined with design suction profiles to estimate movement. Improvements in the test approach and refinement of design suctions will be needed as knowledge is improved. Estimates of soil reactivity from soil index tests should look to indices that are readily evaluated and that indicate clay mineralogy, as recommended by Covar and Lytton (2001) and Kariuki and van der Meer (2004).

Predictions of the cohesion or shear strength afforded by the roots of plants are needed for calculations of soil stability by geotechnical engineers. However, inadequate knowledge of specific tree species and their root distribution in a soil profile limit what can be done. Hubble et al. (2013) developed a model for the influence of roots on soil stability, which is of interest to the geotechnical engineering fraternity. The model was applied successfully to the assessment of slope stability in the eastern Australian highlands, where it was found that mature native Australian forest trees reduced the likelihood of slope mass failure. Further research is needed to validate or improve the model in a range of situations.

The findings presented in this section outline situations with focus on practical implementation and a pathway forward to develop existing understanding of current knowledge. Lack of certainty about the interaction of soil and vegetation has meant that vegetation tends not to be the preferred design measure because there are concerns regarding consistency of performance. Development of techniques, construction methodology, along with data collation and categorization to inform and eventually upgrade standards are the necessary measures required to encourage stabilization of clay-based soil profiles in a long-term, cost-effective manner.

22.3 TREES AND CLAY SOILS

Trees and their roots growing close to buildings and pavements (e.g., roads) can cause direct or indirect structural damage (Satriani et al., 2010). This damage is thought to be exacerbated when trees are growing in expansive clay soils, as soil moisture is extracted through trees roots (Cameron, 2001). Due to the current lack of integrated, conclusive research on trees and their root systems growing in suburban environments, effective suburban designs are unable to be proposed that are complementary to specific trees and growth of root systems over time. With the development of non-invasive techniques to observe the root architecture of suburban trees, whereby an invasive technique is not required, observations in areas previously unable to be observed may occur.

22.3.1 Tree root growth in built environments

The tree root architecture system is the spatial structure of the tree's root system, encompassing all roots that branch, interlock, cross over, and grow in all three dimensions (Butnor et al., 2012). The unpredictable growth and interaction of roots and barriers in a suburban environment can be seen in the exhumation of part of a root growing under a driveway in Fig. 22.4.

Currently, the basic guide for a tree's root spread is the diameter of the canopy. However, as summarized by Day et al. (2010), Tubbs (1977) and Gilman (1988), the relationship between canopy and roots is independent of each other. Therefore, further tree root architecture mapping is required, along with understanding moisture movement through trees.

Growth of roots within a suburban environment is considered unpredictable because there is a lack of knowledge about how various plant species grow in suburban environments (O'Malley and Cameron, 2001;

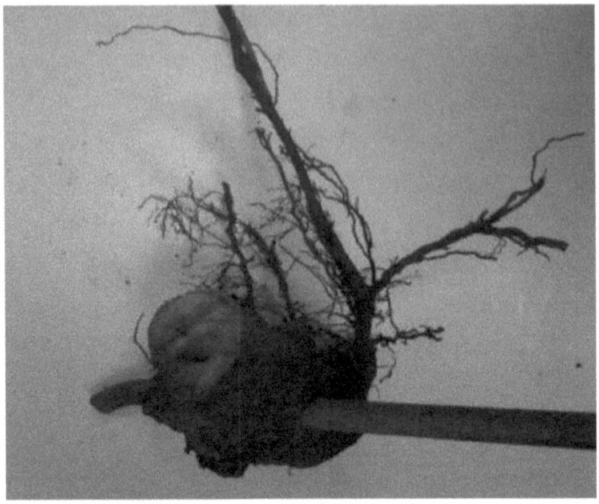

Figure 22.4 A large tree root growing under a driveway concrete slab and around a galvanized steel pipe (25 mm in diameter) in sandy soil, Adelaide.

Standards Australia, 2011), where barriers to root growth exist, including the following:

- Roads, residential footings, concrete footpaths, and buried pipes
- Material trenched in place to prevent root growth in a certain direction (Wilson and Lester, 2002)
- Vegetation planted close together (root competition)
- Species' mechanisms to prevent other species growing nearby by emitting a natural toxin (Jensen, 1907)

Reduced evaporation of soil moisture through semi-impermeable surfaces (asphalt and/or concrete) causes condensation on the underside of the pavement with temperature differences between the soil and pavement (Randrup et al., 2001). The condensation effect of moisture gathering under either a concrete or an asphalt road allows roots to grow within this zone in the soil profile due to heat and moisture.

Tree roots (in general) growing in the soil profile extract water. Diametric increase in specific tree species' roots can directly disturb infrastructure if located at shallow depths.

22.3.2 Exploration of tree root systems

In order to care for and ensure trees are correctly placed in suburban designed environments, observations of plants growing in suburban environments are required. To practically observe tree roots at depth in a suburban environment, tree root architecture observations may be made with noninvasive techniques, although an invasive technique is required to validate results (Butnor et al., 2012). Development of reliable noninvasive techniques to observe suburban trees' root architecture, which forgoes the need for an invasive technique, will allow for greater numbers of observations of tree roots. A variety of methods to observe tree root systems are summarized in the following sections.

Surface ground-penetrating radar

The ground-penetrating radar (GPR) technique requires an electromagnetic signal (transmitted and received) that is characterized by a given frequency. The GPR method utilizes dielectric permittivity, which is the ratio of the permittivity of a substance to the permittivity of free space (Malmberg and Maryott, 1956). Nondestructive monitoring of sub-asphalt pavement soils as well as tree roots (Ow and Sim, 2012) and soil moisture content (Tosti and Benedetto, 2012) can be undertaken using GPR. Coarse roots with diameters greater than 0.05 m are clearly detected by the GPR system in clay soil (Ow and Sim, 2012), although finer ones are commonly not. However, this is an area of rapid development, and techniques may soon be available to provide better detection.

Borehole ground-penetrating radar

Vertical roots cannot be accurately assessed by surface-based GPR. The borehole ground-penetrating radar (BGPR) technique requires the equipment to be lowered vertically into the ground in an access tube, allowing vertically oriented roots to be captured (Butnor et al., 2012). The placement of access tubes approximately 50 mm in diameter, 3 m deep, and 400 mm apart was recommended by Mala (2014).

Electrical resistivity tomography

Tree roots and trunks have been observed using electrical resistivity tomography (EST; al Hagrey, 2007). ERT uses electrical resistivity of the subsurface from measurements made at the surface to produce a cross section of the tree trunk. Spatial images showing whole root zones and moisture variations around the vegetation root zones (al Hagrey, 2007) can be produced with this method. Resistive woody (water transporting) and conductive soft (absorbing) roots in the root zones of old trees have both been identified with geoelectrical imaging. Laboratory experiments conducted by Satriani et al. (2010) integrated GPR data and ERT data to obtain high-resolution electrical images of the subsurface occupied by tree roots.

Seismic refraction tomography

Seismic refraction tomography is a geophysical investigation method for mapping shallow subsurface ground (Azwin et al., 2013). Leucci (2010) used seismic refraction tomography to identify tree roots. The three-dimensional images developed for the total root volume of soil within a suburban environment were completed in combination with data obtained from GPR and ERT.

Excavation

Even if roots are excavated carefully, there is still a chance of loss of roots due to roots breaking during the excavation process. Fine roots are generally assessed using area-based sampling techniques such as soil coring and soil pits (Niiyama et al., 2010).

Water and vacuum system

The water and vacuum system is a destructive process that utilizes pressurized water (hydro) and a vacuum system (vac) to remove soil cover to expose tree roots. The use of this method provides significantly reduced risk of damage to existing plants during construction (Hydrovac Excavations, 2000).

Root–soil separation system

The root–soil separation system (Smucker et al., 1982) separates roots in the laboratory from compacted soils (including heavy clay soils), ensuring small lateral roots, nodules, and other fragile roots remain intact. This process is completed on soil samples taken from the field (potentially from soil cores) in an in-laboratory system that includes washing (with appropriately pressurized water and air jetting) and separation (by sieve) of tree roots from the soil (Smucker et al., 1982).

22.3.3 Moisture movement

In order to understand water movement through trees, a tree's water requirements, the moisture in soil in which the tree grows, and the physiology of the tree (Salisbury and Ross, 1992) need to be studied. Prevailing weather patterns and potential irrigation in the area where the tree is growing play a key role in moisture availability. Evapotranspiration includes the evaporation of moisture from soil and plant surfaces and the transpiration of moisture through plant surfaces (Burman and Pochop, 1994). The Australian Bureau of Meteorology (BOM) expresses the rate of evapotranspiration in millimeters per day.

Data presented on the BOM website are calculated using the Food and Agriculture Organization (FAO) Penman–Monteith method equation, using meteorological data such as solar radiation, air temperature, humidity, and wind speed (Webb, 2010). The FAO Penman–Monteith method is the standard method for the computation of potential evapotranspiration from meteorological data (Allen et al., 1998).

In built environments, soil moisture content is affected by a variety of elements acting synergistically: seasonal regional weather influences, tree species (Dawson, 1996), the design and permeability of materials used to build infrastructure, and the percentage of soil surface covered. The built suburban environment can impact on soil moisture evaporation. Water loss is increased where asphalt covers the base surface of a tree compared to non-covered soils (Kjelgren and Montague, 1998).

O'Malley and Cameron (2005) used the following instruments to obtain information on water requirements of individual trees in a suburban environment: the neutron moisture meter (NMM) for monitoring soil moisture changes and the xylem pressure apparatus and Wescor leaf suction chamber to measure suctions of stems and leaves, respectively. Other instruments applicable for tree physiological observations include the sap flow meter and canopy analyzer. Practical placement and/or use of these instruments are outlined in Fig. 22.5.

Tree Canopy Porometer	The porometer is able to measure live time exchange rates of leaf conductance to water vapor from leaves on a growing tree. The open gas exchange system within the monitor measures concentration differences of CO2 and H2O entering and leaving a cuvette, clamped to a leaf (Schulze 1982)
Xylem Pressure Apparatus	The xylem pressure apparatus technique measure xylem pressure, which approximates the water potential of plant tissues. A leaf stem, removed from a tree, is partly placed within a sealed chamber in which pressurised gas is slowly added (Tyree & Hammel 1972). With increased pressure, the sap is forced form the xylem or stem; the amount of pressure represents the negative water potential of the leaf (Scholander et al. 1965).
Under Tree Canopy Plant Canopy Analyser	The Plant Canopy Analyser (Brand: Licor, Product no. LAI-2200) measures Leaf Area Index (LAI). Watson (1947) defined LAI as the total one-sided leaf tissue area per unit ground surface area. This ration of foliage area to ground area allows for an indirect canopy structure measurement (radiative transfer through a canopy) to be obtained. The gap fraction of a canopy is the fraction of view that is unobstructed by canopy in any particular direction. The Plant Canopy Analyser is used to determine canopy light (sun flecks) interception at five angles, through the gaps in the canopy (Welles & Cohen 1996).
Sap Flow Meter	The sap flow meter measure the movement of sap within a tree and utilises the Heat Ratio Method (HRM), whereby a low grade heat is applied (Nouri et al. 2012). The instrument typically consists of three 35mm needles connected to a microprocessor, placed parallel into the truck/branch of the tree. The probes (top and bottom) contain two sets of matches high precision temperature sensors, located usually at 7.5mm and 22.5mm from the tip of each probe. The third and more centrally located needle is the line heater (full length of the needle), which provides a uniform heat pulse through the sapwood, every 30minutes (ICT International 2014). Direct measure of plant water use can be obtained through knowing the sap flow rate and sap flow volume of which is calculated via the thermal changes recorded by the three needle system (ICT International 2014)
Ground Soil Profile Neutron Moisture Meter (NMM)	The neutron moisture meter probe measure a specific volume of the moisture within the soil, through emitting radioactive material (high energy neutrons). The probe is lowered into a pre-drilled and cased hole and low-energy moderated neutrons are recorded. The emitted radioactive material is reflected back to the probe, in proportion to the hydrogen atoms (predominately in water molecules), within soils (APSIM 2014). Wet soil provides a high count per time of test and a dry soil provides a low count for the same time period (CPN International 2013).

Figure 22.5 Instruments to measure an individual tree's water moisture requirements.

22.4 QUANTIFYING THE IMPACT OF VEGETATION ON INFRASTRUCTURE

Plants have been used to stabilize soils since 28 BC China (Redfield, 2000). A quantified design method to ensure soils are indeed stabilized using vegetation is yet to be completed. For effective management of suburban environments, further information is required with regard to trees growing successfully in those environments.

22.4.1 Measuring ground movement

Potential ground movement within a suburban environment with mature growing trees of specific species is an area yet to be explored adequately through monitoring and modeling (Standards Australia, 2011). Monitoring ground movements through the soil profile can be completed with a magnetic extensometer (Kirby et al., 2003), although reported cases of the use of extensometers in expansive clay suburban environments are rare. Magnetic extensometers can indicate soil settlement zones, as well as the total displacement of soil (Slope Indicator, 2014).

Extensometer readings need to be supplemented with level surveys of the surrounding ground surface.

22.4.2 Research being conducted in an Adelaide suburb

Currently, research is being undertaken with the assistance of the Australian Research Council in Walkley Heights, a northern suburb of Adelaide (Vorwerk et al., 2014). The primary aim of this 3-year project is to uncover ground movement patterns within an urban environment with expansive clays through

- observations of ground movement in a suburban environment and distortion of asphalt roads/concrete house footings;
- characterization of different street tree species, in terms of moisture drawn from soils in different seasons;
- modeling the water balance to describe the exchange of water among soil, plants, and the atmosphere;
- evaluating the moisture loss from the soil profiles near four specific tree species and the associated ground movements; and
- evaluating the impact of the prevailing climate on tree water demand for selected tree species.

The total on-site assets that have been set up in seven residential streets in research sites at Walkley Heights include nine extensometers, 25 NMM

Figure 22.6 Site set up at Walkley Heights with a prickly paperbark tree.

access holes, and 207 pins for surface-level surveys, and three deep benchmarks (~6 m deep). An example of one site in the research area where ground movement is being monitored at Walkley Heights is depicted in Fig. 22.6. Five different species of trees are being closely monitored and will be fitted with sap flow meters. Two of the species are native trees.

This ongoing research will prove extremely beneficial to understanding the interaction of soil–plant–atmosphere and the long-term stability of the ground and the sustainability of the natural built environment.

22.5 INFRASTRUCTURE DAMAGE PREVENTION

Movement of expansive clays is a costly worldwide problem (Rajeev et al., 2012), and through stabilizing these soils during construction, this cost could be reduced (Petry and Armstrong, 1989). This is particularly important for structures such as roads, railways, and canals, which cannot be readily designed (with available knowledge) to mitigate the influence of expansive soil movement in situations either with or without vegetation. Roads and pavements have been neglected during the past four decades of research into damage to buildings caused by volumetric expansion (swelling) and shrinkage of expansive clay soils (Dafalla and Shamrani, 2011). Deformation, ruts, and/or cracking damage to pavements can occur due to either wetting or drying of clay when asphalt-sealed roads are founded on clay soil (Tosti and Benedetto, 2013). Pavement cracking can be minimized by reducing

Figure 22.7 Relatively impermeable asphalt on clay soil with *Pyrus calleryana* (ornamental pear) as the selected tree species planted in the street corridor of Walkley Heights, Adelaide, South Australia.

water infiltration into the pavement and completing ongoing asphalt maintenance measures. An example of such an area where prevention measures have been implemented is outlined in Fig. 22.7; minimal cracking and/or deformation is shown in the asphalt road profile.

The following sections introduce some published case studies of interactions between expansive clay and buildings and/or infrastructure, with or without vegetation.

22.5.1 Building case studies and building standards

Lightweight structures with shallow footings are most susceptible to expansive soil movement. Expansive soils exerting uplift pressures of 250 kPa potentially cause considerable damage to lightly loaded wood-frame structures (Rogers et al., 1993). In Australia, there is awareness of expansive soils and of the need to mitigate movement through footing design, and building standards are available to help deal with the potential problems. The following are specific items taken into account with the Australian Standard for residential footings (AS2870; Standards Australia, 2011):

- Swelling and/or shrinkage movements of reactive clay soils due to moisture change
- Site classification based on soil profile and prevailing climate
- Suppression of swell from applied loads from buildings
- Movement tolerance of the superstructure

AS2870 provides guidelines for classification of sites for potential movement and the design of footings for houses and similar structures on expansive clays in editions of the Standard published in 1986, 1996, and 2011. Nonetheless, failures can still occur if the scope of the Standard is breached—that is, in a highly variable soil profile or if mistakes are made.

AS2870 introduced guidelines for design of footings for trees in 2011. Previous editions made it clear that designs for tree drying effects were outside the scope of the Standard. In this edition, it is acknowledged that there are few data currently available to guide the drafting committee (Standards Australia, 1995). The tree guidelines were based on a South Australian approach to design for trees, modified to suit a range of climates and for the observation that suction changes from tree water uptake were deep rather than shallow (Cameron and Beal, 2011). The South Australian approach had proven successful in protecting buildings from damage over a period of at least 15 years. Observations of deep suction changes arose from forensic investigations of damaged structures with determination of suction profiles around the building (Cameron and Beal, 2011).

Two case studies of building damage are summarized in the following sections.

Case study 1: House with overly wet soil conditions

Li et al. (2014) completed a field investigation on a house built on a stiffened raft slab complying with the Australian Standard (Standards Association of Australia, 1986). It was found, however, that leaking pipes and excessive watering of the garden provided nonuniform soil moisture conditions under the residential housing slab, causing the house slab and vertical walls in the house to crack.

In performing the back-analysis of the distressed raft slab, a three-dimensional investigation of the expansive soil was able to be completed. The movement of the raft slab footing was 100 mm over a 6-year period (Li et al., 2014). Because of the expansive soil movement and the previously described "abnormal" moisture conditions, this site was not considered a "normal site" by the Australian Standard (Standards Australia, 2011).

To design for these abnormal conditions, much stiffer and stronger footings would be required, as defined by Li et al. (2014), which would add significantly to the initial building costs. Another available option is to ensure the owners perform management of houses built on these expansive clays, through site maintenance, to ensure leaks are not occurring and to minimize of garden watering (Li et al., 2014).

Case study 2: House and soil drying by trees

In the United Kingdom, subsidence has been reported to occur where buildings were constructed too close to existing trees and/or vegetation (Hill, 1983). Vegetation was thought to have removed moisture from clay soils, causing subsidence of low-rise building foundations. Heave (upward movement), like subsidence, is difficult to predict, especially in the rare case that drains are blocked and/or broken by tree roots, releasing water into the surrounding clays causing swelling (Hill, 1983). Recommended depths of strip footings in the UK near isolated trees depend on the distance of the tree to the building and may vary in depth between 1.5 and 2.8 m for Poplar, Elm and Willow, and 1.0 to 2.4 m for other species (Hill, 1983).

Hill (1983) outlined that 90% of damage incidents reported in buildings occurred near trees that were a distance of less than the mature height of the tree species. These data were captured as observations of situations in which a tree caused damage to a house through recording the species, position, and size of the tree.

As for plants growing on slopes, the literature is currently missing case studies that compare areas that have heaved and/or subsided with similar areas in which stability occurs.

22.5.2 Asphalt road case studies

Unlike concrete raft slabs for houses, asphalt roads in residential streets are not strong structurally and are much more susceptible to deformation and cracking as clay subgrades move. Examples of ways to prevent such deformation and cracking of roads and to provide stabilization are outlined in the following case studies.

Case study 1: Trees and asphalt roads, Australia

Jameson (1986) found in the State of Victoria, Australia, that certain plantations of tree species growing adjacent to medians and roadsides were thought to be the cause of deformation of pavements on expansive soil, reducing the ride quality of the road. The following specific tree species were identified as unsuitable in such situations:

- *Eucalyptus cladocalyx*—sugar gum
- *Eucalyptus cladocalyx nan*—dwarf sugar gum
- *Eucalyptus spathulata*—swamp mallet
- *Eucalyptus occidentalis*—flat-topped yate

Observations on pavement movement by Jameson (1986) indicated the following:

1. Sections bordering areas without plantations had no substantial change in pavement levels.
2. Seasonal variations of pavement levels occurred closest to the plantations. Generally, pavement levels were affected by areas with plantation at a general distance of approximately twice the height of the plantation, causing decline of the road ride quality.
3. Plantations thinned by 50% had a slight effect on pavement levels, although the limited data indicated that thinning the plantation produced considerable pavement level reduction along the shoulder edge.
4. Twelve months after two plantations were removed, there was substantially less variation in levels than for untreated plantation sections.

Case study 2: Trees and roads, Oklahoma, USA

Snethen (2001) described the central Oklahoma region as being well-known for volumetric shrinkage and swelling. The Oklahoma expansive soil is a combination of clayey silts, silty clays, and clay, with low to moderate plasticity. Pavements and lightly loaded structures sitting on these soils have had measurable deformations, which have exceeded 50 mm.

During the summer of 1999, arc-shaped hairline cracks formed on an asphalt road in Stillwater, Oklahoma, matching the outer limit of the canopy of a Chinese elm tree. The Chinese elm tree was thought to be the cause of the cracks due to moisture extraction from the soil formation. The road was initially built with several layers of crusher-run limestone gravel to a 75- to 100-mm profile depth and compacted with local traffic running over the surface. A 75-mm layer of hot mix asphalt was placed over the road base.

Case study 3: Stabilization trial

Quality road base material is virtually nonexistent in the western areas of New South Wales, Australia (Walter et al., 2012), leading to the use of local expansive clay and/or "black soil" as road subgrade material. A trial road constructed in Bourke involved the stabilization of the in situ black soil through the addition of lime in the subgrade, sub-base, and base layer (1, 3, and 6%, respectively) with a final crumb rubber bitumen seal placed over the road base (Walter et al., 2012). A design life of 20 years was targeted. After 30 years of service, the road had no observable difference in overall operational performance, as a rural road with low traffic

(Walter et al., 2012). Normal road resealing maintenance was required over the life of the road.

22.5.3 Rail track case studies

Similar to roads, rail tracks are not strong structurally and are much more susceptible to deformation and cracking as clay subgrades move. Ballasted rail tracks tend to capture moisture especially on poorly drained sites. Examples of stabilization of clay are outlined in the following two cases.

Case study 1: Subgrade stabilization

At the Seaford Rail Extension Project in South Australia, chemical lime treatment of expansive clays was implemented (Spriggs and Drechsler, 2011). The treatment occurred during construction of the rail line extension as a precaution against ground movement. The addition of 2 or 3% hydrated lime to the soil profile to a depth of 300 mm in the rail track subgrade reduced erodibility and increased durability of the soil. The Emerson Class and pin-hole dispersion tests (Standards Australia, 1997) were used to determine erodibility and durability of the lime-treated soil to meet design specifications.

Case study 2: Rail track formation improvement with trees

Train operators noted better aligned track alongside trees than in sections of the rail corridor devoid of trees in the western plains of Victoria, Australia. Therefore, a field study was initiated in rail corridors in Victoria (Miram, Horsham, and Wal Wal) and Queensland (Emerald). Subsequently, Potter and Cameron (2005) demonstrated that vegetated sites had soils with greater strengths and stiffness than those of the same or similar soil at sites without significant trees. The tracks were laid on ballast over relatively flat ground; hence, the sites were poorly drained.

Over time, moisture infiltrating the ballast builds up because the ballast forms a capillary break that inhibits evaporation. Consequently, the subgrade becomes saturated and prone to settlement with further rail traffic. Thus, it was conjectured that in the long term, trees (in general) were useful for removing excess pore water from the rail corridor and setting up unsaturated soil conditions, thereby improving the engineering properties of the soil. Soil suction profiles determined across the various trees and corresponding non-treed sites confirmed this hypothesis.

Potter and Cameron (2005) further analyzed the data with respect to the influence of solute suction on the benefits of vegetation by comparing two of the sites. It was found that the stabilizing influence of vegetation was much more pronounced at the less saline site in Horsham, Victoria, than in the saline site in Emerald in central Queensland. The drying influence of a red gum (*Eucalyptus camaldulensis*) at Horsham, located just 4 m from the toe of the ballast, was evident below the track to at least a depth of 4 m.

22.6 CONCLUSION

Expansive clay soils are known to cause damage to buildings and roads within urban environments through heave (soil swelling) or settlement (soil desiccation) movements. Experienced practitioners need to be able to accurately predict volumetric movements in expansive clays, although testing methods are still being developed.

General observations of trees in urban environments have led to indirect evidence that trees desiccate the clay soil at depth, increasing shrinkage settlement (Cameron, 2001). Paradoxically, rail tracks and sloped embankments built on expansive clay soils may be stabilized using vegetation. The soil's stability and/or movement due to tree roots extracting moisture in suburban environments has yet to be quantified over a sufficiently wide range of soil profiles, climates, and tree species. The distribution of tree roots within clay soil profiles is not well understood and remains unpredictable, particularly along residential streets. Reliable estimation of ground movements and soil improvement due to matric suction requires knowledge of where tree roots are extracting moisture. Development of noninvasive methods to observe tree roots more reliably is needed.

Models based on climatic data, the moisture in soil, ground movement, and plant physiology need to be developed to predict soil improvement due to partial saturation or damage to houses and roads in communities in areas with expansive clay soils. Deliberate design of residential buildings, roads, rail lines, and sloped embankments may provide considerable savings by minimizing rectification during operation through the employment of various combinations of vegetation, the use of appropriate building materials, and the addition of chemical stabilizers as appropriate.

Through complementary infrastructure design and construction processes and ongoing operational management, expansive clay stabilization may be possible.

REFERENCES

Abe, K., Ziemer, R.R., 1991. Effect of tree roots on shear zone: Modeling reinforced shear stress. Can. J. Forest Res. 21 (7), 1012–1019.

al Hagrey, S.A., 2007. Geophysical imaging of root-zone, trunk, and moisture heterogeneity. J. Exp. Bot. 58 (4), 839–854.

Allen, R.G., Pereira, L.S., Raes, D., Smith, M., 1998. Crop evapotranspiration: Guidelines for computing crop water requirements, Irrigation and Drainage Paper 56. Food and Agriculture Organization of the United Nations, Rome. accessed January 2014, http://www.fao.org/docrep/X0490E/X0490E00.htm.

Al-Mukhtar, M., Lasledj, A., Alcover, J., 2010. Behaviour and mineralogy changes in lime-treated expansive soil at 20 °C. Appl. Clay Sci. 50, 191–198.

Al-Rawas, A.A., Goosen, M.F.A., Al-Rawa, G.A., 2006. Geology, classification, and distribution of expansive soils and rocks. In: Al-Rawas, A.A., Goosen, M.F.A. (Eds.), Expansive Soils: Recent Advances in Characterization and Treatment. Taylor & Francis, London, pp. xi and 3.

APSIM, 2014. Module 4: Determining plant available water capacity—The neutron moisture meter (NMM). http://www.apsim.info/Portals/0/APSoil/SoilMatters/pdf/Mod4pt2a.pdf; accessed May 2014.

Azwin, I., Saad, R., Nordiana, M., 2013. Applying the seismic refraction tomography for site characterization. APCBEE Procedia 5, 227–231.

Burman, R., Pochop, L.O., 1994. Evaporation, Evapotranspiration and Climatic Data. Elsevier, Amsterdam.

Butnor, J.R., Barton, C., Day, F.P., Kurt, H., Johnsen, K.H., Mucciardi, A.N., et al., 2012. Using ground-penetrating radar to detect tree roots and estimate biomass. In: Mancuso, S. (Ed.), Measuring Roots. Springer-Verlag, Berlin, pp. 213–238.

Cameron, D.A., 1989. Tests for reactivity and prediction of ground movement. Civ. Eng. Trans., I.E Australia 31 (3), 121–132.

Cameron, D.A., 2001. The extent of soil desiccation near trees in a semi-arid environment. Geotech. Geol. Eng. 19, 357–370.

Cameron, D.A., Beal, N.S., 2011. Estimation of foundation movement and design of footing systems on reactive soils for the effects of trees. Aust. Geomech. J. 46 (3), 97–113.

Chabrillat, S., Goetz, A.F.H., 1999. The search for swelling clays along the Colorado Front Range: The role of AVIRIS resolution in detection. Proc. Airborne Earth Science Workshop. JPL Publication.

Chen, F.H., 1988. Foundation on Expansive Soils. Elsevier, Amsterdam.

Chen, L., Yin, Z., Zhang, P., 2007. Relationship of resistivity with water content and fissures of unsaturated expansive soils. J. China Univ. Mining Technol. 17 (4), 537–540.

Covar, A.P., Lytton, R.L., 2001. Estimating soil swelling behavior using soil classification properties. Geotech. Spec. Publ. 115, 44–62, Expansive Clay Soils and Vegetative Influence on Shallow Foundations, ASCE.

CPN International, 2013. 503 Dr Hydroprobe operating manual. http://www.instrotek.com/pdfs/503%20Op%20manual.pdf; accessed September 2014.

CSIRO, 2012. Mineralogical services. http://www.clw.csiro.au/services/mineral/x-ray.html; accessed October 2014.

Dafalla, M.A., Shamrani, M.A., 2011. Road damage due to expansive soils: Survey of the phenomenon and measures for improvement. Geotechnical Special Publication No. 219, American Society of Civil Engineers. http://wendang.baidu.com/view/0013943b31126edb6f1a1021.html, accessed February 2014.

Dawson, T.E., 1996. Determining water use by trees and forests from isotopic, energy balance and transpiration analyses: The roles of tree size and hydraulic lift. In: Tree Physiology. Heron Publishing, Victoria, Canada, 16, pp. 263–272.

Day, S.D., Wiseman, E., Dickinson, S.B., Harris, J.R., 2010. Contemporary concepts of root system architecture of urban trees. Arboriculture Urban Forestry 36 (4), 149–157.

De Baets, S., Poesen, J., Reubens, B., Wemans, K., De Baerdemeker, J., Muys, B., 2008. Root tensile strength and root distribution of typical Mediterranean plant species and their contribution to soil shear strength. Plant Soil 305, 207–226.

Deakin, N., 2005. Repair of subsidence damage: An insurer's perspective. J. Building Appraisal 1 (3), 225–243.

Dorioz, J.M., Robert, M., Chenu, C., 1993. The role of roots, fungi and bacteria on clay particle organization: An experimental approach. Geoderma 56, 179–194.

Erguler, Z.A., Ulusay, R., 2003. A simple test and predictive models for assessing swell potential of Ankara (Turkey) clay. Eng. Geol. 67, 331–352.

Fatahi, B., 2007. Modelling of influence of matric suction induced by native vegetation on sub-soil improvement. PhD thesis, School of Civil, Mining and Environmental Engineering, University of Wollongong, Australia.

Fityus, S.G., Smith, D.W., Allman, M.A., 2004. Expansive soil test site near Newcastle. J. Geotech. Geoenviron. Eng. 130, 686–695.

Fityus, S.G., Cameron, D.A., Walsh, P.F., 2005. The shrink swell test. ASTM Geotech. Testing J. 28, 1–10.

Fortin, S., 2014. Permeability testing in unconsolidated materials. http://www.robertsongeoconsultants.com/index.php?page=pageandid=56; accessed March 2014.

Fredlund, D.G., Xing, Z., 1994. Equations for the soil–water characteristic curve. Can. Geotech. J. 31 (3), 521–531.

Fredlund, D.G., Xing, A., Huang, S., 1994. Predicting the permeability function for unsaturated soils using the soil–water characteristic curve. Can. Geotech. J. 31 (3), 521–531.

Gilman, E., 1988. Predicting root spread from trunk diameter and branch spread. J. Arboriculture 14 (4), 85–89.

Gray, D.H., 1995. Influence of vegetation on the stability of slopes. In: Vegetation and Slopes: Stabilisation, Protection and Ecology. In: Proceedings of the international conference held at the University Museum, Oxford, September 29–30, 1994.

Gromko, G.J., 1974. June. Review of expansive soils. J. Geotech. Eng. Div., ASCE 667–689.

Hengchaovanich, D., 1996. Use of vetiver grass for engineering purposes in Malaysia with particular reference to slope stabilisation and erosion control. In: Proc. Vetiver Workshop on the Research & Development and Application of the Vetiver Grass System (VGS), Toowoomba, Queensland, Australia, pp. 6–7.

Hill, W., 1983. Tree root damage to low rise buildings. Struct. Surv. 1 (3), 254–261.

Hillel, D., 1998. Environmental Soil Physics. Academic Press, San Diego.

Hubble, T.C.T., Airey, D.W., Sealey, H.K., De Carli, E.V., Clarke, S.L., 2013. A little cohesion goes a long way: Estimating appropriate values of additional root cohesion for evaluating slope stability in the Eastern Australian highlands. Ecol. Eng. 61, 621–632.

Hydrovac Excavations, 2000. Hydrovac Excavations vacuum excavation, direction drilling and civil construction. http://www.hydrovac.com.au/--!services/ca4p#!services/ca4p; accessed January 2014.

ICT International, 2014. Heat ratio method—Sap flow meter. http://www.ictinternational.com; accessed January 2014.

Jameson, G.W., 1986. Trees and pavement shape loss. In: 13th ARRB Conference. Australia Road Research Board, Adelaide, pp. 93–106.

Jensen, G.H., 1907. Toxic limits and stimulation effects of some salts and poisons on wheat. Bot. Gaz. 11–44.

Jotisankasa, A., Mairaing, W., Tansamrit, S., 2014. Infiltration and stability of soil slope with vetiver grass subjected to rainfall from numerical modeling. In: Khalili, N., Russell, A., Khoshghalb, A. (Eds.), In: Unsaturated soils: Research & applications, 2. Taylor & Francis, London, pp. 1241–1247.

Kalantari, B., 2012. Engineering significance of swelling soils. Res. J. Appl. Sci. Eng. Technol. 4 (17), 2874–2878.

Kariuki, P.C., van der Meer, F., 2004. A unified swelling potential index for expansive soils. Eng. Geol. 72, 1–8.

Kirby, J.M., Bernardi, A.L., Ringrose-Voase, A.J., Young, R., Rose, H., 2003. Field swelling, shrinking, and water content change in a heavy clay soil. Aust. J. Soil Res. 41, 963–978.

Kjelgren, R., Montague, T., 1998. Urban tree transpiration over turf and asphalt surfaces. Atmospheric Environ. 32, 35–41.

Krahn, J., Fredlund, D.G., 1972. On total, matric and osmotic suction. Soil Sci. 114 (5), 339–348.

Leucci, G., 2010. The use of three geophysical methods for 3D images of total root volume of soil in urban environments. Exploration Geophys. 41, 268–278.

Li, J., Cameron, D.A., Ren, G., 2014. Case study and back analysis of a residential building damaged by expansive soils. Comput. Geotech. 56, 89–99.

Mala, 2014. Mala ground explorer. http://www.malagpr.com.au/ground-explorer-hdr.html; accessed March 2014.

Malmberg, C.G., Maryott, A.A., 1956. Dielectric constant of water from 0 °C to 100 °C. J. Res. Natl. 56 (1), Bureau Standards, Research Paper 2641.

Mielenz, R.C., King, M.E., 1955. Physical chemical properties and engineering performance of clays. Calif., Div. Mines Geol Bull. 169, 196–254.

Mitchell, J.K., 1993. Fundamentals of Soil Behavior, second ed. University of California Press, Berkeley.

Mitchell, P.W., Avalle, D.L., 1984. A technique to predict expansive soil movements. Proceedings. In: Fifth International Conference on Expansive Soils Institution of Engineers, Australia. pp. 124–130.

Muttuvel, T., 2008. Erosion rate of chemically stabilised soils incorporating tensile stress—Deformation behaviour. PhD thesis, University of Wollongong, Australia.

Navarro, V., Candel, M., Yustres, A., Sánchez, J., Alonso, J., 2009. Trees, soil moisture and foundation movements. Comput. Geotech. 36, 810–818.

Nelson, J.D., Miller, D.J., 1992. Expansive Soils: Problems and Practice in Foundation and Pavement Engineering. Wiley, New York.

Niiyama, K., Kajimoto, T., Matsuura, Y., Yamashita, T., Matsuo, N., Yashiro, Y., et al., 2010. Estimation of root biomass based on excavation of individual root systems in a primary dipterocarp forest in Pasoh Forest Reserve, Peninsular Malaysia. J. Trop. Ecol. 26, 271–284.

Nimmo, J.R., Landa, E.R., 2005. The soil physics contributions of Edgar Buckingham. Soil Sci. Am. J. 69 (2), 328–342.

Nouri, H., Beecham, S., Kazemi, F., Hassanli, A.M., 2012. A review of ET measurement techniques for estimating the water requirements of urban landscape vegetation. Urban Water J. 10 (4), 247–259.

O'Malley, A., Cameron, D., 2001. Street trees influence on soil moisture, urban dwellings and pavements and Walkley Heights as a UniSA–Treenet research site, Paper presented at the Treenet Proc. 2nd National Street Tree Symposium, Adelaide, Australia, September 6–7. http://treenetmedia.com/?120.

O'Malley, A.P.K., Cameron, D.A., 2005. The Influence of Trees on Soil Moisture, Dwellings and Pavements in an Urban Environment. University of South Australia, School of Natural and Built Environments, Mawson Lakes Campus, Sponsors, Local Government Association and Salisbury City Council http://www.hbc-consolidamenti.it/userfiles/files/The_Influence_of_Trees_on_Soil_Moisture_Dwellings_and_Pavements_in_an_Urban_Environment_-_2005_Report_-_Section_1.pdf.

Ow, L.F., Sim, E.K., 2012. Detection of urban tree roots with the ground penetrating radar. Plant Biosystems 146, 288–297.

Petry, T.M., Armstrong, J.C., 1989. Stabilization of expansive clay soils. Transport. Res. Rec. 1219.

Potter, W., Cameron, D.A., 2005. Potential remediation of rail track foundations in poorly drained clay sites with native vegetation. In: Proc. 7th Int. Conf. on the Bearing Capacity of Roads, Railways and Airfields, June, Trondheim.

Puppala, A.J., Manosuthikij, T., Chittoori, B.C.C., 2014. Swell and shrinkage strain prediction models for expansive clays. Eng. Geol. 169, 1–8.

Rajeev, P., Chan, D., Kodikara, J., 2012. Ground–atmosphere interaction modelling for long-term prediction of soil moisture and temperature. Can. Geotech. J. 49 (9), 1059–1073.

Randrup, T.B., McPherson, E.G., Costello, L.R., 2001. A review of tree root conflicts with sidewalks, curbs and roads. Urban Ecosyst. 5, 209–225.

Redfield, E., 2000. Soil bioengineering and biotechnical stabilization. Renewable Resources, Vol. 575: Advanced Revegetation. University of Alberta, Alberta, Canada.

Rhodes, G., 2000. Crystallography Made Crystal Clear. Academic Press, San Diego.

Richards, B.G., Peter, P., Emerson, W.W., 1983. The effect of vegetation on the swelling and shrinking of soils in Australia. Geotechnique 33, 39–127.

Rillig, M.C., Mardatin, N.F., Leifheit, E.F., Antunes, P.M., 2010. Mycelium of arbuscular mycorrhizal fungi increases soil water repellency and is sufficient to maintain water-stable soil aggregates. Soil Biol. Biochem. 42 (7), 1189–1191.

Rogers, J.D., Olshansky, R., Rogers, R.B., 1993. Damage to foundations from expansive soils. Claims People 3 (4), 1–4.

Salisbury, F.B., Ross, C.W., 1992. Plant Physiology, fourth ed. Wadsworth, Belmont, CA.

Samudre, M.P., Mangulkar, M.N., Saptarshi, S.D., 2014. A review of emerging way to enhance the durability and strength of concrete structures: Microbial concrete. Int. J. Innov. Res. Sci. Eng. Technol. 3(2).

Satriani, A., Loperte, A., Proto, M., Bavusi, M., 2010. Building damage caused by tree roots: Laboratory experiments of GPR and ERT surveys. Adv. Geosci. 24, 133–137.

Scholander, P., Bradstreet, E., Hemmingsen, E., Hammel, H., 1965. Sap pressure in vascular plants: Negative hydrostatic pressure can be measured in plants. Science 148 (3668), 339–346.

Schulze, E.D., 1982. Plant life forms and their carbon, water and nutrient relations. In: Lange, O.L., Nobel, P.S., Osmond, C.B., Ziegler, H. (Eds.), Physiological Plant Ecology: II. Encyclopedia of Plant Physiology, New Series, 12B. Springer-Verlag, Berlin, pp. 615–676.

Slope Indicator, 2014. Magnet Extensometer (51817199). http://www.slopeindicator.com/pdf/manuals/magnet-extensometer.pdf; accessed October 2014.

Smucker, A.J.M., McBurney, S.L., Srivastava, A.K., 1982. Quantitative separation of roots from compacted soil profiles by the hydropneumatic elutriation system. Agron. J. 74 (3), 500–503.

Snethen, D., 2001. Influence of local tree species on shrink/swell behavior of Permian clays in central Oklahoma. In: Vipulanandan, C., Addison, M.B., Hasen, M. (Eds.), Expansive clay soils and vegetative influence on shallow foundations. Geotechnical Special Publication No. 115. American Society of Civil Engineers, Reston, Va, pp. 158–171.

Spohn, M., Rillig, M.C., 2012. Temperature- and moisture-dependent soil water repellency induced by the basidiomycete Agaricus bisporus, Pedobiologia. Int. J. Soil Biol. 55, 59–61.

Spriggs, L., Drechsler, M., 2011. Geology and geotechnical design for seaford rail extension earthworks in Adelaide. South Australia. Unpublished manuscript.

Standards Association of Australia, 1986. Residential slabs and footings, AS 2870–1986. SAI Global database.

Standards Australia, 1997. Methods of testing soils for engineering purposes: Soil classification tests, AS1289.5.4.1-1997. SAI Global database.
Standards Australia, 2011. Residential slabs and footings, AS 2870-2011. SAI Global database.
Stokes, A., Atger, C., Bengough, A.G., Fourcaud, T., Sidle, R.C., 2009. Desirable plant root traits for protecting natural and engineered slopes against landslides. Plant Soil 324, 1–30.
Tayabji, S.D., Nussbaum, P.J., Ciolko, A.T., 1982. Evaluation of heavily loaded cement-stabilized bases. Transport. Res. Rec. 839, 6–11.
Tingle, J.S., Newman, J.K., Larson, S.L., Weiss, C.A., Rushing, J.F., 2007. Stabilization mechanisms of nontraditional additives. Transport. Res. Rec. 1989, 59–67.
Tosti, F., Benedetto, A., 2012. Pavement pumping prediction using ground penetrating radar. Procedia Soc. Behav. Sci. 53, 1045–1054.
Tubbs, C.H., 1977. Root–crown relations of young sugar maple and yellow birch, Research Note NC-225. USDA Forest Service, North Central Forest Experiment Station.
Tyree, M.T., Hammel, H.T., 1972. The measurement of the turgor pressure and the water relations of plants by the pressure-bomb technique. J. Exp. Bot. 23, 267–282.
Vanapalli, S.K., Adem, H.H., 2013. A simple modeling approach for estimation of soil deformation behaviour of natural expansive soils using the modulus of elasticity as a tool. In: Hellmich, C., Pichler, B., Adam, D. (Eds.), Poromechanics V: Proceedings of the Fifth Biot Conference on Poromechanics. American Society for Civil Engineers, Reston, VA.
Vorwerk, S.J., Cameron, D., Keppel, G., 2014. Modelling moisture movement: Trees growing in clay soil. University of South Australia, Project proposal.
Walter, P., Anwar, S., Sutherland, M., 2012. Performance of pavements constructed with expansive clay. In: Paper presented at the 25th ARRB Conference—Shaping the future: Linking policy, research and outcomes, Perth, Australian Road Research Board (ARRB).
Watson, D.J., 1947. Comparative physiological studies in the growth of field crops: I. Variation in net assimilation rate and leaf area between species and varieties and within and between years. Ann. Bot. 11, 41–76.
Watts, C.W., Whalley, W.R., Brookes, P.C., Devonshire, B.J., Whitmore, A.P., 2005. Biological and physical processes that mediate micro-aggregation of clays. Soil Sci. 170 (8), 573–583.
Webb, C.P., 2010. Bureau of Meteorology reference evapotranspiration calculations. http://www.bom.gov.au/watl/eto/reference-evapotranspiration-report.pdf; accessed June 2014.
Welles, J.M., Cohen, S., 1996. Canopy structure measurement by gap fraction analysis using commercial instrumentation. J. Exp. Bot. 47 (302), 1335–1342.
Wilson, A.D., Lester, D.G., 2002. Trench inserts as long-term barriers to root transmission for control of oak wilt. Plant Dis. 1067–1074.
Wu, Q., He, X., Zou, Y., He, K., Sun, Y., Cao, M., 2012. Spatial distribution of glomalin-related soil protein and its relationships with root mycorrhization, soil aggregates, carbohydrates, activity of protease and β-glucosidase in the rhizosphere of *Citrus unshiu*. Soil Biol. Biochem. 45, 181–183.
Yitagesu, F.A., van, d.M., van, d.W., Seged, H., 2011. Evaluation of Soil Expansion Index from routinely determined geotechnical parameters. Soil Sci. Soc. Am. J 75 (5), 1640–1651.
Yerima, B.P.K., van Ranst, E., 2005. Introduction to Soil Science: Soils of the Tropics. Trafford, Victoria, British Columbia.

INDEX

Note: Page numbers followed by *f* indicate figures and *t* indicate tables.

D

F

FSC
www.fsc.org
MIX
Papier | Fördert gute Waldnutzung
FSC® C083411

Kolibri 360 GmbH
Ferdinand-Jühlke-Straße 7
99095 Erfurt, Deutschland
produktsicherheit@kolibri360.de